Applications to Regular and Bang-Bang Control

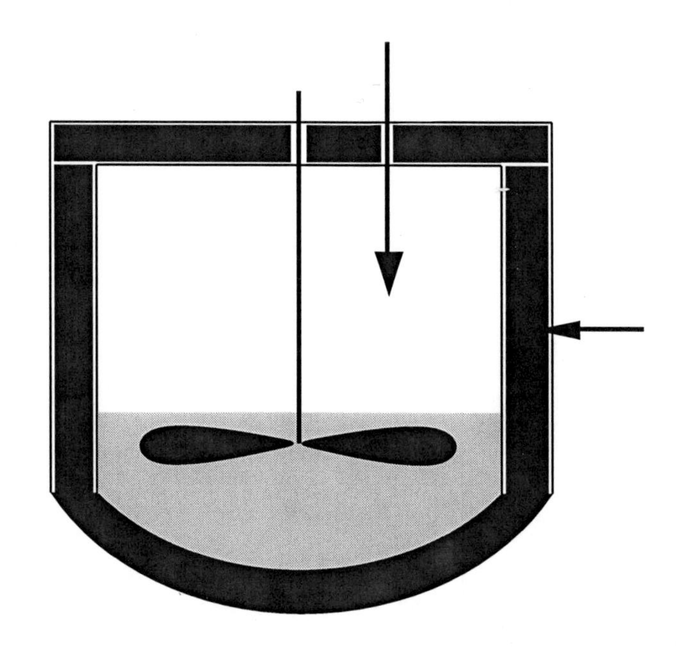

Advances in Design and Control

SIAM's Advances in Design and Control series consists of texts and monographs dealing with all areas of design and control and their applications. Topics of interest include shape optimization, multidisciplinary design, trajectory optimization, feedback, and optimal control. The series focuses on the mathematical and computational aspects of engineering design and control that are usable in a wide variety of scientific and engineering disciplines.

Series Volumes

Osmolovskii, Nikolai P. and Maurer, Helmut, *Applications to Regular and Bang-Bang Control: Second-Order Necessary and Sufficient Optimality Conditions in Calculus of Variations and Optimal Control*

Biegler, Lorenz T., Campbell, Stephen L., and Mehrmann, Volker, eds., *Control and Optimization with Differential-Algebraic Constraints*

Delfour, M. C. and Zolésio, J.-P., *Shapes and Geometries: Metrics, Analysis, Differential Calculus, and Optimization, Second Edition*

Hovakimyan, Naira and Cao, Chengyu, *$\mathcal{L}_1$ Adaptive Control Theory: Guaranteed Robustness with Fast Adaptation*

Speyer, Jason L. and Jacobson, David H., *Primer on Optimal Control Theory*

Betts, John T., *Practical Methods for Optimal Control and Estimation Using Nonlinear Programming, Second Edition*

Shima, Tal and Rasmussen, Steven, eds., *UAV Cooperative Decision and Control: Challenges and Practical Approaches*

Speyer, Jason L. and Chung, Walter H., *Stochastic Processes, Estimation, and Control*

Krstic, Miroslav and Smyshlyaev, Andrey, *Boundary Control of PDEs: A Course on Backstepping Designs*

Ito, Kazufumi and Kunisch, Karl, *Lagrange Multiplier Approach to Variational Problems and Applications*

Xue, Dingyü, Chen, YangQuan, and Atherton, Derek P., *Linear Feedback Control: Analysis and Design with MATLAB*

Hanson, Floyd B., *Applied Stochastic Processes and Control for Jump-Diffusions: Modeling, Analysis, and Computation*

Michiels, Wim and Niculescu, Silviu-Iulian, *Stability and Stabilization of Time-Delay Systems: An Eigenvalue-Based Approach*

Ioannou, Petros and Fidan, Barış, *Adaptive Control Tutorial*

Bhaya, Amit and Kaszkurewicz, Eugenius, *Control Perspectives on Numerical Algorithms and Matrix Problems*

Robinett III, Rush D., Wilson, David G., Eisler, G. Richard, and Hurtado, John E., *Applied Dynamic Programming for Optimization of Dynamical Systems*

Huang, J., *Nonlinear Output Regulation: Theory and Applications*

Haslinger, J. and Mäkinen, R. A. E., *Introduction to Shape Optimization: Theory, Approximation, and Computation*

Antoulas, Athanasios C., *Approximation of Large-Scale Dynamical Systems*

Gunzburger, Max D., *Perspectives in Flow Control and Optimization*

Delfour, M. C. and Zolésio, J.-P., *Shapes and Geometries: Analysis, Differential Calculus, and Optimization*

Betts, John T., *Practical Methods for Optimal Control Using Nonlinear Programming*

El Ghaoui, Laurent and Niculescu, Silviu-Iulian, eds., *Advances in Linear Matrix Inequality Methods in Control*

Helton, J. William and James, Matthew R., *Extending H^∞ Control to Nonlinear Systems: Control of Nonlinear Systems to Achieve Performance Objectives*

Applications to Regular and Bang-Bang Control

Second-Order Necessary and Sufficient Optimality Conditions in Calculus of Variations and Optimal Control

Nikolai P. Osmolovskii
Systems Research Institute
Warszawa, Poland
University of Technology and Humanities in Radom
Radom, Poland
University of Natural Sciences and Humanities in Siedlce
Siedlce, Poland
Moscow State University
Moscow, Russia

Helmut Maurer
Institute of Computational and Applied Mathematics
Westfälische Wilhelms-Universität Münster
Münster, Germany

Society for Industrial and Applied Mathematics
Philadelphia

10 9 8 7 6 5 4 3 2 1

Library of Congress Cataloging-in-Publication Data

Osmolovskii, N. P. (Nikolai Pavlovich), 1948-
Applications to regular and bang-bang control : second-order necessary and sufficient optimality conditions in calculus of variations and optimal control / Nikolai P. Osmolovskii, Helmut Maurer.
p. cm. -- (Advances in design and control ; 24)
Includes bibliographical references and index.
ISBN 978-1-611972-35-1
1. Calculus of variations. 2. Control theory. 3. Mathematical optimization. 4. Switching theory. I. Maurer, Helmut. II. Title.
QA315.O86 2012
515'.64--dc23 2012025629

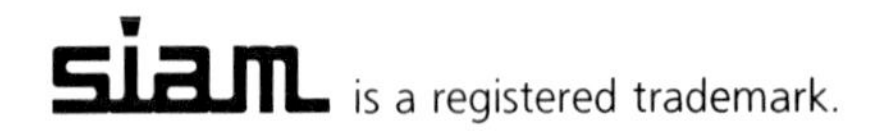

For our wives, Alla and Gisela

Contents

List of Figures

Notation

$\{x \mid P(x)\}$	:	set of elements x with the property P.
$\mathbb{R}$	:	set of real numbers.
$\xi_+ := \max\{\xi, 0\}$	:	positive part of $\xi \in \mathbb{R}$.
$x \in \mathbb{R}^n$	:	$x = \begin{pmatrix} x_1 \\ \vdots \\ x_n \end{pmatrix}$, $x_i \in \mathbb{R}$.
$y \in \mathbb{R}^{n*}$	:	$\Leftrightarrow\ y = (y_1, \dots, y_n),\ y_i \in \mathbb{R}$.
$x^* \in \mathbb{R}^{n*}$	:	$x^* = (x_1, \dots, x_n)$ for $x \in \mathbb{R}^n$.
$yx = \sum_{i=1}^n y_i x_i$	:	$x \in \mathbb{R}^n$, $y \in \mathbb{R}^{n*}$.
$\langle x, y \rangle = \sum_{i=1}^n x_i y_i$	:	$x, y \in \mathbb{R}^n$.
$d(a)$	:	dimension of the vector a.
$\|\cdot\|_X$	:	norm in the normed space X.
$[a,b] \subset X$	:	$[a,b] := \{x \in X \mid x = \alpha a + (1-\alpha)b$, where $\alpha \in [0,1]\}$, closed interval in the linear space X with endpoints a, b.
$\overline{A}$	:	closure of the set A.
X^*	:	dual space to the normed space X.
$\langle x^*, x \rangle$	:	value of the linear functional $x^* \in X^*$ at the point $x \in X$.
K^*	:	$K^* := \{x^* \in X^* \mid \langle x^*, x \rangle \geq 0$ for all $x \in K\}$, dual cone of K.
$g'(x_0)$	:	Fréchet derivative of the mapping $g : X \to Y$ at x_0.
$J(x) \to \min$	:	$J(x) \to \min$, $f_i(x) \leq 0, i = 1, \dots, k$, $g(x) = 0$ ($g : X \to Y$); abstract optimization problem in the space X.
$\sigma(\delta x)$	:	$\sigma(\delta x) = \max\{m(\delta x), \|g(x_0 + \delta x)\|\}$,
	:	$m(\delta x) = \max_{i=0,\dots,k} f_i(x_0 + \delta x)$, $f_0(x) := J(x) - J(x_0)$, violation function in the abstract optimization problem at x_0.
$\lambda = (\alpha_0, \dots, \alpha_k, y^*)$	:	Lagrange multipliers in the abstract optimization problem.

$L(\lambda,x)$	:	$L(\lambda,x) = \alpha_0 J(x) + \sum_{i=1}^k \alpha_i f_i(x) + \langle y^*, g(x)\rangle$, Lagrange function in the abstract optimization problem.
Λ_0	:	$\Lambda_0 = \{\lambda \mid \alpha_i \geq 0,\ (i = 0,\dots,m),\ \alpha_i f_i(\hat{x}) = 0\ (i = 1,\dots,k),$
	:	$\sum_{i=0}^k \alpha_i + \|y^*\| = 1,\ L_x(\lambda,\hat{x}) = 0\}$, $L_x = \partial L/\partial x$, set of the normed tuples of Lagrange multipliers at $\hat{x}$.
$\mathcal{K}$	:	$\mathcal{K} = \{\bar{x} \in X \mid \langle f_i'(x_0),\bar{x}\rangle \leq 0,\ i \in I \cup \{0\};\ g'(x_0)\bar{x} = 0\}$, cone of critical directions (critical cone) at the point x_0, $I = \{i \in \{1,\dots,k\} \mid f_i(x_0) = 0\}$, set of active indices at x_0.
$z(\cdot)$	:	element of a function space; $z(t)$ is the value of $z(\cdot)$ at t.
$x(t)$	:	state variable, $x \in \mathbb{R}^{d(x)}$.
$u(t)$	:	control variable, $u \in \mathbb{R}^{d(u)}$.
$\Delta = [t_0, t_f]$	:	time interval.
$w := (x,u)$	:	state x and control u.
$p = (x_0, x_f)$	:	$\in \mathbb{R}^{2d(x)}$, if Δ is a fixed time interval, $x_0 := x(t_0)$, $x_f := x(t_f)$.
$p := (t_0, x_0, t_f, x_f)$	:	$\in \mathbb{R}^{2+2d(x)}$, if Δ is a variable time interval.
$\psi(t)$	:	adjoint variable, $\psi \in \mathbb{R}^{d(x)*}$.
$\dot{x} = f(t,x,u)$	:	control system, where $\dot{x} := \frac{dx}{dt}$.
$H(t,x,u,\psi)$	:	$= \psi f(t,x,u)$, Pontryagin function or Hamiltonian
H_x, H_u	:	partial derivatives of H with respect to x and u, e.g., $H_x := \frac{\partial H}{\partial x} = \left(\frac{\partial H}{\partial x_1},\dots,\frac{\partial H}{\partial x_n}\right) \in \mathbb{R}^{n*}$, $n = d(x)$.
H_{ux}	:	second partial derivative of H with respect to x and u, $H_{ux} = \begin{pmatrix} \frac{\partial^2 H}{\partial u_1 \partial x_1} & \dots & \frac{\partial^2 H}{\partial u_1 \partial x_n} \\ \dots & \dots & \dots \\ \frac{\partial^2 H}{\partial u_m \partial x_1} & \dots & \frac{\partial^2 H}{\partial u_m \partial x_n} \end{pmatrix}$, $m = d(u), n = d(x)$.
$w^0(t) = (x^0(t), u^0(t))$	:	pair satisfying the constraints of an optimal control problem.
$\delta x(t)$ or $\bar{x}(t)$	:	variation of the state $x^0(t)$.
$\delta u(t)$ or $\bar{u}(t)$	:	variation of the control $u^0(t)$.
$\langle H_{ux}\bar{x}, \bar{u}\rangle$	:	$= \sum_{i=1}^m \sum_{j=1}^n \frac{\partial^2 H}{\partial u_i \partial x_j} \bar{u}_i \bar{x}_j$, $m = d(u), n = d(x)$.

$C([t_0,t_f],\mathbb{R}^n)$	:	space of continuous vector functions $x\colon [t_0,t_f] \to \mathbb{R}^n$ with norm $\|x(\cdot)\|_\infty = \|x(\cdot)\|_C = \max_{t\in[t_0,t_f]} \lvert x(t)\rvert$.
$PC([t_0,t_f],\mathbb{R}^n)$	:	class of piecewise continuous functions $u\colon [t_0,t_f] \to \mathbb{R}^n$.
$\Theta = \{t_1,\dots,t_s\}$	:	set of discontinuity points of $u^0(\cdot) \in PC([t_0,t_f],\mathbb{R}^n)$, $u^{0k-} = u^0(t_k-)$, $u^{0k+} = u^0(t_k+)$, $[u^0]^k = u^{0k+} - u^{0k-}$.
$C^1([t_0,t_f],\mathbb{R}^n)$	:	space of continuously differentiable functions $x\colon [t_0,t_f] \to \mathbb{R}^n$ endowed with the norm $\|x(\cdot)\|_{C^1} = \max\{\|x(\cdot)\|_\infty, \|\dot{x}(\cdot)\|_\infty\}$.
$PC^1([t_0,t_f],\mathbb{R}^n)$	:	class of continuous functions $x\colon [t_0,t_f] \to \mathbb{R}^n$ with a piecewise continuous derivative.
$L^1([t_0,t_f],\mathbb{R}^m)$	:	space of Lebesgue integrable functions $u : [t_0,t_f] \to \mathbb{R}^m$ endowed with the norm $\|u(\cdot)\|_1 = \int_{t_0}^{t_f} \lvert u(t)\rvert\, dt$.
$L^2([t_0,t_f],\mathbb{R}^m)$	:	Hilbert space of square Lebesgue integrable functions $u : [t_0,t_f] \to \mathbb{R}^m$ with the inner product $\langle u(\cdot), v(\cdot)\rangle = \int_{t_0}^{t_f} \langle u(t), v(t)\rangle\, dt$.
$L^\infty([t_0,t_f],\mathbb{R}^m)$	:	space of bounded measurable functions $u : [t_0,t_f] \to \mathbb{R}^m$ endowed with the norm $\|u(\cdot)\|_\infty = \operatorname{ess\,sup}_{t\in[t_0,t_f]} \lvert u(t)\rvert$.
$W^{1,1}([t_0,t_f],\mathbb{R}^n)$	:	space of absolutely continuous functions $x : [t_0,t_f] \to \mathbb{R}^n$ endowed with the norm $\|x(\cdot)\|_{1,1} = \lvert x(t_0)\rvert + \int_{t_0}^{t_f} \lvert\dot{x}(t)\rvert\, dt$.
$W^{1,2}([t_0,t_f],\mathbb{R}^n)$	:	Hilbert space of absolutely continuous functions $x : [t_0,t_f] \to \mathbb{R}^n$ with square integrable derivative and inner product $\langle x(\cdot), y(\cdot)\rangle = \langle x(t_0), y(t_0)\rangle + \int_{t_0}^{t_f} \langle \dot{x}(t), \dot{y}(t)\rangle\, dt$.
$W^{1,\infty}([t_0,t_f],\mathbb{R}^n)$	:	space of Lipschitz continuous functions $x : [t_0,t_f] \to \mathbb{R}^n$ endowed with the norm $\|x(\cdot)\|_{1,\infty} = \lvert x(t_0)\rvert + \|\dot{x}(\cdot)\|_\infty$.
$P_\Theta W^{1,1}([t_0,t_f],\mathbb{R}^{d(x)})$	:	space of piecewise continuous functions $\bar{x} : [t_0,t_f] \mapsto \mathbb{R}^{d(x)}$ that are absolutely continuous on each of the intervals of the set $(t_0,t_f)\setminus\Theta$.
$P_\Theta W^{1,2}([t_0,t_f],\mathbb{R}^{d(x)})$	:	Hilbert space of functions $\bar{x}(\cdot) \in P_\Theta W^{1,1}([t_0,t_f],\mathbb{R}^{d(x)})$ such that the first derivative $\dot{\bar{x}}$ is Lebesgue square integrable; with $[\bar{x}]^k = \bar{x}^{k+} - \bar{x}^{k-} = \bar{x}(t_k+) - \bar{x}(t_k-)$ and inner product $\langle \bar{x}, \bar{y}\rangle = \langle \bar{x}(t_0), \bar{y}(t_0)\rangle + \sum_{k=1}^{s} \langle [\bar{x}]^k, [\bar{y}]^k\rangle + \int_{t_0}^{t_f} \langle \dot{\bar{x}}(t), \dot{\bar{y}}(t)\rangle\, dt$.

Preface

The book is devoted to the theory and application of second-order necessary and sufficient optimality conditions in the Calculus of Variations and Optimal Control. The theory is developed for control problems with ordinary differential equations subject to boundary conditions of equality and inequality type and mixed control-state constraints of equality type. The book exhibits two distinctive features: (a) necessary and sufficient conditions are given in the form of no-gap conditions, and (b) the theory covers broken extremals, where the control has finitely many points of discontinuity. Sufficient conditions for regular controls that satisfy the strict Legendre condition can be checked either via the classical Jacobi condition or through the existence of solutions to an associated Riccati equation.

Particular emphasis is given to the study of bang-bang control problems. Bang-bang controls induce an optimization problem with respect to the switching times of the control. It is shown that the classical second-order sufficient condition for the Induced Optimization Problem (IOP), together with the so-called strict bang-bang property, ensures second-order sufficient conditions (SSC) for the bang-bang control problem. Numerical examples in different areas of application illustrate the verification of SSC for both regular controls and bang-bang controls.

SSC are crucial for exploring the sensitivity analysis of parametric optimal control problems. It is well known in the literature that for regular controls satisfying the strict Legendre condition, SSC allow us to prove the parametric solution differentiability of optimal solutions and to compute parametric sensitivity derivatives. This property has lead to efficient real-time control techniques. Recently, similar results have been obtained for bang-bang controls via SSC for the IOP. Though the discussion of sensitivity analysis and the ensuing real-time control technques are an immediate consequence of the material presented in this book, a systematic treatment of these issues is beyond the scope of this book.

The results of Sections 1.1–1.3 are due to Levitin, Milyutin, and Osmolovskii. The results of Section 6.8 were obtained by Milyutin and Osmolovskii. The results of Sections 2.1–3.4, 5.1, 5.2, 6.1, 6.2, and 6.5 were obtained by Osmolovskii; some important ideas used in these sections are due to Milyutin. The results of Sections 4.1 and 4.2 (except for Section 4.1.5) are due to Lempio and Osmolovskii. The results of Sections 5.3, 6.3, 6.6, and 7.1–7.5 were obtained by Maurer and Osmolovskii. All numerical examples in Sections 4.1, 5.4, 6.4, and Chapter 8 were collected and investigated by Maurer, who is grateful for the numerical assistance provided by Christof Büskens, Laurenz Göllmann, Jang-Ho Robert Kim, and Georg Vossen. Together we solved a lot more bang-bang and singular control problems than could be included in this book. H. Maurer is indebted to Yalçin Kaya for drawing his attention to the arc-parametrization method presented in Section 8.1.2.

Acknowledgments. We are thankful to our colleagues at Moscow State University and University of Münster for their support. A considerable part of the book was written by the first author during his work in Poland (System Research Institute of Polish Academy of Science in Warszawa, Politechnika Radomska in Radom, Siedlce University of Natural Sciences and Humanities), and during his stay in France (Ecole Polytechnique, INRIA Futurs, Palaiseau). We are grateful to J. Frédéric Bonnans from laboratory CMAP (INRIA) at Ecole Polytechnique for his support and for fruitful discussions. Many important ideas used in this book are due to A.A. Milyutin. N.P. Osmolovskii was supported by the grant RFBR 11-01-00795; H. Maurer was supported by the grant MA 691/18 of the Deutsche Forschungsgemeinschaft.

Finally, we wish to thank Elizabeth Greenspan, Lisa Briggeman, and the SIAM compositors for their helpful, patient, and excellent work on our book.

Introduction

By *quadratic conditions*, we mean second-order extremality conditions formulated for a given extremal in the form of positive (semi)definiteness of the corresponding quadratic form. So, in the *simplest problem of the calculus of variations*

$$\int_{t_0}^{t_f} F(t,x,\dot{x})\,dt \to \min, \quad x(t_0)=a, \quad x(t_f)=b,$$

considered on the space C^1 of continuously differentiable vector-valued functions $x(\cdot)$ on the given closed interval $[t_0,t_f]$, the *quadratic form* is as follows:

$$\omega = \int_{t_0}^{t_f} \Big(\langle F_{xx}\bar{x},\bar{x}\rangle + 2\langle F_{\dot{x}x}\bar{x},\dot{\bar{x}}\rangle + \langle F_{\dot{x}\dot{x}}\dot{\bar{x}},\dot{\bar{x}}\rangle\Big)\,dt, \quad \bar{x}(t_0)=\bar{x}(t_f)=0,$$

where the second derivatives $F_{xx}, F_{\dot{x}x}$, and $F_{\dot{x}\dot{x}}$ are calculated along the extremal $x(\cdot)$ that is of interest to us. It is considered in the space $W^{1,2}$ of absolutely continuous functions $\bar{x}$ with square integrable derivative $\dot{\bar{x}}$. By using ω for a given extremal, one formulates a necessary second-order condition for a weak minimum (the positive semidefiniteness of the form), as well as a sufficient second-order condition for a weak minimum (the positive definiteness of the form). As is well known, these quadratic conditions are equivalent (under the strengthened Legendre condition) to the corresponding Jacobi conditions.

The simplest problem of the calculus of variations can also be considered in the space $W^{1,\infty}$ of Lipschitz-continuous functions $x(\cdot)$, and then, in particular, there arises the problem of studying the extremality of broken extremals $x(\cdot)$, i.e., extremals such that the derivative $\dot{x}(\cdot)$ has finitely many points of discontinuity of the first kind. What are second-order conditions for broken extremals, and what is the corresponding quadratic form for them? A detailed study of this problem for the simplest problem was performed by the first author in the book [79], where quadratic necessary and sufficient conditions for the so-called "Pontryagin minimum" (corresponding to L^1-small variations of the control $u=\dot{x}$ under the condition of their uniform L^∞-boundedness) were obtained, and also the relation between the obtained conditions and the conditions for the strong minimum (and also the so-called "Θ-weak" and "bounded strong" minima) was established. For an extremal $x(\cdot)$ with one break at a point t_*, the corresponding form becomes (cf. [39, 85, 86, 107]):

$$\Omega = a\bar{\xi}^2 + 2[F_x]\bar{x}_{\text{av}}\bar{\xi} + \int_{t_0}^{t_f} \Big(\langle F_{xx}\bar{x},\bar{x}\rangle + 2\langle F_{\dot{x}x}\bar{x},\dot{\bar{x}}\rangle + \langle F_{\dot{x}\dot{x}}\dot{\bar{x}},\dot{\bar{x}}\rangle\Big)\,dt,$$

where $\bar{\xi}$ is a numerical parameter, $\bar{x}(\cdot)$ is a function that can have a nonzero jump $[\bar{x}] := \bar{x}(t_*+) - \bar{x}(t_*-)$ at the point t_* and is absolutely continuous on the semiopen intervals

$[t_0,t_*)$ and $(t_*,t_f]$, and the derivative $\dot{\bar{x}}(\cdot)$ is square integrable, and, moreover, the following conditions hold:

$$[\bar{x}] = [\dot{x}]\bar{\xi}, \quad \bar{x}(t_0) = \bar{x}(t_f) = 0.$$

Here, $[F_x]$ and $[\dot{x}]$ denote the *jumps* of the gradient $F_x(t,x(t),\dot{x}(t))$ and the derivative $\dot{x}(t)$ of the extremal at the point t_*, respectively (e.g., $[\dot{x}] = \dot{x}(t_*+) - \dot{x}(t_*-)$); a is the derivative in t of the function

$$F_{\dot{x}}(t,x(t),\dot{x}(t))[\dot{x}] - F(t,x(t),\dot{x}(t_*+)) + F(t,x(t),\dot{x}(t_*-))$$

at the same point (its existence is proved); and $\bar{x}_{\text{av}}$ is the average value of the left-hand and the right-hand values of $\bar{x}$ at t_*, i.e., $\bar{x}_{\text{av}} = \frac{1}{2}(\bar{x}(t_*-) + \bar{x}(t_*+))$. The Weierstrass condition (the minimum principle) implies the inequality $a \geq 0$, which complements the well-known Weierstrass–Erdmann conditions for broken extremals.

In [96] and [97] it was shown (see also [90]) that the problem of the "sign" of the quadratic form Ω can be studied by using methods analogous to the classical methods. The Jacobi conditions and criteria formulated by using the corresponding Riccati equation are extended to the case of a broken extremal. In all these conditions, a new aspect consists only of the fact that the solutions of the corresponding differential equations should have completely certain jumps at the point of break of the extremal. Moreover, it was shown in [97] that, as in the classical case, the quadratic form Ω reduces to a sum of squares solving, by the corresponding Riccati equation satisfying (this is the difference from the classical case), a definite jump condition at the point t_*, which also gives a criterion for positive definiteness of the form Ω and, therefore, a sufficient extremality condition for a given extremal.

In book [79], quadratic extremality conditions for discontinuous controls were also presented in the following problem on a fixed time interval $[t_0,t_f]$:

$$\begin{aligned}
&\mathcal{J}(x(\cdot),u(\cdot)) = J(x(t_0),x(t_f)) \to \min,\\
&F(x(t_0),x(t_f)) \leq 0, \quad K(x(t_0),x(t_f)) = 0,\\
&\dot{x} = f(t,x,u), \quad g(t,x,u) = 0, \quad (t,x,u) \in \mathcal{Q},
\end{aligned}$$

where $\mathcal{Q}$ is an open set, x, u, F, K, and g are vector-valued functions, and J is a scalar-valued function. The functions J, F, K, f, and g belong to the class C^2, and, moreover, the derivative g_u has the maximum rank on the surface $g = 0$ (the nondegeneracy condition for the relation $g = 0$). We seek the minimum among pairs $(x(\cdot),u(\cdot))$ admissible by the constraints such that the function $x(\cdot)$ is absolutely continuous and $u(\cdot)$ is bounded and measurable. This statement corresponds to the general canonical optimal control problem in the Dubovitskii–Milyutin form, but, in contrast to the latter, it is considered on a fixed interval of time, and, which is of special importance, it does not contain pointwise (or *local*, in the Dubovitskii–Milyutin terminology) mixed inequality-type constraints $\varphi(t,x,u) \leq 0$. Precisely, these constraints caused the major difficulties in the study of quadratic conditions [85, 88]. Also, due to the absence of local inequalities, we refer this problem to the calculus of variations (rather than to optimal control) and call it the *general problem of the calculus of variations* (with mixed equality-type constraints, on a fixed time interval). Its statement is close to the Mayer problem, but the existence of endpoint inequality-type constraints determines its specifics.

On the other hand, this problem, even being referred to as the calculus of variations, is sufficiently general and its statement is close to optimal control problems, especially owing to the local relation $g(t,x,u)=0$. In [79], it was shown how, by using quadratic conditions for the general problem of calculus of variations, one can obtain quadratic (necessary and sufficient) conditions in optimal control problems in which the controls enter linearly and the constraint on the control is given in the form of a convex polyhedron under the assumption that the optimal control is piecewise-constant and (outside the switching points) belongs to vertices of the polyhedron (the so-called bang-bang control). To show this, in [79], we first used the property that the set V of vertices of a polyhedron U can be given by a nondegenerate relation $g(u)=0$ on an open set $\mathcal{Q}$ consisting of disjoint open neighborhoods of vertices. This allows us to write quadratic necessary conditions for bang-bang controls. Further, in [79], it was shown that a sufficient minimality condition on V guarantees (when the control enters linearly) the minimum on its convexification $U = \text{co } V$. In this way, the quadratic sufficient conditions were obtained for bang-bang controls.

However, in [79], there is a substantial gap stemming from the fact that, to avoid making the book too long, the authors decided to omit the proofs of quadratic conditions for the general problem of the calculus of variations and restricted themselves to their formulation and the presentation of proofs only for the simplest problem. Although the latter gives the idea of the proofs in the general case, there are no formal proofs of quadratic conditions for the general problem of the calculus of variations in [79]. Part I of the present book is devoted to removing this gap. Therefore, Part I can be considered as a necessary supplement to the book [79]. At the same time, the material contained in Part I is independent and is a complete theory of quadratic conditions for smooth problems of the calculus of variations and optimal control that are covered by the statement presented above.

Part I is organized as follows. First, in Chapter 1, we present a fragment of the abstract theory of higher-order conditions of Levitin, Milyutin, and Osmolovskii [54, 55, 56], more precisely, a modification of this theory for smooth problems on the set of sequences Π determining one or another concept of minimum. In this theory, by a *higher order*, we mean a nonnegative functional determining a growth estimate of the objective function on admissible sequences of variations from Π. The main result of the abstract theory is that for a given class of problems, a given higher order γ, and a given set of sequences Π, one defines a constant C_γ (by using the Lagrange function) such that $C_\gamma \geq 0$ is a necessary minimality condition (corresponding to Π), and $C_\gamma > 0$ is a sufficient condition. The constant C_γ is said to be *basic*. In each concrete class of problems (for given Π and γ), there arises the problem of "decoding" the basic constant. By decoding we mean the simplest method for calculating its sign. We illustrate the decoding of the basic constant by two simple examples, obtaining conditions of the order γ, where γ is a certain quadratic functional.

In Chapter 2, on the basis of the results of Chapter 1, we create a quadratic theory of conditions for Pontryagin minimum in the general problem of the calculus of variations without local mixed constraint $g(t,x,u)=0$. We perform the decoding of the basic constant for the set of the so-called "Pontryagin sequences" and a special higher order γ, which is characteristic for extremals with finitely many discontinuities of the first kind of control. We first estimate the basic constant from above, thus obtaining quadratic necessary conditions for Pontryagin minimum, and then we estimate it from below, thus obtaining sufficient conditions. After that (in Section 2.7), we establish the relation of the obtained sufficient conditions for Pontryagin minimum with conditions for strong minimum.

In Chapter 3, we extend the quadratic conditions obtained in Chapter 2 to the general problem with the local relation $g(t,x,u)=0$ using a special method of projection contained in [79]. Moreover, we extend these conditions to the problem on a variable interval of time using a simple change of time variable. We also formulate, without proofs, quadratic conditions in an optimal control problem with local relations $g(t,x,u)=0$ and $\varphi(t,x,u)\leq 0$. The proofs are set forth in [94, 95].

In Chapter 4, following the results of papers [96] and [97], we derive the tests for positive semidefiniteness and that for positive definiteness of the quadratic form Ω on the critical cone $\mathcal{K}$ (obtained in Chapter 2 for extremals with jumps of the control); these are necessary and sufficient conditions for local minimum, respectively. First, we derive such tests for the simplest problem of the calculus of variations and for extremal with only one corner point. In these tests we exploit the classical Jacobi and Riccati equations, but, as it was said, we use the discontinuous solutions to these equations satisfying specific jump conditions at the corner point of extremal. In the proofs we use the ideas in [25] and [26]. Namely, we consider a one parameter family of the auxiliary minimization problems and reduce the question of the "sign" of our quadratic form on the critical subspace to the condition of the existence of a nonzero point of minimum in the auxiliary problem for a certain value of the parameter. This condition was called in [25] the "passage of quadratic form through zero". Then we obtain a dual test for minimum in the auxiliary problem. As a result we arrive at a generalization of the concept of conjugate point. Such a point is called a Θ-*conjugate*, where Θ is a singleton consisting of one corner point t_* of extremal at hand. This generalization allows us to formulate both necessary and sufficient second-order optimality conditions for broken extremal. Next, we concentrate only on sufficient conditions for positive definiteness of the quadratic form Ω in the auxiliary problem. Following [97], we show that if there exists a solution to the Riccati matrix equation satisfying a definite jump condition, then the quadratic form Ω can be transformed into a perfect square, just as in the classical case. This gives the possibility of proving a sufficient condition for positive definiteness of the quadratic form in the auxiliary problem and thus to obtain one more sufficient condition for optimality of broken extremal. First we prove this result for extremal with one corner point in the simplest problem of the calculus of variations, and then for extremal with finitely many points of discontinuity of the control in the general problem of the calculus of variations.

Part II is devoted to optimal control problems. In Chapter 5, we derive quadratic optimality conditions for optimal control problems with a vector control variable having two components: a continuous unconstrained control appearing nonlinearly in the control system and a bang-bang control appearing linearly and belonging to a convex polyhedron. Such type of control problem arises in many applications. The proofs of quadratic optimality conditions for the mixed continuous-bang case are very similar to the proofs given in [79] for the pure bang-bang case, but some modifications were inevitable. We demonstrate these modifications. In the proofs we use the optimality conditions obtained in Chapter 3 for extremal with jumps of the control in the general problem of calculus of variations. Further, we show that, also for the mixed case, there is a techniques for checking positive definiteness of the quadratic form on the critical cone via a discontinuous solution of an associated Riccati equation with corresponding jump conditions (for this solution) at the points of discontinuity of bang-bang control. This technique is applied to an economic control problem in optimal production and maintenance which was introduced by Cho, Abad, and Parlar [22]. We show that the numerical solution obtained in Maurer, Kim, and

Vossen [67] satisfies the second-order test derived in Chapter 5 while existing sufficiency results fail to hold.

In Chapter 6, we investigate the pure bang-bang case. We obtain second-order necessary and sufficient optimality conditions for this case as a consequence of the conditions obtained in Chapter 5. In the pure bang-bang case, the conditions amount to testing the positive (semi)definiteness of a quadratic form on a *finite-dimensional* critical cone. Nevertheless, the assumptions are appropriate for numerical verification only in some special cases. Therefore, again we study various transformations of the quadratic form and the critical cone which will be tailored to different types of control problems in practice. In particular, by means of a solution to a linear matrix differential equation, the quadratic form can be converted to perfect squares. We demonstrate by practical examples that the obtained conditions can be verified numerically.

We also study second-order optimality conditions for time optimal control problems with control appearing linearly. More specifically, we consider the special case of time optimal bang-bang controls with a given initial and terminal state. We aim at showing that an approach similar to the above-mentioned Riccati equation approach works as well for such problems. Again, the test requires us to find a solution of a linear matrix differential equation which satisfies certain jump conditions at the switching points. We discuss three numerical examples that illustrate the numerical procedure of verifying positive definiteness of the corresponding quadratic forms. Finally, following [79], we study second-order optimality conditions in a simple, but important, class of time optimal control problems for linear systems with constant entries.

Second-order optimality conditions in bang-bang control problems have been derived in the literature in two different forms. The first form was discussed above. The second form belongs to Agrachev, Stefani, and Zezza [1], who first reduce the bang-bang control problem to a finite-dimensional Induced Optimization Problem (IOP) and then show that well-known sufficient optimality conditions for the induced problem supplemented by the strict bang-bang property furnish sufficient conditions for the bang-bang control problem.

In Chapter 7, we establish the equivalence of both forms of sufficient conditions. The proof of this equivalence make extensive use of explicit formulas for first- and second-order derivatives of the trajectory with respect to variations of the optimization variable ζ comprising the switching times, the free initial and final time, and the free initial state. We formulate the IOP with optimization variable ζ which is associated with the bang-bang control problem. We give formulas for the first- and second-order derivatives of trajectories with respect to ζ which follow from elementary properties of ordinary differential equations (ODEs). The formulas are used to establish the explicit relations between the multipliers of Pontryagin's minimum principle and the Lagrange multipliers, critical cones and quadratic forms of the original and IOPs. In our opinion, the resulting formulas seem to have been mostly unknown in the literature. These formulas provide the main technical tools to obtain explicit representations of the second-order derivatives of the Lagrangian. The remarkable fact to be noted here is that by using a suitable transformation, these derivatives are seen to involve only *first-order* variations of the trajectory with respect to ζ. This property facilitates considerably the numerical computation of the Hessian of the Lagrangian. Thus, we arrive at a representation of the quadratic form associated with the Hessian of the Lagrangian.

Finally, Chapter 8 is devoted to numerical methods for solving the IOP and testing the second-order sufficient conditions in Theorem 7.10. After a brief survey on numerical methods for solving optimal control problems, we present in Section 8.1.2 the arc-parametrization method for computing bang-bang controls [44, 45, 66] and its extension to piecewise feedback controls [111, 112, 113]. Arc parametrization can be efficiently implemented using the code NUDOCCCS developed by Büskens [13, 14]. Several numerical examples illustrate the arc-parametrization method and the verification of second-order conditions.

Part I

Second-Order Optimality Conditions for Broken Extremals in the Calculus of Variations

Chapter 1

Abstract Scheme for Obtaining Higher-Order Conditions in Smooth Extremal Problems with Constraints

Here, we present the general theory of higher-order conditions [78] which will be used in what follows in obtaining quadratic optimality conditions in the canonical problem of the calculus of variations. In Section 1.1, we formulate the main result of the general theory in the smooth case. Section 1.2 is devoted to its proof. In Section 1.3, we present two simple applications of the general theory.

1.1 Main Concepts and Main Theorem

1.1.1 Minimum on a Set of Sequences

Let X and Y be Banach spaces. Let a set $\Omega \subset X$, functionals $J : \Omega \to \mathbb{R}^1$, $f_i : \Omega \to \mathbb{R}^1$, $i = 1,\dots,k$, and an operator $g : \Omega \to Y$ be given. Consider the problem

$$J(x) \to \min; \quad f_i(x) \leq 0, \quad i = 1,\dots,k; \quad g(x) = 0; \quad x \in \Omega. \tag{1.1}$$

Let a point $x_0 \in \Omega$ satisfy the constraints, and let us study its optimality. By $\{\delta x_n\}$, and also by $\{\bar{x}_n\}$, we denote countable sequences in X. Denote by Π_0 the set of sequences $\{\bar{x}_n\}$ converging in norm to zero in X. Let us introduce the set of sequences determining the type of the minimum at the point x_0, which will be used. Let Π be an arbitrary set of sequences $\{\delta x_n\}$ satisfying the following conditions:

(a) Π is closed with respect to passing to a subsequence;

(b) $\Pi + \Pi_0 \subset \Pi$; i.e., the conditions $\{\delta x_n\} \in \Pi$ and $\{\bar{x}_n\} \in \Pi_0$ imply $\{\delta x_n + \bar{x}_n\} \in \Pi$ (in this case, we say that Π *sustains a* Π_0*-extension*).

Moreover, it is assumed that the following condition holds for Π, Ω, and x_0:

(c) for any sequence $\{\delta x_n\} \in \Pi$, we have $x_0 + \delta x_n \in \Omega$ for sufficiently large n (in this case, we say that the set Ω is *absorbing for* Π *at the point* x_0).

We give the following definition for problem (1.1).

Definition 1.1. We say that *the minimum is attained at a point* x_0 *on* Π (or x_0 is *a point of* Π*-minimum*) if there is no sequence $\{\delta x_n\} \in \Pi$ such that for all n,

$$J(x_0 + \delta x_n) - J(x_0) < 0, \quad f_i(x_0 + \delta x_n) \leq 0 \; (i = 1,\dots,k), \quad g(x_0 + \delta x_n) = 0.$$

In a similar way, the *strict minimum on* Π *at* x_0 is defined. We need only replace the strict inequality $J(x_0+\delta x_n)-J(x_0)<0$ in the previous definition with nonstrict inequality and require additionally that the sequence $\{\delta x_n\}$ contains nonzero terms.

Obviously, the minimum on Π_0 is a local minimum. If $\Pi_0 \subset \Pi$, then the minimum on Π is not weaker than a local minimum. The inclusion $\Pi_0 \subset \Pi$ holds iff Π contains a zero sequence. This condition holds in all applications of the general theory.

In what follows, the point x_0 is fixed, and, therefore, as a rule, it will be omitted in the definitions and notation. By $\delta\Omega$, we denote the set of variations $\delta x \in X$ such that $x_0+\delta x \in \Omega$. Note that $0 \in \delta\Omega$. We set $f_0(x) = J(x)-J(x_0)$ for $x \in \Omega$. Denote by S the *system* consisting of the functionals $f_0, f_1, \dots, f_k$ and the operator g. The concepts of minimum and strict minimum on Π are naturally extended to the system S. The concepts introduced below can also be related to problem (1.1), as well as to the system S. Sometimes, it is more convenient to speak about the system and not about the problem.

For $\delta x \in \delta\Omega$, we set $m(\delta x) = \max_{0\le i\le k} f_i(x_0+\delta x)$. Let Π_g be the set of sequences $\{\delta x_n\} \in \Pi$ such that $g(x_0+\delta x_n)=0$ for all sufficiently large n. Consider the condition $m \ge 0 \mid \Pi_g$. By definition, this condition means that for any sequence $\{\delta x_n\} \in \Pi_g$, there exists a number starting from which $m(\delta x_n) \ge 0$. In what follows, such a notation will be used without additional explanations. The following proposition follows from the definitions directly.

Proposition 1.2. *If the minimum is attained on* Π, *then* $m \ge 0 \mid \Pi_g$.

Therefore, the condition $m \ge 0 \mid \Pi_g$ is necessary for the minimum on Π. It will serve as a source of other, coarser necessary conditions.

Now let us consider two obvious sufficient conditions. Let Π^+ be the set of all sequences from Π that do not vanish. Define Π_g^+ analogously. The following proposition follows directly from the definitions.

Proposition 1.3. *The condition* $m > 0 \mid \Pi_g^+$ *is equivalent to the strict minimum on* Π.

For $\delta x \in \delta\Omega$, we set $\sigma(\delta x) = \max\{m(\delta x), \| g(x_0+\delta x) \|\}$. We say that σ is the *violation function*. If, in problem (1.1), there is no equality-type constraint $g(x)=0$, then we set $\sigma(\delta x) = m^+(\delta x)$, where $m^+ = \max\{m, 0\}$. The following proposition is elementary.

Proposition 1.4. *Condition* $\sigma > 0 \mid \Pi^+$ *is equivalent to the strict minimum on* Π.

1.1.2 Smooth Problem

Let us formulate the assumptions in problem (1.1) that define it as a smooth problem. These assumptions are related not only to the functionals $J, f_1, \dots, f_k$, the operator g, and the point x_0, but also to the set of sequences Π. First let us give several definitions.

Let Z be a Banach space. A mapping $h : \Omega \to Z$ is said to be Π-*continuous at* x_0 if the condition $\|h(x_0+\delta x_n) - h(x_0)\| \to 0 \quad (n \to \infty)$ holds for any $\{\delta x_n\} \in \Pi$. A Π-continuous mapping is said to be *strictly* Π-*differentiable at* x_0 if there exists a linear operator $H : X \to Z$ such that for any sequences $\{\delta x_n\} \in \Pi$ and $\{\bar{x}_n\} \in \Pi_0$, there exists a sequence $\{z_n\}$ in Z such that $\|z_n\| \to 0$, and for all sufficiently large n, we have the relation

$$h(x_0+\delta x_n+\bar{x}_n)=h(x_0+\delta x_n)+H\bar{x}_n+z_n\|\bar{x}_n\|.$$

The definition easily implies the uniqueness of the operator H. If Π contains the zero sequence (and hence $\Pi_0\subset\Pi$), then H is the Frechét derivative of the operator h at the point x_0, and the strict Π-differentiability implies the strict differentiability. In what follows, we set $H=h'(x_0)$. The function of two variables, $(\delta x,\bar{x})\longmapsto h(x_0+\delta x)+h'(x_0)\bar{x}$, that maps $\delta\Omega\times X$ into Z, is called a *fine linear approximation of the operator h at the point x_0 on Π*. These concepts are used for operators, as well as for functionals.

It is assumed that all functionals J, $f_1,\dots,f_k$ and the operator g in problem (1.1) are Π-continuous at x_0. Introduce the *set of active indices*:

$$I=\{i\in\{0,1,\dots,k\}\bigm| f_i(x_0)=0\},\text{ where } f_0(x)=J(x)-J(x_0).\tag{1.2}$$

Obviously, $0\in I$. It is assumed that the functionals f_i, $i\in I$, and the operator g are strictly Π-differentiable at x_0. Also, it is assumed that either $g'(x_0)X=Y$ (in this case, we say that for g at x_0 on Π, the *Lyusternik condition holds*) or the image $g'(x_0)X$ is closed in Y and has a direct complement which is a closed subspace in Y. Precisely, these assumptions define a *smooth problem (smooth system) on Π at the point x_0*. In this chapter we will consider only this type of problem.

1.1.3 Conditions of Order γ

As usual, by the order of an extremality condition, one means the order of the highest derivative entering this condition. We give another definition of the order. A functional $\gamma:\delta\Omega\to\mathbb{R}^1$ is called an *order on* Π if it is nonnegative on $\delta\Omega$, Π-continuous at zero, and $\gamma(0)=0$. An order γ on Π is said to be *higher* if it is strictly Π-differentiable at zero, and hence $\gamma'(0)=0$. An order γ on Π is said to be *strict* if $\gamma>0\bigm|\Pi^+$.

Let γ be strict higher order on Π. Define the following two conditions on Π: the γ-necessity and the γ-sufficiency. To this end, we set

$$C_\gamma(m,\Pi_g)=\inf_{\Pi_g}\left\{\liminf\frac{m}{\gamma}\right\}.$$

Let us explain that, in calculating this quantity, for each sequence $\{\delta x_n\}\in\Pi_g$ that does not vanish, we first calculate the limit inferior (lim inf) of the ratio $m(\delta x_n)/\gamma(\delta x_n)$ as $n\to\infty$, and then we take the greatest lower bound of the limits inferior over the whole set of sequences from Π_g that do not vanish. An analogous notation will be used for other functions and other sets of sequences. Proposition 1.2 implies the following assertion.

Proposition 1.5. *If we have the minimum on Π, then $C_\gamma(m,\Pi_g)\ge 0$.*

The condition $C_\gamma(m,\Pi_g)\ge 0$ is called the *γ-necessity on Π*. It is easy to see that the γ-necessity on Π is equivalent to the following condition: there are no $\varepsilon>0$ and sequence $\{\delta x_n\}\in\Pi^+$ such that

$$f_i(x_0+\delta x_n)\le-\varepsilon\gamma(\delta x_n)\ (i=0,\dots,k),\quad g(x_0+\delta x_n)=0.$$

It is convenient to compare the concept of γ-necessity in this form with the concept of minimum on Π. Further, we set

$$C_\gamma(\sigma,\Pi) = \inf_{\Pi}\left\{\liminf \frac{\sigma}{\gamma}\right\}.$$

Propositions 1.3 and 1.4 imply the following assertion.

Proposition 1.6. *Each of the two conditions $C_\gamma(m,\Pi_g) > 0$ and $C_\gamma(\sigma,\Pi) > 0$ is sufficient for the strict minimum on Π.*

The condition $C_\gamma(\sigma,\Pi) > 0$ is said to be *γ-sufficiency on* Π. It is equivalent to the following condition: there exists $C > 0$ such that $\sigma \geq C\gamma \mid \Pi^+$. Since, obviously,

$$C_\gamma(m,\Pi_g)^+ = C_\gamma(m^+,\Pi_g) = C_\gamma(\sigma,\Pi_g) \geq C_\gamma(\sigma,\Pi),$$

the inequality $C_\gamma(\sigma,\Pi) > 0$ implies the inequality $C_\gamma(m,\Pi_g) > 0$. Therefore, the inequality $C_\gamma(\sigma,\Pi) > 0$ is not a weaker sufficient condition for the strict minimum on Π than the inequality $C_\gamma(m,\Pi_g) > 0$. In what follows, we will show that if the Lyusternik condition holds, then these two sufficient conditions are equivalent. As the main sufficient condition, we will consider the inequality $C_\gamma(\sigma,\Pi) > 0$, i.e., the γ-sufficiency on Π.

Therefore, the γ-necessity and the γ-sufficiency on Π are obvious weakening and strengthening of the concept of minimum on Π, respectively. We aim at obtaining the criteria for the *γ-conditions*, i.e., the γ-necessity and the γ-sufficiency, that are formulated by using the Lagrange function.

1.1.4 Lagrange Function. Main Result

By $\lambda = (\alpha, y^*)$, we denote an arbitrary tuple of multipliers, where $\alpha = (\alpha_0,\alpha_1,\dots,\alpha_k) \in \mathbb{R}^{k+1}$, $y^* \in Y^*$. Denote by Λ_0 the set of tuples λ such that

$$\begin{aligned} &\alpha_i \geq 0 \quad (i = 0,\dots,k), \quad \sum_{i=0}^{k} \alpha_i + \|y^*\| = 1, \\ &\alpha_i f_i(x_0) = 0 \quad (i = 1,\dots,k), \quad \sum_{i=0}^{k} \alpha_i f_i'(x_0) + y^* g'(x_0) = 0, \end{aligned} \tag{1.3}$$

where $\langle y^* g'(x_0), x\rangle = \langle y^*, g'(x_0)x\rangle$ for all $x \in X$ by definition (here, we prefer not to use the notation $g'(x_0)^*$ for the adjoint operator). Therefore, Λ_0 is the set of normalized tuples of Lagrange multipliers. The relation $\sum \alpha_i + \|y^*\| = 1$ is the *normalization condition* here. Such a normalization is said to be *standard.* Introduce the *Lagrange function*

$$L(\lambda, x) = \sum_{i=0}^{k} \alpha_i f_i(x) + \langle y^*, g(x)\rangle, \quad x \in \Omega,$$

and the functions

$$\begin{aligned} &\Phi(\lambda,\delta x) = L(\lambda, x_0 + \delta x) = \sum_{i=0}^{k} \alpha_i f_i(x_0 + \delta x) + \langle y^*, g(x_0 + \delta x)\rangle, \\ &\Phi_0(\delta x) = \max_{\lambda\in\Lambda_0} \Phi(\lambda,\delta x), \quad \delta x \in \delta\Omega. \end{aligned}$$

Here and in what follows, we set $\max_{\emptyset}(\cdot) = -\infty$.

Denote by $\Pi_{\sigma\gamma}$ the set of sequences $\{\delta x_n\} \in \Pi$ satisfying the condition $\sigma(\delta x_n) \leq O(\gamma(\delta x_n))$. The latter means that there exists $C > 0$ depending on the sequence such that $\sigma(\delta x_n) \leq C\gamma(\delta x_n)$ for all n. We set

$$C_\gamma(\Phi_0, \Pi_{\sigma\gamma}) = \inf_{\Pi_{\sigma\gamma}} \left\{ \liminf \frac{\Phi_0}{\gamma} \right\}.$$

The constant $C_\gamma(\Phi_0, \Pi_{\sigma\gamma})$ is said to be *basic*. It turns out that for an arbitrary higher strict order γ on Π, the constant $C_\gamma(\Phi_0, \Pi_{\sigma\gamma})$ allows us to formulate the following pair of *adjacent conditions* for Π: the inequality $C_\gamma(\Phi_0, \Pi_{\sigma\gamma}) \geq 0$ is necessary for the minimum on Π, and the strict inequality $C_\gamma(\Phi_0, \Pi_{\sigma\gamma}) > 0$ is sufficient for the strict minimum on Π. Moreover, the following assertion holds.

Theorem 1.7. (a) *If $g'(x_0)X = Y$, then the inequality $C_\gamma(\Phi_0, \Pi_{\sigma\gamma}) \geq 0$ is equivalent to the inequality $C_\gamma(m, \Pi_g) \geq 0$. If $g'(x_0)X \neq Y$, then $\Phi_0 \geq 0$, and, therefore, $C_\gamma(\Phi_0, \Pi_{\sigma\gamma}) \geq 0$.*
(b) *The inequality $C_\gamma(\Phi_0, \Pi_{\sigma\gamma}) > 0$ is always equivalent to the inequality $C_\gamma(\sigma, \Pi) > 0$. In the case where $g'(x_0)X = Y$, the following three inequalities are pairwise equivalent to each other*:

$$C_\gamma(\Phi_0, \Pi_{\sigma\gamma}) > 0, \qquad C_\gamma(\sigma, \Pi) > 0, \quad \text{and} \quad C_\gamma(m, \Pi_g) > 0.$$

Therefore, the γ-necessity on Π always implies the inequality $C_\gamma(\Phi_0, \Pi_{\sigma\gamma}) \geq 0$, and the γ-sufficiency on Π is always equivalent to the inequality $C_\gamma(\Phi_0, \Pi_{\sigma\gamma}) > 0$. Theorem 1.7 is the main result of the abstract theory of higher-order conditions for smooth problems. Note that it remains valid if, in the definition of the set Λ_0, we replace the *standard normalization* $\sum \alpha_i + \|y^*\| = 1$ by any equivalent normalization. Let us make more precise what we mean by an equivalent normalization.

1.1.5 Equivalent Normalizations

Let $\nu(\lambda)$ be a positively homogeneous function of the first degree. A normalization $\nu(\lambda) = 1$ is said to be *equivalent* to the standard normalization if the condition that Λ_0 is nonempty implies the inequalities

$$0 < \inf_{\Lambda_0} \nu(\lambda) \leq \sup_{\Lambda_0} \nu(\lambda) < +\infty.$$

The following assertion holds.

Proposition 1.8. *Let $g'(x_0)X = Y$. Then the condition $\sum_{i=0}^k \alpha_i = 1$ defines an equivalent normalization.*

Proof. Let Λ_0 be nonempty, and let $\lambda \in \Lambda_0$. Then $\sum_I \alpha_i f_i'(x_0) + y^* g'(x_0) = 0$, which implies $\|y^* g'(x_0)\| \leq (\sum \alpha_i) \max_I \|f_i'(x_0)\|$. Therefore,

$$\sum \alpha_i \leq \sum \alpha_i + \|y^* g'(x_0)\| \leq \left(\sum \alpha_i\right)\left(1 + \max_I \|f_i'(x_0)\|\right).$$

It remains to note that $\|y^* g'(x_0)\||$ and $\|y^*\|$ are two equivalent norms on Y^*, since $g'(x_0)X = Y$. ∎

Therefore, in the case where the Lyusternik condition $g'(x_0)X = Y$ holds, we can use the normalization $\sum \alpha_i = 1$ in the definition of Λ_0. This normalization is called the *Lyusternik normalization*. We need to compare the functions Φ_0 for the standard and Lyusternik normalizations. Therefore, in the case of Lyusternik normalization, let us agree to equip the set Λ_0 and the function Φ_0 with the subscript L; i.e., we write Λ_0^L and Φ_0^L. The following assertion holds.

Proposition 1.9. *Let $g'(x_0)X = Y$. Then there exists a number a, $0 < a \le 1$, such that*

$$\Phi_0 \le \max\{a\Phi_0^L, \Phi_0^L\}, \tag{1.4}$$

$$\Phi_0^L \le \max\left\{\Phi_0, \frac{1}{a}\Phi_0\right\}. \tag{1.5}$$

Proof. We first prove inequality (1.4). If Λ_0 is empty, then $\Phi_0 = -\infty$, and hence inequality (1.4) holds. Suppose that Λ_0 is not empty. By Proposition 1.8, there exists $a, 0 < a \le 1$, such that for any $\lambda \in \Lambda_0$, the inequality $a \le \sum \alpha_i$ holds. Moreover, the condition $\sum \alpha_i + \|y^*\| = 1$ implies $\sum \alpha_i \le 1$. Let $\lambda = (\alpha, y^*) \in \Lambda_0$. We set $\nu := \sum \alpha_i$. Then $\hat{\lambda} := \lambda/\nu \in \Lambda_0^L$, $a \le \nu \le 1$. Therefore, for any $\delta x \in \delta\Omega$,

$$\Phi(\lambda, \delta x) = \Phi(\nu\hat{\lambda}, \delta x) = \nu\Phi(\hat{\lambda}, \delta x) \le \max\{a\Phi_0^L(\delta x), \Phi_0^L(\delta x)\}.$$

This implies estimate (1.4).

Now let us prove (1.5). If Λ_0^L is empty, then $\Phi_0^L = -\infty$, and hence (1.5) holds. Now let Λ_0^L be nonempty, and let $\hat{\lambda} = (\hat{\alpha}, \hat{y}^*) \in \Lambda_0^L$. We set $\mu := 1 + \|\hat{y}^*\|$, $\lambda = (\alpha, y^*) := \hat{\lambda}/\mu$. Then $\lambda \in \Lambda_0$. Moreover, $a \le \sum \alpha_i = \frac{1}{\mu}\sum \hat{\alpha}_i = \frac{1}{\mu} \le 1$, which implies $1 \le \mu \le 1/a$. Therefore, for any $\delta x \in \delta\Omega$, we have

$$\Phi(\hat{\lambda}, \delta x) = \mu\Phi(\lambda, \delta x) \le \max\left\{\Phi_0(\delta x), \frac{1}{a}\Phi_0(\delta x)\right\}.$$

This implies estimate (1.5). ∎

Proposition 1.9 immediately implies the following assertion.

Proposition 1.10. *Let $g'(x_0)X = Y$. Then there exists a number a, $0 < a \le 1$, such that*

$$C_\gamma(\Phi_0, \Pi_{\sigma\gamma}) \le \max\{aC_\gamma(\Phi_0^L, \Pi_{\sigma\gamma}), C_\gamma(\Phi_0^L, \Pi_{\sigma\gamma})\}, \tag{1.6}$$

$$C_\gamma(\Phi_0^L, \Pi_{\sigma\gamma}) \le \max\{C_\gamma(\Phi_0, \Pi_{\sigma\gamma}), \frac{1}{a}C_\gamma(\Phi_0, \Pi_{\sigma\gamma})\}. \tag{1.7}$$

Therefore, if the Lyusternik condition holds, constants $C_\gamma(\Phi_0, \Pi_{\sigma\gamma})$ and $C_\gamma(\Phi_0^L, \Pi_{\sigma\gamma})$ have the same signs, which in this case allows us to replace the first constant by the second in Theorem 1.7.

1.1.6 Sufficient Conditions

As was already noted, the inequalities $C_\gamma(\Phi_0, \Pi_{\sigma\gamma}) \geq 0$ and $C_\gamma(\Phi_0, \Pi_{\sigma\gamma}) > 0$ are a pair of adjacent conditions for a minimum on Π at the point x_0. The nonstrict inequality is a necessary condition, and the strict inequality is a sufficient condition. This is implied by Theorem 1.7. The necessary condition is not trivial as we will verify below in proving Theorem 1.7. As for the sufficient condition $C_\gamma(\Phi_0, \Pi_{\sigma\gamma}) > 0$, its sufficiency for the minimum on Π is simple (this is characteristic for sufficient conditions in general: their sources are simple as a rule). Let us prove the sufficiency of this condition. The following estimate easily follows from the definitions of the functions Φ_0 and σ: $\Phi_0 \leq \sigma$. Hence

$$C_\gamma(\Phi_0, \Pi_{\sigma\gamma}) \leq C_\gamma(\sigma, \Pi_{\sigma\gamma}). \tag{1.8}$$

Let us show that

$$C_\gamma(\sigma, \Pi_{\sigma\gamma}) = C_\gamma(\sigma, \Pi). \tag{1.9}$$

Indeed, the inclusion $\Pi_{\sigma\gamma} \subset \Pi$ implies the inequality $C_\gamma(\sigma, \Pi_{\sigma\gamma}) \geq C_\gamma(\sigma, \Pi)$. Let us prove the converse inequality

$$C_\gamma(\sigma, \Pi_{\sigma\gamma}) \leq C_\gamma(\sigma, \Pi). \tag{1.10}$$

If $C_\gamma(\sigma, \Pi) = \infty$, then inequality (1.10) holds. Let $C_\gamma(\sigma, \Pi) < \infty$, and let a number C be such that $C_\gamma(\sigma, \Pi) < C$, i.e.,

$$\inf_{\Pi} \liminf \frac{\sigma}{\gamma} < C.$$

Then there exists a sequence $\{\delta x_n\} \in \Pi^+$ such that $\sigma(\delta x_n)/\gamma(\delta x_n) < C$ for all n. This implies $\{\delta x_n\} \in \Pi_{\sigma\gamma}$ and

$$C_\gamma(\sigma, \Pi_{\sigma\gamma}) := \inf_{\Pi_{\sigma\gamma}} \liminf \frac{\sigma}{\gamma} \leq C.$$

We have shown that the inequality $C_\gamma(\sigma, \Pi) < C$ always implies inequality $C_\gamma(\sigma, \Pi_{\sigma\gamma}) \leq C$. This implies inequality (1.10) and, therefore, relation (1.9). From (1.8) and (1.9) we obtain

$$C_\gamma(\Phi_0, \Pi_{\sigma\gamma}) \leq C_\gamma(\sigma, \Pi). \tag{1.11}$$

Thus, the inequality $C_\gamma(\Phi_0, \Pi_{\sigma\gamma}) > 0$ implies the inequality $C_\gamma(\sigma, \Pi) > 0$, i.e., the γ-sufficiency on Π. Therefore, the inequality $C_\gamma(\Phi_0, \Pi_{\sigma\gamma}) > 0$ is sufficient for the strict minimum on Π.

The latter assertion contained in Theorem 1.7 turns out to be very simple. However, a complete proof of Theorem 1.7 requires considerably greater effort. Before passing directly to its proof, we present all necessary auxiliary assertions. The next section is devoted to this.

1.2 Proof of the Main Theorem

We will need the main lemma for the proof of Theorem 1.7. In turn, the proof of the main lemma is based on the following three important properties used in the extremum theory: the compatibility condition of a set of linear inequalities and equations, the Hoffman lemma, and the Lyusternik theorem. These three properties compose the basis of the abstract theory of higher order for smooth problems, and for the reader's convenience they are formulated in this section.

1.2.1 Basis of the Abstract Theory

As above, let X and Y be Banach spaces, and let X^* and Y^* be their duals. Let a tuple $l = \{l_1, \ldots, l_m\}$, $l_i \in X^*$, $i = 1, \ldots, m$, and a linear surjective operator $A : X \xrightarrow{\text{on}} Y$ be given. With the tuple l and the operator A we associate the set $\Lambda = \Lambda(l, A)$ consisting of the tuples of multipliers $\lambda = (\alpha, y^*)$, $\alpha = (\alpha_1, \ldots, \alpha_m) \in \mathbb{R}^{m*}$, $y^* \in Y^*$, satisfying the conditions

$$\alpha_i \geq 0 \ (i = 1, \ldots, m), \quad \sum_{i=1}^{m} \alpha_i = 1, \quad \sum_{i=1}^{m} \alpha_i l_i + y^* A = 0.$$

The compatibility criterion of a set of linear inequalities and equations has the following form.

Lemma 1.11. *Let $\xi = (\xi_1, \ldots, \xi_m)^* \in \mathbb{R}^m$, $y \in Y$. The set of conditions*

$$\langle l_i, x \rangle + \xi_i < 0 \quad (i = 1, \ldots, m), \qquad Ax + y = 0$$

is compatible iff

$$\sup_{\lambda \in \Lambda} \left\{ \sum_{i=1}^{m} \alpha_i \xi_i + \langle y^*, y \rangle \right\} < 0.$$

(By definition, $\sup_{\emptyset} = -\infty$.) Along with this lemma, in studying problems with inequality-type constraints, an important role is played by the estimate of the distance to the set of solutions of a set of linear inequalities and equations [41], which is presented below.

Lemma 1.12 (Hoffman). *There exists a constant $C = C(l, A)$ with the following property: if for certain $\xi \in \mathbb{R}^m$ and $y \in Y$ the system*

$$\langle l_i, x \rangle + \xi_i \leq 0 \quad (i = 1, \ldots, m), \quad Ax + y = 0$$

is compatible, then there exists its solution x satisfying the estimate

$$\|x\| \leq C \max\{\xi_1, \ldots, \xi_m, \|y\|\}.$$

In the case where there is no equation $Ax + y = 0$, Lemma 1.12 holds with the estimate $\|x\| \leq C \max\{\xi_1^+, \ldots, \xi_m^+\}$.

Finally, we present the Lyusternik-type theorem on the estimate of the distance to the level of the equality operator in the form which is convenient for us (see [78, Theorem 2]). Let a set of sequences Π satisfy the same conditions as in Section 1.1. The following theorem holds for an operator $g : X \to Y$ strictly Π-differentiable at a point x_0.

Theorem 1.13. *Let $g(x_0) = 0$ and let $g'(x_0)X = Y$. Then there exists $C > 0$ such that for any $\{\delta x_n\} \in \Pi$, there exists $\{\bar{x}_n\} \in \Pi_0$ satisfying the following conditions for all sufficiently large n: $g(x_0 + \delta x_n + \bar{x}_n) = 0$, $\|\bar{x}_n\| \leq C \|g(x_0 + \delta x_n)\|$.*

1.2.2 Main Lemma

We now turn to problem (1.1). Let all assumptions of Section 1.1 hold. Let $g'(x_0)X = Y$. In the definition of the set Λ_0, we choose the normalization $\sum \alpha_i = 1$. According to Proposition 1.8, it is equivalent to the standard normalization. The following assertion holds.

Lemma 1.14 (Main Lemma). *Let a sequence $\{\delta x_n\}$ and a sequence of numbers $\{\zeta_n\}$ be such that $\delta x_n \in \delta\Omega$ for all n, $\zeta_n^+ \to 0$, and $\Phi_0^L(\delta x_n) + \zeta_n < 0$ for all n. Then there exists a sequence $\{\bar{x}_n\} \in \Pi_0$ such that the following conditions hold:*

(1) $\|\bar{x}_n\| \le O(\sigma(\delta x_n) + \zeta_n^+)$;

(2) $f_i(x_0 + \delta x_n + \bar{x}_n) + \zeta_n \le o(\|\bar{x}_n\|)$, $i \in I$;

(3) $g(x_0 + \delta x_n + \bar{x}_n) = 0$ for all sufficiently large n.

Proof. For an arbitrary n, let us consider the following set of conditions on $\bar{x}$:

$$\langle f_i'(x_0), \bar{x}\rangle + f_i(x_0 + \delta x_n) + \zeta_n < 0,\ i \in I; \quad g'(x_0)\bar{x} + g(x_0 + \delta x_n) = 0. \tag{1.12}$$

Let $(\alpha_i)_{i\in I}$ and y^* be a tuple from the set Λ of system (1.12). We set $\alpha_i = 0$ for $i \notin I$. Then $\lambda = (\alpha_0, \ldots, \alpha_k, y^*) \in \Lambda_0^L$. The converse is also true: if $\lambda = (\alpha_0, \ldots, \alpha_k, y^*) \in \Lambda_0^L$, then $\alpha_i = 0$ for $i \notin I$ and the tuple $((\alpha_i)_{i\in I}, y^*)$ belong to the set Λ of system (1.12). Therefore,

$$\begin{aligned} &\max_{\Lambda} \left\{ \sum_I \alpha_i (f_i(x_0 + \delta x_n) + \zeta_n) + \langle y^*, g(x_0 + \delta x_n)\rangle \right\} \\ &= \max_{\Lambda_0^L} \left\{ \sum_{i=0}^k \alpha_i (f_i(x_0 + \delta x_n) + \zeta_n) + \langle y^*, g(x_0 + \delta x_n)\rangle \right\} \\ &= \Phi_0^L(\delta x_n) + \zeta_n < 0. \end{aligned}$$

The latter relation is implied by the definition of the function Φ_0^L and the normalization $\Sigma\alpha_i = 1$. According to Lemma 1.11, system (1.12) is compatible. Then by Hoffman's lemma (Lemma 1.12), there exist $C > 0$ and a sequence $\{\bar{x}_n'\}$ such that for all n,

$$\langle f_i'(x_0), \bar{x}_n'\rangle + f_i(x_0 + \delta x_n) + \zeta_n \le 0, \quad i \in I; \tag{1.13}$$

$$g'(x_0)\bar{x}_n' + g(x_0 + \delta x_n) = 0; \tag{1.14}$$

$$\|\bar{x}_n'\| \le C \max\left\{ \max_{i\in I}\{f_i(x_0 + \delta x_n) + \zeta_n\}, \|g(x_0 + \delta x_n)\| \right\}. \tag{1.15}$$

It follows from (1.15) that for all sufficiently large n,

$$\|\bar{x}_n'\| \le C(\sigma(\delta x_n) + \zeta_n^+) \to 0, \tag{1.16}$$

since $f_i(x_0 + \delta x_n) \to f_i(x_0) < 0$ for $i \notin I$. Therefore, $\{\bar{x}_n'\} \in \Pi_0$. Since g is strictly Π-differentiable at the point x_0, condition (1.14) implies $g(x_0 + \delta x_n + \bar{x}_n') = o(\|\bar{x}_n'\|)$. Then by the Lyusternik theorem (see Theorem 1.13), there exists $\{\bar{x}_n''\} \in \Pi_0$ such that for all sufficiently large n, we have

$$g(x_0 + \delta x_n + \bar{x}_n' + \bar{x}_n'') = 0, \tag{1.17}$$

$$\|\bar{x}_n''\| = o(\|\bar{x}_n'\|). \tag{1.18}$$

We set $\{\bar{x}_n\} = \{\bar{x}'_n + \bar{x}''_n\}$. Condition (1.17) implies $g(x_0 + \delta x_n + \bar{x}_n) = 0$ for all sufficiently large n, and conditions (1.16) and (1.18) imply $\|\bar{x}_n\| \le O(\sigma(\delta x_n) + \zeta_n^+)$. Further, we obtain from conditions (1.13) and the property of strong Π-differentiability of the functionals f_i at the point x_0 that

$$\begin{aligned} & f_i(x_0 + \delta x_n + \bar{x}_n) + \zeta_n \\ &= f_i(x_0 + \delta x_n) + \langle f_i'(x_0), \bar{x}'_n \rangle + \langle f_i'(x_0), \bar{x}''_n \rangle + \zeta_n + o(\|\bar{x}_n\|) \\ &\le \langle f_i'(x_0), \bar{x}''_n \rangle + o(\|\bar{x}_n\|) = o_1(\|\bar{x}_n\|), \quad i \in I. \end{aligned}$$

The latter relation holds because of (1.18). The lemma is proved. ∎

Now, we prove a number of assertions from which the main result (Theorem 1.7) will follow. Below we assume that the order γ is strict and higher on Π, and all assumptions of Section 1.1 hold for the set of sequences Π and problem (1.1) at the point x_0.

1.2.3 Case Where the Lyusternik Condition Holds

We have the following theorem.

Theorem 1.15. *Let $g'(x_0)X = Y$. Then $C_\gamma(\Phi_0^L, \Pi_{\sigma\gamma}) = C_\gamma(m, \Pi_g)$.*

Proof. We first show that $C_\gamma(\Phi_0^L, \Pi_{\sigma\gamma}) \le C_\gamma(m, \Pi_g)$. Indeed,

$$C_\gamma(\Phi_0^L, \Pi_{\sigma\gamma}) \le C_\gamma(\Phi_0^L, \Pi_{\sigma\gamma} \cap \Pi_g) \le C_\gamma(m, \Pi_{\sigma\gamma} \cap \Pi_g) = C_\gamma(m, \Pi_g).$$

Here, the first inequality is obvious, and the second inequality follows from the obvious estimate $\Phi_0^L \le m \mid \Pi_g$. The equality is proved in the same way as relation (1.9).

Now let us prove inequality $C_\gamma(m, \Pi_g) \le C_\gamma(\Phi_0^L, \Pi_{\sigma\gamma})$, which will finish the proof of the theorem. If $C_\gamma(\Phi_0^L, \Pi_{\sigma\gamma}) = +\infty$, the inequality holds. Let $C_\gamma(\Phi_0^L, \Pi_{\sigma\gamma}) < +\infty$, and let C be such that

$$C_\gamma(\Phi_0^L, \Pi_{\sigma\gamma}) := \inf_{\Pi_{\sigma\gamma}} \liminf \frac{\Phi_0^L}{\gamma} < -C.$$

Then there exists a sequence $\{\delta x_n\} \in \Pi_{\sigma\gamma}^+$ at which $\Phi_0^L(\delta x_n) + C\gamma(\delta x_n) < 0$, and, moreover, $\delta x_n \in \delta\Omega$ for all n. We set $\zeta_n = C\gamma(\delta x_n)$. According to the main lemma, there exists a sequence $\{\bar{x}_n\}$ such that the following conditions hold:

(α) $\|\bar{x}_n\| \le O(\sigma(\delta x_n) + C^+\gamma(\delta x_n))$;

(β) $f_i(x_0 + \delta x_n + \bar{x}_n) + C\gamma(\delta x_n) \le o(\|\bar{x}_n\|)$, $i \in I$;

(γ) $g(x_0 + \delta x_n + \bar{x}_n) = 0$ for all sufficiently large n.

Since $\{\delta x_n\} \in \Pi_{\sigma\gamma}$, the first condition implies $\|\bar{x}_n\| \le O(\gamma(\delta x_n))$. We set $\{\delta x'_n\} = \{\delta x_n + \bar{x}_n\}$. Then condition ($\gamma$) implies $\{\delta x'_n\} \in \Pi_g$, and condition ($\beta$) implies

$$f_i(x_0 + \delta x'_n) + C\gamma(\delta x_n) \le o(\gamma(\delta x_n)), \quad i \in I.$$

From this we obtain

$$\liminf \frac{m(\delta x'_n)}{\gamma(\delta x_n)} \le -C.$$

Since γ is a higher order, we have

$$\gamma(\delta x'_n) = \gamma(\delta x_n + \bar{x}_n) = \gamma(\delta x_n) + o(\|\bar{x}_n\|) = \gamma(\delta x_n) + o(\gamma(\delta x_n)).$$

Therefore,

$$\liminf \frac{m(\delta x'_n)}{\gamma(\delta x'_n)} \le -C.$$

Taking into account that $\{\delta x'_n\} \in \Pi_g$, we obtain from this that $C_\gamma(m, \Pi_g) \le -C$. Therefore, we have proved that the inequality $C_\gamma(\Phi_0^L, \Pi_{\sigma\gamma}) < -C$ always implies the inequality $C_\gamma(m, \Pi_g) \le -C$. Therefore, $C_\gamma(m, \Pi_g) \le C_\gamma(\Phi_0^L, \Pi_{\sigma\gamma})$. The theorem is completely proved. ∎

Theorem 1.15 and Proposition 1.10 imply the following theorem.

Theorem 1.16. *Let $g'(x_0)X = Y$. Then the following three inequalities are pairwise equivalent*:

$$C_\gamma(m, \Pi_g) \ge 0, \quad C_\gamma(\Phi_0^L, \Pi_{\sigma\gamma}) \ge 0, \quad \textit{and} \quad C_\gamma(\Phi_0, \Pi_{\sigma\gamma}) \ge 0.$$

Now, consider the sequence of relations

$$C_\gamma(\Phi_0, \Pi_{\sigma\gamma}) \le C_\gamma(\sigma, \Pi) \le C_\gamma(\sigma, \Pi_g) = C_\gamma(m^+, \Pi_g) = C_\gamma(m, \Pi_g)^+. \tag{1.19}$$

The first of these relations was proved in Section 1.1 (inequality (1.11)), and the other relations are obvious. The following assertion follows from (1.19), Theorem 1.15, and inequality (1.7).

Corollary 1.17. *Let $g'(x_0)X = Y$. Then for $0 < a \le 1$ the following inequalities hold*:

$$C_\gamma(\Phi_0, \Pi_{\sigma\gamma}) \le C_\gamma(\sigma, \Pi) \le C_\gamma(m, \Pi_g)^+ \le C_\gamma(\Phi_0^L, \Pi_{\sigma\gamma})^+ \le \frac{1}{a} C_\gamma(\Phi_0, \Pi_{\sigma\gamma})^+.$$

This implies the following theorem.

Theorem 1.18. *Let $g'(x_0)X = Y$. Then the following four inequalities are pairwise equivalent*:

$$C_\gamma(\sigma, \Pi) > 0, \quad C_\gamma(m, \Pi_g) > 0, \quad C_\gamma(\Phi_0, \Pi_{\sigma\gamma}) > 0, \quad \textit{and} \quad C_\gamma(\Phi_0^L, \Pi_{\sigma\gamma}) > 0.$$

Therefore, in the case where the Lyusternik condition holds, we have proved all the assertions of Theorem 1.7.

1.2.4 Case Where the Lyusternik Condition Is Violated

To complete the proof of Theorem 1.7, we need to prove the following: if $g'(x_0)X \ne Y$, then (a) $\Phi_0 \ge 0$, and (b) the inequality $C_\gamma(\sigma, \Pi) > 0$ is equivalent to the inequality $C_\gamma(\Phi_0, \Pi_{\sigma\gamma}) > 0$. We begin with the proof of (a).

Proposition 1.19. *If $g'(x_0)X \ne Y$, then $\Phi_0(\delta x) \ge 0$ for all $\delta x \in \delta\Omega$.*

Proof. Since the image $Y_1 := g'(x_0)X$ is closed in Y, the condition $Y_1 \neq Y$ implies the existence of $y^* \in Y$, $\|y^*\| = 1$ such that $\langle y^*, y\rangle = 0$ for all $y \in Y_1$, and hence $\lambda' = (0, y^*) \in \Lambda_0$ and $\lambda'' = (0, -y^*) \in \Lambda_0$. From this we obtain that for any $\delta x \in \delta\Omega$,

$$\max_{\Lambda_0} \Phi(\lambda, \delta x) \geq \max\{\Phi(\lambda', \delta x), \Phi(\lambda'', \delta x)\} = |\langle y^*, g(x_0 + \delta x)\rangle| \geq 0.$$

The proposition is proved. ∎

Now let us prove assertion (b). Since the inequality $C_\gamma(\Phi_0, \Pi_{\sigma\gamma}) \leq C_\gamma(\sigma, \Pi)$ always holds by (1.11), in order to prove (b), we need to prove the following lemma.

Lemma 1.20. *Let $g'(x_0)X \neq Y$. Then there exists a constant $b = b(g'(x_0)) > 0$ such that*

$$C_\gamma(\sigma, \Pi) \leq b\, C_\gamma(\Phi_0, \Pi_{\sigma\gamma})^+. \tag{1.20}$$

Proof. The proof uses a special method for passing from the system $S = \{f_0, \dots, f_k, g\}$ to a certain auxiliary system $\hat{S}$. We set $Y_1 = g'(x_0)X$. According to the definition of the smooth problem, $Y = Y_1 \oplus Y_2$, where Y_2 is a closed subspace in Y. Then $Y^* = W_1 \oplus W_2$, where W_1 and W_2 are such that any functional from W_1 is annihilated on Y_2, and any functional from W_2 is annihilated on Y_1. Without loss of generality, we assume that if $y = y_1 + y_2$, $y_1 \in Y_1$, and $y_2 \in Y_2$, then $\|y\| = \max\{\|y_1\|, \|y_2\|\}$. Then for $y^* = y_1^* + y_2^*, y_1^* \in W_1, y_2^* \in W_2$, we have $\|y^*\| = \|y_1^*\| + \|y_2^*\|$.

Let $P_1 : Y \to Y_1$ and $P_2 : Y \to Y_2$ be projections compatible with the decomposition of Y into a direct sum $Y = Y_1 \oplus Y_2$. Then $P_1 + P_2 = I$, $P_1 P_2 = 0$, and $P_2 P_1 = 0$. We set $g_1 = P_1 g$ and $g_2 = P_2 g$. Then $g = g_1 + g_2$, $g_1'(x_0)X = Y_1$, and $g_2'(x_0)X = \{0\}$. Introduce the functional $f_g(x) = \|g_2(x)\|$. The condition $g_2'(x_0)X = \{0\}$ implies that f_g is strictly Π-differentiable at the point x_0 and $f_g'(x_0) = 0$. Consider the system $\hat{S}$ consisting of the functionals $f_0, \dots, f_k, f_g$ and the operator g_1. All subjects related to this system will be endowed with the sign $\wedge$. Since $g = g_1 + g_2$, we have $\|g\| = \max\{\|g_1\|, \|g_2\|\}$. Therefore,

$$\begin{aligned}\sigma(\delta x) &:= \max\{f_0(x_0 + \delta x), \dots, f_k(x + \delta x), \|g(x_0 + \delta x)\|\} \\ &= \max\{f_0(x_0 + \delta x), \dots, f_k(x_0 + \delta x), f_g(x_0 + \delta x), \|g_1(x_0 + \delta x)\|\} =: \hat{\sigma}(\delta x).\end{aligned}$$

This implies

$$C_\gamma(\sigma, \Pi) = C_\gamma(\hat{\sigma}, \Pi). \tag{1.21}$$

Further, since the Lyusternik condition $g_1(x_0)X = Y_1$ holds for the system $\hat{S}$, by Corollary 1.17, there exists $\hat{a} > 0$ such that

$$C_\gamma(\hat{\sigma}, \Pi) \leq \frac{1}{\hat{a}} C_\gamma(\hat{\Phi}_0, \Pi_{\hat{\sigma}\gamma})^+. \tag{1.22}$$

Now let us show that $\hat{\Phi}_0 \leq \Phi_0$. If $\hat{\Lambda}_0$ is empty, then $\hat{\Phi}_0 = -\infty$, and hence the inequality holds. Let $\hat{\Lambda}_0$ be nonempty, and let $\hat{\lambda} = (\alpha_0, \dots, \alpha_k, \alpha_g, y_1^*)$ be an arbitrary element of the set $\hat{\Lambda}_0$. Then

$$\alpha_i \geq 0 \ (i = 0, \dots, k), \qquad \alpha_i f_i(x_0) = 0 \ (i = 1, \dots k), \quad \alpha_g \geq 0; \quad y_1^* \in W_1,$$
$$\sum_{i=0}^{k} \alpha_i + \alpha_g + \|y_1^*\| = 1, \quad \sum_{i=0}^{k} \alpha_i f_i'(x_0) + \alpha_g f_g'(x_0) + y_1^* g_1'(x_0) = 0.$$

Moreover, $f_g'(x_0) = 0$. Let $\delta x \in \delta\Omega$ be an arbitrary element. Choose $y_2^* \in W_2$ so that the following conditions hold:

$$\|y_2^*\| = \alpha_g, \quad \langle y_2^*, g_2(x_0+\delta x)\rangle = \alpha_g \|g_2(x_0+\delta x)\|.$$

We set $y^* = y_1^* + y_2^*$, and $\lambda = (\alpha_0, \dots, \alpha_k, y^*)$. As is easily seen, then we have $\lambda \in \Lambda_0$ and $\hat{\Phi}(\hat{\lambda}, \delta x) = \Phi(\lambda, \delta x)$. Therefore, for arbitrary δx and $\hat{\lambda} \in \hat{\Lambda}_0$, there exists $\lambda \in \Lambda_0$ such that the indicated relation holds. This implies $\hat{\Phi}_0(\delta x) \leq \Phi_0(\delta x)$. Also, taking into account that $\hat{\sigma} = \sigma$, we obtain

$$C_\gamma(\hat{\Phi}_0, \Pi_{\hat{\sigma}\gamma}) \leq C_\gamma(\Phi_0, \Pi_{\sigma\gamma}). \tag{1.23}$$

It follows from (1.21)–(1.23) that

$$C_\gamma(\sigma, \Pi) \leq \frac{1}{\hat{a}} C_\gamma(\Phi_0, \Pi_{\sigma\gamma})^+. \tag{1.24}$$

It remains to set $b = 1/\hat{a}$. The lemma is proved. ∎

Therefore, we have shown that in the case of violation of the Lyusternik condition, the inequalities $C_\gamma(\sigma, \Pi) > 0$ and $C_\gamma(\Phi_0, \Pi_{\sigma\gamma}) > 0$ are equivalent. Thus, we have completed the proof of Theorem 1.7.

1.3 Simple Applications of the Abstract Scheme

In this section, following [55], we shall obtain quadratic conditions in a smooth problem in $\mathbb{R}^n$ and in the problem of Bliss with endpoint inequalities.

1.3.1 A Smooth Problem in $\mathbb{R}^n$

Let $X = \mathbb{R}^n$, $Y = \mathbb{R}^m$, and $\Omega = X$. Consider the problem

$$J(x) \to \min; \quad f_i(x) \leq 0 \quad (i = 1, \dots, k), \quad g(x) = 0. \tag{1.25}$$

We assume that the functions $J : \mathbb{R}^n \to \mathbb{R}^1$, $f_i : \mathbb{R}^n \to \mathbb{R}^1$, $i = 1, \dots, k$, and the operator $g : \mathbb{R}^n \to \mathbb{R}^m$ are twice differentiable at each point. Let a point x_0 satisfy the constraints, and let us study its optimality. We define f_0 and I as in relation (1.2). Let $\Pi = \Pi_0 := \{\{\delta x_n\} \mid \delta x_n \to 0 \ (n \to \infty)\}$. Obviously, (1.25) is a smooth problem on Π_0 at the point x_0, and the minimum on Π_0 is a local minimum. For an order we take a unique (up to a nonsingular transformation) quadratic positive definite functional $\gamma(\delta x) = \langle \delta x, \delta x \rangle$ in $\mathbb{R}^n$. Obviously, γ is a strict higher order (see the definition in Section 1.1.3). Let us define the violation function σ as in Section 1.1.1. Denote by $\Pi_{\sigma\gamma}$ the set of sequences $\{\delta x_n\} \in \Pi$ satisfying the condition $\sigma(\delta x_n) \leq O(\gamma(\delta x_n))$. Define the following necessary condition for a local minimum at a point x_0.

Condition $\aleph_\gamma$: $g'(x_0)X \neq Y$ or γ-necessity holds.

By results in Section 1.1.4, the inequality $C_\gamma(\Phi_0, \Pi_{\sigma\gamma}) \geq 0$ is a necessary condition for a local minimum and is equivalent to Condition $\aleph_\gamma$, and $C_\gamma(\Phi_0, \Pi_{\sigma\gamma}) > 0$ is a sufficient condition for a local minimum and is equivalent to γ-sufficiency. Below we transform the expression for

$$C_\gamma(\Phi_0,\Pi_{\sigma\gamma}) := \inf_{\{\delta x_n\}\in\Pi_{\sigma\gamma}} \liminf_{n\to\infty} \frac{\Phi_0(\delta x_n)}{\gamma(\delta x_n)}$$

into an equivalent simpler form. We recall that

$$\Phi(\lambda,\delta x) := \sum_{i=0}^{k} \alpha_i f_i(x_0+\delta x) + \langle y^*, g(x_0+\delta x)\rangle = L(\lambda, x_0+\delta x),$$
$$\Phi_0(\delta x) := \max_{\lambda\in\Lambda_0} \Phi(\lambda,\delta x),$$

where the set Λ_0 is defined by (1.3). We set

$$\begin{aligned}
\Omega^\lambda(\bar{x}) &:= \frac{1}{2}\langle L_{xx}(\lambda,x_0)\bar{x},\bar{x}\rangle, \quad \Omega_0(\bar{x}) := \max_{\lambda\in\Lambda_0}\Omega^\lambda(\bar{x}),\\
\mathcal{K} &:= \{\bar{x}\in X \mid \langle f_i'(x_0),\bar{x}\rangle \le 0,\ i\in I;\ g'(x_0)\bar{x}=0\},\\
\sigma'(\bar{x}) &:= \sum_{i\in I}\langle f_i'(x_0),\bar{x}\rangle^+ + \|g'(x_0)\bar{x}\|,\\
C_\gamma(\Omega_0,\mathcal{K}) &:= \inf\{\Omega_0(\bar{x}) \mid \bar{x}\in\mathcal{K},\ \gamma(\bar{x})=1\}.
\end{aligned}$$

If $\mathcal{K}=\{0\}$, then as usual we set $C_\gamma(\Omega_0,\mathcal{K})=+\infty$. Obviously, $\mathcal{K}=\{\bar{x}\in X \mid \sigma'(\bar{x})=0\}$. We call $\mathcal{K}$ the *critical cone*.

Theorem 1.21. *The following equality holds:* $C_\gamma(\Phi_0,\Pi_{\sigma\gamma}) = C_\gamma(\Omega_0,\mathcal{K})$.

Proof. Evidently,

$$\Phi_0(\delta x) = \Omega_0(\delta x) + o(\gamma(\delta x)) \quad \text{as } \delta x\to 0, \tag{1.26}$$
$$\sigma(\delta x) = \sigma'(\delta x) + O(\gamma(\delta x)) \quad \text{as } \delta x\to 0. \tag{1.27}$$

We set

$$\Pi_{\sigma'\gamma} := \big\{\{\delta x_n\}\in\Pi_0 \mid \sigma'(\delta x)\le O(\gamma(\delta x))\big\}.$$

It follows from (1.27) that $\Pi_{\sigma'\gamma}=\Pi_{\sigma\gamma}$; hence, by taking into account (1.26), we immediately obtain $C_\gamma(\Phi_0,\Pi_{\sigma\gamma}) = C_\gamma(\Omega_0,\Pi_{\sigma'\gamma})$, where

$$C_\gamma(\Omega_0,\Pi_{\sigma'\gamma}) := \inf_{\{\delta x_n\}\in\Pi_{\sigma'\gamma}} \liminf_{n\to\infty}\frac{\Omega_0(\delta x_n)}{\gamma(\delta x_n)}.$$

Then we set

$$\Pi_{\sigma'} := \big\{\{\delta x_n\}\in\Pi_0 \mid \sigma'(\delta x)=0\big\}, \quad C_\gamma(\Omega_0,\Pi_{\sigma'}) := \inf_{\{\delta x_n\}\in\Pi_{\sigma'}} \liminf_{n\to\infty}\frac{\Omega_0(\delta x_n)}{\gamma(\delta x_n)}.$$

Since, obviously, $\Pi_{\sigma'}\subset\Pi_{\sigma'\gamma}$, we have $C_\gamma(\Omega_0,\Pi_{\sigma'}) \ge C_\gamma(\Omega_0,\Pi_{\sigma'\gamma})$. We show that, in fact, equality holds. Suppose that $C_\gamma(\Omega_0,\Pi_{\sigma'\gamma}) \ne +\infty$. Take any $\varepsilon>0$. Let $\{\delta x_n\}\in\Pi_{\sigma'\gamma}$ be a nonvanishing sequence such that

$$\lim_{n\to\infty}\frac{\Omega_0(\delta x_n)}{\gamma(\delta x_n)} \le C_\gamma(\Omega_0,\Pi_{\sigma'\gamma})+\varepsilon.$$

By applying the Hoffman lemma (see Lemma 1.12) for each δx_n $(n = 1,2,\dots)$ to the system

$$\langle f'(x_0), \delta x_n + \bar{x}\rangle \leq 0,\ i \in I, \quad g'(x_0)(\delta x_n + \bar{x}) = 0,$$

regarded as a system in the unknown $\bar{x}$, and by bearing in mind that $\sigma'(\delta x_n) \leq O(\gamma(\delta x_n))$, we obtain the following assertion: we can find an $\{\bar{x}_n\}$ such that $\sigma'(\delta x_n + \bar{x}_n) = 0$ and $\|\bar{x}_n\| \leq O(\gamma(\delta x_n))$. Consequently,

$$\lim_{n\to\infty} \frac{\Omega_0(\delta x_n + \bar{x}_n)}{\gamma(\delta x_n + \bar{x}_n)} = \lim_{n\to\infty} \frac{\Omega_0(\delta x_n)}{\gamma(\delta x_n)} \leq C_\gamma(\Omega_0, \Pi_{\sigma'\gamma}) + \varepsilon,$$

and $\{\delta x_n + \bar{x}_n\} \in \Pi_{\sigma'}$. This implies that $C_\gamma(\Omega_0, \Pi_{\sigma'}) \leq C_\gamma(\Omega_0, \Pi_{\sigma'\gamma})$. Consequently, the equality $C_\gamma(\Omega_0, \Pi_{\sigma'}) = C_\gamma(\Omega_0, \Pi_{\sigma'\gamma})$ holds, from which we also obtain $C_\gamma(\Phi_0, \Pi_{\sigma\gamma}) = C_\gamma(\Omega_0, \Pi_{\sigma'})$. But since Ω_0 and γ are positively homogeneous of degree 2, by applying the definition of the cone $\mathcal{K}$, in an obvious way we obtain that, in turn, $C_\gamma(\Omega_0, \Pi_{\sigma'}) = C_\gamma(\Omega_0, \mathcal{K})$. Thus, $C_\gamma(\Phi_0, \Pi_{\sigma\gamma}) = C_\gamma(\Omega_0, \mathcal{K})$, and the theorem is proved. ∎

Corollary 1.22. *The condition $C_\gamma(\Omega_0, \mathcal{K}) \geq 0$ is equivalent to Condition $\aleph_\gamma$ and so is necessary for a local minimum. The condition $C_\gamma(\Omega_0, \mathcal{K}) > 0$ is equivalent to γ-sufficiency and so is sufficient for a strict local minimum. In particular, $\mathcal{K} = \{0\}$ is sufficient for a strict local minimum.*

It is obvious, that $C_\gamma(\Omega_0, \mathcal{K}) \geq 0$ is equivalent to the condition $\Omega_0 \geq 0$ on $\mathcal{K}$, and since $\mathcal{K}$ is finite-dimensional, $C_\gamma(\Omega_0, \mathcal{K}) > 0$ is equivalent to $\Omega_0 > 0$ on $\mathcal{K} \setminus \{0\}$. We also remark that these conditions are stated by means of the maximum of the quadratic forms, and they cannot be reduced in an equivalent way to a condition on one of these forms. Here is a relevant example.

Example 1.23 (Milyutin). Let $X = \mathbb{R}^2$, and let φ and ρ be polar coordinates in $\mathbb{R}^2$. Let the four quadratic forms Q_i, $i = 1,2,3,4$, be defined by their traces q_i on a circle of unit radius:

$$q_1(\varphi) = \sin 2\varphi - \varepsilon, \quad q_2(\varphi) = -\sin 2\varphi - \varepsilon,$$
$$q_3(\varphi) = \cos 2\varphi - \varepsilon, \quad q_4(\varphi) = -\cos 2\varphi - \varepsilon.$$

We choose the constant $\varepsilon > 0$ so that $\max_{1\leq i\leq 4} q_i(\varphi) > 0$, $0 \leq \varphi < 2\pi$. We consider the system S formed from the functionals $f_i(x) = Q_i(x)$, $i = 1,2,3,4$, in a neighborhood of $x_0 = 0$. For $\gamma(x) = \langle x, x\rangle$ this system has γ-sufficiency. In fact, since $f_i'(0) = 0$, $i = 1,2,3,4$, we have $\mathcal{K} = \mathbb{R}^2$, $\Lambda_0 = \{\alpha \in \mathbb{R}^4 \mid \alpha_i \geq 0, \sum \alpha_i = 1\}$ and $\Omega_0(x) = \max_{1\leq i\leq 4} Q_i(x)$, and so $\Omega_0(x) \geq \varepsilon\gamma(x)$ for all $x \in X$. But no form $\sum_{i=1}^4 \alpha_i Q_i$, where $\alpha_i \geq 0$ and $\sum \alpha_i = 1$, is nonnegative, since its trace q on a circle of unit radius has the form $q(\varphi) = A\sin 2\varphi + B\cos 2\varphi - \varepsilon$. ∎

1.3.2 The Problem of Bliss with Endpoint Inequalities

We consider the following problem. It is required to minimize the function of the initial and final states

$$J(x(t_0), x(t_f)) \to \min \tag{1.28}$$

under the constraints

$$F_i(x(t_0), x(t_f)) \leq 0 \ (i = 1,\dots,k), \quad K(x(t_0), x(t_f)) = 0, \quad \dot{x} = f(t,x,u), \tag{1.29}$$

where $x \in \mathbb{R}^n$, $u \in \mathbb{R}^r$, $J \in \mathbb{R}$, $F_i \in \mathbb{R}$, $K \in \mathbb{R}^s$, $f \in \mathbb{R}^n$, and the interval $[t_0, t_f]$ is fixed. Strictly speaking, the problem of Bliss [6] includes the local equation $g(t,x,u) = 0$, where $g \in \mathbb{R}^q$, and it is traditional to require that the matrix g_u has maximal rank at points (t,x,u) such that $g(t,x,u) = 0$. This statement will be considered later. Here, for simplicity, we consider the problem without the local equation.

We set $W = W^{1,1}([t_0,t_f],\mathbb{R}^n) \times L^\infty([t_0,t_f],\mathbb{R}^r)$, where $W^{1,1}([t_0,t_f],\mathbb{R}^n)$ is the space of n-dimensional absolutely continuous functions, and $L^\infty([t_0,t_f],\mathbb{R}^r)$ is the space of r-dimensional bounded measurable functions. We consider problem (1.28)–(1.29) in W. We denote the pair (x,u) by w, and we define the norm in W by

$$\|w\| = \|x\|_{1,1} + \|u\|_\infty = |x(t_0)| + \int_{t_0}^{t_f} \dot{x}(t)\,dt + \operatorname*{ess\,sup}_{t\in[t_0,t_f]} |u(t)|.$$

Clearly, a local minimum in this space is weak. This is the minimum on the set of sequences $\Pi_0 := \{\{\delta w_n\} \mid \|\delta w_n\| \to 0\ (n \to \infty)\}$. Again we set $\Pi = \Pi_0$.

We denote the argument of the functions J, F_i, and K by $p = (x_0, x_f)$, where $x_0 \in \mathbb{R}^n$ and $x_f \in \mathbb{R}^n$. All relations containing measurable sets and functions are understood with accuracy up to a set of measure zero. We assume the following.

Assumption 1.24. The functions J, F_i, and K are twice continuously differentiable with respect to p; the function f is twice differentiable with respect to w; the function f and its second derivative f_{ww} are uniformly bounded and equicontinuous with respect to w on any bounded set of values (t,w) and are measurable in t for any fixed w.

Evidently, f satisfies these conditions if f and f_{ww} are continuous jointly in both variables. Let $w^0(\cdot) \in W$ be a trajectory satisfying all constraints that is being investigated for an optimal situation. We set $p^0 = (x^0(t_0), x^0(t_f))$, $F_0(p) = J(p) - J(p^0)$, $I = \{i \in \{0,1,\dots,k\} \mid F_i(p^0) = 0\}$. Obviously, problem (1.28)–(1.29) is smooth on Π_0 at the point w^0. The set Λ_0 consists of aggregates $\lambda = (\alpha,\beta,\psi)$ for which the local form of Pontryagin's minimum principle holds:

$$\alpha \in \mathbb{R}^{(k+1)*}, \quad \beta \in \mathbb{R}^{s*}, \quad \psi(\cdot) \in W^{1,1}([t_0,t_f],\mathbb{R}^{n*}), \tag{1.30}$$

$$\alpha \ge 0, \quad \alpha_i F_i(p^0) = 0 \ (i = 1,\dots,k), \quad \sum_{i=0}^{k} \alpha_i + \sum_{j=1}^{s} |\beta_j| = 1, \tag{1.31}$$

$$\dot{\psi} = -H_x(t,w^0,\psi), \quad \psi(t_0) = -l_{x_0}(p^0,\alpha,\beta), \quad \psi(t_f) = l_{x_f}(p^0,\alpha,\beta), \tag{1.32}$$

$$H_u(t,w^0,\psi) = 0, \tag{1.33}$$

where

$$H = \psi f, \quad l = \alpha F + \beta K. \tag{1.34}$$

The notation $\mathbb{R}^{n*}$ stands for the space of n-dimensional row vectors. We emphasize the dependence of the *Pontryagin function* H and the *endpoint Lagrange function* l on λ that is defined by (1.34) by writing $H = H^\lambda(t,w)$ and $l = l^\lambda(p)$. Under our assumptions Λ_0 is a finite-dimensional compact set, each point of which is uniquely determined by its projection (α,β).

For any $\delta w \in W$ and $\lambda \in \Lambda_0$ the Lagrange function Φ has the form

$$\begin{aligned}\Phi(\lambda, \delta w) &= l^\lambda(p^0 + \delta p) + \textstyle\int_{t_0}^{t_f} \left(H^\lambda(t, w^0 + \delta w) - H^\lambda(t, w^0)\right) dt \\ &\quad - \textstyle\int_{t_0}^{t_f} \psi \delta \dot{x}\, dt,\end{aligned} \tag{1.35}$$

where $\delta p = (\delta x(t_0), \delta x(t_f))$. For arbitrary $\bar{w} \in W$ and $\lambda \in \Lambda_0$ we set

$$\omega^\lambda(\bar{w}) := \frac{1}{2}\langle l^\lambda_{pp}(p^0)\bar{p}, \bar{p}\rangle + \frac{1}{2}\int_{t_0}^{t_f} \langle H^\lambda_{ww}(t, w^0)\bar{w}, \bar{w}\rangle\, dt, \quad \omega_0(\bar{w}) = \max_{\lambda \in \Lambda_0} \omega^\lambda(\bar{w}), \tag{1.36}$$

where $\bar{p} = (\bar{x}(t_0), \bar{x}(t_f))$. Set

$$\gamma(\delta w) = \langle \delta x(t_0), \delta x(t_0)\rangle + \int_{t_0}^{t_f} \langle \delta \dot{x}(t), \delta \dot{x}(t)\rangle\, dt + \int_{t_0}^{t_f} \langle \delta u(t), \delta u(t)\rangle\, dt. \tag{1.37}$$

Obviously, γ is a strict higher order. We define the cone of critical variations

$$\mathcal{K} = \left\{\bar{w} \in W \mid F_i'(p^0)\bar{p} \le 0,\ i \in I,\ K'(p^0)\bar{p} = 0,\ \dot{\bar{x}} = f_w(t, w^0)\bar{w}\right\} \tag{1.38}$$

and the constant

$$C_\gamma(\omega_0, \mathcal{K}) = \inf\left\{\omega_0(\bar{w}) \mid \bar{w} \in \mathcal{K},\ \gamma(\bar{w}) = 1\right\}. \tag{1.39}$$

(Let us note that the sign of the constant $C_\gamma(\omega_0, \mathcal{K})$ will not change if we replace, in its definition, the functional γ with the functional $\bar{\gamma}(\bar{w}) = \langle \bar{x}(t_0), \bar{x}(t_0)\rangle + \int_{t_0}^{t_f} \langle \bar{u}(t), \bar{u}(t)\rangle\, dt$.) We define Condition $\aleph_\gamma$ as in Section 1.3.1.

Theorem 1.25. *The condition $C_\gamma(\omega_0, \mathcal{K}) \ge 0$ is equivalent to the condition $\aleph_\gamma$ and so is necessary for a local minimum; the condition $C_\gamma(\omega_0, \mathcal{K}) > 0$ is equivalent to γ-sufficiency and so is sufficient for a strict local minimum.*

This result among others, is given, in [84]. But it is not difficult to derive it directly by following the scheme indicated in our discussion of the finite-dimensional case. Because the problem is smooth, Theorem 1.25 follows from $C_\gamma(\Phi_0, \Pi_{\sigma\gamma}) = C_\gamma(\omega_0, \mathcal{K})$, which is established in the same way as in Section 1.3.1.

Notes on SSC for abstract optimization problems. Maurer and Zowe [75] considered optimization problems in Banach spaces with fully infinite-dimensional equality and inequality constraints defined by cone constraints and derived SSC for quadratic functionals γ. Maurer [62] showed that the SSC in [75] can be applied to optimal control problems by taking into account the so-called "two-norm discrepancy."

Chapter 2

Quadratic Conditions in the General Problem of the Calculus of Variations

In this chapter, on the basis of the results of Chapter 1, we create the quadratic theory of conditions for a Pontryagin minimum in the general problem of the calculus of variations without local mixed constraint $g(t,x,u)=0$. Following [92], we perform the decoding of the basic constant for the set of the so-called "Pontryagin sequences" and a special higher order γ, which is characteristic for extremals with finitely many discontinuities of the first kind of control. In Section 2.1, we formulate both necessary and sufficient quadratic conditions for a Pontryagin minimum, which will be obtained as a result of the decoding. In Sections 2.2 and 2.3, we perform some preparations for the decoding. In Section 2.4, we estimate the basic constant from above, thus obtaining quadratic necessary conditions for Pontryagin minimum, and in Section 2.5, we estimate it from below, thus obtaining sufficient conditions. In Section 2.7, we establish the relation of the obtained sufficient conditions for Pontryagin minimum with conditions for strong minimum.

2.1 Statements of Quadratic Conditions for a Pontryagin Minimum

2.1.1 Statement of the Problem and Assumptions

In this chapter, we consider the following *general problem of the calculus of variations* on a fixed time interval $\Delta := [t_0, t_f]$:

$$J(x(t_0), x(t_f)) \to \min, \tag{2.1}$$

$$F(x(t_0), x(t_f)) \le 0, \quad K(x(t_0), x(t_f)) = 0, \tag{2.2}$$

$$\dot{x} = f(t,x,u), \tag{2.3}$$

$$(x(t_0), x(t_f)) \in \mathcal{P}, \quad (t,x,u) \in \mathcal{Q}, \tag{2.4}$$

where $\mathcal{P} \subset \mathbb{R}^{2d(x)}$ and $\mathcal{Q} \subset \mathbb{R}^{1+d(x)+d(u)}$ are open sets. By $d(a)$ we denote the dimension of vector a. Problem (2.1)–(2.4) also will be called the *canonical problem*. For the sake of brevity, we set

$$x(t_0) = x_0, \quad x(t_f) = x_f, \quad (x_0, x_f) = p, \quad (x,u) = w.$$

We seek the minimum among pairs of functions $w = (x,u)$ such that $x(t)$ is an absolutely continuous function on $\Delta = [t_0,t_f]$ and $u(t)$ is a bounded measurable function on Δ. Recall that by $W^{1,1}(\Delta,\mathbb{R}^{d(x)})$ we denote the space of absolutely continuous functions $x : \Delta \to \mathbb{R}^{d(x)}$, endowed with the norm $\|x\|_{1,1} := |x(t_0)| + \int_{t_0}^{t_f} |\dot{x}(t)|\,dt$, and $L^\infty(\Delta,\mathbb{R}^{d(u)})$ denotes the space of bounded measurable functions $u : \Delta \to \mathbb{R}^{d(u)}$, endowed with the norm $\|u\|_\infty := \operatorname{ess\,sup}_{t\in[t_0,t_f]} |u(t)|$. We set

$$W = W^{1,1}(\Delta,\mathbb{R}^{d(x)}) \times L^\infty(\Delta,\mathbb{R}^{d(u)}).$$

We define the norm in the space W as the sum of the norms in the spaces $W^{1,1}(\Delta,\mathbb{R}^{d(x)})$ and $L^\infty(\Delta,\mathbb{R}^{d(u)})$: $\|w\| = \|x\|_{1,1} + \|u\|_\infty$. The space W with this norm is a Banach space. Therefore, we seek the minimum in the space W. A pair $w = (x,u)$ is said to be *admissible*, if $w \in W$ and constraints (2.2)–(2.4) are satisfied by w.

We assume that the functions $J(p)$, $F(p)$, and $K(p)$ are defined and twice continuously differentiable on the open set $\mathcal{P}$, and the function $f(t,w)$ is defined and twice continuously differentiable on the open set $\mathcal{Q}$. These are the assumptions on the functions of the problem. Before formulating the assumptions on the point $w^0 \in W$ being studied, we give the following definition.

Definition 2.1. We say that $t_* \in (t_0,t_f)$ is an *L-point* (or *Lipschitz point*) of a function $\varphi : [t_0,t_f] \to \mathbb{R}^n$ if at t_*, there exist the left and right limit values

$$\lim_{\substack{t\to t_*\\ t<t_*}} \varphi(t) = \varphi(t_*-), \qquad \lim_{\substack{t\to t_*\\ t>t_*}} \varphi(t) = \varphi(t_*+)$$

and there exist $L > 0$ and $\varepsilon > 0$ such that

$$|\varphi(t) - \varphi(t_*-)| \le L|t - t_*| \quad \forall\, t \in (t_* - \varepsilon, t_*) \cap [t_0,t_f],$$
$$|\varphi(t) - \varphi(t_*+)| \le L|t - t_*| \quad \forall\, t \in (t_*, t_* + \varepsilon) \cap [t_0,t_f].$$

A point of discontinuity of the first kind that is an L-point will be called a *point of L-discontinuity*.

Let $w^0 = (x^0,u^0)$ be a pair satisfying the constraints of the problem whose optimality is studied. We assume that the control $u^0(\cdot)$ is piecewise continuous. Denote by $\Theta = \{t_1,\dots,t_s\}$ the set of points of discontinuity for the control u^0, where $t_0 < t_1 < \cdots < t_s < t_f$. We assume that Θ is nonempty (in the case where Θ is empty, all the results remain valid and are obviously simplified). We assume that each $t_k \in \Theta$ is a point of L-discontinuity. By $u^{0k-} = u^0(t_k-)$ and $u^{0k+} = u^0(t_k+)$ we denote the left and right limit values of the function $u^0(t)$ at the point $t_k \in \Theta$, respectively. For a piecewise continuous function $u^0(t)$, the condition $(t,x^0,u^0) \in \mathcal{Q}$ means that $(t,x^0(t),u^0(t)) \in \mathcal{Q}$ for all $t \in [t_0,t_f]\backslash\Theta$. We also assume that $(t_k,x^0(t_k),u^{0k-}) \in Q$ and $(t_k,x^0(t_k),u^{0k+}) \in \mathcal{Q}$ for all $t_k \in \Theta$. As above, all relations and conditions involving measurable functions are assumed to be valid with accuracy up to a set of zero measure even if this is not specified.

2.1.2 Minimum on the Set of Sequences

Let S be an arbitrary set of sequences $\{\delta w_n\}$ in the space W invariant with respect to the operation of passing to a subsequence. According to Definition 1.1, w^0 is a *minimum point on* S if there is no sequence $\{\delta w_n\} \in S$ such that the following conditions hold for all its members:

$$J(p^0+\delta p_n) < J(p^0), \quad F(p^0+\delta p_n) \le 0, \quad K(p^0+\delta p_n) = 0,$$
$$\dot{x}^0+\delta\dot{x}_n = f(t,w^0+\delta w_n), \quad (p^0+\delta p_n) \in \mathcal{P}, \quad (t,w^0+\delta w_n) \in \mathcal{Q},$$

where $p^0 = (x^0(t_0),x^0(t_f))$, $\delta w_n = (\delta x_n,\delta u_n)$, and $\delta p_n = (\delta x_n(t_0),\delta x_n(t_f))$. In a similar way, we define the *strict minimum on* S: it is necessary to only replace the strict inequality $J(p^0+\delta p_n) < J(p^0)$ by nonstrict in the previous definition and additionally assume that the sequence $\{\delta w_n\}$ contains only nonzero members.

We can define any local (in the sense of a certain topology) minimum as a minimum on the corresponding set of sequences. For example, a *weak minimum* is a minimum on the set of sequences $\{\delta w_n\}$ in W such that $\|\delta x_n\|_C + \|\delta u_n\|_\infty \to 0$, where $\|x\|_C = \max_{t\in[t_0,t_f]}|x(t)|$ is the norm in the space of continuous functions. Let Π^0 be the set of sequences $\{\delta w_n\}$ in W such that $\|\delta w_n\| = \|\delta x_n\|_{1,1} + \|\delta u_n\|_\infty \to 0$. (We note that Π^0 corresponds to the set of sequences Π_0 introduced in Section 1.1.) It is easy to see that a minimum on Π^0 is also a weak minimum. Therefore, we can define the same type of minimum using various sets of sequences. We often use this property when choosing a set of sequences for the type of minimum considered which is most convenient for studying. In particular, this refers to the definition of Pontryagin minimum studied in this chapter.

2.1.3 Pontryagin Minimum

Let $\|u\|_1 := \int_{t_0}^{t_f}|u(t)|\,dt$ be the norm in the space $L^1(\Delta,\mathbb{R}^{d(u)})$ of functions $u : [t_0,t_f] \to \mathbb{R}^{d(u)}$ Lebesque integrable with first degree. Denote by Π the set of sequences $\{\delta w_n\}$ in W satisfying the following two conditions:

(a) $\|\delta x_n\|_{1,1} + \|\delta u_n\|_1 \to 0$;

(b) there exists a compact set $\mathcal{C} \subset \mathcal{Q}$ (for each sequence) such that, starting from a certain number, the following condition holds: $(t,w^0(t)+\delta w_n(t)) \in \mathcal{C}$ a.e. on $[t_0,t_f]$.

A minimum on Π is called a *Pontryagin minimum*. For convenience, let us formulate an equivalent definition of the Pontryagin minimum.

The pair $w^0 = (x^0,u^0)$ is a point of *Pontryagin minimum* iff for each compact set $\mathcal{C} \subset \mathcal{Q}$ there exists $\varepsilon > 0$ such that $J(p) \ge J(p^0)$ (where $p^0 = (x^0(t_0),x^0(t_f))$, $p = (x(t_0),x(t_f))$) for all admissible pairs $w = (x,u)$ such that

(a) $\max_{t\in\Delta}|x(t)-x^0(t)| < \varepsilon$,

(b) $\int_\Delta |u(t)-u^0(t)|\,dt < \varepsilon$,

(c) $(t,x(t),u(t)) \in \mathcal{C}$ a.e. on $[t_0,t_f]$.

We can show that it is impossible to define a Pontryagin minimum as a local minimum with respect to a certain topology. Therefore the concept of minimum on a set of sequences is more general than the concept of local minimum. Since $\Pi \supset \Pi^0$, a Pontryagin minimum implies a weak minimum.

2.1.4 Pontryagin Minimum Principle

Define two sets Λ_0 and M_0 of tuples of Lagrange multipliers. They are related to the first-order necessary conditions for the weak and Pontryagin minimum, respectively. We set

$l = \alpha_0 J + \alpha F + \beta K$, $H = \psi f$, where α_0 is a number and α, β, and ψ are row vectors of the same dimension as F, K, and f, respectively (note that x, u, w, F, K, and f are column vectors). Denote by $(\mathbb{R}^n)^*$ the space of row vectors of the dimension n. The functions l and H depend on the following variables: $l = l(p,\alpha_0,\alpha,\beta)$, $H = H(t,w,\psi)$. Denote by λ an arbitrary tuple $(\alpha_0,\alpha,\beta,\psi(\cdot))$ such that $\alpha_0 \in \mathbb{R}^1$, $\alpha \in (\mathbb{R}^{d(F)})^*$, $\beta \in (\mathbb{R}^{d(K)})^*$, $\psi(\cdot) \in W^{1,\infty}(\Delta,(\mathbb{R}^{d(x)})^*)$, where $W^{1,\infty}(\Delta,(\mathbb{R}^{d(x)})^*)$ is the space of Lipschitz continuous functions mapping $[t_0,t_f]$ into $(\mathbb{R}^{d(x)})^*$. For arbitrary λ, w, and p, we set

$$l^\lambda(p) = l(p,\alpha_0,\alpha,\beta), \quad H^\lambda(t,w) = H(t,w,\psi(t)).$$

We introduce an analogous notation for partial derivatives (except for the derivative with respect to t) $l^\lambda_{x_0} = \frac{\partial l}{\partial x_0}(p,\alpha_0,\alpha,\beta)$, $H^\lambda_x(t,w) = \frac{\partial H}{\partial x}(t,w,\psi(t))$, etc. Denote by Λ_0 the set of tuples λ such that

$$\alpha_0 \geq 0, \quad \alpha \geq 0, \quad \alpha F(p^0) = 0, \quad \alpha_0 + \sum_{i=1}^{d(F)} \alpha_i + \sum_{j=1}^{d(K)} |\beta_j| = 1, \tag{2.5}$$

$$\dot{\psi} = -H^\lambda_x(t,w^0), \quad \psi(t_0) = -l^\lambda_{x_0}(p^0), \quad \psi(t_f) = l^\lambda_{x_f}(p^0), \quad H^\lambda_u(t,w^0) = 0. \tag{2.6}$$

Here, α_i are components of the row vector α and β_j are components of the row vector β. If a point w^0 yields a weak minimum, then Λ_0 is nonempty. This was shown in [79, Part 1].

We set $\mathcal{U}(t,x) = \{u \in \mathbb{R}^{d(u)} \mid (t,x,u) \in \mathcal{Q}\}$. Denote by M_0 the set of tuples $\lambda \in \Lambda_0$ such that for all $t \in [t_0,t_f]\backslash\Theta$, the condition $u \in \mathcal{U}(t,x^0(t))$ implies the inequality

$$H(t,x^0(t),u,\psi(t)) \geq H(t,x^0(t),u^0(t),\psi(t)). \tag{2.7}$$

If w^0 is a point of Pontryagin minimum, then M_0 is nonempty; i.e., the *Pontryagin minimum principle* holds. This also was shown in [79, Part 1].

The sets Λ_0 and M_0 are finite-dimensional compact sets, and, moreover, the projection $\lambda \mapsto (\alpha_0,\alpha,\beta)$ is injective on the largest set Λ_0 and, therefore, on M_0. Denote by $\operatorname{co}\Lambda_0$ the convex hull of the set Λ_0, and let M_0^{co} be the set of all $\lambda \in \operatorname{co}\Lambda_0$ such that for all $t \in [t_0,t_f]\backslash\Theta$, the condition $u \in \mathcal{U}(t,x^0(t))$ implies inequality (2.7).

We now formulate a quadratic necessary condition for the Pontryagin minimum. For this purpose, along with the set M_0, we need to define a critical cone $\mathcal{K}$ and a quadratic form on it.

2.1.5 Critical Cone

Denote by $P_\Theta W^{1,2}(\Delta,\mathbb{R}^{d(x)})$ the space of piecewise continuous functions $\bar{x}(t) : [t_0,t_f] \to \mathbb{R}^{d(x)}$ absolutely continuous on each of the intervals of the set $(t_0,t_f)\backslash\Theta$ such that their first derivative is square Lebesgue integrable. We note that all points of discontinuity of functions in $P_\Theta W^{1,2}(\Delta,\mathbb{R}^{d(x)})$ are contained in Θ. Below, for $t_k \in \Theta$ and $\bar{x} \in P_\Theta W^{1,2}(\Delta,\mathbb{R}^{d(x)})$, we set $\bar{x}^{k-} = \bar{x}(t_k-)$, $\bar{x}^{k+} = \bar{x}(t_k+)$, and $[\bar{x}]^k = \bar{x}^{k+} - \bar{x}^{k-}$. Let $Z_2(\Theta)$ be the space of triples $\bar{z} = (\bar{\xi},\bar{x},\bar{u})$ such that

$$\bar{\xi} = (\bar{\xi}_1,\ldots,\bar{\xi}_s) \in \mathbb{R}^s, \quad \bar{x} \in P_\Theta W^{1,2}(\Delta,\mathbb{R}^{d(x)}), \quad \bar{u} \in L^2(\Delta,\mathbb{R}^{d(u)}),$$

where $L^2(\Delta,\mathbb{R}^{d(u)})$ is the space of Lebesgue square integrable functions $\bar{u}(t):[t_0,t_f]\to\mathbb{R}^{d(u)}$. Let $I_F(w^0)=\{i\in\{1,\dots,d(F)\}\mid F_i(p^0)=0\}$ be the set of active subscripts of the constraints $F_i(p)\le 0$ at the point w^0. Denote by $\mathcal{K}$ the set of $\bar{z}=(\bar{\xi},\bar{x},\bar{u})\in Z_2(\Theta)$ such that

$$J_p(p^0)\bar{p}\le 0,\quad F_{ip}(p^0)\bar{p}\le 0\quad \forall\, i\in I_F(w^0),\quad K_p(p^0)\bar{p}=0, \tag{2.8}$$

$$\dot{\bar{x}}=f_w(t,w^0)\bar{w},\quad [\bar{x}]^k=[\dot{x}^0]^k\bar{\xi}_k\quad \forall\, t_k\in\Theta, \tag{2.9}$$

where $\bar{p}=(\bar{x}(t_0),\bar{x}(t_f))$, $\bar{w}=(\bar{x},\bar{u})$, and $[\dot{x}^0]^k$ is the jump of the function $\dot{x}^0(t)$ at the point t_k, i.e.,

$$[\dot{x}^0]^k=\dot{x}^{0k+}-\dot{x}^{0k-}=\dot{x}^0(t_k+)-\dot{x}^0(t_k-).$$

Clearly, $\mathcal{K}$ is a convex polyhedral cone. It will be called the *critical cone*.

The following question is of interest: which inequalities in the definition of $\mathcal{K}$ can be replaced by equalities not changing $\mathcal{K}$? An answer to this question follows from the next proposition. For $\lambda=(\alpha_0,\alpha,\beta,\psi)\in\Lambda_0$, denote by $[H^\lambda]^k$ the jump of the function $H(t,x^0(t),u^0(t),\psi(t))$ at the point $t_k\in\Theta$, i.e.,

$$[H^\lambda]^k=H^{\lambda k+}-H^{\lambda k-},$$

where

$$H^{\lambda k+}=H(t_k,x^0(t_k),u^{0k+},\psi(t_k)),\qquad H^{\lambda k-}=H(t_k,x^0(t_k),u^{0k-},\psi(t_k)).$$

Let Λ_0^Θ be the subset of $\lambda\in\Lambda_0$ such that $[H^\lambda]^k=0$ for all $t_k\in\Theta$. Then Λ_0^Θ is a finite-dimensional compact set, and, moreover, $M_0\subset\Lambda_0^\Theta\subset\Lambda_0$.

Proposition 2.2. *The following conditions hold for any* $\lambda\in\Lambda_0^\Theta$ *and any* $\bar{z}\in\mathcal{K}$*:*

$$\alpha_0 J_p(p^0)\bar{p}=0,\qquad \alpha_i F_{ip}(p^0)\bar{p}=0\quad \forall\, i\in I_F(w^0). \tag{2.10}$$

Also, the following question is of interest: in which case can one of the inequalities in the definition of $\mathcal{K}$ be omitted not changing $\mathcal{K}$? For example, in which case can we omit the inequality $J_p(p^0)\bar{p}\le 0$?

Proposition 2.3. *If there exist* $\lambda\in\Lambda_0^\Theta$ *such that* $\alpha_0>0$, *then the conditions*

$$F_{ip}(p^0)\bar{p}\le 0,\quad \alpha_i F_i(p^0)\bar{p}=0\quad \forall\, i\in I_F(w^0),\quad K_p(p^0)\bar{p}=0, \tag{2.11}$$

$$\dot{\bar{x}}=f_w(t,w^0)\bar{w},\quad [\bar{x}]^k=[\dot{x}^0]^k\bar{\xi}_k\quad \forall\, t_k\in\Theta \tag{2.12}$$

imply $J_p(p^0)\bar{p}=0$, *i.e., conditions* (2.11) *and* (2.12) *determine* $\mathcal{K}$ *as before.*

An analogous assertion holds for any other inequality $F_{ip}(p^0)\bar{p}\le 0$, $i\in I_F(w^0)$, in the definition of $\mathcal{K}$.

2.1.6 Quadratic Form

For $\lambda \in \Lambda_0$, $t_k \in \Theta$, we set

$$(\Delta_k H^\lambda)(t) := H(t, x^0(t), u^{0k+}, \psi(t)) - H(t, x^0(t), u^{0k-}, \psi(t)).$$

The following assertion holds: for any $\lambda \in \Lambda_0$, $t_k \in \Theta$, the function $(\Delta_k H^\lambda)(t)$ has the derivative at the point t_k. We set[1]

$$D^k(H^\lambda) = -\frac{d}{dt}(\Delta_k H^\lambda)(t)|_{t=t_k}.$$

The quantity $D^k(H^\lambda)$ can be calculated by the formula

$$D^k(H^\lambda) = -H_x^{\lambda k+} H_\psi^{k-} + H_x^{\lambda k-} H_\psi^{k+} - [H_t^\lambda]^k,$$

where

$$\begin{aligned}
&H_x^{\lambda k+} = \psi(t_k) f_x(t_k, x^0(t_k), u^{0k+}), \quad H_x^{\lambda k-} = \psi(t_k) f_x(t_k, x^0(t_k), u^{0k-}), \\
&H_\psi^{k+} = f(t_k, x^0(t_k), u^{0k+}) = \dot{x}^{0k+}, \quad H_\psi^{k-} = f(t_k, x^0(t_k), u^{0k-}) = \dot{x}^{0k-}, \\
&\dot{x}^{0k+} = \dot{x}(t_k+), \quad \dot{x}^{0k-} = \dot{x}(t_k-),
\end{aligned}$$

and $[H_t^\lambda]^k$ is the jump of the function $H_t^\lambda = \psi(t) f_t(t, x^0(t), u^0(t))$ at the point t_k, i.e.,

$$[H_t^\lambda]^k = H_t^{\lambda k+} - H_t^{\lambda k-} = \psi(t_k) f_t(t_k, x^0(t_k), u^{0k+}) - \psi(t_k) f_t(t_k, x^0(t_k), u^{0k-}).$$

We note that $D^k(H^\lambda)$ depends on λ linearly, and $D^k(H^\lambda) \geq 0$ for any $\lambda \in M_0$ and any $t_k \in \Theta$. Let $[H_x^\lambda]^k$ be the jump of the function $H_x^\lambda = \psi(t) f_x(t, x^0(t), u^0(t)) = -\dot{\psi}(t)$ at the point $t_k \in \Theta$, i.e., $[H_x^\lambda]^k = H_x^{\lambda k+} - H_x^{\lambda k-}$. For $\lambda \in M_0$, $\bar{z} \in Z_2(\Theta)$, we set (see also Henrion [39])

$$\begin{aligned}
\Omega^\lambda(\bar{z}) \quad = \quad & \frac{1}{2} \sum_{k=1}^{s} \left(D^k(H^\lambda) \bar{\xi}_k^2 + 2[H_x^\lambda]^k \bar{x}_{\text{av}}^k \bar{\xi}_k \right) \\
& + \frac{1}{2} \left(\langle l_{pp}^\lambda(p^0) \bar{p}, \bar{p} \rangle + \int_{t_0}^{t_f} \langle H_{ww}^\lambda(t, w^0) \bar{w}, \bar{w} \rangle \, dt \right),
\end{aligned} \tag{2.13}$$

where $\bar{x}_{\text{av}}^k = \frac{1}{2}(\bar{x}^{k-} + \bar{x}^{k+})$, $\bar{p} = (\bar{x}(t_0), \bar{x}(t_f))$. Obviously, $\Omega^\lambda(\bar{z})$ is a *quadratic form* in $\bar{z}$ and a linear form in λ.

2.1.7 Necessary Quadratic Condition for a Pontryagin Minimum

In the case where the set M_0 is nonempty, we set

$$\Omega_0(\bar{z}) = \max_{\lambda \in M_0} \Omega^\lambda(\bar{z}). \tag{2.14}$$

If M_0 is empty, we set $\Omega_0(\cdot) = -\infty$.

Theorem 2.4. *If w^0 is a Pontryagin minimum point, then M_0 is nonempty and the function $\Omega_0(\cdot)$ is nonnegative on the cone $\mathcal{K}$.*

[1] In the book [79], the value $D^k(H^\lambda)$ was defined by the formula $D^k(H^\lambda) = \frac{d}{dt}(\Delta_k H^\lambda)(t_k)$, since there the Pontryagin *maximum* principle was used.

We say that *Condition* $\mathfrak{A}$ holds for the point w^0 if the set M_0 is nonempty and the function Ω_0 is nonnegative on the cone $\mathcal{K}$. According to Theorem 2.4, the condition $\mathfrak{A}$ is necessary for a Pontryagin minimum.

2.1.8 Sufficient Quadratic Condition for a Pontryagin Minimum

A natural strengthening of Condition $\mathfrak{A}$ is sufficient for the strict Pontryagin minimum. (We will show in Section 2.7 that it is also sufficient for the so-called "bounded strong minimum".) We now formulate this strengthening. For this purpose, we define the set M_0^+ of elements $\lambda \in M_0$ satisfying the so-called "strict minimum principle" and its subset $\mathrm{Leg}_+(M_0^+)$ of "strictly Legendre elements." We begin with the definition of the concept of strictly Legendre element. An element $\lambda = (\alpha_0, \alpha, \beta, \psi) \in \Lambda_0$ is said to be *strictly Legendre* if the following conditions hold:

(a) $[H^\lambda]^k = 0, \quad D^k(H^\lambda) > 0$ for all $t_k \in \Theta$;

(b) for any $t \in [t_0, t_f]\backslash\Theta$, the inequality $\langle H_{uu}(t, x^0(t), u^0(t), \psi(t))\bar{u}, \bar{u}\rangle > 0$ holds for all $\bar{u} \in \mathbb{R}^{d(u)}$, $\bar{u} \neq 0$;

(c) for each point $t_k \in \Theta$, the inequalities $\langle H_{uu}(t_k, x^0(t_k), u^{0k-}, \psi(t_k))\bar{u}, \bar{u}\rangle > 0$ and $\langle H_{uu}(t_k, x^0(t_k), u^{0k+}, \psi(t_k))\bar{u}, \bar{u}\rangle > 0$ hold for all $\bar{u} \in \mathbb{R}^{d(u)}$, $\bar{u} \neq 0$.

We note that each element $\lambda \in M_0$ is *Legendre in nonstrict sense*, i.e., $[H^\lambda]^k = 0$, $D^k(H^\lambda) \geq 0$ for all $t_k \in \Theta$, and in conditions (b) and (c) nonstrict inequalities hold for the quadratic form $\langle H_{uu}\bar{u}, \bar{u}\rangle$. In other words, $\mathrm{Leg}(M_0) = M_0$, where $\mathrm{Leg}(M)$ is a subset of Legendre (in nonstrict sense) elements of the set $M \subset \Lambda_0$. Further, denote by M_0^+ the set of $\lambda \in M_0$ such that

(a) $H(t, x^0(t), u, \psi(t)) > H(t, x^0(t), u^0(t), \psi(t))$ if $t \in [t_0, t_f]\backslash\Theta$, $u \in \mathcal{U}(t, x^0(t))$, $u \neq u^0(t)$;

(b) $H(t_k, x^0(t_k), u, \psi(t_k)) > H^{\lambda k-} = H^{\lambda k+}$ if $t_k \in \Theta$, $u \in \mathcal{U}(t_k, x^0(t_k))$, $u \notin \{u^{0k-}, u^{0k+}\}$.

Denote by $\mathrm{Leg}_+(M_0^+)$ the subset of all strictly Legendre elements $\lambda \in M_0^+$. We set

$$\bar{\gamma}(\bar{z}) = \langle \bar{\xi}, \bar{\xi}\rangle + \langle \bar{x}(t_0), \bar{x}(t_0)\rangle + \int_{t_0}^{t_f} \langle \bar{u}(t), \bar{u}(t)\rangle\, dt. \tag{2.15}$$

We say that *Condition* $\mathfrak{B}$ holds for the point w^0 if there exist a nonempty compact set $M \subset \mathrm{Leg}_+(M_0^+)$ and a constant $C > 0$ such that

$$\max_{\lambda \in M} \Omega^\lambda(\bar{z}) \geq C\bar{\gamma}(\bar{z}) \quad \forall\, \bar{z} \in \mathcal{K}.$$

Theorem 2.5. *If Condition* $\mathfrak{B}$ *holds, then* w^0 *is a strict Pontryagin minimum point.*

In conclusion, we note that the space $P_\Theta W^{1,2}(\Delta, \mathbb{R}^{d(x)})$ with the inner product

$$\langle \bar{x}, \bar{\bar{x}}\rangle = \langle \bar{x}(t_0), \bar{\bar{x}}(t_0)\rangle + \sum_{k=1}^{s} \langle [\bar{x}]^k, [\bar{\bar{x}}]^k\rangle + \int_{t_0}^{t_f} \langle \dot{\bar{x}}(t), \dot{\bar{\bar{x}}}(t)\rangle\, dt,$$

and also the space $Z_2(\Theta)$ with the inner product

$$\langle \bar{z}, \bar{\bar{z}}\rangle = \langle \bar{\xi}, \bar{\bar{\xi}}\rangle + \langle \bar{x}, \bar{\bar{x}}\rangle + \int_{t_0}^{t_f} \langle \bar{u}(t), \bar{\bar{u}}(t)\rangle\, dt,$$

are Hilbert spaces. Moreover, the functional $\sqrt{\bar{\gamma}(\bar{z})}$ is equivalent to the norm $\|\bar{z}\|_{Z_2(\Theta)} = \sqrt{\langle \bar{z}, \bar{z} \rangle}$ on the subspace $\{\bar{z} \in Z_2(\Theta) \mid \dot{\bar{x}} = f_w \bar{w},\ [\bar{x}]^k = [\dot{x}^0]^k \bar{\xi}_k,\ k \in I^*\}$, where $I^* = \{1, \ldots, s\}$.

2.2 Basic Constant and the Problem of Its Decoding

2.2.1 Verification of Assumptions in the Abstract Scheme

The derivation of quadratic conditions for the Pontryagin minimum in the problem (2.1)–(2.4) is based on the abstract scheme presented in Chapter 1. We begin with the verification of the property that the problem (2.1)–(2.4), the set of sequences Π, and the point w^0 satisfy all the assumptions of the abstract scheme. As the Banach space X, we consider the space $W = W^{1,1}(\Delta, \mathbb{R}^{d(x)}) \times L^\infty(\Delta, \mathbb{R}^{d(u)})$ of pairs of functions $w = (x, u)$. As the set Ω entering the statement of the abstract problem (1.1), we consider the set $\mathcal{W}$ of pairs $w = (x, u) \in W$ such that $(x(t_0), x(t_f)) \in \mathcal{P}$ and there exists a compact set $\mathcal{C} \subset \mathcal{Q}$ such that $(t, x(t), u(t)) \in \mathcal{C}$ a.e. on $[t_0, t_f]$.

Let $w^0 = (x^0, u^0) \in W$ be a point satisfying the constraints of the canonical problem (2.1)–(2.4) and the assumptions of Section 2.1. As the set of sequences defining the type of minimum at this point, we take the set of Pontryagin sequences Π in the space W. Obviously, Π is invariant with respect to the operation of passing to a subsequence. Also, it is elementary verified that Π is invariant with respect to the Π^0-extension, where $\Pi^0 = \{\{\bar{w}_n\} \mid \|\bar{w}_n\| = \|\bar{x}_n\|_{1,1} + \|\bar{u}_n\|_\infty \to 0\}$. As was already mentioned, Π^0 corresponds to the set of sequences Π_0 introduced in Section 1.1. (Here, the notation Π^0 is slightly more convenient.) Finally, it is easy to see that the set $\mathcal{W}$ is absorbing for Π at the point w^0. Therefore, all the assumptions on w^0, $\mathcal{W}$, and Π from Section 1.1 hold.

Furthermore, consider the functional

$$w(\cdot) = (x(\cdot), u(\cdot)) \longmapsto J(x(t_0), x(t_f)). \tag{2.16}$$

Since the function $J(p)$ is defined on $\mathcal{P}$, we can consider this functional as a functional given on $\mathcal{W}$. We denote it by $\hat{J}$. Therefore, the functional $\hat{J} : \mathcal{W} \to \mathbb{R}^1$ is given by formula (2.16). Analogously, we define the following functionals on $\mathcal{W}$:

$$\hat{F}_i : w(\cdot) = (x(\cdot), u(\cdot)) \in \mathcal{W} \longmapsto F_i(x(t_0), x(t_f)), \quad i = 1, \ldots, d(F), \tag{2.17}$$

and the operator

$$\hat{g} : w(\cdot) = (x(\cdot), u(\cdot)) \in \mathcal{W} \longmapsto (\dot{x} - f(t, w), K(p)) \in Y, \tag{2.18}$$

where $Y = L^1(\Delta, \mathbb{R}^{d(x)}) \times \mathbb{R}^{d(K)}$. We omit the verification of the property that all the functionals $\hat{J}(w)$, $\hat{F}_i(w)$, $i = 1, \ldots, d(F)$ and the operator $\hat{g}(w) : \mathcal{W} \longmapsto Y$ are Π-continuous and strictly Π-differentiable at the point w^0, and, moreover, their Fréchet derivatives at this point are linear functionals

$$\hat{J}'(w^0) : \bar{w} = (\bar{x}, \bar{u}) \in W \longmapsto J_p(p^0)\bar{p},$$
$$\hat{F}_i'(w^0) : \bar{w} = (\bar{x}, \bar{u}) \in W \longmapsto F_{ip}(p^0)\bar{p}, \quad i = 1, \ldots, d(F),$$

and a linear operator

$$\hat{g}'(w^0) : \bar{w} = (\bar{x}, \bar{u}) \in W \longmapsto (\dot{\bar{x}} - f_w(t, w^0)\bar{w},\ K_p(p^0)\bar{p}) \in Y, \tag{2.19}$$

respectively; moreover, here, $\bar{p} = (\bar{x}(t_0), \bar{x}(t_f))$. The verification of this assertion is also very elementary.

We prove only that if $\hat{g}'(w^0)W \neq Y$, then the image $\hat{g}'(w^0)W$ is closed in Y and has a direct complement that is a closed subspace in Y. Note that the operator

$$\bar{w} = (\bar{x}, \bar{u}) \in W \longmapsto \dot{\bar{x}} - f_w(t, w^0)\bar{w} \in L^1(\Delta, \mathbb{R}^{d(x)}) \tag{2.20}$$

is surjective. Indeed, for an arbitrary function $\bar{v} \in L^1(\Delta, \mathbb{R}^{d(x)})$, there exists a function $\bar{x}$ satisfying $\dot{\bar{x}} - f_x(t, w^0)\bar{x} = \bar{v}$ and $\bar{x}(t_0) = 0$. Then under the mapping given by this operator, the pair $\bar{w} = (\bar{x}, \bar{u})$ yields $\bar{v}$ in the image. Further, we note that the operator

$$\bar{w} = (\bar{x}, \bar{u}) \in W \longmapsto K_p(p^0)\bar{p} \in \mathbb{R}^{d(K)} \tag{2.21}$$

is finite-dimensional. The surjectivity of operator (2.20) and the finite-dimensional property of operator (2.21) imply the closedness of the image of operator (2.19). The following more general assertion holds.

Lemma 2.6. *Let X, Y, and Z be Banach spaces, let $A : X \to Y$ be a linear operator with closed range, and let $B : X \to Z$ be a linear operator such that the image $B(\operatorname{Ker} A)$ of the kernel of the operator A under the mapping of the operator B is closed in Z. Then the operator $T : X \to Y \times Z$ defined by the relation $Tx = (Ax, Bx)$ for all $x \in X$ has a closed range.*

Proof. Let a sequence $\{x_n\}$ in X be such that $Tx_n = (Ax_n, Bx_n) \to (y, z) \in Y \times Z$. Since the range of AX is closed in Y, by the Banach theorem on the inverse operator there exists a convergent subsequence $x_n' \to x$ in X such that $Ax_n' = Ax_n \to y$ and, therefore, $Ax = y$. Then $A(x_n - x_n') = 0$ for all n and $B(x_n - x_n') \to z - Bx$. By the closedness of the range $B(\operatorname{Ker} A)$, there exists $x'' \in X$ such that $Ax'' = 0$ and $Bx'' = z - Bx$. Then $B(x + x'') = z$ and $A(x + x'') = y$ as required. The lemma is proved. ∎

Since the image of any subspace under the mapping defined by a finite-dimensional operator is a finite-dimensional and, therefore, closed subspace, Lemma 2.6 implies the following assertion.

Corollary 2.7. *Let X, Y, and Z be Banach spaces, let $A : X \to Y$ be a surjective linear operator, and let $B : X \to Z$ be a finite-dimensional linear operator. Then the operator $T : X \to Y \times Z$ defined by the relation $Tx = (Ax, Bx)$ for all $x \in X$ has a closed range.*

This implies that operator (2.19) has a closed range. Since the range of operator (2.19) is of finite codimension, this range has a direct complement that is finite-dimensional, and, therefore, it is a closed subspace.

Therefore, all the conditions defining a smooth problem on Π at the point w^0 hold for the problem (2.1)–(2.4). Hence we can apply the main result of the abstract theory of higher-order conditions, Theorem 1.7.

2.2.2 Set Λ_0

In what follows, for the sake of brevity we make the following agreement: if a point at which the derivative of a given function is taken is not indicated, then for the functions J, F, K, and l^λ, p^0 is such a point, and for the functions f and H^λ, it is the point $(t, w^0(t))$. Using the definition given by relations (1.3), we denote by Λ_0 the set of tuples $\lambda = (\alpha_0, \alpha, \beta, \psi)$ such that

$$\alpha_0 \in \mathbb{R}, \quad \alpha \in (\mathbb{R}^{d(F)})^*, \quad \beta \in (\mathbb{R}^{d(K)})^*, \quad \psi(\cdot) \in L^\infty(\Delta, (\mathbb{R}^{d(x)})^*)$$

and the following conditions hold:

$$\alpha_0 \geq 0, \quad \alpha_i \geq 0 \ (i = 1, \dots, d(F)), \quad \alpha_i F_i(p^0) = 0 \ (i = 1, \dots, d(F)), \tag{2.22}$$

$$\alpha_0 + \sum_{i=1}^{d(F)} \alpha_i + \sum_{j=1}^{d(K)} |\beta_j| + \|\psi\| = 1, \tag{2.23}$$

$$\alpha_0 J_p \bar{p} + \sum_{i=1}^{d(F)} \alpha_i F_{ip} \bar{p} + \beta K_p \bar{p} - \int_{t_0}^{t_f} \psi(\dot{\bar{x}} - f_w \bar{w})\, dt = 0 \tag{2.24}$$

for all $\bar{w} = (\bar{x}, \bar{u}) \in W$. Here, $\bar{p} = (\bar{x}(t_0), \bar{x}(t_f))$.

Let us show that the normalization (2.23) is equivalent to the normalization

$$\alpha_0 + \sum_{i=1}^{d(F)} \alpha_i + \sum_{j=1}^{d(K)} |\beta_j| = 1 \tag{2.25}$$

(the definition of equivalent normalizations was given in Section 1.1.5). The upper estimate is obvious: $\alpha_0 + \sum \alpha_i + \sum |\beta_j| \leq 1$ for any $\lambda \in \Lambda_0$. It is required to establish the lower estimate: there exists $\varepsilon > 0$ such that

$$\alpha_0 + \sum \alpha_i + \sum |\beta_j| \geq \varepsilon \tag{2.26}$$

for any $\lambda \in \Lambda_0$. Suppose that this is not true. Then there exists a sequence

$$\lambda_n = (\alpha_{0n}, \alpha_n, \beta_n, \psi_n) \in \Lambda_0$$

such that $\alpha_{0n} \to 0$, $\alpha_n \to 0$, and $\beta_n \to 0$. By (2.24), this implies that the functional $\bar{w} \in W \longmapsto \int_{t_0}^{t_f} \psi_n(\dot{\bar{x}} - f_w \bar{w})\, dt$ is norm-convergent to zero. Since the operator $\bar{w} \in W \longmapsto (\dot{\bar{x}} - f_w \bar{w}) \in L^1(\Delta, \mathbb{R}^{d(x)})$ is surjective, this implies $\|\psi_n\|_\infty \to 0$. Therefore, $\alpha_{0n} + |\alpha_n| + |\beta_n| + \|\psi_n\| \to 0$, which contradicts the condition $\alpha_{0n} + \sum \alpha_{in} + \sum |\beta_{jn}| + \|\psi_n\| = 1$, which follows from $\lambda_n \in \Lambda_0$. Therefore, estimate (2.26) also holds with some $\varepsilon > 0$. The equivalence of normalizations (2.23) and (2.25) is proved. In what follows, we will use normalization (2.25) preserving the old notation for the new Λ_0.

Let us show that the set Λ_0 with normalization (2.25) coincides with the set Λ_0 defined by (1.3) in Section 1.1.4. For $\lambda = (\alpha_0, \alpha, \beta, \psi)$, let conditions (2.22), (2.24), and (2.25) hold. We rewrite condition (2.24) in the form

$$l_p \bar{p} - \int_{t_0}^{t_f} \psi(\dot{\bar{x}} - f_w \bar{w})\, dt = 0 \quad \forall\, \bar{w} \in W, \tag{2.27}$$

where $l = \alpha_0 J + \alpha F + \beta K$. We set $\bar{x} = 0$ in (2.27). Then $\int_{t_0}^{t_f} \psi f_u \bar{u}\, dt = 0$ for all $\bar{u} \in L^\infty$. Therefore, $\psi f_u = 0$ or $H_u = 0$, where $H = \psi f$. Then we obtain from (2.27) that

$$l_p \bar{p} - \int_{t_0}^{t_f} \psi \dot{\bar{x}}\, dt + \int_{t_0}^{t_f} \psi f_x \bar{x}\, dt = 0 \tag{2.28}$$

for all $\bar{x} \in W^{1,1}$. Let us show that this implies the conditions

$$\psi \in W^{1,\infty}, \quad -\dot{\psi} = H_x, \quad \psi(t_0) = -l_{x_0}, \quad \psi(t_f) = l_{x_f}. \tag{2.29}$$

Let ψ' satisfy the conditions $\psi' \in W^{1,\infty}$, $-\dot{\psi}' = \psi' f_x$, and $\psi'(t_0) = -l_{x_0}$. Integrating by parts, we obtain

$$\int_{t_0}^{t_f} \psi' \dot{\bar{x}}\, dt = \psi' \bar{x} \,|_{t_0}^{t_f} - \int_{t_0}^{t_f} \dot{\psi}' \bar{x}\, dt = l_{x_0} \bar{x}_0 + \psi'(t_f) \bar{x}_f + \int_{t_0}^{t_f} \psi' f_x \bar{x}\, dt$$

for any $\bar{x} \in W^{1,1}$. (Hereafter $\bar{x}_0 := \bar{x}(t_0)$, $\bar{x}_f := \bar{x}(t_f)$). Therefore,

$$\int_{t_0}^{t_f} \psi' \dot{\bar{x}}\, dt - \int_{t_0}^{t_f} \psi' f_x \bar{x}\, dt - l_{x_0} \bar{x}_0 - \psi'(t_f) \bar{x}_f = 0. \tag{2.30}$$

Adding (2.30) to (2.28), we obtain

$$(l_{x_f} - \psi'(t_f)) \bar{x}_f + \int_{t_0}^{t_f} (\psi' - \psi) \dot{\bar{x}}\, dt + \int_{t_0}^{t_f} (\psi - \psi') f_x \bar{x}\, dt = 0$$

for all $\bar{x} \in W^{1,1}$. Let $\bar{c} = l_{x_f} - \psi'(t_f)$ and $\bar{\psi} = \psi' - \psi$. Then

$$\bar{c} \bar{x}_f + \int_{t_0}^{t_f} \bar{\psi} (\dot{\bar{x}} - f_x \bar{x})\, dt = 0. \tag{2.31}$$

This is true for any $\bar{x} \in W^{1,1}$. Let $\bar{a} \in \mathbb{R}^n$, $\bar{v} \in L^1$. Let us find $\bar{x} \in W^{1,1}$ such that $\dot{\bar{x}} - f_x \bar{x} = \bar{v}$ and $\bar{x}(t_f) = \bar{a}$. Then $\bar{c}\bar{a} + \int_{t_0}^{t_f} \bar{\psi} \bar{v}\, dt = 0$. Since this relation holds for an arbitrary $\bar{a} \in \mathbb{R}^n$, $\bar{v} \in L^1$, we obtain that $\bar{c} = 0$ and $\bar{\psi} = 0$, i.e., $\psi = \psi'$ and $\psi'(t_f) = l_{x_f}$. Therefore, conditions (2.29) hold. Conversely, if conditions (2.29) hold for ψ, then applying the integration-by-parts formula $\int_{t_0}^{t_f} \psi \dot{\bar{x}}\, dt = \psi \bar{x} \,|_{t_0}^{t_f} - \int_{t_0}^{t_f} \dot{\psi} \bar{x}\, dt$, we obtain condition (2.28). From (2.28) and the condition $H_u = 0$, condition (2.27), and therefore, condition (2.24) follow. Therefore, we have shown that the set Λ_0 defined in this section coincides with the set Λ_0 of Section 1.1.4.

We note that the set Λ_0 is a finite-dimensional compact set, and the projection $\lambda = (\alpha_0, \alpha, \beta, \psi) \longmapsto (\alpha_0, \alpha, \beta)$ is injective on Λ_0. Indeed, according to $(\alpha_0, \alpha, \beta)$, the vector $l_{x_0}^{\lambda}$ is uniquely defined, and from the conditions $-\dot{\psi} = \psi f_x$ and $\psi(t_0) = l_{x_0}$ the function ψ is uniquely found. The same is also true for the set $\operatorname{co} \Lambda_0$.

2.2.3 Lagrange Function

We set $F_0(p) = J(p) - J(p^0)$. Introduce the *Lagrange function* of the problem (2.1)–(2.4). Let $\delta\mathcal{W}$ be the set of variations $\delta w \in W$ such that $(w^0 + \delta w) \in \mathcal{W}$, i.e., $(p^0 + \delta p) \in \mathcal{P}$, where $\delta p = (\delta x(t_0), \delta x(t_f))$, and there exists a compact set $\mathcal{C} \subset \mathcal{Q}$ such that $(t, w^0(t) + \delta w(t)) \in \mathcal{C}$ a.e. on $[t_0, t_f]$. For $\lambda = (\alpha_0, \alpha, \beta, \psi) \in \operatorname{co} \Lambda_0$ and $\delta w = (\delta x, \delta u) \in \delta\mathcal{W}$, we set

$$\begin{aligned} \Phi(\lambda, \delta w) &= \alpha_0 F_0(p^0 + \delta p) + \alpha F(p^0 + \delta p) + \beta K(p^0 + \delta p) \\ &\quad - \int_{t_0}^{t_f} \psi(\dot{x}^0 + \delta \dot{x} - f(t, w^0 + \delta w))\, dt. \end{aligned}$$

We set

$$\delta F_i = F_i(p^0+\delta p) - F_i(p^0), \qquad i = 1,\dots,d(F),$$
$$\delta K = K(p^0+\delta p) - K(p^0) = K(p^0+\delta p), \quad \delta f = f(t,w^0+\delta w) - f(t,w^0).$$

Then

$$\begin{aligned}\Phi(\lambda,\delta w) &= \sum_{i=0}^{d(F)} \alpha_i \delta F_i + \beta\delta K - \int_{t_0}^{t_f} \psi(\delta\dot{x} - \delta f)\,dt \\ &= \delta l^\lambda - \int_{t_0}^{t_f} \psi\delta\dot{x}\,dt + \int_{t_0}^{t_f} \delta H^\lambda\,dt,\end{aligned}$$

where

$$l^\lambda = \sum_{i=0}^{d(F)} \alpha_i F_i + \beta K, \quad H^\lambda = \psi f, \quad \delta l^\lambda = l^\lambda(p^0+\delta p) - l^\lambda(p^0),$$
$$\delta H^\lambda = H(t,w^0+\delta w,\psi) - H(t,w^0,\psi) = \psi\delta f.$$

Note that in contrast to the classical calculus of variations, where δJ stands for the first variation of the functional, we denote by δJ, δf, etc. the full increments corresponding to the variation δw.

2.2.4 Basic Constant

For $\delta w \in \delta\mathcal{W}$, we set

$$\Phi_0(\delta w) = \max_{\Lambda_0} \Phi(\lambda,\delta w).$$

Let $\gamma = \gamma(\delta w) : \delta\mathcal{W} \longmapsto \mathbb{R}^1$ be an arbitrary strict higher order on Π whose definition is given in Section 1.1.3. We set

$$\Pi_{\sigma\gamma} = \{\{\delta w_n\} \in \Pi \mid \sigma(\delta w_n) \le O(\gamma(\delta w_n))\},$$

where $\sigma(\delta w) = \max\{F_0(p^0+\delta p),\dots,F_{d(F)}(p^0+\delta p), |\delta K|, \ \|\delta\dot{x} - \delta f\|_1\}$ is the violation function of the problem (4.1)–(4.4). In what follows, we shall use the shorter notation $\underline{\lim}$ for the limit inferior instead of lim inf. We set

$$C_\gamma(\Phi_0,\Pi_{\sigma\gamma}) = \inf_{\Pi_{\sigma\gamma}} \underline{\lim}\, \frac{\Phi_0}{\gamma}.$$

Theorem 1.7 implies the following theorem.

Theorem 2.8. *The condition $C_\gamma(\Phi_0,\Pi_{\sigma\gamma}) \ge 0$ is necessary for a Pontryagin minimum at the point w^0, and the condition $C_\gamma(\Phi_0,\Pi_{\sigma\gamma}) > 0$ is sufficient for a strict Pontryagin minimum at this point.*

Further, there arises the problem of the choice of a higher order γ and decoding the constant $C_\gamma(\Phi_0,\Pi_{\sigma\gamma})$ corresponding to the chosen order. The constant $C_\gamma(\Phi_0,\Pi_{\sigma\gamma})$ is said to be *basic*, and by decoding of the basic constant, we mean the simplest method for calculating its sign. In what follows, we will deal with the choice of γ and decoding of the basic constant. As a result, we will obtain theorems on quadratic conditions for the Pontryagin minimum formulated in Section 2.1.

2.3 Local Sequences, Higher Order γ, Representation of the Lagrange Function on Local Sequences with Accuracy up to $o(\gamma)$

2.3.1 Local Sequences and Their Structure

As before, let $w^0 \in W$ be a point satisfying the constraints of the canonical problem and the assumptions of Section 2.1. For convenience, we assume that u^0 is left continuous at each point of discontinuity $t_k \in \Theta$. Denote by $\overline{u^0}$ the closure of the graph of $u^0(t)$. Let Π^{loc} be the set of sequences $\{\delta w_n\}$ in the space W satisfying the following two conditions:

(a) $\|\delta x_n\|_{1,1} \to 0$;

(b) for any neighborhood V of the compact set $\overline{u^0}$ there exists $n_0 \in \mathbb{N}$ with

$$(t, u^0(t) + \delta u_n(t)) \in V \quad \text{a.e. on } [t_0, t_f] \ \forall\, n \geq n_0. \tag{2.32}$$

Sequences from Π^{loc} are said to be *local*. Obviously, $\Pi^0 \subset \Pi^{\mathrm{loc}} \subset \Pi$. Although the set of local sequences Π^{loc} is only a part of the set Π of Pontryagin sequences, all main considerations in obtaining quadratic conditions for the Pontryagin minimum are related namely to the set Π^{loc}.

Let us consider the structure of local sequences. Denote by Π_u^{loc} the set of sequences $\{\delta u_n\}$ in $L^\infty(\Delta, \mathbb{R}^{d(u)})$ such that for any neighborhood V of the compact set $\overline{u^0}$ there exists a number starting from which condition (2.32) holds. We briefly write the condition defining the sequences from Π_u^{loc} in the form $(t, u^0 + \delta u_n) \to \overline{u^0}$. Therefore, $\Pi^{\mathrm{loc}} = \{\{\delta w_n\} \mid \|\delta x_n\|_{1,1} \to 0,\ (t, u^0 + \delta u_n) \to \overline{u^0}\}$. In what follows, in order not to abuse the notation, we will omit the number n in sequences.

Let $\mathcal{Q}_{tu}$ be the projection of $\mathcal{Q}$ under the mapping $(t,x,u) \to (t,u)$. Then $\mathcal{Q}_{tu}$ is an open set in $\mathbb{R}^{d(u)+1}$ containing the compact set $\overline{u^0}$. Denote by $\overline{u^0}(t_{k-1}, t_k)$ the closure in $\mathbb{R}^{d(u)+1}$ of the intersection of the compact set $\overline{u^0}$ with the layer $\{(t,u) \mid u \in \mathbb{R}^{d(u)}, t \in (t_{k-1}, t_k)\}$, where $k = 1, \ldots, s+1$, and $t_{s+1} = t_f$. In other words, $\overline{u^0}(t_{k-1}, t_k)$ is the closure of the graph of the restriction of the function $u^0(\cdot)$ to the interval (t_{k-1}, t_k). Obviously, $\overline{u^0}$ is the union of disjoint compact sets $\overline{u^0}(t_{k-1}, t_k)$. For brevity, we set $\overline{u^0}(t_{k-1}, t_k) = \overline{u^0}^k$.

Let $\mathcal{V}_k \subset \mathcal{Q}_{ut}$ be fixed disjoint bounded neighborhoods of the compact sets $\overline{u^0}^k$, $k = 1, \ldots, s+1$. We set

$$\mathcal{V} = \bigcup_{k=1}^{s+1} \mathcal{V}_k. \tag{2.33}$$

Then $\mathcal{V}$ is a neighborhood of $\overline{u^0}$. Without loss of generality, we assume that $\mathcal{V}$, together with its closure, is contained in $\mathcal{Q}_{ut}$. Recall that for brevity, we set $I^* = \{1, \ldots, s\}$. By the superscript "star" we denote the functions and sets related to the set Θ of points of discontinuity of the function u^0. Define the following subsets of the neighborhood $\mathcal{V}$; cf. the illustration in Figure 2.1 for $k = 1$:

$$\mathcal{V}_{k-}^* = \{(t,u) \in \mathcal{V}_{k+1} \mid t < t_k\}; \quad \mathcal{V}_{k+}^* = \{(t,u) \in \mathcal{V}_k \mid t > t_k\};$$
$$\mathcal{V}_k^* = \mathcal{V}_{k-}^* \cup \mathcal{V}_{k+}^*, \quad k \in I^*; \quad \mathcal{V}^* = \bigcup_{k \in I^*} \mathcal{V}_k^*; \quad \mathcal{V}^0 = \mathcal{V} \backslash \mathcal{V}^*.$$

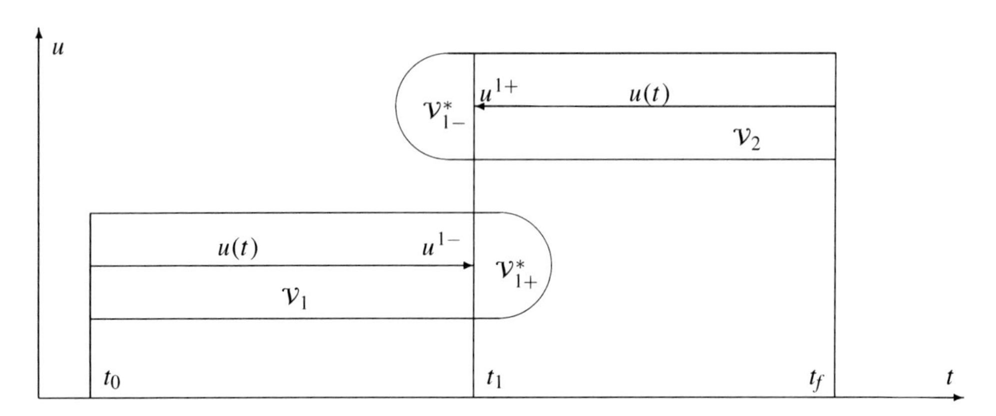

Figure 2.1. *Neighborhoods of the control at a point t_1 of discontinuity.*

While the superscripts $k-$, $k+$, and k were used for designation of left and right limit values and ordinary values of functions at the point $t_k \in \Theta$, the same subscripts will be used for enumeration of sets and functions related to the point $t_k \in \Theta$. The notation $\text{vraimax}_{t\in\mathcal{M}} u(t)$ will be often used to denote the essential supremum (earlier denoted also by ess sup) of a function $u(\cdot)$ on a set $\mathcal{M}$.

Further, let $\{\delta u\} \in \Pi_u^{\text{loc}}$, i.e., $(t, u^0 + \delta u) \to \overline{u^0}$, and let $k \in I^*$. For a sequence $\{\delta u\}$, introduce the sequence of sets $\mathcal{M}^*_{k-} = \{t \mid (t, u^0(t) + \delta u(t)) \in \mathcal{V}^*_{k-}\}$ and assume that χ^*_{k-} is the characteristic function of the set $\mathcal{M}^*_{k-}$. We set $\{\delta u^*_{k-}\} = \{\delta u \chi^*_{k-}\}$. Then

$$\underset{t\in\mathcal{M}^*_{k-}}{\text{vraimax}} |u^0(t) + \delta u^*_{k-}(t) - u^{0k+}| \to 0, \quad \text{where} \quad u^{0k+} = u^0(t_k+).$$

In short, we write this fact as $(u^0 + \delta u^*_{k-})\,|_{\mathcal{M}^*_{k-}} \to u^{0k+}$. Analogously, for the sequence $\{\delta u\}$, we define

$$\mathcal{M}^*_{k+} = \{t \mid (t, u^0(t) + \delta u(t)) \in \mathcal{V}^*_{k+}\} \quad \text{and} \quad \{\delta u^*_{k+}\} = \{\delta u \chi^*_{k+}\},$$

where χ^*_{k+} is the characteristic function of the set $\mathcal{M}^*_{k+}$. Then $(u^0 + \delta u^*_{k+})\,|_{\mathcal{M}^*_{k+}} \to u^{0k-}$, i.e.,

$$\underset{t\in\mathcal{M}^*_{k+}}{\text{vraimax}} |u^0(t) + \delta u^*_{k+}(t) - u^{0k-}| \to 0, \quad \text{where} \quad u^{0k-} = u^0(t_k-).$$

The sequence $\{\delta u^*_{k-}\}$ belongs to the set of sequences Π^*_{uk-} defined as follows: Π^*_{uk-} consists of sequences $\{\delta u\}$ in $L^\infty(\Delta, \mathbb{R}^{d(u)})$ such that

(a) $\text{vraimax}_{t\in\mathcal{M}}\, t \le t_k$, where $\mathcal{M} = \{t \mid \delta u(t) \neq 0\}$, i.e., the support $\mathcal{M}$ of each member δu of the sequence $\{\delta u\}$ is located to the left from t_k (here and in what follows, by the *support* of a measurable function, we mean the set of points at which it is different from zero); for brevity, we write this fact in the form $\mathcal{M} \le t_k$;

(b) $\text{vraimax}_{t\in\mathcal{M}} |t - t_k| \to 0$, i.e., the support $\mathcal{M}$ tends to t_k; for brevity, we write this fact as $\mathcal{M} \to t_k$;

(c) $\text{vraimax}_{t\in\mathcal{M}} |u^0(t) + \delta u(t) - u^{0k+}| \to 0$, i.e., the values of the function $u^0 + \delta u$ on the support $\mathcal{M}$ tend to u^{0k+}; in short, we write this fact in the form

$$(u^0 + \delta u)\,|_{\mathcal{M}} \to u^{0k+}.$$

Analogously, we define the set Π^*_{uk+} consisting of sequences $\{\delta u\}$ such that

$$\mathcal{M} \geq t_k, \qquad \mathcal{M} \to t_k, \qquad (u^0 + \delta u)\,|_{\mathcal{M}} \to u^{0k-},$$

where $\mathcal{M} = \{t \mid \delta u(t) \neq 0\}$. Clearly, $\{\delta u^*_{k+}\} \in \Pi^*_{uk+}$. We set

$$\Pi^*_{uk-} + \Pi^*_{uk+} = \Pi^*_{uk}, \quad \sum_{k \in I^*} \Pi^*_{uk} = \Pi^*_u.$$

By definition, the sum of sets of sequences consists of all sums of sequences belonging to these sets.

As before, let $\{\delta u\} \in \Pi^{\rm loc}_u$. Let the sets $\mathcal{M}^*_{k-}$, $\mathcal{M}^*_{k+}$ and the functions χ^*_{k-}, χ^*_{k+} correspond to the members of this sequence. We set

$$\mathcal{M}^*_k = \mathcal{M}^*_{k-} \cup \mathcal{M}^*_{k+}, \quad \mathcal{M}^* = \bigcup_{k \in I^*} \mathcal{M}^*_k,$$
$$\chi^*_k = \chi^*_{k-} + \chi^*_{k+}, \quad \chi^* = \sum_{k \in I^*} \chi^*_k,$$
$$\delta u^*_k = \delta u \chi^*_k = \delta u^*_{k-} + \delta u^*_{k+}, \quad \delta u^* = \delta u \chi^* = \sum_{k \in I^*} \delta u^*_k.$$

Then the sequence $\{\delta u^*\}$ belongs to the set Π^*_u.

Further, for members δu of the sequence $\{\delta u\}$, we set $\delta u^0 = \delta u - \delta u^*$. Then starting from a certain number, we have for the sequence $\{\delta u^0\}$ that $(t, u^0(t) + \delta u^0(t)) \in \mathcal{V}^0$ a.e. on $[t_0, t_f]$. Moreover, $\delta u^0 \chi^* = 0$. Here, we assume that members of the sequences $\{\delta u^0\}$ and $\{\chi^*\}$ having the same number are multiplied. This remark refers to all relations which contain members of distinct sequences. In what follows, we do not make such stipulations. Clearly, $\|\delta u^0\|_\infty \to 0$. We denote the set of sequence in $L^\infty(\Delta, \mathbb{R}^{d(u)})$ having this property by Π^0_u. Therefore, we have shown that an arbitrary sequence $\{\delta u\} \in \Pi^{\rm loc}_u$ admits the representation

$$\{\delta u\} = \{\delta u^0\} + \{\delta u^*\}, \quad \{\delta u^0\} \in \Pi^0_u, \quad \{\delta u^*\} \in \Pi^*_u, \quad \{\delta u^0 \chi^*\} = \{0\},$$

where χ^* is the characteristic function of the set $\mathcal{M}^* = \{t \mid \delta u^*(t) \neq 0\}$. Such a representation is said to be *canonical.* It is easy to see that the canonical representation is unique. In particular, the existence of the canonical representation implies $\Pi^{\rm loc}_u \subset \Pi^0_u + \Pi^*_u$. Obviously, the converse inclusion holds. Therefore, the relation $\Pi^{\rm loc}_u = \Pi^0_u + \Pi^*_u$ holds.

We now introduce the set Π^* consisting of sequences $\{\delta w^*\} = \{(0, \delta u^*)\}$ in the space W such that $\{\delta u^*\} \in \Pi^*_u$ (the component δx of these sequence vanishes identically).

Proposition 2.9. *We have the relation $\Pi^{\rm loc} = \Pi^0 + \Pi^*$. Moreover, we can represent any sequence $\{\delta w\} = \{(\delta x, \delta u)\} \in \Pi^{\rm loc}$ in the form $\{\delta w\} = \{\delta w^0\} + \{\delta w^*\}$, where $\{\delta w^0\} = \{(\delta x, \delta u^0)\} \in \Pi^0$, $\{\delta w^*\} = \{(0, \delta u^*)\} \in \Pi^*$, and, moreover, for all members of sequences with the same numbers, the condition $\delta u^0 \chi^* = 0$ holds; here, χ^* is the characteristic function of the set $\{t \mid \delta u^*(t) \neq 0\}$.*

Recall that by definition, $\Pi^0 = \{\{\delta w\} \mid \|\delta x\|_{1,1} + \|\delta u\|_\infty \to 0\}$. Proposition 2.9 follows from the relation $\Pi^{\rm loc}_u = \Pi^0_u + \Pi^*_u$ and the existence of the canonical representation for sequences in $\Pi^{\rm loc}_u$.

The representation of a sequence from Π^{loc} in the form of a sum of sequences from Π^0 and Π^* with the condition $\{\delta u^0 \chi^*\} = \{0\}$, which was indicated in Proposition 2.9, will also be called *canonical*. Obviously, the canonical representation is unique. It will play an important role in what follows.

We introduce one more notation related to an arbitrary sequence $\{\delta w\} \in \Pi^{\text{loc}}$. For such a sequence, we set

$$\begin{aligned}
&\delta v_{k-} = (u^0 + \delta u - u^{0k+})\chi^*_{k-} = (u^0 + \delta u^*_{k-} - u^{0k+})\chi^*_{k-},\\
&\delta v_{k+} = (u^0 + \delta u - u^{0k-})\chi^*_{k+} = (u^0 + \delta u^*_{k+} - u^{0k-})\chi^*_{k+},\\
&\{\delta v_k\} = \{\delta v_{k-}\} + \{\delta v_{k+}\}, \quad \{\delta v\} = \sum_{k \in I^*} \{\delta v_k\}.
\end{aligned}$$

Then the supports of δv_{k-}, δv_{k+}, δv_k, and δv are the sets $\mathcal{M}^*_{k-}$, $\mathcal{M}^*_{k+}$, $\mathcal{M}^*_k$, and $\mathcal{M}^*$, respectively. Moreover, it is obvious that $\|\delta v^*_{k-}\|_\infty \to 0$, $\|\delta v^*_{k+}\|_\infty \to 0$, $\|\delta v_k\|_\infty \to 0$, and $\|\delta v\|_\infty \to 0$, i.e., the sequences $\{\delta v^*_{k-}\}$, $\{\delta v^*_{k+}\}$, $\{\delta v_k\}$, and $\{\delta v\}$ belong to Π^0_u.

2.3.2 Representation of the Function $f(t,x,u)$ on Local Sequences

Let $\{\delta w\} \in \Pi^{\text{loc}}$. Recall that

$$\Phi_0(\delta w) := \max_{\lambda \in \Lambda_0} \Phi(\lambda, \delta w) = \max_{\lambda \in \text{co}\,\Lambda_0} \Phi(\lambda, \delta w),$$

where $\Phi(\lambda, \delta w) = \delta l^\lambda - \int_{t_0}^{t_f} \psi \delta \dot{x}\, dt + \int_{t_0}^{t_f} \delta H^\lambda\, dt$, $\delta H^\lambda = \psi \delta f$, and $\text{co}\,\Lambda_0$ is the convex hull of Λ_0.

Consider δf on the sequences $\{\delta w\}$. Represent $\{\delta w\}$ in the canonical form:

$$\{\delta w\} = \{\delta w^0\} + \{\delta w^*\}, \quad \{\delta w^0\} \in \Pi^0, \quad \{\delta w^*\} \in \Pi^*, \quad \{\delta u^0 \chi^*\} = \{0\}.$$

Then

$$\begin{aligned}
\delta f &= f(t, w^0 + \delta w) - f(t, w^0)\\
&= f(t, w^0 + \delta w) - f(t, w^0 + \delta w^*) + f(t, w^0 + \delta w^*) - f(t, w^0)\\
&= f(t, w^0 + \delta w^* + \delta w^0) - f(t, w^0 + \delta w^*) + \delta^* f\\
&= f_w(t, w^0 + \delta w^*)\delta w^0 + \frac{1}{2}\langle f_{ww}(t, w^0 + \delta w^*)\delta w^0, \delta w^0\rangle + r + \delta^* f,
\end{aligned} \tag{2.34}$$

where $\delta^* f = f(t, w^0 + \delta w^*) - f(t, w^0)$, and the residue term r with the components r_i, $i = 1, \dots, d(f)$, is defined by the mean value theorem as follows:

$$\begin{aligned}
&r_i = \frac{1}{2}\langle (f_{iww}(t, w^0 + \delta w^* + \zeta_i \delta w^0) - f_{iww}(t, w^0 + \delta w^*))\delta w^0, \delta w^0\rangle,\\
&\zeta_i = \zeta_i(t), \quad 0 \le \zeta_i(t) \le 1, \quad i = 1, \dots, d(f).
\end{aligned}$$

Therefore,

$$\|r\|_1 = o\left(\|\delta x\|_C^2 + \int_{t_0}^{t_f} |\delta u^0|^2\, dt\right). \tag{2.35}$$

Further, $f_w(t, w^0 + \delta w^*)\delta w^0 = f_w(t, w^0)\delta w^0 + (\delta^* f_w)\delta w^0$, where $\delta^* f_w = f_w(t, w^0 + \delta w^*) - f_w(t, w^0)$. Since $(\delta^* f_w)\delta w^0 = (\delta^* f_x)\delta x + (\delta^* f_u)\delta u^0 = (\delta^* f_x)\delta x$ (because $\delta u^0 \chi^* = 0$), we have

$$f_w(t, w^0 + \delta w^*)\delta w^0 = f_w \delta w^0 + (\delta^* f_x)\delta x. \tag{2.36}$$

Here and in what follows, we set $f_w = f_w(t,w^0)$ for brevity. Analogously,

$$\begin{aligned}\langle f_{ww}(t,w^0+\delta w^*)\delta w^0,\delta w^0\rangle &= \langle f_{ww}(t,w^0)\delta w^0,\delta w^0\rangle + \langle(\delta^* f_{ww})\delta w^0,\delta w^0\rangle \\ &= \langle f_{ww}(t,w^0)\delta w^0,\delta w^0\rangle + \langle(\delta^* f_{xx})\delta x,\delta x\rangle.\end{aligned}$$

Setting $f_{ww}(t,w^0) = f_{ww}$ for brevity, we obtain

$$\langle f_{ww}(t,w^0+\delta w^*)\delta w^0,\delta w^0\rangle = \langle f_{ww}\delta w^0,\delta w^0\rangle + \langle(\delta^* f_{xx})\delta x,\delta x\rangle, \tag{2.37}$$

where $\delta^* f_{xx} = f_{xx}(t,w^0+\delta w^*) - f_{xx}(t,w^0)$. Moreover, $\|\delta^* f_{xx}\|_1 \le \text{const}\cdot \text{meas}\,\mathcal{M}^* \to 0$. This implies

$$\|\langle(\delta^* f_{xx})\delta x,\delta x\rangle\|_1 = o(\|\delta x\|_C^2). \tag{2.38}$$

Substituting (2.36) and (2.37) in (2.34) and taking into account estimates (2.35) and (2.38), we obtain the following assertion.

Proposition 2.10. *Let $\{\delta w\} \in \Pi^{\text{loc}}$. Then the following formula holds for the canonical representation $\{\delta w\} = \{\delta w^0\} + \{\delta w^*\}$:*

$$\delta f = f_w\delta w^0 + \frac{1}{2}\langle f_{ww}\delta w^0,\delta w^0\rangle + \delta^* f + (\delta^* f_x)\delta x + \tilde{r},$$

where

$$\|\tilde{r}\|_1 = o\left(\|\delta x\|_C^2 + \int_{t_0}^{t_f} |\delta u^0|^2 dt\right), \quad \delta^* f = f(t,w^0+\delta w^*) - f(t,w^0),$$

$$\delta^* f_x = f_x(t,w^0+\delta w^*) - f_x(t,w^0), \quad f_w = f_w(t,w^0), \quad f_{ww} = f_{ww}(t,w^0).$$

Therefore, for any $\lambda \in \text{co}\,\Lambda_0$, we have

$$\begin{aligned}\int_{t_0}^{t_f} \delta H^\lambda\, dt &= \int_{t_0}^{t_f} H_x^\lambda \delta x\, dt + \frac{1}{2}\int_{t_0}^{t_f} \langle H_{ww}^\lambda \delta w^0,\delta w^0\rangle\, dt \\ &\quad + \int_{t_0}^{t_f} \delta^* H^\lambda\, dt + \int_{t_0}^{t_f} (\delta^* H_x^\lambda \delta x)\, dt + \rho^\lambda,\end{aligned} \tag{2.39}$$

where $\sup_{\lambda\in\text{co}\,\Lambda_0} |\rho^\lambda| = o(\|\delta x\|_C^2 + \int_{t_0}^{t_f} |\delta u^0|^2 dt)$, $\delta^* H^\lambda = \psi\delta^* f$, and $\delta^* H_x^\lambda = \psi\delta^* f_x$.

Here, we have used the relation $H_u^\lambda = 0$ for all $\lambda \in \text{co}\,\Lambda_0$. As above, all derivatives whose argument is not indicated are taken for $w = w^0(t)$.

2.3.3 Representation of the Integral $\int_{t_0}^{t_f} (\delta^* H_x^\lambda)\delta x\, dt$ on Local Sequences

Proposition 2.11. *Let two sequences $\{\delta x\}$ and $\{\delta u^*\}$ such that $\|\delta x\|_C \to 0$ and $\{\delta u\} \in \Pi_u^*$ be given. Let $\lambda \in \text{co}\,\Lambda_0$. Then*

$$\int_{t_0}^{t_f} (\delta^* H_x^\lambda)\delta x\, dt = \sum_{k\in I^*} \int_{t_0}^{t_f} [H_x^\lambda]^k \delta x(\chi_{k-}^* - \chi_{k+}^*)\, dt + \rho^{*\lambda}, \tag{2.40}$$

where $\sup_{\lambda\in\operatorname{co}\Lambda_0}|\rho^{*\lambda}| = o(\|\delta x\|_C^2 + \sum_{I^*}\int_{\mathcal{M}_k^*}|\delta t_k|\,dt)$, $\delta t_k = t_k - t$, $[H_x^\lambda]^k = H_x^{\lambda k+} - H_x^{\lambda k-} = \psi(t_k)(f_x^{k+} - f_x^{k-}) = \psi(t_k)(f_x(t_k, x^0(t_k), u^{0k+}) - f_x(t_k, x^0(t_k), u^{0k-}))$.

Proof. Since $\chi^* = \sum \chi_k^*$, we have $\delta^* H_x^\lambda = \sum \delta_k^* H_x^\lambda$, and, therefore,

$$\int_{t_0}^{t_f} \delta^* H_x^\lambda \delta x\,dt = \sum_{I^*} \int_{t_0}^{t_f} \delta_k^* H_x^\lambda \delta x\,dt, \quad \delta_k^* H_x^\lambda = \psi \delta_k^* f = \psi(f(t, w^0 + \delta w_k^*) - f(t, w^0)).$$

Further, for $\psi = \psi(t)$ we have $\psi(t) = \psi(t_k) + \Delta_k\psi$, where $\Delta_k\psi = \psi(t_k + \delta t_k) - \psi(t_k)$, and, moreover, $\sup_{\operatorname{co}\Lambda_0}|\Delta_k\psi| \le \text{const}\,|\delta t_k|$, since $-\dot{\psi} = \psi f_x$, and $\sup_{\operatorname{co}\Lambda_0}\|\psi\|_\infty < +\infty$. Consider $\delta_k^* f_x$. Since $\chi_k^* = \chi_{k-}^* + \chi_{k+}^*$, we have $\delta_k^* f_x = \delta_{k-}^* f_x + \delta_{k+}^* f_x$, where the increments $\delta_{k-}^* f_x$ and $\delta_{k+}^* f_x$ correspond to the variations δu_{k-}^* and δu_{k+}^*, respectively. For $\delta_{k-}^* f_x$, we have

$$\delta_{k-}^* f_x = [f_x]^k \chi_{k-}^* + (\delta_{k-}^* f_x - [f_x]^k)\chi_{k-}^*,$$

where

$$[f_x]^k = f_x^{k+} - f_x^{k-}, \quad f_x^{k-} = f_x(t_k, x^0(t_k), u^{0k-}), \quad f_x^{k+} = f_x(t_k, x^0(t_k), u^{0k+}).$$

Further, let $\eta_{k-} := (\delta_{k-}^* f_x - [f_x]^k)\chi_{k-}^* = (f_x(t, x^0, u^0 + \delta u_{k-}^*) - f_x^{k+})\chi_{k-}^* - (f_x(t, x^0, u^0) - f_x^{k-})\chi_{k-}^*$. Then $\|\eta_{k-}\|_\infty \to 0$, since $u^0 + \delta u_{k-}^* \mid_{\mathcal{M}_{k-}^*} \to u^{0k+}$ and $u^0 \mid_{\mathcal{M}_{k-}^*} \to u^{0k-}$. Therefore, $\delta_{k-}^* f_x = [f_x]^k \chi_{k-}^* + \eta_{k-}$, $\eta_{k-}\chi_{k-}^* = \eta_{k-}$, $\|\eta_{k-}\|_\infty \to 0$. Therefore,

$$\begin{aligned} \delta_{k-}^* H_x^\lambda &= \psi \delta_{k-}^* f_x = (\psi(t_k) + \Delta_k\psi)([f_x]^k + \eta_{k-})\chi_{k-}^* \\ &= \psi(t_k)[f_x]^k \chi_{k-}^* + \eta_{k-}^\lambda = [H_x^\lambda]^k \chi_{k-}^* + \eta_{k-}^\lambda, \end{aligned}$$

where $\sup_{\operatorname{co}\Lambda_0}\|\eta_{k-}^\lambda\|_\infty \to 0$, $\eta_{k-}^\lambda \chi_{k-}^* = \eta_{k-}^\lambda$. This implies

$$\int_{t_0}^{t_f} \delta_{k-}^* H_x^\lambda \delta x\,dt = \int_{t_0}^{t_f} [H_x^\lambda]^k \delta x \chi_{k-}^*\,dt + \rho_{k-}^\lambda, \tag{2.41}$$

where $\rho_{k-}^\lambda = \int_{t_0}^{t_f} \eta_{k-}^\lambda \delta x\,dt$. Let us estimate ρ_{k-}^λ. We have

$$|\rho_{k-}^\lambda| \le \sup_{\operatorname{co}\Lambda_0}\|\eta_{k-}^\lambda\|_\infty \|\delta x\|_C \operatorname{meas}\mathcal{M}_{k-}^* \le \frac{1}{2}\sup_{\operatorname{co}\Lambda_0}\|\eta_{k-}^\lambda\|_\infty \left(\|\delta x\|_C^2 + (\operatorname{meas}\mathcal{M}_{k-}^*)^2\right).$$

We use the obvious estimate $(\operatorname{meas}\mathcal{M}_{k-}^*)^2 \le 2\int_{\mathcal{M}_{k-}^*}|\delta t_k|\,dt$. Then

$$\sup_{\operatorname{co}\Lambda_0}|\rho_{k-}^\lambda| = o\left(\|\delta x\|_C^2 + \int_{\mathcal{M}_{k-}^*}|\delta t_k|\,dt\right). \tag{2.42}$$

Analogously, we prove that

$$\int_{t_0}^{t_f} \delta_{k+}^* H_x^\lambda \delta x\,dt = -\int_{t_0}^{t_f} [H_x^\lambda]^k \delta x \chi_{k+}^*\,dt + \rho_{k+}^\lambda, \tag{2.43}$$

where

$$\sup_{\mathrm{co}\,\Lambda_0} |\rho_{k+}^{\lambda}| = o\left(\|\delta x\|_C^2 + \int_{\mathcal{M}_{k+}^*} |\delta t_k \mid dt \right). \tag{2.44}$$

Taking into account that $\delta_k^* H_x^\lambda = \delta_{k-}^* H_x^\lambda + \delta_{k+}^* H_x^\lambda$, we obtain from (2.41)–(2.43) that

$$\int_{t_0}^{t_f} \delta_k^* H_x^\lambda \delta x \, dt = \int_{t_0}^{t_f} [H_x^\lambda]^k \delta x (\chi_{k-}^* - \chi_{k+}^*) dt + \rho_k^\lambda,$$

where $\sup_{\mathrm{co}\,\Lambda_0} |\rho_k^\lambda| = o\left(\|\delta x\|_C^2 + \int_{\mathcal{M}_k^*} |\delta t_k| \, dt \right)$. This and the relation $\delta^* H_x^\lambda = \sum \delta_k^* H_x^\lambda$ imply (2.40). The proposition is proved. ∎

2.3.4 Representation of the Integral $\int_{t_0}^{t_f} \delta^* H^\lambda \, dt$ on Local Sequences

We consider the term $\int_{t_0}^{t_f} \delta^* H^\lambda \, dt$ on an arbitrary sequence $\{\delta u^*\} \in \Pi_u^*$. Since $\chi^* = \sum \chi_k^*$, we have $\delta^* H^\lambda = \sum \delta_k^* H^\lambda$, where $\delta_k^* H^\lambda = \psi \delta_k^* f$. In turn, $\delta_k^* H^\lambda = \delta_{k-}^* H^\lambda + \delta_{k+}^* H^\lambda$, which corresponds to the representation $\delta u_k^* = \delta u_{k-}^* + \delta u_{k+}^*$. Consider the increment $\delta_{k-}^* H^\lambda = \psi \delta_{k-}^* f$. By definition, $\delta_{k-}^* f = f(t, x^0, u^0 + \delta u_{k-}^*) - f(t, x^0, u^0)$, and, moreover, $\delta_{k-}^* f = \delta_{k-}^* f \chi_{k-}^*$. Recall that we have introduced the function

$$\delta v_{k-}^* = (u^0 + \delta u_{k-}^* - u^{0k+}) \chi_{k-}^*.$$

On $\mathcal{M}_{k-}^*$, we have $t = t_k + \delta t_k$, $u^0 + \delta u_{k-}^* = u^{0k+} + \delta v_{k-}$, $x^0(t) = x^0(t_k + \delta t_k) = x^0(t_k) + \Delta_k x^0$, where $\Delta_k x^0 = x^0(t_k + \delta t_k) - x^0(t_k)$. But $\Delta_k x^0 = \dot{x}^{0k-} \delta t_k + o(\delta t_k)$, where $\dot{x}^{0k-} = \dot{x}^0(t_k-)$. Therefore, $|\Delta_k x^0| = O(\delta t_k)$. This implies that on $\mathcal{M}_{k-}^*$,

$$\begin{aligned}
f(t, x^0, u^0 + \delta u_{k-}^*) &= f(t_k + \delta t_k, x^0(t_k) + \Delta_k x^0, u^{0k+} + \delta v_{k-}) \\
&= f^{k+} + f_t^{k+} \delta t_k + f_x^{k+} \Delta_k x^0 + f_u^{k+} \delta v_{k-} \\
&\quad + \frac{1}{2} \langle (f'')^{k+} (\delta t_k, \Delta_k x^0, \delta v_{k-}), (\delta t_k, \Delta_k x^0, \delta v_{k-}) \rangle \\
&\quad + o\left(|\delta t_k|^2 + |\Delta_k x^0|^2 + |\delta v_{k-}|^2 \right),
\end{aligned}$$

where f^{k+}, f_t^{k+}, f_x^{k+}, f_u^{k+}, and $(f'')^{k+}$ are values of the function f and its derivatives at the point $(t_k, x^0(t_k), u^{0k+})$. Taking into account that $\Delta_k x^0 = \dot{x}^{0k-} \delta t_k + o(\delta t_k)$, we obtain from this that on $\mathcal{M}_{k-}^*$,

$$\begin{aligned}
f(t, x^0, u^0 + \delta u_{k-}^*) &= f^{k+} + f_t^{k+} \delta t_k + f_x^{k+} \dot{x}^{0k-} \delta t_k + f_u^{k+} \delta v_{k-} \\
&\quad + \frac{1}{2} \langle f_{uu}^{k+} \delta v_{k-}, \delta v_{k-} \rangle + o(|\delta t_k| + |\delta v_{k-}|^2).
\end{aligned} \tag{2.45}$$

Further, consider $f(t, x^o, u^0)$ on $\mathcal{M}_{k-}^*$. For $t < t_k$, we set $\Delta_k u^0 = u^0(t) - u^{0k-}$. Note that $|\Delta_k u^0| = O(\delta t_k)$ by the assumption that t_k is an L-point of the function u^0. Moreover, $\Delta_k x^0 = \dot{x}^{0k-} \delta t_k + o(\delta t_k)$. Therefore,

$$\begin{aligned}
f(t, x^0, u^0) &= f(t_k + \delta t_k, x^0(t_k) + \Delta_k x^0, u^{0k-} + \Delta_k u^0) \\
&= f^{k-} + f_t^{k-} \delta t_k + f_x^{k-} \dot{x}^{0k-} \delta t_k + f_u^{k-} \Delta_k u^0 + o(\delta t_k),
\end{aligned} \tag{2.46}$$

where f^{k-}, f_t^{k-}, f_x^{k-}, and f_u^{k-} are the values of the function f and its derivatives at the point $(t_k, x^0(t_k), u^{0k-})$. Subtracting (2.46) from (2.45), we obtain the following on $\mathcal{M}_{k-}$:

$$\begin{aligned} \delta^*{}_{k-}f &= [f]^k + [f_t]^k\delta t_k + [f_x]^k\dot{x}^{0k-}\delta t_k + f_u^{k+}\delta v_{k-} - f_u^{k-}\Delta_k u^0 \\ &+\frac{1}{2}\langle f_{uu}^{k+}\delta v_{k-}, \delta v_{k-}\rangle + o(|\delta t_k| + |\delta v_{k-}|^2). \end{aligned} \tag{2.47}$$

Taking into account that $\delta^*_{k-}f$ and δv_{k-} are concentrated on $\mathcal{M}^*_{k-}$ and $\delta t_k = -|\delta t_k|$ on $\mathcal{M}^*_{k-}$, we obtain from this that

$$\begin{aligned} \delta^*{}_{k-}f &= [f]^k\chi^*_{k-} - \left([f_t]^k + [f_x]^k\dot{x}^{0k-}\right)|\delta t_k|\chi^*_{k-} + f_u^{k+}\delta v_{k-} \\ &-f_u^{k-}\Delta_k u^0\chi^*_{k-} + \frac{1}{2}\langle f_{uu}^{k+}\delta v_{k-}, \delta v_{k-}\rangle + o(|\delta t_k| + |\delta v_{k-}|^2)\chi^*_{k-}. \end{aligned} \tag{2.48}$$

Analogously, for $\delta^*_{k+}f$, we have the formula

$$\begin{aligned} \delta^*{}_{k+}f &= -[f]^k\chi^*_{k+} - \left([f_t]^k + [f_x]^k\dot{x}^{0k+}\right)|\delta t_k|\chi^*_{k+} + f_u^{k-}\delta v_{k+} \\ &-f_u^{k+}\Delta_k u^0\chi^*_{k+} + \frac{1}{2}\langle f_{uu}^{k-}\delta v_{k+}, \delta v_{k+}\rangle + o(|\delta t_k| + |\delta v_{k+}|^2)\chi^*_{k+}, \end{aligned} \tag{2.49}$$

where $\Delta_k u^0 = u^0(t) - u^{0k+}$ for $t > t_k$, and $\dot{x}^{0k+} = \dot{x}^0(t_k+)$.

We mention a consequence of formulas (2.48) and (2.49), which will be needed in what follows. Adding (2.48) and (2.49) and taking into account that $|\Delta_k u^0| = O(\delta t_k)$, we obtain

$$\delta^*_k f = [f]^k(\chi^*_{k-} - \chi^*_{k+}) + f_u^{k+}\delta v_{k-} + f_u^{k-}\delta v_{k+} + O(|\delta t_k| + |\delta v_k|^2)\chi^*_k. \tag{2.50}$$

This formula holds for an arbitrary sequence $\{\delta u^*_k\} \in \Pi^*_{uk}$.

Let us return to formula (2.48) and use it to obtain the expression for $\delta^*_{k-}H^\lambda = \psi\delta^*_{k-}f$. On $\mathcal{M}^*_{k-}$, we have

$$\begin{aligned} \psi(t) &= \psi(t_k + \delta t_k) = \psi(t_k) + \dot{\psi}^{k-}\delta t_k + \eta^\psi_{k-}|\delta t_k| \\ &= \psi(t_k) - \dot{\psi}^{k-}|\delta t_k| + \eta^\psi_{k-}|\delta t_k|, \end{aligned} \tag{2.51}$$

where $\dot{\psi}^{k-} = \dot{\psi}(t_k-)$ and $\sup_{\operatorname{co}\Lambda_0}|\eta^\psi_{k-}| \to 0$ as $|\delta t_k| \to 0$, since $\dot{\psi} = -H^\lambda_x$ is left continuous at the point t_k. We obtain from (2.48) and (2.51) that

$$\begin{aligned} \delta^*_{k-}H^\lambda = [H^\lambda]^k\chi^*_{k-} - \left([H^\lambda_t]^k + [H^\lambda_x]^k\dot{x}^{0k-} + \dot{\psi}^{k-}[H_\psi]^k\right)|\delta t_k|\chi^*_{k-} \\ +\frac{1}{2}\langle H^{\lambda k+}_{uu}\delta v_{k-}, \delta v_{k-}\rangle + \tilde{\eta}^\lambda_{k-}(|\delta t_k| + |\delta v_{k-}|^2), \end{aligned} \tag{2.52}$$

where $\tilde{\eta}^\lambda_{k-}\chi^*_{k-} = \tilde{\eta}^\lambda_{k-}$ and $\sup_{\lambda\in\operatorname{co}\Lambda_0}\|\tilde{\eta}^\lambda_{k-}\|_\infty \to 0$. Here, we have taken into account that $H^{\lambda k+}_u = H^{\lambda k-}_u = 0$ for all $\lambda \in \operatorname{co}\Lambda_0$ and $[f]^k = [H_\psi]^k$.

We now turn to formula (2.49) and use it to obtain the expression for $\delta^*_{k+}H^\lambda = \psi\delta^*_{k+}f$. On $\mathcal{M}^*_{k+}$, we have

$$\begin{aligned} \psi(t) &= \psi(t_k + \delta t_k) = \psi(t_k) + \dot{\psi}^{k+}\delta t_k + \eta^\psi_{k+}\delta t_k \\ &= \psi(t_k) + \dot{\psi}^{k+}|\delta t_k| + \eta^\psi_{k+}|\delta t_k|, \end{aligned} \tag{2.53}$$

where $\dot{\psi}^{k+} = \dot{\psi}(t_k+)$, and $\sup_{\operatorname{co}\Lambda_0} |\eta^{\psi}_{k+}| \to 0$. Analogously to formula (2.52), we obtain from (2.49) and (2.53) that

$$\delta^*_{k+} H^\lambda = -[H^\lambda]^k \chi^*_{k+} - \left([H^\lambda_t]^k + [H^\lambda_x]^k \dot{x}^{0k+} + \dot{\psi}^{k+}[H_\psi]^k\right)|\delta t_k|\chi^*_{k+}$$
$$+\frac{1}{2}\langle H^{\lambda k-}_{uu}\delta v_{k+}, \delta v_{k+}\rangle + \tilde{\eta}^\lambda_{k+}(|\delta t_k| + |\delta v_{k+}|^2), \tag{2.54}$$

where $\tilde{\eta}^\lambda_{k+}\chi^*_{k+} = \tilde{\eta}^\lambda_{k+}$ and $\sup_{\lambda\in\operatorname{co}\Lambda_0} \|\tilde{\eta}^\lambda_{k+}\|_\infty \to 0$.

A remarkable fact is that the coefficients of $|\delta t_k|$ in formulas (2.52) and (2.54) coincide and are the derivative at the point $t = t_k$ of the function $\{\Delta_k H^\lambda\}(t)$ introduced in Section 2.1.6. Let us show this. Let $k \in I^*$ and $\lambda \in \operatorname{co}\Lambda_0$. Recall that by definition,

$$(\Delta_k H^\lambda)(t) = H(t, x^0(t), u^{0k+}, \psi(t)) - H(t, x^0(t), u^{0k-}, \psi(t)) = \psi(t)(\Delta_k f)(t),$$

where $(\Delta_k f)(t) = f(t, x^0(t), u^{0k+}) - f(t, x^0(t), u^{0k-})$. In what follows, we will omit the superscript λ of H. The conditions $\dot{\psi}(t) = -\psi(t) f_x(t, x^0(t), u^0(t))$, $\dot{x}^0(t) = f(t, x^0(t), u^0(t))$, and the property of the function $u^0(t)$ implies the existence of the left and right derivatives of the functions $\psi(t)$ and $x^0(t)$ at the point t_k, and, moreover, the left derivative is the left limit of the derivatives, and the right derivative is the right limit of the derivatives. On each of the intervals of the set $[t_0, t_f]\backslash\Theta$, the derivatives $\dot{\psi}$, $\dot{x}^0$ are continuous. Analogous assertions hold for the function $(\Delta_k H)(t)$. Its left derivative at the point t_k can be calculated by the formula

$$\frac{d}{dt_-}(\Delta_k H)(t_k) = \frac{d}{dt_-}\left(H(t, x^0(t), u^{0k+}, \psi(t)) - H(t, x^0(t), u^{0k-}, \psi(t))\right)\Bigg|_{t=t_k}$$
$$= [H_t]^k + [H_x]^k \dot{x}^{0k-} + \dot{\psi}^{k-}[H_\psi]^k. \tag{2.55}$$

This derivative is the coefficient of $|\delta t_k|$ in expression (2.52) for $\delta^*_{k-}H$. The right derivative of the function $\{H\}^k(t)$ at the point t_k can be calculated by the formula

$$\frac{d}{dt_+}(\Delta_k H)(t_k) = \frac{d}{dt_+}\left(H(t, x^0(t), u^{0k+}, \psi(t)) - H(t, x^0(t), u^{0k-}, \psi(t))\right)\Bigg|_{t=t_k}$$
$$= [H_t]^k + [H_x]^k \dot{x}^{0k+} + \dot{\psi}^{k+}[H_\psi]^k. \tag{2.56}$$

This derivative is the coefficient of $|\delta t_k|$ in expression (2.54) for $\delta^*_{k+}H$. We show that

$$\frac{d}{dt_-}(\Delta_k H)(t_k) = \frac{d}{dt_+}(\Delta_k H)(t_k); \tag{2.57}$$

i.e., the function $(\Delta_k H)(t_k)$ is differentiable at the point t_k. Indeed,

$$\frac{d}{dt_-}(\Delta_k H)(t_k) = [H_t]^k + [H_x]^k \dot{x}^{0k-} + \dot{\psi}^{k-}[H_\psi]^k$$
$$= [H_t]^k + (H^{k+}_x - H^{k-}_x)H^{k-}_\psi - H^{k-}_x(H^{k+}_\psi - H^{k-}_\psi)$$
$$= [H_t]^k + H^{k+}_x H^{k-}_\psi - H^{k-}_x H^{k+}_\psi. \tag{2.58}$$

But the right derivative has the same form:

$$\begin{aligned}\frac{d}{dt_+}(\Delta_k H)(t_k) &= [H_t]^k + [H_x]^k \dot{x}^{0k+} + \dot{\psi}^{k+}[H_\psi]^k \\ &= [H_t]^k + (H_x^{k+} - H_x^{k-})H_\psi^{k+} - H_x^{k+}(H_\psi^{k+} - H_\psi^{k-}) \\ &= [H_t]^k + H_x^{k+}H_\psi^{k-} - H_x^{k-}H_\psi^{k+}.\end{aligned}$$

We denote the derivative of the function $-(\Delta_k H^\lambda)(t)$ at the point t_k by $D^k(H^\lambda)$. Therefore, we have proved the following assertion.

Lemma 2.12. *The function $(\Delta_k H)(t)$ is differentiable at each point $t_k \in \Theta$. Its derivative at this point can be calculated by the formulas*

$$\begin{aligned}\frac{d}{dt}(\Delta_k H)(t_k) &= [H_t^\lambda]^k + H_x^{\lambda k+}H_\psi^{\lambda k-} - H_x^{\lambda k-}H_\psi^{\lambda k+} \\ &= [H_t^\lambda]^k + [H_x^\lambda]^k \dot{x}^{k-} + \dot{\psi}^{k-}[H_\psi^\lambda]^k = [H_t^\lambda]^k + [H_x^\lambda]^k \dot{x}^{k+} + \dot{\psi}^{k+}[H_\psi^\lambda]^k.\end{aligned}$$

In particular, these formulas imply

$$\frac{d}{dt}(\Delta_k H)(t_k-) = \frac{d}{dt}(\Delta_k H)(t_k+),$$

and, therefore, $(\Delta_k H)(t)$ is continuously differentiable at each point $t_k \in \Theta$. By definition

$$-\frac{d}{dt}(\Delta_k H)(t_k) = D^k(H).$$

We now turn to formulas (2.52) and (2.54). Since $\delta_k^* H^\lambda = \delta_{k-}^* H^\lambda + \delta_{k+}^* H^\lambda$ and $\delta^* H^\lambda = \sum_k \delta_k^* H^\lambda$, the following assertions follows from these formulas and Lemma 2.12.

Proposition 2.13. *Let $\{\delta w^*\} = \{(0, \delta u^*)\} \in \Pi^*$ and $\lambda = (\alpha_0, \alpha, \beta, \psi) \in \operatorname{co} \Lambda_0$. Then,*

$$\begin{aligned}\delta^* H^\lambda = \sum_{k \in I^*} \Big(&[H^\lambda]^k (\chi_{k-}^* - \chi_{k+}^*) + D^k(H^\lambda)|\delta t_k| \chi_k^* + \frac{1}{2}\langle H_{uu}^{\lambda k+} \delta v_{k-}, \delta v_{k-} \rangle \\ &+ \frac{1}{2}\langle H_{uu}^{\lambda k-} \delta v_{k+}, \delta v_{k+} \rangle + \tilde{\eta}_k^\lambda (|\delta t_k| + |\delta v_k|^2) \Big),\end{aligned}$$

where $\sup_{\lambda \in \operatorname{co} \Lambda_0} \|\tilde{\eta}_k^\lambda\|_\infty \to 0$, $\tilde{\eta}_k^\lambda \chi_k^* = \tilde{\eta}_k^\lambda$, $k \in I^*$. *Therefore,*

$$\begin{aligned}\int_{t_0}^{t_f} \delta^* H^\lambda \, dt = &\sum_{k \in I^*} \Big([H^\lambda]^k (\operatorname{meas} \mathcal{M}_{k-}^* - \operatorname{meas} \mathcal{M}_{k+}^*) + D^k(H^\lambda) \int_{\mathcal{M}_k^*} |\delta t_k| \, dt \\ &+ \frac{1}{2} \int_{t_0}^{t_f} \Big(\langle H_{uu}^{\lambda k+} \delta v_{k-}, \delta v_{k-} \rangle + \langle H_{uu}^{\lambda k-} \delta v_{k+}, \delta v_{k+} \rangle \Big) dt \Big) \\ &+ \sum_{k \in I^*} \varepsilon_k^\lambda \int_{\mathcal{M}_k^*} (|\delta t_k| + |\delta v_k|^2) \, dt,\end{aligned}$$

where $\sup_{\lambda \in \operatorname{co} \Lambda_0} |\varepsilon_k^\lambda| \to 0$, $k \in I^*$.

2.3.5 Representation of the Integral $\int_{t_0}^{t_f} \delta H^\lambda\, dt$ on Local Sequences

Propositions 2.10, 2.11, and 2.13 imply the following assertion.

Proposition 2.14. *Let $\{\delta w\} \in \Pi^{\rm loc}$ be represented in the canonical form: $\{\delta w\} = \{\delta w^0\} + \{\delta w^*\}$, where $\{\delta w^0\} = \{(\delta x, \delta u^0)\}$, $\{\delta w^*\} = \{(0, \delta u^*)\}$, $\{\delta u^0 \chi^*\} = \{0\}$. Then the following formula holds for any $\lambda \in \operatorname{co}\Lambda_0$:*

$$\begin{aligned}\int_{t_0}^{t_f} \delta H^\lambda\, dt \;=\; & \int_{t_0}^{t_f} H_x^\lambda \delta x\, dt + \frac{1}{2}\int_{t_0}^{t_f} \langle H_{ww}^\lambda \delta w^0, \delta w^0\rangle\, dt \\ & + \sum_{k\in I^*}\Bigg([H^\lambda]^k(\operatorname{meas}\mathcal{M}_{k-}^* - \operatorname{meas}\mathcal{M}_{k+}^*) + D^k(H^\lambda)\int_{\mathcal{M}_k^*} |\delta t_k|\, dt \\ & \quad + \frac{1}{2}\int_{t_0}^{t_f}\Big(\langle H_{uu}^{\lambda k+}\delta v_{k-}, \delta v_{k-}\rangle + \langle H_{uu}^{\lambda k-}\delta v_{k+}, \delta v_{k+}\rangle\Big)\, dt \\ & \quad + [H_x^\lambda]^k \int_{t_0}^{t_f} \delta x(\chi_{k-}^* - \chi_{k+}^*)\, dt\Bigg) + \rho_H^\lambda,\end{aligned}$$

where $\rho_H^\lambda = \varepsilon_H^\lambda\big(\sum_{k\in I^*}\int_{\mathcal{M}_k^*}(|\delta t_k| + |\delta v_k|^2)\, dt + \int_{t_0}^{t_f}|\delta u^0|^2\, dt + \|\delta x\|_C^2\big)$, $\sup_{\lambda\in\operatorname{co}\Lambda_0}|\varepsilon_H^\lambda| \to 0$.

2.3.6 Expansion of the Lagrange Function on Local Sequences

Now recall that $\Phi(\lambda, \delta w) = \delta l^\lambda - \int_{t_0}^{t_f} \psi\delta\dot{x}\, dt + \int_{t_0}^{t_f} \delta H^\lambda\, dt$ for $\{\delta w\} \in \Pi^{\rm loc}$. For δl^λ, we have the decomposition

$$\delta l^\lambda = l^\lambda(p^0 + \delta p) = l_p^\lambda \delta p + \frac{1}{2}\langle l_{pp}^\lambda \delta p, \delta p\rangle + \varepsilon_l^\lambda |\delta p|^2,$$

where $\sup_{\lambda\in\operatorname{co}\Lambda_0}|\varepsilon_l^\lambda| \to 0$. We integrate the term $\int_{t_0}^{t_f} \psi\delta\dot{x}\, dt$ by parts and use the transversality conditions and also the adjoint equation:

$$\begin{aligned}\int_{t_0}^{t_f} \psi\delta\dot{x}\, dt &= -\psi(t_0)\delta x(t_0) + \psi(t_f)\delta x(t_f) - \int_{t_0}^{t_f} \dot{\psi}\delta x\, dt \\ &= l_{x_0}^\lambda \delta x(t_0) + l_{x_f}^\lambda \delta x(t_f) + \int_{t_0}^{t_f} H_x^\lambda \delta x\, dt = l_p^\lambda \delta p + \int_{t_0}^{t_f} H_x^\lambda \delta x\, dt.\end{aligned}$$

From this we obtain

$$\delta l^\lambda - \int_{t_0}^{t_f} \psi\delta\dot{x}\, dt = \frac{1}{2}\langle l_{pp}^\lambda \delta p, \delta p\rangle - \int_{t_0}^{t_f} H_x^\lambda \delta x\, dt + \varepsilon_l^\lambda |\delta p|^2. \tag{2.59}$$

For the sequence $\{\delta w\} \in \Pi^{\rm loc}$ represented in the canonical form, we set

$$\gamma(\delta w) = \|\delta x\|_C^2 + \int_{t_0}^{t_f} |\delta u^0|^2\, dt + 2\sum_{k=1}^{s}\int_{\mathcal{M}_k^*} |\delta t_k|\, dt + \int_{t_0}^{t_f} |\delta v|^2\, dt. \tag{2.60}$$

Formula (2.59) and Proposition 2.14 imply the following assertion.

Proposition 2.15. *For any sequence* $\{\delta w\} \in \Pi^{\mathrm{loc}}$ *represented in the canonical form,* $\{\delta w\} = \{\delta w^0\} + \{\delta w^*\}$, $\{\delta w^0\} = \{(\delta x, \delta u^0)\} \in \Pi^0$, $\{\delta w^*\} = \{(0, \delta u^*)\} \in \Pi^*$, $\{\delta u^0 \chi^*\} = \{0\}$, *and for any* $\lambda \in \operatorname{co}\Lambda_0$, *we have the formula*

$$\Phi(\lambda, \delta w) = \Phi^{1\lambda}(\delta w) + \varepsilon_\Phi^\lambda \gamma(\delta w),$$

where

$$\begin{aligned}\Phi^{1\lambda}(\delta w) = &\frac{1}{2}\langle l_{pp}^\lambda \delta p, \delta p\rangle + \frac{1}{2}\int_{t_0}^{t_f} \langle H_{ww}^\lambda \delta w^0, \delta w^0\rangle\, dt \\ &+ \sum_{k=1}^{s} \Big([H^\lambda]^k (\operatorname{meas} \mathcal{M}_{k-}^* - \operatorname{meas} \mathcal{M}_{k+}^*) \\ &+ D^k(H^\lambda) \int_{\mathcal{M}_k^*} |\delta t_k|\, dt + [H_x^\lambda]^k \int_{t_0}^{t_f} \delta x (\chi_{k-}^* - \chi_{k+}^*)\, dt \\ &+ \frac{1}{2}\int_{t_0}^{t_f} \Big(\langle H_{uu}^{\lambda k+} \delta v_{k-}, \delta v_{k-}\rangle + \langle H_{uu}^{\lambda k-} \delta v_{k+}, \delta v_{k+}\rangle \Big)\, dt \Big) \end{aligned} \tag{2.61}$$

and $\sup_{\lambda \in \operatorname{co}\Lambda_0} |\varepsilon_\Phi^\lambda| \to 0$.

In expression (2.61) for $\Phi^{1\lambda}(\delta w)$, all terms, except for $\sum_{k=1}^{s} [H^\lambda]^k (\operatorname{meas} \mathcal{M}_{k-}^* - \operatorname{meas} \mathcal{M}_{k+}^*)$, are estimated through γ on any sequence $\{\delta w\} \in \Pi^{\mathrm{loc}}$ starting from a certain number. For example,

$$\left| \int_{t_0}^{t_f} \delta x (\chi_{k-}^* - \chi_{k+}^*)\, dt \right| \le \|\delta x\|_C \operatorname{meas} \mathcal{M}_k^* \le \frac{1}{2}(\|\delta x\|_C^2 + (\operatorname{meas} \mathcal{M}_k^*)^2) \le \gamma(\delta w),$$

since $(\operatorname{meas} \mathcal{M}_k^*)^2 \le 4 \int_{\mathcal{M}_k^*} |\delta t_k|\, dt$. (This estimate follows from the estimates $\frac{1}{2}(\operatorname{meas} \mathcal{M}_{k-}^*)^2 \le \int_{\mathcal{M}_{k-}^*} |\delta t_k|\, dt$, $\frac{1}{2}(\operatorname{meas} \mathcal{M}_{k+}^*)^2 \le \int_{\mathcal{M}_{k+}^*} |\delta t_k|\, dt$, and the equality $\operatorname{meas} \mathcal{M}_k^* = \operatorname{meas} \mathcal{M}_{k-}^* + \operatorname{meas} \mathcal{M}_{k+}^*$.)

Recall that by Λ_0^Θ we denote the set consisting of those $\lambda \in \Lambda_0$ for which the conditions $[H^\lambda]^k = 0$ for all $k \in I^*$ hold. Proposition 2.15 implies the following assertion.

Proposition 2.16. *Let the set* Λ_0^Θ *be nonempty. Then there exists a constant* $C_\Theta > 0$ *such that the following estimate holds at any sequence* $\{\delta w\} \in \Pi^{\mathrm{loc}}$ *represented in the canonical form, starting from a certain number*:

$$\max_{\lambda \in \operatorname{co}\Lambda_0^\Theta} |\Phi(\lambda, \delta w)| \le C_\Theta \gamma(\delta w).$$

We will need this estimate later, in Section 2.5.

We have made an important step in the way of distinguishing the quadratic form. Also, we have defined the functional γ on local sequences. We must extend this functional to Pontryagin sequences. Precisely, this functional will define the higher order which we will use to obtain quadratic conditions in the problem considered. Note that $\Phi''(\lambda, \delta w^0) := \frac{1}{2}\langle l_{pp}^\lambda \delta p, \delta p\rangle + \frac{1}{2}\int_{t_0}^{t_f} \langle H_{ww}^\lambda \delta w^0, \delta w^0\rangle\, dt$ is the second variation of the Lagrange functional.

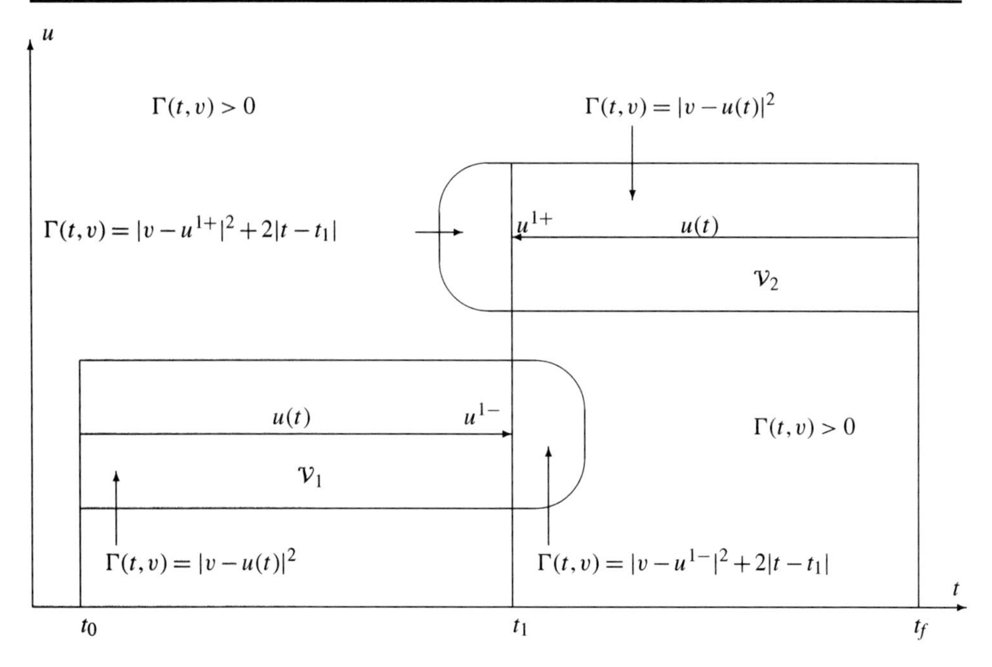

Figure 2.2. *Definition of functions* $\Gamma(t,v)$ *on neighborhoods of discontinuity points.*

2.3.7 Higher Order γ

We first define the concept of admissible function $\Gamma(t,u)$; cf. the illustration in Figure 2.2 for $k=1$.

Definition 2.17. A function $\Gamma(t,u): \mathcal{Q}_{tu} \to \mathbb{R}$ is said to be *admissible* (or an *order function*) if it is continuous on $\mathcal{Q}_{tu}$ and there exist disjoint neighborhoods $\mathcal{V}_k \subset \mathcal{Q}_{tu}$ of the compact sets $\overline{u^0}(t_{k-1},t_k)$ such that the following five conditions hold:

(1) $\Gamma(t,u) = |u-u^0(t)|^2$ if $(t,u)\in\mathcal{V}_k$, $t\in(t_{k-1},t_k)$, $k=1,\dots,s+1$;

(2) $\Gamma(t,u) = 2|t-t_k|+|u-u^{0k-}|^2$ if $(t,u)\in\mathcal{V}_k$, $t>t_k$, $k=1,\dots,s$;

(3) $\Gamma(t,u) = 2|t-t_k|+|u-u^{0k+}|^2$ if $(t,u)\in\mathcal{V}_{k+1}$, $t<t_k$, $k=1,\dots,s$;

(4) $\Gamma(t,u)>0$ on $\mathcal{Q}_{tu}\setminus\mathcal{V}$, where $\mathcal{V}=\bigcup_{k=1}^{s+1}\mathcal{V}_k$;

(5) for any compact set $\mathcal{F}\subset\mathcal{Q}_{tu}\setminus\mathcal{V}$, there exists a constant $L>0$ such that $|\Gamma(t,u')-\Gamma(t,u'')|\le L|u'-u''|$ if (t,u') and (t,u'') belong to $\mathcal{F}$.

Let us show that there exists at least one admissible function Γ. Fix arbitrary disjoint neighborhoods $\mathcal{V}_k\subset\mathcal{Q}_{tu}$ of the compact sets $\overline{u^0}(t_{k-1},t_k)$ and define Γ on $\mathcal{V}=\bigcup\mathcal{V}_k$ by conditions (1)–(3). We set $\mathcal{V}_\varepsilon=\{(t,u)\in\mathcal{V}\mid\Gamma(t,u)<\varepsilon\}$. For a sufficiently small $\varepsilon=\varepsilon_0>0$, the set $\mathcal{V}_{\varepsilon_0}$ is a neighborhood of $\overline{u^0}$ contained in $\mathcal{V}$ together with its closure. For the above ε_0, we set

$$\Gamma_0(t,u)=\begin{cases}\Gamma(t,u) & \text{if } (t,u)\in\mathcal{V}_{\varepsilon_0},\\ \varepsilon_0 & \text{if } (t,u)\in\mathcal{Q}_{tu}\setminus\mathcal{V}_{\varepsilon_0}.\end{cases}$$

Then the function Γ_0 is admissible. An admissible function Γ is not uniquely defined, but any two of them coincide in a sufficiently small neighborhood of the compact set $\overline{u^0}$.

Let $\Gamma(t,u)$ be a certain admissible function. We set

$$\gamma(\delta w) = \|\delta x\|_C^2 + \int_{t_0}^{t_f} \Gamma(t, u^0 + \delta u)\, dt. \tag{2.62}$$

This functional is defined for pairs $\delta w = (\delta x, \delta u) \in W$ such that $(t, u^0 + \delta u) \in \mathcal{Q}_{tu}$ a.e. on $[t_0, t_f]$. Such pairs are said to be *admissible with respect to* $\mathcal{Q}_{tu}$ (also, in this case, the variation δu is said to be *admissible with respect to* $\mathcal{Q}_{tu}$). It is easy to see that for any local sequence $\{\delta w\} \in \Pi^{\text{loc}}$, the values of $\gamma(\delta w)$ can be calculated by formula (2.60) starting from a certain number, and, therefore in the definition of γ and in formula (2.60), we have used the same notation.

Let us verify that γ is a strict higher order on Π, where Π is the set of Pontryagin sequences. Obviously, $\gamma \geq 0$, $\gamma(0) = 0$, and for any variation δw admissible on $\mathcal{Q}_{tu}$, the condition $\gamma(\delta w) = 0$ implies $\delta w = 0$.

Let us show that the functional γ is Π-continuous at zero. It is required to show that $\gamma(\delta w) \to 0$ for any Pontryagin sequence $\{\delta w\}$. Since the condition $\|\delta x\|_{1,1} \to 0$ holds for $\{\delta w\} \in \Pi$ and $\|\delta x\|_C \leq \|\delta x\|_{1,1}$, it suffices to show that for $\{\delta w\} \in \Pi$, the condition $\int_{t_0}^{t_f} \Gamma(t, u^0 + \delta u)\, dt \to 0$ holds. Let $U_\varepsilon(\overline{u^0})$ be an ε-neighborhood of the set $\overline{u^0}$ in $\mathbb{R}^{1+d(u)}$. Assume that $\varepsilon > 0$ is chosen so that $U_\varepsilon(\overline{u^0}) \subset \mathcal{Q}_{tu}$. Represent δu in the form $\delta u = \delta u_\varepsilon + \delta u^\varepsilon$, where

$$\delta u_\varepsilon(t) = \begin{cases} \delta u(t) & \text{if } (t, u^0(t) + \delta u(t)) \in U_\varepsilon(\overline{u^0}), \\ 0 & \text{otherwise.} \end{cases}$$

We set $\mathcal{M}^\varepsilon = \{t \mid \delta u^\varepsilon(t) \neq 0\}$, $\mathcal{M}_\varepsilon = [t_0, t_f] \backslash \mathcal{M}^\varepsilon$. Since $|\delta u^\varepsilon| \geq \varepsilon$ on $\mathcal{M}^\varepsilon$, we have

$$\varepsilon \operatorname{meas} \mathcal{M}^\varepsilon \leq \int_{t_0}^{t_f} |\delta u^\varepsilon|\, dt \leq \int_{t_0}^{t_f} |\delta u|\, dt \to 0.$$

Therefore, for any fixed ε, we have $\operatorname{meas} \mathcal{M}^\varepsilon \to 0$. But then we can choose a subsequence $\varepsilon \to +0$ such that $\operatorname{meas} \mathcal{M}^\varepsilon \to 0$. (Recall that $\mathcal{M}^\varepsilon$ is defined by a member of the sequence $\{\delta u\}$ and the corresponding member of the sequence $\{\varepsilon\}$; when defining $\mathcal{M}^\varepsilon$, we take the members of the sequences $\{\delta u\}$ and $\{\varepsilon\}$ with the same numbers.) Fix such a sequence $\{\varepsilon\}$. Since $\|\delta u^\varepsilon\|_\infty \leq O(1)$, we have $\|\Gamma(t, u^0 + \delta u^\varepsilon)\|_\infty \leq O(1)$. Therefore, $\int_{t_0}^{t_f} \Gamma(t, u^0 + \delta u^\varepsilon)\, dt \leq \|\Gamma(t, u^0 + \delta u^\varepsilon)\|_\infty \operatorname{meas} \mathcal{M}^\varepsilon \to 0$. Moreover, the condition $\varepsilon \to +0$ implies $\{\delta u_\varepsilon\} \in \Pi_u^{\text{loc}}$, and therefore, $\|\Gamma(t, u^0 + \delta u_\varepsilon)\|_\infty \to 0$, which implies $\int_{t_0}^{t_f} \Gamma(t, u^0 + \delta u_\varepsilon)\, dt \to 0$. Then

$$\int_{t_0}^{t_f} \Gamma(t, u^0 + \delta u)\, dt = \int_{t_0}^{t_f} \Gamma(t, u^0 + \delta u_\varepsilon)\, dt + \int_{t_0}^{t_f} \Gamma(t, u^0 + \delta u^\varepsilon)\, dt \to 0;$$

this is what was required to be proved. Therefore, we have shown that the functional γ is Π-continuous at zero. Therefore, γ is an order. Moreover, γ is a strict order.

Let us verify that γ is a higher order. Let $\{\delta w\} \in \Pi$, and let $\{\bar{w}\} \in \Pi^0$. We need to show that $\gamma(\delta w + \bar{w}) = \gamma(\delta w) + o(\|\bar{w}\|)$, where $\|\bar{w}\| = \|\bar{x}\|_{1,1} + \|\bar{u}\|_\infty$. Since $\|\delta x + \bar{x}\|_C^2 = \|\delta x\|_C^2 + o(\|\bar{x}\|_C)$ and $\|\bar{x}\|_C \leq \|\bar{x}\|_{1,1}$, it suffices to show that

$$\int_{t_0}^{t_f} \Gamma(t, u^0 + \delta u + \bar{u})\, dt = \int_{t_0}^{t_f} \Gamma(t, u^0 + \delta u)\, dt + o(\|\bar{u}\|_\infty).$$

As above, represent $\{\delta u\}$ in the form $\{\delta u\} = \{\delta u_\varepsilon\} + \{\delta u^\varepsilon\}$, where $\varepsilon \to 0$, $\operatorname{meas} \mathcal{M}^\varepsilon \to 0$.

Then

$$\int_{t_0}^{t_f} \Gamma(t,u^0+\delta u+\bar{u})dt = \int_{\mathcal{M}_\varepsilon} \Gamma(t,u^0+\delta u_\varepsilon+\bar{u})dt + \int_{\mathcal{M}^\varepsilon} \Gamma(t,u^0+\delta u^\varepsilon+\bar{u})dt$$

$$= \int_{t_0}^{t_f} \Gamma(t,u^0+\delta u)dt + \int_{\mathcal{M}_\varepsilon} (\Gamma(t,u^0+\delta u_\varepsilon+\bar{u})$$

$$-\Gamma(t,u^0+\delta u_\varepsilon))dt$$

$$+\int_{\mathcal{M}^\varepsilon} (\Gamma(t,u^0+\delta u^\varepsilon+\bar{u}) - \Gamma(t,u^0+\delta u^\varepsilon))dt.$$

Here, we have used the relations

$$\Gamma(t,u^0+\delta u) = \Gamma(t,u^0+\delta u)(\chi_\varepsilon+\chi^\varepsilon) = \Gamma(t,u^0+\delta u_\varepsilon)+\Gamma(t,u^0+\delta u^\varepsilon),$$

where χ_ε and χ^ε are the characteristic functions of the sets $\mathcal{M}_\varepsilon$ and $\mathcal{M}^\varepsilon$, respectively. By property (5) of Definition 2.17, we have

$$\left|\int_{\mathcal{M}^\varepsilon} \left(\Gamma(t,u^0+\delta u^\varepsilon+\bar{u}) - \Gamma(t,u^0+\delta u^\varepsilon)\right) dt\right| \le \text{const}(\text{meas}\,\mathcal{M}^\varepsilon)\|\bar{u}\|_\infty = o(\|\bar{u}\|_\infty).$$

Therefore, it suffices to show that

$$\int_{\mathcal{M}_\varepsilon} (\Gamma(t,u^0+\delta u_\varepsilon+\bar{u}) - \Gamma(t,u^0+\delta u_\varepsilon))dt = o(\|\bar{u}\|_\infty) \tag{2.63}$$

or, which is the same,

$$\int_{t_0}^{t_f} (\Gamma(t,u^0+\delta u_\varepsilon+\bar{u}_\varepsilon) - \Gamma(t,u^0+\delta u_\varepsilon))dt = o(\|\bar{u}\|_\infty),$$

where $\bar{u}_\varepsilon = \bar{u}\chi_\varepsilon$. As was already noted, $\{\delta u_\varepsilon\} \in \Pi_u^{\text{loc}}$. Moreover, $\|\bar{u}_\varepsilon\|_\infty \to 0$, i.e., $\{\bar{u}_\varepsilon\} \in \Pi_u^0$. Therefore, it suffices to prove the following assertion.

Proposition 2.18. *The following estimate holds for any* $\{\delta u\} \in \Pi_u^{\text{loc}}$ *and* $\{\bar{u}\} \in \Pi_u^0$:

$$\int_{t_0}^{t_f} \Gamma(t,u^0+\delta u+\bar{u})dt = \int_{t_0}^{t_f} \Gamma(t,u^0+\delta u)dt + o(\|\bar{u}\|_\infty).$$

Proof. Represent $\{\delta u\}$ in the canonical form $\{\delta u\} = \{\delta u^0\}+\{\delta u^*\}$, $\{\delta u^0\} \in \Pi_u^0$, $\{\delta u^*\} \in \Pi_u^*$, $|\delta u^0|\cdot|\delta u^*| = 0$. The latter property holds for all members of the sequences $\{\delta u^0\}$ and $\{\delta u^*\}$ with the same numbers. According to the definition of the function $\Gamma(t,u)$, we have

$$\int_{t_0}^{t_f} \Gamma(t,u^0+\delta u)dt = \int_{t_0}^{t_f} |\delta u^0|^2\,dt + \sum_k \int_{\mathcal{M}_k^*} (2|\delta t_k| + |\delta v_k^*|^2)dt.$$

Let $\mathcal{M} = [t_0,t_f]\backslash\mathcal{M}^*$. Then

$$\int_{t_0}^{t_f} \Gamma(t,u^0+\delta u+\bar{u})dt = \int_{\mathcal{M}} |\delta u^0+\bar{u}|^2\,dt + \sum_k \int_{\mathcal{M}_k^*} (2|\delta t_k| + |\delta v_k^*+\bar{u}|^2)dt$$

$$= \int_{t_0}^{t_f} \Gamma(t,u^0+\delta u)dt + o(\|\bar{u}\|_\infty).$$

The proposition is proved. ∎

According to Proposition 2.18, formula (2.63) holds. Therefore, γ is a higher order on Π. Obviously, γ is a strict order on Π. We will perform decoding of the constant $C_\gamma := C_\gamma(\Phi_0, \Pi_{\sigma\gamma})$ precisely with this order. According to Theorem 2.8, the inequality $C_\gamma \geq 0$ is necessary for the Pontryagin minimum at the point w^0, and the strict inequality $C_\gamma > 0$ is sufficient for the strict Pontryagin minimum at this point. As was already mentioned, the decoding of the basic constant C_γ consists of the following two stages: estimating C_γ from above and estimating C_γ from below.

2.4 Estimation of the Basic Constant from Above

2.4.1 Passing to Local Sequences and Needles

Recall that

$$C_\gamma = C_\gamma(\Phi_0, \Pi_{\sigma\gamma}) = \inf_{\Pi_{\sigma\gamma}} \varliminf \frac{\Phi_0}{\gamma}, \qquad \Pi_{\sigma\gamma} = \{\{\delta w\} \in \Pi \mid \sigma \leq O(\gamma)\},$$
$$\sigma = \max\{F_0(p^0+\delta p), \ldots, F_{d(F)}(p^0+\delta p),\ |\delta K|,\ \|\delta\dot{x} - \delta f\|_1\}.$$

We will estimate C_γ from above. Since $C_\gamma \geq 0$ is a necessary condition for the Pontryagin minimum, the nonnegativity of any upper estimate for C_γ is also a necessary condition for the Pontryagin minimum. Therefore, this stage of decoding can be considered as obtaining a necessary condition for the Pontryagin minimum.

Let $t' \in (t_0, t_f)\backslash\Theta$, $\varepsilon > 0$, and let $[t'-\varepsilon, t'+\varepsilon]$ be entirely contained in one of the intervals of the set $(t_0, t_f)\backslash\Theta$. Let a point $u' \in \mathbb{R}^{d(u)}$ be such that $(t', x^0(t'), u') \in \mathcal{Q}$, $u' \neq u^0(t')$. Define the needle-shaped variation

$$\delta u' = \delta u'(t; t', \varepsilon, u') = \begin{cases} u' - u^0(t), & t \in [t'-\varepsilon, t'+\varepsilon], \\ 0 & \text{otherwise.} \end{cases}$$

Consider a sequence of needle-shaped variations $\{\delta u'\} := \{\delta u'(\cdot, t', \frac{1}{n}, u')\}$, enumerated by the parameter ε so that $\varepsilon = \varepsilon_n = 1/n$. Clearly, $\{\delta w'\} := \{(0, \delta u')\}$ is a Pontryagin sequence. Obviously, $\gamma' := \gamma(\delta w') = \int_{[t'-\varepsilon, t'+\varepsilon]} \Gamma(t, u')\,dt$ is of order ε. We denote by Π' the set of sequences $\{\delta w'\} = \{(0, \delta u')\}$ such that $\{\delta u'\}$ is a sequence of needle-shaped variations. We set $\Pi^{\text{loc}}_{\sigma\gamma} = \Pi^{\text{loc}} \cap \Pi_{\sigma\gamma}$. Therefore,

$$\Pi^{\text{loc}}_{\sigma\gamma} = \{\{\delta w\} \in \Pi^{\text{loc}} \mid \sigma \leq O(\gamma)\}.$$

We have the following assertion.

Lemma 2.19. *Let $\{\delta w^{\text{loc}}\} \in \Pi^{\text{loc}}_{\sigma\gamma}$ and $\{\delta w'\} \in \Pi'$ be such that $\gamma' \leq O(\gamma^{\text{loc}})$, where $\gamma' = \gamma(\delta w')$ and $\gamma^{\text{loc}} = \gamma(\delta w^{\text{loc}})$. Then*

$$\varliminf \max_{\Lambda_0} \frac{\Phi(\lambda, \delta w^{\text{loc}}) + \int_{t_0}^{t_f} \delta' H^\lambda\,dt}{\gamma^{\text{loc}} + \gamma'} \geq C_\gamma,$$

where $\delta' H^\lambda = H(t, x^0, u^0 + \delta u', \psi) - H(t, x^0, u^0, \psi)$.

To prove Lemma 2.19, we need the following proposition.

Proposition 2.20. *Let $\varphi(t,w) : \mathcal{Q} \mapsto \mathbb{R}^{d(\varphi)}$ be a continuous function. Let $\{\delta w^{\rm loc}\} \in \Pi^{\rm loc}$, $\{\delta w'\} \in \Pi'$, $\{\delta w\} = \{\delta w^{\rm loc} + \delta w'\}$. Then $\delta\varphi = \delta^{\rm loc}\varphi + \delta'\varphi + r_\varphi$, where $\|r_\varphi\|_1 = o(\gamma')$, $\|r_\varphi\|_\infty \to 0$, $\delta\varphi = \varphi(t,w^0+\delta w) - \varphi(t,w^0)$, $\delta^{\rm loc}\varphi = \varphi(t,w^0+\delta w^{\rm loc}) - \varphi(t,w^0)$, $\delta'\varphi = \varphi(t,w^0+\delta w') - \varphi(t,w^0)$.*

Proof. The following relations hold:

$$\begin{aligned}\delta\varphi &= \varphi(t,w^0+\delta w^{\rm loc}+\delta w') - \varphi(t,w^0+\delta w^{\rm loc}) + \delta^{\rm loc}\varphi \\ &= \bar\delta'\varphi + \delta^{\rm loc}\varphi = \bar\delta'\varphi\chi' + \delta^{\rm loc}\varphi,\end{aligned}$$

where $\bar\delta'\varphi = \varphi(t,w^0+\delta w^{\rm loc}+\delta w') - \varphi(t,w^0+\delta w^{\rm loc})$ and χ' is the characteristic function of the set $\mathcal{M}' = \{t \mid \delta u' \neq 0\}$. Further, let $\{\delta w^{\rm loc}\} = \{\delta w^0\} + \{\delta w^*\}$ be the canonical representation, where $\{\delta w^0\} \in \Pi^0$, $\{\delta w^*\} \in \Pi^*$, and $|\delta u^0|\cdot|\delta u^*| = 0$. It follows from the definitions of sequences $\{\delta w'\}$ and $\{\delta w^*\}$ that $|\delta u'|\cdot|\delta u^*| = 0$ starting from a certain number. Therefore,

$$\begin{aligned}(\bar\delta'\varphi)\chi' &= \Big(\varphi(t,w^0+\delta w^0+\delta w') - \varphi(t,w^0+\delta w^0)\Big)\chi' \\ &= \Big(\varphi(t,w^0+\delta w^0+\delta w') - \varphi(t,w^0+\delta w') - \delta^0\varphi + \delta'\varphi\Big)\chi' = r_\varphi + \delta'\varphi,\end{aligned}$$

where

$$\begin{aligned}r_\varphi &= (\bar\delta^0\varphi - \delta^0\varphi)\chi', \\ \bar\delta^0\varphi &= \varphi(t,w^0+\delta w'+\delta w^0) - \varphi(t,w^0+\delta w'), \\ \delta^0\varphi &= \varphi(t,w^0+\delta w^0) - \varphi(t,w^0).\end{aligned}$$

Therefore, $\delta\varphi = \delta^{\rm loc}\varphi + \delta'\varphi + r_\varphi$. Since $\|\bar\delta^0\varphi\|_\infty \to 0$, $\|\delta^0\varphi\|_\infty \to 0$, $\operatorname{meas}\mathcal{M}' = O(\gamma')$, we have $\|r_\varphi\|_\infty \le \|\bar\delta^0\varphi\|_\infty + \|\delta^0\varphi\|_\infty \to 0$, $\|r_\varphi\|_1 \le \|r_\varphi\|_\infty \operatorname{meas}\mathcal{M}' = o(\gamma')$. The proposition is proved. ∎

Proposition 2.20 implies the following assertion.

Proposition 2.21. *Let $\{\delta w^{\rm loc}\} \in \Pi^{\rm loc}$, $\{\delta w'\} \in \Pi'$, $\{\delta w\} = \{\delta w^{\rm loc} + \delta w'\}$. Then for any $\lambda \in \Lambda_0$, we have $\Phi(\lambda,\delta w) = \Phi(\lambda,\delta w^{\rm loc}) + \int_{t_0}^{t_f} \delta' H^\lambda\,dt + \rho^\lambda$, where $\sup_{\Lambda_0}|\rho^\lambda| = o(\gamma')$. Moreover, $\gamma = \gamma^{\rm loc} + \gamma' + o(\gamma')$, where $\gamma = \gamma(\delta w)$, $\gamma^{\rm loc} = \gamma(\delta w^{\rm loc})$, and $\gamma' = \gamma(\delta w')$. Finally, $\|\delta\dot x - \delta f\|_1 = \|\delta \dot x - \delta^{\rm loc} f\|_1 + O(\gamma')$.*

Proof. By Proposition 2.20, we have

$$\begin{aligned}\Phi(\lambda,\delta w) &= \delta l^\lambda - \int_{t_0}^{t_f} \psi\delta\dot x\,dt + \int_{t_0}^{t_f} \psi\delta f\,dt = \delta^{\rm loc} l^\lambda - \int_{t_0}^{t_f} \psi\delta\dot x\,dt + \int_{t_0}^{t_f} \psi\delta^{\rm loc} f\,dt \\ &\quad + \int_{t_0}^{t_f} \psi\delta' f\,dt + \int_{t_0}^{t_f} \psi r_f\,dt = \Phi(\lambda,\delta w^{\rm loc}) + \int_{t_0}^{t_f} \delta' H^\lambda\,dt + \rho^\lambda,\end{aligned}$$

where $\sup_{\Lambda_0}|\rho^\lambda| \le \sup_{\Lambda_0}\|\psi\|_\infty \|r_f\|_1 = o(\gamma')$. Further,

$$\begin{aligned}\gamma(\delta w) &= \|\delta x\|_C^2 + \int_{t_0}^{t_f} \delta\Gamma\,dt = \|\delta x\|_C^2 + \int_{t_0}^{t_f} \delta^{\rm loc}\Gamma\,dt + \int_{t_0}^{t_f} \delta'\Gamma\,dt + o(\gamma') \\ &= \gamma^{\rm loc} + \gamma' + o(\gamma').\end{aligned}$$

Finally, $\|\delta \dot{x} - \delta f\|_1 = \|\delta \dot{x} - \delta^{\mathrm{loc}} f - \delta' f - r_f\|_1 = \|\delta \dot{x} - \delta^{\mathrm{loc}} f\|_1 + O(\gamma')$, since $\|\delta' f\|_1 \leq$ const meas $\mathcal{M}' = O(\gamma')$ and $\|r_f\|_1 = o(\gamma')$. The proposition is proved. ∎

Proof of Lemma 2.19. Let $\{\delta w\} = \{\delta w^{\mathrm{loc}}\} + \{\delta w'\}$, where $\{\delta w^{\mathrm{loc}}\} \in \Pi^{\mathrm{loc}}_{\sigma\gamma}$, $\{\delta w'\} \in \Pi'$ and $\gamma' \leq O(\gamma^{\mathrm{loc}})$. Then, according to Proposition 2.21, $\gamma = \gamma^{\mathrm{loc}} + \gamma' + o(\gamma')$. However, $\gamma' \leq O(\gamma^{\mathrm{loc}})$. Therefore, $\gamma \leq O(\gamma^{\mathrm{loc}})$. On the other hand, since $\gamma = \gamma^{\mathrm{loc}} + (1 + o(1))\gamma'$, $\gamma' \geq 0$, we have $\gamma^{\mathrm{loc}} \leq O(\gamma)$. Therefore, γ and γ^{loc} are of the same order of smallness.

Obviously, $\{\delta w\} \in \Pi$. Let us show that $\{\delta w\} \in \Pi_{\sigma\gamma}$. Indeed, by Proposition 2.21, $\|\delta \dot{x} - \delta f\|_1 = \|\delta \dot{x} - \delta^{\mathrm{loc}} f\|_1 + O(\gamma')$. Therefore,

$$\begin{aligned} \sigma(\delta w) &= \max\{F_i(p^0 + \delta p), |\delta K|, \|\delta \dot{x} - \delta f\|_1\} \\ &\leq \sigma(\delta w^{\mathrm{loc}}) + O(\gamma') \leq O_1(\gamma^{\mathrm{loc}}) \leq O_2(\gamma). \end{aligned}$$

Thus, $\{\delta w\} \in \Pi_{\sigma\gamma}$. Further, according to Proposition 2.21,

$$\Phi(\lambda, \delta w^{\mathrm{loc}}) + \int_{t_0}^{t_f} \delta' H^\lambda \, dt + \rho^\lambda = \Phi(\lambda, \delta w),$$

where $\sup_{\Lambda_0} |\rho^\lambda| = o(\gamma') = o_1(\gamma)$. Therefore,

$$\varliminf \max_{\Lambda_0} \frac{\Phi(\lambda, \delta w^{\mathrm{loc}}) + \int_{t_0}^{t_f} \delta' H^\lambda \, dt}{\gamma^{\mathrm{loc}} + \gamma'} = \varliminf \max_{\Lambda_0} \frac{\Phi(\lambda, \delta w)}{\gamma} = \varliminf \frac{\Phi_0}{\gamma} \geq \inf_{\Pi_{\sigma\gamma}} \varliminf \frac{\Phi_0}{\gamma} = C_\gamma.$$

The inequality holds, since $\{\delta w\} \in \Pi_{\sigma\gamma}$. The lemma is proved. ∎

We can now use the results of Section 2.3. Lemma 2.19 and Proposition 2.15 imply the following assertion.

Lemma 2.22. *Let $\{\delta w\} \in \Pi^{\mathrm{loc}}_{\sigma\gamma}$ and $\{\delta w'\} \in \Pi'$ be such that $\gamma' \leq O(\gamma)$, where $\gamma' = \gamma(\delta w')$ and $\gamma = \gamma(\delta w)$. Then*

$$\varliminf \max_{\Lambda_0} \frac{\Phi^{1\lambda}(\delta w) + \int_{t_0}^{t_f} \delta' H^\lambda \, dt}{\gamma + \gamma'} \geq C_\gamma,$$

where $\delta' H^\lambda = H(t, x^0, u^0 + \delta u', \psi) - H(t, x^0, u^0, \psi)$ and the function $\Phi^{1\lambda}(\delta w)$ is defined by formula (2.61).

2.4.2 Replacement of the Functions in the Definition of the Set $\Pi^{\mathrm{loc}}_{\sigma\gamma}$ by their Decompositions up to First-Order Terms

We represent the sequences $\{\delta w\} \in \Pi^{\mathrm{loc}}$ in the canonical form: $\{\delta w\} = \{\delta w^0\} + \{\delta w^*\}$, where $\{\delta w^0\} = \{(\delta x, \delta u^0)\} \in \Pi^0$, $\{\delta w^*\} = \{0, \delta u^*\} \in \Pi^*$, and, moreover $|\delta u^0| \cdot |\delta u^*| = 0$. We set $I = \{i \in \{0, 1, \ldots, d(F)\} : F_i(p^0) = 0\} = I_F(w^0) \cup \{0\}$. Then $\Pi^{\mathrm{loc}}_{\sigma\gamma}$ consists of sequences $\{\delta w\} \in \Pi^{\mathrm{loc}}$ such that

$$\delta F_i \leq O(\gamma), \quad i \in I; \tag{2.64}$$

$$|\delta K| \leq O(\gamma); \tag{2.65}$$

$$\|\delta \dot{x} - \delta f\|_1 \leq O(\gamma). \tag{2.66}$$

Since

$$\delta F_i = F_{ip}\delta p + O(|\delta p|^2),\ i \in I; \quad \delta K = K_p \delta p + O(|\delta p|^2); \quad |\delta p|^2 \le 2\|\delta x\|_C^2 \le 2\gamma,$$

conditions (2.64) and (2.65) are equivalent to the conditions

$$F_{ip}\delta p \le O(\gamma), \quad i \in I; \tag{2.67}$$

$$|K_p \delta p| \le O(\gamma), \tag{2.68}$$

respectively. Further, consider condition (2.66). By Proposition 2.10,

$$\delta f = f_w \delta w^0 + \frac{1}{2}\langle f_{ww}\delta w^0, \delta w^0\rangle + \delta^* f + \delta^* f_x \delta x + \tilde{r},$$

where $\|\tilde{r}\|_1 = o\big(\|\delta x\|_C^2 + \int_{t_0}^{t_f} |\delta u^0|^2\, dt\big) = o_1(\gamma)$. Here,

$$\|\langle f_{ww}\delta w^0, \delta w^0\rangle\|_1 \le O(\|\delta x\|_C^2 + \int_{t_0}^{t_f} |\delta u^0|^2\, dt) \le O_1(\gamma),$$

$$\|(\delta^* f_x)\delta x\|_1 \le \text{const}\, \|\delta x\|_C \,\text{meas}\, \mathcal{M}^* \le \frac{1}{2}\text{const}(\|\delta x\|_C^2 + (\text{meas}\, \mathcal{M}^*)^2).$$

But, as was already mentioned, $(\text{meas}\, \mathcal{M}_k^*)^2 \le 4\int_{\mathcal{M}_k^*} |\delta t_k|\, dt \le 2\gamma$. Therefore, $\|(\delta^* f_x)\delta x\|_1 \le O(\gamma)$. Further, $\delta^* f = \sum_k \delta_k^* f$. According to formula (2.50), we have

$$\delta_k^* f = [f]^k(\chi_{k-}^* - \chi_{k+}^*) + f_u^{k+}\delta v_{k-} + f_u^{k-}\delta v_{k+} + O(|\delta t_k| + |\delta v_k|^2)\chi_k^*.$$

Therefore, condition (2.66) is equivalent to the following condition:

$$\left\| \delta\dot{x} - f_w \delta w^0 - \sum_k ([f]^k(\chi_{k-}^* - \chi_{k+}^*) - f_u^{k+}\delta v_{k-} - f_u^{k-}\delta v_{k+}) \right\|_1 \le O(\gamma). \tag{2.69}$$

We have shown that $\Pi_{\sigma\gamma}^{\text{loc}}$ consists of sequences $\{\delta w\} \in \Pi^{\text{loc}}$ such that conditions (2.67)–(2.69) hold for their canonical representations.

2.4.3 Narrowing the Set of Sequences $\Pi_{\sigma\gamma}^{\text{loc}}$

In what follows, in the formulation of Lemma 2.22, we narrow the set $\Pi_{\sigma\gamma}^{\text{loc}}$ up to its subset defined by the following conditions:

(a) $\delta v = 0$;

(b) for any $\lambda \in \Lambda_0$,

$$\sum_k [H^\lambda]^k(\text{meas}\, \mathcal{M}_{k-}^* - \text{meas}\, \mathcal{M}_{k+}^*) \le 0. \tag{2.70}$$

These conditions should hold for each member δw of the sequence $\{\delta w\}$. We denote by $\Pi_{\sigma\gamma}^{\text{loc}\,1}$ the set of sequences $\{\delta w\} \in \Pi_{\sigma\gamma}^{\text{loc}}$ satisfying these conditions.

For any sequence $\{\delta w\} \in \Pi_{\sigma\gamma}^{\text{loc}\,1}$, we obviously have

$$\Phi^{1\lambda}(\delta w) \le \Phi^{2\lambda}(\delta w) \quad \forall\, \lambda \in \Lambda_0, \tag{2.71}$$

where $\Phi^{1\lambda}$ is as defined in (2.61) and

$$\begin{aligned}\Phi^{2\lambda}(\delta w) &= \frac{1}{2}\Phi''(\lambda,\delta w^0)\\ &\quad+\sum_k\left(D^k(H^\lambda)\int_{\mathcal{M}_k^*}|\delta t_k|\,dt+[H_x^\lambda]^k\int_{t_0}^{t_f}\delta x(\chi_{k-}^*-\chi_{k+}^*)\,dt\right),\\ \Phi''(\lambda,\delta w^0) &= \langle l_{pp}^\lambda\delta p,\delta p\rangle+\int_{t_0}^{t_f}\langle H_{ww}^\lambda\delta w^0,\delta w^0\rangle\,dt.\end{aligned}$$

Moreover, for any sequence $\{\delta w\}\in\Pi_{\sigma\gamma}^{\mathrm{loc}\,1}$, we have $\gamma(\delta w)=\gamma_1(\delta w)$, where

$$\gamma_1(\delta w)=\|\delta x\|_C^2+\int_{t_0}^{t_f}|\delta u^0|^2\,dt+2\sum_k\int_{\mathcal{M}_k^*}|\delta t_k|\,dt.$$

Finally, condition (2.69) passes to the following condition on these sequences:

$$\left\|\delta\dot{x}-f_w\delta w^0-\sum_k[f]^k(\chi_{k-}^*-\chi_{k+}^*)\right\|_1\le O(\gamma_1).$$

We note that in the definitions of Φ^2, γ_1, and $\Pi_{\sigma\gamma}^{\mathrm{loc}\,1}$, only δw^0 and $\mathcal{M}_k^*$ participate and, moreover, the variation δw is uniquely reconstructed by $\mathcal{M}_k^*$ and δw^0 by using the conditions $\delta u^0\chi^*=0$, $\delta v=0$. We denote the pairs $(\delta w,\mathcal{M}^*)$ by b. Introduce the set of sequences of pairs $\{b\}=\{(\delta w,\mathcal{M}^*)\}$ such that

$$\begin{aligned}&\{\delta w\}=\{(\delta x,\delta u)\}\in\Pi^0,\quad \mathcal{M}^*=\bigcup_{k\in I^*}\mathcal{M}_k^*,\quad \mathcal{M}_k^*\to t_k\ (k\in I^*),\quad \delta u\chi^*=0,\\ &F_{ip}\delta p\le O(\gamma_1)\ (i\in I),\quad |K_p\delta p|\le O(\gamma_1),\\ &\left\|\delta\dot{x}-f_w\delta w-\sum_k[f]^k(\chi_{k-}^*-\chi_{k+}^*)\right\|_1\le O(\gamma_1),\\ &\sum_k[H^\lambda]^k(\operatorname{meas}\mathcal{M}_{k-}^*-\operatorname{meas}\mathcal{M}_{k+}^*)\le 0\quad\forall\,\lambda\in\Lambda_0.\end{aligned}$$

As above, we denote this set of sequences by $\Pi_{\sigma\gamma}^{\mathrm{loc}\,1}$. In what follows, we denote by $\{\delta w\}$ the sequences from Π^0. Lemma 2.22 and inequality (2.71) imply the following assertion.

Lemma 2.23. *Let* $\{b\}=\{(\delta w,\mathcal{M}^*)\}\in\Pi_{\sigma\gamma}^{\mathrm{loc}\,1}$ *and* $\{\delta w'\}\in\Pi'$ *be such that* $\gamma'\le O(\gamma_1)$, *where*

$$\gamma_1=\gamma_1(b):=\|\delta x\|_C^2+\int_{t_0}^{t_f}|\delta u|^2\,dt+\sum_k\int_{\mathcal{M}_k^*}2|\delta t_k|\,dt,\quad \gamma'=\gamma(\delta w').$$

Then,

$$\varliminf\max_{\Lambda_0}\frac{\Phi^{2\lambda}+\int_{t_0}^{t_f}\delta' H^\lambda\,dt}{\gamma_1+\gamma'}\ge C_\gamma,$$

where

$$\begin{aligned}\Phi^{2\lambda} &= \Phi^{2\lambda}(b):=\frac{1}{2}\Phi''(\lambda,\delta w)\\ &\quad+\sum_k\left(D^k(H^\lambda)\int_{\mathcal{M}_k^*}|\delta t_k|\,dt+[H_x^\lambda]^k\int_{t_0}^{t_f}\delta x(\chi_{k-}^*-\chi_{k+}^*)\,dt\right),\\ \delta' H^\lambda &:= H(t,x^0,u^0+\delta u',\psi)-H(t,x^0,u^0,\psi).\end{aligned}$$

2.4.4 Replacement of $\|\delta x\|_C^2$ by $|\delta x(t_0)|^2$ in the Definition of Functional γ_1

We set

$$\gamma_2 = |\delta x(t_0)|^2 + \int_{t_0}^{t_f} |\delta u|^2\,dt + \sum_k \int_{\mathcal{M}_k^*} 2|\delta t_k|\,dt.$$

Since $|\delta x(t_0)| \le \|\delta x\|_C$, we have $\gamma_2 \le \gamma_1$. Let us show that the following estimate also holds on the sequence from $\Pi_{\sigma\gamma}^{\mathrm{loc1}}$: $\gamma_1 \le \mathrm{const}\,\gamma_2$, where const > 0 is independent of the sequence. For this purpose, it suffices to show that $\|\delta x\|_C^2 \le \mathrm{const}\,\gamma_2$ on a sequence from $\Pi_{\sigma\gamma}^{\mathrm{loc1}}$. Let us prove the following assertion.

Proposition 2.24. *There exists a* const > 0 *such that for any sequence* $\{b\} = \{\delta w, \mathcal{M}^*\}$ *satisfying the conditions*

$$\begin{gathered}\{\delta w\} \in \Pi^0, \quad \mathcal{M}^* = \cup_k \mathcal{M}_k^*, \quad \mathcal{M}_k^* \to t_k \quad (k \in I^*),\\ \left\| \delta\dot{x} - f_w\delta w - \sum_k [f]^k(\chi_{k-}^* - \chi_{k+}^*) \right\|_1 = o(\sqrt{\gamma_1})\end{gathered} \tag{2.72}$$

starting from a certain number the following inequality holds: $\|\delta x\|_C^2 \le \mathrm{const}\,\gamma_2$.

Proof. Let $\{b\}$ satisfy conditions (2.72). Then

$$\delta\dot{x} = f_x\delta x + f_u\delta u + \sum_k [f]^k(\chi_{k-}^* - \chi_{k+}^*) + r,$$

where $\|r\|_1 = o(\sqrt{\gamma_1})$. As is known, this implies the estimate

$$\|\delta x\|_{1,1} \le |\delta x(t_0)| + \mathrm{const} \left\| f_u\delta u + \sum_k [f]^k(\chi_{k-}^* - \chi_{k+}^*) + r \right\|_1.$$

Since

$$\begin{gathered}\|\delta u\|_1 \le \sqrt{t_f - t_0}\|\delta u\|_2, \quad \|\chi_{k-}^* - \chi_{k+}^*\|_1 \le \mathrm{meas}\,\mathcal{M}_k^*,\\ (\mathrm{meas}\,\mathcal{M}_{k-}^*)^2 + (\mathrm{meas}\,\mathcal{M}_{k+}^*)^2 \le 2\int_{\mathcal{M}_k^*} |\delta t_k|\,dt \le \gamma_2, \quad k \in I^*,\end{gathered}$$

we have $\|\delta x\|_C^2 \le \|\delta x\|_{1,1}^2 \le \mathrm{const}'\,\gamma_2 + o(\gamma_1)$. This implies what was required. The proposition is proved. ∎

Therefore, γ_1 and γ_2 are equivalent on the set of sequences satisfying conditions (2.72), i.e., on each such sequence, they estimate one another from above and from below:

$$\gamma_2 \le \gamma_1 \le \varkappa\gamma_2 \quad (\varkappa > 1) \tag{2.73}$$

(the constant $\varkappa$ is independent of the sequence). In particular, inequalities (2.73) hold on sequences from $\Pi_{\sigma\gamma}^{\mathrm{loc1}}$. First, this implies that the set $\Pi_{\sigma\gamma}^{\mathrm{loc1}}$ does not change if we replace γ_1 by γ_2 in its definition. Further, inequality (2.73) implies the inequalities $\gamma_2 + \gamma' \le \gamma_1 + \gamma' \le \varkappa(\gamma_2 + \gamma')$, whence

$$1 \le \frac{\gamma_1 + \gamma'}{\gamma_2 + \gamma'} \le \varkappa. \tag{2.74}$$

Let $\{b\} \in \Pi^{\text{loc}\,1}_{\sigma\gamma}$, $\{\delta w'\} \in \Pi'$, and let $\gamma' \le O(\gamma_2)$. Then, we obtain from (2.74) and Lemma 2.23 that

$$\underline{\lim} \max_{\Lambda_0} \frac{\Phi^{2\lambda} + \int_{t_0}^{t_f} \delta' H^\lambda\, dt}{\gamma_2 + \gamma'} = \underline{\lim} \frac{\gamma_1 + \gamma'}{\gamma_2 + \gamma'} \max_{\Lambda_0} \frac{\Phi^{2\lambda} + \int_{t_0}^{t_f} \delta' H^\lambda\, dt}{\gamma_1 + \gamma'}$$

$$\ge \min\{C_\gamma, \varkappa C_\gamma\} = \begin{cases} C_\gamma, & C_\gamma \ge 0, \\ \varkappa C_\gamma, & C_\gamma < 0. \end{cases}$$

We have proved the following assertion.

Lemma 2.25. *The following inequality holds for any* $\{b\} \in \Pi^{\text{loc}\,1}_{\sigma\gamma}$ *and* $\{\delta w'\} \in \Pi'$ *such that* $\gamma' \le O(\gamma_2)$:

$$\underline{\lim} \max_{\Lambda_0} \frac{\Phi^{2\lambda}(b) + \int_{t_0}^{t_f} \delta' H^\lambda\, dt}{\gamma_2(b) + \gamma'} \ge \min\{C_\gamma, \varkappa C_\gamma\}.$$

2.4.5 Passing to Sequences with Discontinuous State Variables

Denote by $P_\Theta W^{1,1}(\Delta, \mathbb{R}^{d(x)})$ the space of functions $\bar{x} : [t_0, t_f] \mapsto \mathbb{R}^{d(x)}$ piecewise continuous on $[t_0, t_f]$ and absolutely continuous on each of the intervals of the set $(t_0, t_f) \setminus \Theta$ (points of discontinuity of such functions are possible only at points of the set Θ). The differential constraint in the set $\Pi^{\text{loc}\,1}_{\sigma\gamma}$ is represented by the condition

$$\left\| \delta\dot{x} - f_w \delta w - \sum_k [f]^k (\chi^*_{k-} - \chi^*_{k+}) \right\|_1 \le O(\gamma_2). \tag{2.75}$$

What is the influence of the terms $\sum_k [f]^k (\chi^*_{k-} - \chi^*_{k+})$ on δx in this condition? We show below that the variations $\delta x \in W^{1,1}(\Delta, \mathbb{R}^{d(x)})$ can be replaced by variations $\bar{x} \in P_\Theta W^{1,1}(\Delta, \mathbb{R}^{d(x)})$ such that $[\bar{x}]^k = [f]^k \xi_k$, where $\xi_k = \operatorname{meas} \mathcal{M}^*_{k-} - \operatorname{meas} \mathcal{M}^*_{k+}$, $k \in I^*$, and, moreover, (2.75) passes to the condition

$$\|\dot{\bar{x}} - f_x \bar{x} - f_u \delta u\|_1 \le O(\gamma_2).$$

We will prove a slightly more general assertion, which will be used later in estimating C_γ from below.

Therefore, we assume that there is a sequence $\{b\} = \{(\delta w, \mathcal{M}^*)\}$ such that $\{\delta w\} \in \Pi^0$, $\mathcal{M}^* = \cup \mathcal{M}^*_k$, $\mathcal{M}^*_k \to t_k$, $k \in I^*$. Moreover, let the following condition (which is weaker than (2.75)) hold:

$$\delta\dot{x} = f_x \delta x + f_u \delta u + \sum_k [f]^k (\chi^*_{k-} - \chi^*_{k+}) + r, \quad \|r\|_1 = o(\sqrt{\gamma_2}). \tag{2.76}$$

For each member $b = (\delta x, \delta u, \mathcal{M}^*)$ of the sequence $\{b\}$, let us define the functions δx^*_k and $\bar{x}^*_k$ by the following conditions:

$$\begin{aligned} &\delta\dot{x}^*_k = [f]^k (\chi^*_{k-} - \chi^*_{k+}), \quad \delta x^*_k(t_0) = 0, \quad \dot{\bar{x}}^*_k = 0, \quad \bar{x}^*_k(t_0) = 0, \\ &[\bar{x}^*_k]^k = [f]^k \xi_k, \quad \xi_k = \operatorname{meas} \mathcal{M}^*_{k-} - \operatorname{meas} \mathcal{M}^*_{k+}. \end{aligned} \tag{2.77}$$

Therefore, $\bar{x}_k^*$ is the jump function: $\bar{x}_k^*(t) = 0$ if $t < t_k$ and $\bar{x}_k^*(t) = [f]^k \xi_k$ if $t > t_k$, and, moreover, the value of the jump is equal to $[f]^k \xi_k$. We set

$$\overline{\delta x}_k^* = \bar{x}_k^* - \delta x_k^*,\ k \in I^*, \quad \overline{\delta x}^* = \sum_k \overline{\delta x}_k^*, \quad \bar{x} = \delta x + \overline{\delta x}^* = \delta x + \sum_k (\bar{x}_k^* - \delta x_k^*).$$

Note that $\bar{x} \in P_\Theta W^{1,1}(\Delta, \mathbb{R}^{d(x)})$. Since the functions $\bar{x}_k^*$ and δx_k^* coincide outside $\mathcal{M}_k^*$, we have $\overline{\delta x}_k^* \chi_k^* = \overline{\delta x}_k^*$ for all k. Hence $\overline{\delta x}^* \chi^* = \overline{\delta x}^*$. Let us estimate $\|\overline{\delta x}^*\|_\infty$ and $\|\overline{\delta x}^*\|_1$. We have

$$\|\overline{\delta x}_k^*\|_\infty \le \|\bar{x}_k^*\|_\infty + \|\delta x_k^*\|_\infty \le \big|[f]^k\big| \cdot |\xi_k| + \big|[f]^k\big| \operatorname{meas} \mathcal{M}_k^*.$$

Moreover,

$$|\xi_k| \le \operatorname{meas} \mathcal{M}_k^* \le \left(4 \int_{\mathcal{M}_k^*} |\delta t_k|\, dt \right)^{\frac{1}{2}} \le \sqrt{2\gamma_2}.$$

Hence $\|\overline{\delta x}^*\|_\infty \le \text{const}\sqrt{\gamma_2}$. Since $\overline{\delta x}^* \chi^* = \overline{\delta x}^*$, we have $\|\overline{\delta x}^*\|_1 \le \|\overline{\delta x}^*\|_\infty \operatorname{meas} \mathcal{M}^* \le \text{const}' \gamma_2$. What equation does $\bar{x}$ satisfy? We obtain from (2.76) and (2.77) that

$$\dot{\bar{x}} = f_x \delta x + f_u \delta u + r = f_x \bar{x} + f_u \delta u - f_x \overline{\delta x}^* + r.$$

Since $\|\overline{\delta x}^*\|_1 \le O(\gamma_2)$, we have $\|\dot{\bar{x}} - f_x \bar{x} - f_u \delta u\|_1 \le O(\gamma_2) + \|r\|_1$. Note that the replacement of δx by $\bar{x}$ does not influence the value of γ_2, since $\bar{x}(t_0) = \delta x(t_0)$ and $\mathcal{M}^*$ and δu are preserved. Now let us show that

$$\int_{t_0}^{t_f} \delta x(\chi_{k-}^* - \chi_{k+}^*)\, dt = \bar{x}_{\text{av}}^k \xi_k + o(\gamma_2), \tag{2.78}$$

where $\xi_k = \operatorname{meas} \mathcal{M}_{k-}^* - \operatorname{meas} \mathcal{M}_{k+}^*$ and $\bar{x}_{\text{av}}^k = \frac{1}{2}(\bar{x}^{k-} + \bar{x}^{k+}) = \frac{1}{2}(\bar{x}(t_k-) + \bar{x}(t_k+))$. Recall that δx satisfies equation (2.76). Represent δx in the form $\delta x = \delta x^0 + \delta x^* + x_r$, where $\dot{x}_r = r$, $x_r(t_0) = 0$, and $\delta x^* = \sum \delta x_k^*$. Then $\delta \dot{x}^0 = f_x \delta x + f_u \delta u$ and $\delta x^0(t_0) = \delta x(t_0)$. This and the conditions $\|\delta x\|_C \to 0$, $\|\delta u\|_\infty \to 0$ imply $\|\delta \dot{x}^0\|_\infty \to 0$.

Now let us consider $\int_{t_0}^{t_f} \delta x(\chi_{k-}^* - \chi_{k+}^*)\, dt$. We set $\delta x_r^0 := \delta x^0 + x_r$. Since

$$\delta x = (\delta x - \delta x^*) + \delta x^* = \delta x^0 + x_r + \delta x^* = \delta x_r^0 + \delta x^*,$$

we have

$$\int_{t_0}^{t_f} \delta x(\chi_{k-}^* - \chi_{k+}^*)\, dt = \int_{t_0}^{t_f} \delta x_r^0(\chi_{k-}^* - \chi_{k+}^*)\, dt + \int_{t_0}^{t_f} \delta x^*(\chi_{k-}^* - \chi_{k+}^*)\, dt. \tag{2.79}$$

Let us estimate each summand separately. We have

$$\begin{aligned} \int_{t_0}^{t_f} \delta x_r^0(\chi_{k-}^* - \chi_{k+}^*)\, dt &= \int_{t_0}^{t_f} \delta x_r^0(t_k)(\chi_{k-}^* - \chi_{k+}^*)\, dt \\ &\quad + \int_{t_0}^{t_f} (\delta x^0(t) - \delta x^0(t_k))(\chi_{k-}^* - \chi_{k+}^*)\, dt \\ &\quad + \int_{t_0}^{t_f} (x_r(t) - x_r(t_k))(\chi_{k-}^* - \chi_{k+}^*)\, dt \\ &= \delta x_r^0(t_k)\xi_k + o(\gamma_2). \end{aligned} \tag{2.80}$$

Here, we have used the following estimates:

(a) $\|x_r\|_\infty \le \|x_r\|_{1,1} = \|\dot{x}_r\|_1 = \|r\|_1 = o(\sqrt{\gamma_2})$, meas $\mathcal{M}_k^* \le \sqrt{2\gamma_2}$, whence

$$\int_{t_0}^{t_f} (x_r(t) - x_r(t_k))(\chi_{k-}^* - \chi_{k+}^*)\,dt = o(\gamma_2); \tag{2.81}$$

(b) $|\delta x^0(t) - \delta x^0(t_k)| \le \|\delta \dot{x}^0\|_\infty |\delta t_k|$, and hence

$$\left|\int_{t_0}^{t_f} (\delta x^0(t) - \delta x^0(t_k))(\chi_{k-}^* - \chi_{k+}^*)\,dt\right| \le \|\delta \dot{x}^0\|_\infty \int_{\mathcal{M}_k^*} |\delta t_k|\,dt = o(\gamma_2), \tag{2.82}$$

since $\|\delta \dot{x}^0\|_\infty \to 0$. Relation (2.80) follows from (2.81) and (2.82). Further, the conditions

$$\begin{cases} \delta \dot{x}_r^0 & = f_x \delta x + f_u \delta u + r, \\ \delta x_r^0(t_0) & = \delta x(t_0), \end{cases} \qquad \begin{cases} \dot{\bar{x}} & = f_x \delta x + f_u \delta u + r, \\ \bar{x}(t_0) & = \delta x(t_0) \end{cases}$$

imply $\delta x_r^0 = \bar{x} - \sum_k \bar{x}_k^*$ outside Θ, and hence $\delta x_r^0(t_k) = \bar{x}^{k-} - \sum_{j<k} [\bar{x}]^j$. We obtain from this and (2.80) that

$$\int_{t_0}^{t_f} \delta x_r^0 (\chi_{k-}^* - \chi_{k+}^*)\,dt = \left(\bar{x}^{k-} - \sum_{j<k} [\bar{x}]^j\right)\xi_k + o(\gamma_2). \tag{2.83}$$

Further, let $y_k^*(t)$ be defined by the conditions $\dot{y}_k^* = \chi_{k-}^* - \chi_{k+}^*$, $y_k^*(t_0) = 0$. Then

$$\int_{t_0}^{t_f} y_k^* \dot{y}_k^*\,dt = \frac{1}{2}(\text{meas}\,\mathcal{M}_{k-}^* - \text{meas}\,\mathcal{M}_{k+}^*)^2 = \frac{1}{2}\xi_k^2.$$

Obviously, $\delta x_k^* = [f]^k y_k^*$. Hence

$$\int_{t_0}^{t_f} \delta x_k^* (\chi_{k-}^* - \chi_{k+}^*)\,dt = \int_{t_0}^{t_f} [f]^k y_k^* \dot{y}_k^*\,dt = [f]^k \frac{1}{2}\xi_k^2 = \frac{1}{2}[\bar{x}_k^*]^k \xi_k = \frac{1}{2}[\bar{x}]^k \xi_k.$$

We obtain from this that

$$\begin{aligned} &\int_{t_0}^{t_f} \delta x^* (\chi_{k-}^* - \chi_{k+}^*)\,dt \\ &= \int_{t_0}^{t_f} \sum_{j<k} \delta x_j^* (\chi_{k-}^* - \chi_{k+}^*)\,dt + \int_{t_0}^{t_f} \delta x_k^* (\chi_{k-}^* - \chi_{k+}^*)\,dt \\ &= \sum_{j<k} [\bar{x}]^j \xi_k + \frac{1}{2}[\bar{x}]^k \xi_k, \end{aligned} \tag{2.84}$$

since we have the following for $j < k$:

$$\delta x_j^* \chi_k^* = \bar{x}_j^* \chi_k^* = [\bar{x}_j^*]^j \chi_k^* = [\bar{x}]^j \chi_k^*.$$

We obtain from (2.79), (2.83), and (2.84) that

$$\int_{t_0}^{t_f} \delta x(\chi^*_{k-} - \chi^*_{k+})\,dt = \left(\bar{x}^{k-} + \frac{1}{2}[\bar{x}]^k\right)\xi_k + o(\gamma_2) = \bar{x}^k_{\text{av}}\xi_k + o(\gamma_2),$$

as required.

Finally, let us show that

$$\int_{t_0}^{t_f} \langle H^\lambda_{ww}\bar{w}, \bar{w}\rangle\,dt = \int_{t_0}^{t_f} \langle H^\lambda_{ww}\delta w, \delta w\rangle\,dt + \rho^\lambda, \tag{2.85}$$

where $\bar{w} = (\bar{x}, \bar{u})$, $\bar{u} = \delta u$, and $\sup_{\Lambda_0}|\rho^\lambda| = o(\gamma_2)$. Indeed, $\bar{x} = \delta x + \overline{\delta x}^*$, where $\|\overline{\delta x}^*\|_\infty \to 0$, $\|\overline{\delta x}^*\|_1 = O(\gamma_2)$. Hence

$$\begin{aligned}
\int_{t_0}^{t_f} \langle H^\lambda_{ww}\bar{w}, \bar{w}\,dt\rangle &= \int_{t_0}^{t_f} \big(\langle H^\lambda_{xx}\bar{x}, \bar{x}\rangle + 2\langle H^\lambda_{ux}\bar{x}, \bar{u}\rangle + \langle H^\lambda_{uu}\bar{u}, \bar{u}\rangle\big)\,dt \\
&= \int_{t_0}^{t_f} \big(\langle H^\lambda_{xx}\delta x, \delta x\rangle + 2\langle H^\lambda_{ux}\delta x, \delta u\rangle + \langle H^\lambda_{uu}\delta u, \delta u\rangle\big)\,dt \\
&\quad + \int_{t_0}^{t_f} \Big(2\langle H^\lambda_{xx}\delta x, \overline{\delta x}^*\rangle + 2\langle H^\lambda_{ux}\overline{\delta x}^*, \delta u\rangle + \langle H^\lambda_{xx}\overline{\delta x}^*, \overline{\delta x}^*\rangle\Big)\,dt \\
&= \int_{t_0}^{t_f} \langle H^\lambda_{ww}\delta w, \delta w\rangle\,dt + \rho^\lambda, \quad \sup_{\Lambda_0}|\rho^\lambda| = o(\gamma_2),
\end{aligned}$$

since

$$\begin{aligned}
\left|\int_{t_0}^{t_f} \langle H^\lambda_{xx}\delta x, \overline{\delta x}^*\rangle\,dt\right| &\le \sup_{\Lambda_0}\|H^\lambda_{xx}\|_\infty\|\delta x\|_C\|\overline{\delta x}^*\|_1 = o(\gamma_2), \\
\left|\int_{t_0}^{t_f} \langle H^\lambda_{ux}\overline{\delta x}^*, \delta u\rangle\,dt\right| &\le \sup_{\Lambda_0}\|H^\lambda_{ux}\|_\infty\|\delta u\|_\infty\|\overline{\delta x}^*\|_1 = o(\gamma_2), \\
\left|\int_{t_0}^{t_f} \langle H^\lambda_{xx}\overline{\delta x}^*, \overline{\delta x}^*\rangle\,dt\right| &\le \sup_{\Lambda_0}\|H^\lambda_{xx}\|_\infty\|\overline{\delta x}^*\|_\infty\|\overline{\delta x}^*\|_1 = o(\gamma_2).
\end{aligned}$$

Therefore, formula (2.85) holds. We set

$$\begin{aligned}
&\bar{b} = (\bar{w}, \mathcal{M}^*); \quad \xi_k = \operatorname{meas}\mathcal{M}^*_{k-} - \operatorname{meas}\mathcal{M}^*_{k+}, \\
&\Phi''(\lambda, \bar{w}) = \langle l^\lambda_{pp}\bar{p}, \bar{p}\rangle + \textstyle\int_{t_0}^{t_f} \langle H^\lambda_{ww}\bar{w}, \bar{w}\rangle\,dt,
\end{aligned}$$

where $\bar{p} = (\bar{x}(t_0), \bar{x}(t_f))$;

$$\Phi^{3\lambda}(\bar{b}) = \frac{1}{2}\Phi''(\lambda, \bar{w}) + \sum_k \left(D^k(H^\lambda)\int_{\mathcal{M}^*_k} |\delta t_k|\,dt + [H^\lambda_x]^k \bar{x}^k_{\text{av}}\xi_k\right); \tag{2.86}$$

$$\gamma_2(\bar{b}) = |\bar{x}(t_0)|^2 + \int_{t_0}^{t_f} |\bar{u}|^2\,dt + 2\sum_k \int_{\mathcal{M}^*_k} |\delta t_k|\,dt.$$

Since $\bar{p} = \delta p$ for the entire sequence $\{b\}$, we obtain from (2.78) and (2.85) that $\Phi^{2\lambda}(b) = \Phi^{3\lambda}(\bar{b}) + \rho^\lambda$, where $\sup_{\Lambda_0} |\rho^\lambda| = o(\gamma_2)$. We have proved the following assertion.

Lemma 2.26. *Let a sequence* $\{b\} = \{(\delta w, \mathcal{M}^*)\}$ *be such that* $\{\delta w\} = \{(\delta x, \delta u)\} \in \Pi^0$, $\mathcal{M}^* = \cup \mathcal{M}_k^*$, $\mathcal{M}_k^* \to t_k$, $k \in I^*$, *and, moreover, let*

$$\delta \dot{x} = f_x \delta x + f_u \delta u + \sum_k [f]^k (\chi_{k-}^* - \chi_{k+}^*) + r,$$

where $\|r\|_1 = o(\sqrt{\gamma_2})$. *Let a sequence* $\{\bar{b}\} = \{(\bar{w}, \mathcal{M}^*)\}$ *be such that* $\bar{w} = (\bar{x}, \bar{u})$, $\bar{x} = \delta x + \overline{\delta x}^*$, *and* $\bar{u} = \delta u$, *where* $\overline{\delta x}^* = \bar{x}^* - \delta x^* = \sum \bar{x}_k^* - \sum \delta x_k^*$, *and* $\bar{x}_k^*$ *and* δx_k^* *are defined by formulas* (2.77). *Let* $\delta p = (\delta x(t_0), \delta x(t_f))$ *and* $\bar{p} = (\bar{x}(t_0), \bar{x}(t_f))$. *Then*

$$\begin{aligned}
&\{\delta p\} = \{\bar{p}\}, \quad \gamma_2(b) = \gamma_2(\bar{b}), \quad \dot{\bar{x}} = f_x \bar{x} + f_u \bar{u} - f_x \overline{\delta x}^* + r,\\
&[\bar{x}]^k = [f]^k \xi_k \quad \forall\, k, \quad \xi_k = \operatorname{meas} \mathcal{M}_{k-}^* - \operatorname{meas} \mathcal{M}_{k+}^*; \quad \|\overline{\delta x}^*\|_1 \le O(\gamma_2);\\
&\Phi^{2\lambda}(b) = \Phi^{3\lambda}(\bar{b}) + \rho^\lambda, \quad \sup_{\Lambda_0} |\rho^\lambda| = o(\gamma_2),
\end{aligned}$$

where $\Phi^{3\lambda}(\bar{b})$ *is defined by formula* (2.86).

We will need this lemma in estimating C_γ from below. We now use the corollary of Lemma 2.26, which is formulated below.

Corollary 2.27. *Let a sequence* $\{\bar{b}\} = \{(\bar{w}, \mathcal{M}^*)\}$ *be such that*

$$\begin{aligned}
&\bar{w} = (\bar{x}, \bar{u}), \quad \bar{x} \in P_\Theta W^{1,1}(\Delta, \mathbb{R}^{d(x)}), \quad \bar{u} \in L^\infty(\Delta, \mathbb{R}^{d(u)}), \quad \|\bar{x}\|_\infty + \|\bar{u}\|_\infty \to 0,\\
&\mathcal{M}^* = \bigcup \mathcal{M}_k^*, \quad \mathcal{M}_k^* \to t_k \quad (k \in I^*), \quad \|\dot{\bar{x}} - f_w \bar{w}\|_1 \le O(\gamma_2).
\end{aligned}$$

Let a sequence $\{b\} = \{(\delta w, \mathcal{M}^*)\}$ *be such that* $\delta w = (\delta x, \delta u)$, $\delta x = \bar{x} - \overline{\delta x}^*$, *and* $\delta u = \bar{u}$, *where* $\overline{\delta x}^* = \bar{x}^* - \delta x^* = \sum \bar{x}_k^* - \sum \delta x_k^*$, *and* $\bar{x}_k^*$ *and* δx_k^* *are defined by formulas* (2.77). *Then*

$$\begin{aligned}
&\{\delta w\} \in \Pi^0, \quad \{\delta p\} = \{\bar{p}\}, \quad \gamma_2(b) = \gamma_2(\bar{b}),\\
&\left\| \delta \dot{x} - f_w \delta w - \sum_k [f]^k (\chi_{k-}^* - \chi_{k+}^*) \right\|_1 \le O(\gamma_2),\\
&\Phi^{2\lambda}(b) = \Phi^{3\lambda}(\bar{b}) + \rho^\lambda, \quad \sup_{\Lambda_0} |\rho^\lambda| = o(\gamma_2).
\end{aligned}$$

Proof. Indeed, by the condition of Corollary 2.27, it follows that $\dot{\bar{x}} = f_x \bar{x} + f_u \bar{u} + \tilde{r}$, $\|\tilde{r}\|_1 \le O(\gamma_2)$. Then $\dot{\bar{x}} = f_x \bar{x} + f_u \bar{u} - f_x \overline{\delta x}^* + r$, where $r = \tilde{r} + f_x \overline{\delta x}^*$, and, moreover, $\|r\|_1 \le \|\tilde{r}\|_1 + \|f_x \overline{\delta x}^*\|_1 \le O(\gamma_2)$. We obtain from this that

$$\delta \dot{x} = f_x \delta x + f_u \delta u + \sum_k [f]^k (\chi_{k-}^* - \chi_{k+}^*) + r.$$

Consequently, $\|\delta \dot{x} - f_w \delta w - \sum_k [f]^k (\chi_{k-}^* - \chi_{k+}^*)\|_1 \le O(\gamma_2)$. The other assertions of Corollary 2.27 follow from Lemma 2.26 directly. ∎

Denote by S^2 the set of sequences $\{\bar{b}\} = \{(\bar{w}, \mathcal{M}^*)\}$ such that

$$\begin{aligned}
&\bar{w} = (\bar{x}, \bar{u}), \quad \bar{u} \in L^\infty(\Delta, \mathbb{R}^{d(u)}), \quad \bar{x} \in P_\Theta W^{1,1}(\Delta, \mathbb{R}^{d(x)}),\\
&\|\bar{x}\|_\infty \to 0, \quad \|\bar{u}\|_\infty \to 0,\\
&\mathcal{M}^* = \bigcup_k \mathcal{M}_k^*, \quad \mathcal{M}_k^* \to t_k \quad (k \in I^*), \quad \bar{u}\chi^* = 0,\\
&F_{ip}\bar{p} \le O(\gamma_2) \quad (i \in I), \quad |K_p\bar{p}| \le O(\gamma_2), \quad \|\dot{\bar{x}} - f_w\bar{w}\|_1 \le O(\gamma_2),\\
&\sum_k [H^\lambda]^k(\operatorname{meas}\mathcal{M}_{k-}^* - \operatorname{meas}\mathcal{M}_{k+}^*) \le 0 \quad \forall\, \lambda \in \Lambda_0.
\end{aligned}$$

Then Lemma 2.25 and also Corollary 2.27 imply the following assertion.

Lemma 2.28. *The following inequality holds for any sequences $\{\bar{b}\} \in S^2$ and $\{\delta w'\} \in \Pi'$ such that $\gamma' \le O(\gamma_2(\bar{b}))$:*

$$\varliminf \max_{\Lambda_0} \frac{\Phi^{3\lambda}(\bar{b}) + \int_{t_0}^{t_f} \delta' H^\lambda\, dt}{\gamma_2(\bar{b}) + \gamma'} \ge \min\{C_\gamma, \varkappa C_\gamma\}.$$

2.4.6 Passing to the Quadratic Form Ω

We remove the condition $\bar{u}\chi^* = 0$ from the definition of S^2. We denote by S^3 the resulting new set of sequences. Let us show that Lemma 2.28 remains valid under the replacement of S^2 by S^3. Indeed, assume that there is a sequence $\{\bar{b}\} = \{(\bar{w}, \mathcal{M}^*)\} \in S^3$, where $\bar{w} = (\bar{x}, \bar{u})$. We set $\bar{\bar{u}} = \bar{u}(1 - \chi^*) = \bar{u} - \bar{u}^*$, where $\bar{u}^* = \bar{u}\chi^*$. Then $\|\bar{u}^*\|_\infty \to 0$ and $\|\bar{u}^*\|_1 = \int_{\mathcal{M}^*} |\bar{u}|\, dt \le \sqrt{\operatorname{meas}\mathcal{M}^*}\, \|\bar{u}\|_2 \le O(\gamma_2)$. Consequently,

$$\begin{aligned}
&\int_{t_0}^{t_f} |\bar{u}^*|^2 dt \le \|\bar{u}^*\|_\infty \|\bar{u}^*\|_1 = o(\gamma_2),\\
&\int_{t_0}^{t_f} \langle H_{uu}^\lambda \bar{u}, \bar{u}\rangle\, dt = \int_{t_0}^{t_f} \langle H_{uu}^\lambda \bar{\bar{u}}, \bar{\bar{u}}\rangle\, dt + \int_{t_0}^{t_f} \langle H_{uu}^\lambda \bar{u}^*, \bar{u}^*\rangle\, dt = \int_{t_0}^{t_f} \langle H_{uu}^\lambda \bar{\bar{u}}, \bar{\bar{u}}\rangle\, dt + \rho_1^\lambda,\\
&\int_{t_0}^{t_f} \langle H_{ux}^\lambda \bar{x}, \bar{u}\rangle\, dt = \int_{t_0}^{t_f} \langle H_{ux}^\lambda \bar{x}, \bar{\bar{u}}\rangle\, dt + \int_{t_0}^{t_f} \langle H_{ux}^\lambda \bar{x}, \bar{u}^*\rangle\, dt = \int_{t_0}^{t_f} \langle H_{ux}^\lambda \bar{x}, \bar{\bar{u}}\rangle\, dt + \rho_2^\lambda,
\end{aligned}$$

where $\sup_{\Lambda_0} |\rho_i^\lambda| = o(\gamma_2)$, $i = 1, 2$. We set $\{\bar{\bar{b}}\} = \{(\bar{x}, \bar{\bar{u}}, \mathcal{M}^*)\}$. Then

$$\gamma_2(\bar{\bar{b}}) = \gamma_2(\bar{b}) + o(\gamma_2), \quad \Phi^{3\lambda}(\bar{\bar{b}}) = \Phi^{3\lambda}(\bar{b}) + \rho^\lambda, \quad \sup_{\Lambda_0} |\rho^\lambda| = o(\gamma_2). \tag{2.87}$$

Moreover, it is easy to see that $\{\bar{\bar{b}}\} \in S^2$. Indeed,

$$\begin{aligned}
\|\dot{\bar{x}} - f_x\bar{x} - f_u\bar{\bar{u}}\|_1 &= \|\dot{\bar{x}} - f_x\bar{x} - f_u\bar{u} - f_u\bar{u}^*\|_1\\
&\le \|\dot{\bar{x}} - f_w\bar{w}\|_1 + \|f_u\bar{u}^*\|_1 \le O(\gamma_2).
\end{aligned}$$

The other conditions of $\{\bar{\bar{b}}\}$ belonging to S^2 obviously hold. Conditions (2.87) and $\{\bar{\bar{b}}\} \in S^2$ imply that Lemma 2.28 remains valid under the replacement of S^2 by S^3.

Further, we narrow the set of sequences S^3 up to the set S^4 by adding the following conditions to the definition of the set S^3:

(i) Each set $\mathcal{M}_k^*$ is a segment adjusting to t_k, i.e., $\mathcal{M}_k^* = [t_k - \varepsilon, t_k]$ or $\mathcal{M}_k^* = [t_k, t_k + \varepsilon]$, where $\varepsilon \to +0$. In this case (see formula (2.13)),

$$\begin{aligned} 2\int_{\mathcal{M}_k^*} |\delta t_k|\,dt &= \xi_k^2, \quad \text{where} \quad \xi_k = \operatorname{meas}\mathcal{M}_{k-}^* - \operatorname{meas}\mathcal{M}_{k+}^* \\ \gamma_2(\bar{b}) &= \sum \xi_k^2 + |\bar{x}(t_0)|^2 + \int_{t_0}^{t_f} |\bar{u}|^2\,dt = \bar{\gamma}; \\ \Phi^{3\lambda}(\bar{b}) &= \frac{1}{2}\Phi''(\lambda,\bar{w}) + \sum_k \left(\frac{1}{2} D^k(H^\lambda)\xi_k^2 + [H_x^\lambda]^k \bar{x}_{\mathrm{av}}^k \xi_k \right) =: \Omega^\lambda. \end{aligned}$$

(ii) Also, the following relations hold:

$$F_{ip}\bar{p} \le 0,\ i \in I, \quad K_p\bar{p} = 0, \quad \dot{\bar{x}} = f_w\bar{w}.$$

We note that for sequences from S^4, each of the quantities $\xi_k = \operatorname{meas}\mathcal{M}_{k-}^* - \operatorname{meas}\mathcal{M}_{k+}^*$ uniquely defines the set $\mathcal{M}_k^* := [t_k - \xi_k^-, t_k + \xi_k^+]$. Here, $\xi_k^- = \max\{0, -\xi_k\}$, and $\xi_k^+ = \max\{0, \xi_k\}$. Moreover, we note that Ω and $\bar{\gamma}$ depend on $\bar{\xi}$, $\bar{x}$, and $\bar{u}$. Therefore, S^4 can be identified with the set of sequences $\{\bar{z}\} = \{(\bar{\xi}, \bar{w})\}$ such that

$$\begin{aligned} &\bar{\xi} \in \mathbb{R}^s, \quad \bar{w} = (\bar{x},\bar{u}), \quad \bar{x} \in P_\Theta W^{1,1}(\Delta, \mathbb{R}^{d(x)}), \quad \bar{u} \in L^\infty(\Delta, \mathbb{R}^{d(u)}), \\ &|\bar{\xi}| \to 0, \quad \|\bar{x}\|_\infty \to 0, \quad \|\bar{u}\|_\infty \to 0, \\ &F_{ip}\bar{p} \le 0\ (i \in I), \quad K_p\bar{p} = 0, \\ &\dot{\bar{x}} = f_w\bar{w}, \quad [\bar{x}]^k = [f]^k\bar{\xi}_k \quad (k \in I^*), \\ &\sum_k [H^\lambda]^k\bar{\xi}_k \le 0 \quad \forall\, \lambda \in \Lambda_0. \end{aligned} \tag{2.88}$$

Therefore, the following assertion holds.

Lemma 2.29. *The following inequality holds for any sequences $\{\bar{z}\} \in S^4$ and $\{\delta w'\} \in \Pi'$ such that $\bar{\gamma}(\bar{z}) \le O(\gamma')$:*

$$\varliminf \max_{\Lambda_0} \frac{\Omega^\lambda(\bar{z}) + \int_{t_0}^{t_f} \delta' H^\lambda\,dt}{\bar{\gamma}(\bar{z}) + \gamma'} \ge \min\{C_\gamma, \varkappa C_\gamma\}.$$

Now let us show that the condition $\sum_k [H^\lambda]^k\bar{\xi}_k \le 0$ for all $\lambda \in \Lambda_0$ holds automatically for the elements of the critical cone $\mathcal{K}$, and therefore, it is extra in the definition of the set of sequences S^4, i.e., it can be removed. We thus will prove that S^4 consists of the sequences of elements of the critical cone $\mathcal{K}$ that satisfy the condition $|\bar{\xi}| + \|\bar{x}\|_\infty + \|\bar{u}\|_\infty \to 0$.

2.4.7 Properties of Elements of the Critical Cone

Proposition 2.30. *Let $\lambda = (\alpha_0, \alpha, \beta, \psi) \in \Lambda_0$, $(\bar{\xi}, \bar{x}, \bar{u}) \in \mathcal{K}$. Then the function $\psi\bar{x}$ is constant on each of the intervals of the set $(t_0, t_f)\backslash\Theta$ and hence is piecewise constant on $[t_0, t_f]$.*

Proof. The conditions $\dot{\bar{x}} = f_w\bar{w}$, $\dot{\psi} = -\psi f_x$, and $\psi f_u = 0$ imply

$$0 = \psi(\dot{\bar{x}} - f_x\bar{x} - f_u\bar{u}) = \psi\dot{\bar{x}} - \psi f_x\bar{x} = \psi\dot{\bar{x}} + \dot{\psi}\bar{x} = \frac{d}{dt}(\psi\bar{x}).$$

Moreover, $\psi \in W^{1,\infty}(\Delta, \mathbb{R}^{d(x)})$ and $\bar{x} \in P_\Theta W^{1,2}(\Delta, \mathbb{R}^{d(x)})$. Therefore, $\psi\bar{x} = \text{const}$ on any interval of the set $(t_0, t_f)\backslash\Theta$. The proposition is proved. ∎

Proposition 2.31. *Let* $\lambda = (\alpha_0, \alpha, \beta, \psi) \in \Lambda_0$, *and let* $(\bar{\xi}, \bar{x}, \bar{u}) \in \mathcal{K}$. *Then*

$$\sum_k [H^\lambda]^k \bar{\xi}_k = l_p^\lambda \bar{p} \le 0.$$

Proof. By Proposition 2.30, $\frac{d}{dt}(\psi\bar{x}) = 0$ a.e. on $[t_0, t_f]$. Hence

$$\begin{aligned} 0 &= \int_{t_0}^{t_f} \frac{d}{dt}(\psi\bar{x})dt = \psi\bar{x}\,\Big|_{t_0}^{t_f} - \sum_k [\psi\bar{x}]^k \\ &= \psi(t_f)\bar{x}(t_f) - \psi(t_0)\bar{x}(t_0) - \sum_k \psi(t_k)[\bar{x}]^k \\ &= l_{x_f}^\lambda \bar{x}(t_f) + l_{x_0}^\lambda \bar{x}(t_0) - \sum_k \psi(t_k)[f]^k \bar{\xi}_k = l_p^\lambda \bar{p} - \sum_k [H^\lambda]^k \bar{\xi}_k. \end{aligned}$$

We obtain from this that $\sum_k [H^\lambda]^k \bar{\xi}_k = l_p^\lambda \bar{p}$. Moreover, the conditions

$$\alpha_i \ge 0\,\forall\, i \in I, \quad \alpha_i = 0\,\forall\, i \notin I, \quad F_{ip}\bar{p} \le 0 \quad \forall\, i \in I, \quad K_p\bar{p} = 0$$

imply $l_p^\lambda \bar{p} \le 0$. The proposition is proved. ∎

Proposition 2.31 implies the following assertion.

Proposition 2.32. *Let* $\lambda = (\alpha_0, \alpha, \beta, \psi) \in \Lambda_0$ *be such that* $[H^\lambda]^k = 0$ *for all* $k \in I^*$. *Let* $\bar{z} = (\bar{\xi}, \bar{x}, \bar{u}) \in \mathcal{K}$. *Then* $\alpha_0(J_p\bar{p}) = 0$, $\alpha_i(F_{ip}\bar{p}) = 0$, $i = 1, \dots, d(F)$.

Proof. By Proposition 2.31, the conditions $\bar{z} \in \mathcal{K}$, $\lambda \in \Lambda_0$, $[H^\lambda]^k = 0$ for all $k \in I^*$ imply $l_p^\lambda \bar{p} = 0$, where $l_p\bar{p} = \alpha_0(J_p\bar{p}) + \sum \alpha_i(F_{ip}\bar{p}) + \sum \beta_j(K_{jp}\bar{p})$. This and the conditions

$$\begin{aligned} &\alpha_0 \ge 0, \quad J_p\bar{p} \le 0, \quad \alpha_i \ge 0, \quad F_{ip}\bar{p} \le 0 \quad \forall\, i \in I_F(w^0), \\ &\alpha_i = 0 \quad \forall\, i \notin I := I_F(w^0) \cup \{0\}, \quad K_p\bar{p} = 0 \end{aligned}$$

imply what was required. The proposition is proved. ∎

In fact, Proposition 2.32 is equivalent to Proposition 2.2. Therefore, using Proposition 2.31, we have proved Proposition 2.2. Proposition 2.3 is proved analogously (we leave this proof to the reader). Further, we use Proposition 2.31.

We set

$$Z(\Theta) = \mathbb{R}^s \times P_\Theta W^{1,1}(\Delta, \mathbb{R}^{d(x)}) \times L^\infty(\Delta, \mathbb{R}^{d(u)}). \tag{2.89}$$

Consider sequences of the form $\{\varepsilon\bar{z}\}$, where $\varepsilon \to +0$ and $\bar{z} \in \mathcal{K} \cap Z(\Theta)$ is a fixed element. According to Proposition 2.31, such a sequence belongs to S^4. Therefore, Lemma 2.29 implies the following assertion.

Lemma 2.33. *Let* $\bar{z} \in \mathcal{K} \cap Z(\Theta)$, $\varepsilon \to +0$, $\{\delta w'\} \in \Pi'$, *and* $\gamma' \le O(\varepsilon^2)$. *Then*

$$\underline{\lim} \max_{\Lambda_0} \frac{\varepsilon^2 \Omega^\lambda(\bar{z}) + \int_{t_0}^{t_f} \delta' H^\lambda \, dt}{\varepsilon^2 \bar{\gamma}(\bar{z}) + \gamma'} \ge \min\{C_\gamma, \varkappa C_\gamma\}.$$

Below, we set $\mathcal{K} \cap Z(\Theta) = \mathcal{K}_Z$.

2.4.8 Cone $\mathfrak{R}_C$ in the Space of Affine Functions on the Compact Set $\mathrm{co}\,\Lambda_0$

To obtain the next upper estimate of the basic constant C_γ, in the estimate of Lemma 2.33, we need to replace the set Λ_0 by the set $M^{\mathrm{co}}(C\Gamma)$ of tuples satisfying the minimum principle of "strictness $C\Gamma$" (the meaning of this will be explained later) and, simultaneously, to remove the sequence of needle-shaped variations. We will attain this using the method of Milyutin, which has become a standard method [27, 86, 92, 95] for this stage of decoding higher-order conditions in problems of the calculus of variations and optimal control. The method of Milyutin is called the "cone technique." The meaning of such a term will become clear after its description. We now give several definitions.

Denote by $L_0 = \mathrm{Lin}(\mathrm{co}\,\Lambda_0)$ the linear span of the set $\mathrm{co}\,\Lambda_0$. Since Λ_0 is a finite-dimensional compact set, L_0 is also of finite dimension. We denote by $l(\lambda)$ an arbitrary linear function $l : L_0 \to \mathbb{R}^1$, and by $\mathfrak{a}(\lambda)$ an arbitrary affine function $\mathfrak{a} : L_0 \to \mathbb{R}^1$, i.e., a function of the form $\mathfrak{a}(\lambda) = l(\lambda) + c$, where c is a number. Since L_0 is of finite dimension, the affine functions $\mathfrak{a}(\lambda) : L_0 \to \mathbb{R}^1$ compose a finite-dimensional space, and each of these functions is uniquely defined by its values on $\mathrm{co}\,\Lambda_0$. In what follows, we consider affine functions $\mathfrak{a}(\lambda)$ on the compact set $\mathrm{co}\,\Lambda_0$ (of course they can be defined directly on $\mathrm{co}\,\Lambda_0$, not passing to the linear span).

With each sequence $\{\delta w'\} \in \Pi'$, we associate the following sequence of linear functions on $\mathrm{co}\,\Lambda_0$:

$$l(\lambda) = \frac{\int_{t_0}^{t_f} \delta' H^\lambda \, dt}{\gamma'},$$

where $\delta' H^\lambda = H^\lambda(t, w^0 + \delta w') - H^\lambda(t, w^0)$, $\gamma' = \int_{t_0}^{t_f} \Gamma' \, dt$, and $\Gamma' = \Gamma(t, u^0 + \delta u')$. It follows from the definition of the sequence $\{\delta w'\} \in \Pi'$ that $\gamma' > 0$ on it, and, therefore, this definition is correct. Let A_0 be the set of all limit points of the sequences $\{l(\lambda)\}$ obtained by the above method from all sequences $\{\delta w'\} \in \Pi'$. Clearly, A_0 is a closed subset in the finite-dimensional space of linear functions $l(\lambda) : \mathrm{co}\,\Lambda_0 \to \mathbb{R}^1$. We note that each convergent sequence $\{l(\lambda)\}$ converges to its limit uniformly on $\mathrm{co}\,\Lambda_0$. Further, to each number C, we associate the set of affine functions A_C obtained from A_0 by means of the shift by $(-C)$:

$$A_C = \{\mathfrak{a}(\lambda) = l(\lambda) - C \mid l(\cdot) \in A_0\}.$$

Denote by $\mathfrak{R}_C := \mathrm{con}\, A_C$ the cone spanned by the set A_C.

Fix an arbitrary element $\bar{z} \in \mathcal{K}_Z$, $\bar{z} \ne 0$ (recall that $\mathcal{K}_Z := \mathcal{K} \cap Z(\Theta)$).

Proposition 2.34. *Let a number* C *be such that*

$$\underline{\lim} \max_{\Lambda_0} \frac{\varepsilon^2 \Omega^\lambda(\bar{z}) + \int_{t_0}^{t_f} \delta' H^\lambda \, dt}{\varepsilon^2 \bar{\gamma}(\bar{z}) + \gamma'} \ge C \tag{2.90}$$

for any pair of sequences $\{\varepsilon\}$, $\{\delta w'\}$ *such that* $\varepsilon \to +0$, $\{\delta w'\} \in \Pi'$, $\gamma' \le O(\varepsilon^2)$. *Then*

$$\inf_{\mathfrak{a}\in\mathfrak{R}_C} \max_{\mathrm{co}\,\Lambda_0} \{\Omega^\lambda(\bar{z}) + \mathfrak{a}(\lambda)\} \ge C\bar{\gamma}(\bar{z}). \tag{2.91}$$

Proof. Let $\mathfrak{a}(\cdot) \in \mathfrak{R}_C$, i.e., $\mathfrak{a}(\lambda) = \rho(l(\lambda) - C)$, where $\rho > 0$, $l(\cdot) \in A_0$. The latter means that there exists $\{\delta w'\} \in \Pi'$ such that

$$\frac{\int_{t_0}^{t_f} \delta' H^\lambda \, dt}{\gamma'} \to l(\lambda) \quad \forall\, \lambda \in \mathrm{co}\,\Lambda_0. \tag{2.92}$$

To the sequence $\{\delta w'\}$, we write the sequence $\{\varepsilon\} = \{\varepsilon(\delta w')\}$ of positive numbers such that

$$\varepsilon^2 = \frac{\gamma'}{\rho}, \quad \text{where } \gamma' = \int_{t_0}^{t_f} \Gamma(t, u^0 + \delta u')\, dt. \tag{2.93}$$

Then $\gamma' = O(\varepsilon^2)$, and, therefore, inequality (2.90) holds. This inequality and conditions (2.92) and (2.93) easily imply

$$\max_{\Lambda_0} \frac{\frac{\Omega^\lambda(\bar{z})}{\rho} + l(\lambda)}{\frac{\bar{\gamma}(\bar{z})}{\rho} + 1} \ge C.$$

Clearly, the maximum over Λ_0 in this inequality can be replaced by the maximum over $\mathrm{co}\,\Lambda_0$, since Ω^λ and $l(\lambda)$ are linear in λ. Therefore, multiplying the inequality by $\bar{\gamma}(\bar{z}) + \rho$, we obtain

$$\max_{\mathrm{co}\,\Lambda_0} (\Omega^\lambda(\bar{z}) + \rho l(\lambda)) \ge C(\bar{\gamma}(\bar{z})) + \rho).$$

Since $\rho(l(\lambda) - C) = \mathfrak{a}(\lambda)$, this implies

$$\max_{\mathrm{co}\,\Lambda_0} (\Omega^\lambda(\bar{z}) + \mathfrak{a}(\lambda)) \ge C\bar{\gamma}(\bar{z}). \tag{2.94}$$

It remains to recall that $\mathfrak{a}(\cdot)$ is an arbitrary element of $\mathfrak{R}_C$, and hence (2.94) implies (2.91). The proposition is proved. ∎

Lemma 2.33 and Proposition 2.34 imply the following assertion.

Lemma 2.35. *Let* $C_\gamma > -\infty$. *Then for any* $C \le \min\{C_\gamma, \varkappa C_\gamma\}$ *and for any* $\bar{z} \in \mathcal{K}_Z$, *we have the inequality*

$$\inf_{\mathfrak{a}\in\mathfrak{R}_C} \max_{\lambda\in\mathrm{co}\,\Lambda_0} \{\Omega^\lambda(\bar{z}) + \mathfrak{a}(\lambda)\} \ge C\bar{\gamma}(\bar{z}). \tag{2.95}$$

In what follows, we will need the convexity property of the cone $\mathfrak{R}_C$. It is implied by the following assertion.

Proposition 2.36. *The set* A_0 *is convex.*

Proof. Let $l_1(\cdot) \in A_0$, $l_2(\cdot) \in A_0$, $p > 0$, $q > 0$, $p + q = 1$. It is required to show that $l(\cdot) = pl_1(\cdot) + ql_2(\cdot) \in A_0$. Let $\{\delta w_i'\} \in \Pi'$, $i = 1, 2$, be two sequences of needle-shaped variations such that

$$\frac{\int_{t_0}^{t_f} \delta_i' H^\lambda \, dt}{\gamma_i'} \to l_i(\lambda), \quad i = 1,2,$$

where $\delta_i' H^\lambda$ and $\gamma_i' = \int_{t_0}^{t_f} \Gamma_i' \, dt$ correspond to the sequences $\{\delta w_i'\}$, $i = 1,2$. Using the sequences $\{\delta w_i'\}$, $i = 1,2$, we construct a sequence $\{\delta w'\} \in \Pi'$ such that

$$\frac{\int_{t_0}^{t_f} \delta' H^\lambda \, dt}{\gamma'} \to l(\lambda). \tag{2.96}$$

Thus, the convexity of A_0 will be proved. For brevity, we set $\int_{t_0}^{t_f} \delta_i' H^\lambda \, dt = \xi_i(\lambda)$, $i = 1,2$. Then

$$\frac{\xi_i(\lambda)}{\gamma_i'} \to l_i(\lambda), \quad i = 1,2.$$

Moreover, we set

$$\frac{p\gamma_2'}{q\gamma_1' + p\gamma_2'} = \alpha', \qquad \frac{q\gamma_1'}{q\gamma_1' + p\gamma_2'} = \beta'.$$

Then $\alpha' > 0$, $\beta' > 0$, $\alpha' + \beta' = 1$, and

$$\frac{\alpha'\gamma_1'}{\alpha'\gamma_1' + \beta'\gamma_2'} = p, \qquad \frac{\beta'\gamma_2'}{\alpha'\gamma_1' + \beta'\gamma_2'} = q.$$

Consequently,

$$\frac{\alpha'\xi_1(\lambda) + \beta'\xi_2(\lambda)}{\alpha'\gamma_1' + \beta'\gamma_2'} = p\frac{\xi_1(\lambda)}{\gamma_1'} + q\frac{\xi_2(\lambda)}{\gamma_2'} \to pl_1(\lambda) + ql_2(\lambda) = l(\lambda).$$

Therefore,

$$\frac{\alpha' \int_{t_0}^{t_f} \delta_1' H^\lambda \, dt + (1-\alpha') \int_{t_0}^{t_f} \delta_2' H^\lambda \, dt}{\alpha' \int_{t_0}^{t_f} \Gamma_1' \, dt + (1-\alpha') \int_{t_0}^{t_f} \Gamma_2' \, dt} \to l(\lambda). \tag{2.97}$$

Assume now that there is a sequence of functions $\{\alpha(t)\}$ in L^∞, each member of which satisfies the following conditions:

(i) $\alpha(t)$ assumes two values, 0 or 1, only;

(ii) $\alpha' \int_{t_0}^{t_f} \delta_1' H^\lambda \, dt = \int_{t_0}^{t_f} \alpha(t)\delta_1' H^\lambda \, dt, \alpha' \int_{t_0}^{t_f} \delta_2' H^\lambda \, dt = \int_{t_0}^{t_f} \alpha(t)\delta_2' H^\lambda \, dt$ for all $\lambda \in \operatorname{co} \Lambda_0$;

(iii) $\alpha' \int_{t_0}^{t_f} \Gamma_1' \, dt = \int_{t_0}^{t_f} \alpha(t)\Gamma_1' \, dt$ and $\alpha' \int_{t_0}^{t_f} \Gamma_2' \, dt = \int_{t_0}^{t_f} \alpha(t)\Gamma_2' \, dt$.

We note that conditions (ii) hold for all elements of $\lambda \in \operatorname{co} \Lambda_0$ whenever they hold for finitely many linearly independent elements of $\operatorname{co} \Lambda_0$ that compose a basis in $L_0 = \operatorname{Lin}(\operatorname{co} \Lambda_0)$. Therefore, in conditions (ii) and (iii) we in essence speak about the preservation of finitely many integrals.

We set $\delta w' = \alpha(t)\delta w_1' + (1-\alpha(t))\delta w_2'$. Then, obviously, $\{\delta w'\} \in \Pi'$, and, moreover,

$$\begin{aligned} \alpha' \int_{t_0}^{t_f} \delta_1' H^\lambda \, dt + (1-\alpha') \int_{t_0}^{t_f} \delta_2' H^\lambda \, dt &= \int_{t_0}^{t_f} \alpha(t)\delta_1' H^\lambda + (1-\alpha(t))\delta_2' H^\lambda \, dt \\ &= \int_{t_0}^{t_f} \delta' H^\lambda \, dt \quad \forall \, \lambda \in \operatorname{co} \Lambda_0, \end{aligned}$$

where $\delta' H^{\lambda}$ corresponds to the sequence $\{\delta w'\}$. Analogously,

$$\alpha' \int_{t_0}^{t_f} \Gamma_1' \, dt + (1-\alpha') \int_{t_0}^{t_f} \Gamma_2' \, dt = \int_{t_0}^{t_f} \Gamma' \, dt,$$

where $\Gamma' = \Gamma(\delta w', t)$. This and (2.97) imply (2.96). Therefore, the convexity of A_0 will be proved if we ensure the existence of a sequence $\{\alpha(t)\}$ satisfying conditions (ii) and (iii). The existence of such a sequence is implied by the Blackwell lemma, which is well known in optimal control theory and is contiguous to the Lyapunov theorem on the convexity of the range of a vector-valued measure. However, we note that we need not satisfy conditions (ii) and (iii) exactly: it suffice to do this with an arbitrary accuracy; i.e., given an arbitrary sequence $\varepsilon \to +0$ in advance, we need to ensure the fulfillment of conditions (ii) and (iii) with accuracy up to ε for each serial number of the sequence. Also, in this case, condition (2.96) certainly holds. With such a weakening of conditions (ii) and (iii), we can refer to Theorem 16.1 in [79, Part 2]. The proposition is proved. ∎

2.4.9 Narrowing of the Set $\operatorname{co}\Lambda_0$ up to the Set $M^{\mathrm{co}}(C\Gamma)$

In what follows, we will deal with the transformation of the left-hand side of inequality (2.95). For this purpose, we need the following abstract assertion.

Lemma 2.37. *Let X be a Banach space, $F : X \to \mathbb{R}^1$ be a sublinear (i.e., convex and positively homogeneous) functional, $K \subset X$ be a nonempty convex cone, and $x_0 \in X$ be a fixed point. Then the following formula holds:*

$$\inf_{x \in K} F(x_0 + x) = \sup_{x^* \in \partial F \cap K^*} \langle x^*, x_0 \rangle,$$

where $\partial F = \{x^* \in X^* \mid \langle x^*, x\rangle \le F(x) \text{ for all } x \in X\}$ is the set of support functionals of F and $K^* = \{x^* \in X^* \mid \langle x^*, x\rangle \ge 0 \text{ for all } x \in K\}$ is the dual cone of K.

We use this lemma in order to transform the expression

$$\inf_{\mathfrak{a} \in \mathfrak{R}_C} \max_{\lambda \in \operatorname{co}\Lambda_0} \{\Omega^{\lambda}(\bar{z}) + \mathfrak{a}(\lambda)\}$$

in the left-hand side of inequality (2.91). Denote by $\mathcal{A}$ the set of all affine functions $\mathfrak{a}(\lambda) : \operatorname{co}\Lambda_0 \to \mathbb{R}^1$. As was already noted, $\mathcal{A}$ is a finite-dimensional space. In this space, we consider the sublinear functional

$$F : \mathfrak{a}(\cdot) \in \mathcal{A} \to \max_{\lambda \in \operatorname{co}\Lambda_0} \mathfrak{a}(\lambda).$$

Since $\operatorname{co}\Lambda_0$ is a convex compact set, we can identify it with the set of support functionals of F; more precisely, there is a one-to-one correspondence between each support functional $\mathfrak{a}^* \in \partial F$ and the element $\lambda \in \operatorname{co}\Lambda_0$ such that $\langle \mathfrak{a}^*, \mathfrak{a}\rangle = \mathfrak{a}(\lambda)$ for all $\mathfrak{a} \in \mathcal{A}$. Moreover, according to this formula, a certain support functional $\mathfrak{a}^* \in \partial F$ corresponds to every element $\lambda \in \operatorname{co}\Lambda_0$.

Further, let C be a number such that the cone $\mathfrak{R}_C$ defined above is nonempty. Then what was said above implies that $\partial F \cap \mathfrak{R}_C^*$ can be identified with the set

$$M(\mathfrak{R}_C; \operatorname{co}\Lambda_0) \stackrel{\text{def}}{=} \left\{\lambda \in \operatorname{co}\Lambda_0 \;\middle|\; \mathfrak{a}(\lambda) \ge 0 \;\forall\, \mathfrak{a} \in \mathfrak{R}_C\right\}.$$

By Lemma 2.37, we obtain from this that

$$\inf_{\mathfrak{a}\in\mathfrak{R}_C}\max_{\lambda\in\operatorname{co}\Lambda_0}\{\Omega^\lambda(\bar z)+\mathfrak{a}(\lambda)\}=\max_{\lambda\in M(\mathfrak{R}_C;\operatorname{co}\Lambda_0)}\Omega^\lambda(\bar z). \tag{2.98}$$

It suffices to make more precise what the set $M(\mathfrak{R}_C;\operatorname{co}\Lambda_0)$ means.

For an arbitrary C, denote by $M^{\mathrm{co}}(C\Gamma)$ the set of $\lambda\in\operatorname{co}\Lambda_0$ such that

$$H(t,x^0(t),u,\psi(t))-H(t,x^0(t),u^0(t),\psi(t))\geq C\Gamma(t,u) \tag{2.99}$$

if $t\in[t_0,t_f]\setminus\Theta$, $u\in\mathcal{U}(t,x^0(t))$, where $\mathcal{U}(t,x)=\{u\in\mathbb{R}^{d(u)}\mid(t,x,u)\in\mathcal{Q}\}$. Namely in this case, we say that the minimum principle "of strictness $C\Gamma$" holds for λ. For a positive C and $\lambda\in\Lambda_0$ it is a strengthening of the usual minimum principle. Also, we note that the set $M^{\mathrm{co}}(C\Gamma)$ for $C=0$ coincides with the set M_0^{co} defined in the same way as the set M_0 from Section 2.1.4, with the only difference being that in the definition of the latter, it is necessary to replace the set Λ_0 by its convex hull $\operatorname{co}\Lambda_0$.

Proposition 2.38. *For any real C, we have*

$$M(\mathfrak{R}_C;\operatorname{co}\Lambda_0)\subset M^{\mathrm{co}}(C\Gamma). \tag{2.100}$$

Proof. Let $C\in\mathbb{R}$. Let $\hat\lambda\in M(\mathfrak{R}_C;\operatorname{co}\Lambda_0)$, i.e., $\hat\lambda\in\operatorname{co}\Lambda_0$ and $\mathfrak{a}(\hat\lambda)\geq 0$ for all $\mathfrak{a}\in\mathfrak{R}_C$. Hence $l(\hat\lambda)\geq C$ for all $l\in A_0$. Using this inequality, we show that $\hat\lambda\in M^{\mathrm{co}}(C\Gamma)$. Fix an arbitrary point t' and a vector u' such that

$$t'\in(t_0,t_f)\setminus\Theta,\quad u'\in\mathbb{R}^{d(u)},\quad (t',x^0(t'),u')\in\mathcal{Q},\quad u'\neq u^0(t'). \tag{2.101}$$

Let $\varepsilon>0$. Define the needle-shaped variation

$$\delta u'(t)=\begin{cases}u'-u^0(t), & t\in[t'-\varepsilon,t'+\varepsilon],\\ 0 & \text{otherwise.}\end{cases}$$

For $\varepsilon\to+0$, we have the sequence $\{\delta w'\}\in\Pi'$, where $\delta w'=(0,\delta u')$. For each $\lambda\in\operatorname{co}\Lambda_0$, there exists the limit

$$\lim_{\varepsilon\to+0}\frac{\int_{t_0}^{t_f}\delta'H^\lambda\,dt}{\int_{t_0}^{t_f}\Gamma'\,dt}=\left.\frac{\delta'H^\lambda}{\Gamma'}\right|_{t=t'}$$

for this sequence. According to the definition of the set A_0, this limit is $l(\lambda)$, where $l(\cdot)\in A_0$. Since $l(\hat\lambda)\geq C$, we have

$$\left.\frac{\delta'H^{\hat\lambda}}{\Gamma'}\right|_{t=t'}\geq C.$$

In other words, for $\lambda=\hat\lambda$, inequality (2.99) holds for arbitrary u' and t' satisfying conditions (2.101). This implies $\hat\lambda\in M^{\mathrm{co}}(C\Gamma)$. The proposition is proved. ∎

(In fact, we have the relation $M(\mathfrak{R}_C;\operatorname{co}\Lambda_0)=M^{\mathrm{co}}(C\Gamma)$, but we need only the inclusion for decoding the constant C_γ.) We obtain from relation (2.98) and inclusion (2.100) that

$$\inf_{\mathfrak{a}\in\mathfrak{R}_C}\max_{\lambda\in\operatorname{co}\Lambda_0}\{\Omega^\lambda(\bar z)+\mathfrak{a}(\lambda)\}\leq\max_{\lambda\in M^{\mathrm{co}}(C\Gamma)}\Omega^\lambda(\bar z).$$

This and Lemma 2.35 imply the following assertion.

Lemma 2.39. *Let* $\min\{C_\gamma, \varkappa C_\gamma\} \geq C > -\infty$. *Then*

$$\max_{\lambda \in M^{\mathrm{co}}(C\Gamma)} \Omega^\lambda(\bar{z}) \geq C\bar{\gamma}(\bar{z}) \quad \forall\, \bar{z} \in \mathcal{K}_Z. \tag{2.102}$$

The distinguishing of the cone $\Re_C$ by using the set Π' of sequences of needle-shaped variations and the "narrowing" of the set $\operatorname{co}\Lambda_0$ up to the set $M^{\mathrm{co}}(C\Gamma)$ by using formula (2.98) referred to the duality theory represents Milyutin's method which is called the "cone technique" bearing in mind the cone $\Re_C$ of affine functions on the convex compact set $\operatorname{co}\Lambda_0$. Also, we note that the use of this method necessarily leads to the convexification of the compact set Λ_0 in all the corresponding formulas.

2.4.10 Closure of the Cone $\mathcal{K}_Z$ in the Space $Z_2(\Theta)$

It remains to prove that Lemma 2.39 remains valid if we replace $\mathcal{K}_Z$ by $\mathcal{K}$ in it. This is implied by the following assertion.

Proposition 2.40. *The closure of the cone* $\mathcal{K}_Z$ *in the space* $Z_2(\Theta)$ *coincides with the cone* $\mathcal{K}$.

The proof of this proposition uses the Hoffman lemma on the estimation of the distance to the solution set of a system of linear inequalities, i.e., Lemma 1.12. More precisely, we use the following consequence of the Hoffman lemma.

Lemma 2.41. *Let* X *and* Y *be two Banach spaces, let* $l_i : X \to \mathbb{R}^1$, $i = 1,\dots,k$, *be linear functionals on* X, *and let* $A : X \to Y$ *be a linear operator with closed range. Then there exists a constant* $N = N(l_1,\dots,l_k,A) > 0$ *such that for any point* $x_0 \in X$, *there exists* $\bar{x} \in X$ *such that*

$$\langle l_i, x_0 + \bar{x}\rangle \leq 0, \quad i = 1,\dots,k; \quad A(x_0 + \bar{x}) = 0, \tag{2.103}$$

$$\|\bar{x}\| \leq N\left(\sum_{i=1}^{k}\langle l_i, x_0\rangle^+ + \|Ax_0\|\right). \tag{2.104}$$

Indeed, system (2.103) is compatible, since it admits the solution $\bar{x} = -x_0$, and then by Lemma 1.12, there exists a solution satisfying estimate (2.104) (clearly, the surjectivity condition for the operator $A : X \to Y$ in Lemma 1.12 can be replaced by the closedness of the range of this operator considering the image AX as Y).

Proof of Proposition 2.40. Since $\mathcal{K}_Z \subset \mathcal{K}$ and $\mathcal{K}$ is closed in $Z_2(\Theta)$, it suffices to only show that $\mathcal{K} \subset [K_Z]_2$, where $[\cdot]_2$ is the closure in $Z_2(\Theta)$. Let $\bar{z} = (\bar{\xi}, \bar{x}, \bar{u}) \in \mathcal{K}$. We show that there exists a sequence in $\mathcal{K}_Z$ that converges to $\bar{z}$ in $Z_2(\Theta)$. Take a sequence $N \to \infty$. For each member of this sequence, we set

$$\bar{u}^N(t) = \begin{cases} \bar{u}(t), & |\bar{u}(t)| \geq N, \\ 0, & |\bar{u}(t)| < N. \end{cases}$$

The sequence $\{\bar{u}^N\}$ satisfies the condition $\int_{t_0}^{t_f} \langle \bar{u}^N, \bar{u}^N\rangle\, dt \to 0$. Let $\bar{x}^N$ be defined from the conditions $\dot{\bar{x}}^N = f_x\bar{x}^N + f_u\bar{u}^N$, $\bar{x}^N(t_0) = 0$, $\bar{x}^N \in W^{1,2}(\Delta, \mathbb{R}^{d(x)})$, where $W^{1,2}(\Delta, \mathbb{R}^{d(x)})$

is the space of absolutely continuous functions $x(t) : [t_0, t_f] \to \mathbb{R}^{d(x)}$ having the Lebesgue square integrable derivatives; it is endowed with the norm

$$\|x\|_{1,2} = \left(\langle x(t_0), x(t_0) \rangle + \int_{t_0}^{t_f} \langle \dot{x}, \dot{x} \rangle \, dt \right)^{1/2}.$$

Then $\|\bar{x}^N\|_{1,2} \to 0$, and hence $\|\bar{x}^N\|_C \to 0$. We set $\bar{z}^N = (0, \bar{x}^N, \bar{u}^N)$ and $\bar{z}_N = \bar{z} - \bar{z}^N$. Then $\|z - \bar{z}_N\|_{Z_2(\Theta)} = \|\bar{z}^N\|_{Z_2(\Theta)} \to 0$. The conditions $\bar{z} \in \mathcal{K}$ and $\|\bar{x}^N\|_C \to 0$ imply

$$\sum_{i \in I} (F_{ip} \bar{p}_N)^+ + |K_p \bar{p}_N| \to 0, \tag{2.105}$$

where $\bar{p}_N$ corresponds to the sequence $\bar{z}_N$. Moreover, $\{\bar{z}_N\}$ belongs to the subspace $\mathcal{T}_2 \subset Z_2(\Theta)$ defined by the conditions

$$\dot{\bar{x}} = f_x \bar{x} + f_u \bar{u}, \quad [\bar{x}]^k = [f]^k \bar{\xi}_k, \quad k = 1, \ldots, s.$$

Applying Lemma 2.41 on $\mathcal{T}_2$, we obtain that for a sequence $\{\bar{z}_N\}$ in $\mathcal{T}_2$ satisfying condition (2.105), there exists a sequence $\{\bar{z}'_N\}$ in $\mathcal{K}_Z$ such that $\|\bar{z}_N - \bar{z}'_N\|_{Z_2(\Theta)} \to 0$. But the following condition also holds for $\{\bar{z}'_N\}$: $\|\bar{z} - \bar{z}'_N\|_{Z_2(\Theta)} \to 0$. Therefore, $\bar{z} \in [\mathcal{K}_Z]_2$. Since $\bar{z}$ is an arbitrary element from $\mathcal{K}$, we have $\mathcal{K} \subset [\mathcal{K}_Z]_2$, and then $\mathcal{K} = [\mathcal{K}_Z]_2$. The proposition is proved. ∎

Lemma 2.39 and Proposition 2.40 imply the following assertion.

Lemma 2.42. *Let* $\min\{C_\gamma, \varkappa C_\gamma\} \geq C > -\infty$. *Then*

$$\max_{\lambda \in M^{\text{co}}(C\Gamma)} \Omega^\lambda(\bar{z}) \geq C \bar{\gamma}(\bar{z}) \quad \forall\, \bar{z} \in \mathcal{K}. \tag{2.106}$$

We now recall that $C_\gamma \geq 0$ is a necessary condition for the Pontryagin minimum. Assume that it holds. Then, setting $C = 0$ in (2.106), we obtain the following result.

Theorem 2.43. *Let* $w^0 = (x^0, u^0)$ *be a Pontryagin minimum point. Then*

$$\max_{\lambda \in M_0^{\text{co}}} \Omega^\lambda(\bar{z}) \geq 0 \quad \forall\, \bar{z} \in \mathcal{K}.$$

Therefore, we have obtained the quadratic necessary condition for the Pontryagin minimum, which is slightly weaker than Condition $\mathfrak{A}$ of Theorem 2.4. It is called Condition $\mathfrak{A}^{\text{co}}$. Using Condition $\mathfrak{A}^{\text{co}}$, we will show below that Condition $\mathfrak{A}$ is also necessary for the Pontryagin minimum. We thus will complete the proof of Theorem 2.4. Section 2.6 is devoted to this purpose. But first we complete (in Section 2.5) the decoding of the constant C_γ.

Denote by $C_{\mathcal{K}}$ the least upper bound of C such that $M^{\text{co}}(C\Gamma)$ is nonempty and condition (2.106) holds. Then Lemma 2.42 implies

$$C_{\mathcal{K}} \geq \min\{C_\gamma, \varkappa C_\gamma\}, \tag{2.107}$$

i.e., the constant $C_{\mathcal{K}}$ estimates the constant C_γ from above with accuracy up to a constant multiplier. We will prove that the constant $C_{\mathcal{K}}$ estimates the constant C_γ from below with accuracy up to a constant multiplier. This will allow us to obtain a sufficient condition for the Pontryagin minimum.

Remark 2.44. Lemma 2.42 also implies the following assertion: if $C_\gamma \geq 0$, then $M_0^{\rm co}$ is nonempty and

$$\max_{\lambda \in M_0^{\rm co}} \Omega^\lambda(\bar{z}) \geq 0 \quad \forall\, \bar{z} \in \mathcal{K}.$$

This assertion serves as an important supplement to inequality (2.107).

2.5 Estimation of the Basic Constant from Below

2.5.1 Method for Obtaining Sufficient Condition for a Pontryagin Minimum

Since $C_\gamma > 0$ is a sufficient condition for the Pontryagin minimum, the positivity requirement for any quantity that estimates C_γ from below is also a sufficient condition. Therefore, the second stage of decoding, the estimation of C_γ from below, can also be considered as a method for obtaining a sufficient condition for the Pontryagin minimum. As was already noted, the sufficiency of the condition $C_\gamma > 0$ for the Pontryagin minimum is a sufficiently elementary fact not requiring the constructions of Chapter 1 for its proof. Therefore a source for obtaining a sufficient condition for the Pontryagin minimum is very simple, which is characteristic for sufficient conditions in general. However, in contrast to other sources and methods for obtaining sufficient conditions, which often are of arbitrary character, in this case, we are familiar with connection of the sufficient condition being used with the necessary condition; it consists of the passage from the strict inequality $C_\gamma > 0$ to the nonstrict inequality. This fact is already not so obvious, it is nontrivial, and it is guaranteed by the abstract scheme.

2.5.2 Extension of the Set $\Pi_{\sigma\gamma}$

For convenience, we recall the following main definitions:

$$C_\gamma = \inf_{\Pi_{\sigma\gamma}} \varliminf \frac{\Phi_0}{\gamma}, \qquad \Phi_0(\delta w) = \max_{\lambda \in \Lambda_0} \Phi(\lambda, \delta w) = \max_{\lambda \in \operatorname{co}\Lambda_0} \Phi(\lambda, \delta w),$$

$$\Phi(\lambda, \delta w) = \delta l^\lambda - \int_{t_0}^{t_f} (\psi \delta \dot{x} - \delta H^\lambda)\, dt, \quad \Pi_{\sigma\gamma} = \{\{\delta w\} \in \Pi \mid \sigma \leq O(\gamma)\},$$

$$\sigma = \max\{F_i(p^0 + \delta p)\ (i \in I), |\delta K|, \|\delta \dot{x} - \delta f\|_1\}.$$

Let $M \subset \operatorname{co}\Lambda_0$ be an arbitrary nonempty compact set. Then we have the following for any variation $\delta w \in \delta \mathcal{W}$:

$$\Phi_0(\delta w) = \max_{\operatorname{co}\Lambda_0} \Phi(\lambda, \delta w) \geq \max_M \Phi(\lambda, \delta w) \stackrel{\rm def}{=} \Phi_M(\delta w).$$

Consequently,

$$C_\gamma := \inf_{\Pi_{\sigma\gamma}} \varliminf \frac{\Phi_0}{\gamma} \geq \inf_{\Pi_{\sigma\gamma}} \varliminf \frac{\Phi_M}{\gamma}. \tag{2.108}$$

We set

$$\Pi_{o(\sqrt{\gamma})} = \{\{\delta w\} \in \Pi \mid \sigma = o(\sqrt{\gamma})\}.$$

Since $\Pi_{\sigma\gamma} \subset \Pi_{o(\sqrt{\gamma})}$, we have

$$\inf_{\Pi_{\sigma\gamma}} \underline{\lim} \frac{\Phi_M}{\gamma} \geq \inf_{\Pi_{o(\sqrt{\gamma})}} \underline{\lim} \frac{\Phi_M}{\gamma}. \tag{2.109}$$

Inequalities (2.108) and (2.109) imply

$$C_\gamma \geq \inf_{\Pi_{o(\sqrt{\gamma})}} \underline{\lim} \frac{\Phi_M}{\gamma}. \tag{2.110}$$

Let $C \in \mathbb{R}^1$ be such that $M^{\text{co}}(C\Gamma)$ is nonempty. We set

$$\Phi_{C\Gamma} = \max_{M^{\text{co}}(C\Gamma)} \Phi(\lambda, \delta w), \quad C_\gamma\Big(\Phi_{C\Gamma}; \Pi_{o(\sqrt{\gamma})}\Big) = \inf_{\Pi_{o(\sqrt{\gamma})}} \underline{\lim} \frac{\Phi_{C\Gamma}}{\gamma}.$$

Then (2.110) implies the following assertion.

Lemma 2.45. *The following inequality holds for an arbitrary C such that $M^{\text{co}}(C\Gamma)$ is nonempty:*

$$C_\gamma \geq C_\gamma(\Phi_{C\Gamma}; \Pi_{o(\sqrt{\gamma})}). \tag{2.111}$$

In what follows, we will fix an arbitrary C such that the set $M^{\text{co}}(C\Gamma)$ is nonempty.

2.5.3 Passing to Local Sequences

We set

$$\Pi^{\text{loc}}_{o(\sqrt{\gamma})} = \{\{\delta w\} \in \Pi^{\text{loc}} \mid \sigma = o(\sqrt{\gamma})\}. \tag{2.112}$$

In other words,

$$\Pi^{\text{loc}}_{o(\sqrt{\gamma})} = \Pi_{o(\sqrt{\gamma})} \cap \Pi^{\text{loc}}.$$

Our further goal consists of passing from the constant $C_\gamma(\Phi_{C\Gamma}, \Pi_{o(\sqrt{\gamma})})$ defined by the set of sequences $\Pi_{o(\sqrt{\gamma})}$ to a constant defined by the set of sequences $\Pi^{\text{loc}}_{o(\sqrt{\gamma})}$. For such a passage, we need the estimate

$$|\Phi_{C\Gamma}| \leq O(\gamma) \mid \Pi^{\text{loc}}, \tag{2.113}$$

i.e., $|\Phi_{C\Gamma}(\delta w)| \leq O(\gamma(\delta w))$ for any sequence $\{\delta w\} \in \Pi^{\text{loc}}$. To prove estimate (2.113), we need a certain property of the set $M^{\text{co}}(C\Gamma)$ analogous to one of the Weierstrass–Erdmann conditions of the classical calculus of variations. Let us formulate this analogue.

Proposition 2.46. *The following conditions hold for any $\lambda \in M^{\text{co}}(C\Gamma)$:*

$$[H^\lambda]^k = 0 \quad \forall\, t_k \in \Theta.$$

Proof. Fix arbitrary $\lambda \in M^{\text{co}}(C\Gamma)$ and $t_k \in \Theta$. We set $t_\varepsilon = t_k - \varepsilon$, $\varepsilon > 0$, and

$$\delta_\varepsilon H^\lambda = H(t_\varepsilon, x^0(t_\varepsilon), u^0(t_\varepsilon) + [u^0]^k, \psi(t_\varepsilon)) - H(t_\varepsilon, x^0(t_\varepsilon), u^0(t_\varepsilon), \psi(t_\varepsilon)).$$

Then for a small $\varepsilon > 0$, the condition $\lambda \in M^{\text{co}}(C\Gamma)$ implies $\delta_\varepsilon H^\lambda \geq C\delta_\varepsilon\Gamma$, where $\delta_\varepsilon\Gamma = \Gamma(t_\varepsilon, u^0(t_\varepsilon) + [u^0]^k) - \Gamma(t_\varepsilon, u^0(t_\varepsilon)) = \Gamma(t_\varepsilon, u^0(t_\varepsilon) + [u^0]^k)$. Taking into account that $\delta_\varepsilon H^\lambda \to [H^\lambda]^k$ and $\delta_\varepsilon\Gamma \to 0$ as $\varepsilon \to +0$, we obtain $[H^\lambda]^k \geq 0$. Constructing an analogous sequence $t^\varepsilon = t_k + \varepsilon$ to the right from the point t_k, we obtain $-[H^\lambda]^k \geq 0$. Therefore, $[H^\lambda]^k = 0$. The proposition is proved. ∎

In Section 2.1.5, we have defined Λ_0^Θ as the set of tuples $\lambda \in \Lambda_0$ such that $[H^\lambda]^k = 0$ for all $t_k \in \Theta$. Proposition 2.46 means that $M^{\text{co}}(C\Gamma) \subset \text{co}\,\Lambda_0^\Theta$. This and Proposition 2.16 imply estimate (2.113). By the way, we note that under the replacement of $M^{\text{co}}(C\Gamma)$ by $\text{co}\,\Lambda_0$, estimate (2.113) does not hold in general.

Recall that by $\overline{u^0}$ we have denoted the closure of the graph of the function $u^0(t)$ assuming that $u^0(t)$ is left continuous. By $\mathcal{Q}_{tu}$ we have denoted the projection of the set $\mathcal{Q}$ under the mapping $(t,x,u) \mapsto (t,u)$. Denote by $\{V\}$ an arbitrary sequence of neighborhoods of the compact set $\overline{u^0}$ contained in $\mathcal{Q}_{tu}$ such that $V \to \overline{u^0}$. The latter means that for any neighborhood $V' \subset \mathcal{Q}_{tu}$, of the compact set $\overline{u^0}$, there exists a number starting from which $V \subset V'$.

Let $\{\delta w\} \in \Pi$ be an arbitrary sequence. For members $\delta w = (\delta x, \delta u)$ and V of the sequences $\{\delta w\}$ and $\{V\}$, respectively, which have the same numbers,
we set

$$\delta u_V(t) = \begin{cases} \delta u(t) & \text{if } (t, u^0(t) + \delta u(t)) \in V, \\ 0 & \text{otherwise,} \end{cases}$$

$$\delta u^V = \delta u(t) - \delta u_V(t), \quad \delta w^{\text{loc}} = (\delta x, \delta u_V), \quad \delta w^V = (0, \delta u^V).$$

Then $\{\delta w\} = \{\delta w^{\text{loc}}\} + \{\delta w^V\}$, $\{\delta w^{\text{loc}}\} \in \Pi^{\text{loc}}$, and $\{\delta w^V\} \in \Pi$.

Proposition 2.47. *The following relation holds for the above representation of the sequence* $\{\delta w\}$:

$$\delta f = \delta^{\text{loc}} f + \delta^V f + r_f^V,$$

where

$$\begin{aligned} \delta f &= f(t, w^0 + \delta w) - f(t, w^0), & \delta^{\text{loc}} f &= f(t, w^0 + \delta w^{\text{loc}}) - f(t, w^0) \\ \delta^V f &= f(t, w^0 + \delta w^V) - f(t, w^0), & r_f^V &= (\bar{\delta}_x f - \delta_x f)\chi^V; \end{aligned}$$

the function χ^V *is the characteristic function of the set* $\mathcal{M}^V = \{t \mid \delta u^V \neq 0\}$; *and*

$$\begin{aligned} \delta_x f &= f(t, x^0 + \delta x, u^0) - f(t, x^0, u^0), \\ \bar{\delta}_x f &= f(t, x^0 + \delta x, u^0 + \delta u) - f(t, x^0, u^0 + \delta u). \end{aligned}$$

Moreover, $\|r_f^V\|_\infty \leq \|\bar{\delta}_x f\|_\infty + \|\delta_x f\|_\infty \overset{\text{def}}{=} \varepsilon_f \to 0$ *and* $\|r_f^V\|_1 \leq \varepsilon_f \operatorname{meas} M^V$.

Proof. We have

$$\begin{aligned}
\delta f &= \delta f(1-\chi^V)+\delta f \chi^V \\
&= \delta^{\text{loc}} f(1-\chi^V)+(f(t,x^0+\delta x,u^0+\delta u)-f(t,x^0,u^0+\delta u) \\
&\quad + f(t,x^0,u^0+\delta u)-f(t,x^0,u^0))\chi^V \\
&= \delta^{\text{loc}} f-\delta^{\text{loc}} f\chi^V+\bar{\delta}_x f\chi^V+\delta^V f \\
&= \delta^{\text{loc}} f+\delta^V f+(\bar{\delta}_x f-\delta_x f)\chi^V \\
&= \delta^{\text{loc}} f+\delta^V f+r_f^V.
\end{aligned}$$

The estimates for r_f^V are obvious. The proposition is proved. ∎

Also, it is obvious that the representation $\gamma=\gamma^{\text{loc}}+\gamma^V$ corresponds to the representation $\{\delta w\}=\{\delta w^{\text{loc}}\}+\{\delta w^V\}$, where $\gamma=\gamma(\delta w)$, $\gamma^{\text{loc}}=\gamma(\delta w^{\text{loc}})$, and $\gamma^V=\gamma(\delta w^V)$. This is implied by the relation $\Gamma(t,u^0+\delta u)=\Gamma(t,u^0+\delta u_V)+\Gamma(t,u^0+\delta u^V)$.

Now let us define a special sequence $\{V\}=\{V(\varepsilon)\}$ of neighborhoods of the compact set $\overline{u^0}$ that converges to this compact set. Namely, for each $\varepsilon>0$, we set

$$V=V(\varepsilon)=\bigcup_{(u,t)\in\overline{u^0}} \mathcal{O}_\varepsilon(t,u),$$

where $\mathcal{O}_\varepsilon(t,u)=\{(t',u')\,|\,|t-t'|<\varepsilon^2,\ |u-u'|<\varepsilon\}$.

Fix a sequence $\{\delta w\}\in\Pi$ and define a sequence $\{\varepsilon\}$ for it such that $\varepsilon\to+0$ sufficiently slowly. The following proposition explains what "sufficiently slowly" means.

Proposition 2.48. *Let $\varepsilon\to+0$ so that*

$$\frac{\varepsilon_f}{\varepsilon^2}\to 0 \quad \textit{and} \quad \frac{\sqrt{\gamma}}{\varepsilon^2}\to 0,$$

where ε_f is the same as in Proposition 2.47. Then $\|r_f^V\|_1\le\varepsilon_f \operatorname{meas}\mathcal{M}^V=o(\gamma^V)$ and $\operatorname{meas}\mathcal{M}^V=o(\sqrt{\gamma^V})$.

Proof. Let there exist the sequences $\{\delta w\}$, $\{\varepsilon\}$, and $\{V\}=\{V(\varepsilon)\}$ defined as above. Since $V\to\overline{u^0}$, we have that, starting from a certain number, $V\subset\mathcal{V}$, where $\mathcal{V}\subset\mathcal{Q}_{tu}$ is the neighborhood of the compact set $\overline{u^0}$ from the definition of the function $\Gamma(t,u)$ given in Section 2.3. Condition $V\subset\mathcal{V}$ implies $(\delta u^V)^{\mathcal{V}}=\delta u^{\mathcal{V}}$ (the meaning of this notation is the same as above). Hence $\delta u^V=\delta u^V_{\mathcal{V}}+\delta u^{\mathcal{V}}$, where $(t,u^0+\delta u^V_{\mathcal{V}})\in\mathcal{V}\setminus V$ on $\mathcal{M}^{\mathcal{V}}_V=\{t\mid\delta u^V_{\mathcal{V}}\ne 0\}$ and $(t,u^0+\delta u^{\mathcal{V}})\notin\mathcal{V}$ on $\mathcal{M}^{\mathcal{V}}=\{t\mid\delta u^{\mathcal{V}}\ne 0\}$. The definitions of $\mathcal{V}$, Γ, and $V=V(\varepsilon)$ imply $\Gamma(t,u^0+\delta u^V_{\mathcal{V}})\ge\varepsilon^2$ on $\mathcal{M}^V_{\mathcal{V}}$. This implies $\gamma^V_{\mathcal{V}}\ge\varepsilon^2\operatorname{meas}\mathcal{M}^V_{\mathcal{V}}$, where $\gamma^V_{\mathcal{V}}:=\int_{t_0}^{t_f}\Gamma(t,u^0+\delta u^V_{\mathcal{V}})\,dt$. Since $\gamma^V_{\mathcal{V}}\le\gamma^V$, we have $\operatorname{meas}\mathcal{M}^V_{\mathcal{V}}\le\frac{\gamma^V}{\varepsilon^2}$. Further, it follows from the definitions of Γ and $\delta u^{\mathcal{V}}$ that $\operatorname{meas}\mathcal{M}^{\mathcal{V}}\le\operatorname{const}\gamma^{\mathcal{V}}$, where $\gamma^{\mathcal{V}}:=\int_{t_0}^{t_f}\Gamma(t,u^0+\delta u^{\mathcal{V}})\,dt$, and const depends on the entire sequence $\{\delta u^{\mathcal{V}}\}$ but not on its specific member. Since $\gamma^{\mathcal{V}}\le\gamma^V$ (because $\mathcal{M}^{\mathcal{V}}\subset\mathcal{M}^V$), we have $\operatorname{meas}\mathcal{M}^{\mathcal{V}}\le\operatorname{const}\gamma^V$. Therefore,

$$\operatorname{meas}\mathcal{M}^V=\operatorname{meas}\mathcal{M}^V_{\mathcal{V}}+\operatorname{meas}\mathcal{M}^{\mathcal{V}}\le\frac{\gamma^V}{\varepsilon^2}+\operatorname{const}\gamma^V=\left(\frac{1}{\varepsilon^2}+\operatorname{const}\right)\gamma^V. \tag{2.114}$$

Taking into account that $\varepsilon_f/\varepsilon^2 \to 0$, we obtain from (2.114) that $\varepsilon_f \operatorname{meas}\mathcal{M}^V = o(\gamma^V)$. Moreover, since $\sqrt{\gamma}/\varepsilon^2 \to 0$ and $\gamma^V \le \gamma$, (2.114) also implies

$$\operatorname{meas}\mathcal{M}^V \le \left(\frac{\sqrt{\gamma^V}}{\varepsilon^2} + \operatorname{const}\sqrt{\gamma^V}\right)\sqrt{\gamma^V} = o\left(\sqrt{\gamma^V}\right).$$

The proposition is proved. ∎

In what follows, for brevity we set

$$\delta\Phi^\lambda = \Phi(\lambda,\delta w), \quad \delta^{\mathrm{loc}}\Phi^\lambda = \Phi(\lambda,\delta w^{\mathrm{loc}}), \quad \delta\Phi_{C\Gamma} = \Phi_{C\Gamma}(\delta w), \quad \delta^{\mathrm{loc}}\Phi_{C\Gamma} = \Phi_{C\Gamma}(\delta w^{\mathrm{loc}}).$$

Assume that there is an arbitrary sequence $\{\delta w\} \in \Pi$. We define a sequence $\{\varepsilon\}$ for it such that the conditions of Proposition 2.48 hold. Also, we define the corresponding sequences $\{V\} = \{V(\varepsilon)\}$, $\{\delta w^{\mathrm{loc}}\}$, and $\{\delta w^V\}$. Then

$$\begin{aligned}
\delta\Phi^\lambda &:= \delta l^\lambda - \int_{t_0}^{t_f} \psi\delta\dot{x}\,dt + \int_{t_0}^{t_f} \delta H^\lambda\,dt \\
&= \delta l^\lambda - \int_{t_0}^{t_f} \psi\delta\dot{x}\,dt + \int_{t_0}^{t_f} \psi(\delta^{\mathrm{loc}} f + \delta^V f + r_f^V)dt \\
&= \delta l^\lambda - \int_{t_0}^{t_f} \psi\delta\dot{x}\,dt + \int_{t_0}^{t_f} \delta^{\mathrm{loc}} H^\lambda\,dt + \int_{t_0}^{t_f} \delta^V H^\lambda\,dt + \int_{t_0}^{t_f} \psi r_f^V\,dt \\
&= \delta^{\mathrm{loc}}\Phi^\lambda + \int_{t_0}^{t_f} \delta^V H^\lambda\,dt + \rho^{V\lambda},
\end{aligned}$$

where $\rho^{V\lambda} = \int_{t_0}^{t_f} \psi r_f^V\,dt$. Since $\sup_{\mathrm{co}\,\Lambda_0} \|\psi\|_\infty < +\infty$ and $\|r_f^V\|_1 = o(\gamma^V)$, we have

$$\sup_{\mathrm{co}\,\Lambda_0} |\rho^{V\lambda}| = o(\gamma^V). \tag{2.115}$$

Finally, we obtain

$$\delta\Phi^\lambda = \delta^{\mathrm{loc}}\Phi^\lambda + \int_{t_0}^{t_f} \delta^V H^\lambda\,dt + \rho^{V\lambda}. \tag{2.116}$$

Moreover, the conditions $\delta f = \delta^{\mathrm{loc}} f + \delta^V f + r_f^V$, $\|\delta^V f\|_1 \le \operatorname{const}\operatorname{meas}\mathcal{M}^V = o\left(\sqrt{\gamma^V}\right)$, and $\|r_f^V\|_1 = o(\gamma^V)$ imply

$$\|\delta\dot{x} - \delta^{\mathrm{loc}} f\|_1 \le \|\delta\dot{x} - \delta f\|_1 + o\left(\sqrt{\gamma^V}\right). \tag{2.117}$$

Therefore, the following assertion holds.

Proposition 2.49. *Let $\{\delta w\} \in \Pi$ be an arbitrary sequence, and let a sequence $\{\varepsilon\}$ satisfy the conditions of Proposition* 2.48. *Then conditions* (2.115)–(2.117) *hold for the corresponding sequences $\{V(\varepsilon)\}$, $\{\delta w^{\mathrm{loc}}\}$, and $\{\delta w^V\}$.*

We now are ready to pass to local sequences. We set

$$C_\gamma\left(\Phi_{C\Gamma}; \Pi^{\mathrm{loc}}_{o(\sqrt{\gamma})}\right) = \inf_{\Pi^{\mathrm{loc}}_{o(\sqrt{\gamma})}} \varliminf \frac{\Phi_{C\Gamma}}{\gamma}.$$

We have the following assertion.

Lemma 2.50. *The following inequality holds:*

$$C_\gamma\Big(\Phi_{C\Gamma};\Pi_{o(\sqrt{\gamma})}\Big) \geq \min\Big\{C, C_\gamma\Big(\Phi_{C\Gamma};\Pi^{\rm loc}_{o(\sqrt{\gamma})}\Big)\Big\}. \tag{2.118}$$

Proof. Let $\{\delta w\} \in \Pi_{o(\sqrt{\gamma})}$ be a sequence with $\gamma > 0$. Choose the sequence $\{\varepsilon\}$ and the corresponding sequence $\{V\}$ as in Proposition 2.48. With the sequences $\{\delta w\}$ and $\{V\}$, we associate the splitting $\{\delta w\} = \{\delta w^{\rm loc}\} + \{\delta w^V\}$. Let us turn to formula (2.116). The definition of the set $M^{\rm co}(C\Gamma)$ implies

$$\delta^V H^\lambda \geq C\Gamma(t, u^0 + \delta u^V) \quad \forall\, \lambda \in M^{\rm co}(C\Gamma)$$

for all sufficiently large numbers. Then (2.115) and (2.116) imply

$$\delta\Phi_{C\Gamma} \geq \delta^{\rm loc}\Phi_{C\Gamma} + C\gamma^V + o(\gamma^V), \tag{2.119}$$

where $\gamma^V = \gamma(\delta w^V)$. We set $\gamma = \gamma(\delta w)$ and $\gamma^{\rm loc} = \gamma(\delta w^{\rm loc})$. Then $\gamma = \gamma^{\rm loc} + \gamma^V$. The following two cases are possible:

Case (a): $\underline{\lim}\, \frac{\gamma^{\rm loc}}{\gamma} = 0$.

Case (b): $\underline{\lim}\, \frac{\gamma^{\rm loc}}{\gamma} > 0$.

Let us consider each of them.

Case (a): Let $\underline{\lim}\, \frac{\gamma^{\rm loc}}{\gamma} = 0$. Choose subsequences such that $\gamma^{\rm loc}/\gamma \to 0$ for them, and, therefore,

$$\gamma^{\rm loc} = o(\gamma) \quad \text{and} \quad \frac{\gamma^V}{\gamma} \to 1. \tag{2.120}$$

We preserve the above notation for the subsequences. It follows from (2.119), (2.120), and estimate (2.113) that $\delta\Phi_{C\Gamma} \geq C\gamma + o(\gamma)$. Hence

$$\underline{\lim}\, \frac{\delta\Phi_{C\Gamma}}{\gamma} \geq C,$$

where the lower limit is taken over the chosen subsequence.

Case (b): Now let $\underline{\lim}\gamma^{\rm loc}/\gamma > 0$, and, therefore, $\gamma \leq O(\gamma^{\rm loc})$ and $\gamma^V \leq O(\gamma^{\rm loc})$. Since $\|\delta\dot{x} - \delta f\|_1 = o(\sqrt{\gamma})$, it follows from (2.117) that $\|\delta\dot{x} - \delta^{\rm loc} f\|_1 = o(\sqrt{\gamma^{\rm loc}})$. Moreover,

$$\delta^{\rm loc} F_i = \delta F_i \leq o(\sqrt{\gamma}) = o_1\Big(\sqrt{\gamma^{\rm loc}}\Big),\ i \in I; \quad |\delta^{\rm loc} K| = |\delta K| = o(\sqrt{\gamma}) = o\Big(\sqrt{\gamma^{\rm loc}}\Big).$$

Hence $\{\delta w^{\rm loc}\} \in \Pi^{\rm loc}_{o(\sqrt{\gamma})}$. Further, we obtain from (2.119) that

$$\begin{aligned}\underline{\lim}\, \frac{\delta\Phi_{C\Gamma}}{\gamma} &\geq \underline{\lim}\, \frac{\delta^{\rm loc}\Phi_{C\Gamma} + C\gamma^V}{\gamma} = \underline{\lim}\left\{\frac{\gamma^{\rm loc}}{\gamma}\frac{\delta^{\rm loc}\Phi_{C\Gamma}}{\gamma^{\rm loc}} + \frac{\gamma^V}{\gamma}C\right\}\\ &\geq \underline{\lim}\min\left\{\frac{\delta^{\rm loc}\Phi_{C\Gamma}}{\gamma^{\rm loc}}, C\right\} = \min\left\{\underline{\lim}\, \frac{\delta^{\rm loc}\Phi_{C\Gamma}}{\gamma^{\rm loc}}, C\right\}.\end{aligned}$$

The second equation follows from the condition

$$\frac{\gamma^{\text{loc}}}{\gamma} + \frac{\gamma^V}{\gamma} = 1.$$

Furthermore,

$$\underline{\lim} \frac{\delta^{\text{loc}} \Phi_{C\Gamma}}{\gamma^{\text{loc}}} \geq C_\gamma \left(\Phi_{C\Gamma}; \Pi^{\text{loc}}_{o(\sqrt{\gamma})} \right),$$

since $\{\delta w^{\text{loc}}\} \in \Pi^{\text{loc}}_{o(\sqrt{\gamma})}$. Hence

$$\underline{\lim} \frac{\delta \Phi_{C\Gamma}}{\gamma} \geq \min \left\{ C_\gamma \left(\Phi_{C\Gamma}; \Pi^{\text{loc}}_{o(\sqrt{\gamma})} \right), C \right\}.$$

Therefore, we have shown that for every sequence $\{\delta w\} \in \Pi_{o(\sqrt{\gamma})}$ on which $\gamma > 0$, there exists a subsequence such that $\underline{\lim}\, \delta \Phi_{C\Gamma}/\gamma$ is not less than the right-hand side of inequality (2.118). This implies inequality (2.118). The lemma is proved. ∎

The method used in this section is very characteristic for decoding higher-order conditions in optimal control in order to obtain sufficient conditions. Now this method is related to the representation of the Pontryagin sequence as the sum $\{\delta w\} = \{\delta w^{\text{loc}}\} + \{\delta w^V\}$ and the use of the maximum principle of strictness $C\Gamma$ for $\{\delta w^V\}$:

$$\delta^V H^\lambda \geq C\Gamma(t, u^0 + \delta u^V), \quad \lambda \in M^{\text{co}}(C\Gamma).$$

In what follows, analogous consideration will be given to various consequences of the minimum principle of strictness $C\Gamma$, the Legendre conditions, and conditions related to them. Lemmas 2.45 and 2.50 imply the following assertion.

Lemma 2.51. *Let the set $M^{\text{co}}(C\Gamma)$ be nonempty. Then*

$$C_\gamma \geq \min \left\{ C_\gamma \left(\Phi_{C\Gamma}; \Pi^{\text{loc}}_{o(\sqrt{\gamma})} \right), C \right\}.$$

2.5.4 Simplifications in the Definition of $C_\gamma(\Phi_{C\Gamma}; \Pi^{\text{loc}}_{o(\sqrt{\gamma})})$

By definition, $\Pi^{\text{loc}}_{o(\sqrt{\gamma})}$ consists of sequences $\{\delta w\} \in \Pi^{\text{loc}}$ satisfying the conditions

$$\delta F_i \leq o(\sqrt{\gamma}) \quad \forall\, i \in I, \quad |\delta K| = o(\sqrt{\gamma}), \tag{2.121}$$

$$\|\delta \dot{x} - \delta f\|_1 = o(\sqrt{\gamma}). \tag{2.122}$$

Obviously, conditions (2.121) are equivalent to the conditions

$$F_{ip} \delta p \leq o(\sqrt{\gamma}) \quad \forall\, i \in I, \quad |K_p \delta p| = o(\sqrt{\gamma}).$$

Let us consider condition (2.122). Assume that the sequence $\{\delta w\}$ is represented in the canonical form (see Proposition 2.9)

$$\{\delta w\} = \{\delta w^0\} + \{\delta w^*\}, \quad \{\delta w^0\} = \{(\delta x, \delta u^0)\} \in \Pi^0,$$
$$\{\delta w^*\} = \{(0, \delta u^*)\} \in \Pi^*, \quad |\delta u^0| \cdot |\delta u^*| = 0.$$

By Proposition 2.10,

$$\delta f = f_w \delta w^0 + \frac{1}{2}\langle f_{ww}\delta w^0, \delta w^0\rangle + \delta^* f + \delta^* f_x \delta x + \tilde{r},$$

where $\|\tilde{r}\|_1 = o(\gamma^0)$ and $\gamma^0 := \|\delta x\|_C^2 + \int_{t_0}^{t_f} |\delta u^0|^2\, dt$. According to formula (2.50),

$$\delta^* f = \sum \delta_k^* f = \sum \Big([f]^k(\chi_{k-}^* - \chi_{k+}^*) + f_u^{k+}\delta v_{k-} + f_u^{k-}\delta v_{k+}\Big) + O(|\delta t_k| + |\delta v_k|^2).$$

As was shown in Section 2.4 in proving the equivalence of conditions (2.66) and (2.69), we have the estimates

$$\|\langle f_{ww}\delta w^0, \delta w^0\rangle\|_1 \le o(\gamma^0), \qquad \left|\int_{t_0}^{t_f} \delta^* f_x \delta x\, dt\right| \le o(\gamma).$$

Moreover,

$$\|f_u^{k+}\delta v_{k-}\|_1 \le |f_u^{k+}|\sqrt{\text{meas}\,\mathcal{M}_{k-}}\,\|\delta v_{k-}\|_2 = o(\sqrt{\gamma^*}),$$
$$\|f_u^{k-}\delta v_{k+}\|_1 \le |f_u^{k-}|\sqrt{\text{meas}\,\mathcal{M}_{k+}}\,\|\delta v_{k+}\|_2 = o(\sqrt{\gamma^*}),$$

where $\gamma^* := \sum_k(\int_{\mathcal{M}_k^*} |\delta t|\, dt + \int_{t_0}^{t_f} |\delta v_k|^2\, dt)$ and $\|v\|_2 = (\int_{t_0}^{t_f} \langle v(t), v(t)\rangle\, dt)^{1/2}$ is the norm of the space $L^2(\Delta, \mathbb{R}^{d(u)})$ of Lebesgue square integrable functions $v(t) : [t_0, t_f] \to \mathbb{R}^{d(u)}$. Therefore, condition (2.122) is equivalent to the condition

$$\left\|\delta\dot{x} - f_w\delta w^0 - \sum [f]^k(\chi_{k-}^* - \chi_{k+}^*)\right\|_1 = o(\sqrt{\gamma}).$$

Finally, we obtain

$$\begin{aligned}\Pi_{o(\sqrt{\gamma})}^{\text{loc}} \;=\; \Big\{&\{\delta w\} \in \Pi^{\text{loc}} \mid F_{ip}\delta p \le o(\sqrt{\gamma})\ \forall\, i \in I; \quad |K_p\delta p| = o(\sqrt{\gamma});\\ &\left\|\delta\dot{x} - f_w\delta w^0 - \sum [f]^k(\chi_{k-}^* - \chi_{k+}^*)\right\|_1 = o(\sqrt{\gamma})\Big\}.\end{aligned}$$

In this relation, we use the canonical representation of the sequence $\{\delta w\} = \{\delta w^0\} + \{\delta w^*\}$, where $\{\delta w^0\} \in \Pi^0$, $\{\delta w^*\} \in \Pi^*$, and $|\delta u^0| \cdot |\delta u^*| = 0$.

By Proposition 2.15, in calculating $C_\gamma(\Phi_{C\Gamma}; \Pi_{o(\sqrt{\gamma})}^{\text{loc}})$, we can use the function

$$\Phi_{C\Gamma}^1(\delta w) := \max_{M^{\text{co}}(C\Gamma)} \Phi^{1\lambda}(\delta w)$$

(see definition (2.61) for $\Phi^{1\lambda}$) instead of the function $\Phi_{C\Gamma} = \max_{M^{\text{co}}(C\Gamma)} \Phi(\lambda, \delta w)$, and since by Proposition 2.46 the conditions $[H^\lambda]^k = 0$ for all $t_k \in \Theta$ hold for any $\lambda \in M^{\text{co}}(C\Gamma)$, we can omit the terms $\sum_k [H^\lambda]^k(\text{meas}\,\mathcal{M}_{k-}^* - \text{meas}\,\mathcal{M}_{k+}^*)$ in the definition of $\Phi^{1\lambda}$, thus passing to the function

$$\begin{aligned}\tilde{\Phi}^{1\lambda}(\delta w) \;:=\; &\frac{1}{2}\Phi''(\lambda, \delta w^0)\\ &+\sum_{k=1}^{s}\Bigg(D^k(H^\lambda)\int_{\mathcal{M}_k^*} |\delta t_k|\, dt + [H_x^\lambda]^k \int_{t_0}^{t_f} \delta x(\chi_{k-}^* - \chi_{k+}^*)\, dt\\ &+\frac{1}{2}\int_{t_0}^{t_f}\Big(\langle H_{uu}^{\lambda k+}\delta v_{k-}, \delta v_{k-}\rangle + \langle H_{uu}^{\lambda k-}\delta v_{k+}, \delta v_{k+}\rangle\Big)\, dt\Bigg),\end{aligned}$$

where

$$\Phi''(\lambda,\delta w) = \langle l^{\lambda}_{pp}\delta p,\delta p\rangle + \int_{t_0}^{t_f} \langle H^{\lambda}_{ww}\delta w,\delta w\rangle\, dt.$$

Therefore, the constant $C_\gamma\big(\Phi_{C\Gamma};\Pi^{\mathrm{loc}}_{o(\sqrt{\gamma})}\big)$ does not change, if we replace the function $\Phi_{C\Gamma}$ by the function $\tilde{\Phi}^1_{C\Gamma}$, where

$$\tilde{\Phi}^1_{C\Gamma}(\delta w) = \max_{M^{\mathrm{co}}(C\Gamma)} \tilde{\Phi}^{1\lambda}(\delta w).$$

Finally, we recall that the functional γ has the following form on local sequences represented in the canonical form:

$$\gamma(\delta w) = \|\delta x\|^2_C + \int_{t_0}^{t_f} |\delta u^0|^2\, dt + \sum_k \left(2\int_{\mathcal{M}^*_k} |\delta t_k|\, dt + \int_{t_0}^{t_f} |\delta v_k|^2\, dt\right)$$

or, in short, $\gamma = \gamma^0 + \gamma^*$, where

$$\gamma^0 = \|\delta x\|^2_C + \int_{t_0}^{t_f} |\delta u^0|^2\, dt, \quad \gamma^* = \sum_k \left(2\int_{\mathcal{M}^*_k} |\delta t_k|\, dt + \int_{t_0}^{t_f} |\delta v_k|^2\, dt\right).$$

We have proved that

$$C_\Gamma\left(\Phi_{C\Gamma};\Pi^{\mathrm{loc}}_{o(\sqrt{\gamma})}\right) = C_\gamma\left(\tilde{\Phi}^1_{C\Gamma};\Pi^{\mathrm{loc}}_{o(\sqrt{\gamma})}\right), \tag{2.123}$$

where

$$C_\gamma\left(\tilde{\Phi}^1_{C\Gamma};\Pi^{\mathrm{loc}}_{o(\sqrt{\gamma})}\right) = \inf_{\Pi^{\mathrm{loc}}_{o(\sqrt{\gamma})}} \varliminf \frac{\tilde{\Phi}^1_{C\Gamma}}{\gamma}.$$

We call attention to the fact that in the definition of this constant, the functions $\tilde{\Phi}^{1\lambda}$ and γ are defined indeed on the set of triples $(\delta w^0,\mathcal{M}^*,\delta v)$ such that

$$\begin{aligned}
&\delta w^0 = (\delta x,\delta u^0) \in W, \quad \mathcal{M}^* = \cup\mathcal{M}^*_k, \quad \mathcal{M}^*_k = \mathcal{M}^*_{k-} \cup \mathcal{M}^*_{k+},\\
&\mathcal{M}^*_{k-} \subset (t_k-\varepsilon,t_k), \quad \mathcal{M}^*_{k+} \subset (t_k,t_k+\varepsilon), \quad k \in I^*, \quad \varepsilon > 0;\\
&\delta v = \sum \delta v_k, \quad \delta v_k = \delta v_{k-} + \delta v_{k+}, \quad k \in I^*;\\
&\{t \mid \delta v_{k-} \neq 0\} \subset \mathcal{M}^*_{k-}, \quad \{t \mid \delta v_{k+} \neq 0\} \subset \mathcal{M}^*_{k+}, \quad k \in I^*.
\end{aligned}$$

Denote by a these triples, and denote by $\{a\}$ an arbitrary sequences of triples a such that

$$\{\delta w^0\} \in \Pi^0; \qquad \mathcal{M}^*_k \to t_k, \quad k \in I^*; \qquad \|\delta v\|_\infty \to 0.$$

To each sequence $\{\delta w\}$ from Π^{loc} represented in canonical form, we naturally associate the sequence of triples $\{a\} = \{a(\delta w)\} = \{(\delta w^0,\mathcal{M}^*,\delta v)\}$. However, note that for an arbitrary sequence $\{a\}$, we do not require the condition $\delta u^0\chi^* = 0$, which holds for the sequences $\{a\}$ corresponding to the sequences $\{\delta w\} \in \Pi^{\mathrm{loc}}$ in their canonical representations (as before, χ^* is the characteristic function of the set $\mathcal{M}^*$). Therefore, on the set of sequences $\{a\}$, we have defined the functions $\tilde{\Phi}^{1\lambda} = \tilde{\Phi}^{1\lambda}(a)$, $\lambda \in M^{\text{co}}(C\Gamma)$ and $\gamma = \gamma(a) = \gamma^0 + \gamma^*$. We set

$$\begin{aligned} S_1 \;=\; & \Big\{\{a\} = \{(\delta w^0, \mathcal{M}^*, \delta v)\} \mid F_{ip}\delta p \le o(\sqrt{\gamma}) \quad \forall\, i \in I, \quad |K_p \delta p| = o(\sqrt{\gamma}), \\ & \Big\| \delta \dot{x} - f_w \delta w^0 - \sum [f]^k (\chi^*_{k-} - \chi^*_{k+}) \Big\|_1 = o(\sqrt{\gamma})\Big\}, \end{aligned}$$

where χ^*_{k-} and χ^*_{k+} are the characteristic functions of the sets $\mathcal{M}^*_{k-}$ and $\mathcal{M}^*_{k+}$, respectively. Also, we set

$$\tilde{\Phi}^1_{C\Gamma}(a) = \max_{M^{\text{co}}(C\Gamma)} \tilde{\Phi}^{1\lambda}(a), \quad C_\gamma(\tilde{\Phi}^1_{C\Gamma}; S_1) = \inf_{S_1} \varliminf \frac{\tilde{\Phi}^1_{C\Gamma}}{\gamma}.$$

The following inequality holds:

$$C_\gamma\left(\tilde{\Phi}^1_{C\Gamma}; \Pi^{\text{loc}}_{o(\sqrt{\gamma})}\right) \ge C_\gamma\left(\tilde{\Phi}^1_{C\Gamma}; S_1\right). \tag{2.124}$$

Indeed, the sequence $\{a\} = \{a(\delta w)\} \in S_1$ corresponds to every sequence $\{\delta w\} \in \Pi^{\text{loc}}_{o(\sqrt{\gamma})}$, and, moreover, the values of $\tilde{\Phi}^1_{C\Gamma}$ and γ are preserved under this correspondence. There is no converse correspondence, since we omit the condition $\delta u^0 \chi^* = 0$ in the definition of S_1. Lemma 2.51 and formulas (2.123) and (2.124) imply the following assertion.

Lemma 2.52. *Let the set $M^{\text{co}}(C\Gamma)$ be nonempty. Then $C_\gamma \ge \min\{C_\gamma(\tilde{\Phi}^1_{C\Gamma}; S_1), C\}$.*

In what follows, we estimate the constant $C_\gamma(\tilde{\Phi}^1_{C\Gamma}; S_1)$ from below.

2.5.5 Use of Legendre Conditions

Now our goal is to pass to sequences with $\delta v = 0$. This will be done by using the Legendre conditions for points $t_k \in \Theta$. Let us formulate the following general assertion on the Legendre conditions.

Proposition 2.53. *Let $\lambda \in M^{\text{co}}(C\Gamma)$. Then we have that*

(a) *the following condition holds for any $t \in [t_0, t_f] \setminus \Theta$:*

$$\frac{1}{2}\langle H_{uu}(t, x^0(t), u^0(t), \psi(t))\bar{u}, \bar{u}\rangle \ge C\langle \bar{u}, \bar{u}\rangle \quad \forall\, \bar{u} \in \mathbb{R}^{d(u)};$$

(b) *the following conditions hold for any $t_k \in \Theta$:*

$$\frac{1}{2}\langle H^{\lambda k+}_{uu}\bar{u}, \bar{u}\rangle \ge C\langle \bar{u}, \bar{u}\rangle \quad \forall\, \bar{u} \in \mathbb{R}^{d(u)};$$

$$\frac{1}{2}\langle H^{\lambda k-}_{uu}\bar{u}, \bar{u}\rangle \ge C\langle \bar{u}, \bar{u}\rangle \quad \forall\, \bar{u} \in \mathbb{R}^{d(u)}.$$

Proof. Let $\lambda \in M^{\text{co}}(C\Gamma)$, $t \in [t_0, t_f]$. Choose $\varepsilon > 0$ so small that the conditions $\tilde{u} \in \mathbb{R}^{d(u)}$ and $|\tilde{u}| < \varepsilon$ imply $(t, x^0(t), u^0(t) + \tilde{u}) \in \mathcal{Q}$ and $\Gamma(t, u^0(t) + \tilde{u}) = |\tilde{u}|^2$, and hence

$$H(t, x^0(t), u^0(t) + \tilde{u}, \psi(t)) - H(t, x^0(t), u^0(t), \psi(t)) \ge C|\tilde{u}|^2.$$

In other words, the function $\varphi(\tilde{u}) := H(t,x^0(t),u^0(t)+\tilde{u},\psi(t)) - C|\tilde{u}|^2$ defined on a neighborhood of the origin of the space $\mathbb{R}^{d(u)}$ has a local minimum at zero. This implies $\varphi'(0) = 0$ and $\langle\varphi''(0)\bar{u},\bar{u}\rangle \geq 0$ for all $\bar{u} \in \mathbb{R}^{d(u)}$. The first condition is equivalent to $H_u(t,x^0(t),u^0(t),\psi(t)) = 0$, and the second is equivalent to

$$\langle H_{uu}(t,x^0(t),u^0(t),\psi(t))\bar{u},\bar{u}\rangle - 2C\langle\bar{u},\bar{u}\rangle \geq 0 \quad \forall\, \bar{u} \in \mathbb{R}^{d(u)}.$$

This implies assertion (a) of the proposition. Assertion (b) is obtained from assertion (a) by passing to the limit as $t \to t_k+0$ and $t \to t_k-0$, $k \in I^*$. The proposition is proved. ∎

We now use assertion (b) only. Denote by b the pair $(\delta w^0, \mathcal{M}^*)$ and by $\{b\}$ the sequence of pairs such that $\{\delta w^0\} \in \Pi^0$, $\mathcal{M}_k^* \to t_k$ for all k, $\mathcal{M}^* = \cup\mathcal{M}_k^*$. On each such sequence, we define the functions

$$\begin{aligned}
\Phi^{2\lambda}(b) &= \frac{1}{2}\Phi''(\lambda,\delta w^0) + \sum_k \left(D^k(H^\lambda)\int_{\mathcal{M}_k^*} |\delta t_k|\,dt + [H_x^\lambda]^k \int_{t_0}^{t_f} \delta x(\chi_{k-}^* - \chi_{k+}^*)\,dt \right),\\
\Phi_{C\Gamma}^2(b) &= \max_{M^{\mathrm{co}}(C\Gamma)} \Phi^{2\lambda}(b),\\
\gamma_1(b) &= \|\delta x\|_C^2 + \int_{t_0}^{t_f} |\delta u^0|^2\,dt + \sum_k \int_{\mathcal{M}_k^*} 2|\delta t_k|\,dt.
\end{aligned}$$

We set

$$\begin{aligned}
S_2 = &\Big\{\{b\} = \{(\delta w^0,\mathcal{M}^*)\} \mid F_{ip}\delta p \leq o(\sqrt{\gamma_1}),\ i \in I; \quad |K_p\delta p| = o(\sqrt{\gamma_1});\\
&\|\delta\dot{x} - f_w\delta w^0 - \textstyle\sum[f]^k(\chi_{k-}^* - \chi_{k+}^*)\|_1 = o(\sqrt{\gamma_1})\Big\}.
\end{aligned}$$

We obtain the definitions of Φ^2, γ_1, and S_2 from the corresponding definitions of $\tilde{\Phi}^1, \gamma$, and S_1 setting $\delta v = 0$ everywhere. We set

$$C_{\gamma_1}(\Phi_{C\Gamma}^2; S_2) = \inf_{S_2} \varliminf \frac{\Phi_{C\Gamma}^2}{\gamma_1}.$$

Lemma 2.54. *The following inequality holds:*

$$C_\gamma(\tilde{\Phi}_{C\Gamma}^1; S_1) \geq \min\Big\{C_{\gamma_1}(\Phi_{C\Gamma}^2; S_2), C\Big\}. \tag{2.125}$$

Proof. Let $\{a\} \in S_1$ be an arbitrary sequence such that $\gamma > 0$ for all its members. For this sequence, we set

$$\{b\} = \{(\delta w^0, \mathcal{M}^*)\}, \quad \hat{\gamma}(\delta v) = \int_{t_0}^{t_f} |\delta v|^2\,dt.$$

Then $\gamma(a) = \gamma_1(b) + \hat{\gamma}(\delta v)$, or, for short, $\gamma = \gamma_1 + \hat{\gamma}$. Let $\lambda \in M^{\mathrm{co}}(C\Gamma)$. Proposition 2.53(b), implies $\tilde{\Phi}^{1\lambda}(\delta w^0) \geq \Phi^{2\lambda}(\delta w^0) + C\hat{\gamma}(\delta v)$. Consequently, $\tilde{\Phi}_{C\Gamma}^1(\delta w^0) \geq \Phi_{C\Gamma}^2(\delta w^0) + C\hat{\gamma}(\delta v)$, or, briefly, $\tilde{\Phi}_{C\Gamma}^1 \geq \Phi_{C\Gamma}^2 + C\hat{\gamma}$. We consider the following two possible cases for the sequence $\{a\}$.

Case (a): Let $\underline{\lim}(\gamma_1/\gamma) = 0$. Extract a subsequence such that $\gamma_1 = o(\gamma)$ on it. Let this condition hold for the sequence $\{a\}$ itself. Then we obtain from the inequality $\tilde{\Phi}^1_{C\Gamma} \geq \Phi^2_{C\Gamma} + C\hat{\gamma}$ and the obvious estimate $|\Phi^2_{C\Gamma}| \leq O(\gamma_1)$ that

$$\underline{\lim}\frac{\tilde{\Phi}^1_{C\Gamma}}{\gamma} \geq \underline{\lim}\frac{\Phi^2_{C\Gamma} + C\hat{\gamma}}{\gamma} = \underline{\lim}\frac{o(\gamma) + C\hat{\gamma}}{\gamma} = C\,\underline{\lim}\frac{\hat{\gamma}}{\gamma} = C,$$

since $\gamma = \gamma_1 + \hat{\gamma} = o(\gamma) + \hat{\gamma}$.

Case (b): Assume now that $\underline{\lim}(\gamma_1/\gamma) > 0$, and hence $\gamma \leq \text{const}\,\gamma_1$ on the subsequence. The inequality $\tilde{\Phi}^1_{C\Gamma} \geq \Phi^2_{C\Gamma} + C\hat{\gamma}$ implies

$$\frac{\tilde{\Phi}^1_{C\Gamma}}{\gamma} \geq \frac{\Phi^2_{C\Gamma} + C\hat{\gamma}}{\gamma} = \frac{\gamma_1}{\gamma}\frac{\Phi^2_{C\Gamma}}{\gamma_1} + C\frac{\hat{\gamma}}{\gamma} \geq \min\left\{\frac{\Phi^2_{C\Gamma}}{\gamma_1}, C\right\},$$

since $\gamma_1 + \hat{\gamma} = \gamma$, $\gamma_1 \geq 0$, and $\hat{\gamma} \geq 0$. Consequently,

$$\underline{\lim}\frac{\tilde{\Phi}^1_{C\Gamma}}{\gamma} \geq \min\left\{\underline{\lim}\frac{\Phi^2_{C\Gamma}}{\gamma_1}, C\right\}.$$

But the conditions $\{a\} \in S$ and $\gamma \leq \text{const}\,\gamma_1$ immediately imply $\{b\} \in S_2$. We obtain from this that

$$\underline{\lim}\frac{\tilde{\Phi}^1_{C\Gamma}}{\gamma_1} \geq C_{\gamma_1}(\Phi^2_{C\Gamma}, S_2).$$

Consequently,

$$\underline{\lim}\frac{\tilde{\Phi}^1_{C\Gamma}}{\gamma} \geq \min\left\{C_{\gamma_1}(\Phi^2_{C\Gamma}; S_2), C\right\}.$$

Therefore, we have shown that from any sequence $\{a\} \in S_1$ at which $\gamma > 0$, it is possible to extract a subsequence such that the lower limit $\underline{\lim}\,\tilde{\Phi}^1_{C\Gamma}/\gamma$ on it is not less than the right-hand side of inequality (2.125). This obviously implies inequality (2.125). The lemma is proved. ∎

Lemmas 2.52 and 2.54 imply the following assertion.

Lemma 2.55. *Let the set $M^{\text{co}}(C\Gamma)$ be nonempty. Then $C_\gamma \geq \min\left\{C_{\gamma_1}(\Phi^2_{C\Gamma}; S_2), C\right\}$.*

In what follows, we will estimate the constant $C_{\gamma_1}(\Phi^2_{C\Gamma}; S_2)$ from below.

2.5.6 Replacement of $\|\delta x\|^2_C$ by $|\delta x(t_0)|^2$ in the Definition of the Functional γ_1

We set

$$\gamma_2 = \gamma_2(b) = |\delta x(t_0)|^2 + \int_{t_0}^{t_f} |\delta u^0|^2\,dt + \sum_k \int_{\mathcal{M}_k^*} 2|\delta t_k|\,dt.$$

Therefore, γ_2 differs from γ_1 by the fact that the term $\|\delta x\|_C^2$ is replaced by $|\delta x(t_0)|^2$. By definition, the sequences $\{b\} = \{(\delta w, \mathcal{M}^*)\}$ from S_2 satisfy the condition

$$\left\| \delta \dot{x} - f_w \delta w - \sum [f]^k (\chi_{k-}^* - \chi_{k+}^*) \right\|_1 = o(\sqrt{\gamma_1}).$$

Therefore, according to Proposition 2.24, there exists a constant $0 < q \le 1$, $q = 1/\varkappa$ (see (2.73)), such that for any sequence $\{b\} \in S_2$, there exists a number starting from which we have $q\gamma_1(b) \le \gamma_2(b)$, or, briefly, $q\gamma_1 \le \gamma_2$. Moreover, since $|\delta x(t_0)| \le \|\delta x\|_C$, we have $\gamma_2 \le \gamma_1$. Hence

$$q \le \frac{\gamma_2}{\gamma_1} \le 1.$$

This implies that the following relations hold for any sequence $\{b\} \in S_2$:

$$\underline{\lim} \frac{\Phi_{C\Gamma}^2}{\gamma_1} = \underline{\lim} \frac{\Phi_{C\Gamma}^2}{\gamma_2} \cdot \frac{\gamma_2}{\gamma_1} \ge \min \left\{ C_{\gamma_2}(\Phi_{C\Gamma}^2; S_2), q C_{\gamma_2}(\Phi_{C\Gamma}^2; S_2) \right\}.$$

Consequently,

$$C_{\gamma_1}(\Phi_{C\Gamma}^2; S_2) \ge \min \left\{ C_{\gamma_2}(\Phi_{C\Gamma}^2; S_2), q C_{\gamma_2}(\Phi_{C\Gamma}^2; S_2) \right\}. \tag{2.126}$$

Here,

$$C_{\gamma_2}(\Phi_{C\Gamma}^2; S_2) = \inf_{S_2} \underline{\lim} \frac{\Phi_{C\Gamma}^2}{\gamma_2}.$$

Inequality (2.126) and Lemma 2.55 imply the following assertion.

Lemma 2.56. *Let the set $M^{\rm co}(C\Gamma)$ be nonempty. Then*

$$C_\gamma \ge \min \left\{ C_{\gamma_2}(\Phi_{C\Gamma}^2; S_2), q C_{\gamma_2}(\Phi_{C\Gamma}^2; S_2), C \right\}, \quad 0 < q \le 1.$$

In what follows, we will estimate the constant $C_{\gamma_2}(\Phi_{C\Gamma}^2; S_2)$ from below.

2.5.7 Passing to Sequences with Discontinuous State Variables

As in Section 2.4, denote by $\{\bar{b}\} = \{(\bar{w}, \mathcal{M}^*)\}$ a sequence such that

$$\begin{aligned} &\mathcal{M}^* = \cup \mathcal{M}_k^*, \quad \mathcal{M}_k^* \to t_k \ (k \in I^*), \\ &\bar{w} = (\bar{x}, \bar{u}), \quad \bar{x} \in P_\Theta W^{1,1}(\Delta, \mathbb{R}^{d(x)}), \quad \bar{u} \in L^\infty(\Delta, \mathbb{R}^{d(u)}), \quad \|\bar{w}\|_\infty \to 0. \end{aligned}$$

Also, we set

$$\Phi^{3\lambda}(\bar{b}) = \frac{1}{2}\Phi''(\lambda, \bar{w}) + \sum_k \left(D^k(H^\lambda) \int_{\mathcal{M}_k^*} |\delta t_k| \, dt + [H_x^\lambda]^k \bar{x}_{\rm av}^k \xi_k \right),$$

where $\xi_k = \operatorname{meas} \mathcal{M}_{k-}^* - \operatorname{meas} \mathcal{M}_{k+}^*$, $\Phi_{C\Gamma}^3(\bar{b}) = \max_{M^{co}(C\Gamma)} \Phi^{3\lambda}(\bar{b})$, and

$$\gamma_2(\bar{b}) = |\bar{x}(t_0)|^2 + \int_{t_0}^{t_f} |\bar{u}|^2 \, dt + 2 \sum_k \int_{\mathcal{M}_k^*} |\delta t_k| \, dt.$$

Let S_3 be the set of sequences $\{\bar{b}\}$ such that

$$F_{ip}\bar{p} \le o(\sqrt{\gamma_2}) \quad (i \in I), \quad |K_p\bar{p}| = o(\sqrt{\gamma_2}),$$
$$\|\dot{\bar{x}} - f_w\bar{w}\|_1 = o(\sqrt{\gamma_2}), \quad [\bar{x}]^k = [f]^k\xi_k \quad \forall\, t_k \in \Theta,$$

where $\xi_k = \operatorname{meas} \mathcal{M}^*_{k-} - \operatorname{meas} \mathcal{M}^*_{k+}$. According to Lemma 2.26, for any sequence $\{b\} \in S_2$ for which $\gamma_2 > 0$, there exists a sequence $\{\bar{b}\}$ such that

$$\delta p = \bar{p}, \quad \|\dot{\bar{x}} - f_w\bar{w}\|_1 = o(\sqrt{\gamma_2}), \quad [\bar{x}]^k = [f]^k\xi_k \quad \forall\, t_k \in \Theta;$$
$$\gamma_2(b) = \gamma_2(\bar{b}), \quad \Phi^2_{C\Gamma}(b) = \Phi^3_{C\Gamma}(\bar{b}) + o(\gamma_2).$$

Consequently, $\{\bar{b}\} \in S_3$ and

$$\underline{\lim} \frac{\Phi^2_{C\Gamma}(b)}{\gamma_2(b)} = \underline{\lim} \frac{\Phi^3_{C\Gamma}(\bar{b})}{\gamma_2(\bar{b})} \ge \inf_{S_3} \underline{\lim} \frac{\Phi^3_{C\Gamma}}{\gamma_2} \stackrel{\text{def}}{=} C_{\gamma_2}(\Phi^3_{C\Gamma}; S_3).$$

Since $\{b\} \in S_2$ is an arbitrary sequence on which $\gamma_2 > 0$, we have

$$C_{\gamma_2}(\Phi^2_{C\Gamma}; S_2) \ge C_{\gamma_2}(\Phi^3_{C\Gamma}; S_3). \tag{2.127}$$

This inequality and Lemma 2.56 imply the following assertion.

Lemma 2.57. *Let the set $M^{\text{co}}(C\Gamma)$ be nonempty. Then*

$$C_\gamma \ge \min\left\{C_{\gamma_2}(\Phi^3_{C\Gamma}; S_3), qC_{\gamma_2}(\Phi^3_{C\Gamma}; S_3), C\right\}.$$

In what follows, we will estimate the constant $C_{\gamma_2}(\Phi^3_{C\Gamma}; S_3)$.

2.5.8 Additional Condition of Legendre, Weierstrass–Erdmann Type Related to Varying Discontinuity Points of the Control

Now our goal is to pass from the sequences $\{(\bar{w}, \mathcal{M}^*)\} \in S_3$ to the sequences $\{(\bar{\xi}, \bar{w})\}$, where $\bar{\xi} = (\bar{\xi}_1, \dots, \bar{\xi}_s)$, $\bar{\xi}_k = \operatorname{meas} \mathcal{M}^*_{k-} - \operatorname{meas} \mathcal{M}^*_{k+}$, $k = 1, \dots, s$.

Proposition 2.58. *The following inequality holds for any $\lambda \in M^{\text{co}}(C\Gamma)$:*

$$D^k(H^\lambda) \ge 2C, \quad k \in I^*.$$

Proof. Fix $\lambda \in M^{\text{co}}(C\Gamma)$, $k \in I^*$. Take a small $\varepsilon > 0$ and construct a variation $\delta u(t)$ in the left neighborhood $(t_k - \varepsilon, t_k)$ of the point t_k such that

$$\delta u(t) = \begin{cases} u^{0k+} - u^0(t), & t \in (t_k - \varepsilon, t_k), \\ 0, & t \notin (t_k - \varepsilon, t_k). \end{cases}$$

For a sufficiently small $\varepsilon > 0$, we have $(t, x^0(t), u^0(t) + \delta u(t)) \in \mathcal{Q}$. Consequently,

$$\delta H^\lambda \ge C\Gamma(t, u^0(t) + \delta u(t)),$$

where $\delta H^\lambda = H^\lambda(t,x^0,u^0+\delta u) - H^\lambda(t,x^0,u^0)$. It follows from the definition of $\Gamma(t,u)$ that the following relation holds for a sufficiently small $\varepsilon > 0$:

$$\Gamma(t,u^0(t)+\delta u(t)) = 2|\delta t_k|\chi^*_{k-},$$

where χ^*_{k-} is the characteristic function of the set $\mathcal{M}^*_{k-} = \{t \mid \delta u(t) \neq 0\} = (t_k - \varepsilon, t_k)$. Moreover, according Proposition 2.13, with the conditions $[H^\lambda]^k = 0$, $\delta v_{k-} = (u^0 + \delta u - u^{0k+})\chi^*_{k-} = 0$ taken into account, we have the following for δu: $\delta H^\lambda = (D^k(H^\lambda)|\delta t_k| + o(|\delta t_k|))\chi^*_{k-}$. Consequently, $D^k(H^\lambda)|\delta t_k| + o(|\delta t_k|) \geq 2C|\delta t_k|$ on $(t_k - \varepsilon, t_k)$. This implies $D^k(H^\lambda) \geq 2C$. The lemma is proved. ∎

Denote by S_4 the set of sequences $\{\bar{z}\} = \{(\bar{\xi},\bar{x},\bar{u})\}$ of elements of the space $Z(\Theta)$ (defined by (2.89)) such that $|\bar{\xi}| + ||\bar{w}||_\infty \to 0$ and, moreover,

$$F_{ip}\bar{p} \leq o(\sqrt{\bar{\gamma}}) \quad (i \in I), \quad |K_p\bar{p}| = o(\sqrt{\bar{\gamma}}),$$
$$\|\dot{\bar{x}} - f_w\bar{w}\|_1 = o(\sqrt{\bar{\gamma}}), \quad [\bar{x}]^k = [f]^k\bar{\xi}_k \qquad \forall\, t_k \in \Theta.$$

Recall that $\bar{\gamma} = \bar{\gamma}(\bar{z}) := |\bar{x}(t_0)|^2 + \int_{t_0}^{t_f} |\bar{u}|^2\,dt + |\bar{\xi}|^2$. Also, recall that (see formula (2.13))

$$\Omega^\lambda(\bar{z}) := \frac{1}{2}\sum_k (D^k(H^\lambda)\bar{\xi}_k^2 + 2[H_x^\lambda]^k \bar{x}^k_{\mathrm{av}}\bar{\xi}_k) + \frac{1}{2}\Phi''(\lambda,\bar{w}),$$

where $\Phi''(\lambda,\bar{w}) = \langle l^\lambda_{pp}\bar{p},\bar{p}\rangle + \int_{t_0}^{t_f}\langle H^\lambda_{ww}\bar{w},\bar{w}\rangle\,dt$. We set

$$\Omega_{C\Gamma}(\bar{z}) = \max_{M^{\mathrm{co}}(C\Gamma)} \Omega^\lambda(\bar{z})$$

and

$$C_{\bar{\gamma}}(\Omega_{C\Gamma}; S_4) = \inf_{S_4} \underline{\lim} \frac{\Omega_{C\Gamma}}{\bar{\gamma}}.$$

Using Proposition 2.58, we prove the following estimate.

Lemma 2.59. *The following inequality holds*:

$$C_{\gamma_2}(\Phi^3_{C\Gamma}; S_3) \geq \min\{C_{\bar{\gamma}}(\Omega_{C\Gamma}; S_4), C\}. \tag{2.128}$$

Proof. Let $\{\bar{b}\} = \{(\bar{w},\mathcal{M}^*)\} \in S_3$ be a sequence on which $\gamma_2 > 0$. The following inequalities hold for each member of this sequence:

$$2\int_{\mathcal{M}_k^*} |\delta t_k|\,dt \geq \bar{\xi}_k^2, \quad k \in I^*,$$

where $\bar{\xi}_k = \operatorname{meas}\mathcal{M}^*_{k-} - \operatorname{meas}\mathcal{M}^*_{k+}$. We set

$$\mu_k = 2\int_{\mathcal{M}_k^*} |\delta t_k|\,dt - \bar{\xi}_k^2, \quad k \in I^*, \quad \mu = \sum \mu_k.$$

Then $\mu_k \geq 0$ for all k, and, therefore, $\mu \geq 0$. To the sequence $\{\bar{b}\} = \{(\bar{w}, \mathcal{M}^*)\}$, we naturally associate the sequence $\{\bar{z}\} = \{\bar{\xi}, \bar{w})\}$ with the same components $\bar{w}$ and with $\bar{\xi}_k = \operatorname{meas} \mathcal{M}^*_{k-} - \operatorname{meas} \mathcal{M}^*_{k+}$, $k \in I^*$. Then $\gamma_2(\bar{b}) = \bar{\gamma}(\bar{z}) + \mu$, or, for short, $\gamma_2 = \bar{\gamma} + \mu$. Moreover,

$$\begin{aligned} \Phi^{3\lambda}(\bar{b}) &= \frac{1}{2}\Phi''(\lambda, \bar{w}) + \sum_k \left(\frac{1}{2} D^k(H^\lambda)(\bar{\xi}_k^2 + \mu_k) + [H_x^\lambda]^k \bar{x}_{\text{av}}^k \bar{\xi}_k \right) \\ &= \Omega^\lambda(\bar{z}) + \frac{1}{2}\sum_k D^k(H^\lambda)\mu_k. \end{aligned}$$

According to Proposition 2.58, $D^k(H^\lambda) \geq 2C$ for all k for any $\lambda \in M^{\text{co}}(C\Gamma)$. Consequently,

$$\Phi^3_{C\Gamma}(\bar{b}) \geq \Omega_{C\Gamma}(\bar{z}) + C\mu$$

or, briefly, $\Phi^3_{C\Gamma} \geq \Omega_{C\Gamma} + C\mu$. Therefore, we have the following for the sequence $\{\bar{b}\}$:

$$\frac{\Phi^3_{C\Gamma}}{\gamma_2} \geq \frac{\Omega_{C\Gamma} + C\mu}{\gamma_2}, \tag{2.129}$$

where $\gamma_2 = \bar{\gamma} + \mu$. Consider the following two cases.

Case (a). Let $\underline{\lim}\, \bar{\gamma}/\gamma_2 = 0$. Choose a subsequence such that $\bar{\gamma} = o(\gamma_2)$ on it. Let this condition hold on the sequence $\{\bar{b}\}$ itself. Since $\gamma_2 = \bar{\gamma} + \mu$, we have $\mu/\gamma_2 \to 1$. Let us show that the following estimate holds in this case:

$$|\Omega_{C\Gamma}| = o(\gamma_2). \tag{2.130}$$

Indeed, the definition of the functional Ω^λ and the boundedness of the set $M^{\text{co}}(C\Gamma)$ imply the existence of a constant $C_\Omega > 0$ such that

$$|\Omega_{C\Gamma}| \leq C_\Omega(\|\bar{x}\|^2_\infty + \|\bar{u}\|^2_2 + |\bar{\xi}|^2). \tag{2.131}$$

Further, since $\{\bar{b}\} \in S_3$, we have

$$\dot{\bar{x}} = f_x\bar{x} + f_u\bar{u} + \bar{r}; \qquad \|\bar{r}\|_1 = o(\sqrt{\gamma_2}); \quad [\bar{x}]^k = [f]^k\bar{\xi}_k, \quad k \in I^*.$$

Consequently, $\|\bar{x}\|_\infty \leq O(|\bar{x}(t_0)| + |\bar{\xi}| + \|\bar{u}\|_1 + \|\bar{r}\|_1)$. Since

$$|\bar{x}(t_0)| + |\bar{\xi}| + \|\bar{u}\|_1 \leq O(\sqrt{\bar{\gamma}}), \quad \|\bar{r}\|_1 = o(\sqrt{\gamma_2}),$$

we have $\|\bar{x}\|^2_\infty \leq O(\bar{\gamma}) + o(\gamma_2)$. This and estimate (2.131) imply $|\Omega_{C\Gamma}| \leq O(\bar{\gamma}) + o(\gamma_2)$. But $\bar{\gamma} = o(\gamma_2)$. Consequently, estimate (2.130) holds.

Taking into account estimate (2.130), we obtain from inequality (2.129) that

$$\underline{\lim}\, \frac{\Phi^3_{C\Gamma}}{\gamma_2} \geq \underline{\lim}\, \frac{\Omega_{C\Gamma} + C\mu}{\gamma_2} = \lim \frac{C\mu}{\gamma_2} = C.$$

Case (b). Let $\underline{\lim}\, \bar{\gamma}/\gamma_2 > 0$. Then $\gamma_2 = O(\bar{\gamma})$. This and the condition $\{\bar{b}\} \in S_3$ easily imply that $\{\bar{z}\} \in S_4$. We obtain from inequality (2.129) that

$$\frac{\Phi^3_{C\Gamma}}{\gamma_2} \geq \frac{\bar{\gamma}}{\gamma_2} \cdot \frac{\Omega_{C\Gamma}}{\bar{\gamma}} + \frac{\mu}{\gamma_2} C \geq \min\left\{ \frac{\Omega_{C\Gamma}}{\bar{\gamma}}, C \right\},$$

since $\bar{\gamma} \geq 0, \mu \geq 0$, and $\gamma_2 = \bar{\gamma} + \mu$. Consequently,

$$\underline{\lim} \frac{\Phi^3_{C\Gamma}}{\gamma_2} \geq \min\left\{\underline{\lim} \frac{\Omega_{C\Gamma}}{\bar{\gamma}}; C\right\} \geq \min\left\{C_{\bar{\gamma}}(\Omega_{C\Gamma}, S_4), C\right\}.$$

The latter inequality holds, since $\{\bar{z}\} \in S_4$.

Therefore, we have proved that it is possible to extract a subsequence from any sequence $\{\bar{b}\} \in S_3$ on which $\gamma_2 > 0$ such that the following inequality holds on it:

$$\underline{\lim} \frac{\Phi^3_{C\Gamma}}{\gamma_2} \geq \min\{C_{\bar{\gamma}}(\Omega_{C\Gamma}; S_4), C\}.$$

This implies inequality (2.128). The lemma is proved. ∎

Lemmas 2.57 and 2.59 imply the following assertion.

Lemma 2.60. *Let the set $M^{\text{co}}(C\Gamma)$ be nonempty. Then*

$$C_\gamma \geq \min\{C_{\bar{\gamma}}(\Omega_{C\Gamma}, S_4), qC_{\bar{\gamma}}(\Omega_{C\Gamma}, S_4), C, qC\}.$$

In what follows, we will estimate the constant $C_{\bar{\gamma}}(\Omega_{C\Gamma}; S_4)$ from below.

2.5.9 Passing to Equality in the Differential Relation

We restrict the set of sequences S_4 to the set of sequences in which each member satisfies the conditions $\dot{\bar{x}} = f_w \bar{w}$, $[\bar{x}]^k = [f]^k \bar{\xi}_k$, $k \in I^*$. We denote by S_5 this new set of sequences. Let us show that under such a restriction, the constant $C_{\bar{\gamma}}$ does not increase. In other words, let us prove the following lemma.

Lemma 2.61. *The following relation holds*: $C_{\bar{\gamma}}(\Omega_{C\Gamma}; S_4) = C_{\bar{\gamma}}(\Omega_{C\Gamma}; S_5)$.

Proof. Let $\{\bar{z}\} \in S_4$ be such that $\bar{\gamma} > 0$ on it. Then

$$\begin{aligned} &F_{ip}\bar{p} \leq o(\sqrt{\bar{\gamma}}) \quad (i \in I), \quad |K_p \bar{p}| = o(\sqrt{\bar{\gamma}}), \\ &\dot{\bar{x}} = f_x \bar{x} + f_u \bar{u} + \bar{r}, \quad \|\bar{r}\| = o(\sqrt{\bar{\gamma}}); \quad [\bar{x}]^k = [f]^k \bar{\xi}_k \quad (k \in I^*). \end{aligned} \tag{2.132}$$

Let a sequence $\{\delta x\}$ satisfy the conditions $\delta\dot{x} = f_x \delta x + \bar{r}$ and $\delta x(t_0) = 0$. Then

$$\|\delta x\|_C \leq \|\delta x\|_{1,1} \leq \text{const} \cdot \|\bar{r}\|_1 = o(\sqrt{\bar{\gamma}}).$$

We set $\delta u = 0$, $\delta w = (\delta x, 0)$, $\bar{w}' = \bar{w} - \delta w$, $\bar{\xi}' = \bar{\xi}$, $\bar{z}' = (\bar{\xi}', \bar{w}')$. For the sequence $\{\bar{z}'\}$, we have $\bar{\gamma}(\bar{z}') = \bar{\gamma}(\bar{z})$, or, for short, $\bar{\gamma}' = \bar{\gamma}$. Moreover,

$$\begin{aligned} &F_{ip}\bar{p}' \leq o(\sqrt{\bar{\gamma}'}) \quad (i \in I), \quad |K_p \bar{p}'| = o(\sqrt{\bar{\gamma}'}), \\ &\dot{\bar{x}}' = f_w \bar{w}', \quad [\bar{x}']^k = [f]^k \bar{\xi}'_k \quad (k \in I^*), \quad \|\bar{w}'\|_\infty + |\bar{\xi}'| \to 0. \end{aligned}$$

Hence $\{\bar{z}'\} \in S_5$. Moreover,

$$\begin{aligned} \Omega^\lambda(\bar{z}) \;&=\; \Omega^\lambda(\bar{z}') + \langle l_{pp}\bar{p}, \delta p\rangle + \frac{1}{2}\langle l_{pp}\delta p, \delta p\rangle \\ &+ \frac{1}{2}\int_{t_0}^{t_f} \left(\langle H^\lambda_{xx}\delta x, \delta x\rangle + 2\langle H^\lambda_{xx}\delta x, \bar{x}\rangle\right) dt + \sum_k [H_x]^k \delta x(t_k)\bar{\xi}_k. \end{aligned} \tag{2.133}$$

Conditions (2.132) easily imply $\|\bar{x}\|_\infty^2 \le O(\bar{\gamma})$. Moreover, $\|\delta x\|_C = o(\sqrt{\bar{\gamma}})$. This and conditions (2.133) imply $\Omega_{C\Gamma}(\bar{z}) = \Omega_{C\Gamma}(\bar{z}') + o(\bar{\gamma})$. Consequently,

$$\underline{\lim} \frac{\Omega_{C\Gamma}(\bar{z})}{\bar{\gamma}(\bar{z})} = \underline{\lim} \frac{\Omega_{C\Gamma}(\bar{z}')}{\bar{\gamma}(\bar{z}')} \ge C_{\bar{\gamma}}(\Omega_{C\Gamma}; S_5).$$

The inequality holds, since $\{\bar{z}'\} \in S_5$. Since $\{\bar{z}\}$ is an arbitrary sequence from S_4 on which $\bar{\gamma} > 0$, we obtain from this that $C_{\bar{\gamma}}(\Omega_{C\Gamma}; S_4) \ge C_{\bar{\gamma}}(\Omega_{C\Gamma}; S_5)$. The inclusion $S_5 \subset S_4$ implies the converse inequality. Therefore, we have an equality here. The lemma is proved. ∎

Lemmas 2.60 and 2.61 imply the following assertion.

Lemma 2.62. *Let the set $M^{\mathrm{co}}(C\Gamma)$ be nonempty. Then*

$$C_\gamma \ge \min\{C_{\bar{\gamma}}(\Omega_{C\Gamma}; S_5), qC_{\bar{\gamma}}(\Omega_{C\Gamma}; S_5), C, qC\}.$$

We now estimate the constant $C_{\bar{\gamma}}(\Omega_{C\Gamma}; S_5)$ from below.

2.5.10 Passing to the Critical Cone

Introduce the constant

$$C_{\bar{\gamma}}(\Omega_{C\Gamma}; S_5) = \inf\left\{ \frac{\Omega_{C\Gamma}(\bar{z})}{\bar{\gamma}(\bar{z})} \,\middle|\, \bar{z} \in \mathcal{K} \setminus \{0\} \right\}.$$

Lemma 2.63. *The following inequality holds*: $C_{\bar{\gamma}}(\Omega_{C\Gamma}; S_5) \ge C_{\bar{\gamma}}(\Omega_{C\Gamma}; \mathcal{K})$.

Proof. Let $\{\bar{z}\}$ be an arbitrary nonvanishing sequence from S_5. For this sequence, we have

$$F_{ip}\bar{p} \le o(\sqrt{\bar{\gamma}}) \quad (i \in I), \quad |K_p\bar{p}| = o(\sqrt{\bar{\gamma}}), \quad \|\bar{w}\|_\infty + |\bar{\xi}| \to 0, \tag{2.134}$$

$$\dot{\bar{x}} = f_w\bar{w}, \quad [\bar{x}]^k = [f]^k\bar{\xi}_k \quad (k \in I^*). \tag{2.135}$$

Moreover, $\bar{\gamma}(\bar{z}) > 0$ on the whole sequence, since it contains nonzero members. Let $\mathcal{T}$ be the subspace in $Z(\Theta)$ defined by conditions (2.135). According to Lemma 2.41 (which follows from the Hoffman lemma), for the sequence $\{\bar{z}\}$, there exists a sequence $\{\bar{\bar{z}}\} = \{(\bar{\bar{\xi}}, \bar{\bar{x}}, \bar{\bar{u}})\}$ in the subspace $\mathcal{T}$ such that

$$F_{ip}(\bar{p} + \bar{\bar{p}}) \le 0 \quad (i \in I), \quad K_p(\bar{p} + \bar{\bar{p}}) = 0,$$

and, moreover, $\|\bar{\bar{x}}\|_\infty + \|\bar{\bar{u}}\|_\infty + |\bar{\bar{\xi}}| = o(\sqrt{\bar{\gamma}})$. We set $\{\bar{z}'\} = \{\bar{z} + \bar{\bar{z}}\}$. As in the proof of Lemma 2.61, for the sequence $\{\bar{z}'\}$, we have

$$\Omega_{C\Gamma}(\bar{z}) = \Omega_{C\Gamma}(\bar{z}') + o(\bar{\gamma}). \tag{2.136}$$

Moreover,

$$\bar{\gamma}(\bar{z}) = \bar{\gamma}(\bar{z}') + o(\bar{\gamma}). \tag{2.137}$$

These conditions follow from the estimate $\|\bar{\bar{x}}\|_\infty + \|\bar{\bar{u}}\|_\infty + |\bar{\bar{\xi}}| = o(\sqrt{\bar{\gamma}})$ and the estimate $\|\bar{x}\|_\infty \le O(\sqrt{\bar{\gamma}})$, which, in turn, follows from (2.135). It follows from (2.137) and the condition $\bar{\gamma}(\bar{z}) > 0$ that $\bar{\gamma}(\bar{z}') > 0$ starting from a certain number. According to (2.136) and (2.137), we have

$$\underline{\lim} \frac{\Omega_{C\Gamma}(\bar{z})}{\bar{\gamma}(\bar{z})} = \underline{\lim} \frac{\Omega_{C\Gamma}(\bar{z}')}{\bar{\gamma}(\bar{z}')} \geq C_{\bar{\gamma}}(\Omega_{C\Gamma};\mathcal{K}). \tag{2.138}$$

The inequality holds, since $\bar{z}' \in \mathcal{K} \setminus \{0\}$. Since this inequality holds for an arbitrary nonvanishing sequence $\{\bar{z}\} \in S_5$, this implies $C_{\bar{\gamma}}(\Omega_{C\Gamma};S_5) \geq C_{\bar{\gamma}}(\Omega_{C\Gamma};\mathcal{K})$. ∎

Lemmas 2.62 and 2.63 imply the following assertion.

Lemma 2.64. *The following inequality holds for any real C such that the set $M^{\text{co}}(C\Gamma)$ is nonempty*:

$$C_\gamma \geq \min\{C_{\bar{\gamma}}(\Omega_{C\Gamma};\mathcal{K}), qC_{\bar{\gamma}}(\Omega_{C\Gamma};\mathcal{K}), C, qC\}. \tag{2.139}$$

Let C be such that $M^{\text{co}}(C\Gamma)$ is nonempty and

$$\Omega_{C\Gamma}(\bar{z}) \geq C\bar{\gamma}(\bar{z}) \quad \forall\, \bar{z} \in \mathcal{K}. \tag{2.140}$$

Then it follows from the definition of the constant $C_{\bar{\gamma}}(\Omega_{C\Gamma};\mathcal{K})$ that $C_{\bar{\gamma}}(\Omega_{C\Gamma};\mathcal{K}) \geq C$. This and (2.139) imply $C_\gamma \geq \min\{C, qC\}$. This inequality holds for all C such that the set $M^{\text{co}}(C\Gamma)$ is nonempty and condition (2.140) holds. Therefore, it also holds for the least upper bound of these C. At the end of Section 2.4, we have denoted this upper bound by $C_{\mathcal{K}}$. Therefore, we have proved the following theorem.

Theorem 2.65. *The following inequality holds*:

$$C_\gamma \geq \min\{C_{\mathcal{K}}, qC_{\mathcal{K}}\}, \quad 0 < q \leq 1. \tag{2.141}$$

Earlier, at the end of Section 2.4, we obtained the following estimate (see (7.43)):

$$C_{\mathcal{K}} \geq \min\{C_\gamma, \varkappa C_\gamma\}, \quad \varkappa = \frac{1}{q} \geq 1,$$

which can be written in the equivalent form:

$$\max\left\{C_{\mathcal{K}}, \frac{1}{\varkappa}C_{\mathcal{K}}\right\} \geq C_\gamma, \quad \varkappa \geq 1. \tag{2.142}$$

2.5.11 Decoding Result

Combining inequalities (2.141) and (2.142), we obtain the following decoding result.

Theorem 2.66. *The following inequalities hold*:

$$\min\{C_{\mathcal{K}}, qC_{\mathcal{K}}\} \leq C_\gamma \leq \max\{C_{\mathcal{K}}, qC_{\mathcal{K}}\}. \tag{2.143}$$

These inequalities are equivalent to

$$\min\{C_\gamma, \varkappa C_\gamma\} \leq C_{\mathcal{K}} \leq \max\{C_\gamma, \varkappa C_\gamma\} \tag{2.144}$$

Recall that $\varkappa \geq 1$, $0 < q \leq 1$, and $\varkappa = 1/q$. Bearing in mind inequalities (2.143) and (2.144), we write

$$C_\gamma \overset{\text{const}}{=} C_{\mathcal{K}}. \tag{2.145}$$

This is the main result of decoding.

We now obtain the following important consequences of this result.

Case (a). Let $C_{\mathcal{K}} \geq 0$. Then $C_\gamma \geq 0$ by (2.143); according to Remark 2.44, this implies the condition

$$M_0^{\text{co}} \neq \emptyset; \qquad \max_{M_0^{\text{co}}} \Omega^\lambda(\bar{z}) \geq 0 \quad \forall\, \bar{z} \in \mathcal{K}. \tag{2.146}$$

Conversely, if condition (2.146) holds, then $C_{\mathcal{K}} \geq 0$. Therefore, condition (2.146) is equivalent to the inequality $C_{\mathcal{K}} \geq 0$; by Theorem 2.66, the latter is equivalent to the inequality $C_\gamma \geq 0$.

Case (b). Let $C_{\mathcal{K}} > 0$. Let the constant C be such that $C_{\mathcal{K}} > C > 0$. Then, by the definition of $C_{\mathcal{K}}$, we have

$$M^{\text{co}}(C\Gamma) \neq \emptyset; \qquad \max_{M^{\text{co}}(C\Gamma)} \Omega^\lambda(\bar{z}) \geq C\bar{\gamma}(\bar{z}) \quad \forall\, \bar{z} \in \mathcal{K}. \tag{2.147}$$

Conversely, if for a certain $C > 0$, condition (2.147) holds, then $C_{\mathcal{K}} \geq C$, and hence $C_{\mathcal{K}} > 0$. Therefore, the existence of $C > 0$ such that (2.147) holds is equivalent to the inequality $C_{\mathcal{K}} > 0$. By Theorem 2.66, the latter is equivalent to the inequality $C_\gamma > 0$.

The following theorem summarizes what was said above.

Theorem 2.67. (a) *The inequality $C_\gamma \geq 0$ is equivalent to condition* (2.146). (b) *The inequality $C_\gamma > 0$ is equivalent to the existence of $C > 0$ such that condition* (2.147) *holds.*

2.5.12 Sufficient Conditions for the Pontryagin Minimum

We give the following definition.

Definition 2.68. Let $\Gamma(t,u)$ be an admissible function. We say that Condition $\mathfrak{B}^{\text{co}}(\Gamma)$ holds at the point w^0 if there exists $C > 0$ such that condition (2.147) holds.

As we already know, the condition $C_\gamma > 0$ is sufficient for the strict Pontryagin minimum at the point w^0, and, according to Theorem 2.67, it is equivalent to Condition $\mathfrak{B}^{\text{co}}(\Gamma)$. Therefore, Condition $\mathfrak{B}^{\text{co}}(\Gamma)$ is also sufficient for the strict Pontryagin minimum.

Let us consider Condition $\mathfrak{B}^{\text{co}}(\Gamma)$. First of all, we show that it is equivalent to Condition $\mathfrak{B}(\Gamma)$ in whose definition we have the set $M(C\Gamma)$ instead of the set $M^{\text{co}}(C\Gamma)$. By definition, the set $M(C\Gamma)$ consists of tuples $\lambda = (\alpha_0, \alpha, \beta, \psi) \in \Lambda_0$ such that $\psi(t)(f(t,x^0(t),u) - f(t,x^0(t),u^0(t))) \geq C\Gamma(t,u)$ if $t \in [t_0,t_f] \setminus \Theta$ and $(t,x^0(t),u) \in \mathcal{Q}$. Therefore, the set $M(C\Gamma)$ differs from the set $M^{\text{co}}(C\Gamma)$ by that $\operatorname{co}\Lambda_0$ is replaced by Λ_0 in the definition of the latter.

Condition $\mathfrak{B}(\Gamma)$ means that the set $M(C\Gamma)$ is nonempty for a certain $C > 0$ and

$$\max_{M(C\Gamma)} \Omega^\lambda(\bar{z}) \geq C\bar{\gamma}(\bar{z}) \quad \forall\, \bar{z} \in \mathcal{K}. \tag{2.148}$$

Since $\Lambda_0 \subset \operatorname{co}\Lambda_0$, and hence $M(C\Gamma) \subset M^{\text{co}}(C\Gamma)$, Condition $\mathfrak{B}(\Gamma)$ implies Condition $\mathfrak{B}^{\text{co}}(\Gamma)$. It is required to prove the converse statement: Condition $\mathfrak{B}^{\text{co}}(\Gamma)$ implies Condition $\mathfrak{B}(\Gamma)$. For this purpose, we prove the following lemma.

Lemma 2.69. *For any $C > 0$ such that $M^{\text{co}}(C\Gamma)$ is nonempty, there exists $0 < \varepsilon < 1$ such that*

$$M^{\text{co}}(C\Gamma) \subset [\varepsilon, 1] \circ M\left(\frac{C}{\varepsilon}\Gamma\right),$$

where $[\varepsilon, 1] \circ M$ is the set of tuples $\lambda = \rho\tilde{\lambda}$ such that $\rho \in [\varepsilon, 1]$ and $\tilde{\lambda} \in M$.

Proof. For an arbitrary $\lambda = (\alpha_0, \alpha, \beta, \psi) \in \operatorname{co}\Lambda_0$, we set $\nu(\lambda) = \alpha_0 + |\alpha| + |\beta|$. Since the function $\nu(\lambda)$ is convex and equals 1 on Λ_0, we have $\nu(\lambda) \le 1$ for all $\lambda \in \operatorname{co}\Lambda_0$. Further, let $C > 0$ be such that $M^{\text{co}}(C\Gamma)$ is nonempty. Then as is easily seen, the compact set $M^{\text{co}}(C\Gamma)$ does not contain zero. This implies

$$\varepsilon = \min_{\lambda \in M^{co}(C\Gamma)} \nu(\lambda) > 0,$$

since the conditions $\lambda \in \operatorname{co}\Lambda_0$ and $\nu(\lambda) = 0$ imply $\lambda = 0$.

Therefore, the inequalities $\varepsilon \le \nu(\lambda) \le 1$ hold for an arbitrary $\lambda \in M^{\text{co}}(C\Gamma)$. We set $\tilde{\lambda} = \lambda/\nu(\lambda)$. Then $\nu(\tilde{\lambda}) = 1$, and hence $\tilde{\lambda} \in \Lambda_0$. Moreover, the condition $\psi\delta_u f \ge C\Gamma$ implies

$$\tilde{\psi}\delta_u f \ge \frac{C}{\nu(\lambda)}\Gamma \ge \frac{C}{\varepsilon}\Gamma,$$

where $\tilde{\psi}$ is a component of $\tilde{\lambda}$, $\delta_u f = f(t, x^0, u) - f(t, x^0, u^0)$, and $\Gamma = \Gamma(t, u)$. Hence, $\tilde{\lambda} \in M(\frac{C}{\varepsilon}\Gamma)$. Therefore, for an arbitrary $\lambda \in M^{\text{co}}(C\Gamma)$, we have found the representation $\lambda = \nu(\lambda)\tilde{\lambda}$, where $\tilde{\lambda} \in M(\frac{C}{\varepsilon}\Gamma)$, $\varepsilon \le \nu(\lambda) \le 1$. This implies

$$M^{\text{co}}(C\Gamma) \subset [\varepsilon, 1] \circ M\left(\frac{C}{\varepsilon}\Gamma\right).$$

The lemma is proved. ■

Lemma 2.69 implies the following theorem.

Theorem 2.70. *Condition $\mathfrak{B}^{\text{co}}(\Gamma)$ is equivalent to Condition $\mathfrak{B}(\Gamma)$.*

Proof. Let Condition $\mathfrak{B}^{\text{co}}(\Gamma)$ hold, i.e., there exists $C > 0$ such that $M^{\text{co}}(C\Gamma)$ is nonempty and

$$\max_{M^{\text{co}}(C\Gamma)} \Omega^\lambda(\bar{z}) \ge C\bar{\gamma}(\bar{z}) \quad \forall\, \bar{z} \in \mathcal{K}.$$

Then, by Lemma 2.69, there exist $\varepsilon > 0$ such that

$$\max_{[\varepsilon,1]\circ M\left(\frac{C}{\varepsilon}\Gamma\right)} \Omega^\lambda(\bar{z}) \ge C\bar{\gamma}(\bar{z}) \quad \forall\, \bar{z} \in \mathcal{K}.$$

By the linearity of Ω^λ in λ and the positivity of C, we obtain from this that

$$\max_{M\left(\frac{C}{\varepsilon}\Gamma\right)} \Omega^\lambda(\bar{z}) \ge C\bar{\gamma}(\bar{z}) \quad \forall\, \bar{z} \in \mathcal{K}.$$

Therefore, Condition $\mathfrak{B}(\Gamma)$ holds. We have shown that Condition $\mathfrak{B}^{\text{co}}(\Gamma)$ implies Condition $\mathfrak{B}(\Gamma)$. As mentioned above, the converse is also true. Therefore, Conditions $\mathfrak{B}^{\text{co}}(\Gamma)$ and $\mathfrak{B}(\Gamma)$ are equivalent. The theorem is proved. ■

Theorems 2.67(b) and 2.70 imply the following theorem.

Theorem 2.71. *The inequality $C_\gamma > 0$ is equivalent to Condition $\mathfrak{B}(\Gamma)$.*

There is a certain inconvenience in Condition $\mathfrak{B}(\Gamma)$, in that it is difficult to verify that $\lambda \in M(C\Gamma)$. Therefore, it is desirable to pass to the sufficient condition, which can be more easily verified. Recall that in Section 2.1.8 we introduced the set M_0^+ consisting of those $\lambda \in M_0$ for which the strict minimum principle holds outside the discontinuity points of the control. In the same section, we have introduced the set $\mathrm{Leg}_+(M_0^+)$ consisting of all strictly Legendrian elements $\lambda \in M_0^+$. The definition of $M(C\Gamma) \subset M^{\mathrm{co}}(C\Gamma)$, its compactness, and Propositions 2.53 and 2.58 imply the following assertion.

Lemma 2.72. *For any admissible function $\Gamma(t,u)$ and any $C > 0$, the set $M(C\Gamma)$ is a compact set contained in the set $\mathrm{Leg}_+(M_0^+)$.*

Also, the following assertion holds.

Lemma 2.73. *For any nonempty compact set $M \subset \mathrm{Leg}_+(M_0^+)$, there exist an admissible function $\Gamma(t,u)$ and a constant $C > 0$ such that $M \subset M(C\Gamma)$.*

Before proving Lemma 2.73, we prove a slightly simpler property. Let $\mathcal{U}$ be an arbitrary neighborhood of the compact set $\overline{u^0}$ containing in Q_{tu}. With the subscript $\mathcal{U}$, we denote all objects referring to the canonical problem complemented by the constraint $(t,u) \in \mathcal{U}$. For example, we write $M_0^{\mathcal{U}}, M^{\mathcal{U}}(C\Gamma)$, etc. Denote by $\mathrm{Leg}_+(\Lambda_0)$ the subset of all strictly Legendre elements $\lambda \in \Lambda_0$.

Lemma 2.74. *Let $M \subset \mathrm{Leg}_+(\Lambda_0)$ be a nonempty compact set, and let $\Gamma(t,u)$ be an admissible function. Then there exist a neighborhood $\mathcal{U}$ of the compact set $\overline{u^0}$ and a constant $C > 0$ such that $M \subset M^{\mathcal{U}}(C\Gamma)$.*

To prove Lemma 2.74, we need several auxiliary assertions.

Proposition 2.75. *Assume that there is a nonempty compact set $M \subset \Lambda_0$ such that the following conditions hold for each of its elements λ:*
(a) *for any $t \in [t_0,t_f] \setminus \Theta$,*

$$\frac{1}{2}\langle H_{uu}(t,x^0(t),u^0(t),\psi(t))\bar{u},\bar{u}\rangle > 0 \quad \forall\, \bar{u} \in \mathbb{R}^{d(u)} \setminus \{0\};$$

(b) *for any $t_k \in \Theta$,*

$$\frac{1}{2}\langle H_{uu}^{\lambda k+}\bar{u},\bar{u}\rangle > 0 \quad \forall\, \bar{u} \in \mathbb{R}^{d(u)} \setminus \{0\} \tag{2.149}$$

and

$$\frac{1}{2}\langle H_{uu}^{\lambda k-}\bar{u},\bar{u}\rangle > 0 \quad \forall\, \bar{u} \in \mathbb{R}^{d(u)} \setminus \{0\}. \tag{2.150}$$

Then there exist $C > 0$ and $\varepsilon > 0$ such that for any $\lambda \in M$, the conditions

$$t \in [t_0, t_f] \setminus \Theta \quad \text{and} \quad |u - u^0(t)| < \varepsilon \tag{2.151}$$

imply

$$H(t, x^0(t), u, \psi(t)) - H(t, x^0(t), u^0(t), \psi(t)) \geq C|u - u^0(t)|^2. \tag{2.152}$$

Proof. Assume the contrary. Let the compact set $M \subset \Lambda_0$ be such that conditions (a) and (b) of the proposition hold for each of its element, but there are no $C > 0$ and $\varepsilon > 0$ such that conditions (2.151) imply inequality (2.152). Then there exist sequences $\{C_n\}, \{t_n\}$, $\{\lambda_n\}$, and $\{\bar{u}_n\}$ such that

$$\begin{aligned} &C_n \to +0, \quad t_n \in [t_0, t_f] \setminus \Theta, \quad \lambda_n \in M, \quad \bar{u}_n \in \mathbb{R}^{d(u)}, \, |\bar{u}_n| \to 0, \\ &H^{\lambda_n}(t_n, x_n^0, u_n^0 + \bar{u}_n) - H^{\lambda_n}(t_n, x_n^0, u_n^0) < C_n |\bar{u}_n|^2, \end{aligned} \tag{2.153}$$

where $x_n^0 = x^0(t_n)$ and $u_n^0 = u^0(t_n)$. Without loss of generality, we assume that

$$t_n \to \hat{t} \in [t_0, t_f], \quad \lambda_n \to \hat{\lambda} \in M, \quad \frac{\bar{u}_n}{|u_n|} \to \bar{u}.$$

Then $\bar{u}_n = \varepsilon_n(\bar{u} + \tilde{u}_n)$, where $\varepsilon_n = |\bar{u}_n| \to 0$, $|\bar{u}_n| \to 0$. In this case, we obtain from (2.153) that

$$\frac{1}{2}\langle H_{uu}^{\lambda_n}(t_n, x_n^0, u_n^0)\bar{u}_n, \bar{u}_n\rangle + o(\varepsilon_n^2) < C_n \varepsilon_n^2. \tag{2.154}$$

Here, we have taken into account that $H_u^{\lambda_n}(t_n, x_n^0, u_n^0) = 0$.

We first assume that $\hat{t} \notin \Theta$. Dividing (2.154) by ε_n^2 and passing to the limit, we obtain

$$\frac{1}{2}\langle H_{uu}^{\hat{\lambda}}(\hat{t}, x^0(\hat{t}), u^0(\hat{t}))\bar{u}, \bar{u}\rangle \leq 0.$$

But this contradicts condition (a), since $\bar{u} \neq 0$. Analogously, in the case where $\hat{t} \in \Theta$, we arrive at a contradiction to one of the conditions in (b). The proposition is proved. ∎

In what follows, we need to use the assumption that each $t_k \in \Theta$ is an L-point of the control u^0.

Proposition 2.76. *Let $M \subset \Lambda_0$ be a nonempty compact set such that the following conditions hold for a fixed point $t_k \in \Theta$ and any $\lambda \in M$:*

$$[H^\lambda]^k = 0, \quad D^k(H^\lambda) > 0, \tag{2.155}$$

$$\frac{1}{2}\langle H_{uu}^{\lambda k-}\bar{u}, \bar{u}\rangle > 0 \quad \forall\, \bar{u} \in \mathbb{R}^{d(u)} \setminus \{0\}. \tag{2.156}$$

Then there exist $C > 0$ and $\varepsilon > 0$ such that for any $\lambda \in M$, the conditions

$$t_k < t < t_k + \varepsilon \quad \text{and} \quad |u - u^{0k-}| < \varepsilon \tag{2.157}$$

imply the inequality

$$H(t, x^0(t), u, \psi(t)) - H(t, x^0(t), u^0(t), \psi(t)) \geq C\left(|t - t_k| + \frac{1}{2}|u - u^{0k-}|^2\right). \tag{2.158}$$

Proof. Let a compact set $M \subset \Lambda_0$ satisfy the condition of the proposition, and let there be no $C > 0$ and $\varepsilon > 0$ such that for any $\lambda \in M$, conditions (2.157) imply inequality (2.158). Then there exist a sequence $\{C\}$ and a sequence of triples $\{(t,u,\lambda)\}$ such that

$$\begin{gathered} C \to +0, \quad t \to t_k + 0, \quad u \to u^{0k+}, \\ H^\lambda(t,x^0(t),u) - H^\lambda(t,x^0(t),u^0(t)) < C\left(|t-t_k| + \frac{1}{2}|u-u^{0k-}|^2\right) \end{gathered} \tag{2.159}$$

(we omit the serial numbers of members). We set $t - t_k = \delta t > 0$, $u - u^{0k-} = \delta v$, $u^0(t) - u^{0k+} = \delta u^0$, $x^0(t_k) = x^{0k}$, $x^0 - x^{0k} = \delta x^0$, etc. Then we get

$$\begin{aligned} H^\lambda(t,x^0,u) &= H(t,x^0,u,\psi) = H(t_k+\delta t, x^{0k}+\delta x^0, u^{0k-}+\delta v, \psi^k+\delta\psi) \\ &= H^{\lambda k-} + \big(H_t^{\lambda k-} + H_x^{\lambda k+}\dot{x}^{0k+} + \dot{\psi}^{k+}H_\psi^{\lambda k-}\big)\delta t + H_u^{\lambda k-}\delta v \\ &\quad + \frac{1}{2}\langle H_{uu}^{\lambda k-}\delta v, \delta v\rangle + o(|\delta t| + |\delta v|^2), \end{aligned}$$

and taking into account that $|\delta u^0| \le L|\delta t|\,(L > 0)$ by assumption, we obtain

$$\begin{aligned} H^\lambda(t,x^0,u^0) &= H(t,x^0,u^0,\psi) = H(t_k+\delta t, x^{0k}+\delta x^0, u^{0k+}+\delta u^0, \psi^k+\delta\psi) \\ &= H^{\lambda k+} + \big(H_t^{\lambda k+} + H_x^{\lambda k+}\dot{x}^{0k+} + \dot{\psi}^{k+}H_\psi^{\lambda k+}\big)\delta t + H_u^{\lambda k+}\delta u^0 + o(\delta t). \end{aligned}$$

Subtracting the latter relation from the previous one, taking into account that $H_u^{\lambda k+} = H_u^{\lambda k-} = 0$ and $[H^\lambda]^k = 0$, and also taking into account inequality (2.159), we obtain

$$\begin{aligned} &-\big([H_t^\lambda]^k + [H_x^\lambda]^k\dot{x}^{0k+} + \dot{\psi}^{k+}[H_\psi^\lambda]^k\big)\delta t + \frac{1}{2}\langle H_{uu}^{\lambda k-}\delta v, \delta v\rangle + o(|\delta t| + |\delta v|^2) \\ &< C\left(|\delta t| + \frac{1}{2}|\delta v|^2\right). \end{aligned}$$

But $[H_t^\lambda]^k + [H_x^\lambda]^k\dot{x}^{0k+} + \dot{\psi}^{k+}[H_\psi^\lambda]^k = -D^k(H^\lambda)$ and $C \to +0$. Hence

$$D^k(H^\lambda)\delta t + \frac{1}{2}\langle H_{uu}^{\lambda k-}\delta v, \delta v\rangle < o_1(|\delta t| + |\delta v|^2). \tag{2.160}$$

Without loss of generality, we assume that $\lambda \to \hat{\lambda} \in M$. We consider the following two possible cases for the sequences $\{\delta t\}$ and $\{\delta v\}$.

Case (a). Assume that there exists a subsequence such that the following relation holds on it: $|\delta v|^2 = o(|\delta t|)$. Then it follows from (2.160) that $D^k(H^\lambda)\delta t < o(\delta t)$. Since $\delta t \to +0$ and $\lambda \to \hat{\lambda}$, we obtain from this that $D^k(H^{\hat{\lambda}}) \le 0$, which contradicts the condition $\hat{\lambda} \in M$.

Case (b). Assume now that

$$\underline{\lim}\, \frac{|\delta v|^2}{|\delta t|} > 0,$$

i.e., $|\delta t| \le O(|\delta v|^2)$. In this case, from (2.160) and the conditions $\delta t > 0$ and $D^k(H^\lambda) > 0$, we obtain

$$\frac{1}{2}\langle H_{uu}^{\lambda k-}\delta v, \delta v\rangle < o(|\delta v|^2). \tag{2.161}$$

Without loss of generality, we assume that

$$\frac{\delta v}{|\delta v|} \to \bar{v}, \quad |\bar{v}| = 1.$$

Dividing inequality (2.161) by $|\delta v|^2$ and passing to the limit, we obtain

$$\frac{1}{2}\langle H_{uu}^{\hat{\lambda}k-}\bar{v},\bar{v}\rangle \leq 0.$$

Since $|\bar{v}| = 1$, this also contradicts the condition $\hat{\lambda} \in M$. Therefore, our assumption that there exist the sequences $\{C\}$ and $\{(t,u,\lambda)\}$ with the above property is not true. We thus have proved the proposition. ∎

The following assertion is proved analogously.

Proposition 2.77. *Let $M \subset \Lambda_0$ be a nonempty compact set such that the following conditions hold for a fixed point $t_k \in \Theta$ and any $\lambda \in M$:*

$$[H^\lambda]^k = 0, \quad D^k(H^\lambda) > 0,$$

and

$$\frac{1}{2}\langle H_{uu}^{\lambda k+}\bar{u},\bar{u}\rangle > 0 \quad \forall\, \bar{u} \in \mathbb{R}^{d(u)} \setminus \{0\}.$$

Then there exist $C > 0$ and $\varepsilon > 0$ such that for any $\lambda \in M$, the conditions $t_k - \varepsilon < t < t_k$, and $|u - u^{0k+}| < \varepsilon$ imply

$$H(t,x^0(t),u,\psi(t)) - H(t,x^0(t),u^0(t),\psi(t)) \geq C\left(|t-t_k| + \frac{1}{2}|u-u^{0k+}|^2\right).$$

Propositions 2.75, 2.76, and 2.77 directly imply Lemma 2.74.

Proof of Lemma 2.73. Assume that there exists a nonempty compact set $M \subset \text{Leg}_+(M_0^+)$. Let $\Gamma_1(t,u)$ be a certain admissible function (as was shown in Section 2.3.7, there exists at least one such function). According to Lemma 2.73, there exist a neighborhood $\mathcal{U} \subset \mathcal{Q}_{tu}$ of the compact set $\overline{u^0}$ and a constant $C > 0$ such that $M \subset M^{\mathcal{U}}(C\Gamma_1)$, i.e., $M \subset \Lambda_0$ and the conditions $t \in [t_0,t_f] \setminus \Theta$, $(t,u) \in \mathcal{U}$, and $(t,x^0(t),u) \in \mathcal{Q}$ imply

$$H^\lambda(t,x^0(t),u) - H^\lambda(t,x^0(t),u^0(t)) \geq C\Gamma_1(t,u).$$

We set

$$\begin{aligned} h(t,u) &= \frac{1}{C}\min_{\lambda \in M}\{H^\lambda(t,x^0(t),u) - H^\lambda(t,x^0(t),u^0(t))\}, \\ \Gamma(t,u) &= \min\{h(t,u),\Gamma_1(t,u)\}. \end{aligned} \tag{2.162}$$

It is easy to see that the function $\Gamma(t,u)$ (defined by (2.162)) is admissible, and, moreover, $M \subset M(C\Gamma)$. The lemma is proved. ∎

We now recall the definition given in Section 2.1.8. We say that Condition $\mathfrak{B}$ holds for the point w^0 if there exist a nonempty compact set $M \subset \text{Leg}_+(M_0^+)$ and a constant $C > 0$

such that

$$\max_M \Omega^\lambda(\bar{z}) \geq C\bar{\gamma}(\bar{z}) \quad \forall\, \bar{z} \in \mathcal{K}. \tag{2.163}$$

The following assertion holds.

Theorem 2.78. *Condition $\mathfrak{B}$ is equivalent to the existence of an admissible function $\Gamma(t,u)$ such that Condition $\mathfrak{B}(\Gamma)$ holds.*

Proof. Let Condition $\mathfrak{B}$ hold; i.e., there exist a nonempty compact set $M \subset \mathrm{Leg}_+(M_0^+)$ and a constant $C > 0$ such that condition (2.163) holds. Then, according to Lemma 2.73, there exist an admissible function $\Gamma(t,u)$ and a constant $C_1 > 0$ such that $M \subset M(C_1\Gamma)$. We set $C_2 = \min\{C, C_1\}$. Then $M(C_1\Gamma) \subset M(C_2\Gamma)$. Consequently, $M \subset M(C_2\Gamma)$ and

$$\max_{M(C_2\Gamma)} \Omega^\lambda(\bar{z}) \geq \max_M \Omega^\lambda(\bar{z}) \geq C\bar{\gamma}(\bar{z}) \geq C_2\bar{\gamma}(\bar{z}) \quad \forall\, \bar{z} \in \mathcal{K}$$

(the second inequality holds by (2.163)). Therefore, Condition $\mathfrak{B}(\Gamma)$ holds.

Conversely, let there exist an admissible function Γ such that Condition $\mathfrak{B}(\Gamma)$ holds. Then there exists $C > 0$ such that $M(C\Gamma)$ is nonempty and condition (2.148) holds. By Lemma 2.72, $M(C\Gamma) \subset \mathrm{Leg}_+(M_0^+)$, and $M(C\Gamma)$ is a compact set. Therefore, Condition $\mathfrak{B}$ also holds. The theorem is proved. ∎

According to Theorem 2.71, the inequality $C_\gamma > 0$ is equivalent to Condition $\mathfrak{B}(\Gamma)$. This is true for every admissible function Γ. This and Theorem 2.78 imply the following theorem.

Theorem 2.79. *Condition $\mathfrak{B}$ is equivalent to the existence of an admissible function Γ such that the condition $C_\gamma > 0$ holds for the order γ corresponding to it.*

In Section 2.2, we have verified all the assumptions of the abstract scheme for the canonical problem, the point w^0, the sets $\mathcal{P}$ and $\mathcal{Q}$ (the absorbing set $\Omega = \mathcal{W}$ corresponds to them in the space W), and the set Π of Pontryagin sequences in the space W. In Section 2.3, the corresponding assumptions of the abstract scheme were also verified for a higher order γ on Π. Therefore, Theorem 1.7 is applicable. According to this theorem, the condition $C_\gamma > 0$ is not only sufficient for the Pontryagin minimum at the point w^0 but is equivalent to the γ-sufficiency on Π. The latter will be also called the Pontryagin γ-sufficiency. For convenience, we define this concept here. Let Γ be an admissible function, and let γ be the higher order corresponding to it.

Definition 2.80. We say that the point w^0 yields the *Pontryagin γ-sufficiency* if there exists $\varepsilon > 0$ such that for any sequence $\{\delta w\} \in \Pi$, there exists a number, starting from which the condition $\sigma(\delta w) \geq \varepsilon\gamma(\delta w)$ holds.

The equivalent condition for the Pontryagin γ-sufficiency consists of the following: There is no sequence $\{\delta w\} \in \Pi$ such that $\sigma = o(\gamma)$ on it.

The violation function σ was already defined in Section 2.2.4. In what follows, it is convenient to use the following expression for σ:

$$\sigma = (\delta J)^+ + \sum F_i^+(p^0 + \delta p) + |\delta K| + \int_{t_0}^{t_f} |\delta\dot{x} - \delta f|\, dt, \tag{2.164}$$

where $\delta J = J(p^0+\delta p) - J(p^0), \delta K = K(p^0+\delta p) - K(p^0), \delta f = f(t,w^0+\delta w) - f(t,w^0)$, and $a^+ = \max\{a,0\}$. This expression differs from the expression from Section 2.2 by only a constant multiplier (more precisely, they estimate each other from above and from below with constant multipliers), and, therefore, it also can be used in all the formulations. Therefore, Theorem 1.7(b) implies the following theorem.

Theorem 2.81. *The condition $C_\gamma > 0$ is equivalent to the Pontryagin γ-sufficiency at the point w^0.*

Theorems 2.79 and 2.81 imply the following theorem.

Theorem 2.82. *Condition $\mathfrak{B}$ is equivalent to the existence of an admissible function Γ such that the Pontryagin γ-sufficiency holds at the point w^0 for the order γ corresponding to it.*

Since the Pontryagin γ-sufficiency implies the strict Pontryagin minimum, Condition $\mathfrak{B}$ is also sufficient for the latter. Therefore, Theorem 2.5 is proved.

However, Theorem 2.82 is a considerably stronger result than Theorem 2.5. It allows us to proceed more efficiently in analyzing sufficient conditions. We will see this in what follows.

2.5.13 An Important Estimate

We devote this section to a certain estimate, which will be needed in Section 2.7 for obtaining the sufficient conditions for the strong minimum.

Let an admissible function Γ and a constant C be such that the set $M(C\Gamma)$ is nonempty. Let M be a nonempty compact set in $M(C\Gamma)$. According to (2.111), we have

$$C_\gamma \geq C_\gamma(\Phi_M; \Pi_{o(\sqrt{\gamma})}), \tag{2.165}$$

where

$$C_\gamma(\Phi_M; \Pi_{o(\sqrt{\gamma})}) = \inf_{\Pi_{o(\sqrt{\gamma})}} \underline{\lim} \frac{\Phi_M}{\gamma}$$

and $\Phi_M(\delta w) = \max_{\lambda \in M} \Phi(\lambda, \delta w)$. We can further estimate $C_\gamma(\Phi_M; \Pi_{o(\sqrt{\gamma})})$ from below exactly in the same way as was done for the constant $C_\gamma(\Phi_{C\Gamma}; \Pi_{o(\sqrt{\gamma})})$ when $M = M^{\mathrm{co}}(C\Gamma)$. All the arguments are repeated literally (see relations (2.118), (2.121), (2.124)–(2.128) and Lemmas 2.61, 2.63). As a result, we arrive at the following estimate:

$$C_\gamma(\Phi_M; \Pi_{o(\sqrt{\gamma})}) \geq \min\{C_{\bar{\gamma}}(\Omega_M; \mathcal{K}), qC_{\bar{\gamma}}(\Omega_M; \mathcal{K}), C, qC\}, \tag{2.166}$$

where $0 < q \leq 1$, $\Omega_M(\bar{z}) = \max_{\lambda \in M} \Omega^\lambda(\bar{z})$, and

$$C_{\bar{\gamma}}(\Omega_M; \mathcal{K}) = \inf \left\{ \frac{\Omega_M(\bar{z})}{\bar{\gamma}(\bar{z})} \;\middle|\; \bar{z} \in \mathcal{K} \setminus \{0\} \right\}.$$

Now let $M \subset \mathrm{Leg}_+(M_0^+)$ be a nonempty compact set, and let there exist a constant $C > 0$ such that

$$\max_M \Omega^\lambda(\bar{z}) \geq C\bar{\gamma}(\bar{z}) \quad \forall\, \bar{z} \in \mathcal{K} \tag{2.167}$$

(i.e., Condition $\mathfrak{B}$ holds). Then by Lemma 2.73, there exist an admissible function Γ and a constant C_1 such that $M \subset M(C_1\Gamma)$. Condition (2.167) implies $C_{\bar{\gamma}}(\Omega_M;\mathcal{K}) \geq C$. Then (2.166) implies $C_\gamma(\Phi_M;\Pi_{o(\sqrt{\gamma})}) \geq q\min\{C,C_1\}$. We set $C_M = \frac{1}{2}q\min\{C,C_1\}$. Then $C_\gamma(\Phi_M;\Pi_{o(\sqrt{\gamma})}) > C_M$. Therefore,

$$\Phi_M \geq C_M \cdot \gamma \mid \Pi_{o(\sqrt{\gamma})}, \tag{2.168}$$

i.e., for any sequence $\{\delta w\} \in \Pi_{o(\sqrt{\gamma})}$, there exists a number starting from which we have $\Phi_M \geq C_M\gamma$. We have obtained the following result.

Lemma 2.83. *Let $M \subset \mathrm{Leg}_+(M_0^+)$ be a nonempty compact set, and let there exist $C > 0$ such that condition* (2.167) *holds. Then there exists a constant $C_M > 0$ such that condition* (2.168) *holds.*

2.6 Completing the Proof of Theorem 2.4

2.6.1 Replacement of M_0^{co} by M_0 in the Necessary Conditions

The purpose of this section is to complete the proof of Theorem 2.4, which we began in Section 2.4. Here we will not use the results of Section 2.5. Instead, we shall need some constructions from [79, Part 1, Chapter 2, Section 7]. Let us note that the proofs in this section are rather technical and could be omitted in a first reading of the book.

We now turn to the following quadratic necessary Condition $\mathfrak{A}^{\mathrm{co}}$ for the Pontryagin minimum obtained in Section 2.4 (see Theorem 2.43):

$$\max_{M_0^{\mathrm{co}}} \Omega^\lambda(\bar{z}) \geq 0 \quad \forall\, \bar{z} \in \mathcal{K}.$$

As was already noted, it is slightly weaker than the necessary condition of Theorem 2.4, since in Condition $\mathfrak{A}$, we have the set M_0, which is more narrow than the set M_0^{co}. However, we will show that the obtained necessary condition remains valid under the replacement of M_0^{co} by M_0, i.e., the necessary Condition $\mathfrak{A}$ holds. We thus will complete the proof of Theorem 2.4.

The passage to the auxiliary problem in [79, Part 1, Chapter 2, Section 7] and the trajectory of this problem corresponding to the index ζ chosen in a special way allows us to do this. For this trajectory, we write the necessary Condition $\mathfrak{A}^{\mathrm{co}}$ in the auxiliary problem with the subsequent transform of this condition into the initial problem. Such a method was already used in [79, Section 7, Part 1] in proving the maximum principle. We use the notation, the concepts, and the results of [79, Section 7, Part 1], briefly mentioning the main constructions. We stress that in contrast to [79, Section 7, Part 1], all the constructions here refer to the problem on a fixed closed interval of time $[t_0,t_f]$. We write the condition that the endpoints of the closed interval of time $[t_0,t_f]$ are fixed as follows: $t_0 = t_0^0$, $t_f = t_f^0$. Therefore, let us consider the problem (2.1)–(2.4) in the form which corresponds to the general problem considered in [79, Section 7, Part 1],

$$\begin{gathered} J(x_0,x_f) \to \min, \quad F(x_0,x_f) \leq 0, \quad K(x_0,x_f) = 0, \\ (x_0,x_f) \in \mathcal{P}, \quad t_0 - t_0^0 = 0, \quad t_f - t_f^0 = 0; \end{gathered} \tag{2.169}$$

$$\frac{dx}{dt} = f(t,x,u), \quad (t,x,u) \in \mathcal{Q}, \tag{2.170}$$

where $x_0 = x(t_0)$, $x_f = x(t_f)$, and let

$$\hat{w}^0 = (\hat{x}^0(t), \hat{u}^0(t) \,|\, t \in [t_0, t_f]) \tag{2.171}$$

be a Pontryagin minimum point in this problem. (Here the components x^0 and u^0 of the pair w^0 are denoted by $\hat{x}^0$ and $\hat{u}^0$, respectively, as in [79, Section 7, Part 1].) Then the minimum principle holds, and hence the set M_0 is nonempty.

2.6.2 Two Cases

We have the following two possibilities:

(a) There exists $\lambda \in M_0$ such that $-\lambda \in M_0$.

(b) There is no $\lambda \in M_0$ such that $-\lambda \in M_0$, i.e., $M_0 \cap (-M_0) = \emptyset$.

In case (a), the necessary Condition $\mathfrak{A}$ holds trivially, since for any $\bar{z}$, at least one of the quadratic forms $\Omega^{\lambda}(\bar{z})$ and $\Omega^{(-\lambda)}(\bar{z})$ is nonnegative. Therefore, we consider case (b).

As in [79, Section 7, Part 1], for a given number N, we denote by $\zeta = (t^i, u^{ik})$ a vector in $\mathbb{R}^N \times \mathbb{R}^{N^2 d(u)}$ with components $t^i \in (t_0^0, t_f^0)$, $i = 1, \dots, N$, and $u^{ik} \in \mathbb{R}^{d(u)}$, $i,k = 1,\dots,N$, such that

$$t^i < t^{i+1}, \quad i = 1,\dots,N-1; \quad (t^i, \hat{x}^0(t^i), u^{ik}) \in \mathcal{Q}, \quad i,k = 1,\dots,N.$$

Here and below in this section, the fixed time interval is denoted by $[t_0^0, t_f^0]$, while all t^i, $i = 1,\dots,N$ are internal points of this interval. Denote by $\mathcal{D}(\Theta)$ the set of all ζ satisfying the condition $t^i \notin \Theta$, $i = 1,\dots,N$.

Further, recall the definition of the set Λ_ζ in [79, Section 7, Part 1]. For the problem (2.169), (2.170), it consists of tuples $\mu = (\alpha_0, \alpha, \beta)$ such that

$$\alpha_0 \geq 0, \quad \alpha \geq 0, \quad \alpha F(\hat{x}_0^0, \hat{x}_f^0) = 0, \quad \alpha_0 + \sum \alpha_i + |\beta| = 1, \tag{2.172}$$

and, moreover, there exist absolutely continuous functions $\hat{\psi}_x(t)$ and $\hat{\psi}_t(t)$ such that

$$\hat{\psi}_x(t_0^0) = -l'_{x_0}, \quad \hat{\psi}_x(t_f^0) = l'_{x_f}, \tag{2.173}$$

$$\begin{aligned} -\frac{d\hat{\psi}_x}{dt} &= \hat{\psi}_x(t) f'_x(t, \hat{x}^0(t), \hat{u}^0(t)), \\ -\frac{d\hat{\psi}_t}{dt} &= \hat{\psi}_x(t) f'_t(t, \hat{x}^0(t), \hat{u}^0(t)), \end{aligned} \tag{2.174}$$

$$\int_{t_0^0}^{t_f^0} \left(\hat{\psi}_x(t) f(t, \hat{x}^0(t), \hat{u}^0(t)) + \hat{\psi}_t(t) \right) dt = 0, \tag{2.175}$$

$$\hat{\psi}_x(t^i) f(t^i, \hat{x}^0(t^i), u^{ik}) + \hat{\psi}_t(t^i) \geq 0, \quad i,k = 1,\dots,N. \tag{2.176}$$

Here,

$$l = l(x_0, x_f, \alpha_0, \alpha, \beta) = \alpha_0 J(x_0, x_f) + \alpha F(x_0, x_f) + \beta K(x_0, x_f) \tag{2.177}$$

and the gradients l'_{x_0} and l'_{x_f} in the transversality conditions (2.173) are taken at the point $(\hat{x}_0^0, \hat{x}_1^0, \alpha_0, \alpha, \beta)$ (note that the components ψ of the tuple λ are denoted by ψ_x here). Let

$$\Lambda = \bigcap_{\zeta} \Lambda_\zeta,$$

where the intersection is taken over all subscripts ζ. At the end of Section 7 in [79, Part 1], we have shown that elements $\mu \in \Lambda$ satisfy the following minimum principle:

$$\hat{\psi}_x(t)f(t,\hat{x}^0(t),u)+\hat{\psi}_t(t)\geq 0 \quad \text{if } t\in[t_0^0,t_f^0], \quad u\in\mathbb{R}^{d(u)}, \quad (t,x^0(t),u)\in\mathcal{Q}; \tag{2.178}$$

$$\hat{\psi}_x(t)f(t,\hat{x}^0(t),\hat{u}^0(t))+\hat{\psi}_t(t)=0 \quad \text{a.e. on } [t_0^0,t_f^0]. \tag{2.179}$$

By continuity, the latter condition extends to all points of the set $[t_0^0,t_f^0]\setminus\Theta$. This implies $\Lambda\subset N_0$, where N_0 is the projection of the set M_0 under the injective mapping $\lambda=(\alpha_0,\alpha,\beta,\psi_x)\mapsto\mu=(\alpha_0,\alpha,\beta)$. We now consider the set

$$\Lambda(\Theta)=\bigcap_{\zeta\in\mathcal{D}(\Theta)}\Lambda_\zeta.$$

Clearly, elements $\mu\in\Lambda(\Theta)$ satisfy condition (2.178) at all points $t\in[t_0^0,t_f^0]\setminus\Theta$; however, by continuity, this condition extends to all points of the interval $[t_0^0,t_f^0]$. Consequently, $\Lambda(\Theta)=\Lambda$, and, therefore,

$$\bigcap_{\zeta\in\mathcal{D}(\Theta)}\Lambda_\zeta\subset N_0$$

(in fact, we have an equality here). In case (b) the following assertion holds.

Proposition 2.84. *There exists a subscript ζ such that $\Lambda_\zeta\cap(-\Lambda_\zeta)=\emptyset$, and, moreover, the following condition holds for all instants of time t^i, $i=1,\dots,N$, entering the definition of the subscript ζ: $t^i\notin\Theta$, $i=1,\dots,N$.*

Proof. Assume that the proposition does not hold. Then each of the sets

$$\mathcal{F}_\zeta:=\Lambda_\zeta\cap(-\Lambda_\zeta), \quad \zeta\in\mathcal{D}(\Theta),$$

is not empty. These sets compose a centered system of nonempty compact sets, and hence their intersection

$$\mathcal{F}:=\bigcap_{\zeta\in\mathcal{D}(\Theta)}\mathcal{F}_\zeta$$

is nonempty. Moreover, $\mathcal{F}\subset\Lambda(\Theta)\subset N_0$, and the condition $\mathcal{F}_\zeta=-\mathcal{F}_\zeta$ for all $\zeta\in\mathcal{D}(\Theta)$ implies $\mathcal{F}=-\mathcal{F}$. Let $\mu=(\alpha_0,\alpha,\beta)\in\mathcal{F}$. Then $\mu\in N_0$ and $(-\mu)\in N_0$; therefore, we have $\lambda\in M_0$ and $(-\lambda)\in M_0$ for the corresponding element λ. But this contradicts case (b) considered. The proposition is proved. ∎

2.6.3 Problem Z_N and Trajectory κ_N

Fix the subscript ζ from Proposition 2.84. For a given N, consider Problem Z_N on a fixed closed interval of time $[\tau_0,\tau_f]$, where $\tau_0=0$, $\tau_f=t_f-t_0+N^2$. Problem Z_N has the form

$$\begin{aligned}
&J(x(\tau_0),x(\tau_f))\to\inf;\\
&F(x(\tau_0),x(\tau_f))\leq 0, \quad K(x(\tau_0),x(\tau_f))=0, \quad (x(\tau_0),x(\tau_f))\in\mathcal{P},\\
&t(\tau_0)-t_0^0=0, \quad t(\tau_f)-t_f^0=0, \quad -z(\tau_0)\leq 0,
\end{aligned} \tag{2.180}$$

$$\frac{dx}{d\tau} = (\varphi(\eta)z)f(t,x,u); \quad \frac{dt}{d\tau} = \varphi(\eta)z; \quad \frac{dz}{d\tau} = 0; \qquad (2.181)$$
$$(t,x,u) \in \mathcal{Q}, \quad \eta \in \mathcal{Q}_1.$$

Here, z and η are of dimension N^2+1 and have the components

$$z_\theta,\ z_{ik} \quad i,k = 1,\dots,N, \quad \text{and} \quad \eta_\theta,\ \eta_{ik}, \quad i,k = 1,\dots,N,$$

respectively.[2] The open set $\mathcal{Q}_1$ is the union of disjoint neighborhoods $\mathcal{Q}^\theta$ and $\mathcal{Q}^{ik}$ of the points e^θ and e^{ik}, $i,k = 1,\dots,N$, respectively, which are the standard basis of $\mathbb{R}^{N^2+1}$ (e^θ has the unit component e_θ, and other components of e^θ are zero, while e^{ik} has the unit component e_{ik}, and other components of e^{ik} are zero), and $\phi(\eta) : \mathcal{Q}_1 \to \mathbb{R}^{d(\eta)}$ is a function mapping each of the mentioned neighborhoods into the element of the basis whose neighborhood it is. Note that the functions $u(\tau)$ and $\eta(\tau)$ are controls, while the functions $z(\tau)$, $x(\tau)$, and $t(\tau)$ are state variables in Problem Z_N.

Recall the definition of the point $\kappa^\zeta = (z^0(\tau), x^0(\tau), t^0(\tau), u^0(\tau), \eta^0(\tau))$ (in Problem Z_N on the closed interval $[\tau_0,\tau_f]$) corresponding to the subscript ζ and the trajectory $(\hat{x}^0(t), \hat{u}^0(t))$, $t \in [t_0^0, t_f^0]$.

We "insert" N closed intervals of unit length adjusting to each other into each point t^i of the closed interval $[t_0^0, t_f^0]$; thus, we enlarge the length of the closed interval by N^2. Place the left endpoint of the new closed interval at zero. We obtain the closed interval $[\tau_0,\tau_f]$ ($\tau_0 = 0$) with N^2 closed intervals (denoted by Δ_{ik}, $i,k = 1,\dots,N$) placed on it; moreover, $\Delta_{i1},\dots,\Delta_{iN}$ are closed intervals adjusted to each other, located in the same order, and corresponding to the point t^i, $i = 1,\dots,N$. We set

$$\mathcal{E} = (\tau_0,\tau_f) \setminus \bigcup_{i,j=1}^{n} \Delta_{ij}.$$

Let $\chi_\mathcal{E}$ and $\chi_{\Delta_{ij}}$ be the characteristic functions of the sets $\mathcal{E}$ and Δ_{ij}, respectively. We set

$$z_\theta^0 = 1, \quad z_{ij}^0 = 0, \quad i,\,j = 1,\dots,N,$$

i.e., $z^0 = e^\theta$. Further, we set

$$\eta^0(\tau) = e^\theta \chi_\mathcal{E}(\tau) + \sum_{i,j} e^{ij} \chi_{\Delta_{ij}}(\tau).$$

Since $\varphi(\eta^0) = \eta^0$, we have $\varphi(\eta^0)z^0 = \eta^0 z^0 = \eta^0 e^\theta = \chi_\mathcal{E}$. Define $t^0(\tau)$ by the conditions

$$\frac{dt^0}{d\tau} = \varphi(\eta^0)z^0, \quad t^0(\tau_0) = t_0^0.$$

Then $t^0(\tau_f) = t_f^0$, since meas $\mathcal{E} = t_f^0 - t_0^0$. We set $x^0(\tau) = \hat{x}^0(t^0(\tau))$, $u^0(\tau) = \hat{u}^0(t^0(\tau))$. As was shown in [79, Part 1, Section 7, Proposition 7.1], the point κ^ζ defined in such a way

[2]We preserve notation accepted in [79, Section 7, Part 1], where z_θ and η_θ were used to denote "zero-components" of vectors z and η, respectively.

that it is a Pontryagin minimum point in Problem Z_N on the fixed closed interval $[\tau_0, \tau_f]$. In what follows, all the functions and sets related to N or ζ are endowed with the indices N or ζ, respectively.

Since κ^ζ yields the Pontryagin minimum in Problem Z_N, the necessary Condition $\mathfrak{A}^{\mathrm{co}\,\zeta}$ holds for it in this problem. We show that the necessary Condition $\mathfrak{A}^\zeta$ also holds for the chosen index ζ. For this purpose, in Problem Z_N, we consider the sets Λ_ζ, Λ_0^ζ, $\mathrm{co}\,\Lambda_0^\zeta$, M_0^ζ, and $M_0^{\mathrm{co}\,\zeta}$ for the trajectory κ^ζ and find the relations between them. The definition of the set Λ_ζ was given in [79, Section 7, Part 1], and the other sets were defined in [79, Part 2].

2.6.4 Condition $\mathfrak{A}^\zeta$

The function l^N has the form

$$\begin{aligned} l^N &= \alpha_0 J + \alpha F + \beta K - \alpha_z z_0 + \beta_{t_0}(t_0 - t_0^0) + \beta_{t_f}(t_f - t_f^0) \\ &= l - \alpha_z z_0 + \beta_{t_0}(t_0 - t_0^0) + \beta_{t_f}(t_f - t_f^0). \end{aligned}$$

The Pontryagin function H^N has the form

$$H^N = \psi_x(\varphi(\eta)z) f(t,x,u) + \psi_t(\varphi(\eta)z) + \psi_z \cdot 0 = (\varphi(\eta)z)(H + \psi_t) + \psi_z \cdot 0,$$

where $H = \psi_x f(t,x,u)$. The set Λ_0^ζ consists of tuples

$$\lambda^N = (\alpha_0, \alpha, \beta, \alpha_z, \beta_{t_0}, \beta_{t_f}, \psi_x(\tau), \psi_t(\tau), \psi_z(\tau)) \tag{2.182}$$

such that

$$\alpha_0 \geq 0, \quad \alpha \geq 0, \quad \alpha_z \geq 0, \quad \alpha F(x_0^0, x_f^0) = 0, \quad \alpha_z z^0(\tau_0) = 0, \tag{2.183}$$

$$\alpha_0 + \sum \alpha_i + \sum \alpha_{zi} + |\beta| + |\beta_{t_0}| + |\beta_{t_f}| = 1, \tag{2.184}$$

$$\psi_x(\tau_0) = -l'_{x_0}, \quad \psi_x(\tau_f) = l'_{x_f}, \tag{2.185}$$

$$\psi_t(\tau_0) = -\beta_{t_0}, \quad \psi_t(\tau_f) = \beta_{t_f}, \tag{2.186}$$

$$\psi_z(\tau_0) = \alpha_z, \quad \psi_z(\tau_f) = 0, \tag{2.187}$$

$$-\frac{d\psi_x}{d\tau} = (\varphi(\eta^0(\tau))z^0)\psi_x(\tau) f'_x(t^0(\tau), x^0(\tau), u^0(\tau)), \tag{2.188}$$

$$-\frac{d\psi_t}{d\tau} = (\varphi(\eta^0(\tau))z^0)\psi_x(\tau) f'_t(t^0(\tau), x^0(\tau), u^0(\tau)), \tag{2.189}$$

$$-\frac{d\psi_z}{d\tau} = \varphi(\eta^0(\tau))\big(\psi_x(\tau) f(t^0(\tau), x^0(\tau), u^0(\tau)) + \psi_t(\tau)\big), \tag{2.190}$$

$$(\varphi(\eta^0(\tau))z^0)\psi_x(\tau) f_u(t^0(\tau), x^0(\tau), u^0(\tau)) = 0. \tag{2.191}$$

The gradients l'_{x_0} and l'_{x_f} are taken at the point $(x_0^0, x_f^0, \alpha_0, \alpha, \beta)$.

In [79, Section 7, Part 1], we have shown that there is the following equivalent normalization for the set Λ_0^ζ:

$$\alpha_0 + \sum \alpha_i + |\beta| = 1 \tag{2.192}$$

(the conditions $\alpha_0 = 0$, $\alpha = 0$, and $\beta = 0$, and also conditions (2.183), (2.185)–(2.191) imply $\alpha_z = 0$, $\beta_{t_0} = 0$, and $\beta_{t_f} = 0$). Therefore, in the definition of Λ_0^ζ, we can replace normalization (2.184) by the equivalent normalization (2.192). In this case, the quadratic Condition $\mathfrak{A}^{\mathrm{co}\,\zeta}$ remains valid. Assume that we have made this replacement. The new set is denoted by Λ_0^ζ as before. In [79, Section 7, Part 1], it was also shown that the element $(\alpha_0, \alpha, \beta) \in \Lambda_\zeta$ corresponds to an element $\lambda^N \in \Lambda_0^\zeta$ and has the same components α_0, α, and β, i.e., the projection $\lambda^N \mapsto (\alpha_0, \alpha, \beta)$ maps Λ_0^ζ into Λ_ζ.

Proposition 2.85. *The convex hull* $\mathrm{co}\,\Lambda_0^\zeta$ *does not contain zero.*

Proof. Assume that this is not true. Then there exist an element $\lambda^N \in \Lambda_0^\zeta$ and a number $\rho > 0$ such that $-\rho\lambda^N \in \Lambda_0^\zeta$. This implies that all nonnegative components α_0, α, and α_z of the element λ^N (see (2.182)) vanish. But then the condition $\lambda^N \in \Lambda_0^\zeta$ implies $-\lambda^N \in \Lambda_0^\zeta$, i.e., we may set $\rho = 1$.

Let $\mu^N = (\alpha_0, \alpha, \beta) = (0, 0, \beta)$ be the projection of the element λ^N. Then μ^N and $-\mu^N$ belong to Λ_ζ. But the existence of such a μ^N contradicts the choice of the index ζ. Therefore, the assumption that $0 \in \mathrm{co}\,\Lambda_\zeta$ is wrong. The proposition is proved. ∎

Proposition 2.85 implies the following assertion.

Corollary 2.86. *For any* $\lambda^N \in \mathrm{co}\,\Lambda_0^\zeta$, *there exists* $\rho > 0$ *such that* $\rho\lambda^N \in \Lambda_0^\zeta$.

Proof. Let $\lambda^N \in \mathrm{co}\,\Lambda_0^\zeta$. Then by Proposition 2.85, $\lambda^N \neq 0$. Obviously, $\mathrm{co}\,\Lambda_0^\zeta$ is contained in the cone $\mathrm{con}\,\Lambda_0^\zeta$ spanned by Λ_0^ζ. The conditions $\lambda^N \in \mathrm{con}\,\Lambda_0^\zeta$, $\lambda^N \neq 0$ imply $\nu(\lambda^N) \stackrel{\mathrm{def}}{=} \alpha_0 + |\alpha| + |\beta| > 0$ (since $\nu = 1$ is a normalization condition). We set

$$\tilde{\lambda}^N = \frac{\lambda^N}{\nu(\lambda^N)}.$$

Then $\tilde{\lambda}^N \in \mathrm{con}\,\Lambda_0^\zeta$ and $\nu(\tilde{\lambda}^N) = 1$. Therefore, $\tilde{\lambda}^N \in \Lambda_0^\zeta$. It remains to set $\rho = 1/\nu(\lambda^N)$. The proposition is proved. ∎

Corollary 2.86 and the definitions of the sets $M_0^{\mathrm{co}\,\zeta}$ and M_0^ζ imply the following assertion.

Corollary 2.87. *Let* $\lambda^N \in M_0^{\mathrm{co}\,\zeta}$. *Then there exists* $\rho > 0$ *such that* $\rho\lambda^N \in M_0^\zeta$.

The condition $\mathfrak{A}^{\mathrm{co}\,\zeta}$ for the point κ^ζ in Problem Z_N has the form

$$\max_{M_0^{\mathrm{co}\,\zeta}} \Omega^\zeta(\lambda^N; \bar{z}^N) \geq 0 \quad \forall\, \bar{z}^N \in \mathcal{K}^\zeta.$$

Here, $\mathcal{K}^\zeta$ is the critical cone and Ω^ζ is the quadratic form of Problem Z_N at the point κ^ζ. Let us show that this implies Condition $\mathfrak{A}^\zeta$:

$$\max_{M_0^\zeta} \Omega^\zeta(\lambda^N; \bar{z}^N) \geq 0 \quad \forall\, \bar{z}^N \in \mathcal{K}^\zeta.$$

Indeed, let $\bar{z}^N \in \mathcal{K}^\zeta$. Condition $\mathfrak{A}^{\text{co}\,\zeta}$ implies the existence of $\lambda^N \in M_0^{\text{co}\,\zeta}$ such that

$$\Omega^\zeta(\lambda^N;\bar{z}^N) \geq 0. \tag{2.193}$$

According to Corollary 2.87, there exists $\rho > 0$ such that $\tilde{\lambda}^N = \rho\lambda^N \in M_0^\zeta$. Multiplying (2.193) by $\rho > 0$, we obtain $\Omega^\zeta(\tilde{\lambda}^N;\bar{z}^N) \geq 0$. Hence, $\max_{M_0^\zeta} \Omega^\zeta(\cdot,\bar{z}^N) \geq 0$. Since $\bar{z}^N$ is an arbitrary element in $\mathcal{K}^\zeta$, this implies Condition $\mathfrak{A}^\zeta$. Thus, we have proved the following lemma.

Lemma 2.88. *Let $\hat{w}^0$ be a Pontryagin minimum point in the problem* (2.169), (2.170), *and let $M_0 \cap (-M_0) = \emptyset$. Then there exists a superscript $\zeta \in \mathcal{D}(\Theta)$ such that Condition $\mathfrak{A}^\zeta$ holds.*

In what follows, we fix a superscript $\zeta \in \mathcal{D}(\Theta)$ such that Condition $\mathfrak{A}^\zeta$ holds. Now our goal is to reveal which information about the trajectory $\hat{w}^0$ can be extracted from Condition $\mathfrak{A}^\zeta$ of superscript ζ. We show that Condition $\mathfrak{A}^\zeta$ implies Condition $\mathfrak{A}$ at the point $\hat{w}^0$ in the initial problem. For this purpose, consider in more detail the definitions of the set M_0^ζ, cone $\mathcal{K}^\zeta$, and quadratic form Ω^ζ at the point κ^ζ in Problem Z_N.

2.6.5 Relation Between the Sets M_0^ζ and M_0

Consider the conditions defining the set M_0^ζ. By definition, M_0^ζ is the set of $\lambda^N \in \Lambda_0^\zeta$ such that the following inequality holds for all τ in the closed interval $[\tau_0,\tau_f]$, except for a finite set of discontinuity points of the controls $u^0(\tau)$ and $\eta^0(\tau)$:

$$\begin{aligned}&(\varphi(\eta)z^0)\big(\psi_x(\tau)f(t^0(\tau),x^0(\tau),u)+\psi_t(\tau)\big)\\ &\quad\geq(\varphi(\eta^0)z^0)\big(\psi_x(\tau)f(t^0(\tau),x^0(\tau),u^0(\tau))+\psi_t(\tau)\big)\end{aligned} \tag{2.194}$$

for all $u \in \mathbb{R}^{d(u)}$ such that $(t^0(\tau),x^0(\tau),u) \in \mathcal{Q}$ and all $\eta \in \mathcal{Q}_1$. Let us analyze condition (2.194). Choose a function $\eta = \eta(\tau) \in \mathcal{Q}_1$ so that the following condition holds:

$$\varphi(\eta(\tau))z^0 = 0. \tag{2.195}$$

Such a choice is possible, since the condition $z^0 = e^\theta$ and the definition of the function φ imply

$$\varphi(\eta)z^0 = \varphi_\theta(\eta) = \begin{cases}1, & \eta \in \mathcal{Q}^\theta,\\ 0, & \eta \notin \mathcal{Q}^\theta,\end{cases}$$

and, therefore, we may set $\eta(\tau) = \eta^*$, where η^* is an arbitrary point in $\mathcal{Q}_1 \setminus \mathcal{Q}^\theta$, for example, $\eta^* = e^{11}$. Therefore, condition (2.195) holds for $\eta(\tau) \equiv e^{11}$. It follows from (2.194) and (2.195) that the right-hand side of inequality (2.194) is nonpositive. But the integral of it over the interval $[\tau_0,\tau_f]$ vanishes (this was shown in [79, Section 7, Part 1]; moreover, this follows from the adjoint equation (2.190), conditions (2.183), and the transversality conditions (2.187) considered for the component ψ_{z_θ} only). But if the integral of a nonpositive function over a closed interval vanishes, then this function equals zero almost everywhere on this closed interval. Hence

$$(\varphi(\eta^0(\tau))z^0)\big(\psi_x(\tau)f(t^0(\tau),x^0(\tau),u^0(\tau))+\psi_t(\tau)\big)=0 \tag{2.196}$$

a.e. on $[\tau_0,\tau_f]$. Further, setting $\eta=\eta^0(\tau)$ in (2.194) and taking into account (2.196), we obtain

$$(\varphi(\eta^0(\tau))z^0)\big(\psi_x(\tau)f(t^0(\tau),x^0(\tau),u)+\psi_t(\tau)\big)\geq 0 \quad \text{if } (t^0(\tau),x^0(\tau),u)\in\mathcal{Q}. \tag{2.197}$$

This condition also holds for almost all $\tau\in[\tau_0,\tau_f]$.

We may rewrite conditions (2.196) and (2.197) for the independent variable t. Recall that in [79, Section 7, Part 1], we have denoted by $\hat{\mathcal{E}}$ the image of the set $\mathcal{E}$ under the mapping $t^0(\tau)$. Also, we note that $t^0(\tau)$ defines a one-to-one and bi-absolutely continuous correspondence between $\mathcal{E}$ and $\hat{\mathcal{E}}$, and, moreover, $[t_0^0,t_f^0]\setminus\hat{\mathcal{E}}$ is a finite set of points $\{t^i\}_{i=1}^N$, and hence $\hat{\mathcal{E}}$ is of full measure in $[t_0^0,t_f^0]$. We have denoted by $\tau^0(t)$ the inverse function mapping $\hat{\mathcal{E}}$ onto $\mathcal{E}$. The function $\tau^0(t)$ monotonically increases on $\hat{\mathcal{E}}$. Let us extend it to the whole closed interval $[t_0^0,t_f^0]$ so that the extended function is left continuous. As before, this function is denoted by $\tau^0(t)$. We set

$$\hat{\psi}_x(t)=\psi_x(\tau^0(t)),\quad \hat{\psi}_t(t)=\psi_t(\tau^0(t)). \tag{2.198}$$

We note that

$$\hat{x}^0(t)=x^0(\tau^0(t)),\quad \hat{u}^0(t)=u^0(\tau^0(t)). \tag{2.199}$$

The first equation holds on $[t_0^0,t_f^0]$, and the second holds at every continuity point of the function $\hat{u}^0(t)$, i.e., on the set $[t_0^0,t_f^0]\setminus\Theta$. Also, recall that $\varphi(\eta^0(\tau))z^0=\chi_{\mathcal{E}}(\tau)$, and hence

$$\varphi(\eta^0(\tau^0(t)))z^0=\chi_{\mathcal{E}}(\tau^0(t))=1 \tag{2.200}$$

a.e. on $[t_0^0,t_f^0]$. Setting $\tau=\tau^0(t)$ in conditions (2.196) and (2.197) and taking into account (2.198)–(2.200), for almost all $t\in[t_0^0,t_f^0]$, we obtain

$$\hat{\psi}_x(t)f(t,\hat{x}^0(t),\hat{u}^0(t))+\hat{\psi}_t(t)=0, \tag{2.201}$$

$$\hat{\psi}_x(t)f(t,\hat{x}^0(t),u)+\hat{\psi}_t(t)\geq 0 \tag{2.202}$$

if $(t,\hat{x}^0(t),u)\in\mathcal{Q}$, $u\in\mathbb{R}^{d(u)}$. Condition (2.201), which holds a.e. on $[t_0^0,t_f^0]$, also holds at every continuity point of the function $\hat{u}^0(t)$, i.e., on $[t_0^0,t_f^0]\setminus\Theta$; condition (2.202) holds for all $t\in[t_0^0,t_f^0]$, since all functions entering this condition are continuous.

In [79, Section 7, Part 1], we have proved that equations (2.188) and (2.189) imply the equations

$$-\frac{d\hat{\psi}_x}{dt}=\hat{\psi}_x(t)f_x'(t,\hat{x}^0(t),\hat{u}^0(t)), \tag{2.203}$$

$$-\frac{d\hat{\psi}_t}{dt}=\hat{\psi}_t(t)f_t'(t,\hat{x}^0(t),\hat{u}^0(t)). \tag{2.204}$$

In proving this, we use the change $\tau = \tau^0(t)$ and the condition

$$\frac{dt^0}{d\tau} = \varphi(\eta^0(\tau))z^0.$$

Finally, the transversality conditions (2.185) imply the following transversality conditions:

$$\hat{\psi}_x(t_0^0) = -l'_{x_0}(\hat{x}_0^0, \hat{x}_1^0), \qquad \hat{\psi}_x(t_f^0) = l'_{x_f}(\hat{x}_0^0, \hat{x}_1^0), \tag{2.205}$$

since $t^0(\tau_0) = t_0^0$ and $t^0(\tau_f) = t_f^0$. Conditions (2.201)–(2.205) and conditions (2.183) and (2.192), which hold for a tuple $(\alpha_0, \alpha, \beta, \psi_x(t), \psi_t(t))$, imply that its projection $(\alpha_0, \alpha, \beta, \psi_x(t))$ belongs to the set M_0 of the problem (2.169), (2.170) at the point $\hat{w}^0(t)$. Therefore, for the superscript ζ indicated in Lemma 2.88 and corresponding function $\tau^0(t)$ (defined above), we have proved the following assertion.

Lemma 2.89. *Let a tuple* $\lambda^N = (\alpha_0, \alpha, \beta, \alpha_z, \beta_{t_0}, \beta_{t_f}, \psi_x(\tau), \psi_t(\tau), \psi_z(\tau))$, *belong to the set* M_0^ζ *of Problem* Z_N *at the point* κ^ζ. *We set* $\hat{\psi}_x(t) = \psi_x(\tau^0(t))$. *Then the tuple* $\lambda = (\alpha_0, \alpha, \beta, \hat{\psi}_x(t))$ *belongs to the set* M_0 *of the problem* (2.169), (2.170) *at the point* $\hat{w}^0$.

2.6.6 Critical Cone $\mathcal{K}^\zeta$ and Its Relation to the Critical Cone $\mathcal{K}$

The discontinuity points $\tau_k = \tau^0(t_k)$, $k = 1, \ldots, s$, of the function $u^0(\tau) = \hat{u}^0(t^0(\tau))$ correspond to the discontinuity points $t_k \in \Theta$, $k = 1, \ldots, s$, of the function $\hat{u}^0(t)$. We set $\Theta^\zeta = \{\tau_k\}_{k=1}^s$. The condition $\zeta \in \mathcal{D}(\Theta)$ implies $\Theta^\zeta \subset \mathcal{E}$. Further, let $\tilde{\Theta}^\zeta = \{\tilde{\tau}_i\}_{i=1}^{\tilde{s}}$ be the set of discontinuity points of the control $\eta^0(\tau)$. The definition of the function $\eta^0(\tau)$ implies that $\tilde{\Theta}^\zeta$ does not intersect the open set $\mathcal{E}$. Therefore, the sets Θ^ζ and $\tilde{\Theta}^\zeta$ are disjoint. We denote their union by $\Theta(\zeta)$.

By definition, the critical cone $\mathcal{K}^\zeta$ for the trajectory κ^ζ in Problem Z_N consists of the tuples

$$\bar{z}^N = (\bar{\xi}, \tilde{\xi}, \bar{t}(\tau), \bar{x}(\tau), \bar{z}(\tau), \bar{u}(\tau), \bar{\eta}(\tau)) \tag{2.206}$$

such that

$$\begin{aligned}
&\bar{x} \in P_{\Theta(\zeta)}W^{1,2}([\tau_0, \tau_f], \mathbb{R}^{d(x)}), \quad \bar{t} \in P_{\Theta(\zeta)}W^{1,2}([\tau_0, \tau_f], \mathbb{R}^1),\\
&\bar{z} \in P_{\Theta(\zeta)}W^{1,2}([\tau_0, \tau_f], \mathbb{R}^{d(z)}), \quad \bar{u} \in L^2([\tau_0, \tau_f], \mathbb{R}^{d(u)}),\\
&\bar{\eta} \in L^2([\tau_0, \tau_f], \mathbb{R}^{d(\eta)}), \quad \bar{\xi} \in \mathbb{R}^s, \quad \tilde{\xi} \in \mathbb{R}^{\tilde{s}}, && (2.207)\\
&J'_p\bar{p} \le 0, \quad F'_{ip}\bar{p} \le 0 \quad (i \in I), \quad K'_p\bar{p} = 0, && (2.208)
\end{aligned}$$

where $\bar{p} = (\bar{x}(\tau_0), \bar{x}(\tau_f))$ and the gradients J'_p, F'_{ip}, and K'_p are taken at the point $(\hat{x}_0^0, \hat{x}_1^0) = \hat{p}^0$,

$$\bar{t}(\tau_0) = 0, \quad \bar{t}(\tau_f) = 0, \tag{2.209}$$

$$\begin{aligned}
\frac{d\bar{x}}{d\tau} = {} & (\varphi(\eta^0(\tau))z^0)\big(f'_t\bar{t}(\tau) + f'_x\bar{x}(\tau) + f'_u\bar{u}(\tau)\big)\\
& + ((\varphi'_\eta(\eta^0(\tau))\bar{\eta}(\tau))\bar{z}^0)f(t^0(\tau), x^0(\tau), u^0(\tau))\\
& + (\varphi(\eta^0(\tau))\bar{z}(\tau))f(t^0(\tau), x^0(\tau), u^0(\tau)),
\end{aligned} \tag{2.210}$$

where the gradients f'_x, f'_u, and f'_t are taken at the trajectory $(t^0(\tau), x^0(\tau), u^0(\tau))$,

$$\frac{d\bar{t}}{d\tau} = (\varphi'_\eta(\eta^0(\tau))\bar{\eta}(\tau))z^0 + \varphi(\eta^0(\tau))\bar{z}, \quad \frac{d\bar{z}}{d\tau} = 0, \tag{2.211}$$

$$\begin{aligned} &[\bar{x}](\tau_k) = [(\varphi(\eta^0)z^0)f(t^0,x^0,u^0)](\tau_k)\bar{\xi}_k, \quad [\bar{t}](\tau_k) = [\varphi(\eta^0)z^0](\tau_k)\bar{\xi}_k, \\ &\bar{z}(\tau_k) = 0, \quad k = 1,\dots,s, \end{aligned} \tag{2.212}$$

$$\begin{aligned} &[\bar{x}](\tilde{\tau}_i) = [(\varphi(\eta^0)z^0)f(t^0,x^0,u^0)](\tilde{\tau}_i)\bar{\tilde{\xi}}_i, \quad [\bar{t}](\tilde{\tau}_i) = [\varphi(\eta^0)z^0](\tilde{\tau}_i)\bar{\tilde{\xi}}_i, \\ &[\bar{z}](\tilde{\tau}_i) = 0, \quad i = 1,\dots,\tilde{s}. \end{aligned} \tag{2.213}$$

Here, $[\cdot](\tau_k)$ is the jump at the point τ_k, and $[\cdot](\tilde{\tau}_i)$ is the jump at the point $\tilde{\tau}_i$. We set

$$\bar{t}(\tau) = 0, \quad \bar{z}(\tau) = 0, \quad \bar{\eta}(\tau) = 0, \quad \bar{\tilde{\xi}} = 0. \tag{2.214}$$

These conditions define the subcone $\mathcal{K}_0^\zeta$ of the cone $\mathcal{K}^\zeta$ such that the following conditions hold:

$$\bar{x}(\tau) \in P_{\Theta^\zeta} W^{1,2}([\tau_0,\tau_f],\mathbb{R}^{d(x)}), \; \bar{u}(\tau) \in L^2([\tau_0,\tau_f],\mathbb{R}^{d(u)}), \quad \bar{\xi} \in \mathbb{R}^s, \tag{2.215}$$

$$J'_p\bar{p} \le 0, \quad F'_{ip}\bar{p} \le 0 \quad (i \in I), \quad K'_p\bar{p} = 0, \tag{2.216}$$

$$\begin{aligned} \frac{d\bar{x}}{d\tau} = (\varphi(\eta^0(\tau))z^0)\big(&f'_x(t^0(\tau),x^0(\tau),u^0(\tau))\bar{x}(\tau) \\ &+ f'_u(t^0(\tau),x^0(\tau),u^0(\tau))\bar{u}(\tau)\big); \end{aligned} \tag{2.217}$$

$$[\bar{x}](\tau_k) = \big[(\varphi(\eta^0)z^0)f(t^0,x^0,u^0)\big](\tau_k)\bar{\xi}_k, \quad k = 1,\dots,s. \tag{2.218}$$

The following assertion holds.

Lemma 2.90. *Let*

$$\hat{\bar{z}} = \big(\hat{\bar{\xi}}, \hat{\bar{x}}(t), \hat{\bar{u}}(t)\big) \tag{2.219}$$

be an arbitrary element of the critical cone $\mathcal{K}$ of the problem (2.169), (2.170) *at the point $\hat{w}^0$. We set*

$$\begin{aligned} &\bar{\xi} = \hat{\bar{\xi}}, \quad \bar{x}(\tau) = \hat{\bar{x}}(t^0(\tau)), \quad \bar{u}(\tau) = \hat{\bar{u}}(t^0(\tau)), \\ &\bar{t}(\tau) = 0, \quad \bar{z}(\tau) = 0, \quad \bar{\eta}(\tau) = 0, \quad \bar{\tilde{\xi}} = 0. \end{aligned} \tag{2.220}$$

Then

$$\bar{z}^N = (\bar{\xi}, \bar{\tilde{\xi}}, \bar{t}(\tau), \bar{x}(\tau), \bar{z}(\tau), \bar{u}(\tau), \bar{\eta}(\tau)) \tag{2.221}$$

is an element of the cone $\mathcal{K}_0^\zeta \subset \mathcal{K}^\zeta$, where $\mathcal{K}^\zeta$ is the critical cone of Problem Z_N at the point κ^ζ, and $\mathcal{K}_0^\zeta$ is defined by conditions (2.214)–(2.218).

Proof. Let $\hat{\bar{z}}$ be an arbitrary element of the critical cone $\mathcal{K}$ of the problem (2.169), (2.170) at the point $\hat{w}^0$ having the form (2.219). Then by the definition of the cone $\mathcal{K}$, we have

$$\hat{\bar{\xi}} \in \mathbb{R}^s, \quad \hat{\bar{x}}(t) \in P_\Theta W^{1,2}([t_0^0,t_f^0],\mathbb{R}^{d(x)}), \quad \hat{\bar{u}}(t) \in L_2([t_0^0,t_f^0],\mathbb{R}^{d(u)}), \tag{2.222}$$

$$J'_p(\hat{p}^0)\hat{\bar{p}} \le 0, \quad F'_{ip}(\hat{p}^0)\hat{\bar{p}} \le 0 \quad (i \in I), \quad K'_p(\hat{p}^0)\hat{\bar{p}} = 0, \tag{2.223}$$

$$\frac{d\hat{\bar{x}}(t)}{dt} = f'_x(t,\hat{x}^0(t),\hat{u}^0(t))\hat{\bar{x}}(t) + f'_u(t,\hat{x}^0(t),\hat{u}^0(t))\hat{\bar{u}}(t), \tag{2.224}$$

$$[\hat{\bar{x}}](t_k) = [f(\cdot,x^0,u^0)](t_k)\hat{\bar{\xi}}_k, \quad k = 1,\dots,s. \tag{2.225}$$

Let conditions (2.220) hold. We show that all conditions (2.214)–(2.218) defining the cone $\mathcal{K}_0^\zeta$ hold for the element $\bar{z}^N$ (having form (2.221)).

Conditions (2.214) follow from (2.220). Conditions (2.215) follow from (2.222). Indeed, the function $\hat{\bar{x}}(t)$ is piecewise absolutely continuous, and the function $t^0(\tau)$ is Lipschitz continuous. Hence $\bar{x}(\tau) = \hat{\bar{x}}(t^0(\tau))$ is a piecewise absolutely continuous function whose set of discontinuity points is contained in Θ^ζ. Further,

$$\frac{d\bar{x}(\tau)}{d\tau} = \frac{d\hat{\bar{x}}(t^0(\tau))}{d\tau} = \left.\frac{d\hat{\bar{x}}}{dt}\right|_{t=t^0(\tau)} \cdot \frac{dt^0(\tau)}{d\tau} = \chi_{\mathcal{E}}(\tau)\left.\frac{d\hat{\bar{x}}}{dt}\right|_{t=t^0(\tau)}. \tag{2.226}$$

Since $\chi_{\mathcal{E}}^2 = \chi_{\mathcal{E}} = \frac{dt^0}{d\tau}$, we have

$$\int_{\tau_0}^{\tau_f} \left(\frac{d\bar{x}(\tau)}{d\tau}\right)^2 d\tau = \int_{\tau_0}^{\tau_f} \left(\frac{d\hat{\bar{x}}}{dt}(t^0(\tau))\right)^2 \frac{dt^0(\tau)}{d\tau} d\tau = \int_{t_0^0}^{t_f^0} \left(\frac{d\hat{\bar{x}}(t)}{dt}\right)^2 dt < +\infty.$$

Hence the derivative $d\bar{x}/d\tau$ is square Lebesgue integrable. Therefore,

$$\bar{x}(\cdot) \in P_{\Theta^\zeta} W^{1,2}([\tau_0, \tau_f], \mathbb{R}^{d(x)}).$$

Further, consider the integral

$$\int_{\tau_0}^{\tau_f} \bar{u}(\tau)^2\, d\tau = \int_{\mathcal{E}} \bar{u}(\tau)^2\, d\tau + \int_{[\tau_0,\tau_f]\setminus\mathcal{E}} \bar{u}(\tau)^2\, d\tau.$$

The function $t^0(\tau)$, and hence the function $\hat{\bar{u}}(t^0(\tau)) = \bar{u}(\tau)$, assumes finitely many values on $[\tau_0, \tau_f] \setminus \mathcal{E}$; hence the second integral in the sum is finite. For the first integral, we have

$$\begin{aligned}\int_{\mathcal{E}} \bar{u}(\tau)^2\, d\tau &= \int_{\mathcal{E}} \bar{u}(\tau)^2 \chi_{\mathcal{E}}\, d\tau = \int_{\mathcal{E}} \bar{u}(\tau)^2 \frac{dt^0(\tau)}{d\tau} d\tau = \int_{\mathcal{E}} \hat{\bar{u}}(t^0(\tau))^2 \frac{dt^0(\tau)}{d\tau} d\tau \\ &= \int_{\hat{\mathcal{E}}} \hat{\bar{u}}(t)^2\, dt = \int_{[t_0^0, t_f^0]} \hat{\bar{u}}(t)^2\, dt < +\infty,\end{aligned}$$

since $\hat{\bar{u}}$ is Lebesgue square integrable. Hence, $\int_{\tau_0}^{\tau_f} \bar{u}(\tau)^2\, d\tau < +\infty$, i.e., $\bar{u}(\cdot) \in L^2([\tau_0, t_f], \mathbb{R}^{d(u)})$. Further, condition (2.223) implies condition (2.216), since $t^0(\tau_0) = t_0^0$, $t^0(\tau_f) = t_f^0$, and, therefore, $\bar{x}(\tau_0) = \hat{\bar{x}}(t_0^0)$, $\bar{x}(\tau_f) = \hat{\bar{x}}(t_f^0)$. Consider the variational equation (2.224). Making the change $t = t^0(\tau)$ in it, multiplying by $\chi_{\mathcal{E}}(\tau)$, and taking into account (2.226), we obtain

$$\frac{d\bar{x}}{d\tau} = \chi_{\mathcal{E}}(\tau)(f_x'(t^0(\tau), x^0(\tau), u^0(\tau))\bar{x}(\tau) + f_u'(t^0(\tau), x^0(\tau), u^0(\tau))\bar{u}(\tau)).$$

But $\chi_{\mathcal{E}}(\tau) = \varphi(\eta^0(\tau))z^0$. Therefore, the variational equation (2.217) holds for $\bar{x}$ and $\bar{u}$.

Finally, we show that the jump conditions (2.218) hold. Note that

$$t_k = t^0(\tau_k), \quad \tau_k \in \mathcal{E}, \quad k = 1, \ldots, s. \tag{2.227}$$

Consequently, each τ_k is a continuity point of the function $\eta^0(\tau)$ and

$$\varphi(\eta^0(\tau_k))z^0 = 1, \quad k = 1, \ldots, s. \tag{2.228}$$

It follows from (2.227) and (2.228) that

$$[(\varphi(\eta^0)z^0)f(t^0,x^0,u^0)](\tau_k) = [f(\cdot,\hat{x}^0,\hat{u}^0)](t_k), \quad k = 1,\dots,s. \tag{2.229}$$

Analogously,

$$[\bar{x}](\tau_k) = [\hat{\bar{x}}](t_k), \quad k = 1,\dots,s. \tag{2.230}$$

Conditions (2.218) follow from (2.225), (2.229), and (2.230) and the relation $\bar{\xi} = \hat{\bar{\xi}}$. Therefore, all the conditions defining the cone $\mathcal{K}_0^\zeta$ hold for the tuple $\bar{z}^N$. ∎

2.6.7 Quadratic Form Ω^ζ and its Relation to the Quadratic Form Ω

Let $\lambda^N \in M_0^\zeta$ and $\bar{z}^N \in \mathcal{K}_0^\zeta$; hence let condition (2.214) hold for $\bar{z}^N$. The value of the quadratic form Ω^ζ (corresponding to the tuple of Lagrange multipliers λ^N, at the point κ^ζ in Problem Z_N) at the element $\bar{z}^N$ is denoted by $\Omega^\zeta(\lambda^N;\bar{z}^N)$. Taking into account conditions (2.214), by definition, we obtain

$$\begin{aligned} 2\Omega^\zeta(\lambda^N,\bar{z}^N) &= \sum_{k=1}^{s}(D^k(H^N)\bar{\xi}_k^2 + 2[H_x^N]^k\bar{x}_{av}^k\bar{\xi}_k) \\ &\quad + \langle l''_{pp}\bar{p},\bar{p}\rangle + \int_{\tau_0}^{\tau_f}\langle H_{ww}^N\bar{w}(\tau),\bar{w}(\tau)\rangle\,d\tau. \end{aligned} \tag{2.231}$$

Here,

$$\bar{p} = (\bar{x}(\tau_0),\bar{x}(\tau_f)), \qquad l''_{pp} = l''_{pp}(\hat{x}_0^0,\hat{x}_f^0;\alpha_0,\alpha,\beta), \tag{2.232}$$

$$\begin{aligned} H_{ww}^N &= (\varphi(\eta^0(\tau))z^0)H''_{ww}(t^0(\tau),x^0(\tau),u^0(\tau),\psi_x(\tau)) \\ &= \chi_\varepsilon(\tau)H''_{ww}(t^0(\tau),x^0(\tau),u^0(\tau),\psi_x(\tau)). \end{aligned} \tag{2.233}$$

Further, $[H_x^N]^k = [H_x^N](\tau_k), k = 1,\dots,s$, where $H_x^N = \chi_\varepsilon(\tau)H_x(t^0(\tau),x^0(\tau),u^0(\tau),\psi_x(\tau))$. Let $\hat{\psi}_x(t) = \psi_x(\tau^0(t))$, $H_x = H_x(t,\hat{x}^0(t),\hat{u}^0(t),\hat{\psi}_x(t))$, $[H_x]^k = [H_x](t_k)$, $k = 1,\dots,s$. Taking into account that

$$\begin{aligned} &\chi_\varepsilon(\tau_k) = 1, \quad \psi_x(\tau_k) = \hat{\psi}_x(t_k), \\ &x^0(\tau_k) = \hat{x}^0(t_k), \quad u^0(\tau_k-) = \hat{u}^0(t_k-), \quad u^0(\tau_k+) = \hat{u}^0(t_k+), \\ &t^0(\tau_k) = t_k, \quad k = 1,\dots,s, \end{aligned}$$

we obtain $[H_x^N](\tau_k) = [H_x](t_k) = [H_x]^k$, $k = 1,\dots,s$. Thus,

$$[H_x^N]^k = [H_x]^k, \quad k = 1,\dots,s. \tag{2.234}$$

Finally, by definition,

$$D^k(H^N) = -\frac{d}{d\tau}(\Delta_k H^N)\bigg|_{\tau=\tau_k}, \quad k = 1,\dots,s.$$

Since τ_k is a continuity point of the function $\eta^0(\tau)$ and $\varphi(\eta^0(\tau_k))z^0 = 1$, we have

$$\begin{aligned}
&(\Delta_k H^N)(\tau)\\
&\quad = (\varphi(\eta^0(\tau_k))z^0)\psi_x(\tau)\big(f(t^0(\tau),x^0(\tau),u^0(\tau_k+)) - f(t^0(\tau),x^0(\tau),u^0(\tau_k-))\big)\\
&\quad = \hat{\psi}_x(t^0(\tau))\big(f(t^0(\tau),\hat{x}^0(t^0(\tau)),\hat{u}^0(t_k+)) - f(t^0(\tau)),\hat{x}^0(t^0(\tau)),\hat{u}^0(t_k-))\big)\\
&\quad = (\Delta_k H)(t^0(\tau)).
\end{aligned}$$

Consequently,

$$\begin{aligned}
D^k(H^N) &= -\frac{d}{d\tau}(\Delta_k H^N)\Big|_{\tau=\tau_k} = -\left(\frac{d}{dt}(\Delta_k H)\right)\Big|_{t=t_k} \cdot \frac{dt^0}{d\tau}\Big|_{\tau=\tau_k}\\
&= D^k(H)\chi_{\mathcal{E}}(\tau_k) = D^k(H).
\end{aligned} \tag{2.235}$$

Let $\hat{\bar{z}} = (\hat{\bar{\xi}},\hat{\bar{w}}) = (\hat{\bar{\xi}},\hat{\bar{x}}(t),\hat{\bar{u}}(t))$ be an arbitrary element of the critical cone $\mathcal{K}$. Let $\bar{z}^N = (\bar{\xi},\tilde{\xi},\bar{t},\bar{x},\bar{z},\bar{u},\bar{\eta})$ be the tuple defined according to $\hat{\bar{z}}$ by using formulas (2.220). Then by Lemma 2.90, $\bar{z}^N \in \mathcal{K}_0^\zeta$. According to (2.233),

$$\begin{aligned}
&\int_{t_0^0}^{t_f^0} \langle H_{ww}(t,\hat{x}^0(t),\hat{u}^0(t),\hat{\psi}(t))\hat{\bar{w}}(t),\hat{\bar{w}}(t)\rangle\,dt\\
&\quad = \int_{\tau_0}^{\tau_f} \langle H_{ww}(t^0(\tau),\hat{x}^0(t^0(\tau)),\hat{u}^0(t^0(\tau)),\hat{\psi}_x(t^0(\tau)))\hat{\bar{w}}(t^0(\tau)),\hat{\bar{w}}(t^0(\tau))\rangle \frac{dt^0(\tau)}{d\tau}\,d\tau\\
&\quad = \int_{\tau_0}^{\tau_f} \langle H_{ww}(t^0(\tau),x^0(\tau),u^0(\tau),\psi_x(\tau))\bar{w}(\tau),\bar{w}(\tau)\rangle \chi_{\mathcal{E}}(\tau)\,d\tau\\
&\quad = \int_{\tau_0}^{\tau_f} \langle H^N_{ww}\bar{w}(\tau),\bar{w}(\tau)\rangle\,d\tau.
\end{aligned} \tag{2.236}$$

Let $\lambda = (\alpha_0,\alpha,\beta,\hat{\psi}_x(t))$ be the element corresponding to the tuple $\lambda^N \in M_0^\zeta$, where $\hat{\psi}_x(t) = \psi_x(\tau^0(t))$. Then $\lambda \in M_0$ according to Lemma 2.89. Recall that by definition, the quadratic form $\Omega^\lambda(\hat{\bar{z}})$ for the problem (2.169), (2.170) at the point $\hat{w}^0$ corresponding to the tuple λ of Lagrange multipliers and calculated at the element $\hat{\bar{z}}$ has the form

$$2\Omega^\lambda(\hat{\bar{z}}) = \sum_{k=1}^{s}(D^k(H)\hat{\bar{\xi}}_k^2 + 2[H_x]^k\hat{\bar{x}}^k_{\mathrm{av}}\hat{\bar{\xi}}_k) + \langle l_{pp}\hat{\bar{p}},\hat{\bar{p}}\rangle + \int_{t_0^0}^{t_f^0}\langle H_{ww}\hat{\bar{w}},\hat{\bar{w}}\rangle\,dt. \tag{2.237}$$

Here, $\hat{\bar{p}} = (\hat{\bar{x}}(t_0^0),\hat{\bar{x}}(t_f^0))$. Note that

$$\langle l''_{pp}\hat{\bar{p}},\hat{\bar{p}}\rangle = \langle l''_{pp}\bar{p},\bar{p}\rangle, \tag{2.238}$$

since $\hat{\bar{x}}(t_0^0) = \hat{\bar{x}}(t^0(\tau_0)) = \bar{x}(\tau_0)$ and $\hat{\bar{x}}(t_f^0) = \hat{\bar{x}}(t^0(\tau_f)) = \bar{x}(\tau_f)$. Formulas (2.231) and (2.234)–(2.238), imply

$$\Omega^\lambda(\hat{\bar{z}}) = \Omega^\zeta(\lambda^N;\bar{z}^N). \tag{2.239}$$

Therefore, we have proved the following assertion.

Lemma 2.91. *Let $\hat{\bar{z}}$ be an arbitrary element* (2.219) *of the critical cone $\mathcal{K}$, and let $\bar{z}^N$ be element* (2.221) *of the cone $\mathcal{K}_0^\zeta \subset \mathcal{K}^\zeta$ obtained by formulas* (2.220). *Let $\lambda^N \in M_0^\zeta$ be an arbitrary tuple* (2.182), *and let $\lambda = (\alpha_0, \alpha, \beta, \hat{\psi}_x(t)) \in M_0$ be the tuple with the same components α_0, α, β and with $\hat{\psi}_x(t) = \psi_x(\tau^0(t))$ corresponding to it by Lemma* 2.89. *Then relation* (2.239) *holds for the quadratic forms.*

2.6.8 Proof of Theorem 2.4

Thus, let $\hat{w}^0$ be a Pontryagin minimum point in the problem (2.169), (2.170). Then the set M_0 is nonempty. If, moreover, $M_0 \cap (-M_0) \neq \emptyset$ (case (a) in Section 2.6.2), then as already mentioned, Condition $\mathfrak{A}$ holds trivially. Otherwise (case (b) in Section 2.6.2), by Lemma 2.88, there exists a superscript ζ such that Condition $\mathfrak{A}^\zeta$ holds, i.e., the set M_0^ζ is nonempty and

$$\max_{M_0^\zeta} \Omega^\zeta(\lambda^N, \bar{z}^N) \geq 0 \quad \forall\, \bar{z}^N \in \mathcal{K}^\zeta. \tag{2.240}$$

Let us show that Condition $\mathfrak{A}$ holds: the set M_0 is nonempty and

$$\max_{M_0} \Omega^\lambda(\bar{z}) \geq 0 \quad \forall\, \bar{z} \in \mathcal{K}. \tag{2.241}$$

Take an arbitrary element $\hat{\bar{z}} \in \mathcal{K}$. According to Lemma 2.90, the element $\bar{z}^N \in \mathcal{K}_0^\zeta \subset \mathcal{K}^\zeta$ corresponds to it by formulas (2.220). By (2.240), for $\bar{z}^N$, there exists $\lambda^N \in M_0^\zeta$ such that

$$\Omega^\zeta(\lambda^N, \bar{z}^N) \geq 0. \tag{2.242}$$

The element $\lambda \in M_0$ corresponds to $\lambda^N \in M_0^\zeta$ by Lemma 2.89, and, moreover, by Lemma 2.91, we have relation (2.239). It follows from (2.239) and (2.242) that

$$\Omega^\lambda(\hat{\bar{z}}) \geq 0, \quad \lambda \in M_0.$$

Since $\hat{\bar{z}}$ is an arbitrary element of $\mathcal{K}$, this implies condition (2.241). Therefore, Condition $\mathfrak{A}$ also holds in case (b). The theorem is completely proved. ∎

2.7 Sufficient Conditions for Bounded Strong and Strong Minima in the Problem on a Fixed Time Interval

2.7.1 Strong Minimum

In [79, Part 1], we considered the strong minimum conditions related to the solutions of the Hamilton-Jacobi equation. We can say that they were obtained as a result of development of the traditional approach to sufficient strong minimum conditions accepted in the calculus of variations. However, it is remarkable that there exists another, nontraditional approach to the strong minimum sufficient conditions using the strengthening of the quadratic sufficient conditions for a Pontryagin minimum. Roughly speaking, the strengthening consists of assuming certain conditions on the behavior of the function H at infinity.

This fact, which had been previously absent in the classical calculus of variations, was first discovered by Milyutin when studying problems of the calculus variations and optimal control. We use this fact in this section.

We first define the concept of strong minimum, which will be considered here. It is slightly different from the usual concept from the viewpoint of strengthening. The usual concept used in the calculus of variations corresponds to the concept of minimum on the set of sequences $\{\delta w\}$ in the space W such that $\|\delta x\|_C \to 0$. It is not fully correct to extend it to the canonical problem without any changes. Indeed, in the classical calculus of variations, it is customary to minimize an integral functional of the form $J = \int_{t_0}^{t_f} F(t,x,u)\,dt$, where $u = \dot{x}$. In passing to the canonical problem, we write the integral functional as the terminal functional: $J = y(t_f) - y(t_0)$, but there arises a new state variable y such that $\dot{y} = F(t,x,u)$. Clearly, the requirement $\|\delta y\|_C \to 0$ must be absent in the canonical problem if we do not want to distort the original concept of strong minimum in rewriting the problem.

How can this be taken into account if we have the canonical form in advance, and it is not known from which problem it originates? It is easy to note that the new state variables y arising in rewriting the integral functionals are characterized by the property that they affinely enter the terminal functionals of the canonical form and are completely absent in the control system of the canonical form. These variables are said to be unessential and the other variables are said to be essential. In defining the strong minimum, we take into account only the essential variables.

Let us give the precise definition. As before, we consider the canonical problem (2.1)–(2.4) on a fixed closed interval of time $[t_0, t_f]$.

Definition 2.92. A state variable x_i (the component x_i of a vector x) is said to be *unessential* if the function f is independent of it and the functions J, F, and K affinely depend on $x_{i0} := x_i(t_0)$, $x_{if} = x_i(t_f)$. The state variables x_i without these properties are said to be *essential.* (One can also use the terms "main" (or "basic") and "complementary" (or "auxiliary") variables.) Respectively, we speak about the essential components of the vector x.

Denote by $\underline{x}$ the vector composed of the essential components of the vector x. Similarly, denote by $\underline{\delta x}$ the vector-valued function composed of the essential components of the variation δx.

Denote by Π^S the set of sequences $\{\delta w\}$ in the space W such that $|\delta x(t_0)| + \|\underline{\delta x}\|_C \to 0$. Let us give the following definition for problem (2.1)–(2.4).

Definition 2.93. We say that w^0 is a *strong minimum point* (with respect to the essential state variables) if it is a minimum point on Π^S.

In what follows, the strong minimum with respect to the essential variables will be called the *strong minimum*, for brevity. By the *strict strong minimum* we mean the strict minimum on Π^S. Since $\Pi \subset \Pi^S$, the strong minimum implies the Pontryagin minimum.

2.7.2 Bounded Strong Minimum, Sufficient Conditions

We now define the concept of bounded strong minimum, which occupies an intermediate place between the strong and Pontryagin minima.

Definition 2.94. We say that w^0 is a *bounded strong minimum point* if it is a minimum point on the set of sequences $\{\delta w\}$ in W satisfying the following conditions:

(a) $|\delta x(t_0)| + \|\underline{\delta x}\|_C \to 0$.

(b) For each sequence there exists a compact set $\mathcal{C} \subset Q$ such that the following condition holds starting from a certain number: $(t, x^0(t), u^0(t) + \delta u(t)) \in \mathcal{C}$ a.e. on $[t_0, t_f]$.

Denote by $\overline{\Pi}^S$ the set of sequences $\{\delta w\}$ in W satisfying conditions (a) and (b) and also the following additional conditions:

(c) starting from a certain number, $(p^0 + \delta p) \in \mathcal{P}$, $(t, w^0 + \delta w) \in \mathcal{Q}$.

(d) $\sigma(\delta w) \to 0$, where σ is the violation function (2.164).

Conditions (c) and (d) hold on every sequence "violating the minimum." Therefore, we may treat the bounded strong minimum as a minimum on $\overline{\Pi}^S$. We will proceed in this way in what follows, since we will need conditions (c) and (d). By the strict bounded strong minimum, we mean the strict minimum on $\overline{\Pi}^S$. Since $\Pi \subset \overline{\Pi}^S \subset \Pi^S$, the strong minimum implies the bounded strong minimum, and the latter implies the Pontryagin minimum.

A remarkable property is that the sufficient conditions obtained in Section 2.5 guarantee not only the Pontryagin minimum but also the bounded strong minimum. This follows from the theorem, which now will be proved. In what follows, w^0 is an admissible point satisfying the standard assumptions of Section 2.1.

Theorem 2.95. *For a point w^0, let there exist an admissible function $\Gamma(t,u)$ and a constant $C > 0$ such that the set $M(C\Gamma)$ is nonempty. Then $\Pi = \overline{\Pi}^S$, and hence the Pontryagin minimum is equivalent to the bounded strong minimum.*

To prove this, we need several auxiliary assertions.

Proposition 2.96. *Let $\lambda \in \Lambda_0$, $\{\delta w\} \in \overline{\Pi}^S$. Then $(\int_{t_0}^{t_f} \delta H^\lambda\, dt)_+ \to 0$, where $a_+ = \max\{a, 0\}$ and $\delta H^\lambda = H(t, x^0 + \delta x, u^0 + \delta u, \psi) - H(t, x^0, u^0, \psi)$.*

Proof. Let $\lambda \in \Lambda_0$, and let δw be an admissible variation with respect to Q, i.e., $(t, w^0 + \delta w) \in \mathcal{Q}$. Then

$$\delta l^\lambda - \int_{t_0}^{t_f} \psi(\delta \dot{x} - \delta f)\, dt \le \text{const}\, \sigma(\delta w). \tag{2.243}$$

On the other hand, we have shown earlier that the conditions $-\dot{\psi} = H_x^\lambda$, $\psi(t_0) = -l_{x_0}^\lambda$, and $\psi(t_f) = l_{x_f}^\lambda$ imply

$$\int_{t_0}^{t_f} \psi \delta \dot{x}\, dt = \psi \delta x \,\Big|_{t_0}^{t_f} - \int_{t_0}^{t_f} \dot{\psi} \delta x\, dt = l_p \delta p + \int_{t_0}^{t_f} H_x^\lambda \delta x\, dt.$$

Taking into account that $\psi \delta f = \delta H^\lambda$, we obtain from inequality (2.243) that

$$\delta l^\lambda - l_p \delta p - \int_{t_0}^{t_f} H_x^\lambda \delta x\, dt + \int_{t_0}^{t_f} \delta H^\lambda\, dt \le \text{const}\, \sigma(\delta w). \tag{2.244}$$

Let $\{\delta w\} \in \overline{\Pi}^S$. The condition $\|\underline{\delta x}\|_C \to 0$ implies $\int_{t_0}^{t_f} H_x^\lambda \delta x\, dt \to 0$ and $(\delta l^\lambda - l_p \delta p) \to 0$. Moreover, $\sigma(\delta w) \to 0$. Therefore, condition (2.244) implies $(\int_{t_0}^{t_f} \delta H^\lambda\, dt)_+ \to 0$. The proposition is proved. ∎

Using Proposition 2.96, we prove the following assertion.

Proposition 2.97. *Let there exist an admissible function $\Gamma(t,u)$ and a constant $C > 0$ such that the set $M(C\Gamma)$ is nonempty. Then the following condition holds for any sequence $\{\delta w\} \in \overline{\Pi}^S$: $\int_{t_0}^{t_f} \Gamma(t, u^0 + \delta u) dt \to 0$.*

Proof. Let $C > 0$, and let $\lambda \in M(C\Gamma)$. According to Proposition 2.96, $(\int_{t_0}^{t_f} \delta H^\lambda\, dt)_+ \to 0$. Represent δH^λ as $\delta H^\lambda = \bar{\delta}_x H^\lambda + \delta_u H^\lambda$, where

$$\begin{aligned} \bar{\delta}_x H^\lambda &= H^\lambda(t, x^0 + \delta x, u^0 + \delta u) - H^\lambda(t, x^0, u^0 + \delta u), \\ \delta_u H^\lambda &= H^\lambda(t, x^0, u^0 + \delta u) - H^\lambda(t, x^0, u^0). \end{aligned}$$

The conditions $\|\underline{\delta x}\|_C \to 0$, $(t, x^0, u^0 + \delta u) \in \mathcal{C}$, where $\mathcal{C} \subset \mathcal{Q}$ is a compact set, imply $\|\bar{\delta}_x H^\lambda\|_\infty \to 0$. Hence $(\int_{t_0}^{t_f} \delta_u H^\lambda\, dt)_+ \to 0$. Further, the condition $\lambda \in M(C\Gamma)$ implies $\delta_u H^\lambda \ge C\Gamma(t, u^0 + \delta u) \ge 0$. Consequently, $\int_{t_0}^{t_f} \delta_u H^\lambda\, dt \ge C \int_{t_0}^{t_f} \Gamma(t, u^0 + \delta u) dt \ge 0$. This and the condition $(\int_{t_0}^{t_f} \delta_u H^\lambda\, dt)_+ \to 0$ imply $\int_{t_0}^{t_f} \Gamma(t, u^0 + \delta u) dt \to 0$. The proposition is proved. ∎

In what follows, $\Gamma(t,u)$ is a function admissible for the point w^0.

Proposition 2.98. *Let $\mathcal{C} \subset \mathcal{Q}$ be a compact set, and let the variation $\delta u \in L^\infty(\Delta, \mathbb{R}^{d(u)})$ be such that $(t, x^0, u^0 + \delta u) \in \mathcal{C}$ a.e. on $[t_0, t_f]$. Then we have the estimate $\|\delta u\|_1 \le$ const$(\int_{t_0}^{t_f} \Gamma(t, u^0 + \delta u) dt)^{1/2}$, where* const *depends only on $\mathcal{C}$.*

Proof. Let $\mathcal{V}$ be the neighborhood of the compact set $\overline{u^0}$ from the definition of the function $\Gamma(t,u)$ in Section 2.3.7, and let $\mathcal{V}^0$ and $\mathcal{V}^*$ be subsets of the neighborhood $\mathcal{V}$ defined in Section 2.3.1. Represent δu as $\delta u = \delta u_v + \delta u^v$, where

$$\delta u_v = \begin{cases} \delta u & \text{if } (t, u^0 + \delta u) \in \mathcal{V}, \\ 0 & \text{otherwise}, \end{cases} \qquad \delta u^v = \delta u - \delta u_v.$$

Further, let the representation $\delta u_v = \delta u^0 + \delta u^*$ correspond to the partition $\mathcal{V} = \mathcal{V}^0 \cup \mathcal{V}^*$:

$$\delta u^0 = \begin{cases} \delta u_v & \text{if } (t, u^0 + \delta u_v) \in \mathcal{V}^0, \\ 0 & \text{otherwise}, \end{cases} \qquad \delta u^* = \begin{cases} \delta u_v & \text{if } (t, u^0 + \delta u_v) \in \mathcal{V}^*, \\ 0 & \text{otherwise}. \end{cases}$$

Then

$$\|\delta u\|_1 = \|\delta u^v\|_1 + \|\delta u^0\|_1 + \|\delta u^*\|_1. \tag{2.245}$$

Let us estimate each of the summands separately.

(1) Since $(t, x^0, u^0 + \delta u^v) \in \mathcal{C}$ and $(t, u^0 + \delta u^v) \notin \mathcal{V}$ for $\delta u^v \ne 0$, by the definition of the function $\Gamma(t,u)$, there exists $\varepsilon = \varepsilon(\mathcal{C}) > 0$ such that $\Gamma(t, u^0 + \delta u^v) \ge \varepsilon$ if $\delta u^v \ne 0$.

Moreover, there exists a constant $N = N(\mathcal{C}) > 0$ such that $\|\delta u^v\|_\infty \le \|\delta u\|_\infty \le N$. Consequently,

$$\|\delta u^v\|_1 \le \|\delta u^v\|_\infty \cdot \text{meas}\{t \mid \delta u^v \ne 0\} \le \frac{N}{\varepsilon}\int_{t_0}^{t_f} \Gamma(t, u^0 + \delta u^v)\,dt.$$

Also, taking into account that $\|\delta u^v\|_1 \le \|\delta u^v\|_\infty (t_f - t_0) \le N(t_f - t_0)$, we obtain from this that

$$\|\delta u^v\|_1^2 \le \|\delta u^v\|_\infty (t_f - t_0)\|\delta u^v\|_1 \le \text{const}\int_{t_0}^{t_f} \Gamma(t, u^0 + \delta u^v)\,dt, \tag{2.246}$$

where const > 0 depends on $\mathcal{C}$ only.

(2) Since $(t, u^0 + \delta u^0) \in \mathcal{V}^0$, we have $\|\delta u^0\|_2^2 = \int_{t_0}^{t_f} \Gamma(t, u^0 + \delta u^0)\,dt$. Consequently,

$$\|\delta u^0\|_1 \le \sqrt{t_f - t_0}\,\|\delta u^0\|_2 = \sqrt{t_f - t_0}\left(\int_{t_0}^{t_f} \Gamma(t, u^0 + \delta u^0)\,dt\right)^{1/2}. \tag{2.247}$$

(3) Obviously, $\|\delta u^*\|_1 = \int_{t_0}^{t_f} |\delta u^*|\,dt \le \|\delta u^*\|_\infty \cdot \text{meas}\,\mathcal{M}^* \le N \cdot \text{meas}\,\mathcal{M}^*$, where $\mathcal{M}^* = \{t \mid \delta u^* \ne 0\}$. Further, as in Section 2.3.1, represent $\mathcal{M}^* = \cup\, \mathcal{M}_k^*$, $\mathcal{M}_k^* = \mathcal{M}_{k-}^* \cup \mathcal{M}_{k+}^*$, $k = 1, \dots, s$. Then

$$(\text{meas}\,\mathcal{M}_{k-}^*)^2 \le 2\int_{\mathcal{M}_{k-}^*} |\delta t_k|\,dt \le \int_{t_0}^{t_f} \Gamma(t, u^0 + \delta u^*)\,dt;$$
$$(\text{meas}\,\mathcal{M}_{k+}^*)^2 \le 2\int_{\mathcal{M}_{k+}^*} |\delta t_k|\,dt \le \int_{t_0}^{t_f} \Gamma(t, u^0 + \delta u^*)\,dt;$$
$$\text{meas}\,\mathcal{M}^* = \sum_k \text{meas}\,\mathcal{M}_{k-}^* + \sum_k \text{meas}\,\mathcal{M}_{k+}^*.$$

Consequently,

$$\|\delta u^*\|_1 \le \text{const}\left(\int_{t_0}^{t_f} \Gamma(t, u^0 + \delta u^*)\,dt\right)^{1/2}, \tag{2.248}$$

where const depends only on $\mathcal{C}$. It follows from (2.245)–(2.248) that

$$\|\delta u\|_1 \le \text{const}\left\{\left(\int_{t_0}^{t_f} \Gamma(t, u^0 + \delta u^v)\,dt\right)^{1/2} + \left(\int_{t_0}^{t_f} \Gamma(t, u^0 + \delta u^0)\,dt\right)^{1/2} + \left(\int_{t_0}^{t_f} \Gamma(t, u^0 + \delta u^*)\,dt\right)^{1/2}\right\}.$$

Now, to obtain the required estimate, it remains to use the inequality

$$a + b + c \le \sqrt{3(a^2 + b^2 + c^2)},$$

which holds for any numbers a, b, and c, and also the relation

$$\int_{t_0}^{t_f} \Gamma(t, u^0 + \delta u^v)\,dt + \int_{t_0}^{t_f} \Gamma(t, u^0 + \delta u^0)\,dt + \int_{t_0}^{t_f} \Gamma(t, u^0 + \delta u^*)\,dt = \int_{t_0}^{t_f} \Gamma(t, u^0 + \delta u)\,dt.$$

The proposition is proved. ∎

Proof of Theorem 2.95. Assume that the conditions of the theorem hold. Let us prove the inclusion $\bar{\Pi}^S \subset \Pi$. Let $\{\delta w\} \in \bar{\Pi}^S$. Then it follows from Propositions 2.97 and 2.98 that $\|\delta u\|_1 \to 0$. Further, the condition $\sigma(\delta w) \to 0$ implies $\|\delta \dot{x} - \delta f\|_1 \to 0$. But $\delta f = \bar{\delta}_x f + \delta_u f$, where

$$\bar{\delta}_x f = f(t,x^0+\delta x,u^0+\delta u) - f(t,x^0,u^0+\delta u),$$
$$\delta_u f = f(t,x^0,u^0+\delta u) - f(t,x^0,u^0).$$

Since $\|\underline{\delta x}\|_C \to 0$ and there exists a compact set $\mathcal{C} \subset \mathcal{Q}$ such that $(t,x^0,u^0+\delta u) \in \mathcal{C}$ starting from a certain number, we have $\|\bar{\delta}_x f\|_\infty \to 0$. The conditions $\|\delta u\|_1 \to 0$ and $(t,x^0,u^0+\delta u) \in \mathcal{C}$ imply $\|\delta_u f\|_1 \to 0$. Consequently,

$$\|\delta \dot{x}\|_1 \le \|\delta \dot{x} - \delta f\|_1 + \|\delta f\|_1 \le \|\delta \dot{x} - \delta f\|_1 + \|\bar{\delta}_x f\|_\infty (t_f - t_0) + \|\delta_u f\|_1 \to 0.$$

The conditions $\|\delta \dot{x}\|_1 \to 0$ and $|\delta x(t_0)| \to 0$ imply $\|\delta x\|_{1,1} \to 0$. Therefore, $\{\delta w\} \in \Pi$. The inclusion $\bar{\Pi}^S \subset \Pi$ is proved. The converse inclusion always holds. Therefore, $\bar{\Pi}^S = \Pi$. The theorem is proved. ∎

Lemma 2.73 and Theorem 2.95 imply the following theorem.

Theorem 2.99. *Let the set* $\mathrm{Leg}_+(M_0^+)$ *be nonempty. Then* $\Pi = \bar{\Pi}^S$, *and hence the Pontryagin minimum is equivalent to the bounded strong minimum.*

Proof. Assume that $\mathrm{Leg}_+(M_0^+)$ is nonempty. Choose an arbitrary compact set $M \subset \mathrm{Leg}_+(\mathrm{M}_0^+)$, e.g., a singleton. According to Lemma 2.73, there exist an admissible function $\Gamma(t,u)$ and a constant $C > 0$ such that $M \subset M(C\Gamma)$. Therefore, $M(C\Gamma)$ is nonempty. Then $\Pi = \bar{\Pi}^S$ by Theorem 2.95. The theorem is proved. ∎

For a point w^0 and the higher order γ corresponding to an admissible function Γ, we give the following definition.

Definition 2.100. We say that the point w^0 is a *point of bounded strong* γ*-sufficiency* if there is no sequence $\{\delta w\} \in \bar{\Pi}^S$ such that $\sigma = o(\gamma)$ on it.

The condition that the set $\mathrm{Leg}_+(M_0^+)$ is nonempty is a counterpart of Condition $\mathfrak{B}$. Therefore, Theorems 2.82 and 2.99 imply the following theorem.

Theorem 2.101. *Condition* $\mathfrak{B}$ *is equivalent to the existence of an admissible function* Γ *such that the bounded strong* γ*-sufficiency holds at the point* w^0 *for the higher order* γ *corresponding to it.*

The bounded strong γ-sufficiency implies the strict bounded strong minimum. Therefore, Theorem 2.101 implies the following theorem.

Theorem 2.102. *Condition* $\mathfrak{B}$ *is sufficient for the strict bounded strong minimum at the point* w^0.

At this point, we complete the consideration of conditions for the bounded strong minimum. Before passing to sufficient conditions for the strong minimum, we prove some estimate for the function Φ, which will be needed in what follows.

2.7.3 Estimate for the Function Φ on Pontryagin Sequences

Recall that in Section 2.1, we introduced the set Λ_0^Θ consisting of those $\lambda \in \Lambda_0$ for which $[H^\lambda]^k = 0$ for all $t_k \in \Theta$, and in Section 2.3, we showed that there exists a constant $C_\Theta > 0$ such that the following estimate holds for any sequence $\{\delta w\} \in \Pi^{\mathrm{loc}}$ starting from a certain number:

$$\max_{\mathrm{co}\,\Lambda_0^\Theta} |\Phi(\lambda, \delta w)| \leq C_\Theta \gamma(\delta w)$$

(see Proposition 2.16). Let us show that the same estimate also holds on any Pontryagin sequence but with a constant depending on the order γ and the sequence.

Lemma 2.103. *Let the set Λ_0^Θ be nonempty. Let $\Gamma(t,u)$ be an admissible function, and let γ be the higher order corresponding to it. Then for any sequence $\{\delta w\} \in \Pi$, there exists a constant $C > 0$ such that*

$$\max_{\mathrm{co}\,\Lambda_0^\Theta} |\Phi(\lambda, \delta w)| \leq C\gamma(\delta w).$$

Briefly, this property will be written as $\max_{\mathrm{co}\,\Lambda_0^\Theta} |\Phi| \leq O(\gamma) \mid \Pi$.

Proof. Proposition 2.16 implies the following assertion: there exist constants $C > 0$ and $\varepsilon > 0$ and a neighborhood V of the compact set $\overline{u^0}$ such that the conditions $\delta w \in W$, $\|\delta x\|_C \leq \varepsilon$, and $(t, u^0 + \delta u) \in V$ imply the estimate $\max_{\mathrm{co}\,\Lambda_0^\Theta} |\Phi(\lambda, \delta w)| \leq C\gamma(\delta w)$. Let us use this estimate. Let $\{\delta w\}$ be an arbitrary sequence from Π. For each member $\delta w = (\delta x, \delta u)$ of the sequence $\{\delta w\}$, represent δu as $\delta u = \delta u_V + \delta u^V$, where

$$\delta u_V = \begin{cases} \delta u & \text{if } (t, u^0 + \delta u) \in V, \\ 0 & \text{otherwise,} \end{cases} \qquad \delta u^V = \delta u - \delta u_V.$$

We set $\delta w_V = (\delta x, \delta u_V)$ and $\delta w^V = (0, \delta u^V)$. Owing to a possible decrease of V, we can assume that both sequences $\{\delta w_V\}$ and $\{\delta w^V\}$ are admissible with respect to $\mathcal{Q}$; i.e., the conditions $(t, x^0 + \delta x, u^0 + \delta u_V) \in \mathcal{Q}$ and $(t, x^0, u^0 + \delta u^V) \in \mathcal{Q}$ hold starting from a certain number (such a possibility follows from the definition of Pontryagin sequence). We assume that this condition holds for all numbers and $\|\delta x\|_C \leq \varepsilon$ holds for all numbers. We set $\mathcal{M}^V = \{t \mid \delta u^V \neq 0\}$. The definitions of admissible function $\Gamma(t,u)$ and Pontryagin sequence imply the existence of constants $0 < a < b$ such that $a \leq \Gamma(t, u^0 + \delta u^V) \leq b \mid \mathcal{M}^V$ for all members of the sequence. This implies $a \cdot \mathrm{meas}\,\mathcal{M}^V \leq \gamma^V \leq b \cdot \mathrm{meas}\,\mathcal{M}^V$, where $\gamma^V = \int_{t_0}^{t_f} \Gamma(t, u^0 + \delta u^V)\,dt = \gamma(\delta w^V)$. Therefore, γ^V and $\mathrm{meas}\,\mathcal{M}^V$ are of the same order of smallness. Moreover, the definitions of γ, δw_V, and δw^V imply $\gamma(\delta w) = \gamma(\delta w_V) + \gamma(\delta w^V)$, or, briefly, $\gamma = \gamma_V + \gamma^V$. In what follows, we will need the formula

$$\delta f = \delta_V f + \bar{\delta}^V f, \tag{2.249}$$

where $\delta f = f(t, w^0 + \delta w) - f(t, w^0)$, $\delta_V f = f(t, w^0 + \delta w_V) - f(t, w^0)$, and $\bar{\delta}^V f = f(t, x^0 + \delta x, u^0 + \delta u^V) - f(t, x^0 + \delta x, u^0)$. The fulfillment of this formula is proved by the following calculation:

$$\begin{aligned}
\delta f &= f(t, x^0 + \delta x, u^0 + \delta u) - f(t, x^0 + \delta x, u^0 + \delta u_V) + \delta_V f \\
&= \big(f(t, x^0 + \delta x, u^0 + \delta u) - f(t, x^0 + \delta x, u^0 + \delta u_V)\big)\chi^V + \delta_V f \\
&= \bar{\delta}^V f \chi^V + \delta_V f = \bar{\delta}^V f + \delta_V f,
\end{aligned}$$

where χ^V is the characteristic function of the set $\mathcal{M}^V$. Formula (2.249) implies the following representation for $\Phi(\lambda,\delta w)$ on the sequence $\{\delta w\}$:

$$\begin{aligned}\Phi(\lambda,\delta w) &:= \delta l^\lambda - \int_{t_0}^{t_f} \psi\delta\dot{x}\,dt + \int_{t_0}^{t_f} \psi\delta f\,dt \\ &= \delta l^\lambda - \int_{t_0}^{t_f} \psi\delta\dot{x}\,dt + \int_{t_0}^{t_f} \psi\delta_V f\,dt + \int_{t_0}^{t_f} \psi\bar{\delta}^V f\,dt \\ &= \Phi(\lambda,\delta w_V) + \int_{t_0}^{t_f} \bar{\delta}^V H^\lambda\,dt,\end{aligned}$$

where $\bar{\delta}^V H^\lambda = \psi\bar{\delta}^V f$ and $\lambda \in \operatorname{co}\Lambda_0$. This implies the estimate

$$\max_{\operatorname{co}\Lambda_0^\Theta} |\Phi(\lambda,\delta w)| \le \max_{\operatorname{co}\Lambda_0^\Theta} |\Phi(\lambda,\delta w_V)| + \max_{\operatorname{co}\Lambda_0^\Theta} \|\psi\|_\infty \int_{t_0}^{t_f} |\bar{\delta}^V f|\,dt.$$

According to the choice of V and ε, the first term of the sum on the right-hand side of the inequality is estimated through $\gamma(\delta w_V)$. The second term of this sum is estimated through meas $\mathcal{M}^V$ and hence through γ^V. Since $\gamma_V \le \gamma$ and $\gamma^V \le \gamma$, the total sum is estimated through γ with a certain positive constant as a multiplier. The lemma is proved. ∎

2.7.4 Sufficient Conditions for the Strong Minimum

In this section, we assume that the set $\mathcal{Q}$ has the form $\mathcal{Q} = \mathcal{Q}_t \times \mathcal{Q}_x \times \mathcal{Q}_u$, where $\mathcal{Q}_t \subset \mathbb{R}$, $\mathcal{Q}_x \subset \mathbb{R}^{d(x)}$, and $\mathcal{Q}_u \subset \mathbb{R}^{d(u)}$ are open sets. Set $\mathcal{Q}_{tu} = \mathcal{Q}_t \times \mathcal{Q}_u$. We now give those additional requirements which, together with Condition $\mathfrak{B}$, turn out to be sufficient for the strong minimum whose definition was given in Section 2.7.1. For $(t,x,u) \in \mathcal{Q}$, we set $\bar{\delta}_u f = f(t,x,u) - f(t,x,u^0(t))$. Further, for $(t,x,u) \in \mathcal{Q}$ and $\lambda = (\alpha_0,\alpha,\beta,\psi) \in \Lambda_0$, we set $\bar{\delta}_u H^\lambda = \psi\bar{\delta}_u f$. The following theorem holds.

Theorem 2.104. *Let the following conditions hold for the point* w^0:

(1) *There exists a nonempty compact set* $M \subset \operatorname{Leg}_+(M_0^+)$ *such that*

  (a) *for a certain* $C > 0$, $\max_{\lambda\in M} \Omega^\lambda(\bar{z}) \ge C\bar{\gamma}(\bar{z})$ *for all* $\bar{z} \in \mathcal{K}$, *i.e., Condition* $\mathfrak{B}$ *holds*;

  (b) *for any* $\varepsilon > 0$, *there exist* $\delta > 0$ *and a compact set* $\mathfrak{C} \subset \mathcal{Q}_{tu}$ *such that for all* $t \in [t_0,t_f]\setminus\Theta$, *the conditions* $(t,w) \in \mathcal{Q}$, $|\underline{x} - \underline{x}^0(t)| \le \delta$, $(t,u) \notin \mathfrak{C}$ *imply* $\min_{\lambda\in M} \bar{\delta}_u H^\lambda \ge -\varepsilon|\bar{\delta}_u f|$;

(2) *there exist* $\delta_0 > 0$, $\varepsilon_0 > 0$, *a compact set* $\mathfrak{C}_0 \subset \mathcal{Q}_{tu}$, *and an element* $\lambda_0 \in M_0$ *such that for all* $t \in [t_0,t_f]\setminus\Theta$ *the conditions* $(t,w) \in \mathcal{Q}$, $|\underline{x} - \underline{x}^0(t)| < \delta_0$, $(t,u) \notin \mathfrak{C}_0$ *imply* $\bar{\delta}_u H^{\lambda_0} \ge \varepsilon_0|\bar{\delta}_u f| > 0$.

Then w^0 *is a strict strong minimum point.*

Remark 2.105. For the point w^0, let there exist δ_0, ε_0, $\mathfrak{C}_0$, and λ_0 satisfying condition (2) of Theorem 2.104. Moreover, let $\lambda_0 \in \operatorname{Leg}_+(M_0^+)$, and for a certain $C > 0$, let $\Omega^{\lambda_0}(\bar{z}) \ge C\bar{\gamma}(\bar{z})$ for all $\bar{z} \in \mathcal{K}$. Then, as is easily seen, all the conditions of Theorem 2.104 hold, and, therefore, w^0 is a strict strong minimum point.

2.7.5 Proof of Theorem 2.104

Assume that for a subset $M \subset \mathrm{Leg}_+(M_0^+)$, $C > 0$, $\delta_0 > 0$, $\varepsilon_0 > 0$, $\mathcal{C}_0 \subset \mathcal{Q}_{tu}$, $\lambda_0 \in M_0$, all conditions of the theorem hold, but there is no strict strong minimum at the point w^0. Let us show that this leads to a contradiction. Since w^0 is not a strict strong minimum point, there exists a sequence $\{\delta w\}$ such that $|\delta x(t_0)| + \|\underline{\delta x}\|_C \to 0$ (i.e., $\{\delta w\} \in \Pi^S$), and the following conditions hold for all members of this sequence: $(p^0+\delta p) \in \mathcal{P}$, $(t, w^0+\delta w) \in \mathcal{Q}$, $\sigma(\delta w) = 0$, $\delta w \neq 0$. The condition $\sigma(\delta w) = 0$ implies

$$\delta \dot{x} - \delta f = 0, \quad \delta K = 0, \quad \delta J \leq 0, \quad F(p^0+\delta p) \leq 0.$$

Hence, for any $\lambda \in \Lambda_0$, we have the following on the sequence $\{\delta w\}$:

$$\Phi(\lambda, \delta w) = \delta l^\lambda \leq \sigma(\delta w) = 0. \tag{2.250}$$

Assume that there is an arbitrary compact set $\mathcal{C}$ satisfying the condition

$$\mathcal{C}_0 \subset \mathcal{C} \subset \mathcal{Q}_{tu}. \tag{2.251}$$

For each member $\delta w = (\delta x, \delta u)$ of the sequence $\{\delta w\}$, we set

$$\delta u^{\mathcal{C}} = \begin{cases} \delta u & \text{if } (t, u^0 + \delta u) \notin \mathcal{C}, \\ 0 & \text{otherwise}, \end{cases}$$

$$\delta w^{\mathcal{C}} = (0, \delta u^{\mathcal{C}}), \qquad \delta u_{\mathcal{C}} = \delta u - \delta u^{\mathcal{C}}, \qquad \delta w_{\mathcal{C}} = (\delta x, \delta u_{\mathcal{C}}).$$

Then $\{\delta w\} = \{\delta w_{\mathcal{C}}\} + \{\delta w^{\mathcal{C}}\}$. The relation $\delta H^\lambda = \delta_{\mathcal{C}} H^\lambda + \bar{\delta}^{\mathcal{C}} H^\lambda$, $\lambda \in \Lambda_0$, where

$$\begin{aligned}
\delta_{\mathcal{C}} H^\lambda &= H^\lambda(t, w^0 + \delta w_{\mathcal{C}}) - H^\lambda(t, w^0), \\
\bar{\delta}^{\mathcal{C}} H^\lambda &= H^\lambda(t, w^0 + \delta w) - H^\lambda(t, w^0 + \delta w_{\mathcal{C}}) \\
&= H^\lambda(t, x^0 + \delta x, u^0 + \delta u) - H^\lambda(t, x^0 + \delta x, u^0 + \delta u_{\mathcal{C}}) \\
&= H^\lambda(t, x^0 + \delta x, u^0 + \delta u^{\mathcal{C}}) - H^\lambda(t, x^0 + \delta x, u^0),
\end{aligned}$$

corresponds to this representation of the sequence $\{\delta w\}$. This and condition (2.250) imply that for any $\lambda \in \Lambda_0$, we have the following inequality on the sequence $\{\delta w\}$:

$$\Phi^\lambda(\delta w_{\mathcal{C}}) + \int_{t_0}^{t_f} \bar{\delta}^{\mathcal{C}} H^\lambda \, dt \leq 0. \tag{2.252}$$

We set $\gamma_{\mathcal{C}} = \gamma(\delta w_{\mathcal{C}})$, $\bar{\delta}^{\mathcal{C}} f = f(t, x^0 + \delta x, u^0 + \delta u^{\mathcal{C}}) - f(t, x^0 + \delta x, u^0)$, $\varphi^{\mathcal{C}} = \int_{t_0}^{t_f} |\bar{\delta}^{\mathcal{C}} f| \, dt$. Since $\mathcal{C} \supset \mathcal{C}_0$ and $\|\underline{\delta x}\|_C \to 0$, condition (2) of the theorem implies $\varphi^{\mathcal{C}} > 0$ for all nonzero members of the sequence $\{\delta w^{\mathcal{C}}\}$ with sufficiently large numbers.

Proposition 2.106. *The following conditions hold*: (a) $\varphi^{\mathcal{C}} \to 0$, (b) $\{\delta w_{\mathcal{C}}\} \in \Pi$, *and hence* $\gamma_{\mathcal{C}} \to 0$.

Proof. By (2.252), we have the following for the sequence $\{\delta w\}$ and the element $\lambda = \lambda_0$:

$$\delta l - \int_{t_0}^{t_f} \psi \delta \dot{x} \, dt + \int_{t_0}^{t_f} \delta_{\mathcal{C}} H \, dt + \int_{t_0}^{t_f} \bar{\delta}^{\mathcal{C}} H \, dt \leq 0 \tag{2.253}$$

(we omit $\lambda = \lambda_0$ in this proof). Represent $\delta_{\mathcal{C}} H$ in the form

$$\delta_{\mathcal{C}} H = H(t, w^0 + \delta w_{\mathcal{C}}) - H(t, w^0) = \hat{\delta}_{\mathcal{C}x} H + \delta_{\mathcal{C}u} H,$$

where

$$\hat{\delta}_{\mathcal{C}x} H = H(t, x^0 + \delta x, u^0 + \delta u_{\mathcal{C}}) - H(t, x^0, u^0 + \delta u_{\mathcal{C}}),$$
$$\delta_{\mathcal{C}u} H = H(t, x^0, u^0 + \delta u_{\mathcal{C}}) - H(t, x^0, u^0).$$

Then we obtain from (2.253) that

$$\delta l - \int_{t_0}^{t_f} \psi \delta \dot{x}\, dt + \int_{t_0}^{t_f} \hat{\delta}_{\mathcal{C}x} H\, dt + \int_{t_0}^{t_f} \delta_{\mathcal{C}u} H\, dt + \int_{t_0}^{t_f} \bar{\delta}^{\mathcal{C}} H\, dt \le 0. \tag{2.254}$$

The conditions $\|\underline{\delta x}\|_C \to 0$ and $(t, u^0 + \delta u_{\mathcal{C}}) \in \mathcal{C}$ imply

$$\int_{t_0}^{t_f} \hat{\delta}_{\mathcal{C}x} H\, dt \to 0. \tag{2.255}$$

Further, the condition $\|\underline{\delta x}\|_C \to 0$ also imply

$$\begin{aligned} \delta l - \int_{t_0}^{t_f} \psi \delta \dot{x}\, dt &= \delta l - \psi \delta x \,\Big|_{t_0}^{t_f} + \int_{t_0}^{t_f} \dot{\psi} \delta x\, dt \\ &= \delta l - l_p \delta p - \int_{t_0}^{t_f} H_x \delta x\, dt \to 0. \end{aligned} \tag{2.256}$$

Conditions (2.254)–(2.256) imply

$$\left(\int_{t_0}^{t_f} \delta_{\mathcal{C}u} H\, dt + \int_{t_0}^{t_f} \bar{\delta}^{\mathcal{C}} H\, dt \right)_+ \to 0. \tag{2.257}$$

Since $\lambda = \lambda_0$ and $\mathcal{C} \supset \mathcal{C}_0$, according to assumption (2) of the theorem, the following inequalities hold for all members of the sequence $\{\delta w^{\mathcal{C}}\}$ with sufficiently large numbers:

$$\int_{t_0}^{t_f} \bar{\delta}^{\mathcal{C}} H\, dt \ge \varepsilon_0 \varphi^{\mathcal{C}} > 0. \tag{2.258}$$

Conditions (2.257) and (2.258) imply

$$\left(\int_{t_0}^{t_f} \delta_{\mathcal{C}u} H\, dt + \varepsilon_0 \varphi^{\mathcal{C}} \right)_+ \to 0. \tag{2.259}$$

Since $\lambda = \lambda_0 \in M_0$ and $(t, x^0, u^0 + \delta u_{\mathcal{C}}) \in \mathcal{Q}$, we have $\delta_{\mathcal{C}u} H \ge 0$. Therefore, both terms in (2.259) are nonnegative for all sufficiently large numbers of the sequence $\{\delta w\}$. But then (2.259) implies $\int_{t_0}^{t_f} \delta_{\mathcal{C}u} H\, dt \to 0$, $\varphi^{\mathcal{C}} \to 0$.

We now show that $\{\delta w_{\mathcal{C}}\} \in \bar{\Pi}^S$. For this purpose, we prove that $\sigma(\delta w_{\mathcal{C}}) \to 0$ (the other conditions of $\{\delta w_{\mathcal{C}}\}$ belonging to the set of sequences $\bar{\Pi}^S$ obviously hold). Since $\sigma(\delta w) = 0$, we need show only that $\|\delta \dot{x} - \delta_{\mathcal{C}} f\|_1 \to 0$. We have the following for the sequence $\{\delta w\}$:

$$\delta \dot{x} - \delta f = 0, \qquad \delta f = \delta_{\mathcal{C}} f + \bar{\delta}^{\mathcal{C}} f. \tag{2.260}$$

The condition $\varphi^{\mathcal{C}} \to 0$ means that $\|\bar{\delta}^{\mathcal{C}} f\|_1 \to 0$. This and (2.260) imply $\|\delta\dot{x} - \delta_{\mathcal{C}} f\|_1 \to 0$. Therefore, $\{\delta w_{\mathcal{C}}\} \in \bar{\Pi}^S$.

We now recall that, by condition (1) of the theorem, the set $\mathrm{Leg}_+(M_0^+)$ is nonempty. By Theorem 2.99, it follows from this that $\bar{\Pi}^S = \Pi$. Therefore, $\{\delta w_{\mathcal{C}}\} \in \Pi$. The proposition is proved. ∎

We continue the proof of the theorem. Consider the following two possible cases for the sequence $\{\delta w\}$.

Case (a). Assume that there exist a compact set $\mathcal{C}$ satisfying conditions (2.251) and a subsequence of the sequence $\{\delta w\}$ such that the following conditions hold on this subsequence:

$$\varphi^{\mathcal{C}} > 0, \qquad \gamma_{\mathcal{C}} = o(\varphi^{\mathcal{C}}). \tag{2.261}$$

Assume that these conditions hold for the sequence $\{\delta w\}$ itself. Inequality (2.252) and the conditions $\lambda_0 \in M_0 \subset \Lambda_0$ imply that the following inequality holds on the sequence $\{\delta w\}$:

$$\int_{t_0}^{t_f} \bar{\delta}^{\mathcal{C}} H^{\lambda_0}\, dt \le -\Phi^{\lambda_0}(\delta w_{\mathcal{C}}). \tag{2.262}$$

As was already mentioned in the proof of Proposition 2.106, the following inequalities hold for all members of the sequence $\{\delta w\}$ having sufficiently large numbers:

$$\varepsilon_0 \varphi^{\mathcal{C}} \le \int_{t_0}^{t_f} \bar{\delta}^{\mathcal{C}} H^{\lambda_0}\, dt. \tag{2.263}$$

On the other hand, according to Lemma 2.103, the conditions $\{\delta w_{\mathcal{C}}\} \in \Pi$, $\lambda_0 \in M_0 \subset \Lambda_0^{\Theta}$ imply the estimate

$$-\Phi^{\lambda_0}(\delta w_{\mathcal{C}}) \le O(\gamma_{\mathcal{C}}). \tag{2.264}$$

We obtain from (2.261)–(2.264) that $0 < \varphi^{\mathcal{C}} \le O(\gamma_{\mathcal{C}}) = o(\varphi^{\mathcal{C}})$. This is a contradiction.

Case (b). Consider the second possibility. Assume that for any compact set $\mathcal{C}$ satisfying conditions (2.251), there exists a constant $N > 0$ such that the following estimate holds on the sequence $\{\delta w\}$:

$$\varphi^{\mathcal{C}} \le N\gamma_{\mathcal{C}}. \tag{2.265}$$

We show that this also leads to a contradiction. We will thus prove the theorem. First of all, we note that the constant N in (2.265) can be chosen common for all compact sets $\mathcal{C}$ satisfying conditions (2.251). Indeed, let N_0 correspond to a compact set $\mathcal{C}_0$, i.e., the estimate $\varphi^{\mathcal{C}_0} \le N_0\gamma_{\mathcal{C}_0}$ holds on the sequence $\{\delta w\}$. Let $\mathcal{C}$ be an arbitrary compact set such that (2.251) hold. Then we have the following on the sequence $\{\delta w\}$ for all sufficiently large numbers: $\varphi^{\mathcal{C}} \le \varphi^{\mathcal{C}_0} \le N_0\gamma_{\mathcal{C}_0} \le N_0\gamma_{\mathcal{C}}$. Therefore, $N = N_0$ is also appropriate for $\mathcal{C}$. Also, we note that for any $\mathcal{C}$ satisfying (2.251), there exists a (serial) number of the sequence starting from which $\gamma_{\mathcal{C}} > 0$. Indeed, otherwise, there exist a compact set $\mathcal{C}$, satisfying (2.251) and a subsequence of the sequence $\{\delta w\}$ such that $\gamma_{\mathcal{C}} = 0$ on the subsequence, and then $\varphi^{\mathcal{C}} = 0$ by (2.265). By assumption (2) of the theorem, this implies that all members of the subsequence vanish. The latter is impossible, since the sequence $\{\delta w\}$ contains nonzero members by assumption.

Now let the compact set $\mathcal{C}$ satisfy conditions (2.251). Inequality (2.252) and the inclusion $M \subset \Lambda_0$ imply that the following inequality holds on the sequence $\{\delta w\}$:

$$\max_M \Phi(\lambda, \delta w_{\mathcal{C}}) \le -\min_M \int_{t_0}^{t_f} \bar{\delta}^{\mathcal{C}} H^{\lambda}\, dt. \tag{2.266}$$

Obviously,

$$\min_M \int_{t_0}^{t_f} \bar{\delta}^{\mathcal{C}} H^{\lambda}\, dt \ge \int_{t_0}^{t_f} \min_M \bar{\delta}^{\mathcal{C}} H^{\lambda}\, dt. \tag{2.267}$$

Condition 1(b) of the theorem implies that, for any $\varepsilon > 0$, there exists a compact set $\mathcal{C}$ satisfying (2.251) such that

$$\int_{t_0}^{t_f} \min_M \bar{\delta}^{\mathcal{C}} H^{\lambda}\, dt \ge -\varepsilon \cdot \varphi^{\mathcal{C}} \tag{2.268}$$

for all sufficiently large numbers of the sequence. We obtain from (2.265)–(2.268) that for any $\varepsilon > 0$, there exists a compact set $\mathcal{C}$ satisfying (2.251) such that following estimate holds starting from a certain number:

$$\max_M \Phi(\lambda, \delta w_{\mathcal{C}}) \le \varepsilon N \gamma_{\mathcal{C}}, \tag{2.269}$$

where $N = N_0$ is independent of $\mathcal{C}$.

We now estimate the left-hand side of inequality (2.269) from below. For this purpose, we show that for any compact set $\mathcal{C}$ satisfying (2.251), the sequence $\{\delta w_{\mathcal{C}}\}$ belongs to $\Pi_{o(\sqrt{\gamma})}$. Let $\mathcal{C}$ be an arbitrary compact set satisfying (2.251). According to Proposition 2.106, $\{\delta w_{\mathcal{C}}\} \in \Pi$. Since $\sigma(\delta w) = 0$, we have $(\delta J)^+ + \sum F_i^+(p^0 + \delta p) + |\delta K| = 0$ on the whole sequence. Moreover, the conditions $\delta \dot{x} = \delta f$, $\delta f = \delta_{\mathcal{C}} f + \bar{\delta}^{\mathcal{C}} f$, $\|\bar{\delta}^{\mathcal{C}} f\|_1 = \varphi^{\mathcal{C}} \le O(\gamma_{\mathcal{C}})$ imply $\|\delta \dot{x} - \delta_{\mathcal{C}} f\|_1 \le O(\gamma_{\mathcal{C}})$. Therefore, $\{\delta w_{\mathcal{C}}\} \in \Pi_{\sigma\gamma} \subset \Pi_{o(\sqrt{\gamma})}$. But then, by Lemma 2.83, condition 1(a) of the theorem implies that there exists a constant $C_M > 0$ such that, starting from a certain number,

$$\max_M \Phi(\lambda, \delta w_{\mathcal{C}}) \ge C_M \gamma_{\mathcal{C}} > 0. \tag{2.270}$$

Moreover, the constant C_M is independent of the sequence from $\Pi_{o(\sqrt{\gamma})}$ and hence is independent of $\mathcal{C}$. Comparing estimates (2.269) and (2.270), we obtain the following result: For any $\varepsilon > 0$, there exists a compact set $\mathcal{C}$ satisfying (2.251) such that, starting from a certain number, $0 < C_M \gamma_{\mathcal{C}} \le \varepsilon N \gamma_{\mathcal{C}}$. Choosing $0 < \varepsilon < C_M/N$, we obtain a contradiction. The theorem is proved. ∎

Chapter 3

Quadratic Conditions for Optimal Control Problems with Mixed Control-State Constraints

In Sections 3.1 and 3.2 of this chapter, following [92], we extend the quadratic conditions obtained in Chapter 2 to the general problem with the local relation $g(t,x,u) = 0$ using a special method of projection contained in [79]. In Section 3.3, we extend these conditions to the problem on a variable interval of time using a simple change of time variable. In Section 3.4, we formulate (without proofs) quadratic conditions in an optimal control problem with the local relations $g(t,x,u) = 0$ and $\varphi(t,x,u) \leq 0$.

3.1 Quadratic Necessary Conditions in the Problem with Mixed Control-State Equality Constraints on a Fixed Time Interval

3.1.1 Statement of the Problem with a Local Equality and Passage to an Auxiliary Problem without Local Constraints

We consider the following problem on a fixed interval $[t_0, t_f]$ with a local equality-type constraint:

$$\begin{gathered} J(x(t_0), x(t_f)) \to \min, \\ F(x(t_0), x(t_f)) \leq 0, \quad K(x(t_0), x(t_f)) = 0, \quad (x(t_0), x(t_f)) \in \mathcal{P}, \end{gathered} \tag{3.1}$$

$$\dot{x} = f(t,x,u), \quad g(t,x,u) = 0, \quad (t,x,u) \in \mathcal{Q}. \tag{3.2}$$

It is assumed that the functions J, F, and K are twice continuously differentiable on the open set $\mathcal{P} \subset \mathbb{R}^{2d(x)}$, and f and g are twice continuously differentiable on the open set $\mathcal{Q} \subset \mathbb{R}^{1+d(x)+d(u)}$. Moreover, the following full-rank condition is assumed for the local equality:

$$\operatorname{rank} g_u(t,x,u) = d(g) \tag{3.3}$$

for all $(t,x,u) \in \mathcal{Q}$ such that $g(t,x,u) = 0$.

As in Section 2.1.2, we define a (strict) minimum on a set of sequences S: w^0 is a *(strict) minimum point on* S in problem (3.1), (3.2) if there is no sequence $\{\delta w\} \in S$ such that the following conditions hold for all its members:

$$J(p^0+\delta p) < J(p^0) \quad (J(p^0+\delta p) \le J(p^0),\ \delta w \ne 0),$$
$$F(p^0+\delta p) \le 0, \quad K(p^0+\delta p) = 0, \quad \dot{x}^0+\delta\dot{x} = f(t, w^0+\delta w),$$
$$g(t, w^0+\delta w) = 0, \quad (p^0+\delta p) \in \mathcal{P}, \quad (t, w^0+\delta w) \in \mathcal{Q},$$

where $p^0 = (x^0(t_0), x^0(t_f))$, $\delta w = (\delta x, \delta u)$, and $\delta p = (\delta x(t_0), \delta x(t_f))$. A (strict) minimum on Π (see Section 2.1.3) is said to be a *(strict) Pontryagin minimum.*

Our goal is to obtain quadratic conditions in problem (3.1), (3.2) using the quadratic conditions obtained in problem (2.1)–(2.4). In this case, we will use the same method for passing to a problem without local constraints, which was already used in [79, Section 17, Part 1] for obtaining first-order conditions. Recall that in [79, Section 17, Part 1], we have introduced the set

$$\mathcal{G} = \{(t,x,u) \in \mathcal{Q} \mid g(t,x,u) = 0\}.$$

We have shown that there exist a neighborhood $\mathcal{Q}_1 \subset \mathcal{Q}$ of the set $\mathcal{G}$ and a continuously differentiable function $U(t,x,u) : \mathcal{Q}_1 \to \mathbb{R}^{d(x)}$ such that

$$\begin{aligned} &\text{(i)} \quad (t,x,U(t,x,u)) \in \mathcal{G} \quad \forall\, (t,x,u) \in \mathcal{Q}_1, \\ &\text{(ii)} \quad U(t,x,u) = u \quad \forall\, (t,x,u) \in \mathcal{G}. \end{aligned} \tag{3.4}$$

Owing to these properties, U is called a *projection.* Since g is a twice continuously differentiable function on $\mathcal{Q}$, we can choose the function U, together with the neighborhood $\mathcal{Q}_1$, so that U is a twice continuously differentiable function on $\mathcal{Q}_1$. This can be easily verified by analyzing the scheme for proving the existence of a projection presented in [79, Section 17, Part 1]. We fix certain $\mathcal{Q}_1$ and U with the above properties. Instead of system (3.2) with the local equality-type constraint $g(t,x,u) = 0$, consider the following system without local constraint:

$$\dot{x} = f(t,x,U(t,x,u)), \quad (t,x,u) \in \mathcal{Q}_1. \tag{3.5}$$

We find a connection between the necessary conditions in the problem (3.1), (3.2) and those in the problem (3.1), (3.5). Preparatorily, we prove the following assertion.

Proposition 3.1. *Let* $(x^0,u^0) = w^0$ *be a Pontryagin minimum point in the problem* (3.1), (3.2). *Then* w^0 *is a Pontryagin minimum point in the problem* (3.1), (3.5).

Proof. The property that w^0 is a Pontryagin minimum point in the problem (3.1), (3.2) implies that w^0 is an admissible point in the problem (3.1), (3.2), and then by the second property (3.4) of the projection, w^0 is also admissible in the problem (3.1), (3.5). Suppose that w^0 is not a Pontryagin minimum point in the problem (3.1), (3.5). Then there exist $\mathcal{C} \in \mathcal{Q}_1$ and a sequence $\{\delta w\} = \{(\delta x, \delta u)\}$ such that

$$\|\delta x\|_{1,1} \to 0, \quad \|\delta u\|_1 \to 0, \tag{3.6}$$

and the following conditions hold for all members of the sequence:

$$(t, w^0+\delta w) \in \mathcal{C}, \tag{3.7}$$
$$\delta J < 0, \quad F(p^0+\delta p) \le 0, \quad K(p^0+\delta p) = 0, \tag{3.8}$$
$$\dot{x}^0+\delta\dot{x} = f(t, x^0+\delta x,\ U(t, x^0+\delta x, u^0+\delta u)), \tag{3.9}$$

where $p^0 = (x^0(t_0), x^0(t_f))$ and $\delta p = (\delta x(t_0), \delta x(t_f))$. Therefore, $\{\delta w\}$ is a Pontryagin sequence "violating the Pontryagin minimum" at the point w^0 in the problem (3.1), (3.5). We set $\{\delta w^1\} = \{(\delta x, \delta u^1)\}$, where

$$\delta u^1 = U(t, x^0 + \delta x, u^0 + \delta u) - U(t, x^0, u^0) = U(t, x^0 + \delta x, u^0 + \delta u) - u^0.$$

We show that $\{\delta w^1\}$ is a Pontryagin sequence "violating the Pontryagin minimum" at w^0 in the problem (3.1), (3.2).

First of all, we show that $\{\delta w^1\}$ is a Pontryagin sequence for system (3.2) at w^0. For this purpose, we represent δu^1 in the form

$$\begin{aligned} \delta u^1 \quad = \quad & U(t, x^0 + \delta x, u^0 + \delta u) - U(t, x^0, u^0 + \delta u) \\ & + U(t, x^0, u^0 + \delta u) - U(t, x^0, u^0). \end{aligned} \tag{3.10}$$

This representation is correct, since the conditions

$$\|\delta x\|_C \to 0 \quad \text{and} \quad (t, x^0 + \delta x, u^0 + \delta u) \in \mathcal{C} \subset \mathcal{Q}_1 \tag{3.11}$$

imply $(t, x^0, u^0 + \delta u) \in \mathcal{Q}_1$ for all sufficiently large serial numbers (here, we use the compactness of $\mathcal{C}$ and the openness of $\mathcal{Q}_1$). Representation (3.10) implies

$$\begin{aligned} \|\delta u^1\|_1 \quad \le \quad & \|U(t, x^0 + \delta x, u^0 + \delta u) - U(t, x^0, u^0 + \delta u)\|_\infty (t_f - t_0) \\ & + \|U(t, x^0, u^0 + \delta u) - U(t, x^0, u^0)\|_1. \end{aligned}$$

This, condition (3.11), and also the condition $\|\delta u\|_1 \to 0$ imply $\|\delta u^1\|_1 \to 0$. Further, denote by $\mathcal{C}_1$ the image of the compact set $\mathcal{C}$ under the mapping $(t, x, u) \mapsto (t, x, U(t, x, u))$. Then $\mathcal{C} \subset \mathcal{G}$, and hence $\mathcal{C}_1 \subset \mathcal{Q}_1$. Moreover, $\mathcal{C}_1$ is a compact set and

$$(t, x^0 + \delta x, u^0 + \delta u^1) = (t, x^0 + \delta x, U(t, x^0 + \delta x, u^0 + \delta u)) \in \mathcal{C}_1.$$

This implies $g(t, x^0 + \delta x, u^0 + \delta u^1) = 0$. Finally, we note that the sequence $\{w^0 + \delta w^1\}$ satisfies the differential equation of system (3.2):

$$\dot{x}^0 + \delta \dot{x} = f(t, x^0 + \delta x, U(t, x^0 + \delta x, u^0 + \delta u)) = f(t, x^0 + \delta x, u^0 + \delta u^1)$$

and the local equality constraint

$$g(t, x^0 + \delta x, u^0 + \delta u^1) = g(t, x^0 + \delta x, U(t, x^0 + \delta x, u^0 + \delta u)) = 0.$$

Therefore, we have shown that the mapping

$$(\delta x, \delta u) \mapsto \big(\delta x, U(t, x^0 + \delta x, u^0 + \delta u) - u^0\big)$$

transforms the Pontryagin sequence $\{\delta w\}$ of the system (3.1), (3.5) at the point w^0 into the Pontryagin sequence $\{\delta w^1\}$ of the system (3.1), (3.2) at the same point. Since the sequence $\{\delta w\}$ "violates" the Pontryagin minimum in the problem (3.1), (3.5) at the point w^0, conditions (3.8) hold. But these conditions can be also referred to as the sequence $\{\delta w^1\}$, since the members δx of these two sequence coincide. Therefore, $\{\delta w^1\}$ "violates" the Pontryagin minimum at the point w^0 in the problem (3.1), (3.2). Therefore, the absence of the Pontryagin minimum at the point w^0 in the problem (3.1), (3.5) implies the same case in the problem (3.1), (3.2). This implies what was required. ∎

Let $w^0 = (x^0, u^0)$ be a fixed point of the Pontryagin minimum in the problem (3.1), (3.2). Then w^0 is a Pontryagin minimum point in the problem (3.1), (3.5). We now write the quadratic necessary conditions for the Pontryagin minimum at the point w^0 in the problem (3.1), (3.5) (which contains no local constraints) so that the projection U can be excluded from these conditions. Then we will obtain the quadratic necessary conditions for the Pontryagin minimum in the problem (3.1), (3.2).

3.1.2 Set M_0

For the problem (3.1), (3.2), we set $l = \alpha_0 J + \alpha K$, $H = \psi f$, and $\bar{H} = H + \nu g$, where $\nu \in (\mathbb{R}^{d(g)})^*$. Therefore, $\bar{H} = \bar{H}(t,x,u,\psi,\nu)$. We also set

$$\mathcal{H}(t,x,\psi) = \min_{\{u \mid (t,x,u)\in \mathcal{G}\}} \psi f(t,x,u).$$

For problem (3.1), (3.2) and the point w^0, we introduce the set M_0 consisting of tuples $\lambda = (\alpha_0, \alpha, \beta, \psi(t), \nu(t))$ such that

$$\alpha_0 \geq 0, \quad \alpha \geq 0, \quad \alpha F(p^0) = 0, \quad \alpha_0 + |\alpha| + |\beta| = 1, \tag{3.12}$$

$$\psi(t_0) = -l_{x_0}, \quad \psi(t_f) = l_{x_f}, \tag{3.13}$$

$$-\dot{\psi} = \bar{H}_x, \quad \bar{H}_u = 0, \tag{3.14}$$

$$H(t,x^0,u^0,\psi) = \mathcal{H}(t,x^0,\psi). \tag{3.15}$$

Here, $\psi(t)$ is an absolutely continuous function and $\nu(t)$ is a bounded measurable function. All the derivatives are taken for $p = p^0$ and $w = w^0(t)$. The results of [79, Section 17, Part 1] imply that the set M_0 of problem (3.1), (3.5) at the point w^0 can be represented in this form. More precisely, the linear projection $(\alpha_0,\alpha,\beta,\psi,\nu) \mapsto (\alpha_0,\alpha,\beta,\psi)$ yields a one-to-one correspondence between the elements of the set M_0 of the problem (3.1), (3.2) at the point w^0 and the elements of the set M_0 of the problem (3.1), (3.5) at the same point. To differentiate these two sets from one another, we denote the latter set by M_0^U. We will equip all objects referring to the problem (3.1), (3.5) with the superscript U.

In what follows, we will assume that all assumptions of Section 2.1 hold for the point w^0, i.e., $u^0(t)$ is a piecewise continuous function whose set of discontinuity points is $\Theta = \{t_1,\dots,t_s\} \subset (t_0,t_f)$, and each point of the set Θ is an L-point. The condition

$$-f_u^*(t,x^0(t),u^0(t))\psi^*(t) = g_u^*(t,x^0(t),u^0(t))\nu^*(t), \tag{3.16}$$

which is equivalent to the condition $\bar{H}_u = 0$, and also the full-rank condition (3.3) imply that $\nu(t)$ has the same properties as $u^0(t)$: the function $\nu(t)$ is piecewise continuous and each of its point of discontinuity is an L-point which belongs to Θ. To verify this, it suffices to premultiply the above relation by the matrix $g_u(t,x^0(t),u^0(t))$,

$$-g_u(t,x^0(t),u^0(t))f_u^*(t,x^0(t),u^0(t))\psi^*(t) = g_u(t,x^0(t),u^0(t))g_u^*(t,x^0,(t),u^0(t))\nu^*(t),$$

and use the properties of the functions g, f, x^0, and u^0; in particular, the property

$$|\det g_u(t,x^0,u^0)g_u^*(t,x^0,u^0)| \geq \text{const} > 0,$$

which is implied by the full-rank condition (3.3). Therefore, the basic properties of the function $\nu(t)$ are proved exactly in the same way as in [79, Section 17, Part 1] for the bounded measurable control $u^0(t)$.

3.1.3 Critical Cone

We now consider the conditions defining critical cone at the point w^0 in the problem (3.1), (3.5). The variational equation has the form

$$\dot{\bar{x}} = (f_x + f_u U_x)\bar{x} + f_u U_u \bar{u}. \tag{3.17}$$

All the derivatives are taken for $w = w^0(t)$. Setting $\tilde{u} = U_x\bar{x} + U_u\bar{u}$, we obtain

$$\dot{\bar{x}} = f_x\bar{x} + f_u\tilde{u}. \tag{3.18}$$

This is the usual variational equation, but for the pair $\tilde{w} = (\bar{x}, \tilde{u})$. Let us show that the pair $(\bar{x}, \tilde{u})$ also satisfies the condition

$$g_x\bar{x} + g_u\tilde{u} = 0. \tag{3.19}$$

By the first condition in (3.4), we have

$$g(t,x,U(t,x,u)) = 0 \quad \forall\,(t,x,u) \in \mathcal{Q}_1. \tag{3.20}$$

Differentiating this relation in x, u, and t, as in [79, Section 17, Part 1], we obtain

$$g_x + g_u U_x = 0, \tag{3.21}$$

$$g_u U_u = 0, \tag{3.22}$$

$$g_t + g_u U_t = 0. \tag{3.23}$$

These relations hold on $\mathcal{Q}_1$, but it suffices to consider them only on the trajectory $(t, x^0(t), u^0(t))$, and, moreover, we now need only the first two conditions. By (3.21) and (3.22), we have

$$g_x\bar{x} + g_u\tilde{u} = g_x\bar{x} + g_u(U_x\bar{x} + U_u\bar{u}) = (g_x + g_u U_x)\bar{x} + g_u U_u\bar{u} = 0.$$

Therefore, we have proved the following proposition.

Proposition 3.2. *Let a pair of functions $(\bar{x}, \bar{u})$ satisfy the variational equation* (3.17) *of system* (3.5). *We set $\tilde{u} = U_x\bar{x} + U_u\bar{u}$. Then conditions* (3.18) *and* (3.19) *hold for $(\bar{x}, \tilde{u})$.*

In this proposition, $\bar{x} \in P_\Theta W^{1,2}(\Delta, \mathbb{R}^{d(x)})$, $\bar{u} \in L^2(\Delta, \mathbb{R}^{d(u)})$, and $\tilde{u} \in L^2(\Delta, \mathbb{R}^{d(u)})$, where $\Delta = [t_0, t_f]$. Also, we are interested in the possibility of the converse passage from conditions (3.18) and (3.19) to condition (3.17). For this purpose, we prove the following proposition.

Proposition 3.3. *Let a pair of functions $(\bar{x}, \tilde{u})$ be such that $g_x\bar{x} + g_u\tilde{u} = 0$. Then setting $\bar{u} = \tilde{u} - U_x\bar{x}$, we obtain $U_u\bar{u} = \bar{u}$, and hence $\tilde{u} = U_x\bar{x} + U_u\bar{u}$. Here, as above, $\bar{x} \in P_\Theta W^{1,2}(\Delta, \mathbb{R}^{d(x)})$, $\bar{u} \in L^2(\Delta, \mathbb{R}^{d(u)})$, and $\tilde{u} \in L^2(\Delta, \mathbb{R}^{d(u)})$; all the derivatives are taken for $x = x^0(t)$ and $u = u^0(t)$.*

Proof. First of all, we note that properties (3.4) of the function $U(t,x,u)$ imply the following assertion: at each point $(t,x,u) \in \mathcal{G}$, the finite-dimensional linear operator

$$\bar{u} \in \mathbb{R}^{d(u)} \mapsto U_u(t,x,u)\bar{u} \in \mathbb{R}^{d(u)} \tag{3.24}$$

is the linear projection of the space $\mathbb{R}^{d(u)}$ on the subspace

$$L_g(t,x,u) = \{\bar{u} \in \mathbb{R}^{d(u)} \mid g_u(t,x,u)\bar{u} = 0\}. \tag{3.25}$$

Indeed, let $(t,x,u) \in \mathcal{G}$, and then condition (3.22) implies that the image of operator (3.24) is contained in subspace (3.25). The condition

$$U_u(t,x,u)\bar{u} = \bar{u} \quad \forall\, \bar{u} \in L_g(t,x,u) \tag{3.26}$$

is easily proved by using the Lyusternik theorem [28]. Indeed, let $\bar{u} \in L_g(t,x,u)$, i.e., $g_u(t,x,u)\bar{u} = 0$. Let $\varepsilon \to +0$. Then

$$g(t,x,u+\varepsilon\bar{u}) = g(t,x,u) + g_u(t,x,u)\varepsilon\bar{u} + r_g(\varepsilon) = r_g(\varepsilon),$$

where $|r_g(\varepsilon)| = o(\varepsilon)$. By the Lyusternik theorem [28], there exists a "sequence" $\{u_g(\varepsilon)\}$ such that $u_g(\varepsilon) \in \mathbb{R}^{d(u)}$, $|u_g(\varepsilon)| = o(\varepsilon)$, and, moreover, $g(t,x,u+\varepsilon\bar{u}+u_g(\varepsilon)) = 0$, i.e., $(t,x,u+\varepsilon\bar{u}+u_g(\varepsilon)) \in \mathcal{G}$. Hence $U(t,x,u+\varepsilon\bar{u}+u_g(\varepsilon)) = u+\varepsilon\bar{u}+u_g(\varepsilon)$. But $U(t,x,u+\varepsilon\bar{u}+u_g(\varepsilon)) = U(t,x,u)+U_u(t,x,u)(\varepsilon\bar{u}+u_g(\varepsilon))+r_U(\varepsilon)$, where $|r_U(\varepsilon)| = o(\varepsilon)$. We obtain from the latter two conditions and the condition $U(t,x,u) = u$ that $\varepsilon\bar{u}+u_g(\varepsilon) = U_u(t,x,u)(\varepsilon\bar{u}+u_g(\varepsilon))+r_U(\varepsilon)$. Dividing this relation by ε and passing to the limit as $\varepsilon \to +0$, we obtain $\bar{u} = U_u(t,x,u)\bar{u}$. Condition (3.26) is proved. Condition (3.26) holds at each point $(t,x,u) \in \mathcal{G}$, but we use it only at the trajectory $(t,x^0(t),u^0(t)) \mid t \in [t_0,t_f]$.

If a pair of functions $\bar{x}(t), \tilde{u}(t)$ satisfies the condition $g_x\bar{x}+g_u\tilde{u} = 0$, then by (3.21), we have $-g_uU_x\bar{x}+g_u\tilde{u} = 0$, i.e., $-U_x\bar{x}+\tilde{u} \in L_g(t,x^0,u^0)$. Then the condition $U_u\bar{u} = \bar{u}$ also holds for $\bar{u} = -U_x\bar{x}+\tilde{u}$, and hence $\tilde{u} = U_x\bar{x}+U_u\bar{u}$. The proposition is proved. ∎

Proposition 3.3 implies the following assertion.

Proposition 3.4. *Let a pair of functions* $(\bar{x},\tilde{u})$ *be such that conditions* (3.18) *and* (3.19) *hold*: $\dot{\bar{x}} = f_x\bar{x}+f_u\tilde{u}$ *and* $g_x\bar{x}+g_u\tilde{u} = 0$. *We set* $\bar{u} = -U_x\bar{x}+\tilde{u}$. *Then* $U_u\bar{u} = \bar{u}$, *and the variational equation* (3.17) *of system* (3.5) *holds for the pair of functions* $(\bar{x},\bar{u})$ *at the point* w^0.

Proof. Indeed,

$$\dot{\bar{x}} = f_x\bar{x}+f_u\tilde{u} = f_x\bar{x}+f_u(U_x\bar{x}+\bar{u}) = f_x\bar{x}+f_u(U_x\bar{x}+U_u\bar{u}) = (f_x+f_uU_x)\bar{x}+f_uU_u\bar{u}$$

as required. The proposition is proved. ∎

We now give the following definition.

Definition 3.5. The *critical cone* $\mathcal{K}$ of problem (3.1), (3.2) at the point w^0 is the set of triples $\bar{z} = (\bar{\xi},\bar{x},\bar{u})$ satisfying the following conditions:

$$\bar{\xi} \in \mathbb{R}^s, \quad \bar{x} \in P_\Theta W^{1,2}(\Delta, \mathbb{R}^{d(x)}), \quad \bar{u} \in L^2(\Delta, \mathbb{R}^{d(u)}), \tag{3.27}$$
$$J_p \bar{p} \le 0, \quad F_{ip} \bar{p} \le 0 \quad \forall\, i \in I, \quad K_p \bar{p} = 0, \tag{3.28}$$
$$\dot{\bar{x}} = f_x \bar{x} + f_u \bar{u}, \tag{3.29}$$
$$[\bar{x}]^k = [f]^k \bar{\xi}_k \quad \forall\, t_k \in \Theta, \tag{3.30}$$
$$g_x \bar{x} + g_u \bar{u} = 0. \tag{3.31}$$

Let us compare this definition with the definition of the critical cone $\mathcal{K}^U$ of the problem (3.1), (3.5) at the point w^0. According to Section 2.1, the latter is defined by the same conditions (3.27) and (3.28), the variational equation (3.17), and the jump condition (3.30). Thus, all the conditions referring to the components $\bar{\xi}$ and $\bar{x}$ in the definitions of two critical cones coincide. This and Proposition 3.2 imply the following assertion.

Lemma 3.6. *Let $\bar{z} = (\bar{\xi}, \bar{x}, \bar{u})$ be an arbitrary element of the critical cone $\mathcal{K}^U$ of the problem* (3.1), (3.5) *at the point w^0. We set $\tilde{u} = U_x \bar{x} + U_u \bar{u}$. Then $\tilde{z} = (\bar{\xi}, \bar{x}, \tilde{u})$ is an element of the critical cone $\mathcal{K}$ of the problem* (3.1), (3.2) *at the point w^0.*

Respectively, Proposition 3.4 implies the following lemma.

Lemma 3.7. *Let $\tilde{z} = (\bar{\xi}, \bar{x}, \tilde{u})$ be an arbitrary element of the critical cone $\mathcal{K}$ of the problem* (3.1), (3.2) *at the point w^0. We set $\bar{u} = \tilde{u} - U_x \bar{x}$. Then $\bar{z} = (\bar{\xi}, \bar{x}, \bar{u})$ is an element of the critical cone $\mathcal{K}^U$ of the problem* (3.1), (3.5) *at the point w^0, and, moreover, $U_u \bar{u} = \bar{u}$, which implies $\tilde{u} = U_x \bar{x} + U_u \bar{u}$.*

This is the connection between the critical cones at the point w^0 in the problems (3.1), (3.2) and (3.1), (3.5). We will need Lemma 3.6 later in deducing quadratic sufficient conditions in the problem with a local equality; now, in deducing necessary conditions, we use Lemma 3.7. Preparatorily, we find the connection between the corresponding quadratic forms.

3.1.4 Quadratic Form

We write the quadratic form $\Omega^U(\bar{z})$ for the point w^0 in the problem (3.1), (3.5) in accordance with its definition in Section 2.1. Let $\lambda^U = (\alpha_0, \alpha, \beta, \psi)$ be an element of the set M_0^U of the problem (3.1), (3.5) at the point w^0, and let $\lambda = (\alpha_0, \alpha, \beta, \psi, \nu)$ be the corresponding element of the set M_0 of the problem (3.1), (3.2) at the same point. As above, we set

$$H(t,x,u,\psi) = \psi f(t,x,u); \quad \bar{H}(t,x,u,\psi) = \psi f(t,x,u) + \nu g(t,x,u).$$

Also, we introduce the notation

$$f^U(t,x,u) = f(t,x,U(t,x,u)), \tag{3.32}$$
$$H^U(t,x,u,\psi) = \psi f^U(t,x,u) = H(t,x,U(t,x,u),\psi). \tag{3.33}$$

We omit the superscripts λ and λ^U in the notation. For each $t \in [t_0, t_f]$, let us calculate the quadratic form

$$\langle H^U_{ww} \bar{w}, \bar{w} \rangle = \langle H^U_{xx} \bar{x}, \bar{x} \rangle + 2\langle H^U_{xu} \bar{u}, \bar{x} \rangle + \langle H^U_{uu} \bar{u}, \bar{u} \rangle, \tag{3.34}$$

where $\bar{x} \in \mathbb{R}^{d(x)}$, $\bar{u} \in \mathbb{R}^{d(u)}$, and all the second derivatives are taken at the point $(t,x,u) = (t,x^0(t),u^0(t))$.

Let us calculate $\langle H^U_{xx}\bar{x},\bar{x}\rangle$. It follows from (3.33) that

$$H^U_x = H_x + H_u U_x. \tag{3.35}$$

Differentiating this equation in x and twice multiplying it by $\bar{x}$, we obtain

$$\begin{aligned}\langle H^U_{xx}\bar{x},\bar{x}\rangle &= \langle H_{xx}\bar{x},\bar{x}\rangle + 2\langle H_{xu}(U_x\bar{x}),\bar{x}\rangle + \langle H_{uu}(U_x\bar{x}),(U_x\bar{x})\rangle \\ &\quad + \langle (H_u U_{xx})\bar{x},\bar{x}\rangle,\end{aligned} \tag{3.36}$$

where $\langle (H_u U_{xx})\bar{x},\bar{x}\rangle = \langle (\sum_{i=1}^{d(u)} H_{u_i} U_{ixx})\bar{x},\bar{x}\rangle$ by definition. Further, let us calculate $\langle H^U_{xu}\bar{u},\bar{x}\rangle$. Differentiating (3.35) in u and multiplying it by $\bar{u}$ and $\bar{x}$, we obtain

$$\langle H^U_{xu}\bar{u},\bar{x}\rangle = \langle H_{xu}(U_u\bar{u}),\bar{x}\rangle + \langle H_{uu}(U_u\bar{u}),(U_x\bar{x})\rangle + \langle (H_u U_{xu})\bar{u},\bar{x}\rangle, \tag{3.37}$$

where $\langle (H_u U_{xu})\bar{u},\bar{x}\rangle = \langle (\sum_{i=1}^{d(u)} H_{u_i} U_{ixu})\bar{u},\bar{x}\rangle$. Finally, let us calculate $\langle H^U_{uu}\bar{u},\bar{u}\rangle$. It follows from (3.33) that $H^U_u = H_u U_u$. Differentiating this equation in u and twice multiplying it by $\bar{u}$, we obtain

$$\langle H^U_{uu}\bar{u},\bar{u}\rangle = \langle H_{uu}(U_u\bar{u}),(U_u\bar{u})\rangle + \langle (H_u U_{uu})\bar{u},\bar{u}\rangle, \tag{3.38}$$

where $\langle (H_u U_{uu})\bar{u},\bar{u}\rangle = \langle (\sum_{i=1}^{d(u)} H_{u_i} U_{iuu})\bar{u},\bar{u}\rangle$. Formulas (3.34), (3.36)–(3.38) imply

$$\begin{aligned}\langle H^U_{ww}\bar{w},\bar{w}\rangle &= \langle H_{xx}\bar{x},\bar{x}\rangle + 2\langle H_{xu}(U_x\bar{x}),\bar{x}\rangle + \langle H_{uu}(U_x\bar{x}),(U_x\bar{x})\rangle \\ &\quad + \langle (H_u U_{xx})\bar{x},\bar{x}\rangle + 2\langle H_{xu}(U_u\bar{u}),\bar{x}\rangle + 2\langle H_{uu}(U_u\bar{u}),(U_x\bar{x})\rangle \\ &\quad + 2\langle (H_u U_{xu})\bar{u},\bar{x}\rangle + \langle H_{uu}(U_u\bar{u}),(U_u\bar{u})\rangle + \langle (H_u U_{uu})\bar{u},\bar{u}\rangle \\ &= \langle H_{xx}\bar{x},\bar{x}\rangle + 2\langle H_{xu}\tilde{u},\bar{x}\rangle + \langle H_{uu}\tilde{u},\tilde{u}\rangle \\ &\quad + \langle (H_u U_{xx}\bar{x},\bar{x}\rangle + 2\langle (H_u U_{xu})\bar{u},\bar{x}\rangle + \langle (H_u U_{uu})\bar{u},\bar{u}\rangle,\end{aligned} \tag{3.39}$$

where $\tilde{u} = U_x\bar{x} + U_u\bar{u}$. Further, differentiating in x the relations

$$g_{ix} + g_{iu}U_x = 0, \qquad i = 1,\dots,d(g), \tag{3.40}$$

which hold on $\mathcal{Q}_1$, and twice multiplying the result by $\bar{x}$, we obtain

$$\langle g_{ixx}\bar{x},\bar{x}\rangle + 2\langle g_{ixu}(U_x\bar{x}),\bar{x}\rangle + \langle g_{iuu}(U_x\bar{x}),(U_x\bar{x})\rangle + \langle g_{iu}U_{xx}\bar{x},\bar{x}\rangle = 0,$$

$i = 1,\dots,d(g)$. Multiplying each of these relations by the ith component ν_i of the vector-valued function $\nu(t)$, summing with respect to i, and using the relation $H_u + \nu g_u = 0$, we obtain

$$\langle \nu g_{xx}\bar{x},\bar{x}\rangle + 2\langle \nu g_{xu}(U_x\bar{x}),\bar{x}\rangle + \langle \nu g_{uu}(U_x\bar{x}),(U_x\bar{x})\rangle - \langle H_u U_{xx}\bar{x},\bar{x}\rangle = 0, \tag{3.41}$$

where $\nu g_{xx} = \sum \nu_i g_{ixx}$, $\nu g_{xu} = \sum \nu_i g_{ixu}$, and $\nu g_{uu} = \sum \nu_i g_{iuu}$. Differentiating the same relations (3.40) in u and multiplying by $\bar{u}$ and $\bar{x}$, we obtain

$$\langle g_{ixu}(U_u\bar{u}),\bar{x}\rangle + \langle g_{iuu}(U_u\bar{u}),(U_x\bar{x})\rangle + \langle (g_{iu}U_{xu})\bar{u},\bar{x}\rangle = 0.$$

Multiplying each of these relations by $2\nu_i$, summing with respect to i, and using the property that $H_u + \nu g_u = 0$, we obtain

$$2\langle \nu g_{xu}(U_u\bar{u}), \bar{x}\rangle + 2\langle \nu g_{uu}(U_u\bar{u}), (U_x\bar{x})\rangle - 2\langle (H_u U_{xu})\bar{u}, \bar{x}\rangle = 0, \tag{3.42}$$

where $\nu g_{xu} = \sum \nu_i g_{ixu}$ and $\nu g_{uu} = \sum \nu_i g_{iuu}$. Finally, differentiating in u the relations $g_{iu}U_u = 0$, $i = 1,\dots,d(g)$, which hold on $\mathcal{Q}_1$, and twice multiplying the result by $\bar{u}$, we obtain $\langle g_{iuu}(U_u\bar{u}), (U_u\bar{u})\rangle + \langle (g_{iu}U_{uu})\bar{u}, \bar{u}\rangle = 0$. Multiplying each of these equations by ν_i and using the property that $H_u + \nu g_u = 0$, we obtain

$$\langle \nu g_{uu}(U_u\bar{u}), (U_u\bar{u})\rangle - \langle H_u U_{uu}\bar{u}, \bar{u}\rangle = 0. \tag{3.43}$$

This and (3.38) imply

$$\langle H^U_{uu}\bar{u}, \bar{u}\rangle = \langle H_{uu}(U_u\bar{u}), (U_u\bar{u})\rangle + \langle \nu g_{uu}(U_u\bar{u}), (U_u\bar{u})\rangle. \tag{3.44}$$

Using the notation $\bar{H} = H + \nu g$, we present this relation in the form

$$\langle H^U_{uu}\bar{u}, \bar{u}\rangle = \langle \bar{H}_{uu}(U_u\bar{u}), (U_u\bar{u})\rangle. \tag{3.45}$$

We will use this relation later in Section 3.2. Summing relations (3.41)–(3.43), we obtain

$$\begin{aligned}&\langle \nu g_{xx}\bar{x}, \bar{x}\rangle + 2\langle \nu g_{xu}\tilde{u}, \bar{x}\rangle + 2\langle \nu g_{uu}\tilde{u}, \tilde{u}\rangle \\ &\quad - \langle H_u U_{xx}\bar{x}, \bar{x}\rangle - 2\langle H_u U_{xu}\bar{u}, \bar{x}\rangle - \langle H_u U_{uu}\bar{u}, \bar{u}\rangle = 0,\end{aligned} \tag{3.46}$$

where $\tilde{u} = U_x\bar{x} + U_u\bar{u}$. It follows from (3.39) and (3.46) that $\langle H^U_{ww}\bar{w}, \bar{w}\rangle = \langle H_{ww}\tilde{w}, \tilde{w}\rangle + \langle \nu g_{ww}\tilde{w}, \tilde{w}\rangle$ or

$$\langle H^U_{ww}\bar{w}, \bar{w}\rangle = \langle \bar{H}_{ww}\tilde{w}, \tilde{w}\rangle, \tag{3.47}$$

where $\tilde{w} = (\bar{x}, \tilde{u}) = (\bar{x}, U_x\bar{x} + U_u\bar{u})$.

We now consider the terms referring to the points of discontinuity of the control u^0. We set

$$D^k(\bar{H}) = -\bar{H}^{k+}_x \bar{H}^{k-}_\psi + \bar{H}^{k-}_x \bar{H}^{k+}_\psi - [\bar{H}_t]^k, \quad k = 1,\dots,s, \tag{3.48}$$

where

$$\begin{aligned}
\bar{H}^{k+}_x &= \bar{H}_x(t_k, x^0(t_k), u^{0k+}, \psi(t_k), \nu^{k+}), \quad \bar{H}^{k-}_x = \bar{H}_x(t_k, x^0(t_k), u^{0k-}, \psi(t_k), \nu^{k-}),\\
\bar{H}^{k+}_\psi &= f(t_k, x^0(t_k), u^{0k+}) = H^{k+}_\psi, \quad \bar{H}^{k-}_\psi = f(t_k, x^0(t_k), u^{0k-}) = H^{k-}_\psi,\\
[\bar{H}_t]^k &= \bar{H}^{k+}_t - \bar{H}^{k-}_t = \psi(t_k)[f_t]^k + [\nu g_t]^k\\
&= \psi(t_k)\big(f_t(t_k, x^0(t_k), u^{0k+}) - f_t(t_k, x^0(t_k), u^{0k-})\big)\\
&\quad + \nu^{k+} g_t(t_k, x^0(t_k), u^{0k+}) - \nu^{k-} g_t(t_k, x^0(t_k), u^{0k-}),\\
\nu^{k+} &= \nu(t_k+), \quad \nu^{k-} = \nu(t_k-).
\end{aligned}$$

Therefore, the definition of $D^k(\bar{H})$ is analogous to that of $D^k(H)$.

We can define $D^k(\bar{H})$ using another method, namely, as the derivative of the "jump of $\bar{H}$" at the point t_k. Introduce the function

$$\begin{aligned}
(\Delta_k\bar{H})(t) &= (\Delta_k H)(t) + (\Delta_k(\nu g))(t)\\
&= \psi(t)\big(f(t, x^0(t), u^{0k+}) - f(t, x^0(t), u^{0k-})\big)\\
&\quad + \big(\nu^{k+} g(t, x^0(t), u^{0k+}) - \nu^{k-} g(t, x^0(t), u^{0k-})\big).
\end{aligned} \tag{3.49}$$

Similarly to what was done for $(\Delta_k H)(t)$ in Section 2.3 (see Lemma 2.12), we can show that the function $(\Delta_k \bar{H})(t)$ is continuously differentiable at the point $t_k \in \Theta$, and its derivative at this point coincides with $-D^k(\bar{H})$. Therefore, we can obtain the value of $D^k(\bar{H})$ calculating the left or right limit of the derivatives of the function $(\Delta_k \bar{H})(t)$ defined by formula (3.49):

$$D^k(\bar{H}) = -\frac{d}{dt}(\Delta_k \bar{H})(t_k).$$

We now show that

$$D^k(\bar{H}) = D^k(H^U), \quad k = 1, \dots, s. \tag{3.50}$$

Indeed, by definition,

$$-D^k(H^U) = H_x^{Uk+} H_\psi^{Uk-} - H_x^{Uk-} H_\psi^{Uk+} + [H_t^U]^k. \tag{3.51}$$

Furthermore,

$$H_x^U = H_x + H_u U_x = H_x - \nu g_u U_x = H_x + \nu g_x = \bar{H}_x. \tag{3.52}$$

Here, we have used the formulas $H_u + \nu g_u = 0$, $g_x + g_u U_x = 0$, and $\bar{H} = H + \nu g$. Also, it is obvious that for $(t, w) = (t, w^0(t))$,

$$H_\psi^U = f^U = f = H_\psi = \bar{H}_\psi. \tag{3.53}$$

Finally,

$$H_t^U = H_t + H_u U_t = H_t - \nu g_u U_t = H_t + \nu g_t = \bar{H}_t, \tag{3.54}$$

since $g_t + g_u U_t = 0$. The formulas (3.48), (3.51)–(3.54) imply relation (3.50). Further, note that relations (3.52) imply

$$[H_x^U]^k = [\bar{H}_x]^k, \quad k = 1, \dots, s, \tag{3.55}$$

where

$$[\bar{H}_x]^k = \bar{H}_x(t_k, x^0(t_k), u^{0k+}, \psi(t_k), \nu^{k+}) - \bar{H}_x(t_k, x^0(t_k), u^{0k-}, \psi(t_k), \nu^{k-}) \tag{3.56}$$

is the jump of the function $\bar{H}_x(t, x^0(t), u^0(t), \psi(t), \nu(t))$ at the point $t_k \in \Theta$.

For the problem (3.1), (3.2) and the point w^0, we define the following quadratic form in $\bar{z} = (\bar{\xi}, \bar{x}, \bar{u})$ for each $\lambda = (\alpha_0, \alpha, \beta, \psi, \nu) \in M_0$:

$$2\Omega^\lambda(\bar{z}) = \sum_{k=1}^{s} \left(D^k(\bar{H}^\lambda)\bar{\xi}_k^2 + 2[\bar{H}_x^\lambda]^k \bar{x}_{\mathrm{av}}^k \bar{\xi}_k \right) + \langle l_{pp}^\lambda \bar{p}, \bar{p} \rangle + \int_{t_0}^{t_f} \langle \bar{H}_{ww}^\lambda \bar{w}, \bar{w} \rangle \, dt. \tag{3.57}$$

Therefore, the quadratic form in the problem with local equality-type constraints is defined in the same way as in the problem without local constraints; the only difference is that instead of the function $H = \psi f$ in the definition of the new quadratic form, we must use the function $\bar{H} = H + \nu g$.

According to Section 2.1, the quadratic form takes the following form for the problem (3.1), (3.5) and the point w^0:

$$2\Omega^{U\lambda^U}(\bar{z}) = \sum_{k=1}^{s}\left(D^k\left(H^{U\lambda^U}\right)\bar{\xi}_k^2+2\left[H_x^{U\lambda^U}\right]\bar{x}_{\text{av}}^k\bar{\xi}_k\right) + \langle l_{pp}^{\lambda^U}\bar{p},\bar{p}\rangle + \int_{t_0}^{t_f}\langle H^{U\lambda^U}\bar{w},\bar{w}\rangle\,dt. \tag{3.58}$$

Here, as above, we have used the superscript U in the notation Ω^U of the quadratic form in order to stress that this quadratic form corresponds to the problem (3.1), (3.5) being considered for a given projection $U(t,x,u)$. We have denoted by λ^U the tuple $(\alpha_0,\alpha,\beta,\psi)$, which uniquely defines the tuple $\lambda=(\alpha_0,\alpha,\beta,\psi,\nu)$ by the condition $\psi f_u+\nu g_u=0$. In what follows, these tuples correspond to one another and belong to the sets M_0^U and M_0 of the problems (3.1), (3.5) and (3.1), (3.2), respectively. Formulas (3.47), (3.50), (3.55), (3.57), and (3.58) imply the following assertion.

Lemma 3.8. *Let $\bar{z}=(\bar{\xi},\bar{x},\bar{u})$ be an arbitrary element of the space $Z_2(\Theta)$, and let $\tilde{z}=(\bar{\xi},\bar{x},\tilde{u})=(\bar{\xi},\tilde{w})$, where $\tilde{u}=U_x\bar{x}+U_u\bar{u}$. Let λ^U be an arbitrary element of M_0^U, and let λ be the corresponding element of M_0. Then $\Omega^\lambda(\tilde{z})=\Omega^{U\lambda^U}(\bar{z})$.*

3.1.5 Necessary Quadratic Conditions

The following theorem holds.

Theorem 3.9. *If w^0 is a Pontryagin minimum point in the problem* (3.1), (3.2), *then the following Condition $\mathfrak{A}$ holds: the set M_0 is nonempty and*

$$\max_{\lambda\in M_0}\Omega^\lambda(\bar{z})\geq 0\quad\forall\,\bar{z}\in\mathcal{K}, \tag{3.59}$$

where $\mathcal{K}$ is the critical cone at the point w^0 defined by conditions (3.27)–(3.31), *$\Omega^\lambda(\bar{z})$ is the quadratic form at the same point defined by* (3.57), *and M_0 is the set of tuples of Lagrange multipliers satisfying the minimum principle defined by* (3.12)–(3.15).

Proof. Let w^0 be a Pontryagin minimum point in the problem (3.1), (3.2). Then according to Proposition 3.1, w^0 is a Pontryagin minimum point in the problem (3.1), (3.5). Hence, by Theorem 2.4, the following necessary Condition $\mathfrak{A}^U$ holds at the point w^0 in the problem (3.1), (3.5): the set M_0^U is nonempty and

$$\max_{M_0^U}\Omega^{U\lambda^U}(\bar{z})\geq 0\quad\forall\,\bar{z}\in\mathcal{K}^U. \tag{3.60}$$

Let us show that the necessary Condition $\mathfrak{A}$ holds at the point w^0 in the problem (3.1), (3.2). Let $\tilde{z}=(\bar{\xi},\bar{x},\tilde{u})$ be an arbitrary element of the critical cone $\mathcal{K}$ at the point w^0 in the problem (3.1), (3.2). According to Lemma 3.7, there exists a function $\bar{u}(t)$ such that $\tilde{u}=U_x\bar{x}+U_u\bar{u}$, and, moreover, $\bar{z}=(\bar{\xi},\bar{x},\bar{u})$ is an element of the critical cone $\mathcal{K}^U$ at the point w^0 in the problem (3.1), (3.5) (we can set $\bar{u}=\tilde{u}-U_x\bar{x}$). Since the necessary

Condition $\mathfrak{A}^U$ holds, there exists an element $\lambda^U \in M_0^U$ such that $\Omega^{U\lambda^U}(\bar{z}) \geq 0$. According to Section 3.1.2, an element $\lambda \in M_0$ corresponds to the element $\lambda^U \in M_0^U$. Moreover, by Lemma 3.8, $\Omega^\lambda(\bar{z}) = \Omega^{U\lambda^U}(\bar{z})$. Therefore, $\Omega^\lambda(\bar{z}) \geq 0$. Since $\bar{z}$ is an arbitrary element of $\mathcal{K}$, Condition $\mathfrak{A}$ holds. The theorem is proved. ∎

Therefore, we have obtained the final form of the quadratic necessary condition for the Pontryagin minimum in the problem with local equality-type constraints, Condition $\mathfrak{A}$, in which there is no projection U. This condition is a natural generalization of the quadratic necessary Condition $\mathfrak{A}$ in the problem without local constraints.

3.2 Quadratic Sufficient Conditions in the Problem with Mixed Control-State Equality Constraints on a Fixed Time Interval

3.2.1 Auxiliary Problem V

Since there is the projection $U(t,x,u)$, the problem (3.1), (3.5) has the property that the strict minimum is not attained at any point. For this reason, the problem (3.1), (3.5) cannot be directly used for obtaining quadratic sufficient conditions that guarantee the strict minimum. To overcome this difficulty, we consider a new auxiliary problem adding the additional constraint

$$\int_{t_0}^{t_f} (u - U(t,x,u))^2 \, dt \leq 0,$$

where $(u-U)^2 = \langle u-U, u-U \rangle$. Representing this constraint as an endpoint constraint by introducing a new state variable y, we arrive at the following problem on a fixed interval $[t_0, t_f]$:

$$\begin{gathered} J(x_0, x_f) \to \min, \\ F(x_0,x_f) \leq 0, \quad K(x_0,x_f) = 0, \quad (x_0,x_f) \in \mathcal{P}, \end{gathered} \tag{3.61}$$

$$y_0 = 0, \qquad y_f \leq 0, \tag{3.62}$$

$$\dot{y} = \frac{1}{2}(u - U(t,x,u))^2, \tag{3.63}$$

$$\dot{x} = f(t,x,U(t,x,u)), \quad (t,x,u) \in \mathcal{Q}_1, \tag{3.64}$$

where $x_0 = x(t_0)$, $x_f = x(t_f)$, $y_0 = y(t_0)$, and $y_f = y(t_f)$.

Problem (3.61)–(3.64) is called the *auxiliary problem* V. By the superscript V we denote all objects referring to this problem. Therefore, the auxiliary problem V differs from the auxiliary problem (3.1), (3.5) or problem U by the existence of the additional constraints (3.62) and (3.63).

If (y,x,u) is an admissible triple in problem (3.61)–(3.64), then $y = 0$, and the pair $w = (x,u)$ satisfies the constraints of problem (3.1), (3.2). Indeed, (3.62) and (3.63) imply

$$y(t) = 0, \quad U(t,x(t),u(t)) = u(t), \tag{3.65}$$

and then

$$\dot{x}(t) - f(t,x(t),u(t)) = \dot{x}(t) - f(t,x(t),U(t,x(t),u(t))) = 0,$$
$$g(t,x(t),u(t)) = g(t,x(t),U(t,x(t),u(t))) = 0,$$

since $(t,x(t),u(t)) \in \mathcal{Q}_1$. The converse is also true: if $w = (x,u)$ is an admissible pair in problem (3.1), (3.2) and $y = 0$, then the triple (y,x,u) is admissible in Problem V, since conditions (3.65) hold for it.

3.2.2 Bounded Strong γ-Sufficiency

Fix an admissible point $w^0 = (x^0,u^0) \in W$ in problem (3.1), (3.2) satisfying the assumption of Section 2.1. Let $y^0 = 0$. Then $(y^0,w^0) = (0,w^0)$ is an admissible point in Problem V. Let $\Gamma(t,u)$ be the admissible function defined in Section 2.3 (see Definition 2.17). The higher order at the point $(0,w^0)$ in Problem V is defined by the relation

$$\gamma^V(\delta y,\delta w) = \|\delta y\|_C^2 + \|\delta x\|_C^2 + \int_{t_0}^{t_f} \Gamma(t,u^0+\delta u)\,dt = \|\delta y\|_C^2 + \gamma(\delta w). \tag{3.66}$$

The violation function is defined by the relation

$$\begin{aligned}\sigma^V(\delta y,\delta w) &= (\delta J)^+ + |F(p^0+\delta p)^+| + |\delta K| + \|\delta\dot{x} - \delta f^U\|_1 \\ &\quad + \|\delta\dot{y} - \frac{1}{2}(u^0+\delta u - U(t,x^0+\delta x,u^0+\delta u))^2\|_1 + (\delta y_f)^+ + |\delta y_0|,\end{aligned} \tag{3.67}$$

where

$$\begin{aligned}\delta f^U &= f^U(t,w^0+\delta w) - f^U(t,w^0) \\ &= f(t,x^0+\delta x,U(t,x^0+\delta x,u^0+\delta u)) - f(t,x^0,u^0).\end{aligned} \tag{3.68}$$

We will use the concept of a bounded strong minimum and also that of a bounded strong γ^V-sufficiency at the point $(0,w^0)$ in the auxiliary problem V. Let us introduce analogous concepts for problem (3.1), (3.2) at the point w^0.

Definition 3.10. We say that a point $w^0 = (x^0,u^0)$ is a point of *strict bounded strong minimum* in problem (3.1), (3.2) if there is no sequence $\{\delta w\}$ in the space W that does not contain zero members, there is no compact set $\mathcal{C}$ such that $|\delta x(t_0)| \to 0$, $\|\underline{\delta x}\|_C \to 0$, and the following conditions hold for all members of the sequence $\{\delta w\}$:

$$\begin{aligned}&(t,w^0+\delta w) \in \mathcal{C}, \quad \delta J = J(p^0+\delta p) - J(p^0) \le 0,\\ &F(p^0+\delta p) \le 0, \quad K(p^0+\delta p) = 0,\\ &\dot{x}^0+\delta\dot{x} = f(t,w^0+\delta w), \quad g(t,w^0+\delta w) = 0.\end{aligned}$$

As in Section 2.7 (see Definition 2.92), we denote by $\underline{\delta x}$ the tuple of essential components of the variation δx. (The definition of unessential components in problem (3.1), (3.2) is the same as in problem (2.1)–(2.4), but now neither functions f nor g depends on these components.)

Definition 3.11. We say that $w^0 = (x^0,u^0)$ is a point of *bounded strong γ-sufficiency* in problem (3.1), (3.2) if there is no sequence $\{\delta w\}$ in the space W without zero members and there is no compact set $\mathcal{C} \subset \mathcal{Q}$ such that

$$|\delta x(t_0)| \to 0, \quad \|\underline{\delta x}\|_C \to 0, \quad \sigma(\delta w) = o(\gamma(\delta w)), \tag{3.69}$$

and the following conditions hold for all members of the sequence $\{\delta w\}$:

$$g(t, w^0 + \delta w) = 0 \quad \text{and} \quad (t, w^0 + \delta w) \in \mathcal{C}. \tag{3.70}$$

Here,

$$\sigma(\delta w) = (\delta J)^+ + |F(p^0 + \delta p)^+| + |\delta K| + \|\delta \dot{x} - \delta f\|_1. \tag{3.71}$$

Therefore, the violation function $\sigma(\delta w)$ in problem (3.1), (3.2) contains no term related to the local constraint $g(t,x,u) = 0$; however, in the definition of the bounded strong γ-sufficiency of this problem, it is required that the sequence $\{w^0 + \delta w\}$ satisfies this constraint. The local constraint does not have the same rights as the other constraints. Obviously, the following assertion holds.

Proposition 3.12. *The bounded strong γ-sufficiency at the point w^0 in problem* (3.1), (3.2) *implies the strict bounded strong minimum.*

Our goal is to obtain a sufficient condition for the bounded strong γ-sufficiency in problem (3.1), (3.2) using the sufficient condition for the bounded strong γ^V-sufficiency in the auxiliary problem V without local constraints. For this purpose, we prove the following assertion.

Proposition 3.13. *Let a point $(0, w^0)$ be a point of bounded strong γ^V-sufficiency in problem V. Then w^0 is a point of bounded strong γ-sufficiency in problem* (3.1), (3.2).

Proof. Suppose that w^0 is not a point of the bounded strong γ-sufficiency in problem (3.1), (3.2). Then there exist a sequence $\{\delta w\}$ containing nonzero members and a compact set $\mathcal{C} \subset \mathcal{Q}$ such that conditions (3.69) and (3.70) hold. We show that in this case, $(0, w^0)$ is not a point of bounded strong γ^V-sufficiency in Problem V. Consider the sequence $\{(0, \delta w)\}$ with $\delta y = 0$. Condition (3.70) implies

$$U(t, x^0 + \delta x, u^0 + \delta u) = u^0 + \delta u, \tag{3.72}$$

and then

$$\begin{aligned} \delta f &= f(t, x^0 + \delta x, u^0 + \delta u) - f(t, x^0, u^0) \\ &= f(t, x^0 + \delta x, U(t, x^0 + \delta x, u^0 + \delta u)) - f(t, x^0, u^0) = \delta f^U. \end{aligned} \tag{3.73}$$

It follows from (3.66)–(3.73) that

$$\sigma^V(0, \delta w) = \sigma(\delta w) = o(\gamma(\delta w)) = o(\gamma^V(0, \delta w)).$$

Therefore, $(0, w^0)$ is not a point of the bounded strong γ^V-sufficiency in Problem V. The proposition is proved. ∎

Next, we formulate the main result of this section: the quadratic sufficient conditions for the bounded strong γ-sufficiency at the point w^0 in problem (3.1), (3.2). Then we show that these conditions guarantee the bounded strong γ^V-sufficiency in Problem V, and hence, by Proposition 3.13, this implies the bounded strong γ-sufficiency at the point w^0 in problem (3.1), (3.2). This is our program. We now formulate the main result.

3.2.3 Quadratic Sufficient Conditions

For an admissible point $w^0 = (x^0, u^0)$ in problem (3.1), (3.2), we give the following definition.

Definition 3.14. An element $\lambda = (\alpha_0, \alpha, \beta, \psi, \nu) \in M_0$ is said to be *strictly Legendre* if the following conditions hold:

(1) $D^k(\bar{H}^\lambda) > 0$ for all $t_k \in \Theta$;

(2) for any $t \in [t_0, t_f] \setminus \Theta$, the form

$$\langle \bar{H}_{uu}(t, w^0(t), \psi(t), \nu(t))\bar{u}, \bar{u}\rangle \tag{3.74}$$

quadratic in $\bar{u}$ is positive definite on the subspace of vectors $\bar{u} \in \mathbb{R}^{d(u)}$ such that

$$g_u(t, w^0(t))\bar{u} = 0; \tag{3.75}$$

(3) the following condition C^{k-} holds for each point $t_k \in \Theta$: the form

$$\langle \bar{H}_{uu}(t_k, x^0(t_k), u^{0k-}, \psi(t_k), \nu^{k-})\bar{u}, \bar{u}\rangle \tag{3.76}$$

quadratic in $\bar{u}$ is positive definite on the subspace of vectors $\bar{u} \in \mathbb{R}^{d(u)}$ such that

$$g_u(t_k, x^0(t_k), u^{0k-})\bar{u} = 0 \tag{3.77}$$

(4) the following condition C^{k+} holds for each point $t_k \in \Theta$: the quadratic form

$$\langle \bar{H}_{uu}(t_k, x^0(t_k), u^{0k+}, \psi(t_k), \nu^{k+})\bar{u}, \bar{u}\rangle \tag{3.78}$$

is positive definite on the subset of vectors $\bar{u} \in \mathbb{R}^{d(u)}$ such that

$$g_u(t_k, x^0(t_k), u^{0k+})\bar{u} = 0. \tag{3.79}$$

Further, denote by M_0^+ the set of $\lambda \in M_0$ such that the following conditions hold:

$$H(t, x^0(t), u, \psi(t)) > H(t, x^0(t), u^0(t), \psi(t)), \tag{3.80}$$
$$\text{if} \quad t \in [t_0, t_f]\setminus\Theta, \quad u \in \mathcal{U}(t, x^0(t)), \quad u \neq u^0(t), \tag{3.81}$$

where $\mathcal{U}(t,x) = \{u \in \mathbb{R}^{d(u)} \mid (t,x,u) \in \mathcal{Q},\ g(t,x,u) = 0\}$;

$$H(t_k, x^0(t_k), u, \psi(t_k)) > H(t_k, x^0(t_k), u^{0k-}, \psi(t_k)) = H(t_k, x^0(t_k), u^{0k+}, \psi(t_k)) \tag{3.82}$$
$$\text{if} \quad t_k \in \Theta, \quad u \in \mathcal{U}(t_k, x^0(t_k)), \quad u \notin \{u^{0k-}, u^{0k+}\}. \tag{3.83}$$

Denote by $\mathrm{Leg}_+(M_0^+)$ the set of all strictly Legendrian elements $\lambda \in M_0^+$.

Definition 3.15. We say that *Condition* $\mathfrak{B}$ holds at the point w^0 in problem (3.1), (3.2) if the set $\mathrm{Leg}_+(M_0^+)$ is nonempty and there exist a compact set $M \subset \mathrm{Leg}_+(M_0^+)$ and a constant $C > 0$ such that

$$\max_{\lambda \in M} \Omega^\lambda \geq C\bar{\gamma}(\bar{z}) \quad \forall\, \bar{z} \in \mathcal{K}, \tag{3.84}$$

where the quadratic form $\Omega^\lambda(\bar{z})$, the critical cone $\mathcal{K}$ for problem (3.1), (3.2), and the point w^0 were defined by relations (3.57) and (3.27)–(3.31), respectively, and

$$\bar{\gamma}(\bar{z}) = \langle \bar{\xi}, \bar{\xi} \rangle + \langle \bar{x}(t_0), \bar{x}(t_0) \rangle + \int_{t_0}^{t_f} \langle \bar{u}(t), \bar{u}(t) \rangle \, dt \tag{3.85}$$

(as in (2.15)). We have the following theorem.

Theorem 3.16. *If Condition* $\mathfrak{B}$ *holds for the point* w^0 *in the problem* (3.1), (3.2), *then we have the bounded strong* γ*-sufficiency at this point.*

We now prove this theorem. As was said, by Proposition 3.13, it suffices to show that Condition $\mathfrak{B}$ guarantees the bounded strong γ^V-sufficiency at the point $(0, w^0)$ in the problem without local constraints. For this purpose, we write the quadratic sufficient condition of Section 2.1 for the auxiliary Problem V at the point $(0, w^0)$.

3.2.4 Proofs of Sufficient Quadratic Conditions

We first write the set M_0^V of normalized Lagrange multipliers satisfying the maximum principle for the point $(0, w^9)$ in Problem V. The Pontryagin function in Problem V has the form

$$\begin{aligned} H^V(t, y, x, u, \psi_y, \psi_x) &= \psi_x f(t, x, U(t, x, u)) + \frac{1}{2}\psi_y (u - U(t, x, u))^2 \\ &= H^U(t, x, u, \psi_x) + \frac{1}{2}\psi_y (u - U(t, x, u))^2, \end{aligned} \tag{3.86}$$

and the endpoint Lagrange function is defined by the relation

$$\begin{aligned} &l^V(y_0, x_0, y_f, x_f, \alpha_0, \alpha_y, \alpha, \beta_y, \beta) \\ &\quad = \alpha_0 J(x_0, x_f) + \alpha F(x_0, x_f) + \beta K(x_0, x_f) + \alpha_y y_f + \beta_y y_0 \\ &\quad = l(x_0, x_f, \alpha_0, \alpha, \beta) + \alpha_y y_f + \beta_y y_0. \end{aligned} \tag{3.87}$$

The set M_0^V consists of tuples $(\alpha_0, \alpha_y, \alpha, \beta_y, \beta, \psi_y, \psi_x)$ such that

$$\alpha_0 \geq 0, \quad \alpha \geq 0, \quad \alpha_y \geq 0, \tag{3.88}$$

$$\alpha F(p^0) = 0, \tag{3.89}$$

$$\alpha_y y_f^0 = 0, \tag{3.90}$$

$$\alpha_0 + |\alpha| + \alpha_y + |\beta_y| + |\beta| = 1, \tag{3.91}$$

$$\dot{\psi}_y = 0, \qquad \psi_y(t_0) = -\beta_y, \qquad \psi_y(t_f) = \alpha_y, \tag{3.92}$$

$$\dot{\psi}_x = -H_x^V, \tag{3.93}$$

$$\psi_x(t_0) = -l_{x_0}, \qquad \psi_x(t_f) = l_{x_f}, \tag{3.94}$$

$$H^V(t,y^0(t),x^0(t),u,\psi_y(t),\psi_x(t)) \\ \geq H^V(t,y^0(t),x^0(t),u^0(t),\psi_y(t),\psi_x(t)) \tag{3.95}$$

if $t \in [t_0,t_f] \setminus \Theta$ and $(t,x^0(t),u) \in \mathcal{Q}_1$. Let us analyze these conditions. Since $y_f^0 = y^0(t_f) = 0$, condition (3.90) holds automatically, and we can exclude it from consideration. Further, since $H_x^V = H_x^U - \psi_y(u^0 - U(t,w^0))U_x(t,w^0) = H_x^U$, condition (3.93) is equivalent to the condition

$$\dot{\psi}_x = -H_x^U. \tag{3.96}$$

It follows from (3.92) that

$$\psi_y = \text{const} = -\beta_y = \alpha_y. \tag{3.97}$$

Therefore, the normalization condition (3.91) is equivalent to the condition

$$\alpha_0 + |\alpha| + |\beta| + \alpha_y = 1. \tag{3.98}$$

We now turn to the minimum condition (3.95). It follows from (3.86) and (3.97) that it is equivalent to the condition

$$H^U(t,x^0(t),u,\psi_x(t)) + \frac{1}{2}\alpha_y(u - U(t,x^0(t),u))^2 \\ \geq H^U(t,x^0(t),u^0(t),\psi_x(t)) \tag{3.99}$$

whenever $t \in [t_0,t_f] \setminus \Theta$ and $(t,x^0(t),u) \in \mathcal{Q}_1$. Therefore, we can identify the set M_0^V with the set of tuples $\lambda^V = (\alpha_0,\alpha,\beta,\psi_x,\alpha_y)$ such that conditions (3.88), (3.89), (3.98), (3.96), (3.94), and (3.99) hold.

Let there exist an element $\lambda = (\alpha_0,\alpha,\beta,\psi,\nu)$ of the set M_0 of problem (3.1), (3.2) at the point w^0. Then its projection $\lambda^U = (\alpha_0,\alpha,\beta,\psi)$ is an element of the set M_0^U of problem (3.1), (3.5) (of problem U) at the same point. Let $0 \leq \alpha_y \leq 1$. We set

$$\lambda^V = ((1-\alpha_y)\lambda^U, \alpha_y). \tag{3.100}$$

Let us show that $\lambda^V \in M_0^V$. Indeed,

$$(1-\alpha_y)\alpha_0 + (1-\alpha_y)|\alpha| + (1-\alpha_y)|\beta| + \alpha_y = (1-\alpha_y)(\alpha_0 + |\alpha| + |\beta|) + \alpha_y = 1;$$

i.e., the normalization condition (3.98) holds for λ^V. Also conditions (3.88), (3.89), (3.96), and (3.94) hold.

Let us verify the minimum condition (3.99). Since $\lambda^U \in M_0^U$, the conditions

$$t \in [t_0,t_f] \setminus \Theta, \quad (t,x^0(t),u) \in \mathcal{Q}_1 \tag{3.101}$$

imply

$$H^U(t,x^0(t),u,\psi_x(t)) \geq H^U(t,x^0(t),u^0(t),\psi_x(t)). \tag{3.102}$$

Moreover, the condition $\alpha_y \geq 0$ implies

$$\frac{1}{2}\alpha_y(u - U(t,x^0(t),u))^2 \geq 0. \tag{3.103}$$

Adding inequalities (3.102) and (3.103), we obtain the minimum condition (3.99). Therefore, we have proved the following assertion.

Proposition 3.17. *The conditions* $\lambda = (\alpha_0,\alpha,\beta,\psi,\nu) \in M_0$, $0 \leq \alpha_y \leq 1$, $\lambda^U = (\alpha_0,\alpha,\beta,\psi)$, *and* $\lambda^V = ((1-\alpha_y)\lambda^U,\alpha_y)$ *imply* $\lambda^V \in M_0^V$.

Further, let $\lambda = (\alpha_0,\alpha,\beta,\psi,\nu) \in M_0^+$, i.e., $\lambda \in M_0$, and the following conditions of the strict minimum principle hold: (3.81) implies (3.80) and (3.83) implies (3.82). We set $\lambda^U = (\alpha_0,\alpha,\beta,\psi)$. It follows from the condition $\lambda \in M_0$ that $\lambda^U \in M_0^U$. Let

$$0 < \alpha_y < 1 \quad \text{and} \quad \lambda^V = ((1-\alpha_y)\lambda^U,\alpha_y). \tag{3.104}$$

Then according to Proposition 3.17, $\lambda^V \in M_0^V$. Let us show that $\lambda^V \in M_0^{V+}$, i.e., the strict minimum principle at the point $(0, w^0)$ in Problem V holds for λ^V. First, let $t \in [t_0,t_f]\setminus\Theta$, and let

$$u \in \mathbb{R}^{d(u)}, \quad (t,x^0(t),u) \in \mathcal{Q}_1, \quad u \neq u^0(t). \tag{3.105}$$

If $g(t,x^0(t),u) = 0$ in this case, then

$$U(t,x^0(t),u) = u \tag{3.106}$$

and the strict inequality (3.80) holds. It follows from (3.106) and (3.80) that

$$H^U(t,x^0(t),u,\psi_x(t)) > H^U(t,x^0(t),u^0(t),\psi_x(t)). \tag{3.107}$$

Taking into account that $\alpha_y > 0$ and $(u-U)^2 \geq 0$, we obtain

$$\begin{aligned} &H^U(t,x^0(t),u,\psi_x(t)) + \frac{1}{2}\alpha_y(u - U(t,x^0(t),u))^2 \\ &> H^U(t,x^0(t),u^0(t),\psi_x(t)), \end{aligned} \tag{3.108}$$

i.e., the strict minimum condition holds for the function H^V defined by relation (3.86). If, along with conditions (3.105), the condition $g(t,x^0(t),u) \neq 0$ holds, then this implies

$$U(t,x^0(t),u) \neq u, \tag{3.109}$$

and then

$$\frac{1}{2}\alpha_y(u - U(t,x^0(t),u))^2 > 0, \tag{3.110}$$

since $\alpha_y > 0$. Since $\lambda^U \in M_0^U$, (3.105) implies the nonstrict inequality

$$H^U(t,x^0(t),u,\psi_x(t)) \geq H^U(t,x^0(t),u^0(t),\psi_x(t)). \tag{3.111}$$

Again, inequalities (3.110) and (3.111) imply the strict inequality (3.108). Therefore, for $t \notin \Theta$, we have the strict minimum in u at the point $u^0(t)$ for $H^V(t,0,x^0(t),u,\psi_y,\psi_x(t))$. The case $t = t_k \in \Theta$ is considered analogously. Therefore, we have proved the following assertion.

Proposition 3.18. *The conditions* $\lambda = (\alpha_0,\alpha,\beta,\psi,\nu) \in M_0^+$, $0 < \alpha_y < 1$, $\lambda^U = (\alpha_0,\alpha,\beta,\psi)$, *and* $\lambda^V = ((1-\alpha_y)\lambda^U,\alpha_y)$ *imply* $\lambda^V \in M_0^{V+}$.

Let $\mathrm{Leg}_+(M_0)$ be the set of all strictly Legendre elements $\lambda \in M_0$ in the problem (3.1), (3.2) at the point w^0. The definition of these elements was given in Section 3.2.3. Also, denote by $\mathrm{Leg}_+(M_0^V)$ the set of all strictly Legendre elements $\lambda^V \in M_0^V$ of Problem V at the point $(0,w^0)$. The definition of these elements was given in Section 2.1. Let us prove the following assertion.

Proposition 3.19. *The conditions* $\lambda = (\alpha_0,\alpha,\beta,\psi,\nu) \in \mathrm{Leg}_+(M_0)$, $0 < \alpha_y < 1$, $\lambda^U = (\alpha_0,\alpha,\beta,\psi)$, *and* $\lambda^V = ((1-\alpha_y)\lambda^U,\alpha_y)$ *imply* $\lambda^V \in \mathrm{Leg}_+(M_0^V)$.

Proof. The definitions of the element λ^V and the function H^V imply

$$H^V = (1-\alpha_y)H^U + \frac{1}{2}\alpha_y(u-U)^2, \tag{3.112}$$

where H^V corresponds to the element λ^V and H^U corresponds to the element λ^U. It follows from (3.112) that the relation $H_x^V = (1-\alpha_y)H_x^U$ holds on the trajectory $(t,w^0(t))$. But, according to (3.52), $H_x^U = \bar{H}_x$, where $\bar{H}_x = \bar{H}_x^\lambda$ corresponds to the element λ. Therefore,

$$H_x^V = (1-\alpha_y)\bar{H}_x. \tag{3.113}$$

Further, since H^V is independent of y, we have

$$H_y^V = 0. \tag{3.114}$$

Also, (3.112) implies that $H_t^V = (1-\alpha_y)H_t^U$ on the trajectory $(t,w^0(t))$. But, according to (3.54), $H_t^U = \bar{H}_t$. Hence

$$H_t^V = (1-\alpha_y)\bar{H}_t. \tag{3.115}$$

Finally, the definitions of the functions H^V and $\bar{H}$ imply that the following relations hold on the trajectory $(t,w^0(t))$:

$$H_{\psi_x}^V = f^U(t,w^0) = f(t,w^0) = \bar{H}_{\psi_x}, \tag{3.116}$$

$$H_{\psi_y}^V = \frac{1}{2}(u^0 - U(t,w^0))^2 = 0. \tag{3.117}$$

We obtain from the definitions of $D^k(H^V)$ and $D^k(\bar{H})$ and also from conditions (3.113)–(3.117) that

$$\begin{aligned} -D^k(H^V) &= H_x^{Vk+}H_{\psi_x}^{Vk-} + H_y^{Vk+}H_{\psi_y}^{Vk-} \\ &\quad -(H_x^{Vk-}H_{\psi_x}^{Vk+} + H_y^{Vk-}H_{\psi_y}^{Vk+}) + [H_t^V]^k \\ &= H_x^{Vk+}H_{\psi_x}^{Vk-} - H_x^{Vk-}H_{\psi_x}^{Vk+} + [H_t^V]^k \\ &= (1-\alpha_y)\bar{H}_x^{k+}\bar{H}_{\psi_x}^{k-} - (1-\alpha_y)\bar{H}_x^{k-}\bar{H}_{\psi_x}^{k+} + (1-\alpha_y)[\bar{H}_t]^k \\ &= -(1-\alpha_y)D^k(\bar{H}) \quad \forall\, t_k \in \Theta. \end{aligned} \tag{3.118}$$

Since $\lambda \in \text{Leg}_+(M_0)$ by condition, $D^k(\bar{H}) > 0$ for all $t_k \in \Theta$. This and (3.118) together with the inequality $1-\alpha_y > 0$ imply

$$D^k(H^V) > 0 \quad \forall\, t_k \in \Theta. \tag{3.119}$$

Let us verify the conditions for the strict Legendre property of the element λ^V. For this purpose, we calculate the quadratic form $\langle H_{uu}^V\bar{u},\bar{u}\rangle$, where $\bar{u} \in \mathbb{R}^{d(u)}$, for this element on the trajectory $(t, w^0(t))$. Differentiating relation (3.112) in u and multiplying it by $\bar{u}$, we obtain

$$H_u^V\bar{u} = (1-\alpha_y)H_u^U\bar{u} + \alpha_y\langle(u-U),(\bar{u}-U_u\bar{u})\rangle.$$

The repeated differentiation in u and the multiplication by $\bar{u}$ yield

$$\langle H_{uu}^V\bar{u},\bar{u}\rangle = (1-\alpha_y)\langle H_{uu}^U\bar{u},\bar{u}\rangle + \alpha_y(\bar{u}-U_u\bar{u})^2 - \alpha_y\langle(u-U)U_{uu}\bar{u},\bar{u}\rangle.$$

Substituting $(t,w) = (t,w^0(t))$ and, moreover, taking into account that $u^0 = U(t,w^0)$, we obtain $\langle H_{uu}^V\bar{u},\bar{u}\rangle = (1-\alpha_y)\langle H_{uu}^U\bar{u},\bar{u}\rangle + \alpha_y(\bar{u}-U_u\bar{u})^2$. Finally, according to (3.45), $\langle H_{uu}^U\bar{u},\bar{u}\rangle = \langle\bar{H}_{uu}(U_u\bar{u}),(U_u\bar{u})\rangle$, where $\bar{H}_{uu} = \bar{H}_{uu}^\lambda$ corresponds to the element λ. Hence

$$\langle H_{uu}^V\bar{u},\bar{u}\rangle = (1-\alpha_y)\langle\bar{H}_{uu}(U_u\bar{u}),(U_u\bar{u})\rangle + \alpha_y(\bar{u}-U_u\bar{u})^2. \tag{3.120}$$

The values of the derivatives are taken for $(t,w) = (t,w^0(t))$.

It is easy to verify that for each $t \in [t_0,t_f]$, form (3.120) quadratic in $\bar{u}$ is positive-definite on $\mathbb{R}^{d(u)}$. Indeed, suppose first that $t \in [t_0,t_f]\setminus\Theta$. Recall that the mapping $\bar{u} \in \mathbb{R}^{d(u)} \mapsto U_u(t,w^0(t))\bar{u} \in \mathbb{R}^{d(u)}$ is the projection on the subspace

$$\{\bar{u} \in \mathbb{R}^{d(u)} \mid g_u(t,w^0(t))\bar{u} = 0\}. \tag{3.121}$$

This and the condition $\lambda \in \text{Leg}_+(M_0)$ imply that the quadratic form $\langle\bar{H}_{uu}(U_u\bar{u}),(U_u\bar{u})\rangle$ is positive semidefinite on $\mathbb{R}^{d(u)}$ and positive definite on subspace (3.121). Furthermore, the quadratic form $(\bar{u}-U_u\bar{u})^2$ is positive semidefinite on $\mathbb{R}^{d(u)}$ and positive outside subspace (3.121). This and the conditions $0 < \alpha_y < 1$ imply the positivity of the quadratic form $\langle H_{uu}^V\bar{u},\bar{u}\rangle$ outside the origin of the space $\mathbb{R}^{d(u)}$ and, therefore, the positive definiteness on $\mathbb{R}^{d(u)}$. The case $t \in [t_0,t_f]\setminus\Theta$ has been considered. The case $t = t_k \in \Theta$ is considered similarly. Therefore, all the conditions needed for element λ^V to belong to the set $\text{Leg}_+(M_0^V)$ hold. The proposition is proved. ∎

Propositions 3.18 and 3.19 imply the following assertion.

Lemma 3.20. *The conditions* $\lambda = (\alpha_0, \alpha, \beta, \psi, \nu) \in \mathrm{Leg}_+(M_0^+)$, $0 < \alpha_y < 1$, $\lambda^U = (\alpha_0, \alpha, \beta, \psi)$, $\lambda^V = ((1-\alpha_y)\lambda^U, \alpha_y)$ *imply* $\lambda^V \in \mathrm{Leg}_+(M_0^{V+})$.

Fix a number α_y such that $0 < \alpha_y < 1$, e.g., $\alpha_y = 1/2$, and consider the linear operator

$$\lambda = (\alpha_0, \alpha, \beta, \psi, \nu) \mapsto \lambda^V = ((1-\alpha_y)\lambda^U, \alpha_y), \tag{3.122}$$

where $\lambda^U = (\alpha_0, \alpha, \beta, \psi)$ is the projection of the element λ. Lemma 3.20 implies the following assertion.

Lemma 3.21. *Operator* (3.122) *transforms an arbitrary nonempty compact set* $M \subset \mathrm{Leg}_+(M_0^+)$ *into a nonempty compact set* $M^V \subset \mathrm{Leg}_+(M_0^{V+})$.

Now let us consider the critical cone $\mathcal{K}^V$ of Problem V at the point $(0, w^0)$. According to the definition given in Section 2.1, it consists of elements $\bar{z}^V = (\bar{\xi}, \bar{y}, \bar{w}) = (\bar{\xi}, \bar{y}, \bar{x}, \bar{u})$ such that the following conditions hold:

$$\begin{aligned}
&\bar{z} = (\bar{\xi}, \bar{x}, \bar{u}) = (\bar{\xi}, \bar{w}) \in Z_2(\Theta), \quad \bar{y} \in P_\Theta W^{1,2}(\Delta, \mathbb{R}),\\
&J_p\bar{p} \le 0, \quad F_{ip}\bar{p} \le 0 \quad \forall\, i \in I, \quad K_p\bar{p} = 0,\\
&\bar{y}_0 = 0, \quad \bar{y}_f \le 0,\\
&\dot{\bar{y}} = 0, \quad [\bar{y}]^k = 0 \quad \forall\, t_k \in \Theta,\\
&\dot{\bar{x}} = f_x\bar{x} + f_u(U_w\bar{w}),\\
&[\bar{x}]^k = [f]^k\bar{\xi}_k \quad \forall\, t_k \in \Theta.
\end{aligned}$$

These conditions imply $\bar{y} = 0$, $\bar{z} = (\bar{\xi}, \bar{x}, \bar{u}) \in \mathcal{K}^U$, where $\mathcal{K}^U$ is the critical cone of problem U, i.e., the problem (3.1), (3.5), at the point w^0. Then, according to Lemma 3.6, $\tilde{z} = (\bar{\xi}, \bar{x}, U_w\bar{w})$ is an element of the critical cone $\mathcal{K}$ of the problem (3.1), (3.2) at the point w^0. Therefore, we have proved the following assertion.

Lemma 3.22. *Let* $\bar{z}^V = (\bar{\xi}, \bar{y}, \bar{x}, \bar{u}) = (\bar{\xi}, \bar{y}, \bar{w}) \in \mathcal{K}^V$. *Then* $\bar{y} = 0$, $\tilde{z} = (\bar{\xi}, \bar{x}, U_w\bar{w}) \in \mathcal{K}$. *Therefore, the linear operator*

$$(\bar{\xi}, \bar{y}, \bar{x}, \bar{u}) \mapsto (\bar{\xi}, \bar{x}, U_x\bar{x} + U_u\bar{u}) \tag{3.123}$$

transforms the critical cone $\mathcal{K}^V$ *into the critical cone* $\mathcal{K}$.

We now consider the quadratic forms. Let $\lambda \in M_0$ be an arbitrary element, and let λ^V be its image under mapping (3.122). According to Proposition 3.17, $\lambda^V \in M_0^V$. For Problem V and the point $(0, w^0)$, let us consider the quadratic form $\Omega^{V\lambda^V}$ (corresponding to the element λ^V), which was defined in Section 2.1, and let us study its relation with the quadratic form Ω^λ (which corresponds to the element λ) at the point w^0 of the problem (3.1), (3.2). We have already shown that

$$D^k(H^V) = (1-\alpha_y)D^k(\bar{H}), \quad t_k \in \Theta \tag{3.124}$$

(we omit the superscripts λ^V and λ of H^V and $\bar{H}$, respectively). It follows from (3.113) that

$$[H_x^V]^k = (1-\alpha_y)[\bar{H}_x]^k, \quad t_k \in \Theta, \tag{3.125}$$

and (3.114) implies

$$[H_y^V]^k = 0, \quad t_k \in \Theta. \tag{3.126}$$

We obtain from (3.112) that

$$\langle H_{ww}^V \bar{w}, \bar{w} \rangle = (1-\alpha_y)\langle H_{ww}^U \bar{w}, \bar{w} \rangle + \frac{1}{2}\alpha_y \left\langle \left(\frac{\partial^2}{\partial w^2}(u-U)^2 \right) \bar{w}, \bar{w} \right\rangle, \tag{3.127}$$

where $\bar{w} = (\bar{x}, \bar{u})$, $\bar{x} \in P_\Theta W^{1,2}(\Delta, \mathbb{R}^{d(x)})$, $\bar{u} \in L^2(\Delta, \mathbb{R}^{d(u)})$. According to (3.47),

$$\langle H_{ww}^U \bar{w}, \bar{w} \rangle = \langle \bar{H}_{ww} \tilde{w}, \tilde{w} \rangle, \tag{3.128}$$

where $\tilde{w} = (\bar{x}, \tilde{u}) = (\bar{x}, U_w \bar{w})$. Let us calculate the second summand in formula (3.127). We have

$$\frac{1}{2}\left(\frac{\partial}{\partial w}(u-U)^2 \right) \bar{w} = (u-U)(\bar{u} - U_w \bar{w}).$$

Therefore, for $w = w^0(t)$, we have

$$\begin{aligned} \frac{1}{2}\left\langle \left(\frac{\partial^2}{\partial w^2}(u-U)^2 \right) \bar{w}, \bar{w} \right\rangle &= (\bar{u} - U_w \bar{w})^2 - \left\langle \left(u^0 - U(w^0, t) \right) U_{ww} \bar{w}, \bar{w} \right\rangle \\ &= (\bar{u} - U_w \bar{w})^2 = (\bar{u} - \tilde{u})^2. \end{aligned} \tag{3.129}$$

We obtain from (3.127)–(3.129) that

$$\langle H_{ww}^V \bar{w}, \bar{w} \rangle = (1-\alpha_y)\langle \bar{H}_{ww} \tilde{w}, \tilde{w} \rangle + \alpha_y (\bar{u} - \tilde{u})^2, \tag{3.130}$$

where $\tilde{u} = U_w \bar{w}$ and $\tilde{w} = (\bar{x}, \tilde{u})$. Since H^V is independent of y, we have

$$H_{yy}^V = 0 \quad \text{and} \quad H_{yw}^V = 0. \tag{3.131}$$

Let $z^V = (\bar{\xi}, \bar{y}, \bar{x}, \bar{u})$ be an arbitrary tuple such that

$$\bar{z} = (\bar{\xi}, \bar{x}, \bar{u}) = (\bar{\xi}, \bar{w}) \in Z_2(\Theta), \quad \bar{y} \in P_\Theta W^{1,2}. \tag{3.132}$$

The definitions of the quadratic forms $\Omega^{V\lambda^V}$ and Ω^λ and also relations (3.124)–(3.126), (3.130), and (3.131) imply

$$\Omega^{V\lambda^V}(\bar{z}^V) = (1-\alpha_y)\Omega^\lambda(\tilde{z}) + \alpha_y \int_{t_0}^{t_f} (\bar{u} - \tilde{u})^2 \, dt, \tag{3.133}$$

where $\tilde{u} = U_x \bar{x} + U_u \bar{u}$ and $\tilde{z} = (\bar{\xi}, \bar{x}, \tilde{u})$. Therefore, we have proved the following assertion.

Lemma 3.23. *Let an element $\lambda \in M_0$ and a tuple $\bar{z}^V = (\bar{\xi}, \bar{y}, \bar{x}, \bar{u})$ satisfying conditions* (3.132) *be given. Let λ^V be the image of λ under mapping* (3.122). *Then formula* (3.133) *holds for the quadratic form $\Omega^{V\lambda^V}$ calculated for Problem V, the point $(0, w^0)$, and the element λ^V and also holds for the quadratic form Ω^λ calculated for the problem* (3.1), (3.2), *the point w^0, and the element λ.*

We now assume that the following sufficient Condition $\mathfrak{B}$ holds at the point w^0 in problem (3.1), (3.2): there exist a nonempty compact set $M \subset \mathrm{Leg}_+(M_0^+)$ and a constant $C > 0$ such that

$$\max_M \Omega^\lambda(\bar{z}) \geq C\bar{\gamma}(\bar{z}) \quad \forall\, \bar{z} \in \mathcal{K}. \tag{3.134}$$

Let us show that in this case, the sufficient condition of Section 2.1 (denoted by $\mathfrak{B}^V$) holds at the point $(0, w^0)$ in Problem V.

Let $\bar{z}^V = (\bar{\xi}, \bar{y}, \bar{x}, \bar{u})$ be an arbitrary element of the critical cone $\mathcal{K}^V$ of Problem V at the point $(0, w^0)$. Then $\bar{y} = 0$, and, by Lemma 3.22, $\tilde{z} = (\bar{\xi}, \bar{x}, \tilde{u}) \in \mathcal{K}$, where $\tilde{u} = U_w \bar{w}$. Condition (3.134) implies the existence of $\lambda \in M$ such that

$$\Omega^\lambda(\tilde{z}) \geq C\bar{\gamma}(\tilde{z}). \tag{3.135}$$

Let λ^V be the image of λ under the mapping defined by operator (3.122). Then, by Lemma 3.23, formula (3.133) holds. It follows from (3.133) and (3.135) that

$$\Omega^{V\lambda^V}(\bar{z}^V) \geq (1-\alpha_y)C\bar{\gamma}(\tilde{z}) + \alpha_y \int_{t_0}^{t_f} (\bar{u} - \tilde{u})^2\, dt. \tag{3.136}$$

Therefore,

$$\max_{\lambda^V \in M^V} \Omega^{V\lambda^V}(\bar{z}^V) \geq (1-\alpha_y)C\bar{\gamma}(\tilde{z}) + \alpha_y \int_{t_0}^{t_f} (\bar{u} - \tilde{u})^2\, dt, \tag{3.137}$$

where M^V is the image of the compact set M under the mapping defined by operator (3.122). By Lemma 3.21,

$$M^V \subset \mathrm{Leg}_+(M_0^{V+}). \tag{3.138}$$

Conditions (3.137) and (3.138) imply that the sufficient Condition $\mathfrak{B}^V$ holds at the point $(0, w^0)$ in Problem V. To verify this, it suffices to show that the right-hand side of inequality (3.137) on the cone $\mathcal{K}^V$ is estimated from below by the functional

$$\bar{\gamma}^V(\bar{z}^V) := \bar{\xi}^2 + \bar{y}_0^2 + \bar{x}_0^2 + \int_{t_0}^{t_f} \bar{u}^2\, dt = \bar{\gamma}(\bar{z}) + \bar{y}_0^2$$

with a small coefficient $\varepsilon > 0$. Since $\bar{y} = 0$ for all $\bar{z} \in \mathcal{K}^V$, it suffices to prove that there exists $\varepsilon > 0$ such that $(1-\alpha_y)C\bar{\gamma}(\tilde{z}) + \alpha_y \int (\bar{u} - \tilde{u})^2\, dt \geq \varepsilon \bar{\gamma}(\bar{z})$ or

$$(1-\alpha_y)C\Big(\bar{\xi}^2 + \bar{x}_0^2 + \int_{t_0}^{t_f} \tilde{u}^2\, dt\Big) + \alpha_y \int_{t_0}^{t_f} (\bar{u} - \tilde{u})^2\, dt \geq \varepsilon\Big(\bar{\xi}^2 + \bar{x}_0^2 + \int_{t_0}^{t_f} \bar{u}^2\, dt\Big). \tag{3.139}$$

We set $\bar{u} - \tilde{u} = \hat{u}$. Then $\bar{u}^2 = \hat{u}^2 + 2\hat{u}\tilde{u} + \tilde{u}^2 \leq 2\hat{u}^2 + 2\tilde{u}^2$. This obviously implies the estimate required. Therefore, we have proved the following assertion.

Lemma 3.24. *Let the sufficient Condition $\mathfrak{B}$ hold at the point w^0 in problem* (3.1), (3.2). *Then the sufficient Condition $\mathfrak{B}^V$ of Section* 2.1 *holds at the point* $(0, w^0)$ *in Problem V.*

By Theorem 2.101, Condition $\mathfrak{B}^V$ implies the existence of the bounded strong γ^V-sufficiency at the point $(0, w^0)$ in Problem V; by Proposition 3.13, this implies the existence of the bounded strong γ-sufficiency at the point w^0 in the problem (3.1), (3.2). Also, taking into account Lemma 3.24, we obtain that Condition $\mathfrak{B}$ is sufficient for the bounded strong γ-sufficiency at the point w^0 in problem (3.1), (3.2). Therefore, we have proved Theorem 3.16, to which this section is devoted.

3.3 Quadratic Conditions in the Problem with Mixed Control-State Equality Constraints on a Variable Time Interval

3.3.1 Statement of the Problem

Here, quadratic optimality conditions, both necessary and sufficient, are presented (as in [93]) in the following canonical Dubovitskii–Milyutin problem on a variable time interval. Let $\mathcal{T}$ denote a trajectory $(x(t),u(t) \mid t \in [t_0,t_f])$, where the state variable $x(\cdot)$ is a Lipschitz-continuous function, and the control variable $u(\cdot)$ is a bounded measurable function on a time interval $\Delta = [t_0,t_f]$. The interval Δ is not fixed. For each trajectory $\mathcal{T}$ we denote by $p = (t_0,x(t_0),t_f,x(t_f))$ the vector of the endpoints of time-state variable (t,x). It is required to find $\mathcal{T}$ minimizing the functional

$$\mathcal{J}(\mathcal{T}) := J(p) \to \min \tag{3.140}$$

subject to the constraints

$$F(p) \le 0, \quad K(p) = 0, \tag{3.141}$$
$$\dot{x}(t) = f(t,x(t),u(t)), \tag{3.142}$$
$$g(t,x(t),u(t)) = 0, \tag{3.143}$$
$$p \in \mathcal{P}, \quad (t,x(t),u(t)) \in \mathcal{Q}, \tag{3.144}$$

where $\mathcal{P}$ and $\mathcal{Q}$ are open sets, and x, u, F, K, f, and g are vector functions.

We assume that the functions J, F, and K are defined and twice continuously differentiable on $\mathcal{P}$, and that the functions f and g are defined and twice continuously differentiable on $\mathcal{Q}$. It is also assumed that the gradients with respect to the control $g_{iu}(t,x,u)$, $i = 1,\dots,d(g)$ are linearly independent at each point $(t,x,u) \in \mathcal{Q}$ such that $g(t,x,u) = 0$. Here $d(g)$ is a dimension of the vector g.

3.3.2 Necessary Conditions for a Pontryagin Minimum

Let $\mathcal{T}$ be a fixed admissible trajectory such that the control $u(\cdot)$ is a piecewise Lipschitz-continuous function on the interval Δ with the set of discontinuity points $\Theta = \{t_1,\dots,t_s\}$, $t_0 < t_1 < \cdots < t_s < t_f$. Let us formulate a first-order necessary condition for optimality of the trajectory $\mathcal{T}$. We introduce the Pontryagin function

$$H(t,x,u,\psi) = \psi f(t,x,u) \tag{3.145}$$

and the augmented Pontryagin function

$$\bar{H}(t,x,u,\psi,\nu) = H(t,x,u,\psi) + \nu g(t,x,u), \tag{3.146}$$

where ψ and ν are row vectors of the dimensions $d(x)$ and $d(g)$, respectively. Let us define the endpoint Lagrange function

$$l(p,\alpha_0,\alpha,\beta) = \alpha_0 J(p) + \alpha F(p) + \beta K(p), \tag{3.147}$$

where $p=(t_0,x_0,t_f,x_f)$, $x_0=x(t_0)$, $x_f=x(t_f)$, $\alpha_0\in\mathbb{R}$, $\alpha\in(\mathbb{R}^{d(F)})^*$, $\beta\in(\mathbb{R}^{d(K)})^*$. Also we introduce a tuple of Lagrange multipliers

$$\lambda=(\alpha_0,\alpha,\beta,\psi(\cdot),\psi_0(\cdot),\nu(\cdot)) \tag{3.148}$$

such that $\psi(\cdot):\Delta\to(\mathbb{R}^{d(x)})^*$ and $\psi_0(\cdot):\Delta\to\mathbb{R}^1$ are piecewise smooth functions, continuously differentiable on each interval of the set $\Delta\setminus\Theta$, and $\nu(\cdot):\Delta\to(\mathbb{R}^{d(g)})^*$ is a piecewise continuous function, Lipschitz continuous on each interval of the set $\Delta\setminus\Theta$.

Denote by M_0 the set of the normed tuples λ satisfying the conditions of the minimum principle for the trajectory $\mathcal{T}$:

$$\begin{aligned}
&\alpha_0\ge 0,\quad \alpha\ge 0,\quad \alpha F(p)=0,\quad \alpha_0+\sum\alpha_i+\sum|\beta_j|=1,\\
&\dot\psi=-\bar H_x,\quad \dot\psi_0=-\bar H_t,\quad \bar H_u=0,\quad t\in\Delta\setminus\Theta,\\
&\psi(t_0)=-l_{x_0},\quad \psi(t_f)=l_{x_f},\quad \psi_0(t_0)=-l_{t_0},\quad \psi_0(t_f)=l_{t_f},\\
&\min_{u\in\mathcal{U}(t,x(t))} H(t,x(t),u,\psi(t))=H(t,x(t),u(t),\psi(t)),\quad t\in\Delta\setminus\Theta,\\
&H(t,x(t),u(t),\psi(t))+\psi_0(t)=0,\quad t\in\Delta\setminus\Theta,
\end{aligned} \tag{3.149}$$

where $\mathcal{U}(t,x)=\{u\in\mathbb{R}^{d(u)}\mid g(t,x,u)=0,\ (t,x,u)\in\mathcal{Q}\}$. The derivatives l_{x_0} and l_{x_f} are at $(p,\alpha_0,\alpha,\beta)$, where $p=(t_0,x(t_0),t_f,x(t_f))$, and the derivatives $\bar H_x$, $\bar H_u$, and $\bar H_t$ are at $(t,x(t),u(t),\psi(t),\nu(t))$, where $t\in\Delta\setminus\Theta$. (Condition $\bar H_u=0$ follows from the others conditions in this definition, and therefore could be excluded; yet we need to use it later.)

Let us give the definition of Pontryagin minimum in problem (3.140)–(3.144) on a variable interval $[t_0,t_f]$.

Definition 3.25. The trajectory $\mathcal{T}$ affords a *Pontryagin minimum* if there is no sequence of admissible trajectories $\mathcal{T}^n=(x^n(t),u^n(t)\mid t\in[t_0^n,t_f^n])$, $n=1,2,\dots$ such that

(a) $\mathcal{J}(\mathcal{T}^n)<\mathcal{J}(\mathcal{T})$ for all n;
(b) $t_0^n\to t_0$, $t_f^n\to t_f$ $(n\to\infty)$;
(c) $\max_{\Delta^n\cap\Delta}|x^n(t)-x(t)|\to 0$ $(n\to\infty)$, where $\Delta^n=[t_0^n,t_f^n]$;
(d) $\int_{\Delta^n\cap\Delta}|u^n(t)-u(t)|\,dt\to 0$ $(n\to\infty)$;
(e) there exists a compact set $\mathcal{C}\subset\mathcal{Q}$ such that $(t,x^n(t),u^n(t))\in\mathcal{C}$ a.e. on Δ^n for all n.

For convenience, let us give an equivalent definition of the Pontryagin minimum.

Definition 3.26. The trajectory $\mathcal{T}$ affords a *Pontryagin minimum* if for each compact set $\mathcal{C}\subset\mathcal{Q}$ there exists $\varepsilon>0$ such that $\mathcal{J}(\tilde{\mathcal{T}})\ge\mathcal{J}(\mathcal{T})$ for all admissible trajectories $\tilde{\mathcal{T}}=(\tilde x(t),\tilde u(t)\mid t\in[\tilde t_0,\tilde t_f])$ satisfying the following conditions:

(a) $|\tilde t_0-t_0|<\varepsilon$, $|\tilde t_f-t_f|<\varepsilon$;
(b) $\max_{\tilde\Delta\cap\Delta}|\tilde x(t)-x(t)|<\varepsilon$, where $\tilde\Delta=[\tilde t_0,\tilde t_f]$;
(c) $\int_{\tilde\Delta\cap\Delta}|\tilde u(t)-u(t)|\,dt<\varepsilon$;
(d) $(t,\tilde x(t),\tilde u(t))\in\mathcal{C}$ a.e. on $\tilde\Delta$.

The condition $M_0\ne\emptyset$ is equivalent to the Pontryagin minimum principle. It is a first-order necessary condition of Pontryagin minimum for the trajectory $\mathcal{T}$. Thus, the following theorem holds.

Theorem 3.27. *If the trajectory $\mathcal{T}$ affords a Pontryagin minimum, then the set M_0 is nonempty.*

Assume that M_0 is nonempty. Using the definition of the set M_0 and the full rank condition of the matrix g_u on the surface $g = 0$, one can easily prove the following statement.

Proposition 3.28. *The set M_0 is a finite-dimensional compact set, and the mapping $\lambda \mapsto (\alpha_0, \alpha, \beta)$ is injective on M_0.*

As in Section 3.1, for each $\lambda \in M_0, t_k \in \Theta$, we set

$$D^k(\bar{H}) = -\bar{H}_x^{k+}\bar{H}_\psi^{k-} + \bar{H}_x^{k-}\bar{H}_\psi^{k+} - [\bar{H}_t]^k, \tag{3.150}$$

where $\bar{H}_x^{k-} = \bar{H}_x(t_k, x(t_k), u(t_k-), \psi(t_k), \nu(t_k-))$, $\bar{H}_x^{k+} = \bar{H}_x(t_k, x(t_k), u(t_k+), \psi(t_k), \nu(t_k+))$, $[\bar{H}_t]^k = \bar{H}_t^{k+} - \bar{H}_t^{k-}$, etc.

Theorem 3.29. *For each $\lambda \in M_0$ the following conditions hold:*

$$D^k(\bar{H}) \geq 0, \quad k = 1, \ldots, s. \tag{3.151}$$

Thus, conditions (3.151) follows from the minimum principle conditions (3.149). The following is an alternative method for calculating $D^k(\bar{H})$: For $\lambda \in M_0$, $t_k \in \Theta$, consider the function

$$(\Delta_k \bar{H})(t) = \bar{H}(t_k, x(t), u(t_k+), \psi(t), \nu(t_k+)) - \bar{H}(t_k, x(t), u(t_k-), \psi(t), \nu(t_k-)).$$

Proposition 3.30. *For each $\lambda \in M_0$ the following equalities hold:*

$$\frac{d}{dt}(\Delta_k \bar{H})\big|_{t=t_k-} = \frac{d}{dt}(\Delta_k \bar{H})\big|_{t=t_k+} = -D^k(\bar{H}), \quad k = 1, \ldots, s. \tag{3.152}$$

Hence, for $\lambda \in M_0$ the function $(\Delta_k \bar{H})(t)$ has a derivative at the point $t_k \in \Theta$ equal to $-D^k(\bar{H})$, $k = 1, \ldots, s$. Let us formulate a *quadratic necessary condition* of a Pontryagin minimum for the trajectory $\mathcal{T}$. First, for this trajectory, we introduce a Hilbert space $\mathcal{Z}_2(\Theta)$ and the critical cone $\mathcal{K} \subset \mathcal{Z}_2(\Theta)$. We denote by $P_\Theta W^{1,2}(\Delta, \mathbb{R}^{d(x)})$ the Hilbert space of piecewise continuous functions $\bar{x}(\cdot) : \Delta \to \mathbb{R}^{d(x)}$, absolutely continuous on each interval of the set $\Delta \setminus \Theta$ and such that their first derivative is square integrable. For each $\bar{x} \in P_\Theta W^{1,2}(\Delta, \mathbb{R}^{d(x)})$, $t_k \in \Theta$, we set

$$\bar{x}^{k-} = \bar{x}(t_k-), \quad \bar{x}^{k+} = \bar{x}(t_k+), \quad [\bar{x}]^k = \bar{x}^{k+} - \bar{x}^{k-}.$$

Further, we denote $\bar{z} = (\bar{t}_0, \bar{t}_f, \bar{\xi}, \bar{x}, \bar{u})$, where

$$\bar{t}_0 \in \mathbb{R}^1, \quad \bar{t}_f \in \mathbb{R}^1, \quad \bar{\xi} \in \mathbb{R}^s, \quad \bar{x} \in P_\Theta W^{1,2}(\Delta, \mathbb{R}^{d(x)}), \quad \bar{u} \in L^2(\Delta, \mathbb{R}^{d(u)}).$$

Thus,

$$\bar{z} \in \mathcal{Z}_2(\Theta) := \mathbb{R}^2 \times \mathbb{R}^s \times P_\Theta W^{1,2}(\Delta, \mathbb{R}^{d(x)}) \times L^2(\Delta, \mathbb{R}^{d(u)}).$$

Moreover, for given $\bar{z}$ we set

$$\bar{w} = (\bar{x}, \bar{u}), \quad \bar{x}_0 = \bar{x}(t_0), \quad \bar{x}_f = \bar{x}(t_f), \tag{3.153}$$

$$\bar{\bar{x}}_0 = \bar{x}(t_0) + \bar{t}_0 \dot{x}(t_0), \quad \bar{\bar{x}}_f = \bar{x}(t_f) + \bar{t}_f \dot{x}(t_f), \quad \bar{\bar{p}} = (\bar{t}_0, \bar{\bar{x}}_0, \bar{t}_f, \bar{\bar{x}}_f). \tag{3.154}$$

By $I_F(p) = \{i \in \{1,\dots,d(F)\} \mid F_i(p) = 0\}$, we denote the set of active indices of the constraints $F_i(p) \le 0$.

Let $\mathcal{K}$ be the set of all $\bar{z} \in \mathcal{Z}_2(\Theta)$ satisfying the following conditions:

$$\begin{aligned} & J'(p)\bar{\bar{p}} \le 0, \quad F_i'(p)\bar{\bar{p}} \le 0 \quad \forall\, i \in I_F(p), \quad K'(p)\bar{\bar{p}} = 0, \\ & \dot{\bar{x}}(t) = f_w(t, w(t))\bar{w}(t) \text{ for a.a. } t \in [t_0, t_f], \\ & [\bar{x}]^k = [\dot{x}]^k \bar{\xi}_k, \quad k = 1, \dots, s, \\ & g_w(t, w(t))\bar{w}(t) = 0 \text{ for a.a. } t \in [t_0, t_f], \end{aligned} \tag{3.155}$$

where $p = (t_0, x(t_0), t_f, x(t_f))$, $w = (x,u)$. It is obvious that $\mathcal{K}$ is a convex cone in the Hilbert space $Z_2(\Theta)$, and we call it the *critical cone*. If the interval Δ is fixed, then we set $p := (x_0, x_f) = (x(t_0), x(t_f))$, and in the definition of $\mathcal{K}$ we have $\bar{t}_0 = \bar{t}_f = 0$, $\bar{\bar{x}}_0 = \bar{x}_0$, $\bar{\bar{x}}_f = \bar{x}_f$, and $\bar{\bar{p}} = \bar{p} := (\bar{x}_0, \bar{x}_f)$.

Let us introduce a quadratic form on $\mathcal{Z}_2(\Theta)$. For $\lambda \in M_0$ and $\bar{z} \in \mathcal{K}$, we set

$$\begin{aligned} \omega_e(\lambda, \bar{z}) = {} & \langle l_{pp}\bar{\bar{p}}, \bar{\bar{p}} \rangle - 2\dot{\psi}(t_f)\bar{x}(t_f)\bar{t}_f - \Big(\dot{\psi}(t_f)\dot{x}(t_f) + \dot{\psi}_0(t_f)\Big)\bar{t}_f^2 \\ & + 2\dot{\psi}(t_0)\bar{x}(t_0)\bar{t}_0 + \Big(\dot{\psi}(t_0)\dot{x}(t_0) + \dot{\psi}_0(t_0)\Big)\bar{t}_0^2, \end{aligned} \tag{3.156}$$

where $l_{pp} = l_{pp}(p, \alpha_0, \alpha, \beta)$, $p = (t_0, x(t_0), t_f, x(t_f))$. We also set

$$\omega(\lambda, \bar{z}) = \omega_e(\lambda, \bar{z}) + \int_{t_0}^{t_f} \langle \bar{H}_{ww}\bar{w}(t), \bar{w}(t) \rangle \, dt, \tag{3.157}$$

where $\bar{H}_{ww} = \bar{H}_{ww}(t, x(t), u(t), \psi(t), \nu(t))$. Finally, we set

$$2\Omega(\lambda, \bar{z}) = \omega(\lambda, \bar{z}) + \sum_{k=1}^{s} \Big(D^k(\bar{H})\bar{\xi}_k^2 - 2[\dot{\psi}]^k \bar{x}_{\mathrm{av}}^k \bar{\xi}_k \Big), \tag{3.158}$$

where $\bar{x}_{\mathrm{av}}^k = \frac{1}{2}(\bar{x}^{k-} + \bar{x}^{k+})$, $[\dot{\psi}]^k = \dot{\psi}^{k+} - \dot{\psi}^{k-}$.

Now, we formulate the main necessary quadratic condition of Pontryagin minimum in the problem on a variable time interval.

Theorem 3.31. *If the trajectory $\mathcal{T}$ yields a Pontryagin minimum, then the following Condition $\mathfrak{A}$ holds: the set M_0 is nonempty and*

$$\max_{\lambda \in M_0} \Omega(\lambda, \bar{z}) \ge 0 \qquad \forall\, \bar{z} \in \mathcal{K}.$$

3.3.3 Sufficient Conditions for a Bounded Strong Minimum

Next, we give the definition of a bounded strong minimum in problem (3.140)–(3.144) on a variable interval $[t_0,t_f]$. To this end, let us give the definition of essential component of vector x in this problem: the ith component x_i of vector x is called *unessential* if the functions f and g do not depend on this component and the functions J, F, and K are affine in $x_{i0}=x_i(t_0)$, $x_{if}=x(t_f)$; otherwise the component x_i is called *essential*. We denote by $\underline{x}$ a vector composed of all essential components of vector x.

Definition 3.32. We say that the trajectory $\mathcal{T}$ affords a *bounded strong minimum* if there is no sequence of admissible trajectories $\mathcal{T}^n=(x^n(t),u^n(t)\mid t\in[t_0^n,t_f^n])$, $n=1,2,\dots$ such that

(a) $\mathcal{J}(\mathcal{T}^n)<\mathcal{J}(\mathcal{T})$;

(b) $t_0^n\to t_0$, $t_f^n\to t_f$, $x^n(t_0^n)\to x(t_0)$ $(n\to\infty)$;

(c) $\max_{\Delta^n\cap\Delta}|\underline{x}^n(t)-\underline{x}(t)|\to 0$ $(n\to\infty)$, where $\Delta^n=[t_0^n,t_f^n]$;

(d) there exists a compact set $\mathcal{C}\subset\mathcal{Q}$ such that $(t,x^n(t),u^n(t))\in\mathcal{C}$ a.e. on Δ^n for all n.

An equivalent definition has the following form.

Definition 3.33. The trajectory $\mathcal{T}$ affords *a bounded strong minimum* if for each compact set $\mathcal{C}\subset\mathcal{Q}$ there exists $\varepsilon>0$ such that $\mathcal{J}(\tilde{\mathcal{T}})\ge\mathcal{J}(\mathcal{T})$ for all admissible trajectories $\tilde{\mathcal{T}}=(\tilde{x}(t),\tilde{u}(t)\mid t\in[\tilde{t}_0,\tilde{t}_f])$ satisfying the following conditions:

(a) $|\tilde{t}_0-t_0|<\varepsilon$, $|\tilde{t}_f-t_f|<\varepsilon$, $|\tilde{x}(\tilde{t}_0)-x(t_0)|<\varepsilon$;

(b) $\max_{\tilde{\Delta}\cap\Delta}|\underline{\tilde{x}}(t)-\underline{x}(t)|<\varepsilon$, where $\tilde{\Delta}=[\tilde{t}_0,\tilde{t}_f]$;

(c) $(t,\tilde{x}(t),\tilde{u}(t))\in\mathcal{C}$ a.e. on $\tilde{\Delta}$.

The *strict* bounded strong minimum is defined in a similar way, with the nonstrict inequality $\mathcal{J}(\tilde{\mathcal{T}})\ge\mathcal{J}(\mathcal{T})$ replaced by the strict one and the trajectory $\tilde{\mathcal{T}}$ required to be different from $\mathcal{T}$. Finally, we define a *(strict) strong minimum* in the same way but omit condition (c) in the last definition. The following statement is quite obvious.

Proposition 3.34. *If there exists a compact set $\mathcal{C}\subset\mathcal{Q}$ such that $\{(t,x,u)\in\mathcal{Q}\mid g(t,x,u)=0\}\subset\mathcal{C}$, then a (strict) strong minimum is equivalent to a (strict) bounded strong minimum.*

Let us formulate a sufficient optimality Condition $\mathfrak{B}$, which is a natural strengthening of the necessary Condition $\mathfrak{A}$. The condition $\mathfrak{B}$ is sufficient not only for a Pontryagin minimum, but also for a strict bounded strong minimum.

To formulate the condition $\mathfrak{B}$, we introduce, for $\lambda\in M_0$, the following conditions of the *strict minimum principle*:

$$(\mathrm{MP}^+_{\Delta\setminus\Theta})\qquad H(t,x(t),u,\psi(t))>H(t,x(t),u(t),\psi(t))$$

for all $t\in\Delta\setminus\Theta$, $u\ne u(t)$, $u\in\mathcal{U}(t,x(t))$, and

$$(\mathrm{MP}^+_{\Theta})\qquad H(t_k,x(t_k),u,\psi(t_k))>H^k$$

for all $t_k\in\Theta$, $u\in\mathcal{U}(t_k,x(t_k))$, $u\ne u(t_k-)$, $u\ne u(t_k+)$, where $H^k:=H^{k-}=H^{k+}$, $H^{k-}=H(t_k,x(t_k),u(t_k-),\psi(t_k))$, $H^{k+}=H(t_k,x(t_k),u(t_k+),\psi(t_k))$. We denote by M_0^+ the set of all $\lambda\in M_0$ satisfying conditions $(\mathrm{MP}^+_{\Delta\setminus\Theta})$ and (MP^+_{Θ}).

For $\lambda \in M_0$ we also introduce the strengthened Legendre–Clebsch conditions:
($\mathrm{SLC}_{\Delta\setminus\Theta}$): For each $t \in \Delta \setminus \Theta$, the quadratic form

$$\langle \bar{H}_{uu}(t, x(t), u(t), \psi(t), \nu(t))\bar{u}, \bar{u}\rangle$$

is positive definite on the subspace of vectors $\bar{u} \in \mathbb{R}^{d(u)}$ such that

$$g_u(t, x(t), u(t))\bar{u} = 0.$$

(SLC^{k-}): For each $t_k \in \Theta$, the quadratic form

$$\langle \bar{H}_{uu}(t_k, x(t_k), u(t_k-), \psi(t_k), \nu(t_k-))\bar{u}, \bar{u}\rangle$$

is positive definite on the subspace of vectors $\bar{u} \in \mathbb{R}^{d(u)}$ such that

$$g_u(t_k, x(t_k), u(t_k-))\bar{u} = 0.$$

(SLC^{k+}): this condition is symmetric to condition (SLC^{k-}) by replacing (t_k-) everywhere by (t_k+).

Note that for each $\lambda \in M_0$ the nonstrengthened Legendre–Clebsch conditions hold; i.e., the same quadratic forms are *nonnegative* on the corresponding subspaces.

We denote by $\mathrm{Leg}_+(M_0^+)$ the set of all $\lambda \in M_0^+$ satisfying the strengthened Legendre–Clebsch conditions ($\mathrm{SLC}_{\Delta\setminus\Theta}$), ($\mathrm{SLC}^{k-}$), ($\mathrm{SLC}^{k+}$), $k = 1, \ldots, s$, and also the conditions

$$D^k(\bar{H}) > 0 \qquad \forall\, k = 1, \ldots, s. \tag{3.159}$$

Let us introduce the functional

$$\bar{\gamma}(\bar{z}) = \bar{t}_0^2 + \bar{t}_f^2 + \langle \bar{\xi}, \bar{\xi}\rangle + \langle \bar{x}(t_0), \bar{x}(t_0)\rangle + \int_{t_0}^{t_f} \langle \bar{u}(t), \bar{u}(t)\rangle\, dt, \tag{3.160}$$

which is equivalent to the norm squared on the subspace

$$\dot{\bar{x}} = f_w(t, x(t), u(t))\bar{w}; \quad [\bar{x}]^k = [\dot{x}]^k \bar{\xi}_k, \quad k = 1, \ldots, s, \tag{3.161}$$

of Hilbert space $Z_2(\Theta)$. Recall that the critical cone $\mathcal{K}$ is contained in the subspace (3.161).

Theorem 3.35. *For the trajectory $\mathcal{T}$, assume that the following Condition $\mathfrak{B}$ holds: The set $\mathrm{Leg}_+(M_0^+)$ is nonempty and there exist a nonempty compact set $M \subset \mathrm{Leg}_+(M_0^+)$ and a number $C > 0$ such that*

$$\max_{\lambda \in M} \Omega(\lambda, \bar{z}) \geq C\bar{\gamma}(\bar{z}) \tag{3.162}$$

for all $\bar{z} \in \mathcal{K}$. Then the trajectory $\mathcal{T}$ affords a strict bounded strong minimum.

3.3.4 Proofs

The proofs are based on the quadratic optimality conditions, obtained in this chapter for problems on a fixed interval of time. We will give the proofs but omit some details. In order to extend the proofs to the case of a variable interval $[t_0, t_f]$ we use a simple change of the time variable. Namely, we associate the fixed admissible trajectory $\mathcal{T} = (x(t), u(t) \mid t \in [t_0, t_f])$ in the problem on a variable time interval (3.140)–(3.144) with a trajectory $\mathcal{T}^\tau = (v(\tau), t(\tau), x(\tau), u(\tau) \mid \tau \in [\tau_0, \tau_f])$, considered on a fixed interval $[\tau_0, \tau_f]$, where $\tau_0 = t_0$, $\tau_f = t_f$, $t(\tau) \equiv \tau$, $v(\tau) \equiv 1$. This is an admissible trajectory in the following problem on a fixed interval $[\tau_0, \tau_f]$: Minimize the cost function

$$\mathcal{J}(\mathcal{T}^\tau) := J(t(\tau_0), x(\tau_0), t(\tau_f), x(\tau_f)) \to \min \tag{3.163}$$

subject to the constraints

$$F(t(\tau_0), x(\tau_0), t(\tau_f), x(\tau_f)) \le 0, \quad K(t(\tau_0), x(\tau_0), t(\tau_f), x(\tau_f)) = 0, \tag{3.164}$$

$$\frac{dx(\tau)}{d\tau} = v(\tau) f(t(\tau), x(\tau), u(\tau)), \quad \frac{dt(\tau)}{d\tau} = v(\tau), \quad \frac{dv(\tau)}{d\tau} = 0, \tag{3.165}$$

$$g(t(\tau), x(\tau), u(\tau)) = 0, \tag{3.166}$$

$$(t(\tau_0), x(\tau_0), t(\tau_f), x(\tau_f)) \in \mathcal{P}, \quad (t(\tau), x(\tau), u(\tau)) \in \mathcal{Q}. \tag{3.167}$$

In this problem, $x(\tau)$, $t(\tau)$, and $v(\tau)$ are state variables, and $u(\tau)$ is a control variable. For brevity, we refer to problem (3.140)–(3.144) as problem P (on a variable interval $\Delta = [t_0, t_f]$), and to problem (3.163)–(3.167) as problem P^τ (on a fixed interval $[\tau_0, \tau_f]$). We denote by $\mathfrak{A}^\tau$ the necessary quadratic Condition $\mathfrak{A}$ for problem P^τ on a fixed interval $[\tau_0, \tau_f]$. Similarly, we denote by $\mathfrak{B}^\tau$ the sufficient quadratic Condition $\mathfrak{B}$ for problem P^τ on a fixed interval $[\tau_0, \tau_f]$.

Recall that the control $u(\cdot)$ is a piecewise Lipschitz continuous function on the interval $\Delta = [t_0, t_f]$ with the set of discontinuity points $\Theta = \{t_1, \dots, t_s\}$, where $t_0 < t_1 < \cdots < t_s < t_f$. Hence, for each $\lambda \in M_0$, the function $\nu(t)$ is also piecewise Lipschitz continuous on the interval Δ, and, moreover, all discontinuity points of ν belong to Θ. This easily follows from the equation $\bar{H}_u = 0$ and the full-rank condition for matrix g_u. Consequently, $\dot{u}$ and $\dot{\nu}$ are bounded measurable functions on Δ. The proof of Theorem 3.31 is composed of the following chain of implications:

(i) A Pontryagin minimum is attained on the trajectory $\mathcal{T}$ in problem $P \Longrightarrow$
(ii) A Pontryagin minimum is attained on the trajectory $\mathcal{T}^\tau$ in problem $P^\tau \Longrightarrow$
(iii) Condition $\mathfrak{A}^\tau$ holds for the trajectory $\mathcal{T}^\tau$ in problem $P^\tau \Longrightarrow$
(iv) Condition $\mathfrak{A}$ holds for the trajectory $\mathcal{T}$ in problem P.

The first implication is readily verified, the second follows from Theorem 3.9. The verification of the third implication (iii) $\Rightarrow$ (iv) is not short and rather technical: we have to compare the sets of Lagrange multipliers, the critical cones, and the quadratic forms in both problems. This will be done below.

In order to prove the sufficient conditions in problem P, given by Theorem 3.35, we have to check the following chain of implications:

(v) Condition $\mathfrak{B}$ holds for the trajectory $\mathcal{T}$ in problem $P \Longrightarrow$
(vi) Condition $\mathfrak{B}^\tau$ holds for the trajectory $\mathcal{T}^\tau$ in problem $P^\tau \Longrightarrow$

(vii) A bounded strong minimum is attained on the trajectory $\mathcal{T}^\tau$ in problem P^τ $\Longrightarrow$
(viii) A bounded strong minimum is attained on the trajectory $\mathcal{T}$ in problem P.

The verification of the first implication (v) $\Rightarrow$ (vi) is similar to the verification of the third implication (iii) $\Rightarrow$ (iv) in the proof of the necessary conditions, the second implication (vi) $\Rightarrow$ (vii) follows from Theorem 3.16, and the third (vii) $\Rightarrow$ (viii) is readily verified.

Thus, it remains to compare the sets of Lagrange multipliers, the critical cones, and the quadratic forms in problems P and P^τ for the trajectories $\mathcal{T}$ and $\mathcal{T}^\tau$, respectively.

Comparison of the sets of Lagrange multipliers. Let us formulate the Pontryagin minimum principle in problem P^τ for the trajectory $\mathcal{T}^\tau$. The endpoint Lagrange function l, the Pontryagin function H, and the augmented Pontryagin function $\bar{H}$ (all of them are equipped with the superscript τ) have the form

$$l^\tau = \alpha_0 J + \alpha F + \beta K = l,$$
$$H^\tau = \psi v f + \psi_0 v + \psi_v \cdot 0 = v(\psi f + \psi_0), \quad \bar{H}^\tau = H^\tau + \nu g.$$

The set M_0^τ in problem P^τ for the trajectory $\mathcal{T}^\tau$ consists of all tuples of Lagrange multipliers $\lambda^\tau = (\alpha_0, \alpha, \beta, \psi, \psi_0, \psi_v, \nu)$ such that the following conditions hold:

$$\begin{aligned}
&\alpha_0 + |\alpha| + |\beta| = 1,\\
&-\frac{d\psi}{d\tau} = v\psi f_x + \nu g_x, \quad -\frac{d\psi_0}{d\tau} = v\psi f_t + \nu g_t, \quad -\frac{d\psi_v}{d\tau} = \psi f + \psi_0\\
&\psi(\tau_0) = -l_{x_0}, \quad \psi(\tau_f) = l_{x_f}, \quad \psi_0(\tau_0) = -l_{t_0}, \quad \psi_0(\tau_f) = l_{t_f},\\
&\psi_v(\tau_0) = \psi_v(\tau_f) = 0, \quad v\psi f_u + \nu g_u = 0,\\
&v(\tau)\big(\psi(\tau) f(t(\tau), x(\tau), u) + \psi_0(\tau)\big)\\
&\quad \geq v(\tau)\big(\psi(\tau) f(t(\tau), x(\tau), u(\tau)) + \psi_0(\tau)\big).
\end{aligned} \tag{3.168}$$

The last inequality holds for all $u \in \mathbb{R}^{d(u)}$ such that $g(t(\tau), x(\tau), u) = 0$, $(t(\tau), x(\tau), u) \in \mathcal{Q}$. Recall that $v(\tau) \equiv 1$, $t(\tau) \equiv \tau$, $\tau_0 = t_0$, and $\tau_f = t_f$. In (3.168), the function f and its derivatives f_x, f_u, f_t, g_x g_u, and g_t are taken at $(t(\tau), x(\tau), u(\tau))$, $\tau \in [\tau_0, \tau_f] \setminus \Theta$, while the derivatives l_{t_0}, l_{x_0}, l_{t_f}, and l_{x_f} are calculated at $(t(\tau_0), x(\tau_0), t(\tau_f), x(\tau_f)) = (t_0, x(t_0), t_f, x(t_f))$.

Conditions $-d\psi_v/d\tau = \psi f + \psi_0$ and $\psi_v(\tau_0) = \psi_v(\tau_f) = 0$ imply that $\int_{\tau_0}^{\tau_f} (\psi f + \psi_0)\, d\tau = 0$. As is well known, conditions (3.168) of the minimum principle also imply that $\psi f + \psi_0 = \text{const}$, whence $\psi f + \psi_0 = 0$ and $\psi_v = 0$. Taking this fact into account and comparing the definitions of the sets M_0^τ (3.168) and M_0 (3.149), we see that the projector

$$\big(\alpha_0, \alpha, \beta, \psi, \psi_0, \psi_v, \nu\big) \to \big(\alpha_0, \alpha, \beta, \psi, \psi_0, \nu\big) \tag{3.169}$$

realizes a one-to-one correspondence between these two sets. (Moreover, in the definition of the set M_0^τ one could replace the relations $-d\psi_v/d\tau = \psi f + \psi_0$ and $\psi_v(\tau_0) = \psi_v(\tau_f) = 0$ with $\psi f + \psi_0 = 0$, and thus identify M_0^τ with M_0.)

We say that an element $\lambda^\tau \in M_0^\tau$ *corresponds* to an element $\lambda \in M_0$ if λ is the projection of λ^τ under the mapping (3.169).

Comparison of the critical cones. For brevity, we set $\varrho = (v, t, x, u) = (v, t, w)$. Let us define the critical cone $\mathcal{K}^\tau$ in problem P^τ for the trajectory $\mathcal{T}^\tau$. It consists of all tuples $(\bar\xi, \bar v, \bar t, \bar x, \bar u) = (\bar\xi, \bar\varrho)$ satisfying the relations

$$J_{t_0}\bar t(\tau_0) + J_{x_0}\bar x(\tau_0) + J_{t_f}\bar t(\tau_f) + J_{x_f}\bar x(\tau_f) \le 0, \tag{3.170}$$

$$F_{it_0}\bar t(\tau_0) + F_{ix_0}\bar x(\tau_0) + F_{it_f}\bar t(\tau_f) + F_{ix_f}\bar x(\tau_f) \le 0, \quad i \in I_F(p), \tag{3.171}$$

$$K_{t_0}\bar t(\tau_0) + K_{x_0}\bar x(\tau_0) + K_{t_f}\bar t(\tau_f) + K_{x_f}\bar x(\tau_f) = 0, \tag{3.172}$$

$$\frac{d\bar x}{d\tau} = \bar v f + v\big(f_t\bar t + f_x\bar x + f_u\bar u\big), \quad [\bar x]^k = [\dot x]^k\bar\xi_k, \quad k = 1,\dots,s, \tag{3.173}$$

$$\frac{d\bar t}{d\tau} = \bar v, \quad [\bar t]^k = 0, \quad k = 1,\dots,s, \quad \frac{d\bar v}{d\tau} = 0, \quad [\bar v]^k = 0, \quad k = 1,\dots,s, \tag{3.174}$$

$$g_t\bar t + g_x\bar x + g_u\bar u = 0, \tag{3.175}$$

where the derivatives J_{t_0}, J_{x_0}, J_{t_f} J_{x_f}, etc. are calculated at $(t(\tau_0), x(\tau_0), t(\tau_f), x(\tau_f)) = (t_0, x(t_0), t_f, x(t_f))$, while f, f_t, f_x, f_u g_t, g_x, and g_u are taken at $(t(\tau), x(\tau), u(\tau))$, $\tau \in [\tau_0, \tau_f] \setminus \Theta$. Let $(\bar\xi, \bar v, \bar t, \bar x, \bar u)$ be an element of the critical cone $\mathcal{K}^\tau$. We can use the following change of variables:

$$\tilde x = \bar x - \bar t\dot x, \quad \tilde u = \bar u - \bar t\dot u, \tag{3.176}$$

or, briefly,

$$\tilde w = \bar w - \bar t\dot w. \tag{3.177}$$

Since $v = 1$, $\dot x = f$, and $t = \tau$, equation (3.173) is equivalent to

$$\frac{d\bar x}{dt} = \bar v\dot x + f_t\bar t + f_w\bar w. \tag{3.178}$$

Using the relation $\bar x = \tilde x + \bar t\dot x$ in (3.178) along with $\dot{\bar t} = \bar v$, we get

$$\dot{\tilde x} + \bar t\ddot x = \bar t f_t + f_w\bar w. \tag{3.179}$$

By differentiating the equation $\dot x(t) = f(t, w(t))$, we obtain

$$\ddot x = f_t + f_w\dot w. \tag{3.180}$$

Using this relation in (3.179), we get

$$\dot{\tilde x} = f_w\tilde w. \tag{3.181}$$

The relations

$$[\bar x]^k = [\dot x]^k\bar\xi_k, \quad \bar x = \tilde x + \bar t\dot x$$

imply

$$[\tilde x]^k = [\dot x]^k\tilde\xi_k, \tag{3.182}$$

where

$$\tilde\xi_k = \bar\xi_k - \bar t_k, \quad \bar t_k = \bar t(t_k), \quad k = 1,\dots,s. \tag{3.183}$$

Further, relation (3.175) may be written as $g_t\bar t + g_w\bar w = 0$. Differentiating the relation $g(t, w(t)) = 0$, we obtain

$$g_t + g_w\dot w = 0. \tag{3.184}$$

These relations along with (3.177) imply that

$$g_w \tilde{w} = 0. \tag{3.185}$$

Finally, note that since $\bar{x} = \tilde{x} + \bar{t}\dot{x}$ and $\tau_0 = t_0$, $\tau_f = t_f$, we have

$$\bar{p} = \big(\bar{t}_0, \bar{x}(t_0), \bar{t}_f, \bar{x}(t_f)\big) = \big(\bar{t}_0, \tilde{x}(t_0) + \bar{t}_0\dot{x}(t_0), \bar{t}_f, \tilde{x}(t_f) + \bar{t}_f\dot{x}(t_f)\big), \tag{3.186}$$

where $\bar{t}_0 = \bar{t}(t_0)$ and $\bar{t}_f = \bar{t}(t_f)$. The vector in the right-hand side of the last equality has the same form as the vector $\bar{\bar{p}}$ in definition (3.154). Consequently, all relations in definition (3.155) of the critical cone $\mathcal{K}$ in problem P are satisfied for the element $\tilde{z} = (\bar{t}_0, \bar{t}_f, \tilde{\xi}, \tilde{w})$. We have proved that the obtained element $\tilde{z}$ belongs to the critical cone $\mathcal{K}$ in problem P. Conversely, if $(\bar{t}_0, \bar{t}_f, \tilde{\xi}, \tilde{w})$ is an element of the critical cone in problem P, then by setting

$$\bar{v} = \frac{\bar{t}_f - \bar{t}_0}{t_f - t_0}, \quad \bar{t} = \bar{v}(\tau - \tau_0) + \bar{t}_0, \quad \bar{w} = \tilde{w} + \bar{t}\dot{w}, \quad \bar{\xi}_k = \tilde{\xi}_k + \bar{t}(\tau_k),\ k = 1, \dots, s,$$

we obtain an element $(\bar{\xi}, \bar{v}, \bar{t}, \bar{w})$ of the critical cone (3.170)–(3.175) in problem P^τ. Thus, we have proved the following lemma.

Lemma 3.36. *If* $(\bar{\xi}, \bar{v}, \bar{t}, \bar{w})$ *is an element of the critical cone* (3.170)–(3.175) *in problem* P^τ *for the trajectory* $\mathcal{T}^\tau$ *and*

$$\bar{t}_0 = \bar{t}(t_0), \quad \bar{t}_f = \bar{t}(t_f), \quad \tilde{w} = \bar{w} - \bar{t}\dot{w}, \quad \tilde{\xi}_k = \bar{\xi}_k - \bar{t}(t_k),\ k = 1, \dots, s, \tag{3.187}$$

then $(\bar{t}_0, \bar{t}_f, \tilde{\xi}, \tilde{w})$ *is an element of the critical cone* (3.155) *in problem* P *for the trajectory* $\mathcal{T}$. *Moreover, relations* (3.187) *define a one-to-one correspondence between elements of the critical cones in problems* P^τ *and* P.

We say that an element $(\bar{\xi}, \bar{v}, \bar{t}, \bar{w})$ of the critical cone in problem P^τ *corresponds* to an element $(\bar{t}_0, \bar{t}_f, \tilde{\xi}, \tilde{w})$ of the critical cone in problem P if relations (3.187) hold.

Comparison of the quadratic forms. Assume that the element $\lambda^\tau \in M_0^\tau$ corresponds to the element $\lambda \in M_0$. Let us show that the quadratic form $\Omega^\tau(\lambda^\tau, \cdot)$, calculated on the element $(\bar{\xi}, \bar{v}, \bar{t}, \bar{w})$ of the critical cone in problem P^τ for the trajectory $\mathcal{T}^\tau$, can be transformed into the quadratic form $\Omega(\lambda, \cdot)$ calculated on the corresponding element $(\bar{t}_0, \bar{t}_f, \tilde{\xi}, \tilde{w})$ of the critical cone in problem P for the trajectory $\mathcal{T}$.

(i) The relations $\bar{H}^\tau = v(H + \psi_0) + \nu g$, $\bar{H} = H + \nu g$, $v = 1$ imply

$$\langle \bar{H}^\tau_{\varrho\varrho}\bar{\varrho}, \bar{\varrho}\rangle = \langle \bar{H}_{ww}\bar{w}, \bar{w}\rangle + 2\bar{H}_{tw}\bar{w}\bar{t} + \bar{H}_{tt}\bar{t}^2 + 2\bar{v}(H_w\bar{w} + H_t\bar{t}), \tag{3.188}$$

where $\varrho = (v, t, w)$, $\bar{\varrho} = (\bar{v}, \bar{t}, \bar{w})$. Since $\bar{w} = \tilde{w} + \bar{t}\dot{w}$, we have

$$\langle \bar{H}_{ww}\bar{w}, \bar{w}\rangle = \langle \bar{H}_{ww}\tilde{w}, \tilde{w}\rangle + 2\langle \bar{H}_{ww}\dot{w}, \bar{w}\rangle\bar{t} - \langle \bar{H}_{ww}\dot{w}, \dot{w}\rangle\bar{t}^2. \tag{3.189}$$

Moreover, using the relations

$$\begin{aligned} &H_w = \bar{H}_w - \nu g_w, \quad H_t = \bar{H}_t - \nu g_t, \quad g_w\bar{w} + g_t\bar{t} = 0, \\ &-\dot{\psi} = \bar{H}_x, \quad -\dot{\psi}_0 = \bar{H}_t, \quad \bar{H}_u = 0, \end{aligned}$$

we obtain

$$\begin{aligned} H_w\bar{w} + H_t\bar{t} &= \bar{H}_w\bar{w} + \bar{H}_t\bar{t} - \nu(g_w\bar{w} + g_t\bar{t}) \\ &= \bar{H}_w\bar{w} + \bar{H}_t\bar{t} = \bar{H}_x\bar{x} + \bar{H}_t\bar{t} = -\dot{\psi}\bar{x} - \dot{\psi}_0\bar{t}. \end{aligned} \tag{3.190}$$

Relations (3.188)–(3.190) imply

$$\begin{aligned} \langle \bar{H}^{\tau}_{\varrho\varrho}\bar{\varrho}, \bar{\varrho}\rangle &= \langle \bar{H}_{ww}\tilde{w}, \tilde{w}\rangle + 2\langle \bar{H}_{ww}\dot{w}, \bar{w}\rangle\bar{t} + 2\bar{H}_{tw}\bar{w}\bar{t} \\ &\quad - \langle \bar{H}_{ww}\dot{w}, \dot{w}\rangle\bar{t}^2 + \bar{H}_{tt}\bar{t}^2 - 2\bar{v}\big(\dot{\psi}\bar{x} + \dot{\psi}_0\bar{t}\big). \end{aligned} \tag{3.191}$$

(ii) Let us transform the terms $2\langle \bar{H}_{ww}\dot{w}, \bar{w}\rangle\bar{t} + 2\bar{H}_{tw}\bar{w}\bar{t}$ in (3.191). By differentiating $-\dot{\psi} = \bar{H}_x$ with respect to t, we obtain

$$-\ddot{\psi} = \bar{H}_{tx} + (\dot{w})^*\bar{H}_{wx} + \dot{\psi}\bar{H}_{\psi x} + \dot{\nu}\bar{H}_{\nu x}.$$

Here we have $\bar{H}_{\psi x} = f_x$ and $\bar{H}_{\nu x} = g_x$. Therefore

$$-\ddot{\psi} = \bar{H}_{tx} + (\dot{w})^*\bar{H}_{wx} + \dot{\psi}f_x + \dot{\nu}g_x. \tag{3.192}$$

Similarly, by differentiating $\bar{H}_u = 0$ with respect to t, we obtain

$$0 = \bar{H}_{tu} + (\dot{w})^*\bar{H}_{wu} + \dot{\psi}f_u + \dot{\nu}g_u. \tag{3.193}$$

Multiplying (3.192) by $\bar{x}$ and (3.193) by $\bar{u}$ and summing the results, we get

$$-\ddot{\psi}\bar{x} = \bar{H}_{tw}\bar{w} + \langle \bar{H}_{ww}\dot{w}, \bar{w}\rangle + \dot{\psi}f_w\bar{w} + \dot{\nu}g_w\bar{w}. \tag{3.194}$$

Since $(\bar{\xi}, \bar{v}, \bar{t}, \bar{w})$ is an element of the critical cone in problem P^τ, from (3.173) and (3.175) we get $f_w\bar{w} = \dot{\bar{x}} - \bar{v}\dot{x} - f_t\bar{t}$, $g_w\bar{w} = -g_t\bar{t}$. Therefore, equation (3.194) can be represented in the form

$$\bar{H}_{tw}\bar{w} + \langle \bar{H}_{ww}\dot{w}, \bar{w}\rangle = \bar{v}(\dot{\psi}\dot{x}) - \frac{d}{dt}(\dot{\psi}\bar{x}) + \big(\dot{\psi}f_t + \dot{\nu}g_t\big)\bar{t}, \tag{3.195}$$

which implies

$$2\langle \bar{H}_{ww}\dot{w}, \bar{w}\rangle\bar{t} + 2\bar{H}_{tw}\bar{w}\bar{t} = 2\bar{t}\bar{v}(\dot{\psi}\dot{x}) - 2\bar{t}\frac{d}{dt}(\dot{\psi}\bar{x}) + 2\big(\dot{\psi}f_t + \dot{\nu}g_t\big)\bar{t}^2. \tag{3.196}$$

(iii) Let us transform the term $-\langle \bar{H}_{ww}\dot{w}, \dot{w}\rangle\bar{t}^2$ in (3.191). Multiplying (3.192) by $\dot{x}$ and (3.193) by $\dot{u}$ and summing the results, we obtain

$$-\ddot{\psi}\dot{x} = \bar{H}_{tw}\dot{w} + \langle \bar{H}_{ww}\dot{w}, \dot{w}\rangle + \dot{\psi}f_w\dot{w} + \dot{\nu}g_w\dot{w}. \tag{3.197}$$

From (3.180) and (3.184), we get $f_w\dot{w} = \ddot{x} - f_t$, $g_w\dot{w} = -g_t$, respectively. Then (3.197) implies

$$\bar{H}_{tw}\dot{w} + \langle \bar{H}_{ww}\dot{w}, \dot{w}\rangle = -\frac{d}{dt}(\dot{\psi}\dot{x}) + \big(\dot{\psi}f_t + \dot{\nu}g_t\big). \tag{3.198}$$

Multiplying this relation by $-\bar{t}^2$, we get

$$-\langle \bar{H}_{ww}\dot{w}, \dot{w}\rangle\bar{t}^2 = \bar{H}_{tw}\dot{w}\bar{t}^2 + \bar{t}^2\frac{d}{dt}(\dot{\psi}\dot{x}) - \big(\dot{\psi}f_t + \dot{\nu}g_t\big)\bar{t}^2. \tag{3.199}$$

(iv) Finally, let us transform the term $\bar{H}_{tt}\bar{t}^2$ in (3.191). Differentiating $-\dot{\psi}_0 = \bar{H}_t$ with respect to t and using the relations $\bar{H}_{\psi t} = f_t$ and $\bar{H}_{\nu t} = g_t$, we get

$$-\ddot{\psi}_0 = \bar{H}_{tt} + \bar{H}_{tw}\dot{w} + \big(\dot{\psi} f_t + \dot{\nu} g_t\big). \tag{3.200}$$

Consequently,

$$\bar{H}_{tt}\bar{t}^2 = -\ddot{\psi}_0\bar{t}^2 - \bar{H}_{tw}\dot{w}\bar{t}^2 - \big(\dot{\psi} f_t + \dot{\nu} g_t\big)\bar{t}^2. \tag{3.201}$$

(v) Summing (3.199) and (3.201), we obtain

$$-\langle \bar{H}_{ww}\dot{w}, \dot{w}\rangle \bar{t}^2 + \bar{H}_{tt}\bar{t}^2 = -\ddot{\psi}_0\bar{t}^2 - 2\big(\dot{\psi} f_t + \dot{\nu} g_t\big)\bar{t}^2 + \bar{t}^2\frac{d}{dt}(\dot{\psi}\dot{x}). \tag{3.202}$$

Using relations (3.196) and (3.202) in (3.191), we get

$$\begin{aligned}\langle \bar{H}^{\tau}_{\varrho\varrho}\bar{\varrho}, \bar{\varrho}\rangle &= \langle \bar{H}_{ww}\tilde{w}, \tilde{w}\rangle + 2\bar{t}\bar{v}(\dot{\psi}\dot{x}) - 2\bar{t}\frac{d}{dt}(\dot{\psi}\bar{x}) \\ &\quad - \ddot{\psi}_0\bar{t}^2 + \bar{t}^2\frac{d}{dt}(\dot{\psi}\dot{x}) - 2\bar{v}\big(\dot{\psi}\bar{x} + \dot{\psi}_0\bar{t}\big).\end{aligned} \tag{3.203}$$

But

$$\ddot{\psi}_0\bar{t}^2 + 2\bar{v}\bar{t}\dot{\psi}_0 = \frac{d}{dt}\big(\dot{\psi}_0\bar{t}^2\big), \quad \bar{t}\frac{d}{dt}(\dot{\psi}\bar{x}) + \bar{v}(\dot{\psi}\bar{x}) = \frac{d}{dt}\big(\bar{t}\dot{\psi}\bar{x}\big),$$

$$2\bar{t}\bar{v}(\dot{\psi}\dot{x}) + \bar{t}^2\frac{d}{dt}(\dot{\psi}\dot{x}) = \frac{d}{dt}(\dot{\psi}\dot{x}\bar{t}^2).$$

Therefore,

$$\langle \bar{H}^{\tau}_{\varrho\varrho}\bar{\varrho}, \bar{\varrho}\rangle = \langle \bar{H}_{ww}\tilde{w}, \tilde{w}\rangle + \frac{d}{dt}\Big((\dot{\psi}\dot{x})\bar{t}^2 - \dot{\psi}_0\bar{t}^2 - 2\dot{\psi}\bar{x}\bar{t}\Big). \tag{3.204}$$

Finally, using the change of the variable $\bar{x} = \tilde{x} + \bar{t}\dot{x}$ in the right-hand side of this relation, we obtain

$$\langle \bar{H}^{\tau}_{\varrho\varrho}\bar{\varrho}, \bar{\varrho}\rangle = \langle \bar{H}_{ww}\tilde{w}, \tilde{w}\rangle - \frac{d}{dt}\Big((\dot{\psi}_0 + \dot{\psi}\dot{x})\bar{t}^2 + 2\dot{\psi}\tilde{x}\bar{t}\Big). \tag{3.205}$$

We have proved the following lemma.

Lemma 3.37. *Let* $(\bar{\xi}, \bar{v}, \bar{t}, \bar{w}) = (\bar{\xi}, \bar{\varrho})$ *be an element of the critical cone* $\mathcal{K}^{\tau}$ *in problem* P^{τ} *for the trajectory* $\mathcal{T}^{\tau}$. *Set* $\tilde{w} = \bar{w} - \bar{t}\dot{w}$. *Then formula* (3.205) *holds.*

(vi) Recall that λ^{τ} is an arbitrary element of the set M_0^{τ} (consequently $\psi_v = 0$) and λ is the corresponding element of the set M_0, i.e., λ is the projection of λ^{τ} under the mapping (3.169). The quadratic form $\Omega^{\tau}(\lambda^{\tau}, \cdot)$ in problem P^{τ} for the trajectory $\mathcal{T}^{\tau}$ has the following representation:

$$\begin{aligned}\Omega^{\tau}(\lambda^{\tau}; \bar{\xi}, \bar{\varrho}) \;=\; &\sum_{k=1}^{s}\Big(D^k(\bar{H}^{\tau})\bar{\xi}_k^2 - 2[\dot{\psi}]^k\bar{x}^k_{\mathrm{av}}\bar{\xi}_k - 2[\dot{\psi}_0]^k\bar{t}^k_{\mathrm{av}}\bar{\xi}_k\Big) \\ &+ \langle l_{pp}\bar{p}, \bar{p}\rangle + \int_{\tau_0}^{\tau_f}\langle \bar{H}^{\tau}_{\varrho\varrho}\bar{\varrho}, \bar{\varrho}\rangle\, d\tau.\end{aligned} \tag{3.206}$$

Comparing the definitions of $D^k(\bar{H}^\tau)$ and $D^k(\bar{H})$ (see (3.152)) and taking into account that $\bar{H}^\tau = v(\psi f + \psi_0) + \nu g$ and $v = 1$, we get

$$D^k(\bar{H}^\tau) = D^k(\bar{H}). \tag{3.207}$$

Let $\bar{z}^\tau = (\bar{\xi}, \bar{\varrho}) = (\bar{\xi}, \bar{v}, \bar{t}, \bar{x}, \bar{u})$ be an element of the critical cone $\mathcal{K}^\tau$ in the problem P^τ for the trajectory $\mathcal{T}^\tau$, and let $\tilde{z} = (\bar{t}_0, \bar{t}_f, \tilde{\xi}, \tilde{x}, \bar{u})$ be the corresponding element of the critical cone $\mathcal{K}$ in the problem P for the trajectory $\mathcal{T}$; i.e., relations (3.187) hold. Since $[\bar{t}]^k = 0$, $k = 1, \dots, s$, we have

$$\bar{t}^k_{\text{av}} = \bar{t}_k, \quad k = 1, \dots, s \tag{3.208}$$

where $\bar{t}_k = \bar{t}(t_k)$, $k = 1, \dots, s$. Also recall that $\tau_0 = t_0$, $\tau_f = t_f$, $t(\tau) = \tau$, $dt = d\tau$. Since the functions $\dot{\psi}_0$, $\dot{\psi}$, $\dot{x}$, and $\tilde{x}$ may have discontinuities only at the points of the set Θ, the following formula holds:

$$\begin{aligned}\int_{t_0}^{t_f} \frac{d}{dt}\Big((\dot{\psi}_0 + \dot{\psi}\dot{x})\bar{t}^2 + 2\dot{\psi}\tilde{x}\bar{t}\Big)\,dt &= \Big((\dot{\psi}_0 + \dot{\psi}\dot{x})\bar{t}^2 + 2\dot{\psi}\tilde{x}\bar{t}\Big)\Big|_{t_0}^{t_f} \\ &\quad - \sum_{k=1}^{s}\Big([\dot{\psi}_0 + \dot{\psi}\dot{x}]^k \bar{t}(t_k)^2 + 2[\dot{\psi}\tilde{x}]^k \bar{t}(t_k)\Big).\end{aligned} \tag{3.209}$$

Relations (3.205)–(3.209) imply the following representation of the quadratic form Ω^τ on the element $(\bar{\xi}, \bar{\varrho})$ of the critical cone $\mathcal{K}^\tau$:

$$\begin{aligned}\Omega^\tau(\lambda^\tau; \bar{\xi}, \bar{\varrho}) &= \sum_{k=1}^{s}\Big(D^k(\bar{H})\bar{\xi}_k^2 - 2[\dot{\psi}]^k \bar{x}^k_{\text{av}}\bar{\xi}_k - 2[\dot{\psi}_0]^k \bar{t}(t_k)\bar{\xi}_k \\ &\quad + [\dot{\psi}_0 + \dot{\psi}\dot{x}]^k \bar{t}(t_k)^2 + 2[\dot{\psi}\tilde{x}]^k \bar{t}(t_k)\Big) + \langle l_{pp}\bar{p}, \bar{p}\rangle \\ &\quad - \Big((\dot{\psi}_0 + \dot{\psi}\dot{x})\bar{t}^2 + 2\dot{\psi}\tilde{x}\bar{t}\Big)\Big|_{t_0}^{t_f} + \int_{t_0}^{t_f}\langle \bar{H}_{ww}\bar{w}, \bar{w}\rangle\,d\tau.\end{aligned} \tag{3.210}$$

Let us transform the terms related to the discontinuity points t_k of the control $u(\cdot)$, $k = 1, \dots, s$. For any $\lambda \in M_0$, the following lemma holds.

Lemma 3.38. *Let $\bar{z} = (\bar{\xi}, \bar{\varrho}) = (\bar{\xi}, \bar{v}, \bar{t}, \bar{w})$ be an element of the critical cone $\mathcal{K}^\tau$ in the problem P^τ for the trajectory $\mathcal{T}^\tau$. Let the pair $(\tilde{\xi}, \tilde{x})$ be defined by the relations*

$$\tilde{\xi}_k = \bar{\xi}_k - \bar{t}(t_k), \quad k = 1, \dots, s, \quad \tilde{x} = \bar{x} - \bar{t}\dot{x}. \tag{3.211}$$

Then for any $k = 1, \dots, s$ the following formula holds:

$$\begin{aligned}&D^k(\bar{H})\bar{\xi}_k^2 - 2[\dot{\psi}]^k \bar{x}^k_{\text{av}}\bar{\xi}_k - 2[\dot{\psi}_0]^k \bar{t}(t_k)\bar{\xi}_k + [\dot{\psi}_0 + \dot{\psi}\dot{x}]^k \bar{t}(t_k)^2 + 2[\dot{\psi}\tilde{x}]^k \bar{t}(t_k) \\ &= D^k(\bar{H})\tilde{\xi}_k^2 - 2[\dot{\psi}]^k \tilde{x}^k_{\text{av}}\tilde{\xi}_k.\end{aligned} \tag{3.212}$$

Proof. In this proof, we omit the subscript and superscript k. We also write $\bar{t}$ instead of $\bar{t}(t_k)$. Set $a = D(\bar{H})$. Using the relations

$$\bar{\xi} = \tilde{\xi} + \bar{t}, \quad \bar{x}_{\text{av}} = \tilde{x}_{\text{av}} + \bar{t}\dot{x}_{\text{av}}, \tag{3.213}$$

we obtain

$$
\begin{aligned}
&a\bar{\xi}^2 - 2[\dot{\psi}]\bar{x}_{\rm av}\bar{\xi} - 2[\dot{\psi}_0]\bar{t}\bar{\xi} + [\dot{\psi}_0 + \dot{\psi}\dot{x}]\bar{t}^2 + 2[\dot{\psi}\bar{x}]\bar{t} \\
&\quad = a\tilde{\xi}^2 + 2a\tilde{\xi}\bar{t} + a\bar{t}^2 - 2[\dot{\psi}]\tilde{x}_{\rm av}\tilde{\xi} - 2[\dot{\psi}]\dot{x}_{\rm av}\bar{t}\tilde{\xi} - 2[\dot{\psi}_0]\bar{t}\tilde{\xi} + [\dot{\psi}_0 + \dot{\psi}\dot{x}]\bar{t}^2 + 2[\dot{\psi}\bar{x}]\bar{t} \\
&\quad = a\tilde{\xi}^2 - 2[\dot{\psi}]\tilde{x}_{\rm av}\tilde{\xi} + r,
\end{aligned}
\tag{3.214}
$$

where

$$
r = 2a\tilde{\xi}\bar{t} + a\bar{t}^2 - 2[\dot{\psi}]\tilde{x}_{\rm av}\bar{t} - 2[\dot{\psi}]\dot{x}_{\rm av}\bar{t}\tilde{\xi} - 2[\dot{\psi}_0]\bar{t}\tilde{\xi} + [\dot{\psi}_0 + \dot{\psi}\dot{x}]\bar{t}^2 + 2[\dot{\psi}\bar{x}]\bar{t}. \tag{3.215}
$$

It suffices to show that $r = 0$. Using the relations (3.213) in formula (3.215), we get

$$
\begin{aligned}
r &= 2a(\bar{\xi} - \bar{t})\bar{t} + a\bar{t}^2 - 2[\dot{\psi}](\bar{x}_{\rm av} - \bar{t}\dot{x}_{\rm av})\bar{t} - 2[\dot{\psi}]\dot{x}_{\rm av}\bar{t}\tilde{\xi} - 2[\dot{\psi}_0]\bar{t}\tilde{\xi} \\
&\quad + [\dot{\psi}_0 + \dot{\psi}\dot{x}]\bar{t}^2 + 2[\dot{\psi}(\bar{x} - \bar{t}\dot{x})]\bar{t} \\
&= \bar{t}^2\big(-a + 2[\dot{\psi}]\dot{x}_{\rm av} + [\dot{\psi}_0] - [\dot{\psi}\dot{x}]\big) + 2\bar{t}\bar{\xi}\big(a - [\dot{\psi}]\dot{x}_{\rm av} - [\dot{\psi}_0]\big) \\
&\quad + 2\bar{t}\big(-[\dot{\psi}]\bar{x}_{\rm av} + [\dot{\psi}\bar{x}]\big).
\end{aligned}
$$

The coefficient of $\bar{t}^2$ in the right-hand side of the last equality vanishes:

$$
\begin{aligned}
-a + 2[\dot{\psi}]\dot{x}_{\rm av} + [\dot{\psi}_0] - [\dot{\psi}\dot{x}] &= -(\dot{\psi}^+\dot{x}^- - \dot{\psi}^-\dot{x}^+ + [\dot{\psi}_0]) + (\dot{\psi}^+ - \dot{\psi}^-)(\dot{x}^+ + \dot{x}^-) \\
&\quad + [\dot{\psi}_0] - \dot{\psi}^+\dot{x}^+ + \dot{\psi}^-\dot{x}^- = 0.
\end{aligned}
$$

The coefficient of $2\bar{t}\bar{\xi}$ is equal to

$$
\begin{aligned}
a - [\dot{\psi}]\dot{x}_{\rm av} - [\dot{\psi}_0] &= \dot{\psi}^+\dot{x}^- - \dot{\psi}^-\dot{x}^+ + [\dot{\psi}_0] - \frac{1}{2}(\dot{\psi}^+ - \dot{\psi}^-)(\dot{x}^- + \dot{x}^+) - [\dot{\psi}_0] \\
&= \frac{1}{2}(\dot{\psi}^+\dot{x}^- - \dot{\psi}^-\dot{x}^+) - \frac{1}{2}[\dot{\psi}\dot{x}].
\end{aligned}
$$

The coefficient of $2\bar{t}$ is equal to

$$
\begin{aligned}
-[\dot{\psi}]\bar{x}_{\rm av} + [\dot{\psi}\bar{x}] &= -\frac{1}{2}(\dot{\psi}^+ - \dot{\psi}^-)(\bar{x}^- + \bar{x}^+) + (\dot{\psi}^+\bar{x}^+ - \dot{\psi}^-\bar{x}^-) \\
&= \frac{1}{2}\dot{\psi}^+[\bar{x}] + \frac{1}{2}\dot{\psi}^-[\bar{x}] = \dot{\psi}_{\rm av}[\dot{x}]\bar{\xi},
\end{aligned}
$$

since $[\bar{x}] = [\dot{x}]\bar{\xi}$. Consequently,

$$
\begin{aligned}
r &= 2\bar{t}\bar{\xi}\left(\frac{1}{2}(\dot{\psi}^+\dot{x}^- - \dot{\psi}^-\dot{x}^+) - \frac{1}{2}[\dot{\psi}\dot{x}] + \dot{\psi}_{\rm av}[\dot{x}]\right) \\
&= \bar{t}\bar{\xi}\big((\dot{\psi}^+\dot{x}^- - \dot{\psi}^-\dot{x}^+) - (\dot{\psi}^+\dot{x}^+ - \dot{\psi}^-\dot{x}^-) \\
&\quad + (\dot{\psi}^- + \dot{\psi}^+)(\dot{x}^+ - \dot{x}^-)\big) = 0.
\end{aligned}
$$

In view of (3.214) the equality $r = 0$ proves the lemma. ∎

Relation (3.210) along with equality (3.212) gives the following transformation of quadratic form Ω^τ (see (3.206)) on the element $\bar{z}^\tau = (\bar{\xi}, \bar{\varrho})$ of the critical cone $\mathcal{K}^\tau$,

$$
\begin{aligned}
\Omega^\tau(\lambda^\tau; \bar{\xi}, \bar{\varrho}) &= \sum_{k=1}^{s} \big(D^k(\bar{H})\tilde{\xi}_k^2 - 2[\dot{\psi}]^k \tilde{x}_{\rm av}^k \tilde{\xi}_k\big) \\
&\quad + \langle l_{pp}\bar{p}, \bar{p}\rangle - \Big((\dot{\psi}_0 + \dot{\psi}\dot{x})\bar{t}^2 + 2\dot{\psi}\tilde{x}\bar{t}\Big)\Big|_{t_0}^{t_f} + \int_{t_0}^{t_f} \langle \bar{H}_{ww}\tilde{w}, \tilde{w}\rangle\, d\tau.
\end{aligned}
\tag{3.216}
$$

Taking into account (3.186) and definitions (3.156)–(3.158) of quadratic forms ω_e, ω, and Ω, we see that the right-hand side of (3.216) is the quadratic form $\Omega(\lambda,\tilde{z})$ (see (3.158)) in problem P for the trajectory $\mathcal{T}$, where $\tilde{z}=(\bar{t}_0,\bar{t}_f,\tilde{\xi},\tilde{w})$ is the corresponding element of the critical cone $\mathcal{K}$. Thus we have proved the following theorem.

Theorem 3.39. *Let $\bar{z}^\tau=(\bar{\xi},\bar{v},\bar{t},\bar{w})$ be an element of the critical cone $\mathcal{K}^\tau$ in problem P^τ for the trajectory $\mathcal{T}^\tau$. Let $\tilde{z}=(\bar{t}_0,\bar{t}_f,\tilde{\xi},\tilde{w})$ be the corresponding element of the critical cone $\mathcal{K}$ in problem P for the trajectory $\mathcal{T}$, i.e., relations* (3.187) *hold. Then for any $\lambda^\tau\in M_0^\tau$ and the corresponding projection $\lambda\in M_0$ (under the mapping* (3.169)*) the following equality holds: $\Omega^\tau(\lambda^\tau,\bar{z}^\tau)=\Omega(\lambda,\tilde{z})$.*

This theorem proves the implications (iii) $\Rightarrow$ (iv) and (v) $\Rightarrow$ (vi) (see the beginning of this section), and thus completes the proofs of Theorems 3.31 and 3.35.

3.4 Quadratic Conditions for Optimal Control Problems with Mixed Control-State Equality and Inequality Constraints

In this section, we give a statement of the general optimal control problem with mixed control-state equality and inequality constraints on fixed and variable time intervals, recall different concepts of minimum, and formulate optimality conditions.

3.4.1 General Optimal Control Problem on a Fixed Time Interval

Statement of the problem. We consider the following optimal control problem on a fixed interval of time $[t_0, t_f]$:

$$\text{Minimize}\quad \mathcal{J}(x,u)=J\big(x(t_0),x(t_f)\big) \tag{3.217}$$

subject to the constraints

$$F\big(x(t_0),x(t_f)\big)\le 0,\quad K\big(x(t_0),x(t_f)\big)=0,\quad \big(x(t_0),x(t_f)\big)\in\mathcal{P}, \tag{3.218}$$

$$\dot{x}=f(t,x,u),\quad g(t,x,u)=0,\quad \varphi(t,x,u)\le 0,\quad (t,x,u)\in\mathcal{Q}, \tag{3.219}$$

where $\mathcal{P}$ and $\mathcal{Q}$ are open sets and x, u, F, K, f, g, and φ are vector functions, $d(g)\le d(u)$. We use the notation $x(t_0)=x_0$, $x(t_f)=x_f$, $(x_0,x_f)=p$, and $(x,u)=w$. We seek the minimum among the pairs of functions $w(\cdot)=(x(\cdot),u(\cdot))$ such that $x(\cdot)\in W^{1,1}([t_0,t_f],\mathbb{R}^{d(x)})$, $u(\cdot)\in L^\infty([t_0,t_f],\mathbb{R}^{d(u)})$. Therefore, we seek for the minimum in the space

$$W:=W^{1,1}([t_0,t_f],\mathbb{R}^{d(x)})\times L^\infty([t_0,t_f],\mathbb{R}^{d(u)}).$$

A pair $w=(x,u)\in W$ is said to be *admissible* in problem (3.217)–(3.219) if constraints (3.218)–(3.219) are satisfied by w.

Assumption 3.40. (a) The functions $J(p)$, $F(p)$, and $K(p)$ are defined and twice continuously differentiable on the open set $\mathcal{P}\subset\mathbb{R}^{2d(x)}$, and the functions $f(t,w)$, $g(t,w)$, and $\varphi(t,w)$ are defined and twice continuously differentiable on the open set $\mathcal{Q}\subset\mathbb{R}^{d(x)+d(u)+1}$.
(b) The gradients with respect to the control $g_{iu}(t,w)$, $i=1,\dots,d(g)$, $\varphi_{ju}(t,w)$, $j\in I_\varphi(t,w)$

are linearly independent at all points $(t,w) \in \mathcal{Q}$ such that $g(t,w)=0$ and $\varphi(t,w) \le 0$. Here g_i and φ_j are the components of the vector functions g and φ, respectively, and

$$I_\varphi(t,w) = \{j \in \{1,\dots,d(\varphi)\} \mid \varphi_j(t,w)=0\} \tag{3.220}$$

is the *set of indices* of active inequality constraints $\varphi_j(w,t) \le 0$ at $(t,w) \in \mathcal{Q}$.

We refer to (b) as the *linear independence assumption* for the gradients of the active mixed constraints with respect to the control. Let a pair $w^0(\cdot) = (x^0(\cdot), u^0(\cdot)) \in W$ satisfying constraints (3.218)–(3.219) of the problem be the point tested for optimality.

Assumption 3.41. The control $u^0(\cdot)$ is a piecewise continuous function such that all its discontinuity points are L-points (see Definition 2.1). Let $\Theta = \{t_1,\dots,t_s\}$, $t_0 < t_1 < \dots < t_s < t_f$ be the set of all discontinuity points of $u^0(\cdot)$. It is also assumed that $(t_k, x^0(t_k), u^{0k-}) \in \mathcal{Q}$, $(t_k, x^0(t_k), u^{0k+}) \in \mathcal{Q}$, $k=1,\dots,s$, where $u^{0k-} = u^0(t_k - 0)$, $u^{0k+} = u^0(t_k+0)$.

In what follows, we assume for definiteness that the set Θ of discontinuity points of u^0 is nonempty. Whenever this set is empty, all statements admit obvious simplifications.

Minimum on a set of sequences. Weak and Pontryagin minimum. Let S be an arbitrary set of sequences $\{\delta w_n\}$ in the space W closed with respect to the operation of taking subsequences. For problem (3.217)–(3.219), let us define a concept of minimum on S at the admissible point $w^0 = (x^0, u^0)$. Set $p^0 = (x^0(t_0), x^0(t_f))$.

Definition 3.42. We say that w^0 is a *(strict) minimum on S* if there exists no sequence $\{\delta w_n\} \in S$ such that the following conditions hold for all its members:

$$J(p^0+\delta p_n) < J(p^0) \quad (J(p^0+\delta p_n) \le J(p^0),\ \delta w_n \ne 0), \tag{3.221}$$
$$F(p^0+\delta p_n) \le 0, \quad K(p^0+\delta p_n) = 0, \tag{3.222}$$
$$\dot{x}^0 + \delta\dot{x}_n = f(t, w^0+\delta w_n), \quad g(t, w^0+\delta w_n) = 0, \tag{3.223}$$
$$\varphi(t, w^0+\delta w_n) \le 0, \quad (p^0+\delta p_n) \in \mathcal{P}, \quad (t, w^0+\delta w_n) \in \mathcal{Q}, \tag{3.224}$$

where $\delta p_n = (\delta x_n(t_0), \delta x_n(t_f))$, and $\delta w_n = (\delta x_n, \delta u_n)$. Any sequence from S which satisfies conditions (3.221)–(3.224) is said to *violate the (strict) minimality on S*.

Let S^0 be the set of sequences $\{\delta w_n\}$ in W such that $\|\delta w_n\| = \|\delta x_n\|_{1,1} + \|\delta u_n\|_\infty \to 0$. A *weak minimum* is a minimum on S^0.

Definition 3.43. We say that w^0 is *a point of Pontryagin minimum* if this point is a minimum on the set of sequences $\{\delta w_n\}$ in W satisfying the following two conditions:
(a) $\|\delta x_n\|_{1,1} + \|\delta u_n\|_1 \to 0$, where $\|\delta u_n\|_1 = \int_{t_0}^{t_f} |\delta u_n|\,dt$;
(b) there exists a compact set $\mathcal{C} \subset \mathcal{Q}$ (which depends on the choice of the sequence) such that for all sufficiently large n, we have $(t, w^0(t)+\delta w_n(t)) \in \mathcal{C}$ a.e. on $[t_0, t_f]$.

Any sequence satisfying conditions (a) and (b) will be referred to as a *Pontryagin sequence on $\mathcal{Q}$*.

Obviously, every Pontryagin minimum is a weak minimum.

Minimum principle. Let us state the well-known first-order necessary conditions for both a weak and for a Pontryagin minimum. These conditions are often referred to as the local and integral minimum principle, respectively. The local minimum principle, which we give first, is conveniently identified with the nonemptiness of the set Λ_0 defined below. Let

$$l = \alpha_0 J + \alpha F + \beta K, \quad H = \psi f, \quad \bar{H} = H + \nu g + \mu\varphi, \tag{3.225}$$

where α_0 is a scalar, and α, β, ψ, ν, and μ are row vectors of the same dimensions as F, K, f, g, and φ, respectively. The dependence of the functions l, H, and $\bar{H}$ on the variables is as follows: $l = l(p,\alpha_0,\alpha,\beta)$, $H = H(t,w,\psi)$, $\bar{H} = \bar{H}(t,w,\psi,\nu,\mu)$. The function l is said to be the *endpoint Lagrange function*, H is the *Pontryagin function* (or the *Hamiltonian*), and $\bar{H}$ is the *augmented Pontryagin function* (or the *augmented Hamiltonian*).

Denote by $\mathbb{R}^{n*}$ the space of n-dimensional row-vectors. Set

$$\lambda = (\alpha_0,\alpha,\beta,\psi(\cdot),\nu(\cdot),\mu(\cdot)), \tag{3.226}$$

where $\alpha_0 \in \mathbb{R}^1$, $\alpha \in \mathbb{R}^{d(F)*}$, $\beta \in \mathbb{R}^{d(K)*}$, $\psi(\cdot) \in W^{1,1}([t_0,t_f],\mathbb{R}^{d(x)*})$, $\nu(\cdot) \in L^\infty([t_0,t_f],\mathbb{R}^{d(g)*})$, $\mu(\cdot) \in L^\infty([t_0,t_f],\mathbb{R}^{d(\varphi)*})$. Denote by Λ_0 the set of all tuples λ satisfying the conditions

$$\alpha_0 \geq 0, \quad \alpha \geq 0, \quad \alpha F(p^0) = 0, \tag{3.227}$$

$$\alpha_0 + \sum_{i=1}^{d(F)} \alpha_i + \sum_{j=1}^{d(K)} |\beta_j| = 1, \tag{3.228}$$

$$\mu(t) \geq 0, \quad \mu(t)\varphi(t,w^0(t)) = 0, \tag{3.229}$$

$$\dot{\psi} = -\bar{H}_x(t,w^0(t),\psi(t),\nu(t),\mu(t)), \tag{3.230}$$

$$\psi(t_0) = -l_{x_0}(p^0,\alpha_0,\alpha,\beta), \quad \psi(t_f) = l_{x_f}(p^0,\alpha_0,\alpha,\beta), \tag{3.231}$$

$$\bar{H}_u(t,w^0(t),\psi(t),\nu(t),\mu(t)) = 0, \tag{3.232}$$

where α_i and β_j are components of the row vectors α and β, respectively, and $\bar{H}_x$, $\bar{H}_u$, l_{x_0}, and l_{x_f} are gradients with respect to the corresponding variables.

It is well known that if w^0 is a weak minimum, then Λ_0 is nonempty (see, e.g., [30]). The latter condition is just the *local minimum principle*. Note that Λ_0 can consist of more than one element. The following result pertain to this possibility.

Proposition 3.44. *The set Λ_0 is a finite-dimensional compact set, and the projection $\lambda = (\alpha_0,\alpha,\beta,\psi,\nu,\mu) \to (\alpha_0,\alpha,\beta)$ is injective on Λ_0.*

This property of Λ_0 follows from the linear independence assumption for the gradients of the active mixed constraints with respect to the control. This assumption also guarantees the following property.

Proposition 3.45. *Let $\lambda \in \Lambda_0$ be an arbitrary tuple. Then its components $\nu(t)$ and $\mu(t)$ are continuous at each point of continuity of the control $u^0(t)$. Consequently, $\nu(t)$ and $\mu(t)$ are piecewise continuous functions such that all their discontinuity points belong to the set Θ. The adjoint variable $\psi(t)$ is a piecewise smooth function such that all its break points belong to the set Θ.*

In a similar way, the integral minimum principle, which is a first-order necessary condition for a Pontryagin minimum at w^0, can be stated as the nonemptiness of the set M_0 defined below. Let

$$\mathcal{U}(t,x) = \{u \in \mathbb{R}^{d(u)} \mid (t,x,u) \in \mathcal{Q},\ g(t,x,u) = 0,\ \varphi(t,x,u) \leq 0\}. \tag{3.233}$$

Denote by M_0 the set of all tuples $\lambda \in \Lambda_0$ such that for all $t \in [t_0, t_f] \setminus \Theta$, the inclusion $u \in \mathcal{U}(t, x^0(t))$ implies the inequality

$$H(t,x^0(t),u,\psi(t)) \geq H(t,x^0(t),u^0(t),\psi(t)). \tag{3.234}$$

It is known [30] that if w^0 is a Pontryagin minimum, then M_0 is nonempty. The latter condition is just the *integral (or Pontryagin) minimum principle*. Inequality (3.234), satisfied for all $t \in [t_0, t_f] \setminus \Theta$, is called the *minimum condition* of Pontryagin's function H with respect to u. (In the case of a measurable control $u^0(t)$, inequality (3.234) is fulfilled for a.a. $t \in [t_0, t_f]$.)

Note that, just like Λ_0, the set M_0 can contain more than one element. Since this set is closed and $M_0 \subset \Lambda_0$, it follows from Proposition 3.44 that M_0 is also a finite-dimensional compact set. Let us note one more important property of the set M_0 (see [30]).

Proposition 3.46. *Let $\lambda \in M_0$ be an arbitrary tuple. Then there exists an absolutely continuous function $\psi_t(t)$ from $[t_0, t_f]$ into $\mathbb{R}^1$ such that*

$$\dot{\psi}_t = -\bar{H}_t(t, w^0(t), \psi(t), \nu(t), \mu(t)), \tag{3.235}$$

$$H(t, w^0(t), \psi(t)) + \psi_t(t) = 0. \tag{3.236}$$

Consequently, $\psi_t(t)$ is a piecewise smooth function whose break points belong to Θ.

Particularly, this implies the following assertion. Let $\lambda \in M_0$ be an arbitrary tuple. Then the function $H(t, w^0(t), \psi(t))$ satisfies the following condition:

$$[H^\lambda]^k = 0 \quad \forall\, t_k \in \Theta, \tag{3.237}$$

where $[H^\lambda]^k$ is a jump of the function $H(t,x^0(t),u^0(t),\psi(t))$ at the point $t_k \in \Theta$, defined by the relations $[H^\lambda]^k = H^{\lambda k+} - H^{\lambda k-}$, $H^{\lambda k-} = H(t_k, x^0(t_k), u^{0k-}, \psi(t_k))$, and $H^{\lambda k+} = H(t_k, x^0(t_k), u^{0k+}, \psi(t_k))$. Here, by definition, $u^{0k-} = u^0(t_k - 0)$, and $u^{0k+} = u^0(t_k + 0)$.

Let Λ_0^Θ be the set of all tuples $\lambda \in \Lambda_0$ satisfying condition (3.237). From Propositions 3.45 and 3.46 we have the following.

Proposition 3.47. *The set Λ_0^Θ is a finite-dimensional compact set such that $M_0 \subset \Lambda_0^\Theta \subset \Lambda_0$.*

Note that from minimum condition (3.234), the inequality

$$H(t_k, x^0(t_k), u, \psi(t_k)) \geq H^{\lambda k} \quad \forall\, u \in \mathcal{U}(t_k, x^0(t_k)) \tag{3.238}$$

follows by continuity, where by definition $H^{\lambda k} := H^{\lambda k-} = H^{\lambda k+}$. Condition (3.238) holds for any $t_k \in \Theta$ and any $\lambda \in M_0$.

Now we will state two properties of elements of the set M_0 which follow from the minimum principle. The first is a necessary optimality condition related to each discontinuity point of the control u^0. The second is a generalization of the Legendre condition.

The value $D^k(\bar{H}^\lambda)$. Let $\lambda \in \Lambda_0$ and $t_k \in \Theta$. According to Proposition 3.45 the quantities $\mu^{k-} = \mu(t_k - 0)$, $\mu^{k+} = \mu(t_k + 0)$, $\nu^{k-} = \nu(t_k - 0)$, $\nu^{k+} = \nu(t_k + 0)$ are well defined. Set

$$\begin{aligned} \bar{H}_x^{\lambda k-} &:= \bar{H}_x(t_k, x^0(t_k), u^{0k-}, \psi(t_k), \nu^{k-}, \mu^{k-}), \\ \bar{H}_x^{\lambda k+} &:= \bar{H}_x(t_k, x^0(t_k), u^{0k+}, \psi(t_k), \nu^{k+}, \mu^{k+}). \end{aligned} \tag{3.239}$$

Similarly, set

$$\begin{aligned} \bar{H}_\psi^{\lambda k-} &= f^{k-} := f(t_k, x^0(t_k), u^{0k-}), \\ \bar{H}_\psi^{\lambda k+} &= f^{k+} := f(t_k, x^0(t_k), u^{0k+}), \end{aligned} \tag{3.240}$$

$$\begin{aligned} \bar{H}_t^{\lambda k-} &:= \bar{H}_t(t_k, x^0(t_k), u^{0k-}, \psi(t_k), \nu^{k-}, \mu^{k-}), \\ \bar{H}_t^{\lambda k+} &:= \bar{H}_t(t_k, x^0(t_k), u^{0k+}, \psi(t_k), \nu^{k+}, \mu^{k+}). \end{aligned} \tag{3.241}$$

Finally, set

$$D^k(\bar{H}^\lambda) := -\bar{H}_x^{\lambda k+} \bar{H}_\psi^{\lambda k-} + \bar{H}_x^{\lambda k-} \bar{H}_\psi^{\lambda k+} - [\bar{H}_t^\lambda]^k, \tag{3.242}$$

where $[\bar{H}_t^\lambda]^k = \bar{H}_t^{\lambda k+} - \bar{H}_t^{\lambda k-}$ is the jump of $\bar{H}_t(t, x^0(t), u^0(t), \psi(t), \nu(t), \mu(t))$ at t_k. Note that $D^k(\bar{H}^\lambda)$ is linear in λ.

Theorem 3.48. *Let $\lambda \in M_0$. Then $D^k(\bar{H}^\lambda) \geq 0$ for all $t_k \in \Theta$.*

Since conditions $D^k(\bar{H}^\lambda) \geq 0$ for all $t_k \in \Theta$ follow from the minimum principle, they are necessary conditions for the Pontryagin minimum at the point w^0.

As in previous problems, there is another way, convenient for practical use, to calculate the quantities $D^k(\bar{H}^\lambda)$. Given any $\lambda \in \Lambda_0$ and $t_k \in \Theta$, we set

$$\begin{aligned} (\Delta_k \bar{H}^\lambda)(t) \quad = \quad & \bar{H}(t, x^0(t), u^{0k+}, \psi(t), \nu^{k+}, \mu^{k+}) \\ & - \bar{H}(t, x^0(t), u^{0k-}, \psi(t), \nu^{k-}, \mu^{k-}). \end{aligned} \tag{3.243}$$

The function $(\Delta_k \bar{H}^\lambda)(t)$ is continuously differentiable at each point of the set $[t_0, t_f] \setminus \Theta$, since this property hold for $x^0(t)$ and $\psi(t)$. The latter follows from the equation

$$\dot{x}^0(t) = f(t, x^0(t), u^0(t)), \tag{3.244}$$

adjoint equation (3.230), and Assumption 3.41.

Proposition 3.49. *For any $\lambda \in \Lambda_0$ and any $t_k \in \Theta$, the following equalities hold:*

$$D^k(\bar{H}^\lambda) = -\frac{d}{dt}(\Delta_k \bar{H}^\lambda)(t_k - 0) = -\frac{d}{dt}(\Delta_k \bar{H}^\lambda)(t_k + 0). \tag{3.245}$$

Finally, note that equation (3.244) can be written in the form

$$\dot{x}^0(t) = \bar{H}_\psi(t, x^0(t), u^0(t), \psi(t), \nu(t), \mu(t)). \tag{3.246}$$

Relations (3.244), (3.230), (3.235), (3.236), and formula (3.242) can be used for obtaining one more representation of the value $D^k(\bar{H}^\lambda)$.

Proposition 3.50. *For any $\lambda \in \Lambda_0$ and $t_k \in \Theta$, the following equality holds:*

$$D^k(\bar{H}^\lambda) = \dot{\psi}^{k+}\dot{x}^{0k-} - \dot{\psi}^{k-}\dot{x}^{0k+} + [\dot{\psi}_t]^k, \tag{3.247}$$

where the function $\psi_t(t)$ is defined by $\psi_t(t) = -H(t, x^0(t), u^0(t), \psi(t))$, the value $[\dot{\psi}_t]^k = \dot{\psi}_t^{k+} - \dot{\psi}_t^{k-}$ is the jump of the derivative $\dot{\psi}_t(t)$ at the point t_k, and the vectors $\dot{x}^{0k-}, \dot{\psi}^{k-}, \dot{\psi}_t^{k-}$ and $\dot{x}^{0k+}, \dot{\psi}^{k+}, \dot{\psi}_t^{k+}$ are the left and the right limit values of the derivatives $\dot{x}^0(t), \dot{\psi}(t)$ and $\dot{\psi}_t(t)$ at t_k, respectively.

Legendre–Clebsch condition. For any $\lambda = (\alpha_0, \alpha, \beta, \psi, \nu, \mu) \in \Lambda_0$, let us define the following three conditions:

(LC) For any $t \in [t_0, t_f] \setminus \Theta$, the quadratic form

$$\langle \bar{H}_{uu}(t, x^0(t), u^0(t), \psi(t), \nu(t), \mu(t))\bar{u}, \bar{u} \rangle \tag{3.248}$$

of the variable $\bar{u}$ is positive semidefinite on the cone formed by the vectors $\bar{u} \in \mathbb{R}^{d(u)}$ such that

$$\begin{aligned} &g_u(t, x^0(t), u^0(t))\bar{u} = 0, \\ &\varphi_{ju}(t, x^0(t), u^0(t))\bar{u} \le 0 \quad \forall\, j \in I_\varphi(t, x^0(t), u^0(t)), \\ &\mu_j(t)\varphi_{ju}(t, x^0(t), u^0(t))\bar{u} = 0 \quad \forall\, j \in I_\varphi(t, x^0(t), u^0(t)), \end{aligned} \tag{3.249}$$

where $\bar{H}_{uu}$ is the matrix of second derivatives with respect to u of the function $\bar{H}$, and $I_\varphi(t, x, u)$ is the set of indices of active inequality constraints $\varphi_j(t, x, u) \le 0$ at (t, x, u), defined by (3.220).

(LC_Θ^-) For any $t_k \in \Theta$, the quadratic form

$$\langle \bar{H}_{uu}(t_k, x^0(t_k), u^{0k-}, \psi(t_k), \nu^{k-}, \mu^{k-})\bar{u}, \bar{u} \rangle \tag{3.250}$$

of the variable $\bar{u}$ is positive semidefinite on the cone formed by the vectors $\bar{u} \in \mathbb{R}^{d(u)}$ such that

$$\begin{aligned} &g_u(t_k, x^0(t_k), u^{0k-})\bar{u} = 0, \\ &\varphi_{ju}(t_k, x^0(t_k), u^{0k-})\bar{u} \le 0 \quad \forall\, j \in I_\varphi(t_k, x^0(t_k), u^{0k-}), \\ &\mu_j^{k-}\varphi_{ju}(t_k, x^0(t_k), u^{0k-})\bar{u} = 0 \quad \forall\, j \in I_\varphi(t_k, x^0(t_k), u^{0k-}) \end{aligned} \tag{3.251}$$

(LC_Θ^+) For any $t_k \in \Theta$, the quadratic form

$$\langle \bar{H}_{uu}(t_k, x^0(t_k), u^{0k+}, \psi(t_k), \nu^{k+}, \mu^{k+})\bar{u}, \bar{u} \rangle \tag{3.252}$$

of the variable $\bar{u}$ is positive semidefinite on the cone formed by the vectors $\bar{u} \in \mathbb{R}^{d(u)}$ such that

$$\begin{aligned} &g_u(t_k, x^0(t_k), u^{0k+})\bar{u} = 0, \\ &\varphi_{ju}(t_k, x^0(t_k), u^{0k+})\bar{u} \le 0 \quad \forall\, j \in I_\varphi(t_k, x^0(t_k), u^{0k+}), \\ &\mu_j^{k+}\varphi_{ju}(t_k, x^0(t_k), u^{0k+})\bar{u} = 0 \quad \forall\, j \in I_\varphi(t_k, x^0(t_k), u^{0k+}). \end{aligned} \tag{3.253}$$

We say that element $\lambda \in \Lambda_0$ satisfies the *Legendre–Clebsch condition* if conditions (LC), (LC_Θ^-) and (LC_Θ^+) hold. Clearly, these conditions are not independent: conditions (LC_Θ^-) and (LC_Θ^+) follow from condition (LC) by continuity.

Theorem 3.51. *For any $\lambda \in M_0$, the Legendre–Clebsch condition holds.*

Thus, the Legendre–Clebsch condition is also a consequence of the minimum principle.

Legendrian elements. An element $\lambda \in \Lambda_0$ is said to be *Legendrian* if, for this element, the Legendre–Clebsch condition is satisfied and also the following conditions hold:

$$[H^\lambda]^k = 0, \quad D^k(\bar{H}^\lambda) \geq 0 \quad \forall\, t_k \in \Theta. \tag{3.254}$$

Let M be an arbitrary subset of the compact set Λ_0. Denote by $\mathrm{Leg}(M)$ the subset of all Legendrian elements $\lambda \in M$. It follows from Theorems 3.48 and 3.51 and Proposition 3.46 that

$$\mathrm{Leg}(M_0) = M_0. \tag{3.255}$$

Now we introduce the critical cone $\mathcal{K}$ and the quadratic form $\Omega^\lambda(\cdot)$ which will be used for the statement of the quadratic optimality condition.

Critical cone. As above, we denote by $P_\Theta W^{1,2}([t_0,t_f],\mathbb{R}^{d(x)})$ the space of piecewise continuous functions $\bar{x}(\cdot) : [t_0, t_f] \to \mathbb{R}^{d(x)}$ that are absolutely continuous on each interval of the set $[t_0, t_f] \setminus \Theta$ and have a square integrable first derivative. Given $t_k \in \Theta$ and $\bar{x}(\cdot) \in P_\Theta W^{1,2}([t_0,t_f],\mathbb{R}^{d(x)})$, we use the notation $\bar{x}^{k-} = \bar{x}(t_k - 0)$, $\bar{x}^{k+} = \bar{x}(t_k + 0)$, $[\bar{x}]^k = \bar{x}^{k+} - \bar{x}^{k-}$. Denote by $Z_2(\Theta)$ the space of triples $\bar{z} = (\bar{\xi}, \bar{x}, \bar{u})$ such that

$$\bar{\xi} = (\bar{\xi}_1,\ldots,\bar{\xi}_s) \in \mathbb{R}^s, \quad \bar{x} \in P_\Theta W^{1,2}([t_0,t_f],\mathbb{R}^{d(x)}), \quad \bar{u} \in L^2([t_0,t_f],\mathbb{R}^{d(u)}).$$

Thus, $Z_2(\Theta) = \mathbb{R}^s \times P_\Theta W^{1,2}([t_0,t_f],\mathbb{R}^{d(x)}) \times L^2([t_0,t_f],\mathbb{R}^{d(u)})$. Denote by

$$I_F(p^0) = \{i \in \{1,\ldots,d(F)\} \mid F_i(p^0) = 0\}$$

the set of indices of active inequality constraints $F_i(p) \leq 0$ at the point p^0, where F_i are the components of the vector function F.

Let $\mathcal{K}$ denote the set of $\bar{z} = (\bar{\xi}, \bar{x}, \bar{u}) \in Z_2(\Theta)$ such that

$$J'(p^0)\bar{p} \leq 0, \quad F_i'(p^0)\bar{p} \leq 0, \quad i \in I_F(p^0), \quad K'(p^0)\bar{p} = 0, \tag{3.256}$$

$$\dot{\bar{x}}(t) = f_w(t, w^0(t))\bar{w}(t), \quad [\bar{x}]^k = [\dot{x}^0]^k \bar{\xi}_k, \quad t_k \in \Theta, \tag{3.257}$$

$$g_w(t, w^0(t))\bar{w}(t) = 0, \tag{3.258}$$

$$\varphi_{jw}(t, w^0(t))\bar{w}(t) \leq 0 \quad \text{a.e. on } \mathcal{M}_0(\varphi_j^0), \quad j = 1,\ldots,d(\varphi), \tag{3.259}$$

where $\mathcal{M}_0(\varphi_j^0) = \{t \in [t_0,t_f] \mid \varphi_j(t, w^0(t)) = 0\}$, $\bar{p} = (\bar{x}(t_0), \bar{x}(t_f))$, $\bar{w} = (\bar{x}, \bar{u})$, and $[\dot{x}^0]^k$ is the jump of the function $\dot{x}^0(t)$ at the point $t_k \in \Theta$, i.e., $[\dot{x}^0]^k = \dot{x}^{0k+} - \dot{x}^{0k-} = \dot{x}^0(t_k + 0) - \dot{x}^0(t_k - 0)$, $t_k \in \Theta$. Obviously, $\mathcal{K}$ is a closed convex cone in the space $Z_2(\Theta)$. We call it the *critical cone* of problem (3.217)–(3.219) at the point w^0.

The following question is of interest: Which inequalities in the definition of $\mathcal{K}$ can be replaced by equalities without affecting $\mathcal{K}$? This question is answered below.

Proposition 3.52. *For any* $\lambda = (\alpha_0, \alpha, \beta, \psi, \nu, \mu) \in \Lambda_0^\Theta$ *and* $\bar{z} = (\bar{\xi}, \bar{x}, \bar{u}) \in \mathcal{K}$, *we have*

$$\alpha_0 J'(p^0)\bar{p} = 0, \quad \alpha_i F_i'(p^0)\bar{p} = 0, \quad i \in I_F(p^0), \tag{3.260}$$

$$\mu_j(t)\varphi_{jw}(t, w^0(t))\bar{w}(t) = 0, \quad j = 1, \dots, d(\varphi), \tag{3.261}$$

where α_i *and* μ_j *are the components of the vectors* α *and* μ, *respectively.*

Note that conditions (3.260) and (3.261) can be written in brief as $\alpha_0 J'(p^0)\bar{p} = 0$, $\alpha F'(p^0)\bar{p} = 0$, and $\mu(t)\varphi_w(t, w^0(t))\bar{w}(t) = 0$. Proposition 3.52 gives an answer to the question posed above. According to this proposition, for any $\lambda \in \Lambda_0^\Theta$, conditions (3.256)–(3.261) also define $\mathcal{K}$. It follows that if, for some $\lambda = (\alpha_0, \alpha, \beta, \psi, \nu, \mu) \in \Lambda_0^\Theta$, the condition $\alpha_0 > 0$ holds, then, in the definition of $\mathcal{K}$, the inequality $J'(p^0)\bar{p} \le 0$ can be replaced by the equality $J'(p^0)\bar{p} = 0$. If, for some $\lambda \in \Lambda_0^\Theta$ and $i_0 \in \{1, \dots, d(F)\}$, the condition $\alpha_{i_0} > 0$ holds, then the inequality $F_{i_0}'(p^0)\bar{p} \le 0$ can be replaced by the equality $F_{i_0}'(p^0)\bar{p} = 0$. Finally, for any $j \in \{1, \dots, d(\varphi)\}$ and $\lambda \in \Lambda_0^\Theta$, the inequality $\varphi_{jw}(t, w^0(t))\bar{w}(t) \le 0$ can be replaced by the equality $\varphi_{jw}(t, w^0(t))\bar{w}(t) = 0$ a.e. on the set $\{t \in [t_0, t_f] \mid \mu_j(t) > 0\} \subset \mathcal{M}_0(\varphi_j^0)$. Every such change gives an equivalent system of conditions still defining $\mathcal{K}$.

The following question is also of interest: Under what conditions can one of the endpoint inequalities in the definition of $\mathcal{K}$ be omitted without affecting $\mathcal{K}$? In particular, when can the inequality $J'(p^0)\bar{p} \le 0$ be omitted?

Proposition 3.53. *Suppose that there exists* $\lambda \in \Lambda_0^\Theta$ *such that* $\alpha_0 > 0$. *Then the relations*

$$F_i'(p^0)\bar{p} \le 0,\ i \in I_F(p^0), \quad \alpha_i F_i'(p^0)\bar{p} = 0,\ i \in I_F(p^0), \quad K'(p^0)\bar{p} = 0, \tag{3.262}$$

combined with (3.257)–(3.259) *and* (3.261) *imply that* $J'(p^0)\bar{p} = 0$; *i.e.,* $\mathcal{K}$ *can be defined by conditions* (3.257)–(3.259), (3.261), *and* (3.262)) *as well.*

Therefore, if for some $\lambda \in \Lambda_0^\Theta$ all inequalities (3.256) corresponding to positive α_i are replaced by the equalities and each inequality $\varphi_{jw}(t, w^0(t))\bar{w}(t) \le 0$ is replaced by the equality on the set $\{t \in [t_0, t_f] \mid \mu_j(t) > 0\} \subset \mathcal{M}_0(\varphi_j)$, then, after all such changes, the equality $J_p(p^0)\bar{p} = 0$ corresponding to positive α_0 can be excluded, and the obtained new system of conditions still defines $\mathcal{K}$.

Quadratic form. Let us introduce the following notation. Given any

$$\lambda = (\alpha_0, \alpha, \beta, \psi, \nu, \mu) \in \Lambda_0,$$

we set

$$[\bar{H}_x^\lambda]^k = \bar{H}_x^{\lambda k+} - \bar{H}_x^{\lambda k-}, \quad k = 1, \dots, s, \tag{3.263}$$

where $\bar{H}_x^{\lambda k-}$ and $\bar{H}_x^{\lambda k+}$ are defined by (3.239). Thus, $[\bar{H}_x^\lambda]^k$ denotes a jump of the function $\bar{H}_x(t, x^0(t), u^0(t), \psi(t), \nu(t), \mu(t))$ at $t = t_k \in \Theta$. It follows from the adjoint equation (3.230) that

$$[\bar{H}_x^\lambda]^k = -[\dot{\psi}]^k, \quad k = 1, \dots, s, \tag{3.264}$$

where the row vector

$$[\dot{\psi}]^k = \dot{\psi}^{k+} - \dot{\psi}^{k-} = \dot{\psi}(t_k + 0) - \dot{\psi}(t_k - 0)$$

is the jump of the derivative $\dot{\psi}(t)$ at $t_k \in \Theta$. Furthermore, for brevity we set

$$l^\lambda_{pp}(p^0) = \frac{\partial^2 l}{\partial p^2}(p^0, \alpha_0, \alpha, \beta), \; \bar{H}^\lambda_{ww}(w^0) = \frac{\partial^2 \bar{H}}{\partial w^2}(t, x^0(t), u^0(t), \psi(t), \nu(t), \mu(t)). \quad (3.265)$$

Finally, for $\bar{x} \in P_\Theta W^{1,2}([t_0, t_f], \mathbb{R}^{d(x)})$, we set

$$\bar{x}^k_{\text{av}} = \frac{1}{2}(\bar{x}^{k-} + \bar{x}^{k+}), \quad k = 1, \dots, s. \quad (3.266)$$

Here $\bar{x}^k_{\text{av}}$ is an average value of the function $\bar{x}$ at $t_k \in \Theta$.

We are now ready to introduce the quadratic form, which takes into account the discontinuities of the control u^0. For any $\lambda \in \Lambda_0$ and $\bar{z} = (\bar{\xi}, \bar{x}, \bar{u}) \in Z_2(\Theta)$, we set

$$\begin{aligned} \Omega^\lambda(\bar{z}) \;=\; & \frac{1}{2}\sum_{k=1}^{s}\left(D^k(\bar{H}^\lambda)\bar{\xi}_k^2 + 2[\bar{H}^\lambda_x]^k \bar{x}^k_{\text{av}}\bar{\xi}_k\right) \\ & + \frac{1}{2}\langle l^\lambda_{pp}(p^0)\bar{p}, \bar{p}\rangle + \frac{1}{2}\int_{t_0}^{t_f} \langle \bar{H}^\lambda_{ww}(t, w^0)\bar{w}, \bar{w}\rangle \, dt, \end{aligned} \quad (3.267)$$

where $\bar{w} = (\bar{x}, \bar{u})$, $\bar{p} = (\bar{x}(t_0), \bar{x}(t_f))$. Recall that the value $D^k(\bar{H}^\lambda)$ is defined by (3.242), and it is nonpositive for any $\lambda \in M_0$. Obviously, Ω^λ is quadratic in $\bar{z}$ and linear in λ. Set

$$\omega^\lambda_\Theta(\bar{\xi}, \bar{x}) = \frac{1}{2}\sum_{k=1}^{s}\left(D^k(\bar{H}^\lambda)\bar{\xi}_k^2 + 2[\bar{H}^\lambda_x]^k \bar{x}^k_{\text{av}}\bar{\xi}_k\right). \quad (3.268)$$

This quadratic form is related to the discontinuities of control u^0, and we call it the *internal form*. According to (3.247) and (3.264) it can be written as follows:

$$\omega^\lambda_\Theta(\bar{\xi}, \bar{x}) = \frac{1}{2}\sum_{k=1}^{s}\left(\left(\dot{\psi}^{k+}\dot{x}^{0k-} - \dot{\psi}^{k-}\dot{x}^{0k+} + [\dot{\psi}_t]^k\right)\bar{\xi}_k^2 - 2[\dot{\psi}]^k \bar{x}^k_{\text{av}}\bar{\xi}_k\right). \quad (3.269)$$

Furthermore, we set

$$\omega^\lambda(\bar{w}) = \frac{1}{2}\langle l^\lambda_{pp}(p^0)\bar{p}, \bar{p}\rangle + \frac{1}{2}\int_{t_0}^{t_f} \langle \bar{H}^\lambda_{ww}(t, w^0)\bar{w}, \bar{w}\rangle \, dt. \quad (3.270)$$

This quadratic form is the second variation of the Lagrangian of problem (3.217)–(3.219) at the point w^0. We call it the *external form*. Thus, the quadratic form $\Omega^\lambda(\bar{z})$ is a sum of the internal and external forms:

$$\Omega^\lambda(\bar{z}) = \omega^\lambda_\Theta(\bar{\xi}, \bar{x}) + \omega^\lambda(\bar{w}). \quad (3.271)$$

Quadratic necessary condition for Pontryagin minimum. Now, we formulate the main necessary quadratic condition for Pontryagin minimum in problem (3.217)–(3.219) at the point w^0.

Theorem 3.54. *If w^0 is a Pontryagin minimum, then the following Condition $\mathfrak{A}$ holds: The set M_0 is nonempty and*

$$\max_{\lambda \in M_0} \Omega^\lambda(\bar{z}) \geq 0 \quad \forall\, \bar{z} \in \mathcal{K}.$$

The proof of this theorem was given in [86] and published in [95].

Next we formulate the basic *sufficient condition* in problem (3.217)–(3.219). We call it briefly Condition $\mathfrak{B}(\Gamma)$. It is sufficient not only for a Pontryagin minimum, but also for a bounded strong minimum defined below. We give a preliminary definition of a strong minimum which is slightly different from the commonly used definition.

Strong minimum. A state variable x_i (the ith component of x) is said to be *unessential* if the functions f, g, and φ do not depend on this variable and the functions J, F, and K are affine in $p_i = (x_i(t_0), x_i(t_f))$. A state variable which does not possess this property is said to be *essential*. The vector comprised of the essential components x_i of x is denoted by $\underline{x}$. Similarly, $\underline{\delta x}$ denotes the vector consisting of the essential components of a variation δx. Denote by Π^S the set of all sequences $\{\delta w_n\}$ in W such that $|\delta x_n(t_0)| + \|\underline{\delta x}_n\|_C \to 0$. A minimum on Π^S is called *strong*.

Bounded strong minimum. A sequence $\{\delta w_n\}$ in W is said to be *bounded strong on $\mathcal{Q}$* if $\{\delta w_n\} \in \Pi^S$ and there exists a compact set $\mathcal{C} \subset \mathcal{Q}$ such that for all sufficiently large n one has $(t, x^0(t), u^0(t) + \delta u_n(t)) \in \mathcal{C}$ a.e. on $[t_0, t_f]$. By a *(strict) bounded strong minimum* we mean a (strict) minimum on the set of all bounded strong on $\mathcal{Q}$ sequences. Every strong minimum is bounded strong. Hence, it is a Pontryagin minimum.

We know that the bounded strong minimum is equivalent to the strong minimum if there exists a compact set $\mathcal{C} \subset \mathcal{Q}$ such that the conditions $t \in [t_0, t_f]$ and $u \in \mathcal{U}(t, x^0(t))$ imply $(t, x^0(t), u) \in \mathcal{C}$, where $\mathcal{U}(t,x)$ is the set defined by (3.233).

Let us state a quadratic sufficient condition for a point $w^0 = (x^0, u^0)$ to be a bounded strong minimum. Again we assume that u^0 is a piecewise continuous control and $\Theta = \{t_1, \ldots, t_s\}$ is the set of its discontinuity points, every element of Θ being an L-point.

Set $M(C\Gamma)$. Let Γ be an order function (see Definition 2.17). For any $C > 0$, we denote by $M(C\Gamma)$ the set of all $\lambda \in M_0$ such that the following condition holds:

$$H(t, x^0(t), u, \psi(t)) - H(t, x^0(t), u^0(t), \psi(t)) \geq C\Gamma(t, u), \qquad \forall\, t \in [t_0, t_f] \setminus \Theta, \quad u \in \mathcal{U}(t, x^0(t)). \tag{3.272}$$

Condition (3.272) strengthens the minimum condition (3.234), and we call (3.272) the *minimum condition of strictness $C\Gamma$* (or *$C\Gamma$-growth condition for H*). For any $C > 0$, $M(C\Gamma)$ is a closed subset in M_0 and, therefore, a finite dimensional compact set.

Basic sufficient condition. Let

$$\bar{\gamma}(\bar{z}) = \langle \bar{\xi}, \bar{\xi} \rangle + \langle \bar{x}(t_0), \bar{x}(t_0) \rangle + \int_{t_0}^{t_f} \langle \bar{u}(t), \bar{u}(t) \rangle \, dt.$$

On the subspace (3.257)–(3.258) of the space $Z^2(\Theta)$, the value $\sqrt{\bar{\gamma}(\bar{z})}$ is a norm, which is equivalent to the norm of the space $Z^2(\Theta)$. Let Γ be an order function.

Definition 3.55. We say that a point w^0 *satisfies condition* $\mathfrak{B}(\Gamma)$ if there exists $C > 0$ such that the set $M(C\Gamma)$ is nonempty and

$$\max_{\lambda \in M(C\Gamma)} \Omega^{\lambda}(\bar{z}) \geq C\bar{\gamma}(\bar{z}) \quad \forall\, \bar{z} \in \mathcal{K}.$$

Theorem 3.56. *If there exists an order function* $\Gamma(t,u)$ *such that Condition* $\mathfrak{B}(\Gamma)$ *holds, then* w^0 *is a strict bounded strong minimum.*

Condition $\mathfrak{B}(\Gamma)$ obviously holds if for some $C > 0$ the set $M(C\Gamma)$ is nonempty, and if the cone $\mathcal{K}$ consists only of zero. Therefore, Theorem 3.56 implies the following.

Corollary 3.57. *If for some* $C > 0$ *the set* $M(C\Gamma)$ *is nonempty, and if* $\mathcal{K} = \{0\}$, *then* w^0 *is a strict bounded strong minimum.*

Corollary 3.57 states the first-order sufficient condition of a bounded strong minimum.

γ-sufficiency. Quadratic Condition $\mathfrak{B}(\Gamma)$ implies not only a bounded strong minimum, but also a certain strengthening of this concept which is called the γ-sufficiency on the set of bounded strong sequences. Below, we introduce the (higher) order γ and formulate two concepts: γ-sufficiency for Pontryagin minimum and γ-sufficiency for bounded strong minimum. Regarding the point $w^0 = (x^0, u^0)$, tested for optimality, we again use Assumption 3.41. Let $\Gamma(t,u)$ be an order function. Set

$$\gamma(\delta w) = \|\delta x\|_C^2 + \int_{t_0}^{t_f} \Gamma(t, u^0 + \delta u) \, dt.$$

The functional γ is defined on the set of all variations $\delta w = (\delta x, \delta u) \in W$ such that

$$(t, x^0(t), u^0(t) + \delta u(t)) \in \mathcal{Q} \quad \text{a.e. on } [t_0, t_f].$$

The functional γ is the *order associated with the order function* $\Gamma(t,u)$. Following the general theory [55], we also call γ the *higher order*. Thus, with the point w^0 we associate the family of order functions Γ and the family of corresponding orders γ. Let us denote the latter family by $\mathrm{Ord}(w^0)$.

Let us introduce the *violation function of problem* (3.217)–(3.219) at the point w^0:

$$\sigma(\delta w) = \max \left\{ \sigma_{JFK}(\delta p), \sigma_f(\delta w), \sigma_{g\varphi}(\delta w) \right\}, \tag{3.273}$$

where

$$\sigma_{JFK}(\delta p) = \max\left\{J(p^0+\delta p) - J(p^0),\ \max_{i=1,\dots,d(F)} F_i(p^0+\delta p), |K(p^0+\delta p)|\right\},$$
$$\sigma_f(\delta w) = \int_{t_0}^{t_f} |\dot{x}^0 + \delta\dot{x} - f(t,w^0+\delta w)|\,dt,$$
$$\sigma_{g\varphi}(\delta w) = \operatorname{ess\,sup}_{[t_0,t_f]} \max\{|g(t,w^0+\delta w)|,\ \max_{i=1,\dots,d(\varphi)} \varphi_i(t,w^0+\delta w)\},$$
$$\delta w = (\delta x,\delta u) \in W,\ \delta p = (\delta x(t_0),\delta x(t_f)),\ p^0+\delta p \in \mathcal{P},\ (t,w^0(t)+\delta w(t)) \in \mathcal{Q}.$$

Obviously, $\sigma(\delta w) \geq 0$ and $\sigma(0) = 0$.

Let S be a set of sequences $\{\delta w_n\}$ in W, closed with respect to the operation of taking subsequences. Evidently, w^0 is a point of a strict minimum on S iff, for any sequence $\{\delta w_n\} \in S$ containing nonzero terms, one has $\sigma(\delta w_n) > 0$ for all sufficiently large n. A strengthened version of the last condition is suggested by the following definition.

Definition 3.58. We say that w^0 is a point of *γ-sufficiency on S* if there exists an $\varepsilon > 0$ such that, for any sequence $\{\delta w_n\} \in S$, we have $\sigma(\delta w_n) \geq \varepsilon\gamma(\delta w_n)$ for all sufficiently large n.

Let us now introduce a set of sequences related to a bounded strong minimum. Denote by S_{bs} the set of all sequences $\{\delta w_n\}$ which are bounded strong on $\mathcal{Q}$ and satisfy the following conditions:

(a) $(p^0+\delta p_n) \in \mathcal{P}$ for all sufficiently large n,

(b) $\sigma(\delta w_n) \to 0$ as $n \to \infty$.

Conditions (a) and (b) hold for every sequence $\{\delta w_n\}$ that violates minimality, so a bounded strong minimum can be treated as a minimum on S_{bs} (the subscript *bs* means "bounded strong").

Theorem 3.59. *Condition $\mathfrak{B}(\Gamma)$ is equivalent to γ-sufficiency on S_{bs}.*

The proof of this theorem was given in [86] and published in [94, 95]. Theorem 3.59 particularly shows a nontrivial character of minimum guaranteed by condition $\mathfrak{B}(\Gamma)$. Let us explain this in more detail. A sequence $\{\delta w_n\}$ is said to be *admissible* if the sequence $\{w^0+\delta w_n\}$ satisfies all constraints of the canonical problem. We say that w^0 is a point of *γ-minimum on S_{bs}* (or the *γ-growth condition for the cost function holds on S_{bs}*) if there exists $\varepsilon > 0$ such that, for any admissible sequence $\{\delta w_n\} \in S_{\text{bs}}$, we have $J(p^0+\delta p_n) - J(p^0) \geq \varepsilon\gamma(\delta w_n)$ for all sufficiently large n. Clearly, γ-sufficiency on S_{bs} implies *γ-minimum on S_{bs}*. In fact, it is the sufficient Condition $\mathfrak{B}(\Gamma)$ that ensures γ-minimum on S_{bs}. A nontrivial character of γ-minimum on S_{bs} is caused by a nontrivial definition of the order function Γ which specifies the higher-order γ.

Now, let us discuss an important question concerning characterization of Condition $\lambda \in M(C\Gamma)$.

Local quadratic growth condition of the Hamiltonian. Fix an arbitrary tuple $\lambda \in \Lambda_0$. We set

$$\delta H[t,v] := H(t,x^0(t),u^0(t)+v,\psi(t)) - H(t,x^0(t),u^0(t),\psi(t)). \tag{3.274}$$

Definition 3.60. We say that, at the point w^0, the Hamiltonian satisfies a *local quadratic growth condition* if there exist $\varepsilon > 0$ and $\alpha > 0$ such that for all $t \in [t_0, t_f] \setminus \Theta$ the following inequality holds:

$$\begin{cases} \delta H[t,v] \geq \alpha |v|^2 & \text{if } v \in \mathbb{R}^{d(u)}, \quad g(t,x^0(t),u^0(t)+v)=0, \\ \varphi(t,x^0(t),u^0(t)+v) \leq 0, \ |v| < \varepsilon. \end{cases} \tag{3.275}$$

Recall the definition of $\bar{H}$ in (3.225). Let us denote by

$$\begin{cases} \bar{H}_u(t) := \bar{H}_u(t,x^0(t),u^0(t),\psi(t),\nu(t),\mu(t)), \\ \bar{H}_{uu}(t) := \bar{H}_{uu}(t,x^0(t),u^0(t),\psi(t),\nu(t),\mu(t)) \end{cases}$$

the first and second derivative with respect to u of the augmented Hamiltonian, and adopt a similar notation for the Hamiltonian function H. Similarly, we denote $g_u(t) := g_u(t,x^0(t),u^0(t))$, $\varphi_i(t) := \varphi_i(t,x^0(t),u^0(t))$, $\varphi_{iu}(t) := \varphi_{iu}(t,x^0(t),u^0(t))$, $i = 1,\dots,d(\varphi)$. We shall formulate a generalization of the strengthened Legendre condition using the quadratic form $\langle \bar{H}_{uu}(t)v, v\rangle$ complemented by some special nonnegative term $\rho(t,v)$ which will be homogeneous (not quadratic) of the second degree with respect to v. Let us define this additional term.

For any number a, we set $a^+ = \max\{a,0\}$ and $a^- = \max\{-a,0\}$, so that $a^+ \geq 0$, $a^- \geq 0$, and $a = a^+ - a^-$. Denote by

$$\chi_i(t) := \chi_{\{\varphi_i(\tau)<0\}}(t) \tag{3.276}$$

the characteristic function of the set $\{\tau \mid \varphi_i(\tau) < 0\}$, $i = 1,\dots,d(\varphi)$. If $d(\varphi) > 1$, then, for any $t \in [t_0,t_f]$ and any $v \in \mathbb{R}^{d(u)}$, we set

$$\rho(t,v) = \sum_{j=1}^{d(\varphi)} \max_{1\leq i\leq d(\varphi)} \left\{ \frac{\mu_j(t)}{|\varphi_i(t)|} \chi_i(t) \big(\varphi_{ju}(t)v\big)^- \big(\varphi_{iu}(t)v\big)^+ \right\}. \tag{3.277}$$

Here, by definition,

$$\frac{\mu_j(t)}{|\varphi_i(t)|}\chi_i(t) = 0 \quad \text{if } \varphi_i(t) = 0, \quad i,j = 1,\dots,d(\varphi).$$

Particularly, for $d(\varphi) = 2$ the function ρ has the form

$$\begin{aligned} \rho(t,v) \quad &= \quad \frac{\mu_1(t)}{|\varphi_2(t)|}\chi_2(t)\big(\varphi_{1u}(t)v\big)^-\big(\varphi_{2u}(t)v\big)^+ \\ &\quad + \frac{\mu_2(t)}{|\varphi_1(t)|}\chi_1(t)\big(\varphi_{2u}(t)v\big)^-\big(\varphi_{1u}(t)v\big)^+. \end{aligned} \tag{3.278}$$

In the case $d(\varphi) = 1$, we set $\rho(t,v) \equiv 0$.

For any $\Delta > 0$ and any $t \in [t_0,t_f] \setminus \Theta$, denote by $\mathcal{C}_t(\Delta)$ the set of all vectors $v \in \mathbb{R}^{d(u)}$ satisfying

$$\begin{aligned} &g_u(t)v = 0, \quad \varphi_{ju}(t)v \leq 0 \quad \text{if } \varphi_j(t) = 0, \\ &\varphi_{ju}(t)v = 0 \quad \text{if } \mu_j(t) > \Delta, \ j = 1,\dots,d(\varphi). \end{aligned} \tag{3.279}$$

Definition 3.61. We say that the Hamiltonian satisfies the *generalized strengthened Legendre condition* if

$$\left\{\begin{array}{l} \exists\,\alpha > 0,\ \Delta > 0 \text{ such that } \forall\, t \in [t_f, t_f]\setminus\Theta: \\ \frac{1}{2}\langle \bar{H}_{uu}(t)v, v\rangle + \rho(t,v) \geq \alpha|v|^2 \quad \forall\, v \in \mathcal{C}_t(\Delta). \end{array}\right. \tag{3.280}$$

Theorem 3.62. *A local quadratic growth condition for the Hamiltonian is equivalent to the generalized strengthened Legendre condition.*

This theorem was proved in [8] for the control constrained problem (without mixed constraints).

We note that $\mathcal{C}_t(\Delta)$ is in general a larger set than the local cone C_t of critical directions for the Hamiltonian, i.e., the directions $v \in \mathbb{R}^{d(u)}$, such that

$$\begin{array}{l} g_u(t)v = 0, \quad \varphi_{ju}(t)v \leq 0 \quad \text{if } \varphi_j(t) = 0, \\ \varphi_{ju}(t)v = 0 \quad \text{if } \mu_j(t) > 0,\ j = 1,\dots,d(\varphi). \end{array} \tag{3.281}$$

A simple sufficient condition for local quadratic growth of the Hamiltonian. Consider the following second-order condition for the Hamiltonian:

$$\left\{\begin{array}{l} \exists\,\alpha > 0,\ \Delta > 0 \text{ such that } \forall\, t \in [0,T]: \\ \frac{1}{2}\langle \bar{H}_{uu}(t)v, v\rangle \geq \alpha|v|^2 \quad \forall\, v \in \mathcal{C}_t(\Delta). \end{array}\right. \tag{3.282}$$

Let us note that this inequality is stronger than (3.280), since the function $\rho(t,v)$ is nonnegative.

Theorem 3.63. *Condition* (3.282) *implies a local quadratic growth of the Hamiltonian.*

Characterization of condition $\lambda \in M(C\Gamma)$. An element $\lambda \in \Lambda_0$ is said to be *strictly Legendrian* if, for this element, the generalized strengthened Legendre condition (3.280) is satisfied and also the following conditions hold:

$$[H^\lambda]^k = 0, \quad D^k(\bar{H}^\lambda) > 0 \quad \forall\, t_k \in \Theta. \tag{3.283}$$

Denote by M_0^+ the set of $\lambda \in M_0$ such that the following conditions hold:

(a) $H(t,x^0(t),u,\psi(t)) > H(t,x^0(t),u^0(t),\psi(t))$
if $t \in [t_0,t_f]\setminus\Theta$, $u \in \mathcal{U}(t,x^0(t))$, $u \neq u^0(t)$, where
$\mathcal{U}(t,x) = \{u \in \mathbb{R}^{d(u)} \mid (t,x,u) \in \mathcal{Q},\ g(t,x,u) = 0,\ \varphi(t,x,u) \leq 0\}$;

(b) $H(t_k,x^0(t_k),u,\psi(t_k)) > H^k$
if $t_k \in \Theta$, $u \in \mathcal{U}(t_k,x^0(t_k))$, $u \notin \{u^{0k-},u^{0k+}\}$, where
$H^k := H(t_k,x^0(t_k),u^{0k-},\psi(t_k)) = H(t_k,x^0(t_k),u^{0k+},\psi(t_k))$.

Denote by $\mathrm{Leg}_+(M_0^+)$ the set of all strictly Legendrian elements $\lambda \in M_0^+$.

Theorem 3.64. *An element $\lambda \in \mathrm{Leg}_+(M_0^+)$ iff there exists $C > 0$ such that $\lambda \in M(C\Gamma)$.*

The proof will be published elsewhere.

3.4.2 General Optimal Control Problem on a Variable Time Interval

Statement of the problem. Here, quadratic optimality conditions, both necessary and sufficient, are presented in the following optimal control problem on a variable time interval. Let $\mathcal{T}$ denote a trajectory $(x(t),u(t) \mid t \in [t_0,t_f])$, where the state variable $x(\cdot)$ is a Lipschitz continuous function, and the control variable $u(\cdot)$ is a bounded measurable function on a time interval $\Delta = [t_0,t_f]$. The interval Δ is not fixed. For each trajectory $\mathcal{T}$, we denote by $p = (t_0,x(t_0),t_f,x(t_f))$ the vector of the endpoints of time-state variable (t,x). It is required to find $\mathcal{T}$ minimizing the functional

$$\mathcal{J}(\mathcal{T}) := J(t_0,x(t_0),t_f,x(t_f)) \to \min \tag{3.284}$$

subject to the constraints

$$F(t_0,x(t_0),t_f,x(t_f)) \le 0, \quad K(t_0,x(t_0),t_f,x(t_f)) = 0, \tag{3.285}$$

$$\dot{x}(t) = f(t,x(t),u(t)), \tag{3.286}$$

$$g(t,x(t),u(t)) = 0, \quad \varphi(t,x(t),u(t)) \le 0, \tag{3.287}$$

$$p \in \mathcal{P}, \quad (t,x(t),u(t)) \in \mathcal{Q}, \tag{3.288}$$

where $\mathcal{P}$ and $\mathcal{Q}$ are open sets, and x, u, F, K, f, g, and φ are vector functions.

We assume that the functions J, F, and K are defined and twice continuously differentiable on $\mathcal{P}$, and the functions f, g, and φ are defined and twice continuously differentiable on $\mathcal{Q}$; moreover, g and φ satisfy the linear independence assumption (see Assumption 3.40).

Necessary conditions for a Pontryagin minimum. Let $\mathcal{T}$ be a fixed admissible trajectory such that the control $u(\cdot)$ is a piecewise Lipschitz continuous function on the interval Δ with the set of discontinuity points $\Theta = \{t_1,\dots,t_s\}$, $t_0 < t_1 < \dots < t_s < t_f$. Let us formulate a first-order necessary condition for optimality of the trajectory $\mathcal{T}$. We introduce the Pontryagin function

$$H(t,x,u,\psi) = \psi f(t,x,u) \tag{3.289}$$

and the augmented Pontryagin function

$$\bar{H}(t,x,u,\psi,\nu,\mu) = H(t,x,u,\psi) + \nu g(t,x,u) + \mu\varphi(t,x,u), \tag{3.290}$$

where ψ, ν, and μ are row vectors of the dimensions $d(x)$, $d(g)$, and $d(\varphi)$, respectively. Let us define the endpoint Lagrange function

$$l(p,\alpha_0,\alpha,\beta) = \alpha_0 J(p) + \alpha F(p) + \beta K(p), \tag{3.291}$$

where $p = (t_0,x_0,t_f,x_f)$, $x_0 = x(t_0)$, $x_f = x(t_f)$, $\alpha_0 \in \mathbb{R}$, $\alpha \in (\mathbb{R}^{d(F)})^*$, $\beta \in (\mathbb{R}^{d(K)})^*$. Also we introduce a tuple of Lagrange multipliers

$$\lambda = (\alpha_0,\alpha,\beta,\psi(\cdot),\psi_0(\cdot),\nu(\cdot),\mu(\cdot)) \tag{3.292}$$

such that $\psi(\cdot) : \Delta \to (\mathbb{R}^{d(x)})^*$ and $\psi_0(\cdot) : \Delta \to \mathbb{R}^1$ are piecewise smooth functions, continuously differentiable on each interval of the set $\Delta \setminus \Theta$, and $\nu(\cdot) : \Delta \to (\mathbb{R}^{d(g)})^*$ and

$\mu(\cdot) : \Delta \to (\mathbb{R}^{d(\varphi)})^*$ are piecewise continuous functions, Lipschitz continuous on each interval of the set $\Delta \setminus \Theta$.

Denote by M_0 the set of the normed tuples λ satisfying the conditions of the minimum principle for the trajectory $\mathcal{T}$:

$$\begin{gathered}
\alpha_0 \geq 0, \quad \alpha \geq 0, \quad \alpha F(p) = 0, \quad \alpha_0 + \sum_{i=1}^{d(F)} \alpha_i + \sum_{j=1}^{d(K)} |\beta_j| = 1, \\
\dot{\psi} = -\bar{H}_x, \quad \dot{\psi}_0 = -\bar{H}_t, \quad \bar{H}_u = 0, \quad t \in \Delta \setminus \Theta, \\
\psi(t_0) = -l_{x_0}, \quad \psi(t_f) = l_{x_f}, \quad \psi_0(t_0) = -l_{t_0}, \quad \psi_0(t_f) = l_{t_f}, \\
\min_{u \in \mathcal{U}(t,x(t))} H(t,x(t),u,\psi(t)) = H(t,x(t),u(t),\psi(t)), \quad t \in \Delta \setminus \Theta, \\
H(t,x(t),u(t),\psi(t)) + \psi_0(t) = 0, \quad t \in \Delta \setminus \Theta,
\end{gathered} \tag{3.293}$$

where $\mathcal{U}(t,x) = \{u \in \mathbb{R}^{d(u)} \mid g(t,x,u) = 0,\ \varphi(t,x,u) \leq 0,\ (t,x,u) \in \mathcal{Q}\}$. The derivatives l_{x_0} and l_{x_f} are at $(p,\alpha_0,\alpha,\beta)$, where $p = (t_0,x(t_0),t_f,x(t_f))$, and the derivatives $\bar{H}_x$, $\bar{H}_u$, and $\bar{H}_t$ are at $(t,x(t),u(t),\psi(t),\nu(t),\mu(t))$, where $t \in \Delta \setminus \Theta$. (Condition $\bar{H}_u = 0$ follows from the others conditions in this definition, and therefore could be excluded; yet we need to use it later.)

We define the Pontryagin minimum in problem (3.284)–(3.288) on a variable interval $[t_0,t_f]$ as in Section 3.3.2 (see Definition 3.25). The condition $M_0 \neq \emptyset$ is equivalent to the Pontryagin's minimum principle. It is a first-order necessary condition of Pontryagin minimum for the trajectory $\mathcal{T}$. Thus, the following theorem holds (see, e.g., [76]).

Theorem 3.65. *If the trajectory $\mathcal{T}$ affords a Pontryagin minimum, then the set M_0 is nonempty.*

Assume that M_0 is nonempty. Using the definition of the set M_0 and the linear independence assumption for g and φ one can easily prove the following statement.

Proposition 3.66. *The set M_0 is a finite-dimensional compact set, and the mapping $\lambda \mapsto (\alpha_0,\alpha,\beta)$ is injective on M_0.*

As in Section 3.3, for each $\lambda \in M_0$, $t_k \in \Theta$, we set

$$D^k(\bar{H}) = -\bar{H}_x^{k+}\bar{H}_\psi^{k-} + \bar{H}_x^{k-}\bar{H}_\psi^{k+} - [\bar{H}_t]^k, \tag{3.294}$$

where $\bar{H}_x^{k-} = \bar{H}_x(t_k,x(t_k),u(t_k-),\psi(t_k),\nu(t_k-),\mu(t_k-))$, $\bar{H}_x^{k+} = \bar{H}_x(t_k,x(t_k),u(t_k+),\psi(t_k),\nu(t_k+),\mu(t_k+))$, $[\bar{H}_t]^k = \bar{H}_t^{k+} - \bar{H}_t^{k-}$, etc.

Theorem 3.67. *For each $\lambda \in M_0$, the following conditions hold:*

$$D^k(\bar{H}) \geq 0, \quad k = 1,\ldots,s. \tag{3.295}$$

Thus, conditions (3.295) follow from the minimum principle conditions (3.293).

Let us formulate a quadratic necessary condition of a Pontryagin minimum for the trajectory $\mathcal{T}$. First, for this trajectory, we introduce a Hilbert space $\mathcal{Z}_2(\Theta)$ and the critical cone $\mathcal{K} \subset \mathcal{Z}_2(\Theta)$. We denote by $P_\Theta W^{1,2}(\Delta,\mathbb{R}^{d(x)})$ the Hilbert space of piecewise continuous functions $\bar{x}(\cdot) : \Delta \to \mathbb{R}^{d(x)}$, absolutely continuous on each interval of the set $\Delta \setminus \Theta$

and such that their first derivative is square integrable. For each $\bar{x} \in P_\Theta W^{1,2}(\Delta, \mathbb{R}^{d(x)})$, $t_k \in \Theta$, we set $\bar{x}^{k-} = \bar{x}(t_k-)$, $\bar{x}^{k+} = \bar{x}(t_k+)$, $[\bar{x}]^k = \bar{x}^{k+} - \bar{x}^{k-}$. Further, we let $\bar{z} = (\bar{t}_0, \bar{t}_f, \bar{\xi}, \bar{x}, \bar{u})$, where

$$\bar{t}_0 \in \mathbb{R}^1, \quad \bar{t}_f \in \mathbb{R}^1, \quad \bar{\xi} \in \mathbb{R}^s, \quad \bar{x} \in P_\Theta W^{1,2}(\Delta, \mathbb{R}^{d(x)}), \quad \bar{u} \in L^2(\Delta, \mathbb{R}^{d(u)}).$$

Thus,

$$\bar{z} \in Z_2(\Theta) := \mathbb{R}^2 \times \mathbb{R}^s \times P_\Theta W^{1,2}(\Delta, \mathbb{R}^{d(x)}) \times L^2(\Delta, \mathbb{R}^{d(u)}).$$

Moreover, for given $\bar{z}$, we set

$$\bar{w} = (\bar{x}, \bar{u}), \quad \bar{x}_0 = \bar{x}(t_0), \quad \bar{x}_f = \bar{x}(t_f), \tag{3.296}$$

$$\bar{\bar{x}}_0 = \bar{x}(t_0) + \bar{t}_0 \dot{x}(t_0), \quad \bar{\bar{x}}_f = \bar{x}(t_f) + \bar{t}_f \dot{x}(t_f), \quad \bar{\bar{p}} = (\bar{\bar{x}}_0, \bar{t}_0, \bar{\bar{x}}_f, \bar{t}_f). \tag{3.297}$$

By $I_F(p) = \{i \in \{1, \ldots, d(F)\} \mid F_i(p) = 0\}$ we denote the set of active indices of the constraints $F_i(p) \le 0$.

Let $\mathcal{K}$ be the set of all $\bar{z} \in Z_2(\Theta)$ satisfying the following conditions:

$$\begin{aligned}
&J'(p)\bar{\bar{p}} \le 0, \quad F_i'(p)\bar{\bar{p}} \le 0 \quad \forall\, i \in I_F(p), \quad K'(p)\bar{\bar{p}} = 0,\\
&\dot{\bar{x}}(t) = f_w(t, w(t))\bar{w}(t) \quad \text{for a.a. } t \in [t_0, t_f],\\
&[\bar{x}]^k = [\dot{x}]^k \bar{\xi}_k, \quad k = 1, \ldots, s,\\
&g_w(t, w(t))\bar{w}(t) = 0 \quad \text{for a.a. } t \in [t_0, t_f],\\
&\varphi_{jw}(t, w^0(t))\bar{w}(t) \le 0 \quad \text{a.e. on } \mathcal{M}_0(\varphi_j^0), \quad j = 1, \ldots, d(\varphi),
\end{aligned} \tag{3.298}$$

where $\mathcal{M}_0(\varphi_j^0) = \{t \in [t_0, t_f] \mid \varphi_j(t, w^0(t)) = 0\}$, $p = (t_0, x(t_0), t_f, x(t_f))$, $w = (x, u)$. It is obvious that $\mathcal{K}$ is a convex cone in the Hilbert space $Z_2(\Theta)$, and we call it the *critical cone*. If the interval Δ is fixed, then we set $p := (x_0, x_f) = (x(t_0), x(t_f))$, and in the definition of $\mathcal{K}$ we have $\bar{t}_0 = \bar{t}_f = 0$, $\bar{\bar{x}}_0 = \bar{x}_0$, $\bar{\bar{x}}_f = \bar{x}_f$, and $\bar{\bar{p}} = \bar{p} := (\bar{x}_0, \bar{x}_f)$. Define quadratic forms ω_e, ω, and Ω by formulas (3.156), (3.157), and (3.158), respectively. Now, we formulate the main necessary quadratic condition of a Pontryagin minimum in the problem on a variable time interval.

Theorem 3.68. *If the trajectory $\mathcal{T}$ yields a Pontryagin minimum, then the following Condition $\mathfrak{A}$ holds: The set M_0 is nonempty and*

$$\max_{\lambda \in M_0} \Omega(\lambda, \bar{z}) \ge 0 \quad \forall\, \bar{z} \in \mathcal{K}.$$

Sufficient conditions for a bounded strong minimum. The ith component x_i of vector x is called *unessential* if the functions f, g, and φ do not depend on this component and the functions J, F, and K are affine in $x_{i0} = x_i(t_0)$, $x_{if} = x_i(t_f)$; otherwise the component x_i is called *essential*. We denote by $\underline{x}$ a vector composed of all essential components of vector x and we define the (strict) bounded strong minimum as in Definition 3.32. Let Γ be an order function (see Definition 2.17). We formulate a sufficient optimality Condition $\mathfrak{B}(\Gamma)$, which is a natural strengthening of the necessary Condition $\mathfrak{A}$. Let us introduce the functional

$$\bar{\gamma}(\bar{z}) = \bar{t}_0^2 + \bar{t}_f^2 + \langle \bar{\xi}, \bar{\xi} \rangle + \langle \bar{x}(t_0), \bar{x}(t_0) \rangle + \int_{t_0}^{t_f} \langle \bar{u}(t), \bar{u}(t) \rangle \, dt, \tag{3.299}$$

which is equivalent to the norm squared on the subspace

$$\dot{\bar{x}} = f_w(t,x(t),u(t))\bar{w}; \quad [\bar{x}]^k = [\dot{x}]^k \bar{\xi}_k, \quad k = 1,\dots,s, \tag{3.300}$$

of Hilbert space $\mathcal{Z}_2(\Theta)$. Recall that the critical cone $\mathcal{K}$ is contained in the subspace (3.300). For any $C > 0$, we denote by $M(C\Gamma)$ the set of all $\lambda \in M_0$ such that condition (3.272) holds.

Theorem 3.69. *For the trajectory $\mathcal{T}$, assume that the following Condition $\mathfrak{B}(\Gamma)$ holds: There exists $C > 0$ such that the set $M(C\Gamma)$ is nonempty and*

$$\max_{\lambda \in M(C\Gamma)} \Omega(\lambda, \bar{z}) \geq C\bar{\gamma}(\bar{z}) \tag{3.301}$$

for all $\bar{z} \in \mathcal{K}$. Then the trajectory $\mathcal{T}$ affords a strict bounded strong minimum.

Again we can use the characterization of the condition $\lambda \in M(C\Gamma)$ formulated in the previous section.

Chapter 4

Jacobi-Type Conditions and Riccati Equation for Broken Extremals

Here we derive tests for the positive semidefiniteness, respectively, positive definiteness of the quadratic form Ω on the critical cone $\mathcal{K}$ (introduced in Chapter 2 for extremals with jumps of the control). In Section 4.1, we derive such tests for the simplest problem of the calculus of variations and for an extremal with only one corner point. We come to a generalization of the concept of conjugate point which allows us to formulate both necessary and sufficient second-order optimality conditions for broken extremals. Three numerical examples illustrate this generalization. Further, we concentrate on sufficient conditions for positive definiteness of the quadratic form Ω in the auxiliary problem. We show that if there exists a solution to the Riccati matrix equation satisfying a certain jump condition, then the quadratic form Ω can be transformed into a perfect square. This gives a possibility of proving a sufficient condition for positive definiteness of the quadratic form in the auxiliary problem and thus to obtain one more sufficient condition for optimality of broken extremals. At the end of Section 4.1, we obtain such condition for the simplest problem of the calculus of variations, and then, in Section 4.2, we prove it for the general problem (without constraint $g(t,x,u)=0$).

4.1 Jacobi-Type Conditions and Riccati Equation for Broken Extremals in the Simplest Problem of the Calculus of Variations

4.1.1 An Auxiliary Problem for Broken Extremal

Let a closed interval $[t_0,t_f]$, two points $b_0,b_f\in\mathbb{R}^m$, an open set $\mathcal{Q}\subset\mathbb{R}^{2m+1}$, and a function $F:\mathcal{Q}\mapsto\mathbb{R}$ of class C^2 be fixed. The simplest problem of the calculus of variations can be formulated as follows:

$$\text{(SP)}\qquad \text{Minimize } J(x(\cdot),u(\cdot)):=\int_{t_0}^{t_f}F(t,x(t),u(t))dt \tag{4.1}$$

under the constraints

$$\dot{x}(t)=u(t),\quad x(t_0)=b_0,\quad x(t_f)=b_f, \tag{4.2}$$

$$(t,x(t),u(t))\in\mathcal{Q}. \tag{4.3}$$

Here, $x(\cdot):[t_0,t_f]\to\mathbb{R}^m$ is absolutely continuous, and $u(\cdot):[t_0,t_f]\to\mathbb{R}^m$ is bounded and measurable. Set $w(\cdot)=(x(\cdot),u(\cdot))$. Then $w(\cdot)$ is an element of the space

$$W:=W^{1,1}([t_0,t_f],\mathbb{R}^m)\times L^\infty([t_0,t_f],\mathbb{R}^m).$$

We say that $w(\cdot)=(x(\cdot),u(\cdot))$ is an *admissible pair* if $w(\cdot)\in W$ and the constraints (4.2), (4.3) hold for it. Let an admissible pair $w(\cdot)=(x(\cdot),u(\cdot))$ be given. Assume that $u(\cdot)$ is a piecewise continuous function with a unique point of discontinuity $t_*\in(t_0,t_f)$. Denote by Θ the singleton $\{t_*\}$. For t_*, we set $u^-=u(t_*-)$, $u^+=u(t_*+)$, and $[u]=u^+-u^-$. Thus, $[u]$ is a jump of the function $u(\cdot)$ at the point t_*.

In correspondence with (4.3), assume that $(t,x(t),u(t))\in\mathcal{Q}$ for all $t\in[t_0,t_f]\setminus\Theta$, and $(t_*,x(t_*),u^-)\in\mathcal{Q}$, $(t_*,x(t_*),u^+)\in\mathcal{Q}$. Also assume that there exist a constant $C>0$ and a small number $\varepsilon>0$ such that

$$\begin{aligned}|u(t)-u^-| &\le C|t-t_*| \quad \forall\, t\in(t_*-\varepsilon,t_*)\cap[t_0,t_f],\\ |u(t)-u^+| &\le C|t-t_*| \quad \forall\, t\in(t_*,t_*+\varepsilon)\cap[t_0,t_f].\end{aligned}$$

The pair $w(\cdot)=(x(\cdot),u(\cdot))$ is called an *extremal* if

$$\psi(t):=-F_u(t,x(t),u(t)) \tag{4.4}$$

is a Lipschitz-continuous function and a condition equivalent to the Euler equation is satisfied:

$$-\dot\psi(t)=F_x(t,x(t),u(t)) \qquad \forall\, t\in[t_0,t_f]\setminus\Theta. \tag{4.5}$$

Here $\psi\in(\mathbb{R}^m)^*$ is a row vector while $x,u\in R^m$ are column vectors. For the Pontryagin function

$$H(t,x,u,\psi)=\psi u+F(t,x,u), \tag{4.6}$$

we set $H^-=H(t_*,x(t_*),u^-,\psi(t_*))$, $H^+=H(t_*,x(t_*),u^+,\psi(t_*))$, and $[H]=H^+-H^-$. The equalities

$$[H]=0, \quad [\psi]=0 \tag{4.7}$$

are the Weierstrass–Erdmann conditions. They are known as necessary conditions for the strong minimum. However, they are also necessary for the Pontryagin minimum introduced in Chapter 2. For convenience, let us recall the definitions of Pontryagin and bounded strong minima in the simplest problem.

We say that the pair of functions $w=(x,u)$ is a point of *Pontryagin minimum* in problem (4.1)–(4.3) if for each compact set $\mathcal{C}\subset\mathcal{Q}$ there exists $\varepsilon>0$ such that $J(\tilde w)\ge J(w)$ for all admissible pairs $\tilde w=(\tilde x,\tilde u)$ such that

(a) $\max_{[t_0,t_f]}|\tilde x(t)-x(t)|<\varepsilon$,

(b) $\int_{t_0}^{t_f}|\tilde u(t)-u(t)|\,dt<\varepsilon$,

(c) $(t,\tilde x(t),\tilde u(t))\in\mathcal{C}$.

We say that the pair $w = (x,u)$ is a point of *bounded strong minimum* in problem (4.1)–(4.3), if for each compact set $\mathcal{C} \subset \mathcal{Q}$ there exists $\varepsilon > 0$ such that $J(\tilde{w}) \geq J(w)$ for all admissible pairs $\tilde{w} = (\tilde{x},\tilde{u})$, such that

(a) $\max_{[t_0,t_f]} |\tilde{x}(t) - x(t)| < \varepsilon$,
(b) $(t,\tilde{x}(t),\tilde{u}(t)) \in \mathcal{C}$.

Clearly, the following implications hold: *strong minimum* $\Longrightarrow$ *bounded strong minimum* $\Longrightarrow$ *Pontryagin minimum* $\Longrightarrow$ *weak minimum.*

As already mentioned, the Weierstrass–Erdmann conditions are necessary for the Pontryagin minimum. As shown in Chapter 2, they can be supplemented by additional condition of the same type. We set

$$a = D(H) := \dot{\psi}^+ \dot{x}^- - \dot{\psi}^- \dot{x}^+ - [F_t], \tag{4.8}$$

where $\dot{\psi}^- = \dot{\psi}(t_*-)$, $\dot{\psi}^+ = \dot{\psi}(t_*+)$, $\dot{x}^- = \dot{x}(t_*-)$, $\dot{x}^+ = \dot{x}(t_*+)$, and $[F_t] = F_t(t_*,x(t_*),u^+) - F_t(t_*,x(t_*),u^-)$. Then $a \geq 0$ is a necessary condition for the Pontryagin minimum.

As we know, the value $D(H)$ can be computed in a different way. Consider the function

$$\begin{aligned} (\Delta H)(t) &:= H(t,x(t),u^+,\psi(t)) - H(t,x(t),u^-,\psi(t)) \\ &= \psi(t)[\dot{x}] + \left(F(t,x(t),u^+) - F(t,x(t),u^-)\right). \end{aligned} \tag{4.9}$$

Using (4.5), we obtain

$$\frac{d}{dt}(\Delta H)|_{t_*+0} = \dot{\psi}^+[\dot{x}] - [\dot{\psi}]\dot{x}^+ + [F_t], \tag{4.10}$$

$$\frac{d}{dt}(\Delta H)|_{t_*-0} = \dot{\psi}^-[\dot{x}] - [\dot{\psi}]\dot{x}^- + [F_t]. \tag{4.11}$$

Since $\dot{\psi}^+[\dot{x}] - [\dot{\psi}]\dot{x}^+ = \dot{\psi}^- \dot{x}^+ - \dot{\psi}^+ \dot{x}^- = \dot{\psi}^-[\dot{x}] - [\dot{\psi}]\dot{x}^-$, we obtain from this that

$$\frac{d}{dt}(\Delta H)|_{t_*-0} = \frac{d}{dt}(\Delta H)|_{t_*+0} = -D(H). \tag{4.12}$$

Note that the inequality $a \geq 0$ and the Weierstrass–Erdmann conditions are implied by the conditions of the minimum principle, which is equivalent to the Weierstrass condition in this problem. Here, the *minimum principle* has the form: $H(t,x(t),u,\psi(t)) \geq H(t,x(t),u(t),\psi(t))$ if $t \in [t_0,t_f] \setminus \Theta$, $u \in \mathbb{R}^m$, $(t,x(t),u) \in \mathcal{Q}$. Let us also formulate the *strict minimum principle*:

(a) $H(t,x(t),u,\psi(t)) > H(t,x(t),u(t),\psi(t))$
for all $t \in [t_0,t_f] \setminus \Theta$, $u \in \mathbb{R}^m$, $(t,x(t),u) \in \mathcal{Q}$, $u \neq u(t)$,

(b) $H(t_*,x(t_*),u,\psi(t_*)) > H(t_*,x(t_*),u^-,\psi(t_*)) = H(t_*,x(t_*),u^+,\psi(t_*))$
for all $u \in \mathbb{R}^m$, $(t,x(t),u) \in \mathcal{Q}$, $u \neq u^-$, $u \neq u^+$.

Now we define a quadratic form that corresponds to an extremal $w(\cdot)$ with a corner point. As in Section 2.1.5, denote by $P_\Theta W^{1,2}\left([t_0,t_f],\mathbb{R}^m\right)$ the space of all piecewise continuous functions $\bar{x}(\cdot) : [t_0,t_f] \longrightarrow \mathbb{R}^m$ that are absolutely continuous on each of the intervals in $[t_0,t_f] \setminus \Theta$ whose derivatives are square Lebesgue integrable. For t_*, we set

$\bar{x}^- = \bar{x}(t_*-)$, $\bar{x}^+ = \bar{x}(t_*+)$, $[\bar{x}] = \bar{x}^+ - \bar{x}^-$. Recall that the space $P_\Theta W^{1,2}([t_0,t_f],\mathbb{R}^m)$ with the inner product $(\bar{x},\bar{y}) = \langle \bar{x}(0),\bar{y}(0)\rangle + \langle [\bar{x}],[\bar{y}]\rangle + \int_{t_0}^{t_f} \langle \dot{\bar{x}}(t),\dot{\bar{y}}(t)\rangle\, dt$ is a Hilbert space. We set

$$Z_2(\Theta) = \mathbb{R}^1 \times P_\Theta W^{1,2}\left([t_0,t_f],\mathbb{R}^m\right) \times L^2\left([t_0,t_f],\mathbb{R}^m\right)$$

and denote by $\bar{z} = (\bar{\xi},\bar{x},\bar{u})$ an element of the space $Z_2(\Theta)$, where

$$\bar{\xi} \in \mathbb{R}^1, \bar{x}(\cdot) \in P_\Theta W^{1,2}\left([t_0,t_f],\mathbb{R}^m\right), \bar{u}(\cdot) \in L^2\left([t_0,t_f],\mathbb{R}^m\right).$$

In the Hilbert space $Z_2(\Theta)$, we define a subspace and a quadratic form by setting

$$\mathcal{K} = \left\{ \bar{z} \in Z_2(\Theta) \mid \dot{\bar{x}}(t) = \bar{u}(t), \quad \bar{x}(t_0) = \bar{x}(t_f) = 0, \quad [\bar{x}] = [\dot{x}]\bar{\xi} \right\} \tag{4.13}$$

and

$$\Omega(\bar{z}) = \frac{1}{2}\left(a\bar{\xi}^2 - 2[\dot{\psi}]\bar{x}_{\mathrm{av}}\bar{\xi}\right) + \frac{1}{2}\int_{t_0}^{t_f} \langle F_{ww}\bar{w}(t),\bar{w}(t)\rangle\, dt, \tag{4.14}$$

respectively, where

$$\bar{x}_{\mathrm{av}} = \frac{1}{2}\left(\bar{x}^- + \bar{x}^+\right), \quad F_{ww} = F_{ww}(t,x(t),u(t)), \quad w = (x,u),$$

$$F_{ww} = \begin{pmatrix} F_{xx} & F_{xu} \\ F_{ux} & F_{uu} \end{pmatrix}, \qquad \bar{w}(\cdot) = (\bar{x}(\cdot),\bar{u}(\cdot)).$$

The condition of positive semidefiniteness of Ω on $\mathcal{K}$ implies the following *Legendre condition*:

(L) for any $v \in \mathbb{R}^m$,

$$\langle F_{uu}(t,x(t),u(t))v,v\rangle \geq 0 \qquad \forall\, t \in [t_0,t_f]\backslash\Theta, \tag{4.15}$$

$$\langle F_{uu}^- v,v\rangle \geq 0, \quad \langle F_{uu}^+ v,v\rangle \geq 0. \tag{4.16}$$

which is a necessary condition for a weak minimum. Here $F_{uu}^- = F_{uu}(t_*,x(t_*),u^-)$, $F_{uu}^+ = F_{uu}(t_*,x(t_*),u^+)$. The condition of positive definiteness of Ω on K implies the *strengthened Legendre condition*:

(SL) for any $v \in \mathbb{R}^m\backslash\{0\}$,

$$\langle F_{uu}(t,x(t),u(t))v,v\rangle > 0 \quad \forall\, t \in [t_0,t_f]\backslash\Theta, \tag{4.17}$$

$$\langle F_{uu}^- v,v\rangle > 0, \quad \langle F_{uu}^+ v,v\rangle > 0. \tag{4.18}$$

The *auxiliary minimization problem* for an extremal $w(\cdot)$ with a corner point is formulated as follows:

(AP) minimize $\Omega(\bar{z})$ under the constraint $\bar{z} \in \mathcal{K}$.

This setting is stipulated by the following two theorems.

Theorem 4.1. *If $w(\cdot)$ is a Pontryagin minimum, then the following conditions hold:*

(a) *the Euler equation,*

(b) *the minimum principle (the Weierstrass condition),*

(c) *the Legendre condition (L),*
(d) $a \geq 0$,
(e) $\Omega(\bar{z}) \geq 0$ *for all* $\bar{z} \in \mathcal{K}$.

Theorem 4.2. *If the following conditions hold, then* $w(\cdot)$ *is a point of a strict bounded strong minimum:*
(a) *the Euler equation,*
(b) *the strict minimum principle (the strict Weierstrass condition),*
(c) *the strengthened Legendre condition (SL),*
(d) $a > 0$,
(e) *there exists* $\varepsilon > 0$ *such that* $\Omega(\bar{z}) \geq \varepsilon(\bar{\xi}^2 + \int_{t_0}^{t_f} \langle \bar{u}(t), \bar{u}(t)\rangle\, dt)$ *for all* $\bar{z} \in \mathcal{K}$.

Theorems 4.1 and 4.2 follow from Theorems 2.4 and 2.102, respectively. Note that the functional

$$\bar{\gamma}(\bar{z}) = \bar{\xi}^2 + \int_{t_0}^{t_f} \langle \bar{u}(t), \bar{u}(t)\rangle\, dt \tag{4.19}$$

on the subspace $\mathcal{K}$ is equivalent to the squared norm: $\bar{\gamma} \sim (\bar{z}, \bar{z})$.

Now our goal is to derive the tests for positive semidefiniteness and positive definiteness of the quadratic form Ω on the subspace $\mathcal{K}$ in the case of an extremal with a single corner point. We give such tests in the form of the Jacobi-type conditions and in terms of solutions to the Riccati equations.

4.1.2 Jacobi-Type Conditions and the Riccati Equation

In this section, we assume that for an extremal $w(\cdot) = (x(\cdot), u(\cdot))$ with a single corner point $t_* \in (t_0, t_f)$, the condition (SL) holds, and $a > 0$, where a is defined by (4.8). It follows from condition (SL) that at each point $t \in [t_0, t_f] \setminus \Theta$, the matrix $F_{uu} = F_{uu}(t, w(t))$ has the inverse matrix F_{uu}^{-1}. We set

$$\begin{array}{rclrl} A &=& -F_{uu}^{-1} F_{ux}, & B = -F_{uu}^{-1}, \\ C &=& -F_{xu} F_{uu}^{-1} F_{ux} + F_{xx}, & A^* = -F_{xu} F_{uu}^{-1}. \end{array} \tag{4.20}$$

All derivatives are computed along the trajectory $(t, w(t))$. Note that

$$\begin{array}{l} B^* = B, \qquad C^* = C, \\ |\det B| \geq \text{const} > 0 \quad \text{on } [t_0, t_f], \end{array} \tag{4.21}$$

and A, B, and C are matrices with piecewise continuous entries on $[t_0, t_f]$ that are continuous on each of the intervals of the set $(t_0, t_f) \setminus \Theta$.

Further, we formulate Jacobi-type conditions for an extremal $w(\cdot)$ with a single corner point t_*. Denote by $X(t)$ and $\Psi(t)$ two square matrices of order m, where $t \in [t_0, t_f]$. For $X(t)$ and $\Psi(t)$, we consider the set of differential equations

$$\begin{array}{rcl} \dot{X} &=& AX + B\Psi, \\ -\dot{\Psi} &=& CX + A^*\Psi \end{array} \tag{4.22}$$

with the initial conditions

$$X(t_0) = O, \quad \Psi(t_0) = -I, \tag{4.23}$$

where O and I are the zero matrix and the identity matrix, respectively.

Recall that a continuous (and hence a piecewise smooth) solution $X(t), \Psi(t)$ to the Cauchy problem (4.22), (4.23) allows one to formulate the classical concept of the conjugate point. Namely, a point $\tau \in (t_0, t_f]$ is called *conjugate (to the point t_0)* if $\det X(\tau) = 0$. The absence of a conjugate point in (t_0, t_f) is equivalent to the positive semidefiniteness of the quadratic form

$$\omega = \int_{t_0}^{t_f} \langle F_{ww}\bar{w}, \bar{w}\rangle \, dt$$

on the subspace $\mathcal{K}_0$ consisting of pairs $\bar{w} = (\bar{x}, \bar{u})$ such that

$$\dot{\bar{x}} = \bar{u}, \quad \bar{x}(t_0) = \bar{x}(t_f) = 0, \quad \bar{x} \in W^{1,2}([t_0, t_f], \mathbb{R}^m), \quad \bar{u} \in L^2([t_0, t_f], \mathbb{R}^m).$$

The latter condition is necessary for the weak minimum. The absence of a conjugate point in $(t_0, t_f]$ is equivalent to the condition of positive definiteness of ω on $\mathcal{K}_0$, which is a sufficient condition for the strict weak minimum. This is the classical Jacobi condition.

We note that Ω and $\mathcal{K}$ pass into ω and $\mathcal{K}_0$, respectively, if we set $\bar{\xi} = 0$ in the definition of the first pair. In our tests of positive semidefiniteness and positive definiteness of the quadratic form Ω on the subspace $\mathcal{K}$, we use a discontinuous solution X, Ψ to the Cauchy problem (4.22), (4.23) with certain jump conditions at the point t_*. Namely, let a pair $X(t), \Psi(t)$ be a continuous solution to the problem (4.22), (4.23) on the half-interval $[t_0, t_*)$. We set $X^- = X(t_*-)$, and $\Psi^- = \Psi(t_*-)$. The *jumps* $[X]$ and $[\Psi]$ of the matrix-valued functions $X(t)$ and $\Psi(t)$ at the point t_* are uniquely defined by using the relations

$$a[X] = [\dot{x}]\left(-[\dot{x}]^*\Psi^- + [\dot{\psi}]X^-\right), \tag{4.24}$$

$$a[\Psi] = [\dot{\psi}]^*\left(-[\dot{x}]^*\Psi^- + [\dot{\psi}]X^-\right), \tag{4.25}$$

where $[\dot{x}]$ and $[\dot{\psi}]^*$ are column matrices, while $[\dot{x}]^*$ and $[\dot{\psi}]$ are row matrices. Let us define the right limits X^+ and Ψ^+ of the functions X and Ψ at t_* in the following way:

$$X^+ = X^- + [X], \quad \Psi^+ = \Psi^- + [\Psi].$$

Then we continue the process of solution of system (4.22) on $(t_*, T]$ by using the initial conditions for X and Ψ at t_* given by the conditions

$$X(t_*+) = X^+, \quad \Psi(t_*+) = \Psi^+.$$

Thus, on $[t_0, t_f]$, we obtain a piecewise continuous solution $X(t), \Psi(t)$ to system (4.22) with the initial conditions (4.23) at t_0 and the jump conditions (4.24) and (4.25) at t_*. On each of the intervals of the set $[t_0, t_f] \setminus \Theta$, the matrix-valued functions $X(t)$ and $\Psi(t)$ are smooth. Briefly, this pair of functions will be called a *solution to the problem* (4.22)–(4.25) *on* $[t_0, t_f]$.

Theorem 4.3. *The form Ω is positive semidefinite on $\mathcal{K}$ iff the solution $X(t), \Psi(t)$ to the problem* (4.22)–(4.25) *on $[t_0, t_f]$ satisfies the conditions*

$$\det X(t) \neq 0 \quad \forall\, t \in (t_0, t_f) \setminus \Theta, \tag{4.26}$$

$$\det X^- \neq 0, \quad \det X^+ \neq 0, \tag{4.27}$$

$$a - \left([\dot{x}]^* Q^- - [\dot{\psi}]\right)[\dot{x}] > 0, \tag{4.28}$$

where $Q(t) = \Psi(t)X^{-1}(t)$, $Q^- = Q(t_*-)$, and $X^{-1}(t)$ is the inverse matrix to $X(t)$.

Theorem 4.4. *The form* Ω *is positive definite on* $\mathcal{K}$ *iff the solution* $X(t), \Psi(t)$ *to the problem* (4.22)–(4.25) *on* $[t_0, t_f]$ *satisfies conditions* (4.26)–(4.28), *together with the additional condition*

$$\det X(t_f) \neq 0. \tag{4.29}$$

The conditions for positive semidefiniteness, and those for positive definiteness of Ω on K, which are given by these two theorems can easily be reformulated in terms of a solution to the corresponding matrix Riccati equation. Indeed, if $X(t), \Psi(t)$ is a solution to system (4.22) on a certain interval $\Delta \subset [t_0, t_f]$ with $\det X(t) \neq 0$ on Δ, then, as is well known, the matrix-valued function $Q(t) = \Psi(t)X^{-1}(t)$ satisfies the Riccati equation

$$\dot{Q} + QA + A^*Q + QBQ + C = 0 \tag{4.30}$$

on Δ. Let us prove this assertion. Differentiating the equality $\Psi = QX$ and using (4.22), we obtain

$$-CX - A^*\Psi = \dot{\Psi} = \dot{Q}X + Q\dot{X} = \dot{Q}X + Q(AX + B\Psi).$$

Consequently,

$$CX + A^*\Psi + \dot{Q}X + QAX + QB\Psi = 0.$$

Multiplying this equation by X^{-1} from the right, we obtain (4.30). Using (4.20), we can also represent (4.30) as

$$\dot{Q} - (Q + F_{xu})F_{uu}^{-1}(Q + F_{ux}) + F_{xx} = 0. \tag{4.31}$$

The solution $Q = \Psi X^{-1}$ has a singularity at the zero point, since $X(t_0) = O$. The question is: How do we correctly assign the initial condition for Q? We can do this in the following way. In a small half-neighborhood $[t_0, t_0 + \varepsilon)$, $\varepsilon > 0$, we find a solution to the Riccati equation for $R = Q^{-1} = X\Psi^{-1}$ with the initial condition

$$R(t_0) = O, \tag{4.32}$$

which is implied by (4.23). This Riccati equation for R can easily be obtained. Namely, differentiating the equality

$$X = R\Psi \tag{4.33}$$

and using (4.22), we obtain, for small $\varepsilon > 0$,

$$\dot{R} = AR + RA^* + RCR + B, \quad t \in [t_0, t_0 + \varepsilon]. \tag{4.34}$$

Using (4.20), we can transform this Riccati equation into the form

$$\dot{R} + (RF_{xu} + I)F_{uu}^{-1}(F_{ux}R + I) - RF_{xx}R = 0, \quad t \in [t_0, t_0 + \varepsilon]. \tag{4.35}$$

Thus, we solve the Riccati equation (4.34) or (4.35) with initial condition (4.32) in a certain half-neighborhood $[t_0, t_0 + \varepsilon)$ of t_0. Recall that the matrices B and C are symmetric on $[t_0, t_f]$. Consequently, R is also symmetric, and therefore,

$$Q(t_0 + \varepsilon) = R^{-1}(t_0 + \varepsilon) \tag{4.36}$$

is symmetric.

Let $Q(t)$ be the continuous solution of the Riccati equation (4.30) with initial condition (4.36) on the interval (t_0, t_*). The existence of such a solution is a necessary condition for positive semidefiniteness of Ω on $\mathcal{K}$. Since $B(t)$ and $C(t)$ are symmetric on $[t_0, t_f]$ and $Q(\varepsilon)$ is also symmetric, we have that $Q(t)$ is symmetric on (t_0, t_*). Consequently, $Q^- = Q(t_*-)$ is symmetric. Further, we define a *jump condition* for Q at t_* that corresponds to the jump conditions (4.24) and (4.25). This condition has the form

$$\big(a - (q_-)[\dot{x}]\big)[Q] = (q_-)^*(q_-), \tag{4.37}$$

where

$$q_- = [\dot{x}]^* Q^- - [\dot{\psi}]. \tag{4.38}$$

Note that q_- and $[\dot{x}]^*$ are row vectors while $(q_-)^*$ and $[\dot{x}]$ are column vectors. The jump $[Q]$ of the matrix Q at the point t_* is uniquely defined by using (4.37) since, according to Theorem 4.3, the condition

$$a - (q_-)[\dot{x}] > 0$$

is necessary for positive semidefiniteness of Ω on $\mathcal{K}$. Note that $(q_-)^*(q_-)$ is a symmetric positively semidefinite matrix. Hence, $[Q]$ is a symmetric negatively semidefinite matrix.

The right limit $Q^+ = Q(t_*+)$ is defined by the relation

$$Q^+ = Q^- + [Q]. \tag{4.39}$$

It follows that Q^+ is symmetric. Using Q^+ as the initial condition for Q at the point t_*, we continue the solution of the Riccati equation (4.30) for $t > t_*$. The matrix $Q(t)$ is also symmetric for $t > t_*$. Assume that this symmetric solution $Q(t)$ is extended to a certain interval (t_0, τ) or half-interval $(t_0, \tau]$, where $t_* < \tau \le t_f$. It will be called a *solution to problem* (4.30), (4.36), (4.37) *on* (t_0, τ) *or on* $(t_0, \tau]$, respectively.

Theorem 4.5. *The form Ω is positive semidefinite on $\mathcal{K}$ iff there exists a solution Q to the problem* (4.30), (4.36), (4.37) *on (t_0, t_f) that satisfies*

$$a - (q_-)[\dot{x}] > 0, \tag{4.40}$$

where q_- is defined by (4.38).

Theorem 4.6. *The form Ω is positive definite on $\mathcal{K}$ iff there exists a solution Q to the problem* (4.30), (4.36), (4.37) *on $(t_0, t_f]$ that satisfies inequality* (4.40).

We note that the inequality (4.40) is equivalent to condition (4.28). Moreover, set $b_- = a - (q_-)[\dot{x}]$. Then the inequality (4.40) and the jump condition (4.37) obtain the form

$$b_- > 0, \quad (b_-)[Q] = (q_-)^*(q_-),$$

respectively.

4.1.3 Passage of the Quadratic Form through Zero

Now our goal consists of obtaining the tests for positive semidefiniteness and positive definiteness of Ω on $\mathcal{K}$, which were stated in Section 4.1.2. In the present section we shall use some ideas from [25, 26]. Everywhere below, we assume that condition (SL) holds. It follows from condition (SL) that Ω is a *Legendre form* on $\mathcal{K}$ (cf., e.g., [42]) in the abstract sense; i.e., it is a weakly lower semicontinuos functional on $\mathcal{K}$, and the conditions

$$\bar{z}^n \in \mathcal{K} \quad \forall\, n, \quad \bar{z} \in \mathcal{K}, \tag{4.41}$$

$$\bar{\xi}^n \to \bar{\xi}, \quad \bar{u}^n \to \bar{u} \text{ weakly in } L^2, \tag{4.42}$$

$$\Omega(\bar{z}^n) \to \Omega(\bar{z}) \tag{4.43}$$

imply

$$\|\bar{u}^n - \bar{u}\|_2 \to 0, \tag{4.44}$$

and hence $\bar{z}^n \to \bar{z}$ strongly in $Z_2(\Theta)$. Here $\|v\|_2 = (\int_{t_0}^{t_f} \langle v(t), v(t)\rangle\, dt)^{1/2}$ is the norm in L^2. Further, consider a monotonically increasing one-parameter family of subspaces $\mathcal{K}(\tau)$ in $\mathcal{K}$, each of which is defined by the relation

$$\mathcal{K}(\tau) = \left\{\bar{z} = (\bar{\xi}, \bar{x}, \bar{u}) \in \mathcal{K} \mid \bar{u}(t) = 0 \text{ in } [\tau, t_f]\right\}, \tag{4.45}$$

where $\tau \in [t_0, t_f]$. It is clear that

$$\mathcal{K}(t_0) = \{0\}, \quad \mathcal{K}(t_f) = \mathcal{K}. \tag{4.46}$$

We now study the problem of how the property of positive definiteness of Ω on $\mathcal{K}(\tau)$ depends on the change of the parameter τ. This will allow us to obtain Jacobi-type conditions for Ω on $\mathcal{K}$.

The form Ω is *positive* on $\mathcal{K}(\tau)$ if $\Omega(\bar{z}) > 0$ for all $\bar{z} \in \mathcal{K}(\tau)\setminus\{0\}$. As is well known, for Legendrian forms, the positivity of Ω on $\mathcal{K}(\tau)$ is equivalent to its *positive definiteness* on $\mathcal{K}(\tau)$, i.e., to the following property: There exists $\varepsilon > 0$ such that $\Omega(\bar{z}) \geq \varepsilon\bar{\gamma}(\bar{z})$ for all $\bar{z} \in \mathcal{K}(\tau)$, where $\bar{\gamma}$ is defined by relation (4.19).

Obviously, for $\tau = t_0$ the form Ω is positive definite on $\mathcal{K}(t_0) = \{0\}$. Below, we will prove that, due to condition (SL), Ω is also positive definite on $\mathcal{K}(\tau)$ for all sufficiently small $\tau - t_0 > 0$. Let us increase the value of τ. For $\tau = t_f$, we have three possibilities:

Case (1). Ω is positive definite on $\mathcal{K}$.

Case (2). Ω is positive semidefinite on $\mathcal{K}$, but is not positive definite on $\mathcal{K}$.

Case (3). Ω is not positive semidefinite on $\mathcal{K}$.

In Cases (2) and (3) we define

$$\tau_0 := \sup\left\{\tau \in [t_0, t_f] \mid \Omega \text{ is positive definite on } \mathcal{K}(\tau)\right\}. \tag{4.47}$$

We will show that $\Omega(\cdot) \geq 0$ on $\mathcal{K}(\tau_0)$, but Ω is not positive on $\mathcal{K}(\tau_0)$. Consequently, there exists $\bar{z} \in \mathcal{K}(\tau_0)\setminus\{0\}$ such that $\Omega(\bar{z}) = 0$. This fact was called in [25] "the passage of quadratic form through zero." This property of the form Ω follows from condition (SL) and plays a crucial role in our study of the problem of the definiteness of Ω on $\mathcal{K}$. (Note that another possibility is that Ω is still positive on $\mathcal{K}(\tau_0)$. In this case, Ω "does not pass through zero." Such examples are presented in [25] for a quadratic form that does not satisfy the strengthened Legendre condition.) Now, our goal consists of proving the following theorem.

Theorem 4.7. *If Ω is not positive definite on $\mathcal{K}$, then there exists $\tau_0 \in (t_0, t_f]$ such that*

(a) *$\Omega(\cdot)$ is positive definite on $\mathcal{K}(\tau)$ for all $\tau \in (t_0, \tau_0)$ and*

(b) *$\Omega(\cdot) \geq 0$ on $\mathcal{K}(\tau_0)$ and there exists $\bar{z} \in \mathcal{K}(\tau_0)\setminus\{0\}$ such that $\Omega(\bar{z}) = 0$.*

Using this theorem, we can define τ_0 as the minimum value among all $\tau \in (t_0, t_f]$ such that the quadratic form Ω has a nontrivial zero $\bar{z}$ on the subspace $\mathcal{K}(\tau)$. To prove this theorem, we need two auxiliary assertions.

Proposition 4.8. *There exists $\tau \in (t_0, t_f]$ such that Ω is positive definite on $\mathcal{K}(\tau)$.*

Proof. Let $\epsilon_L > 0$ be such that for each $v \in \mathbb{R}^m$,

$$\langle F_{uu}(t, x(t), u(t))v, v\rangle \geq \epsilon_L |v|^2 \qquad \forall\, t \in [t_0, t_f] \backslash \Theta. \tag{4.48}$$

Choose a certain $\tau \in (t_0, t_*)$ and let $\bar{z} = (\bar{\xi}, \bar{x}, \bar{u}) \in \mathcal{K}(\tau) \backslash \{0\}$ be an arbitrary element. Then $\bar{u} \neq 0$ and

$$\begin{aligned} &\bar{u}\chi_{[t_0,\tau]} = \bar{u}, \quad \dot{\bar{x}} = \bar{u}, \quad \bar{x}(t_0) = 0, \\ &\bar{x}\chi_{[t_0,t_*]} = \bar{x}, \quad \bar{x}(\tau) = \bar{x}^-, \quad \bar{x}^- + [\dot{x}]\bar{\xi} = 0, \end{aligned}$$

where $\chi_{\mathcal{M}}$ is the characteristic function of a set $\mathcal{M}$. Consequently,

$$\begin{aligned} &\|\bar{x}\|_\infty \leq \|\bar{u}\|_1 \leq \sqrt{\tau - t_0}\|\bar{u}\|_2, \quad |\bar{\xi}| = \frac{|\bar{x}^-|}{|[\dot{x}]|} \leq \frac{\sqrt{\tau - t_0}}{|[\dot{x}]|}\|\bar{u}\|_2, \\ &|2\bar{x}_{\text{av}}| = |\bar{x}^-| \leq \sqrt{\tau - t_0}\|\bar{u}\|_2. \end{aligned}$$

Therefore,

$$\begin{aligned} &\left|\int_{t_0}^{t_f} \langle F_{xx}\bar{x}, \bar{x}\rangle\, dt\right| \leq \|F_{xx}\|_\infty \|\bar{x}\|_\infty^2 (t_* - t_0) \leq \|F_{xx}\|_\infty (t_* - t_0)(\tau - t_0)\|\bar{u}\|_2^2, \\ &\left|\int_{t_0}^{t_f} \langle 2F_{xu}\bar{u}, \bar{x}\rangle\, dt\right| \leq 2\|F_{xu}\|_\infty \|\bar{x}\|_\infty \|\bar{u}\|_1 \leq 2\|F_{xu}\|_\infty (\tau - t_0)\|\bar{u}\|_2^2, \\ &|a\bar{\xi}^2| \leq \frac{|a|}{|[\dot{x}]|^2}(\tau - t_0)\|\bar{u}\|_2^2, \quad |2[F_x]\bar{x}_{\text{av}}\bar{\xi}| \leq \frac{|[F_x]|}{|[\dot{x}]|}(\tau - t_0)\|\bar{u}\|_2^2. \end{aligned}$$

Consequently,

$$\Omega(\bar{z}) \geq \int_{t_0}^{t_f} \langle F_{uu}\bar{u}, \bar{u}\rangle\, dt - M(\tau - t_0)\|\bar{u}\|_2^2 \geq \epsilon_L \|\bar{u}\|_2^2 - M(\tau - t_0)\|\bar{u}\|_2^2, \tag{4.49}$$

where

$$M = \|F_{xx}\|_\infty (t_* - t_0) + 2\|F_{xu}\|_\infty + \frac{|a|}{|[\dot{x}]|^2} + \frac{|[F_x]|}{|[\dot{x}]|}.$$

Let τ be such that $t_0 < \tau < t_0 + \epsilon_L/M$. Then (4.49) implies that Ω is positive definite on $\mathcal{K}(\tau)$. ∎

Proposition 4.9. *Assume that Ω is not positive definite on $\mathcal{K}$. Then Ω is positive semidefinite on $\mathcal{K}(\tau_0)$, where τ_0 is defined as in* (4.47).

We omit the simple proof of this proposition. Now we are ready to prove the theorem.

Proof of Theorem 4.7. We have already proved that Ω is positive definite on $\mathcal{K}(\tau)$ for all $\tau > t_0$ sufficiently close to t_0 (Proposition 4.8). Consequently, $\tau_0 > t_0$. Further, we consider only the nontrivial case where $\tau_0 < t_f$. We know that $\Omega(\cdot) \geq 0$ on $\mathcal{K}(\tau_0)$ (Proposition 4.9). We have to show that Ω is not positive on $K(\tau_0)$, i.e., there exists $\bar{z} \in \mathcal{K}(\tau_0) \backslash \{0\}$ such that $\Omega(\bar{z}) = 0$ (the passage through zero). Now we follow [26]. For any $\tau > \tau_0$ $(\tau \leq t_f)$, Ω is not positive on $\mathcal{K}(\tau)$. Therefore, for each

$$\tau_n = \tau_0 + \frac{1}{n} < t_f \tag{4.50}$$

there exists

$$\bar{z}^n \in \mathcal{K}(\tau_n) \tag{4.51}$$

such that

$$\Omega(\bar{z}^n) \leq 0, \tag{4.52}$$

$$\bar{\gamma}(\bar{z}^n) = 1. \tag{4.53}$$

The sequence $\{\bar{z}^n\}$ is bounded in $\mathcal{K}$. Therefore, without loss of generality, we assume that

$$\bar{z}^n \longrightarrow \bar{z}^0 \quad \text{weakly}. \tag{4.54}$$

Since each subspace $\mathcal{K}(\tau_n) \subset \mathcal{K}$ is weakly closed, we have

$$\bar{z}^0 \in \mathcal{K}(\tau_n) \quad \forall\, n \tag{4.55}$$

and, therefore,

$$\bar{z}^0 \in \bigcap_n \mathcal{K}(\tau_n) = \mathcal{K}(\tau_0). \tag{4.56}$$

By Proposition 4.9, it follows from (4.56) that

$$\Omega(\bar{z}^0) \geq 0. \tag{4.57}$$

On the other hand, Ω is weakly lower semicontinuous on $\mathcal{K}$. Thus, (4.54) implies

$$\liminf \Omega(\bar{z}^n) \geq \Omega(\bar{z}^0). \tag{4.58}$$

We obtain from (4.52), (4.57), and (4.58) that

$$\Omega(\bar{z}^n) \longrightarrow \Omega(\bar{z}^0) = 0, \tag{4.59}$$

and then

$$\bar{z}^n \longrightarrow \bar{z}^0 \quad \text{strongly} \tag{4.60}$$

since Ω is a Legendre form. Conditions (4.53) and (4.60) imply

$$\bar{\gamma}(\bar{z}^0) = 1. \tag{4.61}$$

Consequently, $\bar{z}^0 \neq 0$, $\Omega(\bar{z}^0) = 0$, i.e., Ω is not positive on $\mathcal{K}(\tau_0)$. ∎

For $\tau \in [t_0, t_f]$, let us consider the problem

(AP$_\tau$) minimize $\Omega(z)$ under the constraint $z \in \mathcal{K}(\tau)$.

Assume that a nonzero element

$$\bar{z} = (\bar{\xi}, \bar{x}, \bar{u}) = (\bar{\xi}, \bar{w}) \in \mathcal{K}(\tau) \tag{4.62}$$

yields the minimum in this problem. Then the following first-order necessary optimality condition holds:

$$\big(\Omega'(\bar{z}), \tilde{z}\big) = 0 \quad \forall\, \tilde{z} \in \mathcal{K}(\tau), \tag{4.63}$$

where $\tilde{z} = (\tilde{\xi}, \tilde{x}, \tilde{u}) = (\tilde{\xi}, \tilde{w})$, $\Omega'(\bar{z})$ is the Fréchet derivative of the functional Ω at the point $\bar{z}$, and $(\cdot,\cdot)$ is the inner product in $Z_2(\Theta)$; in more detail,

$$(\Omega'(\bar{z}), \tilde{z}) = a\bar{\xi}\tilde{\xi} - [\dot{\psi}]\bar{x}_{\mathrm{av}}\tilde{\xi} - [\dot{\psi}]\tilde{x}_{\mathrm{av}}\bar{\xi} + \int_{t_0}^{t_f} \langle F_{ww}\bar{w}, \tilde{w}\rangle\, dt. \tag{4.64}$$

Thus, Theorem 4.7 implies the following.

Corollary 4.10. *Assume that Ω is not positive definite on $\mathcal{K}$. Then, for $\tau = \tau_0 \in (t_0, t_f]$ (given by* (4.47)*), there exists a nonzero element $\bar{z}$ that satisfies* (4.62) *and* (4.63).

On the other hand, we obviously have

$$\frac{1}{2}(\Omega'(\bar{z}), \bar{z}) = \Omega(\bar{z}). \tag{4.65}$$

This implies the following.

Proposition 4.11. *If, for certain $\tau \in [t_0, t_f]$, there exists a nonzero element $\bar{z}$ satisfying* (4.62) *and* (4.63)*, then $\Omega(\bar{z}) = 0$, and hence, Ω is not positive definite on $\mathcal{K}(\tau)$.*

Corollary 4.10 and Proposition 4.11 imply the following.

Theorem 4.12. *Assume that Ω is not positive definite on $\mathcal{K}$. Then τ_0, given by* (4.47)*, is minimal among all $\tau \in (t_0, t_f]$ such that there exists a nonzero element $\bar{z}$ satisfying conditions* (4.62) *and* (4.63).

4.1.4 Θ-Conjugate Point

In this section, we obtain a dual test for condition (4.63), and then we use it to obtain an analogue of a conjugate point for a broken extremal. The most important role is played by the following lemma.

Lemma 4.13. *Let $\tau \in (t_0, t_f]$. A triple $\bar{z} = (\bar{\xi}, \bar{x}, \bar{u})$ satisfies conditions* (4.62) *and* (4.63) *iff there exists a function $\bar{\psi}(\cdot) : [t_0, t_f] \longrightarrow (\mathbb{R}^m)^*$ such that for the tuple $\big(\bar{\xi}, \bar{x}, \bar{u}, \bar{\psi}\big)$,* *the following conditions hold:*

$$\bar{\xi} \in \mathbb{R}^1, \quad \bar{x} \in P_\Theta W^{1,2}, \quad \bar{u} \in L^2, \quad \bar{\psi} \in P_\Theta W^{1,2}, \tag{4.66}$$

$$\bar{x}(t_0) = \bar{x}(t_f) = 0, \tag{4.67}$$

$$\dot{\bar{x}} = \bar{u}, \quad \bar{u}\chi_{[t_0,\tau]} = \bar{u}, \tag{4.68}$$

$$-\bar{\psi} = \bar{x}^* F_{xu} + \bar{u}^* F_{uu} \quad \text{a.e. on } [t_0,\tau], \tag{4.69}$$

$$-\dot{\bar{\psi}} = \bar{x}^* F_{xx} + \bar{u}^* F_{ux} \quad \text{a.e. on } [t_0,t_f], \tag{4.70}$$

$$[\bar{x}] = [\dot{x}]\bar{\xi}, \tag{4.71}$$

$$[\bar{\psi}] = [\dot{\psi}]\bar{\xi}, \tag{4.72}$$

$$a\bar{\xi} = -\bar{\psi}_{\rm av}[\dot{x}] + [\dot{\psi}]\bar{x}_{\rm av}. \tag{4.73}$$

Proof. Let $\bar{z} = (\bar{\xi}, \bar{x}, \bar{u})$ satisfy conditions (4.62) and (4.63). Consider condition (4.63). Define the following subspace:

$$\tilde{L}^2([\tau,t_f],\mathbb{R}^m) := \{\tilde{v} \in L^2([t_0,t_f],\mathbb{R}^m) \mid \tilde{v} = 0 \text{ a.e. on } [t_0,\tau]\}.$$

The operator $\tilde{z} \longmapsto (\dot{\tilde{x}} - \tilde{u}, \tilde{u}\chi_{[\tau,t_f]})$ maps the space $Z_2(\Theta)$ onto the space $L^2([t_0,t_f],\mathbb{R}^m) \times \tilde{L}^2([\tau,t_f],\mathbb{R}^m)$. The operator $\tilde{z} \longmapsto ([\tilde{x}] - [\dot{x}]\tilde{\xi}, \tilde{x}(t_0), \tilde{x}(t_f))$ is finite dimensional. Consequently, the image of the operator

$$\tilde{z} \mapsto \big([\tilde{x}] - [\dot{x}]\tilde{\xi}, \dot{\tilde{x}} - \tilde{u}, \tilde{u}\chi_{[\tau,t_f]}, \tilde{x}(t_0), \tilde{x}(t_f)\big),$$

which maps from $Z_2(\Theta)$ into

$$\mathbb{R}^n \times L^2([t_0,t_f],\mathbb{R}^m) \times \tilde{L}^2([\tau,t_f],\mathbb{R}^m) \times \mathbb{R}^m \times \mathbb{R}^m$$

is closed. The kernel of this operator is equal to $\mathcal{K}(\tau)$. Consequently, an arbitrary linear functional $z^\star$ that vanishes on the kernel of this operator admits the following representation:

$$(z^\star, \tilde{z}) = \bar{\zeta}([\tilde{x}] - [\dot{x}]\tilde{\xi}) - \int_{t_0}^{t_f} \bar{\psi}(\dot{\tilde{x}} - \tilde{u})\,dt + \int_{t_0}^{t_f} \bar{\nu}\tilde{u}\,dt + \bar{c}_0\tilde{x}(t_0) + \bar{c}_f\tilde{x}(t_f), \tag{4.74}$$

where

$$\bar{\zeta} \in (\mathbb{R}^n)^*, \quad \bar{\psi} \in L^2([t_0,t_f],(\mathbb{R}^m)^*), \quad \bar{\nu} \in \tilde{L}^2([\tau,t_f],(\mathbb{R}^m)^*), \quad \bar{c}_0, \bar{c}_f \in (\mathbb{R}^m)^*. \tag{4.75}$$

Consequently, the condition (4.63) is equivalent to the existence of $\bar{\zeta}$, $\bar{\psi}$, $\bar{\nu}$, $\bar{c}_0$, and $\bar{c}_f$ that satisfy (4.75) and are such that for $z^\star$ defined by formula (4.74), we have

$$(\Omega'(\bar{z}), \tilde{z}) + (z^\star, \tilde{z}) = 0 \qquad \forall\, \tilde{z} \in Z_2(\Theta).$$

The exact representation of the latter condition has the form

$$\begin{aligned}
& a\bar{\xi}\tilde{\xi} - [\dot{\psi}]\bar{x}_{\rm av}\tilde{\xi} - [\dot{\psi}]\tilde{x}_{\rm av}\bar{\xi} \\
& + \int_{t_0}^{t_f} (\langle F_{xx}\bar{x}, \tilde{x}\rangle + \langle F_{xu}\bar{u}, \tilde{x}\rangle + \langle F_{ux}\bar{x}, \tilde{u}\rangle + \langle F_{uu}\bar{u}, \tilde{u}\rangle)\,dt \\
& + \bar{\zeta}([\tilde{x}] - [\dot{x}]\tilde{\xi}) - \int_{t_0}^{t_f} \bar{\psi}(\dot{\tilde{x}} - \tilde{u})\,dt + \int_{t_0}^{t_f} \bar{\nu}\tilde{u}\,dt + \bar{c}_0\tilde{x}(t_0) + \bar{c}_f\tilde{x}(t_f) \\
& = 0 \quad \forall\, \tilde{z} = (\tilde{\xi}, \tilde{x}, \tilde{u}) \in Z_2(\Theta).
\end{aligned} \tag{4.76}$$

Let us examine this condition.
(a) We set $\tilde{\xi} = 0$ and $\tilde{x} = 0$ in (4.76). Then

$$\int_{t_0}^{t_f} (\langle F_{ux}\bar{x}, \tilde{u}\rangle + \langle F_{uu}\bar{u}, \tilde{u}\rangle)\, dt + \int_{t_0}^{t_f} \bar{\psi}\tilde{u}\, dt + \int_{t_0}^{t_f} \bar{\nu}\tilde{u}\, dt = 0 \quad \forall\, \tilde{u} \in L^2. \tag{4.77}$$

Consequently,

$$\bar{x}^* F_{xu} + \bar{u}^* F_{uu} + \bar{\psi} = -\bar{\nu}. \tag{4.78}$$

The latter equation is equivalent to condition (4.69).
(b) We set $\tilde{\xi} = 0$ and $\tilde{u} = 0$ in (4.76). Then

$$-[\dot{\psi}]\tilde{x}_{\text{av}}\bar{\xi} + \int_{t_0}^{t_f} (\langle F_{xx}\bar{x}, \tilde{x}\rangle) + \langle F_{xu}\bar{u}, \tilde{x}\rangle)\, dt + \bar{\zeta}[\tilde{x}] - \int_{t_0}^{t_f} \bar{\psi}\dot{\tilde{x}}\, dt$$
$$+ \bar{c}_0\tilde{x}(t_0) + \bar{c}_f\tilde{x}(t_f) = 0 \quad \forall\, \tilde{x} \in P_\Theta W^{1,2}. \tag{4.79}$$

Using (4.79), it is easy to show that $\bar{\psi} \in P_\Theta W^{1,2}([t_0, t_f], (\mathbb{R}^m)^*)$. Consequently,

$$\int_{t_0}^{t_f} \bar{\psi}\dot{\tilde{x}}\, dt = \bar{\psi}\tilde{x}\big|_{t_0}^{t_f} - [\bar{\psi}\tilde{x}] - \int_{t_0}^{t_f} \dot{\bar{\psi}}\tilde{x}\, dt. \tag{4.80}$$

Using (4.80) and the definitions $\tilde{x}_{\text{av}} = \frac{1}{2}(\tilde{x}^- + \tilde{x}^+)$, $[\tilde{x}] = \tilde{x}^+ - \tilde{x}^-$ in (4.79), we obtain

$$-\frac{1}{2}[\dot{\psi}](\tilde{x}^- + \tilde{x}^+)\bar{\xi} + \int_{t_0}^{t_f} \langle F_{xx}\bar{x} + F_{xu}\bar{u}, \tilde{x}\rangle\, dt$$
$$+ \bar{\zeta}(\tilde{x}^+ - \tilde{x}^-) - \bar{\psi}(t_f)\tilde{x}(t_f) + \bar{\psi}(t_0)\tilde{x}(t_0) + \bar{\psi}^+\tilde{x}^+ - \bar{\psi}^-\tilde{x}^-$$
$$+ \int_{t_0}^{t_f} \dot{\bar{\psi}}\tilde{x}\, dt + \bar{c}_0\tilde{x}(t_0) + \bar{c}_f\tilde{x}(t_f) = 0 \qquad \forall\, \tilde{x} \in P_\Theta W^{1,2}. \tag{4.81}$$

Equation (4.81) implies that the coefficients of $\tilde{x}^-$, $\tilde{x}^+$, $\tilde{x}(t_0)$, $\tilde{x}(t_f)$ and the coefficient of $\tilde{x}$ in the integral vanish:

$$-\frac{1}{2}[\dot{\psi}]\bar{\xi} - \bar{\psi}^- - \bar{\zeta} = 0, \tag{4.82}$$
$$-\frac{1}{2}[\dot{\psi}]\bar{\xi} + \bar{\psi}^+ + \bar{\zeta} = 0, \tag{4.83}$$
$$\bar{\psi}(t_0) + \bar{c}_0 = 0, \tag{4.84}$$
$$-\bar{\psi}(t_f) + \bar{c}_f = 0, \tag{4.85}$$
$$\bar{x}^* F_{xx} + \bar{u}^* F_{ux} + \dot{\bar{\psi}} = 0. \tag{4.86}$$

Adding (4.82) and (4.83), we obtain

$$-[\dot{\psi}]\bar{\xi} + [\bar{\psi}] = 0. \tag{4.87}$$

Thus, (4.70) and (4.72) hold. Subtracting (4.82) from (4.83) and dividing the result by two, we obtain

$$\bar{\zeta} = -\bar{\psi}_{\text{av}}. \tag{4.88}$$

(c) We set $\tilde{x} = 0$ and $\tilde{u} = 0$ in (4.76). Then

$$a\bar{\xi}\tilde{\xi} - [\dot{\psi}]\bar{x}_{\text{av}}\tilde{\xi} - \bar{\zeta}[\dot{x}]\tilde{\xi} = 0 \quad \forall\, \tilde{\xi} \in \mathbb{R}^1. \tag{4.89}$$

Consequently, the coefficient of $\tilde{\xi}$ vanishes:

$$a\bar{\xi} - [\dot{\psi}]\bar{x}_{\text{av}} - \bar{\zeta}[\dot{x}] = 0. \tag{4.90}$$

Using (4.88) in (4.90), we obtain (4.73). Conditions (4.67), (4.68), and (4.71) and the first three conditions in (4.66) are implied by (4.62). Thus, all conditions (4.66)–(4.73) hold. Conversely, if a tuple $(\bar{\xi}, \bar{x}, \bar{u}, \bar{\psi})$ satisfies conditions (4.66)–(4.73), then one easily verifies that $\bar{z} = (\bar{\xi}, \bar{x}, \bar{u})$ satisfies (4.62) and (4.63). ■

Let $\bar{z} = (\bar{\xi}, \bar{x}, \bar{u}) \in \mathcal{K}$. Obviously, the condition $\bar{z} \neq 0$ is equivalent to $\bar{x}(\cdot) \neq 0$. Thus, we obtain the following theorem from Theorem 4.12 and Lemma 4.13 under the above condition (SL).

Theorem 4.14. *Assume that Ω is not positive definite on $\mathcal{K}$. Then τ_0 (given by equation* (4.47)) *is a minimal among all $\tau \in (t_0, t_f]$ such that there exists a tuple $(\bar{\xi}, \bar{x}, \bar{u}, \bar{\psi})$ that satisfies* (4.66)–(4.73) *and the condition $\bar{x}(\cdot) \neq 0$.*

In what follows, we will assume that $a > 0$ and, as above, condition (SL) holds.

Assume that $\tau_0 < t_f$. We know that $\Omega \geq 0$ on $\mathcal{K}(\tau_0)$. Theoretically, it is possible that $\Omega \geq 0$ on $\mathcal{K}(\tau_1)$ for a certain $\tau_1 > \tau_0$. In this case, the closed interval $[\tau_0, \tau_1]$ is called a *table* [25]. Tables occur in optimal control problems [25], but, for a smooth extremal, they never arise in the calculus of variations. We now show that for the simplest problem of the calculus of variations, the closed interval $[\tau_0, t_f]$ cannot serve as a table in the case of a broken extremal. To this end, we complete Lemma 4.13 by the following two propositions.

Proposition 4.15. *If the functions $\bar{x}, \bar{\psi} \in P_\Theta W^{1,2}$ and $\bar{u} \in L^2$ satisfy the system*

$$\dot{\bar{x}} = \bar{u}, \quad -\dot{\bar{\psi}} = \bar{x}^* F_{xx} + \bar{u}^* F_{ux}, \quad -\bar{\psi} = \bar{x}^* F_{xu} + \bar{u}^* F_{uu} \tag{4.91}$$

on a certain closed interval $\Delta \subset [t_0, t_f]$, then the functions $\bar{x}$ and $\bar{\psi}$ satisfy the system

$$\dot{\bar{x}} = A\bar{x} + B\bar{\psi}^*, \quad -\dot{\bar{\psi}} = \bar{x}^* C + \bar{\psi} A \tag{4.92}$$

on the same closed interval Δ, where A, B, and C are the same as in (4.20).

Proof. The third equation in (4.91) implies

$$\bar{u} = -F_{uu}^{-1}\bar{\psi}^* - F_{uu}^{-1}F_{ux}\bar{x}, \quad \bar{u}^* = -\bar{\psi}F_{uu}^{-1} - \bar{x}^* F_{xu}F_{uu}^{-1}.$$

Substituting these expressions for $\bar{u}$ and $\bar{u}^*$ into the first and second equations in (4.91), respectively, we obtain (4.92). ■

Proposition 4.16. *Conditions* (4.71)–(4.73) *imply*

$$a\bar{\xi} = -\bar{\psi}^+[\dot{x}] + [\dot{\psi}]\bar{x}^+. \tag{4.93}$$

Proof. Using the expressions $\bar{\psi}_{\rm av} = \bar{\psi}^+ - 0.5[\bar{\psi}]$ and $\bar{x}_{\rm av} = \bar{x}^+ - 0.5[\bar{x}]$, together with (4.71) and (4.72), in (4.73), we obtain (4.93). ∎

Theorem 4.17. *If $\Omega \geq 0$ on $\mathcal{K}$, then Ω is positive definite on $\mathcal{K}(\tau)$ for all $\tau < t_f$.*

Proof. Assume the contrary, i.e., $\Omega \geq 0$ on $\mathcal{K}$ and $\tau_0 < t_f$. Choose an element $\bar{z} = (\bar{\xi}, \bar{x}, \bar{u}) \in \mathcal{K}(\tau_0) \setminus \{0\}$ such that $\Omega(\bar{z}) = 0$. Then $\bar{z}$ yields the minimum in problem (AP$_\tau$) for $\tau = t_f$, and hence, (4.63) holds for $\mathcal{K}(\tau) = \mathcal{K}$. By Lemma 4.13, there exists a function $\bar{\psi}$, defined on $[t_0, t_f]$, such that the conditions (4.66)–(4.73) hold for $\tau = t_f$. Then, according to Propositions 4.15 and 4.16, $\bar{x}$ and $\bar{\psi}$ also satisfy system (4.92) on $[t_0, t_f]$, and relation (4.93) holds for them.

The conditions $\bar{z} \in \mathcal{K}(\tau_0)$, $\tau_0 < t_f$, together with (4.69) and the condition $\tau = t_f$ imply $\bar{x}(t_f) = 0$, $\bar{\psi}(t_f) = 0$ and hence $\bar{x}(t) = 0$, $\bar{\psi}(t) = 0$ on $[t_*, t_f]$ since $\bar{x}, \bar{\psi}$ satisfy (4.92) on the same closed interval. Therefore, $\bar{x}^+ = 0$, $\bar{\psi}^+ = 0$ and then (4.93) and the condition $a > 0$ imply $\bar{\xi} = 0$. By (4.71) and (4.72), we have $[\bar{x}] = 0$, $[\bar{\psi}] = 0$. Consequently, $\bar{x}^- = 0$, $\bar{\psi}^- = 0$, and then $\bar{x}(t) = 0$, $\bar{\psi}(t) = 0$ on $[t_0, t_*]$, since $\bar{x}$ and $\bar{\psi}$ satisfy (4.92) on the same closed interval. Thus, $\bar{x}(t) = 0$ on $[t_0, t_f]$, and hence $\bar{u}(t) = 0$ on $[t_0, t_f]$. We arrive at a contradiction with condition $\bar{z} \neq 0$; this proves the theorem. ∎

Further, we examine conditions (4.66)–(4.73). Using Proposition 4.15, we exclude $\bar{u}$ from these conditions. We obtain the following system:

$$\bar{\xi} \in \mathbb{R}^1, \quad \bar{x}, \bar{\psi} \in P_\Theta W^{1,2}, \tag{4.94}$$

$$\bar{x}(t_0) = \bar{x}(t_f) = 0, \tag{4.95}$$

$$\dot{\bar{x}} = A\bar{x} + B\bar{\psi}^*, \quad -\dot{\bar{\psi}} = \bar{x}^* C + \bar{\psi} A \quad \text{on } [t_0, \tau], \tag{4.96}$$

$$\dot{\bar{x}} = 0, \quad -\dot{\bar{\psi}} = \bar{x}^* F_{xx} \quad \text{on } [\tau, t_f], \tag{4.97}$$

$$[\bar{x}] = [\dot{x}]\bar{\xi}, \quad [\bar{\psi}] = [\dot{\psi}]\bar{\xi}, \tag{4.98}$$

$$a\bar{\xi} = -\bar{\psi}^-[\dot{x}] + [\dot{\psi}]\bar{x}^-. \tag{4.99}$$

The latter condition is obtained from (4.71)–(4.73) similarly to condition (4.93).

If $\bar{\psi}(t_0) = 0$, then (4.94)–(4.99) imply $\bar{x}(\cdot) = 0$. Consequently, we can assign the *nontriviality condition* in the form

$$\bar{\psi}(t_0) \neq 0. \tag{4.100}$$

Definition 4.18. A point $\tau \in (t_0, t_f]$ is called Θ-*conjugate* (*to* t_0) if there exists a triple $(\bar{\xi}, \bar{x}, \bar{\psi})$ that satisfies conditions (4.94)–(4.100).

Obviously, a point $\tau \in (t_0, t_f]$ is Θ-conjugate to t_0 iff, for a given τ, there exists a quadruple $(\bar{\xi}, \bar{x}, \bar{u}, \bar{\psi})$ that satisfies conditions (4.66)–(4.73) and the condition $\bar{x}(\cdot) \neq 0$. Consequently, Theorems 4.14 and 4.17 imply the following.

Theorem 4.19. *The form Ω is positive semidefinite on $\mathcal{K}$ iff there is no point that is Θ-conjugate to t_0 on the interval (t_0, t_f). The form Ω is positive definite on $\mathcal{K}$ iff there is no point that is Θ-conjugate to t_0 on the half-interval $(t_0, t_f]$.*

Now let us examine the condition for positive definiteness of Ω on $\mathcal{K}(t_*)$. This condition implies the positive definiteness of ω on $\mathcal{K}_0(t_*)$ which is defined as a subspace of pairs $\bar{w} = (\bar{x}, \bar{u}) \in \mathcal{K}_0$ (see Section 4.1.2) such that $\bar{u}(t) = 0$ on $[t_*, t_f]$. Let $X(t), \Psi(t)$ be a matrix-valued solution of the Cauchy problem (4.22)–(4.23) on $[t_0, t_*]$. As is well known, the positive definiteness of ω on $\mathcal{K}_0(t_*)$ is equivalent to the condition

$$\det X(t) \neq 0 \quad \forall\, t \in (t_0, t_*]. \tag{4.101}$$

Assume that this condition holds. We set $Q(t) = \Psi(t)X^{-1}(t)$, $t \in (t_0, t_*]$. Consider conditions (4.94)–(4.100) for $\tau \le t_*$. These conditions imply that

$$\begin{aligned} &\bar{x}(\tau) = \bar{x}^- = \bar{x}(t) \quad \forall\, t \in (\tau, t_*), \quad \bar{x}^- + [\dot{x}]\bar{\xi} = \bar{x}^+ = 0, \\ &\bar{\psi}^- = \bar{\psi}(\tau) - \bar{x}(\tau)^*\Phi(\tau), \end{aligned}$$

where $\Phi(\tau) = \int_\tau^{t_*} F_{xx}(t, w(t))\,dt$. Hence, for $\tau \le t_*$ the system (4.94)–(4.100) is equivalent to the system

$$\bar{\xi} \in \mathbb{R}^1, \quad \bar{x}, \bar{\psi} \in P_\Theta W^{1,2}, \tag{4.102}$$

$$\bar{x}(t_0) = 0, \quad \bar{\psi}(t_0) \neq 0, \tag{4.103}$$

$$\dot{\bar{x}} = A\bar{x} + B\bar{\psi}^*, \quad -\dot{\bar{\psi}} = \bar{x}^*C + \bar{\psi}A \quad \text{on } [t_0, \tau], \tag{4.104}$$

$$\bar{x}(\tau) + [\dot{x}]\bar{\xi} = 0, \tag{4.105}$$

$$a\bar{\xi} = -\big(\bar{\psi}(\tau) - \bar{x}(\tau)^*\Phi(\tau)\big)[\dot{x}] + [\dot{\psi}]\bar{x}(\tau). \tag{4.106}$$

Let $(\bar{\xi}, \bar{x}, \bar{\psi})$ be a solution of this system on the closed interval $[t_0, \tau]$, where $\tau \in (t_0, t_*]$. Then there exists $\bar{c} \in \mathbb{R}^m$ such that

$$\bar{x}(t) = X(t)\bar{c}, \quad \bar{\psi}(t) = \bar{c}^*\Psi^*(t). \tag{4.107}$$

Consequently, $\bar{\psi}(t)^* = \Psi(t)\bar{c} = \Psi(t)X^{-1}(t)(X(t)\bar{c}) = Q(t)\bar{x}(t)$. Using this relation, together with (4.105), in (4.106), we obtain

$$a\bar{\xi} - [\dot{x}]^*(Q(\tau) - \Phi(\tau))[\dot{x}]\bar{\xi} + [\dot{\psi}][\dot{x}]\bar{\xi} = 0. \tag{4.108}$$

If $\bar{\xi} = 0$, then from (4.105) and (4.107), we obtain $\bar{x}(\cdot) = 0$, $\bar{\psi}(\cdot) = 0$; this contradicts (4.103). Therefore $\bar{\xi} \neq 0$, and then (4.108) implies

$$a - [\dot{x}]^*(Q(\tau) - \Phi(\tau))[\dot{x}] + [\dot{\psi}][\dot{x}] = 0. \tag{4.109}$$

We have obtained this relation from the system (4.102)–(4.106). Conversely, if (4.109) holds, then, setting $\bar{\xi} = -1$ and $\bar{c} = X^{-1}(\tau)[\dot{x}]$ and defining $\bar{x}, \bar{\psi}$ by formulas (4.107), we obtain a solution of the system (4.102)–(4.106). We have proved the following lemma.

Lemma 4.20. *Assume that condition* (4.101) *holds. Then, $\tau \in (t_0, t_*]$ is a point that is Θ-conjugate to t_0 iff condition* (4.109) *holds.*

We set

$$\mu(t) = a - [\dot{x}]^*(Q(t) - \Phi(t))[\dot{x}] + [\dot{\psi}][\dot{x}]. \tag{4.110}$$

Then, by Lemma 4.20, the absence of a Θ-conjugate point in $(t_0, t_*]$ is equivalent to the condition

$$\mu(t) \neq 0 \quad \forall\, t \in (t_0, t_*]. \tag{4.111}$$

We show further that the function $\mu(t)$ does not increase on $(t_0,t_*]$ and $\mu(t_0+0)=+\infty$. Consequently, condition (4.111) is equivalent to $\mu(t_*)>0$, which is another form of condition (4.28) or condition (4.40).

Proposition 4.21. *Assume that a symmetric matrix $Q(t)$ satisfies the Riccati equation* (4.31) *on (t_0,τ), where $\tau\le t_*$. Then*

$$\dot{\mu}(t)\le 0 \quad \forall\, t\in(t_0,\tau). \tag{4.112}$$

Moreover, if Q satisfies the initial condition (4.36), *then*

$$\mu(t_0+0)=+\infty. \tag{4.113}$$

Proof. For $t\in(t_0,\tau)$, we have

$$\dot{\mu}=-\langle(\dot{Q}+F_{xx})[\dot{x}],[\dot{x}]\rangle. \tag{4.114}$$

Using (4.31) in (4.114), we obtain

$$\begin{aligned}\dot{\mu}&=-\langle(Q+F_{xu})F_{uu}^{-1}(Q+F_{ux})[\dot{x}],[\dot{x}]\rangle\\ &=-\langle F_{uu}^{-1}(Q+F_{ux})[\dot{x}],(Q+F_{ux})[\dot{x}]\rangle\le 0 \quad \text{on } (t_0,\tau),\end{aligned} \tag{4.115}$$

since F_{uu}^{-1} is positive definite. Assume additionally that Q satisfies (4.36). Then $Q=\Psi X^{-1}$, where X and Ψ satisfy (4.22) and (4.23). It follows from (4.22) and (4.23) that

$$X(t_0)=O,\quad \dot{X}(t_0)=-B(t_0)=F_{uu}^{-1}(t_0,w(t_0)),\quad \Psi(t_0)=-I. \tag{4.116}$$

Consequently,

$$X(t)=(t-t_0)F_{uu}^{-1}(t_0,w(t_0))+o(t),\quad \Psi(t)=-I+o(1)\quad \text{as } t\to t_0+0. \tag{4.117}$$

Thus,

$$Q(t)=-\frac{1}{t-t_0}(F_{uu}(t_0,w(t_0))+o(1))\quad \text{as } t\to t_0+0; \tag{4.118}$$

this implies

$$-\langle Q(t)[\dot{x}],[\dot{x}]\rangle=\frac{1}{t-t_0}(\langle F_{uu}(t_0,w(t_0))[\dot{x}],[\dot{x}]\rangle+o(1))\to+\infty\quad \text{as } t\to t_0+0. \tag{4.119}$$

Now (4.110) and (4.119) imply (4.113). ∎

Lemma 4.20 and Proposition 4.21 imply the following.

Theorem 4.22. *Assume that condition* (4.101) *holds. Then the absence of a point τ that is Θ-conjugate to t_0 on $(t_0,t_*]$ is equivalent to condition* (4.28).

Consequently, the positive definiteness of Ω on $\mathcal{K}(t_*)$ is equivalent to the validity of conditions (4.101) and (4.28).

We examine further the system (4.94)–(4.100) for $\tau>t_*$. Using (4.99) and the condition $a>0$, we can exclude ξ from this system. Moreover, for $\bar{x},\bar{\psi}$, we can formulate all conditions on $[t_0,\tau]$ only. As a result, we arrive at the following equivalent system on $[0,\tau]$:

$$\bar{x},\bar{\psi}\in P_\Theta W^{1,2}[t_0,\tau], \tag{4.120}$$

$$\bar{x}(0)=\bar{x}(\tau)=0, \quad \bar{\psi}(0)\neq 0, \tag{4.121}$$

$$\dot{\bar{x}}=A\bar{x}+B\bar{\psi}^*, \quad -\dot{\bar{\psi}}=\bar{x}^*C+\bar{\psi}A \quad \text{on } [t_0,\tau], \tag{4.122}$$

$$a[\bar{x}]=[\dot{x}]\big(-\bar{\psi}^-[\dot{x}]+[\dot{\psi}]\bar{x}^-\big), \tag{4.123}$$

$$a[\bar{\psi}]=[\dot{\psi}]\big(-\bar{\psi}^-[\dot{x}]+[\dot{\psi}]\bar{x}^-\big). \tag{4.124}$$

We have proved the following lemma.

Lemma 4.23. *A point $\tau\in(t_*,t_f]$ is Θ-conjugate to t_0 iff there exists a pair of functions $\bar{x},\bar{\psi}$ that satisfies conditions* (4.120)–(4.124).

Now we can prove the following theorem.

Theorem 4.24. *A point $\tau\in(t_*,t_f]$ is a Θ-conjugate to t_0 iff*

$$\det X(\tau)=0, \tag{4.125}$$

where $X(t),\Psi(t)$ is a solution to the problem (4.22)–(4.25).

Proof. Assume that condition (4.125) holds, in which X,Ψ is a solution to the problem (4.22)–(4.25). Then there exists $\bar{c}\in\mathbb{R}^m$ such that

$$X(\tau)\bar{c}=0, \quad \bar{c}\neq 0. \tag{4.126}$$

We set

$$\bar{x}=X\bar{c}, \quad \bar{\psi}=\bar{c}^*\Psi^*. \tag{4.127}$$

Then $\bar{x}$ and $\bar{\psi}$ satisfy (4.120)–(4.124). Conversely, let $\bar{x}$ and $\bar{\psi}$ satisfy (4.120)–(4.124). We set $\bar{c}=-\bar{\psi}(t_0)^*$. Then conditions (4.126) and (4.127) hold. Conditions (4.126) imply condition (4.125). ∎

Note that Theorems 4.19, 4.22, and 4.24 imply Theorems 4.3 and 4.4.

To complete the proof of the results of Section 4.1.2, we have to consider a jump condition for a solution Q to the Riccati equation (4.30) with initial condition (4.36). In what follows, we assume that Ω is nonnegative on $\mathcal{K}$ and that the pair X,Ψ is a solution to the problem (4.22)–(4.25). Then conditions (4.26) and (4.27) hold. We set $Q=\Psi X^{-1}$. Then $\Psi=QX$. Using the relations $q_-=[\dot{x}]^*Q^- -[\dot{\psi}]$ and $\Psi^-=Q^-X^-$ in the jump conditions (4.24) and (4.25) for X and Ψ, we obtain

$$a[X]=-[\dot{x}](q_-)X^-, \tag{4.128}$$

$$a[\Psi]=-[\dot{\psi}]^*(q_-)X^-. \tag{4.129}$$

Proposition 4.25. *Condition*

$$\det\big(aI-[\dot{x}](q_-)\big)\neq 0 \tag{4.130}$$

holds.

Proof. Relation (4.128) implies

$$aX^{+} = \big(aI - [\dot{x}](q_{-})\big)X^{-}. \tag{4.131}$$

Now (4.130) follows from this relation considered together with (4.27) and the inequality $a > 0$. ∎

Proposition 4.26. *The following equality holds:*

$$(q_{-})^{*}(q_{-})\big(aI - [\dot{x}](q_{-})\big)^{-1} = (q_{-})^{*}(q_{-})\big(a - (q_{-})[\dot{x}]\big)^{-1}, \tag{4.132}$$

where $a - (q_{-})[\dot{x}] > 0$.

Proof. Equality (4.132) is equivalent to

$$(q_{-})^{*}(q_{-})\big(aI - [\dot{x}](q_{-})\big) = (q_{-})^{*}(q_{-})\big(a - (q_{-})[\dot{x}]\big).$$

Let us show that this equality holds. Indeed, we have

$$\begin{aligned}(q_{-})^{*}(q_{-})\big(aI - [\dot{x}](q_{-})\big) &= a(q_{-})^{*}(q_{-}) - \big((q_{-})^{*}(q_{-})\big)[\dot{x}](q_{-})\\ &= a(q_{-})^{*}(q_{-}) - (q_{-})^{*}\big((q_{-})[\dot{x}]\big)(q_{-})\\ &= a(q_{-})^{*}(q_{-}) - \big((q_{-})[\dot{x}]\big)(q_{-})^{*}(q_{-})\\ &= (q_{-})^{*}(q_{-})\big(a - (q_{-})[\dot{x}]\big)\end{aligned}$$

since $(q_{-})[\dot{x}]$ is a number. ∎

Proposition 4.27. *The jump condition* (4.37) *holds.*

Proof. The relation $\Psi = QX$ implies $[\Psi] = Q^{+}X^{+} - Q^{-}X^{-}$. Multiplying it by a and using relations (4.129) and (4.131), we obtain

$$-[\dot{\psi}]^{*}(q_{-})X^{-} = Q^{+}\big(aI - [\dot{x}](q_{-})\big)X^{-} - aQ^{-}X^{-}.$$

Since $\det X^{-} \neq 0$ and $Q^{+} = Q^{-} + [Q]$, we get

$$-[\dot{\psi}]^{*}(q_{-}) = a[Q] - Q^{-}[\dot{x}](q_{-}) - [Q][\dot{x}](q_{-}).$$

This relation and the formula $(q_{-})^{*} = Q^{-}[\dot{x}] - [\dot{\psi}]^{*}$ imply the equality

$$[Q]\big(aI - [\dot{x}](q_{-})\big) = (q_{-})^{*}(q_{-}).$$

By virtue of (4.130) this equality is equivalent to

$$[Q] = (q_{-})^{*}(q_{-})\big(aI - [\dot{x}](q_{-})\big)^{-1}.$$

According to (4.132) this relation can be rewritten as

$$[Q] = (q_{-})^{*}(q_{-})\big(a - (q_{-})[\dot{x}]\big)^{-1}.$$

This implies jump condition (4.37). ∎

4.1.5 Numerical Examples

Rayleigh problem with regulator functional. The following optimal control problem (Rayleigh problem) was thoroughly investigated in [65]. Consider the electric circuit (tunnel-diode oscillator) shown in Figure 4.1. The state variable $x_1(t)$ is taken as the electric current I at time t, and the control variable $u(t)$ is induced by the voltage V_0 at the generator. After a suitable transformation of the voltage $V_0(t)$, we arrive at the following specific Rayleigh equation with a scalar control $u(t)$,

$$\ddot{x}(t) = -x(t) + \dot{x}(t)(1.4 - 0.14\dot{x}(t)^2) + 4u(t).$$

A numerical analysis reveals that the Rayleigh equation with zero control $u(t) \equiv 0$ has a *limit cycle* in the $(x, \dot{x})$-plane. The goal of the control process is to avoid the strong oscillations on the limit cycle by steering the system toward a small neighborhood of the origin $(x, \dot{x}) = (0,0)$ using a suitable control function $u(t)$. With the state variables $x_1 = x$ and $x_2 = \dot{x}$ we arrive at the following control problem with fixed final time $t_f > 0$: Minimize the functional

$$F(x,u) = \int_0^{t_f} (u(t)^2 + x_1(t)^2)\,dt \tag{4.133}$$

subject to

$$\dot{x}_1(t) = x_2(t),\ \dot{x}_2(t) = -x_1(t) + x_2(t)(1.4 - 0.14x_2(t)^2) + 4u(t), \tag{4.134}$$

$$x_1(0) = x_2(0) = -5, \quad x_1(t_f) = x_2(t_f) = 0. \tag{4.135}$$

As in [65], we choose the final time $t_f = 4.5$. The Pontryagin function (Hamiltonian), which corresponds to the minimum principle, becomes

$$H(x_1, x_2, \psi_1, \psi_2, u) = u^2 + x_1^2 + \psi_1 x_2 + \psi_2(-x_1 + x_2(1.4 - 0.14x_2^2) + 4u), \tag{4.136}$$

where ψ_1 and ψ_2 are the adjoint variables associated with x_1 and x_2. The optimal control that minimizes the Pontryagin function is determined by

$$H_u = 2u + 4\psi_2 = 0, \quad \text{i.e., } u(t) = -2\psi_2(t). \tag{4.137}$$

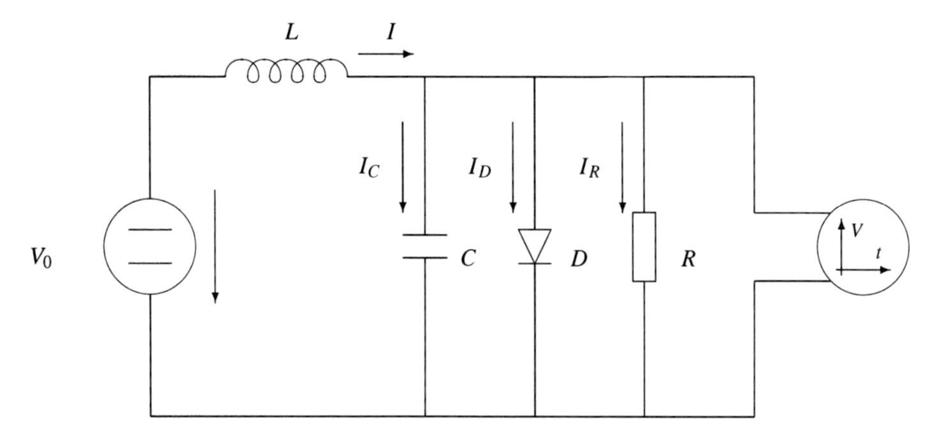

Figure 4.1. *Tunnel-diode oscillator. I denotes inductivity, C capacity, R resistance, I electric current, and D diode.*

The strict Legendre–Clebsch condition holds in view of $H_{uu}(t) \equiv 2 > 0$. The adjoint equation $\dot{\psi} = -H_x$ yields

$$\dot{\psi}_1 = \psi_2 - 2x_1, \quad \dot{\psi}_2 = 0.42\,\psi_2 x_2^2 - 1.4\,\psi_2 - \psi_1. \tag{4.138}$$

Since the final state is specified, there are no boundary conditions for the adjoint variable. The boundary value problem (4.133), (4.134) and (4.138) with control $u = -2\psi_2$ was solved using the multiple shooting code BNDSCO developed by Oberle and Grimm [82]. The optimal state, control, and adjoint variables are shown in Figure 4.2. We get the following initial and final values for the adjoint variables:

$$\begin{array}{rclrcl} \psi_1(0) & = & -9.00247067, & \psi_2(0) & = & -2.67303084, \\ \psi_1(4.5) & = & -0.04456054, & \psi_2(4.5) & = & -0.00010636. \end{array} \tag{4.139}$$

Nearly identical numerical results can be obtained by solving the discretized control problem with a high number of gridpoints.

The optimality of this extremal solution may be checked by producing a finite solution of the Riccati equation (4.30). Since the Rayleigh problem has dimension $n = 2$, we

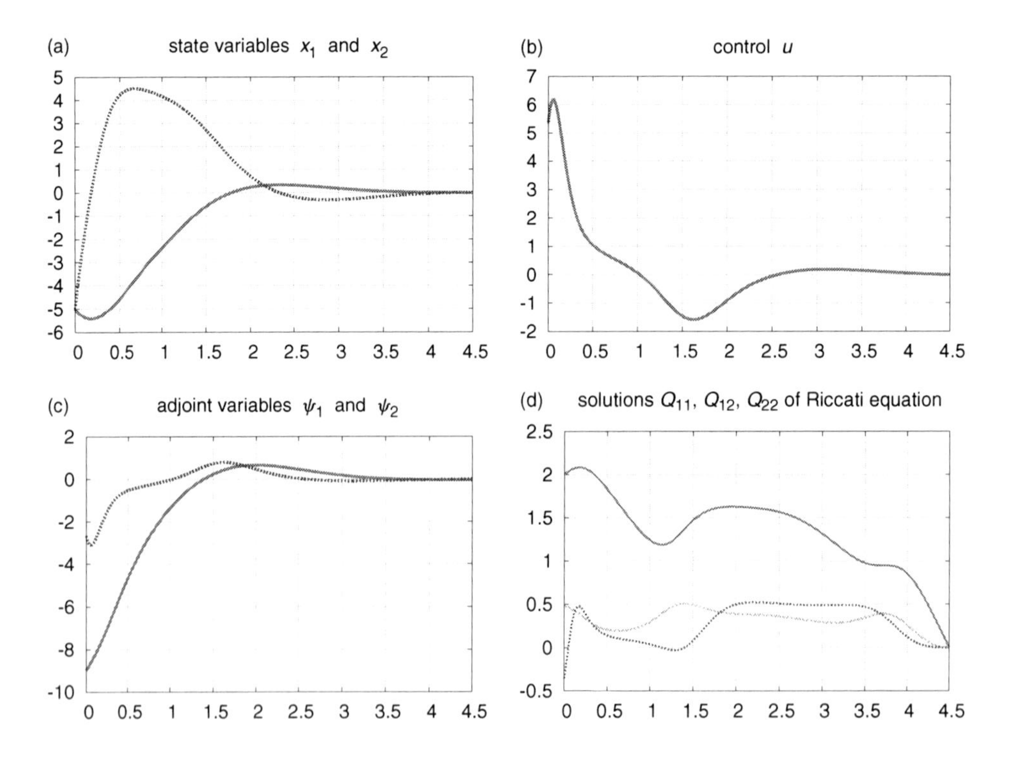

Figure 4.2. *Rayleigh problem with regular control.* (a) *State variables.* (b) *Control.* (c) *Adjoint variables.* (d) *Solutions of the Riccati equation* (4.140).

consider a symmetric 2×2 matrix $Q(t)$ of the form

$$Q(t) = \begin{pmatrix} Q_{11}(t) & Q_{12}(t) \\ Q_{12}(t) & Q_{22}(t) \end{pmatrix}.$$

The Riccati equation (4.30) then reads explicitly as

$$\begin{aligned} \dot{Q}_{11} &= 2Q_{12} + 8Q_{12}^2 - 2, \\ \dot{Q}_{12} &= -Q_{11} + (0.42x_2^2 - 1.4)Q_{12} + Q_{22} + 8\,Q_{12}\,Q_{22}, \\ \dot{Q}_{22} &= -2\,Q_{12} + 2(0.42x_2^2 - 1.4)Q_{22} + 8\,Q_{22}^2 + 0.84\,\psi_2 x_2. \end{aligned} \tag{4.140}$$

Since the final state is fixed, no boundary condition is prescribed for $Q(t_1)$. Choosing for convenience the boundary condition $Q(t_1) = 0$, we obtain the bounded solution of the Riccati equation shown in Figure 4.2. The initial values are computed as

$$Q_{11}(0) = 2.006324, \quad Q_{12}(0) = 0.4705135, \quad Q_{22}(0) = -0.3516654.$$

Thus the unconstrained solution shown in Figure 4.2 provides a local minimum.

To conclude the discussion of the Rayleigh problem, we consider the control problem with free final state $x(t_f)$. The solution is quite similar to that shown in Figure 4.2, with the only difference being that the boundary inequality $Q(t_f) > 0$ should hold. However, it suffices to find a solution $Q(t)$ satisfying the boundary condition $Q(t_f) = 0$ which was imposed earlier. Due to the continuous dependence of solutions on initial or terminal conditions, equation (4.140) then has a solution with $Q(t_f) = \epsilon \cdot I_2 > 0$ for $\epsilon > 0$ small. We obtain

$$\begin{array}{rclrcl} x_1(4.5) &=& -0.0957105, & x_2(4.5) &=& -0.204377, \\ \psi_1(0) &=& -9.00126, & \psi_2(0) &=& -2.67259, \\ Q_{11}(0) &=& 2.00607, & Q_{22}(0) &=& 0.470491, \\ Q_{22}(0) &=& -0.351606. & & & \end{array} \tag{4.141}$$

Variational problem with a conjugate point. The following variational problem was studied in Maurer and Pesch [71]:

$$\text{Minimize} \quad F(x,u) = \frac{1}{2}\int_0^1 (x(t)^3 + \dot{x}(t)^2)\,dt \quad \text{subject to } x(0) = 4,\ x(1) = 1. \tag{4.142}$$

Defining the control variable by $u = \dot{x}$ as usual, the Pontryagin function (Hamiltonian), corresponding to the minimum principle, becomes

$$H(x,u,\psi) = \frac{1}{2}(x^3 + u^2) + \psi\,u.$$

The strict Legendre–Clebsch condition holds in view of $H_{uu} = 1 > 0$. The minimizing control satisfies $H_u = 0$ which gives $u = -\psi$. Using the adjoint equation $\dot{\psi} = -H_x = -3x^2/2$, we get the boundary value problem (Euler–Lagrange equation)

$$\ddot{x} = \frac{3}{2}x^2, \quad x(0) = 4, \quad x(1) = 1. \tag{4.143}$$

The unknown initial value $\dot{x}(0)$ can be determined by shooting methods; cf. Stoer and Bulirsch [106]. The boundary value problem (4.143) has the explicit solution $x^{(1)}(t) = 4/(1+t)^2$ and a second solution $x^{(2)}(t)$ with initial values

$$\dot{x}^{(1)}(0) = -8, \quad \dot{x}^{(2)}(0) = -35.858549. \tag{4.144}$$

Both solutions $x^{(k)}(t)$, $k = 1,2$, may be tested for optimality by the classical Jacobi condition. The variational system (4.22), (4.23), along the two extremals, yields for $k = 1,2$

$$\begin{array}{rclll} \ddot{x}^{(k)}(t) & = & \frac{3}{2}\,x^{(k)}(t), & x^{(k)}(0) = 4, & \dot{x}^{(k)}(0) \quad \text{as in (4.144)}, \\ \ddot{y}^{(k)}(t) & = & 3\,x^{(k)}(t)\,y^{(k)}(t), & y^{(k)}(0) = 0, & \dot{y}^{(k)}(0) = 1. \end{array} \tag{4.145}$$

For $k = 1,2$ the extremals $x^{(k)}$ $(k = 1,2)$ and variational solutions $y^{(k)}$ $(k = 1,2)$ are displayed in Figure 4.3. The extremal $x^{(1)}(t) = 4/(1+t)^2$ is optimal in view of

$$y^{(1)}(t) > 0 \quad \forall\, 0 < t \le 1,$$

whereas the second extremal $x^{(2)}$ is *not* optimal, since it exhibits the conjugate point $t_c = 0.674437$ with $y^{(2)}(t_c) = 0$. The conjugate point t_c has the property that the envelope of

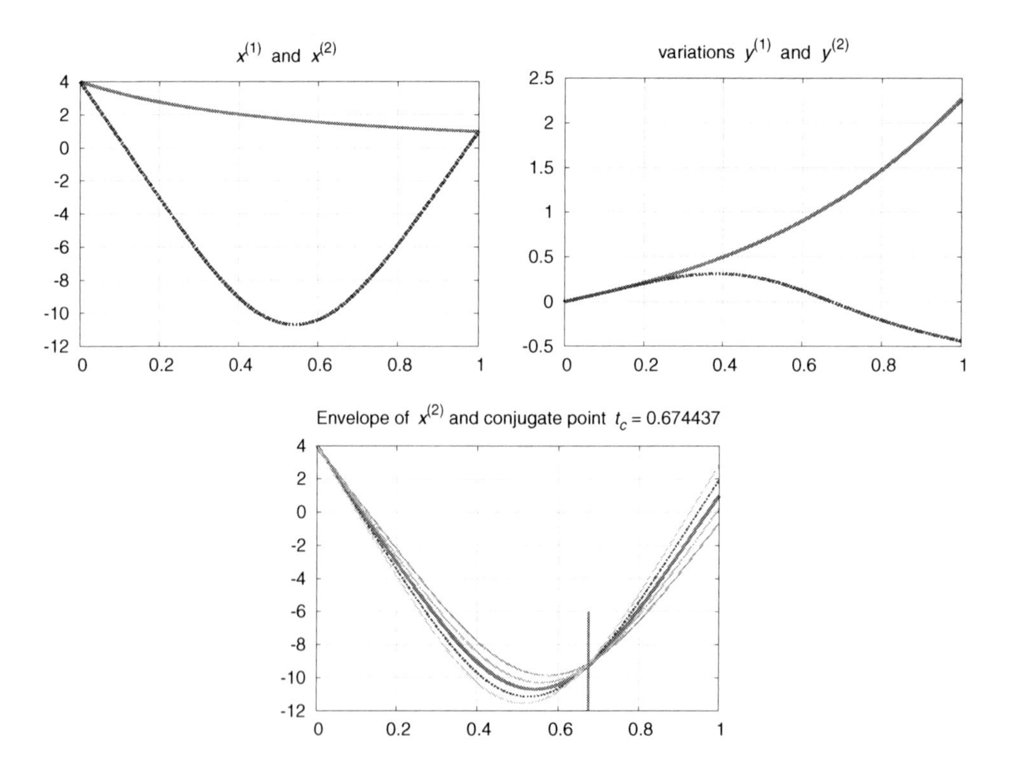

Figure 4.3. *Top left: Extremals $x^{(1)}$, $x^{(2)}$ (lower graph). Top right: Variational solutions $y^{(1)}$ and $y^{(2)}$ (lower graph) to* (4.145). *Bottom: Envelope of neighboring extremals illustrating the conjugate point $t_c = 0.674437$.*

extremals corresponding to neighboring initial slopes $\dot{x}(0)$ touches the extremal $x^{(2)}(t)$ at t_c; cf. Figure 4.3.

Example with a Broken Extremal

Consider the following problem:

$$\text{Minimize } J(x(\cdot),u(\cdot)) := \frac{1}{2}\int_0^{t_f} (f(u(t)) - x^2(t))\,dt \tag{4.146}$$

under the constraints

$$\dot{x}(t) = u(t), \quad x(0) = x(t_f) = 0, \tag{4.147}$$

where $f(u) = \min\{(u+1)^2, (u-1)^2\}$, $t_f > 0$ is fixed, and $u \neq 0$.

This example was examined in [79]. Here, we will study it using the results of this section. In [79] it was shown that for each closed interval $[0,t_f]$ we have an infinite number of extremals. For a given closed interval $[0,t_f]$, each extremal $x(t)$ is a periodic piecewise smooth function with a finite number of switching points of the control $t_k = (2k-1)\varphi$, $k = 1,\dots,s$, where $0 < \varphi < \frac{\pi}{2}$, $t_f = 2s\varphi < s\pi$. In particular, for $s = 1$ we have $t_f = 2\varphi < \pi$. Set $\Theta = \{t_1,\dots,t_s\}$. For each $s \geq 1$ the function $x(t)$ is defined by the conditions

$$\ddot{x} = -x \quad \text{for } t \in [0,t_f]\setminus\Theta, \quad x(0) = x(t_f) = 0, \quad [\dot{x}]^k = (-1)^k 2, \quad k = 1,\dots,s$$

(we identify the extremals differing only by sign). Consequently, $x(t) = \rho\sin t$ for $t \in [0,\varphi)$ and $x(t) = \rho\sin(t-2\varphi)$ for $t \in (\varphi,2\varphi)$, etc., where $\rho = (\cos\varphi)^{-1}$. Further, in [79] it was shown that for $t_f > \pi$ each extremal does not satisfy the necessary second-order condition. The same is true for the case $t_f \leq \pi$ and $s > 1$. Consequently, the only possibility for the quadratic form Ω to be nonnegative on the critical subspace $\mathcal{K}$ is the case $s = 1$. In this case we have $t_f = 2\varphi < \pi$ and

$$\Omega = (\tan\varphi)\bar{\xi}^2 + \frac{1}{2}\int_0^{t_f} (\bar{u}^2 - \bar{x}^2)\,dt.$$

The critical subspace $\mathcal{K}$ is defined by the conditions

$$\dot{\bar{x}} = \bar{u}, \quad \bar{x}(0) = \bar{x}(t_f) = 0, \quad [\bar{x}] = -2\bar{\xi}.$$

The Riccati equation (4.34) has the form (for a small $\varepsilon > 0$)

$$-\dot{R} = 1 + R^2, \quad t \in [0,\varepsilon], \quad R(0) = 0.$$

The solution of this Cauchy problem is $R(t) = -\tan t$. Consequently, $Q(t) = R^{-1}(t) = -\cot t$ is the solution of the Cauchy problem

$$-\dot{Q} + Q^2 + 1 = 0, \quad Q(\varepsilon) = R^{-1}(\varepsilon)$$

on the interval $(0,\varphi)$. It follows that $Q^- = -\cot\varphi$. Using the equalities $[\dot{x}] = [u] = -2$, $[\dot{\psi}] = [x] = 0$, and $a = D(H) = 2\tan\varphi$, we obtain

$$q_- = [\dot{x}]Q^- - [\dot{\psi}] = -2Q^- = 2\cot\varphi,$$
$$b_- = a - q_-[\dot{x}] = 2\tan\varphi + 4\cot\varphi > 0,$$

and then

$$[Q] = (b_-)^{-1}(q_-)^2 = (b_-)^{-1}4\cot^2\varphi,$$

$$Q^+ = Q^- + [Q] = -\cot\varphi + \frac{4\cot^2\varphi}{b_-} = -2(b_-)^{-1} < 0.$$

Now, for $t \in (\varphi, t_f]$, we have to solve the Cauchy problem

$$-\dot{Q} + Q^2 + 1 = 0, \quad Q(\varphi+) = -2(b_-)^{-1}.$$

Since $Q(\varphi+) < 0$, the solution has the form $Q(t) = -\cot(t+\tau)$ with certain $\tau > 0$. It is clear that this solution is defined at least on the closed interval $[\varphi, \varphi + 0.5\pi]$ which contains the closed interval $[\varphi, t_f]$. By Theorem 4.6 the quadratic form Ω is positive definite on the subspace $\mathcal{K}$. It means that the extremal with single switching point satisfies the sufficient second-order optimality condition. By Theorem 4.2, such an extremal is a point of a strict bounded strong minimum in the problem.

4.1.6 Transformation of the Quadratic Form to Perfect Squares for Broken Extremals in the Simplest Problem of the Calculus of Variations

Assume again that $w(\cdot) = (x(\cdot), u(\cdot))$ is an extremal in the simplest problem of the calculus of variations (4.1)–(4.3), with a single corner point $t_* \in (t_0, t_f)$. Assume that condition (SL) is satisfied, and $a > 0$, where a is defined by (4.8). Set $\Theta = \{t_*\}$.

Let $Q(t)$ be a symmetric matrix function on $[t_0, t_f]$ with piecewise continuous entries which are continuously differentiable on each of the two intervals of the set $[t_0, t_f]\backslash\Theta$ (hence, a jump of Q is possible only at the point t_*). We continue to use the following notation

$$\begin{aligned} &Q^- = Q(t_* - 0), \quad Q^+ = Q(t_* + 0), \quad [Q] = Q^+ - Q^-, \\ &q_- = [\dot{x}]^* Q^- - [\dot{\psi}], \quad b_- = a - q_-[\dot{x}], \end{aligned} \tag{4.148}$$

$$\begin{aligned} &A(t) = -F_{uu}^{-1}(t, w(t))F_{ux}(t, w(t)), \quad B(t) = -F_{uu}^{-1}(t, w(t)), \\ &C(t) = -F_{xu}(t, w(t))F_{uu}^{-1}(t, w(t))F_{ux}(t, w(t)) + F_{xx}(t, w(t)). \end{aligned} \tag{4.149}$$

Note that $B(t)$ is negative definite. Assume that Q satisfies
(a) the Riccati equation

$$\dot{Q}(t) + Q(t)A(t) + A(t)^* Q(t) + Q(t)B(t)Q(t) + C(t) = 0 \tag{4.150}$$

for all $t \in [t_0, t_f]\backslash\Theta$,
(b) the inequality

$$b_- > 0, \tag{4.151}$$

(c) the jump condition

$$[Q] = (b_-)^{-1}(q_-)^*(q_-), \tag{4.152}$$

where q_- is a row vector and $(q_-)^*$ is a column vector. (Recall that $(q_-)^*(q_-)$ is a symmetric positive semidefinite matrix.) In this case, we will say that the matrix $Q(t)$ is the *symmetric solution of the problem* (4.150)–(4.152) *on* $[t_0,t_f]$.

For the quadratic form Ω on the subspace $\mathcal{K}$, defined by relations (4.14) and (4.13), respectively, the following theorem holds.

Theorem 4.28. *If there exists a symmetric solution $Q(t)$ of the problem* (4.150)–(4.152) *on $[t_0,t_f]$, then the quadratic form Ω has the following transformation into a perfect square on $\mathcal{K}$:*

$$2\Omega(\bar{z}) = (b_-)^{-1}(a\bar{\xi} + q_-\bar{x}^-)^2 + \int_{t_0}^{t_f} \langle B(t)\bar{v}(t), \bar{v}(t)\rangle\, dt, \tag{4.153}$$

where

$$\bar{v}(t) = \big(Q(t) + F_{ux}(t,w(t))\big)\bar{x}(t) + F_{uu}(t,w(t))\bar{u}(t). \tag{4.154}$$

The proof of this theorem consists of the following four propositions.

Proposition 4.29. *Let $Q(t)$ be a symmetric matrix on $[t_0,t_f]$ with piecewise continuous entries, which are absolutely continuous on each interval of the set $[t_0,t_f]\backslash\Theta$. Then*

$$\begin{aligned} 2\Omega(\bar{z}) \;&=\; \big(b_-\bar{\xi}^2 + 2q_-\bar{x}^+\bar{\xi} + \langle [Q]\bar{x}^+, \bar{x}^+\rangle\big) \\ &\quad + \int_{t_0}^{t_f} \big(\langle (F_{xx}+\dot{Q})\bar{x},\bar{x}\rangle + 2\langle (F_{xu}+Q)\bar{u},\bar{x}\rangle + \langle F_{uu}\bar{u},\bar{u}\rangle\big)\, dt \end{aligned} \tag{4.155}$$

for all $\bar{z} = (\bar{\xi},\bar{x},\bar{u}) \in \mathcal{K}$.

Proof. For $\bar{z} \in \mathcal{K}$, we obviously have

$$\int_{t_0}^{t_f} \frac{d}{dt}\langle Q\bar{x},\bar{x}\rangle\, dt = \langle Q\bar{x},\bar{x}\rangle\Big|_{t_0}^{t_f} - [\langle Q\bar{x},\bar{x}\rangle]. \tag{4.156}$$

Using the conditions $\bar{x}(t_0) = \bar{x}(t_f) = 0$ and $\dot{\bar{x}} = \bar{u}$ in (4.156), we obtain

$$0 = \langle Q^+\bar{x}^+, \bar{x}^+\rangle - \langle Q^-\bar{x}^-, \bar{x}^-\rangle + \int_{t_0}^{t_f} (\langle \dot{Q}\bar{x},\bar{x}\rangle + 2\langle Q\bar{x},\bar{u}\rangle)dt. \tag{4.157}$$

Adding this zero form to the form

$$2\Omega(\bar{z}) = a\bar{\xi}^2 - [\dot{\psi}](\bar{x}^- + \bar{x}^+)\bar{\xi} + \int_{t_0}^{t_f} (\langle F_{xx}\bar{x},\bar{x}\rangle + 2\langle F_{xu}\bar{u},\bar{x}\rangle + \langle F_{uu}\bar{u},\bar{u}\rangle)\, dt,$$

we obtain

$$\begin{aligned} 2\Omega(\bar{z}) &= a\bar{\xi}^2 - [\dot{\psi}](\bar{x}^- + \bar{x}^+)\bar{\xi} - \langle Q^-\bar{x}^-, \bar{x}^-\rangle + \langle Q^+\bar{x}^+, \bar{x}^+\rangle \\ &\quad + \int_{t_0}^{t_f} \big(\langle (F_{xx}+\dot{Q})\bar{x},\bar{x}\rangle + 2\langle (F_{xu}+Q)\bar{u},\bar{x}\rangle + \langle F_{uu}\bar{u},\bar{u}\rangle\big)\, dt. \end{aligned} \tag{4.158}$$

Consider the form

$$\omega_*(\bar{\xi},\bar{x}) := a\bar{\xi}^2 - [\dot{\psi}](\bar{x}^- + \bar{x}^+)\bar{\xi} - \langle Q^-\bar{x}^-,\bar{x}^-\rangle + \langle Q^+\bar{x}^+,\bar{x}^+\rangle \tag{4.159}$$

connected to the jump point $t_* \in \Theta$ of the control $u(\cdot)$, and let us represent it as a function of $\bar{x}^+,\bar{\xi}$,

$$\begin{aligned}\omega_*(\bar{\xi},\bar{x}) &= a\bar{\xi}^2 - [\dot{\psi}](2\bar{x}^+ - [\bar{x}])\bar{\xi} - \langle Q^-(\bar{x}^+ - [\bar{x}]),\bar{x}^+ - [\bar{x}]\rangle + \langle Q^+\bar{x}^+,\bar{x}^+\rangle \\ &= a\bar{\xi}^2 - 2[\dot{\psi}]\bar{x}^+\bar{\xi} + [\dot{\psi}][\bar{x}]\bar{\xi} - \langle Q^-\bar{x}^+,\bar{x}^+\rangle + 2\langle Q^-[\bar{x}],\bar{x}^+\rangle \\ &\quad - \langle Q^-[\bar{x}],[\bar{x}]\rangle + \langle Q^+\bar{x}^+,\bar{x}^+\rangle.\end{aligned}$$

Using the relations $[\bar{x}] = [\dot{x}]\bar{\xi}$ and $Q^+ - Q^- = [Q]$, we obtain

$$\omega_*(\bar{\xi},\bar{x}) = (a - \langle Q[\dot{x}] - [\dot{\psi}]^*,[\dot{x}]\rangle)\bar{\xi}^2 + 2\langle Q^-[\dot{x}] - [\dot{\psi}]^*,\bar{x}^+\rangle\bar{\xi} + \langle [Q]\bar{x}^+,\bar{x}^+\rangle.$$

Now using definitions (4.148) of q_- and b_-, we obtain

$$\omega_*(\bar{\xi},\bar{x}) = b_-\bar{\xi}^2 + 2q_-\bar{x}^+\bar{\xi} + \langle [Q]\bar{x}^+,\bar{x}^+\rangle. \tag{4.160}$$

Formulas (4.158), (4.159), and (4.160) imply formula (4.155). ∎

Proposition 4.30. *If Q satisfies the jump condition $b_-[Q] = (q_-)^*(q_-)$, then*

$$b_-\bar{\xi}^2 + 2q_-\bar{x}^+\bar{\xi} + \langle [Q]\bar{x}^+,\bar{x}^+\rangle = (b_-)^{-1}(a\bar{\xi} + q_-\bar{x}^-)^2. \tag{4.161}$$

Proof. Using the jump conditions for $[Q]$ and $[\bar{x}]$ and the definition (4.148) of b_-, we obtain

$$\begin{aligned}b_-\bar{\xi}^2 + 2q_-\bar{x}^+\bar{\xi} + \langle [Q]\bar{x}^+,\bar{x}^+\rangle &= (b_-)^{-1}\big((b_-)^2\bar{\xi}^2 + 2q_-\bar{x}^+ b_-\bar{\xi} + (q_-\bar{x}^+)^2\big) \\ &= (b_-)^{-1}\big(b_-\bar{\xi} + q_-\bar{x}^+\big)^2 \\ &= (b_-)^{-1}\big((a - q_-[\dot{x}])\bar{\xi} + q_-\bar{x}^+\big)^2 \\ &= (b_-)^{-1}\big(a\bar{\xi} - q_-[\bar{x}] + q_-\bar{x}^+\big)^2 \\ &= (b_-)^{-1}\big(a\bar{\xi} + q_-\bar{x}^-\big)^2.\end{aligned}$$ ∎

Proposition 4.31. *The Riccati equation* (4.150) *is equivalent to the equation*

$$\dot{Q} - (Q + F_{xu})F_{uu}^{-1}(Q + F_{ux}) + F_{xx} = 0 \quad \text{on } [t_0,t_f]\setminus\Theta. \tag{4.162}$$

Proof. Using formulas (4.149), we obtain

$$\begin{aligned}&QA + A^*Q + QBQ + C \\ &\quad = -QF_{uu}^{-1}F_{ux} - F_{xu}F_{uu}^{-1}Q - QF_{uu}^{-1}Q - F_{xu}F_{uu}^{-1}F_{ux} + F_{xx} \\ &\quad = -QF_{uu}^{-1}(Q + F_{ux}) - F_{xu}F_{uu}^{-1}(Q + F_{ux}) + F_{xx} \\ &\quad = -(Q + F_{xu})F_{uu}^{-1}(Q + F_{ux}) + F_{xx}.\end{aligned}$$

Hence, (4.150) is equivalent to (4.162). ∎

Proposition 4.32. *If Q satisfies the Riccati equation* (4.162), *then*

$$\langle (F_{xx}+\dot{Q})\bar{x},\bar{x}\rangle + 2\langle (F_{xu}+Q)\bar{u},\bar{x}\rangle + \langle F_{uu}\bar{u},\bar{u}\rangle = \langle F_{uu}^{-1}\bar{v},\bar{v}\rangle, \tag{4.163}$$

where $\bar{v}$ is defined by (4.154).

Proof. From (4.154), it follows that

$$\begin{aligned}\langle F_{uu}^{-1}\bar{v},\bar{v}\rangle &= \langle F_{uu}^{-1}((Q+F_{ux})\bar{x}+F_{uu}\bar{u}),(Q+F_{ux})\bar{x}+F_{uu}\bar{u})\rangle \\ &= \langle F_{uu}^{-1}(Q+F_{ux})\bar{x},(Q+F_{ux})\bar{x}\rangle + 2\langle F_{uu}^{-1}(Q+F_{ux})\bar{x},F_{uu}\bar{u}\rangle \\ &\quad + \langle F_{uu}^{-1}F_{uu}\bar{u},F_{uu}\bar{u}\rangle \\ &= \langle (Q+F_{xu})F_{uu}^{-1}(Q+F_{ux})\bar{x},\bar{x}\rangle + 2\langle (Q+F_{ux})\bar{x},\bar{u}\rangle + \langle F_{uu}\bar{u},\bar{u}\rangle.\end{aligned} \tag{4.164}$$

Using (4.162), (4.164), we obtain (4.163). ∎

Now, let Q be a symmetric solution of problem (4.150)–(4.152) on $[t_0,t_f]$. Then, Propositions 4.29–4.32 yield the transformation (4.153) of Q to a perfect square on $\mathcal{K}$. This completes the proof of Theorem 4.28.

Now we can easily prove the following sufficient condition for the positive definiteness of the quadratic form Ω on subspace $\mathcal{K}$, which is the main result of this section.

Theorem 4.33. *If there exists a symmetric solution $Q(t)$ of problem* (4.150)–(4.152) *on* $[t_0,t_f]$, *then $\Omega(\bar{z})$ is positive definite on $\mathcal{K}$, i.e., there exists $\epsilon>0$ such that $\Omega(\bar{z})\ge\epsilon\bar{\gamma}(\bar{z})$ for all $\bar{z}\in\mathcal{K}$, where $\bar{\gamma}(\bar{z})=\bar{\xi}^2+\int_{t_0}^{t_f}\langle\bar{u}(t),\bar{u}(t)\rangle\,dt$.*

Proof. By Theorem 4.28, formula (4.153) holds for Ω on $\mathcal{K}$. Hence, the form Ω is nonnegative on $\mathcal{K}$. Let us show that Ω is positive on $\mathcal{K}$. Let $\bar{z}\in\mathcal{K}$ and $\Omega(\bar{z})=0$. Then by (4.153) and (4.154), we have

$$a\bar{\xi}+q_-\bar{x}^-=0, \quad (Q+F_{ux})\bar{x}+F_{uu}\bar{u}=0 \quad \text{on } [t_0,t_f]\backslash\Theta. \tag{4.165}$$

Since $\bar{z}\in\mathcal{K}$, we have $\dot{\bar{x}}=\bar{u}$, and consequently

$$\dot{\bar{x}}=-F_{uu}^{-1}(Q+F_{ux})\bar{x} \quad \text{on } [t_0,t_f]\backslash\Theta. \tag{4.166}$$

Since $\bar{x}(t_0)=0$, we have $\bar{x}(t)=0$ on $[t_0,t_*)$. Hence $\bar{x}^-=0$. Then, by (4.165), we have $\bar{\xi}=0$. Since $[\bar{x}]=[\dot{x}]\bar{\xi}=0$, we also have $\bar{x}^+=0$. Then, by (4.166), $\bar{x}(t)=0$ on $(t_*,t_f]$. Thus, we obtain that $\bar{\xi}=0$, $\bar{x}(t)\equiv 0$ on $[t_0,t_f]$, and then $\bar{u}(t)=0$ a.e. on $[t_0,t_f]$. Thus $\bar{z}=0$, and therefore Ω is positive on $\mathcal{K}$. But Ω is a Legendre form, and therefore positiveness of Ω on $\mathcal{K}$ implies positive definiteness of Ω on $\mathcal{K}$. ∎

Remark. Theorem 4.33 was proved for a single point of discontinuity of the control $u(\cdot)$. A similar result could be proved for finitely many points of discontinuity of the control $u(\cdot)$.

Now, using Theorem 4.33, we can prove, for $s=1$, that the existence of a symmetric solution $Q(t)$ of the problem (4.150)–(4.152) on $[t_0,t_f]$ is not only sufficient (as it was stated by Theorem 4.28) but also necessary for the positive definiteness of Ω on $\mathcal{K}$. This can be done in different ways. We choose the following one. Let Ω be positive definite on $\mathcal{K}$, take a small $\epsilon > 0$, and put $u(t) = u(t_0)$ for $t \in [t_0 - \epsilon, t_0]$.

Thus, $u(t)$ is continued to the left-hand side of the point t_0 by constant value equal to the value at t_0. Now we have $w(t)$ defined on $[t_0 - \epsilon, t_f]$ with the same single discontinuity point t_*, and therefore the "continued" quadratic form

$$\Omega_\epsilon(\bar{z}) = \frac{1}{2}\left(a\bar{\xi}^2 - 2[\dot{\psi}]\bar{x}_{\text{av}}\right) + \frac{1}{2}\int_{t_0-\epsilon}^{t_f} \langle F_{ww}(t, w(t))\bar{w}(t), \bar{w}(t)\rangle\, dt$$

is well defined on the subspace $\mathcal{K}_\epsilon$:

$$\dot{\bar{x}} = \bar{u} \quad \text{a.e. on } [t_0 - \epsilon, t_f], \quad \bar{x}(t_0 - \epsilon) = \bar{x}(t_f) = 0, \quad [\bar{x}] = [\dot{x}]\bar{\xi}.$$

Using the same technique as in [96], one can easily prove that there is $\epsilon > 0$ such that $\Omega_\epsilon(\bar{z})$ is positive definite on $\mathcal{K}_\epsilon$ (note that condition (SL) is satisfied for Ω_ϵ on $[t_0 - \epsilon, t_f]$, which is important for this proof). Then by Theorem 4.33 applied for $[t_0 - \epsilon, t_f]$ there exists a solution Q of the Riccati equation (4.150) on $(t_0 - \epsilon, t_f]$ satisfying inequality (4.151) and jump condition (4.152), and hence we have this solution on the segment $[t_0, t_f]$. Thus, for $s = 1$ the following theorem holds.

Theorem 4.34. *The existence of a symmetric solution $Q(t)$ of problem* (4.150)–(4.152) *together with condition* (SL) *is equivalent to the positive definiteness of Ω on $\mathcal{K}$.*

The formulation of the jump condition (4.152) allows us to solve problem (4.150)–(4.152) in a forward time direction. Certainly it is possible to move in opposite direction. Let us prove the following.

Lemma 4.35. *Conditions* (4.151) *and* (4.152) *imply that*

$$b_+ := a + q_+[\dot{x}] > 0, \tag{4.167}$$

$$(b_+)[Q] = (q_+)^*(q_+), \tag{4.168}$$

where

$$q_+ = [\dot{x}]^* Q^+ - [\dot{\psi}]. \tag{4.169}$$

Proof. By Proposition 4.26, jump condition (4.152) is equivalent to $[Q](aI - [\dot{x}](q_-)) = (q_-)^*(q_-)$. This equality implies that $a[Q] = (q_-)^*(q_-) + [Q][\dot{x}](q_-)$. But

$$[Q][\dot{x}] = (q_+ - q_-)^*. \tag{4.170}$$

Hence

$$a[Q] = (q_+)^*(q_-). \tag{4.171}$$

Using again (4.170) in (4.171), we obtain $a[Q] = (q_+)^*(q_+ - [\dot{x}]^*[Q])$. It follows from this formula that $a[Q] = (q_+)^*(q_+) - (q_+)^*[\dot{x}]^*[Q]$. Consequently

$$(aI + (q_+)^*[\dot{x}]^*)[Q] = (q_+)^*(q_+). \tag{4.172}$$

Let us transform the left-hand side of this equation, using (4.171). Since

$$\begin{aligned}(q_+)^*[\dot{x}]^*[Q] &= (q_+)^*[\dot{x}]^*\left(\frac{1}{a}(q_+)^*(q_-)\right) = \frac{1}{a}(q_+)^*[\dot{x}]^*(q_+)^*(q_-)\\ &= \frac{1}{a}(q_+)^*\big([\dot{x}]^*(q_+)^*\big)(q_-) = \frac{1}{a}(q_+)^*(q_-)\big(q_+[\dot{x}]\big)\\ &= [Q]\big(q_+[\dot{x}]\big),\end{aligned}$$

we have

$$(aI+(q_+)^*[\dot{x}]^*)[Q] = a[Q]+(q_+)^*[\dot{x}]^*[Q] = a[Q]+\big(q_+[\dot{x}]\big)[Q] = (a+q_+[\dot{x}])[Q].$$

Thus, (4.172) is equivalent to

$$b_+[Q] = (q_+)^*(q_+). \tag{4.173}$$

Let us show that $b_+ > 0$. From (4.173) it follows that

$$b_+q_+[Q](q_+)^* = \big(q_+(q_+)^*\big)^2. \tag{4.174}$$

By (4.151) and (4.152), we have

$$q_+[Q](q_+)^* \geq 0. \tag{4.175}$$

If $q_+ = 0$, then $b_+ = a+q_+[\dot{x}] = a > 0$. Assume that $q_+ \neq 0$. Then

$$\big(q_+(q_+)^*\big)^2 > 0. \tag{4.176}$$

Conditions (4.174), (4.175), and (4.176) imply that $b_+ > 0$. ∎

Lemma 4.35 yields the transformation of Ω on $\mathcal{K}$ to a perfect square expressed by q_+ and b_+.

Lemma 4.36. *The following equality holds*:

$$\omega_*(\bar{\xi},\bar{x}) := a\bar{\xi}^2 - [\dot{\psi}](\bar{x}^- + \bar{x}^+)\bar{\xi} - \langle Q^-\bar{x}^-,\bar{x}^-\rangle + \langle Q^+\bar{x}^+,\bar{x}^{k+}\rangle = \frac{1}{b_+}(a\bar{\xi}+q_+\bar{x}^+)^2.$$

Proof. We have

$$\begin{aligned}\omega_*(\bar{\xi},\bar{x}) &= a\bar{\xi}^2 - [\dot{\psi}](\bar{x}^- + \bar{x}^+)\bar{\xi} + \langle Q^+\bar{x}^+,\bar{x}^+\rangle - \langle Q^-\bar{x}^-,\bar{x}^-\rangle\\ &= a\bar{\xi}^2 - [\dot{\psi}](2\bar{x}^- + [\bar{x}])\bar{\xi} + \langle Q^+(\bar{x}^- + [\bar{x}]),\bar{x}^- + [\bar{x}]\rangle - \langle Q^-\bar{x}^-,\bar{x}^-\rangle\\ &= a\bar{\xi}^2 - 2[\dot{\psi}]\bar{x}^-\bar{\xi} - [\dot{\psi}][\dot{x}]\bar{\xi}^2 + \langle Q^+\bar{x}^-,\bar{x}^-\rangle\\ &\quad + 2\langle Q^+[\bar{x}],\bar{x}^+\rangle + \langle Q^+[\bar{x}],[\bar{x}]\rangle - \langle Q^-\bar{x}^-,\bar{x}^-\rangle\\ &= (a - [\dot{\psi}][\dot{x}] + \langle Q^+[\dot{x}],[\dot{x}]\rangle)\bar{\xi}^2 + \langle [Q]\bar{x}^-,\bar{x}^-\rangle + 2([\dot{x}]^*Q^+ - [\dot{\psi}])\bar{x}^-\rangle\bar{\xi}\\ &= (a+q_+[\dot{x}])\bar{\xi}^2 + 2q_+\bar{x}^-\bar{\xi} + \langle [Q]\bar{x}^-,\bar{x}^-\rangle.\end{aligned}$$

Hence $\omega_*(\bar{\xi},\bar{x}) = b_+\bar{\xi}^2 + 2q_+\bar{x}^-\bar{\xi} + \langle [Q]\bar{x}^-,\bar{x}^-\rangle$. Using (4.173), we obtain

$$\begin{aligned}\omega_*(\bar{\xi},\bar{x}) &= \frac{1}{b_+}(b_+^2\bar{\xi}^2 + 2(q_+\bar{x}^-)b_+\bar{\xi} + (q_+,\bar{x}^-)^2)\\ &= \frac{1}{b_+}(b_+\bar{\xi} + q_+\bar{x}^-)^2 = \frac{1}{b_+}(a\bar{\xi} + q_+[\bar{x}] + q_+\bar{x}^-)^2\\ &= \frac{1}{b_+}(a\bar{\xi} + q_+\bar{x}^+)^2. \qquad \blacksquare\end{aligned}$$

Thus, we obtain the following.

Theorem 4.37. *If the assumptions of Theorem* 4.28 *are satisfied, then the quadratic form* Ω *has the following transformation to a perfect square on* $\mathcal{K}$:

$$2\Omega(\bar{z}) = (b_+)^{-1}(a\bar{\xi} + q_+\bar{x}^+)^2 + \int_{t_0}^{t_f} \langle B(t)\bar{v}(t), \bar{v}(t)\rangle\, dt, \tag{4.177}$$

where $\bar{v}$ *is defined by* (4.154).

4.2 Riccati Equation for Broken Extremal in the General Problem of the Calculus of Variations

Now we show that, by analogy with the simplest problem of the calculus of variations, the transformation of the quadratic form to a perfect square may be done in the auxiliary problem connected with the general problem of the calculus of variations (without local equality-type constraints and endpoints inequality-type constraints), and in this way we obtain, for this problem, sufficient optimality conditions for broken extremals in the terms of discontinuous solutions of the corresponding Riccati equation. Consider the problem

$$\mathcal{J}(x,u) = J(x_0,x_f) \longrightarrow \min, \tag{4.178}$$

$$K(x_0,x_f) = 0, \tag{4.179}$$

$$\dot{x} = f(t,x,u), \tag{4.180}$$

$$(x_0,x_f) \in \mathcal{P}, \quad (t,x,u) \in \mathcal{Q}, \tag{4.181}$$

where $x_0 = x(t_0)$, $x_f = x(t_f)$, the interval $[t_0,t_f]$ is fixed, $\mathcal{P}$ and $\mathcal{Q}$ are open sets, $x \in \mathbb{R}^n, u \in \mathbb{R}^r$, J and K are twice continuously differentiable on $\mathcal{P}$, and f is twice continuously differentiable on $\mathcal{Q}$. We set $(x_0,x_f) = p$, $(x,u) = w$. The problem is considered in the space $W^{1,1}([t_0,t_f],\mathbb{R}^n) \times L^\infty([t_0,t_f],\mathbb{R}^r)$. Let an admissible pair $w(\cdot) = (x(\cdot),u(\cdot))$ be given. As in Section 2.1.1, we assume that $u(\cdot)$ is piecewise continuous and each point of discontinuity is a Lipschitz point (see Definition 2.1). We denote by $\Theta = \{t_1,\dots,t_s\}$ the set of all discontinuity points of the control $u(\cdot)$.

For the problem (4.178)–(4.181), let us recall briefly the formulations of the sufficient optimality conditions at the point $w(\cdot)$, given in Chapter 2 for the general problem. Let us introduce the Pontryagin function $H(t,x,u,\psi) = \psi f(t,x,u)$, where $\psi \in (\mathbb{R}^n)^*$. Denote the endpoint Lagrange function by $l(p,\alpha_0,\beta) = \alpha_0 J(p) + \beta K(p)$, where β is a row vector of the same dimension as the vector K. Further, we introduce a collection of Lagrange multipliers $\lambda = (\alpha_0,\beta,\psi(\cdot))$ such that $\psi(\cdot) : [t_0,t_f] \longrightarrow (\mathbb{R}^n)^*$ is an absolutely continuous function continuously differentiable on each interval of the set $[t_0,t_f]\backslash\Theta$.

We denote by Λ_0 the set of the normed collections λ satisfying the conditions of the local minimum principle for an admissible trajectory $w(\cdot) = (x(\cdot), u(\cdot))$,

$$\alpha_0 \geq 0, \quad \alpha_0 + \sum |\beta_j| = 1, \quad \dot{\psi} = -H_x, \quad \psi(t_0) = -l_{x_0}, \quad \psi(t_f) = l_{x_f}, \quad H_u = 0, \tag{4.182}$$

where all derivatives are calculated on the trajectory $w(\cdot)$, respectively, at the endpoints $(x(t_0), x(t_f))$ of this trajectory. The condition $\Lambda_0 \neq \emptyset$ is necessary for a weak minimum at the point $w(\cdot)$.

We set $\mathcal{U}(t,x) = \{u \in \mathbb{R}^{d(u)} \mid (t,x,u) \in \mathcal{Q}\}$. Denote by M_0 the set of tuples $\lambda \in \Lambda_0$ such that for all $t \in [t_0, t_f] \backslash \Theta$, the condition $u \in \mathcal{U}(t, x^0(t))$ implies the inequality

$$H(t, x^0(t), u, \psi(t)) \geq H(t, x^0(t), u^0(t), \psi(t)). \tag{4.183}$$

If w^0 is a point of Pontryagin minimum, then M_0 is nonempty; i.e., the minimum principle holds. Further, denote by M_0^+ the set of $\lambda \in M_0$ such that

(a) $H(t, x^0(t), u, \psi(t)) > H(t, x^0(t), u^0(t), \psi(t))$
for all $t \in [t_0, t_f] \backslash \Theta, \quad u \in \mathcal{U}(t, x^0(t)), \quad u \neq u^0(t)$

(b) $H(t_k, x^0(t_k), u, \psi(t_k)) > H^{\lambda k-} = H^{\lambda k+}$
for all $t_k \in \Theta, \quad u \in \mathcal{U}(t_k, x^0(t_k)), \quad u \notin \{u^{0k-}, u^{0k+}\}$.

According to Definition 2.94, w^0 is a *bounded strong minimum point* if for any compact set $\mathcal{C} \subset Q$ there exists $\varepsilon > 0$ such that $\mathcal{J}(w) \geq \mathcal{J}(w^0)$ for all admissible trajectories w such that

$$|x(t_0) - x^0(t_0)| < \varepsilon, \quad \max_{t \in [t_0, t_f]} |\underline{x}(t) - \underline{x}^0(t)| \leq \varepsilon, \quad (t, w(t)) \in \mathcal{C} \quad \text{a.e. on } [t_0, t_f],$$

where $\underline{x}$ is the vector composed of the essential components of the vector x. Recall that the component x_i of a vector $x = (x_1, \ldots, x_{d(x)})$ is said to be unessential if the function f is independent of x_i and the functions J and K affinely depend on $x_{i0} = x_i(t_0)$, $x_{if} = x_i(t_f)$.

Now we shall formulate quadratic sufficient conditions for a strict bounded strong minimum at the point $w(\cdot)$, which follow from the corresponding conditions in Section 2.7.2; see Theorem 2.102. For $\lambda \in \Lambda_0$ and $t_k \in \Theta$, we put $(\Delta_k H)(t) = H(t, x(t), u^{k+}, \psi(t)) - H(t, x(t), u^{k-}, \psi(t))$, where $u^{k-} = u(t_k - 0)$, $u^{k+} = u(t_k + 0)$, and

$$D^k(H) := -\frac{d}{dt}(\Delta_k H)\Big|_{t_k - 0} = -\frac{d}{dt}(\Delta_k H)\Big|_{t_k + 0}. \tag{4.184}$$

The second equality is a consequence of the minimum principle. As we know, the following formula holds:

$$D^k(H) = -H_x^{k+} H_\psi^{k-} + H_x^{k-} H_\psi^{k+} - [H_t]^k. \tag{4.185}$$

Recall that the conditions $[H]^k = 0$, $D^k(H) \geq 0$, $k = 1, \ldots, s$, also follow from the minimum principle. The same is true for the Legendre condition $\langle H_{uu}(t, w(t), \psi(t))\bar{v}, \bar{v}\rangle \geq 0$ a.e. on $[t_0, t_f]$ for all $\bar{v} \in \mathbb{R}^r$. Let us recall the *strengthened Legendre* (SL) *condition* for $\lambda \in \Lambda_0$:

(a) for any $t \in [t_0, t_f] \backslash \Theta$ the quadratic form $\langle H_{uu}(t, w(t), \psi(t))\bar{v}, \bar{v}\rangle$ is positive definite,

(b) for any $t_k \in \Theta$ the quadratic forms $\langle H_{uu}(t_k, x(t_k), u^{k-}, \psi(t_k))\bar{v}, \bar{v}\rangle$ and $\langle H_{uu}(t_k, x(t_k), u^{k+}, \psi(t_k))\bar{v}, \bar{v}\rangle$ are both positive definite.

An element $\lambda = (\alpha_0, \beta, \psi(\cdot)) \in \Lambda_0$ is said to be *strictly Legendrian* if the following conditions are satisfied:

(i) $[H]^k = 0,\ k = 1, \dots, s$;

(ii) $D^k(H) > 0,\ k = 1, \dots, s$;

(iii) strengthened Legendre (SL) condition.

Assume that there exists a strictly Legendrian element $\lambda \in \Lambda_0$ such that $\alpha_0 > 0$. In this case we put $\alpha_0 = 1$. Denote by $Z_2(\Theta)$ the space of triples $\bar{z} = (\bar{\xi}, \bar{x}(\cdot), \bar{u}(\cdot))$ such that

$$\bar{\xi} \in \mathbb{R}^s, \quad \bar{x}(\cdot) \in P_\Theta W^{1,2}([t_0, t_f], \mathbb{R}^n), \quad \bar{u} \in L^2([t_0, t_f], \mathbb{R}^r).$$

Let $\mathcal{K}$ denote the subspace of all $\bar{z} = (\bar{\xi}, \bar{x}(\cdot), \bar{u}(\cdot)) \in Z_2(\Theta)$ such that

$$K_p \bar{p} = 0, \quad \dot{\bar{x}} = f_w \bar{w}, \quad [\bar{x}]^k = [\dot{x}]^k \bar{\xi}_k, \quad k = 1, \dots, s, \tag{4.186}$$

where $f_w = f_w(t, w(t))$, $K_p = K_p(x(t_0), x(t_f))$, $[\dot{x}]^k$ is the jump of $x(\cdot)$ at t_k, and $[\bar{x}]^k$ is the jump of $\bar{x}(\cdot)$ at t_k. We call $\mathcal{K}$ the *critical subspace*.

As in (2.13), we define the quadratic form, which corresponds to the element $\lambda \in \Lambda_0$, as

$$\Omega^\lambda(\bar{z}) = \frac{1}{2} \sum_{k=1}^{s} \left(D^k(H) \bar{\xi}_k^2 - 2[\dot{\psi}]^k \bar{x}_{\mathrm{av}}^k \bar{\xi}_k \right) + \frac{1}{2} \int_{t_0}^{t_f} \langle H_{ww} \bar{w}, \bar{w} \rangle \, dt + \frac{1}{2} \langle l_{pp} \bar{p}, \bar{p} \rangle, \tag{4.187}$$

where $l_{pp} = l_{pp}(x(t_0), x(t_f), \alpha_0, \beta)$, $H_{ww} = H_{ww}(t, w(t), \psi(t))$. We set

$$\bar{\gamma}(\bar{z}) = \langle \bar{\xi}, \bar{\xi} \rangle + \langle \bar{x}(t_0), \bar{x}(t_0) \rangle + \int_{t_0}^{t_f} \langle \bar{u}, \bar{u} \rangle dt.$$

Theorem 2.102 implies the following assertion.

Theorem 4.38. *Assume that there exists a strictly Legendrian element $\lambda \in M_0^+$ such that $\alpha_0 > 0$ and the quadratic form $\Omega^\lambda(\cdot)$ is positive definite on $\mathcal{K}$, i.e., there exists $\epsilon > 0$ such that*

$$\Omega^\lambda(\bar{z}) \geq \epsilon \bar{\gamma}(\bar{z}) \qquad \forall\, \bar{z} \in \mathcal{K}. \tag{4.188}$$

Then, $w(\cdot)$ is a strict bounded strong minimum.

Now, let us show that the quadratic form Ω could be transformed into a perfect square if the corresponding Riccati equation has a solution $Q(t)$ defined on $[t_0, t_f]$, satisfying certain jump conditions at each point of the set Θ. Define the Riccati equation along $x(t), u(t)$, and $\psi(t)$ by

$$\dot{Q} + Q f_x + f_x^* Q + H_{xx} - (H_{xu} + Q f_u) H_{uu}^{-1} (H_{ux} + f_u^* Q) = 0, \quad t \in [t_0, t_f] \backslash \Theta, \tag{4.189}$$

for a piecewise continuous function $Q(t)$ which is continuously differentiable on each interval of the set $[t_0, t_f] \backslash \Theta$. Set $q_{k-} = \left([\dot{x}]^k\right)^* Q^{k-} - [\dot{\psi}]^k$ and define the conditions

$$b_{k-} := a_k - q_{k-}[\dot{x}]^k > 0, \quad k = 1, \dots, s, \tag{4.190}$$

where $a_k = D^k(H)$, $k = 1, \dots, s$. Define the *jump conditions* for Q at the point $t_k \in \Theta$ as

$$b_{k-}[Q]^k = (q_{k-})^* (q_{k-}), \quad k = 1, \dots, s. \tag{4.191}$$

Theorem 4.39. *Assume that there exists a symmetric solution $Q(t)$ (piecewise continuous on $[t_0,t_f]$ and continuously differentiable on each interval of the set $[t_0,t_f]\backslash\Theta$) of Riccati equation* (4.189) *which satisfies at each point $t_k \in \Theta$ conditions* (4.190) *and jump conditions* (4.191). *Then the quadratic form $\Omega^\lambda(\bar{z})$ (see* (4.187)*) has the following transformation into a perfect square on the subspace $\mathcal{K}$ (see* (4.186)*),*

$$2\Omega^\lambda(\bar{z}) = \sum_{k=1}^{s}(b_{k-})^{-1}\left(a_k\bar{\xi}_k + (q_{k-})\bar{x}^{k-}\right)^2 + \int_{t_0}^{t_f}\langle H_{uu}^{-1}\bar{v},\bar{v}\rangle\,dt + \langle M\bar{p},\bar{p}\rangle, \tag{4.192}$$

where

$$\bar{v} = (H_{ux} + f_u^* Q)\bar{x} + H_{uu}\bar{u}, \tag{4.193}$$

$$M = \begin{pmatrix} l_{x_0x_0} + Q(t_0) & l_{x_0x_f} \\ l_{x_fx_0} & l_{x_fx_f} - Q(t_f) \end{pmatrix}. \tag{4.194}$$

Proof. In perfect analogy with Theorem 4.28 we have to prove statements similar to Propositions 4.29–4.32. They need only small changes. Below we point out these changes. Take a symmetric matrix $Q(t)$ on $[t_0,t_f]$ with piecewise continuous coefficients, which are absolutely continuous on each interval of the set $[t_0,t_f]\backslash\Theta$. Using, for $\bar{z}\in\mathcal{K}$, formula (4.156) and the equalities

$$2\langle Q\dot{\bar{x}},\bar{x}\rangle = 2\langle Qf_x\bar{x},\bar{x}\rangle + 2\langle Qf_u\bar{u},\bar{x}\rangle = \langle (Qf_x + f_x^*Q)\bar{x},\bar{x}\rangle + \langle Qf_u\bar{u},\bar{x}\rangle + \langle f_u^*Q\bar{x},\bar{u}\rangle,$$

we obtain (similar to (4.157)) the following zero form on $\mathcal{K}$:

$$\begin{aligned} 0 = {} & \sum_{k=1}^{s}\left(\langle Q^{k+}\bar{x}^{k+},\bar{x}^{k+}\rangle - \langle Q^{k-}\bar{x}^{k-},\bar{x}^{k-}\rangle\right) \\ & + \int_{t_0}^{t_f}\left\{\langle(\dot{Q} + f_x^*Q + Qf_x)\bar{x},\bar{x}\rangle + \langle f_u^*Q\bar{x},\bar{u}\rangle + \langle Qf_u\bar{u},\bar{x}\rangle\right\}dt \\ & - \langle Q(t_f)\bar{x}_f,\bar{x}_f\rangle + \langle Q(t_0)\bar{x}_0,\bar{x}_0\rangle. \end{aligned} \tag{4.195}$$

Adding this zero form (4.195) to the form $2\Omega^\lambda(\bar{z})$ (see (4.187)) considered for arbitrary $\bar{z}\in\mathcal{K}$, we obtain

$$\begin{aligned} 2\Omega^\lambda(\bar{z}) = {} & \langle M\bar{p},\bar{p}\rangle + \int_{t_0}^{t_f}\left\{\langle(H_{xx} + \dot{Q} + Qf_x + f_x^*Q)\bar{x},\bar{x}\rangle\right. \\ & \left. + \langle(H_{xu} + Qf_u)\bar{u},\bar{x}\rangle + \langle(H_{ux} + f_u^*Q)\bar{x},\bar{u}\rangle + \langle H_{uu}\bar{u},\bar{u}\rangle\right\}dt \\ & + \sum_{k=1}^{s}\omega_k(\bar{\xi},\bar{x}), \end{aligned} \tag{4.196}$$

where M is defined by (4.194) and

$$\omega_k(\bar{\xi},\bar{x}) = a_k\bar{\xi}_k^2 - 2[\dot{\psi}]^k\bar{x}_{\mathrm{av}}^k\bar{\xi}_k + \langle Q^{k+}\bar{x}^{k+},\bar{x}^{k+}\rangle - \langle Q^{k-}\bar{x}^{k-},\bar{x}^{k-}\rangle, \quad k=1,\dots,s. \tag{4.197}$$

According to formula (4.160),

$$\omega_k(\bar{\xi},\bar{x}) = b_{k-}\bar{\xi}_k^2 + 2q_{k-}\bar{x}^{k+}\bar{\xi}_k + \langle [Q]^k\bar{x}^{k+},\bar{x}^{k+}\rangle, \quad k=1,\dots,s. \tag{4.198}$$

Further, by Proposition 4.30,

$$\omega_k = (b_{k-})^{-1}(a_k\bar{\xi}_k + q_{k-}\bar{x}^{k-})^2. \tag{4.199}$$

Using the Riccati equation (4.189) and definition (4.193) for $\bar{v}$, we obtain for the integral part of (4.196),

$$\begin{aligned}\int_{t_0}^{t_f} & \big\{\langle (H_{xx} + \dot{Q} + Qf_x + f_x^* Q)\bar{x}, \bar{x}\rangle + \langle (H_{xu} + Qf_u)\bar{u}, \bar{x}\rangle \\ & + \langle (H_{ux} + f_u^* Q)\bar{x}, \bar{u}\rangle + \langle H_{uu}\bar{u}, \bar{u}\rangle\big\}\, dt \\ & = \int_{t_0}^{t_f} \langle H_{uu}^{-1}\bar{v}, \bar{v}\rangle\, dt,\end{aligned} \tag{4.200}$$

namely,

$$\begin{aligned}\int_{t_0}^{t_f} \langle H_{uu}^{-1}\bar{v}, \bar{v}\rangle\, dt &= \int_{t_0}^{t_f} \big\{\langle H_{uu}^{-1}(H_{ux} + f_u^* Q)\bar{x}, (H_{ux} + f_u^* Q)\bar{x}\rangle \\ &\quad + 2\langle H_{uu}^{-1}(H_{ux} + f_u^* Q)\bar{x}, H_{uu}\bar{u}\rangle + \langle H_{uu}^{-1} H_{uu}\bar{u}, H_{uu}\bar{u}\rangle\big\}\, dt \\ &= \int_{t_0}^{t_f} \big\{\langle (H_{xu} + Qf_u) H_{uu}^{-1}(H_{ux} + f_u^* Q)\bar{x}, \bar{x}\rangle \\ &\quad + 2\langle (H_{ux} + f_u^* Q)\bar{x}, \bar{u}\rangle + \langle H_{uu}\bar{u}, \bar{u}\rangle\big\}\, dt \\ &= \int_{t_0}^{t_f} \big\{\langle (\dot{Q} + Qf_x + f_x^* Q + H_{xx})\bar{x}, \bar{x}\rangle \\ &\quad + \langle (H_{ux} + f_u^* Q)\bar{x}, \bar{u}\rangle + \langle (H_{xu} + Qf_u)\bar{u}, \bar{x}\rangle + \langle H_{uu}\bar{u}, \bar{u}\rangle\big\} dt.\end{aligned}$$

Equalities (4.196)–(4.200) imply the representation (4.192). ∎

Now we can easily prove the following theorem.

Theorem 4.40. *Assume that* (a) *there exists a symmetric solution* $Q(t)$*, piecewise continuous on* $[t_0, t_f]$ *and continuously differentiable on each interval of the set* $[t_0, t_f]\backslash\Theta$*, of Riccati equation* (4.189) *which satisfies at each point* $t_k \in \Theta$ *conditions* (4.190) *and jump conditions* (4.191). *In addition, assume that* (b) $\langle M\bar{p}, \bar{p}\rangle \geq 0$ *for all* $\bar{p} \in \mathbb{R}^{2n}$ *such that* $K_p\bar{p} = 0$. *Also, assume that* (c) *the conditions* $K_p\bar{p} = 0$ *and* $\langle M\bar{p}, \bar{p}\rangle = 0$ *imply that* $\bar{x}_0 = 0$ *or* $\bar{x}_f = 0$. *Then the quadratic form* $\Omega^\lambda(\bar{z})$ *(see* (4.187)*) is positive definite on the subspace* $\mathcal{K}$ *(see* (4.186)*); i.e., condition* (4.188) *holds with some* $\epsilon > 0$.

Proof. From Theorem 4.39 it follows that $\Omega^\lambda(\cdot)$ is nonnegative on $\mathcal{K}$. Let us show that $\Omega^\lambda(\cdot)$ is positive on $\mathcal{K}$. Assume that $\Omega(\bar{z}) = 0$ for some $\bar{z} \in \mathcal{K}$. Then by formula (4.192), we have

$$\langle M\bar{p}, \bar{p}\rangle = 0, \tag{4.201}$$

$$\bar{v} = 0, \tag{4.202}$$

$$a_k\bar{\xi}_k = -q_{k-}\bar{x}^{k-}, \quad k = 1, \ldots, s. \tag{4.203}$$

By assumption (c), we have $\bar{x}_0 = 0$ or $\bar{x}_f = 0$, since $K_p\bar{p} = 0$.

(i) Let $\bar{x}_0 = 0$. By (4.202) and (4.193), we have

$$\bar{u} = -H_{uu}^{-1}(H_{ux} + f_u^* Q)\bar{x}. \tag{4.204}$$

Using this formula in the equality $\dot{\bar{x}} = f_x\bar{x} + f_u\bar{u}$, we obtain

$$\dot{\bar{x}} = (f_x - f_u H_{uu}^{-1}(H_{ux} + f_u^* Q))\bar{x}. \tag{4.205}$$

Together with the initial condition $\bar{x}(t_0) = 0$ this implies that $\bar{x}(t) = 0$ for all $t \in [t_0, t_1)$. Hence $\bar{x}^{1-} = 0$. Then by (4.203), we obtain $\bar{\xi}_1 = 0$. Then $[\bar{x}]^1 = 0$ by the condition $[\bar{x}]^1 = [\dot{x}]^1\bar{\xi}_1$. Hence $\bar{x}^{1+} = \bar{x}^{1-} + [\bar{x}]^1 = 0$ and then again by (4.205) $\bar{x}(t) = 0$ for all $t \in (t_1, t_2)$, etc. By induction, we get $\bar{x}(t) = 0$ on $[t_0, t_f]$, $\bar{\xi} = 0$, and then by (4.204), we get $\bar{u}(t) = 0$ a.e. on $[t_0, t_f]$. Thus $\bar{z} = (\bar{\xi}, \bar{x}, \bar{u}) = 0$. Consequently, $\Omega^\lambda(\cdot)$ is positive on $\mathcal{K}$.

(ii) Consider the case $\bar{x}_f = 0$. Then equation (4.205) and condition $\bar{x}(t_f) = 0$ imply that $\bar{x}(t) = 0$ for all $t \in (t_s, t_f]$. Hence $\bar{x}^{s+} = 0$. Then $[\bar{x}]^s = -x^{s-}$. Using this condition in (4.203), we obtain $a_s\bar{\xi}_s - q_{s-}[\bar{x}]^s = 0$, or

$$(a_s - q_{s-}[\dot{x}]^s)\bar{\xi}_s = 0, \tag{4.206}$$

because $[\bar{x}]^s = [\dot{x}]^s\bar{\xi}_s$. Since $b_{s-} = a_s - q_{s-}[\dot{x}]^s > 0$, condition (4.206) implies that $\bar{\xi}_s = 0$. Hence $[\bar{x}]^s = 0$ and then $\bar{x}^{s-} = 0$. Then, by virtue of (4.205), $\bar{x}(t) = 0$ on (t_{s-1}, t_s), etc. By induction we get $\bar{x}(\cdot) = 0$, $\bar{\xi} = 0$, and then, by (4.204), $\bar{u} = 0$, whence $\bar{z} = 0$. Thus, we have proved that Ω is positive on $\mathcal{K}$. It means that Ω is positive definite on $\mathcal{K}$, since Ω is a Legendre form. ∎

Notes on SSC, Riccati equations, and sensitivity analysis. For regular controls, several authors have used the Riccati equation approach to verify SSC. Maurer and Pickenhain [73] considered optimal control problems with mixed control-state inequality constraints and derived SSC and the associated Riccati matrix equation on the basis of Klötzler's duality theory. Similar results were obtained by Zeidan [116, 117]. Extensions of these results to control problems with free final time are to be found in Maurer and Oberle [68].

It is well known that SSC are fundamental for the stability and sensitivity analysis of parametric optimal control problems; cf., Malanowski and Maurer [58, 59], Augustin and Maurer [3], Maurer and Augustin [64, 65], and Maurer and Pesch [71, 72]. SSC also lay firm theoretical grounds to the method of determining neighboring extremals (Bryson and Ho [12] and Pesch [101]) and to real-time control techniques (see Büskens [13] and Büskens and Maurer [14, 15, 16]).

Part II

Second-Order Optimality Conditions in Optimal Bang-Bang Control Problems

Chapter 5

Second-Order Optimality Conditions in Optimal Control Problems Linear in a Part of Controls

In this chapter, we derive quadratic optimality conditions for optimal control problems with a vector control variable having two components: a *continuous* unconstrained control appearing nonlinearly in the control system and a control appearing linearly and belonging to a convex polyhedron. It is assumed that the control components appearing linearly are of *bang-bang* type. In Section 5.1, we obtain quadratic conditions in the problem with continuous and bang-bang control components on a fixed time interval (that we call the main problem). The case of a nonfixed time interval is considered in Section 5.2. In Section 5.3, we show that, also for the mixed continuous-bang case, there exists a technique to check the positive definiteness of the quadratic form on the critical cone via a discontinuous solution of an associated Riccati equation with appropriate jump conditions at the discontinuity points of the bang-bang control [98]. In Section 5.4, this techniques is applied to an economic control problem in optimal production and maintenance. We show that the numerical solution obtained by Maurer, Kim, and Vossen [67] satisfies the second-order test derived in this chapter, while existing sufficiency results fail to hold.

5.1 Quadratic Optimality Conditions in the Problem on a Fixed Time Interval

In this section, we obtain a necessary quadratic condition of a Pontryagin minimum and then show that a strengthening of this condition yields a sufficient condition of a bounded strong minimum.

5.1.1 The Main Problem

Let $x(t) \in \mathbb{R}^{d(x)}$ denote the state variable, and let $u(t) \in \mathbb{R}^{d(u)}$, $v(t) \in \mathbb{R}^{d(v)}$ denote the control variables in the time interval $t \in [t_0, t_f]$ with fixed initial time t_0 and fixed final time t_f. We shall refer to the following optimal control problem (5.1)–(5.4) as the *main problem*:

$$\text{Minimize} \qquad \mathcal{J}(x(\cdot), u(\cdot), v(\cdot)) = J(x(t_0), x(t_f)) \tag{5.1}$$

subject to the constraints

$$F(x(t_0),x(t_f)) \le 0, \quad K(x(t_0),x(t_f)) = 0, \quad (x(t_0),x(t_f)) \in \mathcal{P}, \tag{5.2}$$

$$\dot{x}(t) = f(t,x(t),u(t),v(t)), \quad u(t) \in U, \quad (t,x(t),v(t)) \in \mathcal{Q}, \tag{5.3}$$

where the control variable u appears linearly in the system dynamics,

$$f(t,x,u,v) = a(t,x,v) + B(t,x,v)u. \tag{5.4}$$

Here, F,K,a are column vector functions, B is a $d(x)\times d(u)$ matrix function, $\mathcal{P} \subset \mathbb{R}^{2d(x)}$, $\mathcal{Q} \subset \mathbb{R}^{1+d(x)+d(v)}$ are open sets, and $U \subset \mathbb{R}^{d(u)}$ is a convex polyhedron. The functions J,F,K are assumed to be twice continuously differentiable on $\mathcal{P}$, and the functions a,B are twice continuously differentiable on $\mathcal{Q}$. The dimensions of F,K are denoted by $d(F)$, $d(K)$. By $\Delta = [t_0,t_f]$ we denote the interval of control and use the abbreviations $x_0 = x(t_0)$, $x_f = x(t_f)$, $p = (x_0,x_f)$.

A process $\mathcal{T} = \{(x(t),u(t),v(t)) \mid t \in [t_0,t_f]\}$ is said to be *admissible* if $x(\cdot)$ is absolutely continuous, $u(\cdot)$, $v(\cdot)$ are measurable bounded on Δ, and the triple of functions $(x(t),u(t),v(t))$, together with the endpoints $p = (x(t_0),x(t_f))$, satisfies the constraints (5.2) and (5.3). Thus, the main problem is considered in the space

$$W := W^{1,1}([t_0,t_f],\mathbb{R}^{d(x)}) \times L^\infty([t_0,t_f],\mathbb{R}^{d(u)}) \times L^\infty([t_0,t_f],\mathbb{R}^{d(v)}).$$

Definition 5.1. An admissible process $\mathcal{T}$ affords a *Pontryagin minimum* if for each compact set $\mathcal{C} \subset \mathcal{Q}$ there exists $\varepsilon > 0$ such that $\mathcal{J}(\tilde{\mathcal{T}}) \ge \mathcal{J}(\mathcal{T})$ for all admissible processes $\tilde{\mathcal{T}} = \{(\tilde{x}(t),\tilde{u}(t),\tilde{v}(t)) \mid t \in [t_0,t_f]\}$ such that (a) $\max_\Delta |\tilde{x}(t) - x(t)| < \varepsilon$, (b) $\int_\Delta |\tilde{u}(t) - u(t)|\,dt < \varepsilon$, (c) $(t,\tilde{x}(t),\tilde{v}(t)) \in \mathcal{C}$ a.e. on Δ.

5.1.2 First-Order Necessary Optimality Conditions

Let $\mathcal{T} = \{(x(t),u(t),v(t) \mid t \in [t_0,t_f]\}$ be a fixed admissible process such that the control $u(t)$ is a *piecewise constant* function taking all its values in the vertices of polyhedron U, and the control $v(t)$ is a *continuous* function on the interval $\Delta = [t_0,t_f]$. Denote by $\Theta = \{t_1,\dots,t_s\}$, $t_0 < t_1 < \cdots < t_s < t_f$ the finite set of all discontinuity points (jump points) of the control $u(t)$. Then $\dot{x}(t)$ is a piecewise continuous function whose discontinuity points belong to Θ, and hence $x(t)$ is a piecewise smooth function on Δ. We use the notation $[u]^k = u^{k+} - u^{k-}$ to denote the jump of the function $u(t)$ at the point $t_k \in \Theta$, where $u^{k-} = u(t_k-)$, $u^{k+} = u(t_k+)$ are the left- and right-hand values of the control $u(t)$ at t_k, respectively. Similarly, we denote by $[\dot{x}]^k$ the jump of the function $\dot{x}(t)$ at the point t_k. In addition, assume that the control $v(t)$ satisfies the following Condition $\mathcal{L}_\theta$: There exist constants $C > 0$ and $\varepsilon > 0$ such that for each point $t_k \in \Theta$ we have $|v(t) - v(t_k)| \le C|t - t_k|$ for all $t \in (t_k - \varepsilon, t_k + \varepsilon)$.

In the literature, one often uses the "Pontryagin minimum principle" instead of the "Pontryagin maximum principle." In order to pass from one principle to the other, one has to change only the signs of the adjoint variables ψ and ψ_0. It leads to obvious changes in the signs of the Hamiltonian, transversality conditions, Legendre conditions, and quadratic forms. Let us introduce the Pontryagin function (or Hamiltonian)

$$H(t,x,u,v,\psi) = \psi f(t,x,u,v) = \psi a(t,x,v) + \psi B(t,x,v)u, \tag{5.5}$$

where ψ is a row vector of dimension $d(\psi) = d(x)$, while x, u, f, F, and K are column vectors. The row vector of dimension $d(u)$,

$$\phi(t,x,v,\psi) = \psi B(t,x,v), \tag{5.6}$$

will be called the *switching function for the u-component* of the control. Denote by l the endpoint Lagrange function

$$l(p,\alpha_0,\alpha,\beta) = \alpha_0 J(p) + \alpha F(p) + \beta K(p),$$

where α and β are row vectors with $d(\alpha) = d(F)$ and $d(\beta) = d(K)$, and α_0 is a number. We introduce a tuple of Lagrange multipliers $\lambda = (\alpha_0, \alpha, \beta, \psi(\cdot))$ such that $\psi(\cdot) : \Delta \to \mathbb{R}^{d(x)}$ is continuous on Δ and continuously differentiable on each interval of the set $\Delta \setminus \Theta$. In what follows, we will denote first- or second-order partial derivatives by subscripts referring to the variables.

Denote by M_0 the set of the normalized tuples λ satisfying the minimum principle conditions for the process $\mathcal{T}$:

$$\alpha_0 \geq 0, \quad \alpha \geq 0, \quad \alpha F(p) = 0, \quad \alpha_0 + \sum_{i=1}^{d(F)} \alpha_i + \sum_{j=1}^{d(K)} |\beta_j| = 1, \tag{5.7}$$

$$\dot{\psi} = -H_x \quad \forall\, t \in \Delta \setminus \Theta, \tag{5.8}$$

$$\psi(t_0) = -l_{x_0}, \quad \psi(t_f) = l_{x_f}, \tag{5.9}$$

$$H(t,x(t),u,v,\psi(t)) \geq H(t,x(t),u(t),v(t),\psi(t))$$
$$\forall\, t \in \Delta \setminus \Theta,\ u \in U,\ v \in \mathbb{R}^{d(v)} \text{ such that } (t,x(t),v) \in \mathcal{Q}. \tag{5.10}$$

The derivatives l_{x_0} and l_{x_f} are taken at the point $(p,\alpha_0,\alpha,\beta)$, where $p = (x(t_0), x(t_f))$, and the derivative H_x is evaluated along the trajectory $(t,x(t),u(t),v(t),\psi(t))$, $t \in \Delta \setminus \Theta$. The condition $M_0 \neq \emptyset$ constitutes the first-order necessary condition of a Pontryagin minimum for the process $\mathcal{T}$ which is called the *Pontryagin minimum principle*, cf. Pontryagin et al. [103], Hestenes [40], and Milyutin and Osmolovskii [79]. The set M_0 is a finite-dimensional compact set and the projector $\lambda \mapsto (\alpha_0,\alpha,\beta)$ is injective on M_0.

In the following, it will be convenient to use the simple abbreviation (t) for indicating all arguments $(t,x(t),u(t),v(t),\psi(t))$, e.g.,

$$H(t) = H(t,x(t),u(t),v(t),\psi(t)), \quad \phi(t) = \phi(t,x(t),v(t),\psi(t)).$$

Let $\lambda = (\alpha_0,\alpha,\beta,\psi(\cdot)) \in M_0$. It is well known that $H(t)$ is a continuous function. In particular, $[H]^k = H^{k+} - H^{k-} = 0$ holds for each $t_k \in \Theta$, where $H^{k-} := H(t_k - 0)$ and $H^{k+} := H(t_k + 0)$. We denote by H^k the common value of H^{k-} and H^{k+}. For $\lambda \in M_0$ and $t_k \in \Theta$ consider the function

$$\begin{aligned}(\Delta_k H)(t) &= H(t,x(t),u^{k+},v(t_k),\psi(t)) - H(t,x(t),u^{k-},v(t_k),\psi(t)) \\ &= \phi(t,x(t),v(t_k),\psi(t))[u]^k.\end{aligned} \tag{5.11}$$

Proposition 5.2. *For each $\lambda \in M_0$, the following equalities hold:*

$$\frac{d}{dt}(\Delta_k H)\big|_{t=t_k-0} = \frac{d}{dt}(\Delta_k H)\big|_{t=t_k+0}, \quad k = 1,\ldots,s.$$

Consequently, for each $\lambda \in M_0$, the function $(\Delta_k H)(t)$ has a derivative at the point $t_k \in \Theta$. In what follows, we will consider the quantities

$$D^k(H) = -\frac{d}{dt}(\Delta_k H)(t_k) = -\dot{\phi}(t_k \pm)[u]^k, \quad k = 1, \dots, s. \tag{5.12}$$

Then the minimum condition (5.10) implies the following property.

Proposition 5.3. *For each $\lambda \in M_0$, the following conditions hold:*

$$D^k(H) \geq 0, \quad k = 1, \dots, s.$$

Note that the value $D^k(H)$ also can be written in the form

$$D^k(H) = -H_x^{k+} H_\psi^{k-} + H_x^{k-} H_\psi^{k+} - [H_t]^k = \dot{\psi}^{k+} \dot{x}^{k-} - \dot{\psi}^{k-} \dot{x}^{k+} + [\dot{\psi}_0]^k,$$

where H_x^{k-} and H_x^{k+} are the left-hand and right-hand values of the function $H_x(t)$ at t_k, respectively, $[H_t]^k$ is the jump of the function $H_t(t) := H_t(t, x(t), u(t), v(t), \psi(t))$ at t_k, etc., and $\psi_0(t) = -H(t)$.

5.1.3 Bounded Strong Minimum

As in Section 10.1, we define essential (or main) and unessential (or complementary) state variables in the problem. The state variable x_i, i.e., the ith component of the state vector x, is called *unessential* if the function f does not depend on x_i and the functions F, J, and K are affine in $x_{i0} = x_i(t_0)$ and $x_{i1} = x_i(t_f)$; otherwise the variable x_i is called *essential.* Let $\underline{x}$ denote the vector of all essential components of state vector x.

Definition 5.4. The process $\mathcal{T}$ affords a *bounded strong minimum* if for each compact set $\mathcal{C} \subset \mathcal{Q}$ there exists $\varepsilon > 0$ such that $\mathcal{J}(\widetilde{\mathcal{T}}) \geq \mathcal{J}(\mathcal{T})$ for all admissible processes $\widetilde{\mathcal{T}} = \{(\tilde{x}(t), \tilde{u}(t), \tilde{v}(t)) \mid t \in [t_0, t_f]\}$ such that (a) $|\tilde{x}(t_0) - x(t_0)| < \varepsilon$, (b) $\max_\Delta |\underline{\tilde{x}}(t) - \underline{x}(t)| < \varepsilon$, (c) $(t, \tilde{x}(t), \tilde{v}(t)) \in \mathcal{C}$ a.e. on Δ.

The *strict* bounded strong minimum is defined in a similar way, with the nonstrict inequality $\mathcal{J}(\widetilde{\mathcal{T}}) \geq \mathcal{J}(\mathcal{T})$ replaced by the strict one and the process $\widetilde{\mathcal{T}}$ required to be different from $\mathcal{T}$.

5.1.4 Critical Cone

For a given process $\mathcal{T}$, we introduce the space $Z_2(\Theta)$ and the *critical cone* $\mathcal{K} \subset Z_2(\Theta)$. As in Section 2.1.5, we denote by $P_\Theta W^{1,2}(\Delta, \mathbb{R}^{d(x)})$ the space of piecewise continuous functions $\bar{x}(\cdot) : \Delta \to \mathbb{R}^{d(x)}$, which are absolutely continuous on each interval of the set $\Delta \setminus \Theta$ and have a square integrable first derivative. For each $\bar{x} \in P_\Theta W^{1,2}(\Delta, \mathbb{R}^{d(x)})$ and for $t_k \in \Theta$, we set

$$\bar{x}^{k-} = \bar{x}(t_k-), \quad \bar{x}^{k+} = \bar{x}(t_k+), \quad [\bar{x}]^k = \bar{x}^{k+} - \bar{x}^{k-}.$$

Let $\bar{z} = (\bar{\xi}, \bar{x}, \bar{v})$, where $\bar{\xi} \in \mathbb{R}^s$, $\bar{x} \in P_\Theta W^{1,2}(\Delta, \mathbb{R}^{d(x)})$, $\bar{v} \in L^2(\Delta, \mathbb{R}^{d(v)})$. Thus,

$$\bar{z} \in Z_2(\Theta) := \mathbb{R}^s \times P_\Theta W^{1,2}(\Delta, \mathbb{R}^{d(x)}) \times L^2(\Delta, \mathbb{R}^{d(v)}).$$

For each $\bar{z}$, we set

$$\bar{x}_0 = \bar{x}(t_0), \quad \bar{x}_f = \bar{x}(t_f), \quad \bar{p} = (\bar{x}_0, \bar{x}_f). \tag{5.13}$$

The vector $\bar{p}$ is considered a column vector. Denote by

$$I_F(p) = \{i \in \{1,\dots,d(F)\} \mid F_i(p) = 0\}$$

the set of indices of all active endpoint inequalities $F_i(p) \le 0$ at the point $p = (x(t_0), x(t_f))$. Denote by $\mathcal{K}$ the set of all $\bar{z} \in Z_2(\Theta)$ satisfying the following conditions:

$$J'(p)\bar{p} \le 0, \quad F_i'(p)\bar{p} \le 0 \ \forall\, i \in I_F(p), \quad K'(p)\bar{p} = 0, \tag{5.14}$$

$$\dot{\bar{x}}(t) = f_x(t)\bar{x}(t) + f_v(t)\bar{v}(t), \tag{5.15}$$

$$[\bar{x}]^k = [\dot{x}]^k \bar{\xi}_k, \quad k = 1,\dots,s, \tag{5.16}$$

where $p = (x(t_0), x(t_f))$ and $[\dot{x}]^k = \dot{x}(t_k + 0) - \dot{x}(t_k - 0)$.

It is obvious that $\mathcal{K}$ is a convex cone in the Hilbert space $Z_2(\Theta)$ with finitely many faces. We call $\mathcal{K}$ the *critical cone*. Note that the variation $\bar{u}(t)$ of the bang-bang control $u(t)$ vanishes in the critical cone.

5.1.5 Necessary Quadratic Optimality Conditions

Let us introduce a quadratic form on the critical cone $\mathcal{K}$ defined by the conditions (5.14)–(5.16). For each $\lambda \in M_0$ and $\bar{z} \in \mathcal{K}$, we set[3]

$$\begin{aligned}\Omega(\lambda, \bar{z}) \;=\;& \langle l_{pp}(p)\bar{p}, \bar{p}\rangle + \sum_{k=1}^{s} \left(D^k(H)\bar{\xi}_k^2 - 2[\dot{\psi}]^k \bar{x}_{\mathrm{av}}^k \bar{\xi}_k\right) \\ &+ \int_{t_0}^{t_f} \big(\langle H_{xx}(t)\bar{x}(t), \bar{x}(t)\rangle + 2\langle H_{xv}(t)\bar{v}(t), \bar{x}(t)\rangle \\ &+ \langle H_{vv}(t)\bar{v}(t), \bar{v}(t)\rangle\big)\, dt,\end{aligned} \tag{5.17}$$

where

$$\begin{aligned} &l_{pp}(p) = l_{pp}(\alpha_0, \alpha, \beta, p), \\ &p = (x(t_0), x(t_f)), \\ &\bar{x}_{\mathrm{av}}^k = \frac{1}{2}(\bar{x}^{k-} + \bar{x}^{k+}), \\ &H_{xx}(t) = H_{xx}(t, x(t), u(t), v(t), \psi(t)),\end{aligned}$$

etc. Note that the functional $\Omega(\lambda, \bar{z})$ is linear in λ and quadratic in $\bar{z}$. The following theorem gives the main second-order necessary condition of optimality.

Theorem 5.5. *If the process $\mathcal{T}$ affords a Pontryagin minimum, then the following Condition $\mathfrak{A}$ holds: The set M_0 is nonempty and $\max_{\lambda \in M_0} \Omega(\lambda, \bar{z}) \ge 0$ for all $\bar{z} \in \mathcal{K}$.*

We call Condition $\mathfrak{A}$ the necessary quadratic condition, although it is truly quadratic only if M_0 is a singleton.

[3] In Part II of the book, we will not use the factor $1/2$ in the definition of the quadratic form Ω.

5.1.6 Sufficient Quadratic Optimality Conditions

A natural strengthening of the necessary Condition $\mathfrak{A}$ turns out to be a sufficient optimality condition not only for a Pontryagin minimum, but also for a bounded strong minimum; cf. Definition 5.4. Denote by M_0^+ the set of all $\lambda \in M_0$ satisfying the following conditions:

(a) $H(t,x(t),u,v,\psi(t)) > H(t,x(t),u(t),v(t),\psi(t))$ for all $t \in \Delta \setminus \Theta$, $u \in U$, $v \in \mathbb{R}^{d(v)}$ such that $(t,x(t),v) \in \mathcal{Q}$ and $(u,v) \neq (u(t),v(t))$;

(b) $H(t_k,x(t_k),u,v,\psi(t_k)) > H^k$ for all $t_k \in \Theta$, $u \in U$, $v \in \mathbb{R}^{d(v)}$ such that $(t_k,x(t_k),v) \in \mathcal{Q}$, $(u,v) \neq (u(t_k-),v(t_k))$, $(u,v) \neq (u(t_k+),v(t_k))$, where $H^k := H^{k-} = H^{k+}$.

Let $\operatorname{Arg\,min}_{\tilde{u}\in U} \phi\tilde{u}$ be the set of points $u \in U$ where the minimum of the linear function $\phi\tilde{u}$ is attained.

Definition 5.6. For a given admissible process $\mathcal{T}$ with a piecewise constant control $u(t)$ and continuous control $v(t)$, we say that $u(t)$ is a *strict bang-bang control* if the set M_0 is nonempty and there exists $\lambda \in M_0$ such that

$$\operatorname*{Arg\,min}_{\tilde{u}\in U} \phi(t)\tilde{u} = [u(t-),u(t+)] \quad \forall\, t \in [t_0,t_f],$$

where $[u(t-),u(t+)]$ denotes the line segment spanned by the vectors $u(t-)$, $u(t+)$.

If $\dim(u) = 1$, then the strict bang-bang property is equivalent to

$$\phi(t) \neq 0 \quad \forall\, t \in \Delta \setminus \Theta.$$

It is easy to show that if the set M_0^+ is nonempty, then $u(t)$ is a strict bang-bang control.

Definition 5.7. An element $\lambda \in M_0$ is said to be *strictly Legendre* if the following conditions are satisfied:

(a) For each $t \in \Delta \setminus \Theta$ the quadratic form $\langle H_{vv}(t,x(t),u(t),v(t),\psi(t))\bar{v},\bar{v}\rangle$ is positive definite on $\mathbb{R}^{d(v)}$;

(b) for each $t_k \in \Theta$ the quadratic form $\langle H_{vv}(t_k,x(t_k),u(t_k-),v(t_k),\psi(t_k))\bar{v},\bar{v}\rangle$ is positive definite on $\mathbb{R}^{d(v)}$;

(c) for each $t_k \in \Theta$ the quadratic form $\langle H_{vv}(t_k,x(t_k),u(t_k+),v(t_k),\psi(t_k))\bar{v},\bar{v}\rangle$ is positive definite on $\mathbb{R}^{d(v)}$;

(d) $D^k(H) > 0$ for all $t_k \in \Theta$.

Denote by $\operatorname{Leg}_+(M_0^+)$ the set of all strictly Legendrian elements $\lambda \in M_0^+$, and set

$$\bar{\gamma}(\bar{z}) = \langle \bar{\xi},\bar{\xi}\rangle + \langle \bar{x}(t_0),\bar{x}(t_0)\rangle + \int_{t_0}^{t_f} \langle \bar{v}(t),\bar{v}(t)\rangle\, dt.$$

Theorem 5.8. *Let the following Condition $\mathfrak{B}$ be fulfilled for the process $\mathcal{T}$:*

(a) *The set $\operatorname{Leg}_+(M_0^+)$ is nonempty;*

(b) *there exists a nonempty compact set* $M \subset \mathrm{Leg}_+(M_0^+)$ *and a number* $C > 0$ *such that* $\max_{\lambda \in M} \Omega(\lambda, \bar{z}) \geq C \bar{\gamma}(\bar{z})$ *for all* $\bar{z} \in \mathcal{K}$.

Then $\mathcal{T}$ *is a strict bounded strong minimum.*

Remark 5.9. If the set $\mathrm{Leg}_+(M_0^+)$ is nonempty and $\mathcal{K}=\{0\}$, then Condition (b) is fulfilled automatically. This case can be considered as a first-order sufficient optimality condition for a strict bounded strong minimum.

As mentioned in the introduction, the proof of Theorem 5.8 is very similar to the proof of the sufficient quadratic optimality condition for the pure bang-bang case given in Milyutin and Osmolovskii [79, Theorem 12.4, p. 302], and based on the sufficient quadratic optimality condition for broken extremals in the general problem of calculus of variations; see Part I of the present book. The proofs of Theorems 5.5 and 5.8 will be given below.

5.1.7 Proofs of Quadratic Conditions in the Problem on a Fixed Time Interval

Problem Z and its convexification with respect to bang-bang control components. Consider the following optimal control problem, which is similar to the main problem (5.1)–(5.4):

$$\text{Minimize} \qquad \mathcal{J}(x(\cdot), u(\cdot), v(\cdot)) = J(x(t_0), x(t_f)) \tag{5.18}$$

subject to the constraints

$$F(x(t_0), x(t_f)) \leq 0, \quad K(x(t_0), x(t_f)) = 0, \quad (x(t_0), x(t_f)) \in \mathcal{P}, \tag{5.19}$$

$$\dot{x} = a(t, x, v) + B(t, x, v)u, \quad u \in \mathcal{U}, \quad (t, x, v) \in \mathcal{Q}, \tag{5.20}$$

where $\mathcal{P} \subset \mathbb{R}^{2d(x)}$, $\mathcal{Q} \subset \mathbb{R}^{1+d(x)+d(v)}$ are open sets, $\mathcal{U} \subset \mathbb{R}^{d(u)}$ is a compact set, and $\Delta = [t_0, t_f]$ is a fixed time interval. The functions J, F, and K are assumed to be twice continuously differentiable on $\mathcal{P}$, and the functions a and B are twice continuously differentiable on $\mathcal{Q}$. The compact set $\mathcal{U}$ is specified by

$$\mathcal{U} = \{u \in \mathcal{Q}_g \mid g(u) = 0\}, \tag{5.21}$$

where $\mathcal{Q}_g \subset \mathbb{R}^{d(u)}$ is an open set and $g : \mathcal{Q}_g \to \mathbb{R}^{d(g)}$ is a twice continuously differentiable function satisfying the full-rank condition

$$\operatorname{rank} g_u(u) = d(g) \tag{5.22}$$

for all $u \in \mathcal{Q}_g$ such that $g(u) = 0$. It follows from (5.22) that $d(g) \leq d(u)$, but it is possible, in particular, that $d(g) = d(u)$. In this latter case for $\mathcal{U}$ we can take any finite set of points in $\mathbb{R}^{d(u)}$, for example, the set of vertices of a convex polyhedron.

For brevity, we will refer to problem (5.18)–(5.20) as the problem Z. Thus, the only difference between problem Z and the main problem (5.1)–(5.4) is that a convex polyhedron U is replaced by an arbitrary compact set $\mathcal{U}$ specified by (5.21). The definitions of the Pontryagin minimum, the bounded strong minimum, and the strict bounded strong minimum in the problem Z are the same as in the main problem.

We will consider the problem (5.18), (5.19) not only for control system (5.20), but also for its convexification with respect to u:

$$\dot{x} = a(t,x,v) + B(t,x,v)u, \quad u \in \operatorname{co}\mathcal{U}, \quad (t,x,v) \in \mathcal{Q}, \tag{5.23}$$

where $\operatorname{co}\mathcal{U}$ is the convex hull of the compact set $\mathcal{U}$. We will refer to the problem (5.18), (5.19), (5.23) as the problem $\operatorname{co} Z$.

We will be interested in the relationships between conditions for a minimum in the problems Z and $\operatorname{co} Z$. Naturally, these conditions concern a process satisfying the constraints of the problem Z. Let $w^0(\cdot) = (x^0(\cdot), u^0(\cdot), v^0(\cdot))$ be such a process. Then it satisfies the constraints of the problem $\operatorname{co} Z$ as well. We assume that w^0 satisfy the following conditions: The function $u^0(t)$ is piecewise continuous with the set $\Theta = \{t_1, \ldots, t_s\}$ of discontinuity points, control $v^0(t)$ is continuous, and each point $t_k \in \Theta$ is an L-point of the controls $u^0(t)$ and $v^0(t)$. The latter means that there exist constants $C > 0$ and $\varepsilon > 0$ such that for each point $t_k \in \Theta$, we have

$$\begin{aligned} |u^0(t) - u^{0k-}| &\leq C|t - t_k| \quad \text{for} \quad t \in (t_k - \varepsilon, t_k), \\ |u^0(t) - u^{0k+}| &\leq C|t - t_k| \quad \text{for} \quad t \in (t_k, t_k + \varepsilon), \\ |v^0(t) - v^0(t_k)| &\leq C|t - t_k| \quad \text{for} \quad t \in (t_k - \varepsilon, t_k + \varepsilon). \end{aligned}$$

We can formulate quadratic conditions for the point w^0 in the problem Z. To what extent can they be carried over to the problem $\operatorname{co} Z$?

Regarding the necessary quadratic conditions (see Condition $\mathfrak{A}$ in Theorem 3.9) this is a simple question. If a point w^0 yields a Pontryagin minimum in the problem $\operatorname{co} Z$, this point, *a fortiori*, affords a Pontryagin minimum in the problem Z. Hence any necessary condition for a Pontryagin minimum at the point w^0 in the problem Z is a necessary condition for a Pontryagin minimum at this point in the problem $\operatorname{co} Z$ as well. Thus we have the following theorem.

Theorem 5.10. *Condition $\mathfrak{A}$ (given by Theorem 3.9) for the point w^0 in the problem Z is a necessary condition for a Pontryagin minimum at this point in the problem* $\operatorname{co} Z$.

Bounded strong γ_1-sufficiency in problem Z and bounded strong minimum in problem $\operatorname{co} Z$. We now turn to derivation of quadratic sufficient conditions in the problem $\operatorname{co} Z$, which is a more complicated task. As we know, Condition $\mathfrak{B}$ for the point w^0 in the problem Z ensures a Pontryagin, and even a bounded strong, minimum in this problem. Does it ensure a bounded strong or at least a Pontryagin minimum at the point w^0 in the convexified problem $\operatorname{co} Z$?

There are examples where the convexification results in the loss of the minimum. The bounded strong minimum is not stable with respect to the convexification operation. However, some stronger property in the problem Z, which is called bounded strong γ_1-sufficiency, ensures a bounded strong minimum in the problem $\operatorname{co} Z$. This property will be defined below. For the problem Z and the point $w^0(\cdot)$ define the violation function

$$\begin{aligned} \sigma(w) = (J(p) - J(p^0))^+ + \sum_{i=1}^{d(F)} (F_i(p))^+ + |K(p)| \\ + \int_{t_0}^{t_f} |\dot{x}(t) - f(t, x(t), u(t), v(t))| \, dt, \end{aligned} \tag{5.24}$$

where $f(t,x,u,v) = a(t,x,v) + B(t,x,v)u$ and $a^+ = \max\{a,0\}$ for $a \in \mathbb{R}^1$. For an arbitrary variation $\delta w = (\delta x, \delta u, \delta v) \in W$, let

$$\gamma_1(\delta w) = (\|\delta u\|_1)^2, \tag{5.25}$$

where $\|\delta u\|_1 = \int_{t_0}^{t_f} |\delta u(t)|\, dt$.

Let $\{w^n\} = \{(x^n, u^n, v^n)\}$ be a bounded sequence in W. In what follows, the notation $\sigma(\delta w^n) = o(\gamma_1(w^n - w^0))$ means that there exists a sequence of numbers $\varepsilon_n \to 0$ such that $\sigma(w^n) = \varepsilon_n \gamma_1(w^n - w^0)$ (even in the case where $\gamma_1(w^n - w^0)$ does not tend to zero).

Definition 5.11. We say that the *bounded strong γ_1-sufficiency* holds at the point w^0 in the problem Z if there are no compact set $\mathcal{C} \subset \mathcal{Q}$ and sequence $\{w^n\} = \{(x^n, u^n, v^n)\}$ in W such that

$$\sigma(w^n) = o(\gamma_1(w^n - w^0)), \quad \max_{t \in \Delta} |\underline{x}^n(t) - \underline{x}^0(t)| \to 0, \quad |x^n(t_0) - x^0(t_0)| \to 0,$$

and for all n, $w^n \neq w^0$, $(t, x^n(t), v^n(t)) \in \mathcal{C}$, $u^n(t) \in \mathcal{U}$ a.e. on Δ, where $\underline{x}^n$ is composed of the essential components of vector x^n.

The following proposition holds for the problem Z.

Proposition 5.12. *A bounded strong γ_1-sufficiency at the point w^0 implies a strict bounded strong minimum at this point.*

Proof. Let w^0 be an admissible point in the problem Z. Assume that w^0 is not a point of a strict bounded strong minimum. Then there exist a compact set $\mathcal{C} \subset \mathcal{Q}$ and a sequence of admissible points $\{w^n\} = \{(x^n, u^n, v^n)\}$ such that $\max_{t \in \Delta} |\underline{x}^n(t) - \underline{x}^0(t)| \to 0$, $|x^n(t_0) - x^0(t_0)| \to 0$ and, for all n,

$$J(p^n) - J(p^0) < 0, \quad (t, x^n(t), v^n(t)) \in \mathcal{C}, \quad u^n(t) \in \mathcal{U} \text{ a.e. on } \Delta, \quad w^n \neq w^0.$$

It follows that $\sigma(w^n) = 0$ for all n. Hence w^0 is not a point of a bounded strong γ_1-sufficiency in the problem Z. ∎

A remarkable fact is that the convexification of the constraint $u \in \mathcal{U}$ turns a bounded strong γ_1-sufficiency into at least a bounded strong minimum.

Theorem 5.13. *Suppose that for an admissible point w^0 in the problem Z the bounded strong γ_1-sufficiency holds. Then w^0 is a point of the strict bounded strong minimum in the problem* co Z.

The proof of Theorem 5.13 is based on the following lemma.

Lemma 5.14. *Let $U \subset \mathbb{R}^{d(u)}$ be a bounded set, and let $u(t)$ be a measurable function on Δ such that $u(t) \in \operatorname{co} U$ a.e. on Δ. Then there exists a sequence $u^n(t)$ of measurable functions on Δ such that for every n we have $u^n(t) \in U$, $t \in \Delta$, and*

$$\int_{t_0}^{t} u^n(\tau)\, d\tau \to \int_{t_0}^{t} u(\tau)\, d\tau \quad \textit{uniformly in } t \in \Delta. \tag{5.26}$$

Moreover, if $u^0(t)$ is a bounded measurable function on Δ such that

$$\operatorname{meas}\{t \in \Delta \mid u(t) \neq u^0(t)\} > 0, \tag{5.27}$$

then (5.26) *implies*

$$\liminf_n \int_\Delta |u^n(t) - u^0(t)|\, dt > 0. \tag{5.28}$$

This lemma is a consequence of Theorem 16.1 in [79, Appendix, p. 361]. In the proof of Theorem 5.13 we will also use the following theorem, which is similar to Theorem 16.2 in [79, Appendix, p. 366].

Theorem 5.15. *Let $w^* = (x^*, u^*, v^*)$ be a triple in W satisfying the control system*

$$\dot{x} = a(t,x,v) + B(t,x,v)u, \quad (t,x,v) \in \mathcal{Q}, \quad x(t_0) = c_0,$$

where $\mathcal{Q}$, a, and B are the same as in the problem Z, and $c_0 \in \mathbb{R}^{d(x)}$. Suppose that there is a sequence $u^n \in L^\infty([t_0,t_f], \mathbb{R}^{d(u)})$ such that

$$\sup_n \|u^n\|_\infty < +\infty \tag{5.29}$$

and, for each $t \in [t_0, t_f]$,

$$\int_{t_0}^t u^n(\tau)\,d\tau \to \int_{t_0}^t u^*(\tau)\,d\tau \qquad (n \to \infty). \tag{5.30}$$

Then, for all sufficiently large n, the system

$$\dot{x} = a(t,x,v^*(t)) + B(t,x,v^*(t))u^n(t), \quad (t,x,v^*(t)) \in \mathcal{Q}, \quad x(t_0) = c_0 \tag{5.31}$$

has the unique solution $x^n(t)$ on $[t_0, t_f]$ and

$$\max_{[t_0,t_f]} |x^n(t) - x^*(t)| \to 0 \qquad (n \to \infty). \tag{5.32}$$

Proof. Consider the equations

$$\dot{x}^n = a(t,x^n,v^*) + B(t,x^n,v^*)u^n,$$
$$\dot{x}^* = a(t,x^*,v^*) + B(t,x^*,v^*)u^*.$$

Their difference can be written as

$$\delta\dot{x} = \delta a + (\delta B)u^n + B(t,x^*,v^*)\delta u, \tag{5.33}$$

where

$$\delta x = x^n - x^*, \quad \delta u = u^n - u^*,$$
$$\delta a = a(t,x^n,v^*) - a(t,x^*,v^*) = a(t,x^* + \delta x, v^*) - a(t,x^*,v^*),$$
$$\delta B = B(t,x^n,v^*) - B(t,x^*,v^*) = B(t,x^* + \delta x, v^*) - B(t,x^*,v^*).$$

We have here

$$\delta x(t_0) = 0. \tag{5.34}$$

It follows from (5.29) and (5.30) that

$$\limsup \|\delta u\|_\infty < +\infty, \qquad \int_{t_0}^{t} \delta u(\tau)d\tau \to 0 \quad \forall\, t \in [t_0, t_f]. \tag{5.35}$$

These properties mean that the sequence $\{\delta u\}$ converges to zero $*$-weakly (L^1-weakly) in L^∞. Therefore the functions

$$\delta y(t) := \int_{t_0}^{t} B(\tau, x^*(\tau), v^*(\tau))\delta u(\tau)d\tau$$

converge to zero pointwise on $[t_0, t_f]$, and hence uniformly on $[t_0, t_f]$ because they possess the common Lipschitz constant. Thus

$$\|\delta y\|_C \to 0. \tag{5.36}$$

Using the relations

$$\delta \dot{y} = B(t, x^*, v^*)\delta u, \quad \delta y(t_0) = 0, \tag{5.37}$$

rewrite (5.33) as

$$\delta \dot{x} - \delta \dot{y} = \delta a + (\delta B)u^n. \tag{5.38}$$

Set $\delta z = \delta x - \delta y$. Then (5.38) can be represented in the form

$$\delta \dot{z} = \delta a + (\delta B)u^n,$$

where

$$\delta a = a(t, x^* + \delta y + \delta z, v^*) - a(t, x^*, v^*), \quad \delta B = B(t, x^* + \delta y + \delta z, v^*) - B(t, x^*, v^*).$$

Thus

$$\begin{aligned} \delta \dot{z} &= a(t, x^* + \delta y + \delta z, v^*) - a(t, x^*, v^*) \\ &\quad + (B(t, x^* + \delta y + \delta z, v^*) - B(t, x^*, v^*))u^n, \\ \delta z(t_0) &= 0, \quad (t, x^* + \delta y + \delta z, v^*) \in \mathcal{Q}. \end{aligned} \tag{5.39}$$

This system is equivalent to the system (5.31) in the following sense: x^n solves (5.31) iff

$$x^n = x^* + \delta y + \delta z, \tag{5.40}$$

where δy satisfies the conditions

$$\delta \dot{y} = B(t, x^*, v^*)(u^n - u^*), \quad \delta y(t_0) = 0, \tag{5.41}$$

and δz solves the system (5.39).

Hence it suffices to show that for all sufficiently large n the system (5.39) has a unique solution, where δy is determined by (5.41). This is so because for $\delta y \equiv 0$ the system (5.39) has a unique solution $\delta z \equiv 0$ on $[t_0, t_f]$ and moreover (5.36) holds. From (5.36) and representation (5.39) it follows also that $\|\delta z\|_C \to 0$ and hence $\|\delta x\|_C \le \|\delta y\|_C + \|\delta z\|_C \to 0$. Therefore (5.32) holds. ∎

Proof of Theorem 5.13. Suppose that w^0 is not a strict bounded strong minimum point in the problem $\operatorname{co} Z$. Then there exist a compact set $\mathcal{C} \subset \mathcal{Q}$ and a sequence $\{w^n\} = \{(x^n, u^n, v^n)\}$ such that for all n one has $w^n \neq w^0$ and

$$(x^n(t_0), x^n(t_f)) \in \mathcal{P}, \quad (t, x^n(t), v^n(t)) \in \mathcal{C} \quad \text{a.e. on } \Delta, \tag{5.42}$$

$$u^n(t) \in \operatorname{co} \mathcal{U} \quad \text{a.e. on } \Delta, \tag{5.43}$$

$$\sigma(w^n) = 0, \tag{5.44}$$

$$\max_{t \in \Delta} |\underline{x}^n(t) - \underline{x}^0(t)| \to 0, \quad |x^n(t_0) - x^0(t_0)| \to 0. \tag{5.45}$$

We will show that w^0 is not a point of the bounded strong γ_1-sufficiency in the problem Z. If $u^n = u^0$ for infinitely many terms, w^0 is not even a strict bounded strong minimum point in the problem Z. Hence, we assume that $u^n \neq u^0$ for all n. Apply Lemma 5.14 to each function u^n. By virtue of this lemma there exists a sequence of measurable functions $\{u^{nk}\}_{k=1}^{\infty}$ such that, for all k,

$$u^{nk}(t) \in \mathcal{U} \quad \forall\, t \in \Delta, \tag{5.46}$$

$$\int_{t_0}^{t} u^{nk}(\tau)\, d\tau \to \int_{t_0}^{t} u^n(\tau)\, d\tau \text{ uniformly in } t \in \Delta \text{ (as } k \to \infty), \tag{5.47}$$

$$\liminf_{k \to \infty} \int_{t_0}^{t_f} |u^{nk}(t) - u^0(t)|\, dt > 0. \tag{5.48}$$

For each k define x^{nk} as the solution of the system

$$\dot{x}^{nk} = a(t, x^{nk}, v^n) + B(t, x^{nk}, v^n) u^{nk}, \quad x^{nk}(t_0) = x^n(t_0). \tag{5.49}$$

According to Theorem 5.15 this system has a solution for all sufficiently large k and

$$\|x^{nk} - x^n\|_C \to 0 \ \text{ as } k \to \infty. \tag{5.50}$$

It follows from (5.44) and (5.50) that

$$(J(p^{nk}) - J(p^0))^+ + \sum_i (F_i(p^{nk}))^+ + |K(p^{nk})| \to 0 \ \text{ as } k \to \infty, \tag{5.51}$$

where $p^{nk} = (x^{nk}(t_0), x^{nk}(t_f)) = (x^n(t_0), x^{nk}(t_f))$. Combined with (5.49) this implies that

$$\sigma(w^{nk}) \to 0 \ \text{ as } k \to \infty, \tag{5.52}$$

where $w^{nk} = (x^{nk}, u^{nk}, v^n)$. It follows from (5.48) and (5.52) that there is a number $k = k(n)$ such that

$$\sigma(w^{nk}) \leq \frac{1}{n} \left(\int_{t_0}^{t_f} |u^{nk}(t) - u^0(t)|\, dt \right)^2 \quad \forall\, k \geq k(n). \tag{5.53}$$

By (5.50), $k(n)$ also can be chosen so that

$$\|x^{nk(n)} - x^n\|_C \leq \frac{1}{n}. \tag{5.54}$$

Then by virtue of (5.53), for the sequence $\{w^{nk(n)}\}$ one has

$$\sigma(w^{nk(n)}) = o(\gamma_1(w^{nk(n)} - w^0)). \tag{5.55}$$

Moreover, we have for this sequence

$$u^{nk(n)}(t) \in \mathcal{U} \quad \forall\, t \in \Delta, \tag{5.56}$$

$$\begin{aligned} \|\underline{x}^{nk(n)} - \underline{x}^0\|_C &\le \|x^{nk(n)} - x^n\|_C + \|\underline{x}^n - \underline{x}^0\|_C \\ &\le \frac{1}{n} + \|\underline{x}^n - \underline{x}^0\|_C \to 0 \quad (n \to \infty), \end{aligned} \tag{5.57}$$

$$|x^{nk(n)}(t_0) - x^0(t_0)| = |x^n(t_0) - x^0(t_0)| \to 0. \tag{5.58}$$

Finally, from (5.42) and (5.54) it follows that there exists a compact set $\mathcal{C}_1$ such that $\mathcal{C} \subset \mathcal{C}_1 \subset \mathcal{Q}$, and for all sufficiently large n we have

$$(x^{nk(n)}(t_0), x^{nk(n)}(t_f)) \in \mathcal{P}, \quad (t, x^{nk(n)}(t), v^n(t)) \in \mathcal{C}_1 \quad \text{a.e. on } \Delta. \tag{5.59}$$

The existence of a compact set $\mathcal{C}_1 \subset \mathcal{Q}$ and a sequence $\{w^{nk(n)}\}$ in the space W satisfying (5.55)–(5.59) means that the bounded strong γ_1-sufficiency fails at the point w^0 in the problem Z. ∎

Now we can prove the following theorem.

Theorem 5.16. *Condition $\mathfrak{B}$ (given in Definition 3.15) for the point w^0 in the problem Z is a sufficient condition for a strict bounded strong minimum at this point in the problem* co Z.

Proof. Assume that Condition $\mathfrak{B}$ for the point w^0 in the problem Z is satisfied. Then by Theorem 3.16 a bounded strong γ-sufficiency holds in problem Z at the point w^0. The latter means (see Definition 3.11) that there are no compact set $\mathcal{C} \subset \mathcal{Q}$ and sequence $\{w^n\} = \{(x^n, u^n, v^n)\}$ in W such that

$$\sigma(w^n) = o(\gamma(w^n - w^0)), \quad \max_{t \in \Delta} |\underline{x}^n(t) - \underline{x}^0(t)| \to 0, \quad |x^n(t_0) - x^0(t_0)| \to 0,$$

and for all n we have $w^n \ne w^0$, $(t, x^n(t), v^n(t)) \in \mathcal{C}$, $g(u^n(t)) = 0$, $u^n(t) \in \mathcal{Q}_g$ a.e. on Δ, where γ is the higher order defined in Definition 2.17.

Note that for each n the conditions $g(u^n(t)) = 0$, $u^n(t) \in \mathcal{Q}_g$ mean that $u^n(t) \in \mathcal{U}$, where $\mathcal{U}$ is a compact set. It follows from Proposition 2.98 that for any compact set $\mathcal{C} \subset \mathcal{Q}$ there exists a constant $C > 0$ such that for any $w = (x, u, v) \in W$ satisfying the conditions $u(t) \in \mathcal{U}$ and $(t, x(t), v(t)) \in \mathcal{C}$ a.e. on Δ, we have $\gamma_1(w - w^0) \le C\gamma(w - w^0)$. By virtue of this inequality, a bounded strong γ-sufficiency at the point w^0 in problem Z implies a bounded strong γ_1-sufficiency at w^0 in the same problem. Then by Theorem 5.13, w^0 is a point of a strict bounded strong minimum in problem co Z. ∎

Proofs of Theorems 5.5 and 5.8. Consider the main problem again:

$$\begin{aligned} &J(p) \to \min, \quad F(p) \le 0, \quad K(p) = 0, \quad p := (x(t_0), x(t_f)) \in \mathcal{P}, \\ &\dot{x} = a(t, x, v) + B(t, x, v)u, \quad u \in U, \quad (t, x, v) \in \mathcal{Q}, \end{aligned}$$

where U is a convex polyhedron. Let $\mathcal{U}$ be the set of vertices of U. Consider an admissible process $w^0 = (x^0, u^0, v^0) \in W$ in the main problem. Assume that the control u^0 is a piecewise constant function on Δ taking all its values in the vertices of U, i.e.,

$$u^0(t) \in \mathcal{U}, \quad t \in \Delta. \tag{5.60}$$

As usual we denote by $\Theta = \{t_1, \ldots, t_s\}$ the set of discontinuity points (switching points) of the control u^0. Assume that the control $v^0(t)$ is a continuous function on Δ, satisfying the Condition $\mathcal{L}_\theta$ (see Section 5.1.2). By virtue of condition (5.60) the process $w^0 = (x^0, u^0, v^0)$ is also admissible in the problem

$$J(p) \to \min, \quad F(p) \le 0, \quad K(p) = 0, \quad p \in \mathcal{P}, \tag{5.61}$$

$$\dot{x}(t) = a(t,x,v) + B(t,x,v)u, \quad u(t) \in \mathcal{U}, \quad (t, x(t), v(t)) \in \mathcal{Q}. \tag{5.62}$$

Since $U = \operatorname{co}\mathcal{U}$, the main problem can be viewed as a convexification of problem (5.61), (5.62) with respect to u.

It is easy to see that, in problem (5.61), (5.62), we can use the results of Sections 3.1.5 and 3.2.3 and formulate both necessary and sufficient optimality conditions for w^0 (see Theorems 3.9 and 3.16, respectively). Indeed, let u^i, $i = 1, \ldots, m$, be the vertices of the polyhedron U, i.e., $\mathcal{U} = \{u^1, \ldots, u^m\}$. Let $\mathcal{Q}(u^i)$, $i = 1, \ldots, m$, be disjoint open neighborhoods of the vertices $u^i \in \mathcal{U}$. Set

$$\mathcal{Q}_g = \bigcup_{i=1}^{m} \mathcal{Q}(u^i).$$

Define the function $g(u) : \mathcal{Q}_g \to \mathbb{R}^{d(u)}$ as follows: On each set $\mathcal{Q}(u^i) \subset \mathcal{Q}_g$, $i = 1, \ldots, m$, let $g(u) = u - u^i$. Then $g(u)$ is a function of class C^∞ on $\mathcal{Q}_g$, specifying the set of vertices of U, i.e., $\mathcal{U} = \{u \in \mathcal{Q}_g \mid g(u) = 0\}$. Moreover, $g'(u) = I$ for all $u \in \mathcal{Q}_g$, where I is the identity matrix of order $d(u)$. Hence, the full-rank condition (3.3) is fulfilled. (This very simple but somewhat unexpected way of using equality constraints $g(u) = 0$ in the problem with a constraint on the control specified by a polyhedron U is due to Milyutin.) Thus, the problem (5.61), (5.62) can be represented as

$$J(p) \to \min, \quad F(p) \le 0, \quad K(p) = 0, \quad p \in \mathcal{P}, \tag{5.63}$$

$$\dot{x}(t) = a(t,x,v) + B(t,x,v)u, \tag{5.64}$$

$$g(u) = 0, \quad u \in \mathcal{Q}_g, \quad (t, x(t), v(t)) \in \mathcal{Q}. \tag{5.65}$$

Let us formulate the quadratic conditions of Sections 3.1.5 and 3.2.3 for the point w^0 in this problem. Put $l = \alpha_0 J + \alpha F + \beta K$, $H = \psi f$, $\bar{H} = H + \nu g$, where $f = a + Bu$. The set M_0 (cf. (3.12)–(3.15)) consists of tuples

$$\lambda = (\alpha_0, \alpha, \beta, \psi, \nu) \tag{5.66}$$

such that

$$\begin{aligned}
&\alpha_0 \in \mathbb{R}^1, \quad \alpha \in \mathbb{R}^{d(F)}, \quad \beta \in \mathbb{R}^{d(K)}, \\
&\psi \in W^{1,1}(\Delta, \mathbb{R}^{d(x)}), \quad \nu \in L^\infty(\Delta, \mathbb{R}^{d(u)}), \\
&\alpha_0 \ge 0, \quad \alpha \ge 0, \quad \alpha F(p^0) = 0, \quad \alpha_0 + |\alpha| + |\beta| = 1, \\
&-\dot{\psi} = \bar{H}_x, \quad \psi(t_0) = -l_{x_0}, \quad \psi(t_f) = l_{x_f}, \\
&H(t, x^0(t), u, v, \psi(t)) \ge H(t, x^0(t), u^0(t), v^0(t), \psi(t)) \\
&\forall\, t \in \Delta \setminus \Theta, \quad u \in \mathcal{U}, \quad v \in \mathbb{R}^{d(v)} \ \text{ such that } (t, x^0(t), v) \in \mathcal{Q}.
\end{aligned} \tag{5.67}$$

The last inequality implies the conditions of a local minimum principle with respect to u and v:

$$\psi f_u(t,x^0(t),u^0(t),v^0(t))+\nu(t)g'(u^0(t))=0, \quad \psi f_v(t,x^0(t),u^0(t),v^0(t))=0.$$

Since $g'(u)=I$, the first equality uniquely determines the multiplier $\nu(t)$.

Note that $\bar{H}_x = H_x$. Therefore conditions (5.67) are equivalent to the minimum principle conditions (5.7)–(5.10). More precisely, the linear projection

$$(\alpha_0,\alpha,\beta,\psi,\nu)\to(\alpha_0,\alpha,\beta,\psi)$$

yields a one-to-one correspondence between the elements of the set (5.67) and the elements of the set (5.7)–(5.10).

Consider the critical cone $\mathcal{K}$ for the point w^0 (see Definition 3.5). To the constraint $g(u)=0$, $u\in\mathcal{Q}_g$ there corresponds the condition $g'(u^0)\bar{u}=0$. But $g'=I$. Hence $\bar{u}=0$. This condition implies that the critical cone $\mathcal{K}$ can be identified with the set of triples $\bar{z}=(\bar{\xi},\bar{x},\bar{v})\in\mathcal{Z}_2(\Theta)$ such that conditions (5.14)–(5.16) are fulfilled.

The definition of the set M_0^+ in Section 5.1.6 corresponds to the definition of this set in Section 3.2.3. The same is true for the set $\mathrm{Leg}_+(M_0^+)$ (again, due to the equality $g'=I$).

Further, for the point w^0 we write the quadratic form Ω (see formula (3.57)), where we set $\bar{u}=0$. Since $\bar{H}_x=H_x$ and $\bar{H}_\psi=H_\psi$, we have $[\bar{H}_x]^k=[H_x]^k$, $D^k(\bar{H})=D^k(H)$, $k=1,\dots,s$. Moreover, $\bar{H}_{xx}=H_{xx}$, $\bar{H}_{xv}=H_{xv}$, $\bar{H}_{vv}=H_{vv}$. Combined with condition $\bar{u}=0$ this implies

$$\langle\bar{H}_{ww}\bar{w},\bar{w}\rangle=\langle H_{xx}\bar{x},\bar{x}\rangle+2\langle H_{xv}\bar{v},\bar{x}\rangle+\langle H_{vv}\bar{v},\bar{v}\rangle. \tag{5.68}$$

Thus, in view of condition $\bar{u}=0$ the quadratic form Ω becomes defined by formula (5.17). Now, Theorem 5.5 easily follows from Theorem 5.10, and similarly Theorem 5.8 becomes a simple consequence of Theorem 5.16.

5.2 Quadratic Optimality Conditions in the Problem on a Variable Time Interval

5.2.1 Optimal Control Problem on a Variable Time Interval

Let $x(t)\in\mathbb{R}^{d(x)}$ denote the state variable, and let $u(t)\in\mathbb{R}^{d(u)}$, $v(t)\in\mathbb{R}^{d(v)}$ be the two types of control variables in the time interval $t\in[t_0,t_f]$ with a nonfixed initial time t_0 and final time t_f. The following optimal control problem (5.69)–(5.72) will be referred to as the *general problem linear in a part of controls*:

$$\text{Minimize}\quad \mathcal{J}(t_0,t_f,x(\cdot),u(\cdot),v(\cdot))=J(t_0,x(t_0),t_f,x(t_f)) \tag{5.69}$$

subject to the constraints

$$\dot{x}(t)=f(t,x(t),u(t),v(t)),\quad u(t)\in U,\quad (t,x(t),v(t))\in\mathcal{Q}, \tag{5.70}$$

$$F(t_0,x(t_0),t_f,x(t_f))\le 0,\quad K(t_0,x(t_0),t_f,x(t_f))=0,\\ (t_0,x(t_0),t_f,x(t_f))\in\mathcal{P}. \tag{5.71}$$

The control variable u appears linearly in the system dynamics,

$$f(t,x,u,v) = a(t,x,v) + B(t,x,v)u, \tag{5.72}$$

whereas the control variable v appears nonlinearly in the dynamics. Here, F, K, and a are column vector functions, B is a $d(x) \times d(u)$ matrix function, $\mathcal{P} \subset \mathbb{R}^{2+2d(x)}$ and $\mathcal{Q} \subset \mathbb{R}^{1+d(x)+d(v)}$ are open sets, and $U \subset \mathbb{R}^{d(u)}$ is a convex polyhedron. The functions J, F, and K are assumed to be twice continuously differentiable on $\mathcal{P}$ and the functions a and B are twice continuously differentiable on $\mathcal{Q}$. The dimensions of F and K are denoted by $d(F)$ and $d(K)$. By $\Delta = [t_0, t_f]$ we denote the interval of control. We shall use the abbreviations $x_0 = x(t_0)$, $x_f = x(t_f)$, and $p = (t_0, x_0, t_f, x_f)$. A process $\mathcal{T} = \{(x(t), u(t), v(t)) \mid t \in [t_0, t_f]\}$ is said to be *admissible*, if $x(\cdot)$ is absolutely continuous, $u(\cdot)$, $v(\cdot)$ are measurable bounded on $\Delta = [t_0, t_f]$, and the triple of functions $(x(t), u(t), v(t))$ together with the endpoints $p = (t_0, x(t_0), t_f, x(t_f))$ satisfies the constraints (5.70) and (5.71).

Definition 5.17. The process $\mathcal{T}$ affords a *Pontryagin minimum* if there is no sequence of admissible processes $\mathcal{T}^n = \{(x^n(t), u^n(t), v^n(t)) \mid t \in [t_0^n, t_f^n]\}$, $n = 1, 2, \ldots$, such that the following properties hold with $\Delta^n = [t_0^n, t_f^n]$:

(a) $\mathcal{J}(\mathcal{T}^n) < \mathcal{J}(\mathcal{T})$ for all n and $t_0^n \to t_0$, $t_f^n \to t_f$ for $n \to \infty$;

(b) $\max_{\Delta^n \cap \Delta} |x^n(t) - x(t)| \to 0$ for $n \to \infty$;

(c) $\int_{\Delta^n \cap \Delta} |u^n(t) - u(t)|\, dt \to 0$, $\int_{\Delta^n \cap \Delta} |v^n(t) - v(t)|\, dt \to 0$ for $n \to \infty$;

(d) there exists a compact set $\mathcal{C} \subset \mathcal{Q}$ (which depends on the choice of the sequence) such that for all sufficiently large n, we have $(t, x^n(t), v^n(t)) \in \mathcal{C}$ a.e. on Δ^n.

For convenience, let us formulate an equivalent definition of the Pontryagin minimum.

Definition 5.18. The process $\mathcal{T}$ affords a *Pontryagin minimum* if for each compact set $\mathcal{C} \subset \mathcal{Q}$ there exists $\varepsilon > 0$ such that $\mathcal{J}(\tilde{\mathcal{T}}) \geq \mathcal{J}(\mathcal{T})$ for all admissible processes $\tilde{\mathcal{T}} = \{(\tilde{x}(t), \tilde{u}(t), \tilde{v}(t)) \mid t \in [\tilde{t}_0, \tilde{t}_f]\}$ such that

(a) $|\tilde{t}_0 - t_0| < \varepsilon$, $|\tilde{t}_f - t_f| < \varepsilon$;

(b) $\max_{\tilde{\Delta} \cap \Delta} |\tilde{x}(t) - x(t)| < \varepsilon$, where $\tilde{\Delta} = [\tilde{t}_0, \tilde{t}_f]$;

(c) $\int_{\tilde{\Delta} \cap \Delta} |\tilde{u}(t) - u(t)|\, dt < \varepsilon$; $\int_{\tilde{\Delta} \cap \Delta} |\tilde{v}(t) - v(t)|\, dt < \varepsilon$;

(d) $(t, \tilde{x}(t), \tilde{v}(t)) \in \mathcal{C}$ a.e. on $\tilde{\Delta}$.

5.2.2 First-Order Necessary Optimality Conditions

Let $\mathcal{T} = \{(x(t), u(t), v(t)) \mid t \in [t_0, t_f]\}$ be a fixed admissible process such that the control $u(t)$ is a piecewise constant function taking all its values in the vertices of the polyhedron U, and the control $v(t)$ is a Lipschitz continuous function on the interval $\Delta = [t_0, t_f]$. Denote by $\Theta = \{t_1, \ldots, t_s\}$, $t_0 < t_1 < \cdots < t_s < t_f$ the finite set of all discontinuity points (jump points) of the control $u(t)$. Then $\dot{x}(t)$ is a piecewise continuous function whose points of discontinuity belong to Θ, and hence $x(t)$ is a piecewise smooth function on Δ.

Let us formulate a first-order necessary condition for optimality of the process $\mathcal{T}$ in the form of the Pontryagin minimum principle. As in Section 5.1.2, we introduce the

Pontryagin function (or Hamiltonian)

$$H(t,x,u,v,\psi) = \psi f(t,x,u,v) = \psi a(t,x,v) + \psi B(t,x,v)u, \tag{5.73}$$

where ψ is a row vector of dimension $d(\psi) = d(x)$, while x, u, f, F, and K are column vectors; the factor of the control u in the Pontryagin function is called the *switching function for the u-component*

$$\phi(t,x,v,\psi) = \psi B(t,x,v) \tag{5.74}$$

which is a row vector of dimension $d(u)$. We also introduce the endpoint Lagrange function

$$l(p,\alpha_0,\alpha,\beta) = \alpha_0 J(p) + \alpha F(p) + \beta K(p), \quad p = (t_0,x_0,t_f,x_f),$$

where α and β are row vectors with $d(\alpha) = d(F)$ and $d(\beta) = d(K)$, and α_0 is a number. We introduce a tuple of Lagrange multipliers $\lambda = (\alpha_0,\alpha,\beta,\psi(\cdot),\psi_0(\cdot))$ such that $\psi(\cdot) : \Delta \to (\mathbb{R}^{d(x)})^*$, $\psi_0(\cdot) : \Delta \to \mathbb{R}^1$ are continuous on Δ and continuously differentiable on each interval of the set $\Delta \setminus \Theta$. As usual, we denote first- or second-order partial derivatives by subscripts referring to the variables.

Denote by M_0 the set of the normalized tuples λ satisfying the minimum principle conditions for the process $\mathcal{T}$:

$$\alpha_0 \geq 0, \quad \alpha \geq 0, \quad \alpha F(p) = 0, \quad \alpha_0 + \sum_{i=1}^{d(F)} \alpha_i + \sum_{j=1}^{d(K)} |\beta_j| = 1, \tag{5.75}$$

$$\dot{\psi} = -H_x, \quad \dot{\psi}_0 = -H_t \quad \forall\, t \in \Delta \setminus \Theta, \tag{5.76}$$

$$\psi(t_0) = -l_{x_0}, \quad \psi(t_f) = l_{x_f}, \quad \psi_0(t_0) = -l_{t_0}, \quad \psi_0(t_f) = l_{t_f}, \tag{5.77}$$

$$H(t,x(t),u,v,\psi(t)) \geq H(t,x(t),u(t),v(t),\psi(t))$$
$$\forall\, t \in \Delta \setminus \Theta,\ u \in U,\ v \in \mathbb{R}^{d(v)} \quad \text{such that } (t,x(t),v) \in \mathcal{Q}, \tag{5.78}$$

$$H(t,x(t),u(t),v(t),\psi(t)) + \psi_0(t) = 0 \quad \forall\, t \in \Delta \setminus \Theta. \tag{5.79}$$

The derivatives l_{x_0} and l_{x_f} are taken at the point $(p,\alpha_0,\alpha,\beta)$, where

$$p = (t_0, x(t_0), t_f, x(t_f)),$$

while the derivatives H_x, H_t are evaluated at the point $(t,x(t),u(t),v(t),\psi(t))$ for $t \in \Delta \setminus \Theta$. The condition $M_0 \neq \emptyset$ constitutes the first-order necessary condition for a Pontryagin minimum of the process $\mathcal{T}$ which is the Pontryagin minimum principle.

Theorem 5.19. *If the process $\mathcal{T}$ affords a Pontryagin minimum, then the set M_0 is nonempty. The set M_0 is a finite-dimensional compact set and the projector $\lambda \mapsto (\alpha_0,\alpha,\beta)$ is injective on M_0.*

Again we use the simple abbreviation (t) for indicating all arguments

$$(t, x(t), u(t), v(t), \psi(t)).$$

Let $\lambda = (\alpha_0,\alpha,\beta,\psi(\cdot),\psi_0(\cdot)) \in M_0$. From condition (5.79), it follows that $H(t)$ is a continuous function. In particular, we have $[H]^k = H^{k+} - H^{k-} = 0$ for each $t_k \in \Theta$, where

$$H^{k-} := H(t_k, x(t_k), u(t_k-), v(t_k), \psi(t_k)), \quad H^{k+} := H(t_k, x(t_k), u(t_k+), v(t_k), \psi(t_k)).$$

We denote by H^k the common value of H^{k-} and H^{k+}. For $\lambda \in M_0$ and $t_k \in \Theta$ we consider the function

$$(\Delta_k H)(t) = H(t,x(t),u^{k+},v(t_k),\psi(t)) - H(t,x(t),u^{k-},v(t_k),\psi(t)) \\ = \phi(t,x(t),v(t_k),\psi(t))[u]^k. \tag{5.80}$$

For this function, Propositions 5.2 and 5.3 hold, so that for each $\lambda \in M_0$ we have $D^k(H) \geq 0$, $k = 1,\ldots,s$, where

$$D^k(H) := -\frac{d}{dt}(\Delta_k H)(t_k) = -\dot{\phi}(t_k\pm)[u]^k \\ = -H_x^{k+}H_\psi^{k-} + H_x^{k-}H_\psi^{k+} - [H_t]^k = \dot{\psi}^{k+}\dot{x}^{k-} - \dot{\psi}^{k-}\dot{x}^{k+} + [\psi_0]^k,$$

where H_x^{k-} and H_x^{k+} are the left- and right-hand values of the function $H_x(t)$ at t_k, respectively, $[H_t]^k$ is the jump of the function $H_t(t)$ at t_k, etc.

5.2.3 Bounded Strong Minimum

As in the case of a fixed time interval we give the following definitions. The state variable x_i, i.e., the ith component of the state vector x, is called *unessential* if the function f does not depend on x_i and if the functions F, J, and K are affine in $x_{i0} = x_i(t_0)$ and $x_{i1} = x_i(t_f)$. We denote by $\underline{x}$ the vector of all essential components of state vector x.

Definition 5.20. We say that the process $\mathcal{T}$ affords a *bounded strong minimum* if there is no sequence of admissible processes $\mathcal{T}^n = \{(x^n(t),u^n(t),v^n(t)) \mid t \in [t_0^n,t_f^n]\}$, $n = 1,2,\ldots$, such that

(a) $\mathcal{J}(\mathcal{T}^n) < \mathcal{J}(\mathcal{T})$;

(b) $t_0^n \to t_0$, $t_f^n \to t_f$, $x^n(t_0) \to x(t_0)$ $(n \to \infty)$;

(c) $\max_{\Delta^n \cap \Delta} |\underline{x}^n(t) - \underline{x}(t)| \to 0$ $(n \to \infty)$, where $\Delta^n = [t_0^n, t_f^n]$;

(d) there exists a compact set $\mathcal{C} \subset \mathcal{Q}$ (which depends on the choice of the sequence) such that for all sufficiently large n we have $(t,x^n(t),v^n(t)) \in \mathcal{C}$ a.e. on Δ^n.

An equivalent definition has the following form.

Definition 5.21. The process $\mathcal{T}$ affords a *bounded strong minimum* if for each compact set $\mathcal{C} \subset \mathcal{Q}$ there exists $\varepsilon > 0$ such that $\mathcal{J}(\tilde{\mathcal{T}}) \geq \mathcal{J}(\mathcal{T})$ for all admissible processes $\tilde{\mathcal{T}} = \{(\tilde{x}(t),\tilde{u}(t),\tilde{v}(t)) \mid t \in [\tilde{t}_0,\tilde{t}_f]\}$ such that

(a) $|\tilde{t}_0 - t_0| < \varepsilon$, $|\tilde{t}_f - t_f| < \varepsilon$, $|\tilde{x}(t_0) - x(t_0)| < \varepsilon$;

(b) $\max_{\tilde{\Delta} \cap \Delta} |\underline{\tilde{x}}(t) - \underline{x}(t)| < \varepsilon$, where $\tilde{\Delta} = [\tilde{t}_0,\tilde{t}_f]$;

(c) $(t,\tilde{x}(t),\tilde{v}(t)) \in \mathcal{C}$ a.e. on $\tilde{\Delta}$.

The *strict* bounded strong minimum is defined in a similar way, with the nonstrict inequality $\mathcal{J}(\tilde{\mathcal{T}}) \geq \mathcal{J}(\mathcal{T})$ replaced by the strict one and the process $\tilde{\mathcal{T}}$ required to be different from $\mathcal{T}$. Below, we shall formulate a quadratic necessary optimality condition of a Pontryagin minimum (Definition 5.17) for given control process $\mathcal{T}$. A strengthening of this quadratic condition yields a quadratic sufficient condition of a bounded strong minimum (Definition 5.20).

5.2.4 Critical Cone

For a given process $\mathcal{T}$ we introduce the space $\mathcal{Z}_2(\Theta)$ and the *critical cone* $\mathcal{K} \subset \mathcal{Z}_2(\Theta)$. Let $\bar{z} = (\bar{t}_0, \bar{t}_f, \bar{\xi}, \bar{x}, \bar{v})$, where $\bar{t}_0, \bar{t}_f \in \mathbb{R}^1$, $\bar{\xi} \in \mathbb{R}^s$, $\bar{x} \in P_\Theta W^{1,2}(\Delta, \mathbb{R}^{d(x)})$, $\bar{v} \in L^2(\Delta, \mathbb{R}^{d(v)})$. Thus,

$$\bar{z} \in \mathcal{Z}_2(\Theta) := \mathbb{R}^2 \times \mathbb{R}^s \times P_\Theta W^{1,2}(\Delta, \mathbb{R}^{d(x)}) \times L^2(\Delta, \mathbb{R}^{d(v)}).$$

For each $\bar{z}$, we set

$$\bar{\bar{x}}_0 = \bar{x}(t_0) + \bar{t}_0 \dot{x}(t_0), \quad \bar{\bar{x}}_f = \bar{x}(t_f) + \bar{t}_f \dot{x}(t_f), \quad \bar{\bar{p}} = (\bar{t}_0, \bar{\bar{x}}_0, \bar{t}_f, \bar{\bar{x}}_f). \tag{5.81}$$

The vector $\bar{\bar{p}}$ is considered a column vector. Note that $\bar{t}_0 = 0$, respectively, $\bar{t}_f = 0$, holds for a *fixed* initial time t_0, respectively, final time t_f. Denote by $I_F(p) = \{i \in \{1, \ldots, d(F)\} \mid F_i(p) = 0\}$ the set of indices of all active endpoint inequalities $F_i(p) \le 0$ at the point $p = (t_0, x(t_0), t_f, x(t_f))$. Denote by $\mathcal{K}$ the set of all $\bar{z} \in \mathcal{Z}_2(\Theta)$ satisfying the following conditions:

$$J'(p)\bar{\bar{p}} \le 0, \quad F_i'(p)\bar{\bar{p}} \le 0 \ \forall\, i \in I_F(p), \quad K'(p)\bar{\bar{p}} = 0, \tag{5.82}$$

$$\dot{\bar{x}}(t) = f_x(t, x(t), u(t), v(t))\bar{x}(t) + f_v(t, x(t), u(t), v(t))\bar{v}(t), \tag{5.83}$$

$$[\bar{x}]^k = [\dot{x}]^k \bar{\xi}_k, \quad k = 1, \ldots, s, \tag{5.84}$$

where $p = (t_0, x(t_0), t_f, x(t_f))$, $[\dot{x}] = \dot{x}(t_k+) - \dot{x}(t_k-)$. It is obvious that $\mathcal{K}$ is a convex cone with finitely many faces in the space $\mathcal{Z}_2(\Theta)$. The cone $\mathcal{K}$ is called the *critical cone.*

5.2.5 Necessary Quadratic Optimality Conditions

Let us introduce a quadratic form on the critical cone $\mathcal{K}$ defined by the conditions (5.82)–(5.84). For each $\lambda \in M_0$ and $\bar{z} \in \mathcal{K}$, we set

$$\begin{aligned} \Omega(\lambda, \bar{z}) = \omega_e(\lambda, \bar{z}) &+ \sum_{k=1}^{s} \left(D^k(H)\bar{\xi}_k^2 - [\dot{\psi}]^k \bar{x}_{\mathrm{av}}^k \bar{\xi}_k\right) \\ &+ \int_{t_0}^{t_f} \big(\langle H_{xx}(t)\bar{x}(t), \bar{x}(t)\rangle + 2\langle H_{xv}(t)\bar{v}(t), \bar{x}(t)\rangle \\ &\qquad + \langle H_{vv}(t)\bar{v}(t), \bar{v}(t)\rangle\big)\, dt, \end{aligned} \tag{5.85}$$

where

$$\begin{aligned} \omega_e(\lambda, \bar{z}) = \langle l_{pp}\bar{\bar{p}}, \bar{\bar{p}}\rangle &- 2\dot{\psi}(t_f)\bar{x}(t_f)\bar{t}_f - \Big(\dot{\psi}(t_f)\dot{x}(t_f) + \dot{\psi}_0(t_f)\Big)\bar{t}_f^2 \\ &+ 2\dot{\psi}(t_0)\bar{x}(t_0)\bar{t}_0 + \Big(\dot{\psi}(t_0)\dot{x}(t_0) + \dot{\psi}_0(t_0)\Big)\bar{t}_0^2, \end{aligned} \tag{5.86}$$

$$l_{pp} = l_{pp}(p, \alpha_0, \alpha, \beta, p), \quad p = (t_0, x(t_0), t_f, x(t_f)), \quad \bar{x}_{\mathrm{av}}^k = \frac{1}{2}(\bar{x}^{k-} + \bar{x}^{k+}),$$

$$H_{xx}(t) = H_{xx}(t, x(t), u(t), v(t), \psi(t)), \quad \text{etc.}$$

Note that for a problem on a fixed time interval $[t_0, t_f]$ we have $\bar{t}_0 = \bar{t}_f = 0$ and, hence, the quadratic form (5.86) reduces to $\langle l_{pp}\bar{\bar{p}}, \bar{\bar{p}}\rangle$. The following theorem gives the main second-order necessary condition of optimality.

Theorem 5.22. *If the process $\mathcal{T}$ affords a Pontryagin minimum, then the following Condition $\mathfrak{A}$ holds: The set M_0 is nonempty and*

$$\max_{\lambda \in M_0} \Omega(\lambda, \bar{z}) \geq 0 \quad \forall\, \bar{z} \in \mathcal{K}.$$

We call Condition $\mathfrak{A}$ the necessary quadratic condition.

5.2.6 Sufficient Quadratic Optimality Conditions

A natural strengthening of the necessary Condition $\mathfrak{A}$ turns out to be a sufficient optimality condition not only for a Pontryagin minimum, but also for a bounded strong minimum; cf. Definition 5.20. Denote by M_0^+ the set of all $\lambda \in M_0$ satisfying the following conditions:

(a) $H(t,x(t),u,v,\psi(t)) > H(t,x(t),u(t),v(t),\psi(t))$ for all $t \in \Delta \setminus \Theta$, $u \in U$, $v \in \mathbb{R}^{d(v)}$ such that $(t,x(t),v) \in \mathcal{Q}$ and $(u,v) \neq (u(t),v(t))$;

(b) $H(t_k,x(t_k),u,v,\psi(t_k)) > H^k$ for all $t_k \in \Theta$, $u \in U$, $v \in \mathbb{R}^{d(v)}$ such that $(t_k,x(t_k),v) \in \mathcal{Q}$, $(u,v) \neq (u(t_k-),v(t_k))$, $(u,v) \neq (u(t_k+),v(t_k))$, where $H^k := H^{k-} = H^{k+}$.

Definition 5.23. An element $\lambda \in M_0$ is said to be *strictly Legendre* if the following conditions are satisfied:

(a) For each $t \in \Delta \setminus \Theta$ the quadratic form $\langle H_{vv}(t,x(t),u(t),v(t),\psi(t))\bar{v},\bar{v}\rangle$ is positive definite in $\mathbb{R}^{d(v)}$;

(b) for each $t_k \in \Theta$ the quadratic form $\langle H_{vv}(t_k,x(t_k),u(t_k-),v(t_k),\psi(t_k))\bar{v},\bar{v}\rangle$ is positive definite in $\mathbb{R}^{d(v)}$;

(c) for each $t_k \in \Theta$ the quadratic form $\langle H_{vv}(t_k,x(t_k),u(t_k+),v(t_k),\psi(t_k))\bar{v},\bar{v}\rangle$ is positive definite in $\mathbb{R}^{d(v)}$;

(d) $D^k(H) > 0$ for all $t_k \in \Theta$.

Denote by $\mathrm{Leg}_+(M_0^+)$ the set of all strictly Legendrian elements $\lambda \in M_0^+$ and set

$$\bar{\gamma}(\bar{z}) = \bar{t}_0^2 + \bar{t}_f^2 + \langle \bar{\xi}, \bar{\xi} \rangle + \langle \bar{x}(t_0), \bar{x}(t_0) \rangle + \int_{t_0}^{t_f} \langle \bar{v}(t), \bar{v}(t) \rangle \, dt.$$

Theorem 5.24. *Let the following Condition $\mathfrak{B}$ be fulfilled for the process Π:*

(a) *The set $\mathrm{Leg}_+(M_0^+)$ is nonempty;*

(b) *there exists a nonempty compact set $M \subset \mathrm{Leg}_+(M_0^+)$ and a number $C > 0$ such that $\max_{\lambda \in M} \Omega(\lambda, \bar{z}) \geq C\bar{\gamma}(\bar{z})$ for all $\bar{z} \in \mathcal{K}$.*

Then $\mathcal{T}$ is a strict bounded strong minimum.

If the set $\mathrm{Leg}_+(M_0^+)$ is nonempty and $\mathcal{K} = \{0\}$, then (b) is fulfilled automatically. This is a first-order sufficient optimality condition of a strict bounded strong minimum. Let us emphasize that there is no gap between the necessary Condition $\mathfrak{A}$ and the sufficient Condition $\mathfrak{B}$.

5.2.7 Proofs

Let us consider the following problem on a variable time interval which is similar to the general problem (5.69)–(5.71):

$$\text{Minimize} \qquad \mathcal{J}(t_0,t_f,x(\cdot),u(\cdot),v(\cdot)) = J(t_0,x(t_0),t_f,x(t_f)) \tag{5.87}$$

subject to the constraints

$$\dot{x}(t) = f(t,x(t),u(t),v(t)), \quad u(t) \in \mathcal{U},\ (t,x(t),v(t)) \in \mathcal{Q}, \tag{5.88}$$

$$\begin{aligned} &F(t_0,x(t_0),t_f,x(t_f)) \le 0, \quad K(t_0,x(t_0),t_f,x(t_f)) = 0, \\ &(t_0,x(t_0),t_f,x(t_f)) \in \mathcal{P}. \end{aligned} \tag{5.89}$$

The control variable u appears linearly in the system dynamics,

$$f(t,x,u,v) = a(t,x,v) + B(t,x,v)u, \tag{5.90}$$

whereas the control variable v appears nonlinearly. Here, F, K, and a are column vector functions, B is a $d(x) \times d(u)$ matrix function, $\mathcal{P} \subset \mathbb{R}^{2+2d(x)}$ and $\mathcal{Q} \subset \mathbb{R}^{1+d(x)+d(v)}$ are open sets, and $\mathcal{U} \subset \mathbb{R}^{d(u)}$ is a compact set. The functions J, F, and K are assumed to be twice continuously differentiable on $\mathcal{P}$ and the functions a and B are twice continuously differentiable on $\mathcal{Q}$. The dimensions of F and K are denoted by $d(F)$ and $d(K)$. By $\Delta = [t_0,t_f]$ we denote the interval of control. The compact set $\mathcal{U}$ is specified by

$$\mathcal{U} = \{u \in \mathcal{Q}_g \mid g(u) = 0\}, \tag{5.91}$$

where $\mathcal{Q}_g \subset \mathbb{R}^{d(u)}$ is an open set and $g : \mathcal{Q}_g \to \mathbb{R}^{d(g)}$ is a twice continuously differentiable function satisfying the full-rank condition

$$\operatorname{rank}\, g_u(u) = d(g) \tag{5.92}$$

for all $u \in \mathcal{Q}_g$ such that $g(u) = 0$.

We refer to the problem (5.87)–(5.89) as the problem A. Along with this problem we treat its convexification $\operatorname{co} A$ with respect to u in which the constraint $u \in \mathcal{U}$ is replaced by the constraint $u \in \operatorname{co}\mathcal{U}$:

$$\dot{x} = a(t,x,v) + B(t,x,v)u, \quad u \in \operatorname{co}\mathcal{U}, \quad (t,x,v) \in \mathcal{Q}. \tag{5.93}$$

Thus $\operatorname{co} A$ is the problem (5.87), (5.89), (5.93).

Let $\mathcal{T} = (x(t),u(t),v(t) \mid t \in [t_0,t_f])$ be an admissible trajectory in the problem A such that $u(t)$ is a piecewise Lipschitz continuous function on the interval $\Delta = [t_0,t_f]$, with the set of discontinuity points $\Theta = \{t_1,\dots,t_s\}$, and the control $v(t)$ is Lipschitz continuous on the same interval. For the trajectory $\mathcal{T}$ we deal with the same question as in Section 5.1.7: What is the relationship between quadratic conditions in the problems A and $\operatorname{co} A$? As in Section 5.1.7, for necessary quadratic conditions this question is simple to answer: A Pontryagin minimum in the problem $\operatorname{co} A$ implies a Pontryagin minimum in the problem A, and hence Theorem 3.31 implies the following assertion.

Theorem 5.25. *Condition $\mathfrak{A}$ for a trajectory $\mathcal{T}$ in the problem A is a necessary condition for a Pontryagin minimum at this trajectory in the problem* $\operatorname{co} A$.

Consider now the same question for sufficient quadratic conditions. It can be solved with the aid of Theorem 5.16 obtained for the problem Z on a fixed time interval, but first we have to make a simple time change. Namely, with the admissible trajectory $\mathcal{T}$ in the problem A we associate the trajectory

$$\mathcal{T}^\tau = \big(z(\tau), t(\tau), x(\tau), u(\tau), v(\tau) \mid \tau \in [\tau_0, \tau_f]\big),$$

where $\tau_0 = t_0$, $\tau_f = t_f$, $t(\tau) \equiv \tau$, $z(\tau) \equiv 1$. This is an admissible trajectory in the problem A^τ specified by conditions

$$\mathcal{J}(\mathcal{T}^\tau) := J(t(\tau_0), x(\tau_0), t(\tau_f), x(\tau_f)) \to \min \tag{5.94}$$

subject to the constraints

$$F(t(\tau_0), x(\tau_0), t(\tau_f), x(\tau_f)) \le 0, \;\; K(t(\tau_0), x(\tau_0), t(\tau_f), x(\tau_f)) = 0, \tag{5.95}$$

$$\frac{dx(\tau)}{d\tau} = z(\tau)\Big(a(t(\tau), x(\tau), v(\tau)) + B(t(\tau), x(\tau), v(\tau))u(\tau)\Big), \tag{5.96}$$

$$\frac{dt(\tau)}{d\tau} = z(\tau), \tag{5.97}$$

$$\frac{dz(\tau)}{d\tau} = 0, \tag{5.98}$$

respectively,[4]

$$(t(\tau_0), x(\tau_0), t(\tau_f), x(\tau_f)) \in \mathcal{P}, \quad (t(\tau), x(\tau)) \in \mathcal{Q}, \quad u(\tau) \in \mathcal{U}. \tag{5.99}$$

The interval $[\tau_0, \tau_f]$ in the problem A^τ is fixed.

Consider also the problem $\operatorname{co} A^\tau$ differing from A^τ by the constraint $u \in \mathcal{U}$ replaced with the constraint $u \in \operatorname{co} \mathcal{U}$. We have the following chain of implications:

Condition $\mathfrak{B}$ for the trajectory $\mathcal{T}$ in the problem A.
$\Longrightarrow$ Condition $\mathfrak{B}$ for the trajectory $\mathcal{T}^\tau$ in the problem A^τ.
$\Longrightarrow$ A strict bounded strong minimum is attained on the trajectory $\mathcal{T}^\tau$ in the problem $\operatorname{co} A^\tau$.
$\Longrightarrow$ A strict bounded strong minimum is attained on the trajectory $\mathcal{T}$ in the problem $\operatorname{co} A$.

The first implication was proved in Section 3.3.4 for problems P and P^τ which are more general than A and A^τ, respectively, the second follows from Theorem 5.8, and the third is readily verified. Thus we obtain the following theorem.

Theorem 5.26. *Condition $\mathfrak{B}$ for an admissible trajectory $\mathcal{T}$ in the problem A is a sufficient condition for a strict strong minimum in the problem* $\operatorname{co} A$.

Now, recall that the representation of the set of vertices of polyhedron U in the form of equality-type constraint $g(u) = 0$, $u \in Q_g$ allowed us to consider the main problem (5.1)–(5.4) as a special case of the problem $\operatorname{co} Z$ (5.18), (5.19), (5.23) and thus to obtain both necessary Condition $\mathfrak{A}$ and sufficient Condition $\mathfrak{B}$ in the main problem as the consequences of these conditions in problem $\operatorname{co} Z$; more precisely, we have shown that Theorems 5.5 and

[4] Note that the function $z(\tau)$ in problem A^τ corresponds to the function $v(\tau)$ in problem P^τ.

5.8 follow from Theorem 5.10 and 5.16, respectively (see the proofs of Theorems 5.5 and 5.8). Similarly, this representation allows us to consider the general problem on a variable time interval (5.69)–(5.72) as a special case of the problem co A and thus to obtain Theorems 5.22 and 5.24 as the consequences of Theorems 5.25 and 5.26, respectively.

5.3 Riccati Approach

The following question suggests itself from a numerical point of view: How does a numerical check of the quadratic sufficient optimality conditions in Theorem 5.8 look? For simplicity, we shall assume that

(a) the initial value $x(t_0)$ is fixed,

(b) there are no endpoint constraints of inequality type, and

(c) the time interval $\Delta = [t_0, t_f]$ is fixed.

Thus, we consider the following control problem:

$$\text{Minimize } J(x(t_f))$$

under the constraints

$$x(t_0) = x_0, \quad K(x(t_f)) = 0, \quad \dot{x} = f(t,x,u,v), \quad u \in U,$$

where

$$f(t,x,u,v) = a(t,x,v) + B(t,x,v)u,$$

$U \subset \mathbb{R}^{d(u)}$ is a convex polyhedron, and J, K, a, and B are C^2-functions.

Let $w = (x,u,v)$ be a fixed admissible process satisfying the assumptions of Section 5.1.2 (consequently, the function $u(t)$ is piecewise constant and the function $v(t)$ is continuous). We also assume, for this process, that the set M_0 is nonempty and there exists $\lambda \in M_0$ such that $\alpha_0 > 0$; let us fix this element λ. Here we set again $\Theta = \{t_1, t_2, \ldots, t_s\}$, where t_k denote the discontinuity points of the bang-bang control $u(t)$. Let $n = d(x)$.

5.3.1 Critical Cone $\mathcal{K}$ and Quadratic Form Ω

In the considered case, the critical cone is a subspace defined by the relations

$$\bar{x}(t_0) = 0, \quad K'(x(t_f))\bar{x}(t_f) = 0,$$
$$\dot{\bar{x}}(t) = f_x(t)\bar{x}(t) + f_v(t)\bar{v}(t), \quad [\bar{x}]^k = [\dot{x}]^k \bar{\xi}_k, \quad k = 1, \ldots, s.$$

These relations imply that $J'(x(t_f))\bar{x}(t_f) = 0$ since $\alpha_0 > 0$. The quadratic form is given by

$$\Omega(\lambda, \bar{z}) = \langle l_{x_f x_f}(x(t_f))\bar{x}_f, \bar{x}_f \rangle + \sum_{k=1}^{s} (D^k(H)\bar{\xi}_k^2 - 2[\dot{\psi}]^k \bar{x}_{\text{av}}^k \bar{\xi}_k)$$
$$+ \int_{t_0}^{t_f} \big(\langle H_{xx}(t)\bar{x}(t), \bar{x}(t) \rangle + 2\langle H_{xv}(t)\bar{v}(t), \bar{x}(t) \rangle + \langle H_{vv}(t)\bar{v}(t), \bar{v}(t) \rangle \big)\, dt,$$

where, by definition, $\bar{x}_f = \bar{x}(t_f)$. We assume that $D^k(H) > 0$, $k = 1, \ldots, s$, and the strengthened Legendre (SL) condition with respect to v is satisfied:

$$\langle H_{vv}(t)\bar{v}, \bar{v} \rangle \geq c\langle \bar{v}, \bar{v} \rangle \quad \forall\, \bar{v} \in \mathbb{R}^{d(v)}, \quad \forall\, t \in [t_0, t_f] \setminus \Theta \quad (c > 0).$$

5.3.2 Q-Transformation of Ω on $\mathcal{K}$

Let $Q(t)$ be a symmetric $n \times n$ matrix on $[t_0, t_f]$ with piecewise continuous entries which are absolutely continuous on each interval of the set $[t_0, t_f] \setminus \Theta$. For each $\bar{z} \in \mathcal{K}$ we obviously have

$$\int_{t_0}^{t_f} \frac{d}{dt} \langle Q\bar{x}, \bar{x} \rangle \, dt = \langle Q\bar{x}, \bar{x} \rangle \Big|_{t_0}^{t_f} - \sum_{k=1}^{s} [\langle Q\bar{x}, \bar{x} \rangle]^k, \tag{5.100}$$

where $[\langle Q\bar{x}, \bar{x} \rangle]^k$ is the jump of the function $\langle Q\bar{x}, \bar{x} \rangle$ at the point $t_k \in \Theta$. Using the equation $\dot{\bar{x}} = f_x \bar{x} + f_v \bar{v}$ and the initial condition $\bar{x}(t_0) = 0$, we obtain

$$\begin{aligned} &-\langle Q(t_f)\bar{x}_f, \bar{x}_f \rangle + \sum_{k=1}^{s} [\langle Q\bar{x}, \bar{x} \rangle]^k \\ &\quad + \int_{t_0}^{t_f} \big(\langle \dot{Q}\bar{x}, \bar{x} \rangle + \langle Q(f_x \bar{x} + f_v \bar{v}), \bar{x} \rangle + \langle Q\bar{x}, f_x \bar{x} + f_v \bar{v} \rangle \big) \, dt = 0. \end{aligned}$$

Adding this zero term to the form $\Omega(\lambda, \bar{z})$, we get

$$\begin{aligned} \Omega(\lambda, \bar{z}) = &\langle (l_{x_f x_f} - Q(t_f))\bar{x}_f, \bar{x}_f \rangle + \sum_{k=1}^{s} \Big(D^k(H)\bar{\xi}_k^2 - 2[\dot{\psi}]^k \bar{x}_{\mathrm{av}}^k \bar{\xi}_k + [\langle Q\bar{x}, \bar{x} \rangle]^k \Big) \\ &+ \int_{t_0}^{t_f} \big(\langle (H_{xx} + \dot{Q} + Qf_x + f_x^* Q)\bar{x}, \bar{x} \rangle + \langle (H_{xv} + Qf_v)\bar{v}, \bar{x} \rangle \\ &\qquad + \langle (H_{vx} + f_v^* Q)\bar{x}, \bar{v} \rangle + \langle H_{vv}(t)\bar{v}(t), \bar{v}(t) \rangle \big) \, dt. \end{aligned}$$

We call this formula the *Q-transformation of Ω on $\mathcal{K}$*.

5.3.3 Transformation of Ω on $\mathcal{K}$ to Perfect Squares

In order to transform the integral term in $\Omega(\lambda, \bar{z})$ into a *perfect square*, we assume that $Q(t)$ satisfies the following matrix Riccati equation (cf. equation (4.189)):

$$\dot{Q} + Qf_x + f_x^* Q + H_{xx} - (H_{xv} + Qf_v)H_{vv}^{-1}(H_{vx} + f_v^* Q) = 0.$$

Then the integral term in Ω can be written as

$$\int_{t_0}^{t_f} \langle H_{vv}^{-1}\bar{h}, \bar{h} \rangle \, dt, \quad \text{where} \quad \bar{h} = (H_{vx} + f_v^* Q)\bar{x} + H_{vv}\bar{v}.$$

As we know, the terms

$$\omega_k := D^k(H)\bar{\xi}_k^2 - 2[\dot{\psi}]^k \bar{x}_{\mathrm{av}}^k \bar{\xi}_k + [\langle Q\bar{x}, \bar{x} \rangle]^k$$

can also be transformed into perfect squares if the matrix $Q(t)$ satisfies a special *jump condition* at each point $t_k \in \Theta$. This jump condition was obtained in Chapter 4. Namely, for each $k = 1, \ldots, s$ put

$$Q^{k-} = Q(t_k-), \quad Q^{k+} = Q(t_k+), \quad [Q]^k = Q^{k+} - Q^{k-}, \tag{5.101}$$

$$q_{k-} = ([\dot{x}]^k)^* Q^{k-} - [\dot{\psi}]^k, \quad b_{k-} = D^k(H) - (q_{k-})[\dot{x}]^k, \tag{5.102}$$

where $[\dot{x}]^k$ is a column vector, while q_{k-}, $([\dot{x}]^k)^*$ and $[\dot{\psi}]^k$ are row vectors, and b_{k-} is a number. We shall assume that

$$b_{k-} > 0, \quad k = 1, \dots, s, \tag{5.103}$$

holds and that Q satisfies the jump conditions

$$[Q]^k = (b_{k-})^{-1}(q_{k-})^*(q_{k-}), \tag{5.104}$$

where (q_{k-}) is a row vector, $(q_{k-})^*$ is a column vector, and hence $(q_{k-})^*(q_{k-})$ is a symmetric $n \times n$ matrix. Then, as it was shown in Section 4.2,

$$\omega_k = (b_{k-})^{-1}((b_{k-})\bar{\xi}_k + (q_{k-})(\bar{x}^{k+}))^2 = (b_{k-})^{-1}(D^k(H)\bar{\xi}_k + (q_{k-})(\bar{x}^{k-}))^2. \tag{5.105}$$

Thus, we obtain the following transformation of the quadratic form $\Omega = \Omega(\lambda, \bar{z})$ to perfect squares on the critical cone $\mathcal{K}$:

$$\Omega = \langle \big(l_{x_f x_f} - Q(t_f)\big)\bar{x}_f, \bar{x}_f\rangle + \sum_{k=1}^{s}(b_{k-})^{-1}\big(D^k(H)\bar{\xi}_k + (q_{k-})(\bar{x}^{k-})\big)^2 + \int_{t_0}^{t_f} \langle H_{vv}^{-1}\bar{h}, \bar{h}\rangle\, dt,$$

where $\bar{h} = (H_{vx} + f_v^* Q)\bar{x} + H_{vv}\bar{v}$. In addition, let us assume that

$$\langle \big(l_{x_f x_f} - Q(t_f)\big)\bar{x}_f, \bar{x}_f\rangle \geq 0$$

for all $\bar{x}_f \in \mathbb{R}^{d(x)} \setminus \{0\}$ such that $K_{x_f}(x(t_f))\bar{x}_f = 0$. Then, obviously, $\Omega(\lambda, \bar{z}) \geq 0$ on $\mathcal{K}$. Now let us show that $\Omega(\lambda, \bar{z}) > 0$ for each nonzero element $\bar{z} \in \mathcal{K}$. This will imply that $\Omega(\lambda, \bar{z})$ is *positive definite* on the critical cone $\mathcal{K}$ since $\Omega(\lambda, \bar{z})$ is a Legendre quadratic form.

Assume that $\Omega(\lambda, \bar{z}) = 0$ for some element $\bar{z} \in \mathcal{K}$. Then, for this element, the following equations hold:

$$\bar{x}(t_0) = 0, \tag{5.106}$$

$$D^k(H)\bar{\xi}_k + (q_{k-})(\bar{x}^{k-}) = 0, \quad k = 1, \dots, s, \tag{5.107}$$

$$\bar{h}(t) = 0 \text{ a.e. in } \Delta. \tag{5.108}$$

From the last equation, we get

$$\bar{v} = -H_{vv}^{-1}(H_{vx} + f_v^* Q)\bar{x}. \tag{5.109}$$

Using this formula in the equation $\dot{\bar{x}} = f_x\bar{x} + f_v\bar{v}$, we see that $\bar{x}$ is a solution to the linear equation

$$\dot{\bar{x}} = (f_x - f_v H_{vv}^{-1}(H_{vx} + f_v^* Q))\bar{x}. \tag{5.110}$$

This equation together with initial condition $\bar{x}(t_0) = 0$ implies that $\bar{x}(t) = 0$ for all $t \in [t_0, t_1)$. Consequently, $\bar{x}^{1-} = 0$, and then, by virtue of (5.107), $\bar{\xi}_1 = 0$. This equality, together with the jump condition $[\bar{x}]^1 = [\dot{x}]^1\bar{\xi}_1$, implies that $[\bar{x}]^1 = 0$, i.e., $\bar{x}$ is continuous at t_1. Consequently, $\bar{x}^{1+} = 0$. From the last condition and equation (5.110) it follows that $\bar{x}(t) = 0$ for all $t \in (t_1, t_2)$. Repeating this argument, we obtain $\bar{\xi}_1 = \bar{\xi}_2 = \cdots = \bar{\xi}_s = 0$, $\bar{x}(t) = 0$ for all $t \in [t_0, t_f]$. Then from (5.109) it follows that $\bar{v} = 0$. Consequently, we have $\bar{z} = 0$ and thus have proved the following theorem; cf. [98].

Theorem 5.27. *Assume that there exists a symmetric matrix $Q(t)$, defined on $[t_0, t_f]$, such that*

(a) *$Q(t)$ is piecewise continuous on $[t_0, t_f]$ and continuously differentiable on each interval of the set $[t_0, t_f] \setminus \Theta$;*

(b) *$Q(t)$ satisfies the Riccati equation*

$$\dot{Q} + Qf_x + f_x^* Q + H_{xx} - (H_{xv} + Qf_v)H_{vv}^{-1}(H_{vx} + f_v^* Q) = 0 \tag{5.111}$$

on each interval of the set $[t_0, t_f] \setminus \Theta$;

(c) *at each point $t_k \in \Theta$ matrix $Q(t)$ satisfies the jump condition*

$$[Q]^k = (b_{k-})^{-1}(q_{k-})^*(q_{k-}),$$

where $q_{k-} = ([\dot{x}]^k)^* Q^{k-} - [\dot{\psi}]^k$, $b_{k-} = D^k(H) - (q_{k-})[\dot{x}]^k > 0$;

(d) *$\langle (l_{x_f x_f} - Q(t_f))\bar{x}_f, \bar{x}_f \rangle \geq 0$ for all $\bar{x}_f \in \mathbb{R}^{d(x)} \setminus \{0\}$ such that $K_{x_f}(x(t_f))\bar{x}_f = 0$.*

Then $\Omega(\lambda, \bar{z})$ is positive definite on the subspace $\mathcal{K}$.

In some problems, it is more convenient to integrate the Riccati equation (5.111) backwards from $t = t_f$. A similar proof shows that we can replace condition (c) in Theorem 4.1 by the following condition:

(c+) at each point $t_k \in \Theta$, the matrix $Q(t)$ satisfies the jump condition

$$\begin{aligned} &[Q]^k = (b_{k+})^{-1}(q_{k+})^*(q_{k+}), \quad \text{where} \\ &q_{k+} = ([\dot{x}]^k)^* Q^{k+} - [\dot{\psi}]^k, \quad b_{k+} = D^k(H) + (q_{k+})[\dot{x}]^k > 0. \end{aligned} \tag{5.112}$$

5.4 Numerical Example: Optimal Control of Production and Maintenance

Cho, Abad, and Parlar [22] introduced an optimal control model where a dynamic maintenance problem is incorporated into a production control problem so as to simultaneously compute optimal production and maintenance policies. In this model, the dynamics is linear with respect to both production and maintenance control, whereas the cost functional is quadratic with respect to production control and linear with respect to maintenance control. Hence, the model fits into the type of control problems considered in (5.1)–(5.4). A detailed numerical analysis of solutions for different final times may be found in Maurer, Kim, and Vossen [67]. For a certain range of final times the maintenance control is bang-bang. We will show that the sufficient conditions in Theorems 5.24 and 5.27 are satisfied for the computed solutions. The notation for the state variables is slightly different from that in [22, 67]. The state and control variables and parameters have the following meaning:

$x_1(t)$: inventory level at time $t \in [0, t_f]$ with fixed final time $t_f > 0$,
$x_2(t)$: proportion of good units of end items produced at time $t \in [0, t_f]$,
$v(t)$: scheduled production rate (*control*),
$m(t)$: preventive maintenance rate to reduce the proportion of defective units produced (*control*),
$\alpha(t)$: obsolescence rate of the process performance in the absence of maintenance,
$s(t)$: demand rate,
$\rho > 0$: discount rate.

The dynamics of the process is given by

$$\begin{aligned} \dot{x}_1(t) &= x_2(t)v(t) - s(t), & x_1(0) = x_{10} > 0, \\ \dot{x}_2(t) &= -\alpha(t)x_2(t) + (1 - x_2(t))m(t), & x_2(0) = x_{20} > 0, \end{aligned} \tag{5.113}$$

with the following bounds on the control variables:

$$0 \le v(t) \le v_{\max}, \quad 0 \le m(t) \le m_{\max} \quad \text{for } 0 \le t \le t_f. \tag{5.114}$$

Since all demands must be satisfied, the following state constraint is imposed:

$$x_1(t) \ge 0 \quad \text{for } 0 \le t \le t_f.$$

Computations show that this state constraint is automatically satisfied if we impose the boundary condition

$$x_1(t_f) = 0. \tag{5.115}$$

The optimal control problem then consists in *maximizing* the total discounted profit plus the salvage value of $x_2(t_f)$,

$$\begin{aligned} J(x_1, x_2, m, v) \;=\; & \int_0^{t_f} [ws - hx_1(t) - rv(t)^2 - cm(t)]e^{-\rho t}\, dt \\ & + b x_2(t_f) e^{-\rho t_f}, \end{aligned} \tag{5.116}$$

under the constraints (5.113)–(5.115). For later computations, the values of constants are chosen as in [22]:

$$\begin{array}{llllll} \alpha \equiv 2, & w = 8, & s(t) \equiv 4, & \rho = 0.1, & h = 1, & c = 2.5, \\ r = 2, & b = 10, & v_{\max} = 3, & m_{\max} = 4, & x_{10} = 3, & x_{20} = 1. \end{array} \tag{5.117}$$

The time horizon t_f will be specified below. In the discussion of the *minimum principle*, we consider the usual Pontryagin function (Hamiltonian), see (5.73), instead of the *current value* Hamiltonian in [22, 67],

$$\begin{aligned} H(t, x_1, x_2, \psi_1, \psi_2, m, v) = \; & e^{-\rho t}(-ws + hx_1 + rv^2 + cm) \\ & + \psi_1(x_2 v - s) + \psi_2(-\alpha x_2 + (1 - x_2)m), \end{aligned} \tag{5.118}$$

where ψ_1, ψ_2 denote the adjoint variables. The adjoint equations and transversality conditions yield, in view of $x_1(t_f) = 0$ and the salvage term in the cost functional,

$$\begin{aligned} \dot{\psi}_1 &= -he^{-\rho t}, & \psi_1(t_f) &= \nu, \\ \dot{\psi}_2 &= -\psi_1 v + \psi_2(\alpha + m), & \psi_2(t_f) &= -be^{-\rho t_f}. \end{aligned} \tag{5.119}$$

The multiplier ν is not known a priori and will be computed later. We will choose a time horizon for which the control constraint $0 \le v(t) \le V = 3$ will not become active. Hence, the minimum condition in the minimum principle leads to the equation $0 = H_v = 2e^{-\rho t}v + \psi_1 x_2$, which yields the control

$$v = -\psi_1 x_2 e^{\rho t}/2r. \tag{5.120}$$

Since the maintenance control enters the Hamiltonian linearly, the control m is determined by the sign of the switching function

$$\phi_m(t) = H_m = e^{-\rho t}c + \psi_2(t)(1 - x_2(t)) \tag{5.121}$$

according to

$$m(t) = \left\{ \begin{array}{lll} m_{\max} & \text{if} & \phi_m(t) > 0 \\ 0 & \text{if} & \phi_m(t) < 0 \\ \text{singular} & \text{if} & \phi_m(t) \equiv 0 \quad \text{for } t \in I_{\text{sing}} \subset [0, t_f] \end{array} \right\}. \tag{5.122}$$

For the final time $t_f = 1$ which was considered in [22] and [67], the maintenance control contains a singular arc. However, the computations in [67] show that for final times $t_f \in [0.15, 0.98]$ the maintenance control is bang-bang and has only one switching time:

$$m(t) = \left\{ \begin{array}{lll} 0 & \text{for} & 0 \le t < t_1 \\ m_{\max} = 4 & \text{for} & t_1 \le t \le t_f \end{array} \right\}. \tag{5.123}$$

Let us study the control problem with final time $t_f = 0.9$ in more detail. To compute a solution candidate, we apply nonlinear programming methods to the discretized control problem with a large number N of grid points $\tau_i = i \cdot t_f / N$, $i = 0, 1, \ldots, N$; cf. [5, 14]. Both the method of Euler and the method of Heun are employed for integrating the differential equation. We use the programming language AMPL developed by Fourer et al. [33] and the interior point optimization code IPOPT of Wächter and Biegler [114]. For $N = 5000$ grid points, the computed state, control and adjoint functions are displayed in Figure 5.1. The following values for the switching time, functional value, and selected state and adjoint

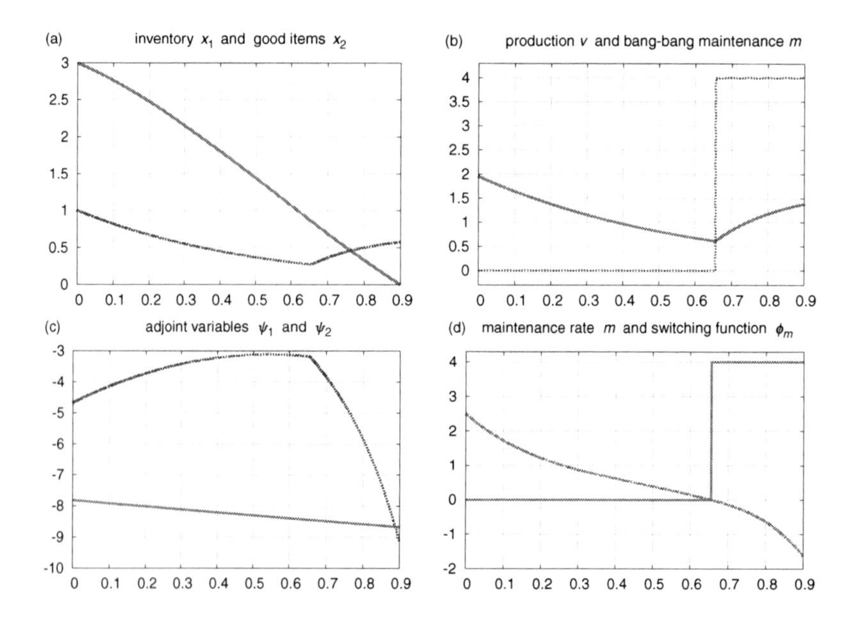

Figure 5.1. *Optimal production and maintenance, final time* $t_f = 0.9$. (a) *State variables* x_1, x_2. (b) *Regular production control* v *and bang-bang maintenance control* m. (c) *Adjoint variables* ψ_1, ψ_2. (d) *Maintenance control* m *with switching function* ϕ_m.

variables are obtained:

$$\begin{array}{rclrcl} t_1 &=& 0.65691, & J &=& 26.705, \\ x_1(t_1) &=& 0.84924, & x_2(t_1) &=& 0.226879, \\ x_1(t_f) &=& 0.0, & x_2(t_f) &=& 0.574104, \\ \psi_1(0) &=& -7.8617, & \psi_2(0) &=& -4.70437, \\ \psi_1(t_1) &=& -8.4975, & \psi_2(t_1) &=& -3.2016, \\ \psi_1(t_f) &=& -8.72313, & \psi_2(t_f) &=& -9.13931. \end{array} \tag{5.124}$$

Now we evaluate the Riccati equation (5.111),

$$\dot{Q} = -Qf_x - f_x Q - H_{xx} + (H_{xv} + Qf_v)(H_{vv})^{-1}(H_{vx} + f_v^* Q), \tag{5.125}$$

for the symmetric 2×2 matrix

$$Q = \begin{pmatrix} Q_{11} & Q_{12} \\ Q_{12} & Q_{22} \end{pmatrix}.$$

Computing the expressions

$$f_x = \begin{pmatrix} 0 & v \\ 0 & -(\alpha + m) \end{pmatrix}, \quad f_v = \begin{pmatrix} x_2 \\ 0 \end{pmatrix}, \quad H_{xx} = \mathbf{0}, \quad H_{xv} = (0, \psi_1),$$

the matrix Riccati equation (5.125) yields the following ODE system:

$$\dot{Q}_{11} = Q_{11}^2 x_2^2 e^{\rho t}/2r, \tag{5.126}$$

$$\dot{Q}_{12} = Q_{11}v - Q_{12}(\alpha + m) + e^{\rho t} Q_{11} x_2(\psi_1 + Q_{12}x_2), \tag{5.127}$$

$$\dot{Q}_{22} = -2Q_{12}v + 2Q_{22}(\alpha + m) + e^{\rho t}(\psi_1 + Q_{12}x_2)^2/2r. \tag{5.128}$$

The equations (5.126) and (5.127) are homogeneous in the variables Q_{11} and Q_{12}. Hence, we can try to find a solution to the Riccati system with $Q_{11}(t) = Q_{12}(t) \equiv 0$ on $[0, t_f]$. Then (5.128) reduces to the *linear equation*

$$\dot{Q}_{22} = 2Q_{22}(\alpha + m) + e^{\rho t}\psi_1^2/2r. \tag{5.129}$$

This linear equation always has a solution. The remaining task is to satisfy the jump and boundary conditions in Theorem 5.27 for the matrix Q. Since we shall integrate equation (5.128) backwards, it is more convenient to evaluate the jump (5.112). Moreover, the boundary conditions for Q in Theorem 5.27(d) show that the initial value $Q(0)$ can be chosen arbitrarily, while the terminal condition imposes the sign condition $Q_{22}(t_f) \le 0$, since $x_2(t_f)$ is free. Therefore, we can choose the terminal value

$$Q_{22}(t_f) = 0. \tag{5.130}$$

Hence, using the computed values in (5.124), we solve the linear equation (5.129) with terminal condition (5.130). At the switching time t_1, we obtain the value

$$Q_{22}(t_1) = -1.5599.$$

Next, we evaluate the jump in the state and adjoint variables and check conditions (5.112). We get

$$([\dot{x}]^1)^* = (0, M(1 - x_2(t_1))), \quad [\dot{\psi}] = (0, M\psi_2(t_1)),$$

which yield the quantities

$$\begin{aligned}
q_{1+} &= ([\dot{x}^1]^T Q^{1+} - [\dot{\psi}]^1 = (0,\, M(1-x_2(t_1))Q_{22}(t_1+) - M\psi_2(t_1)) \\
&= (0, 8.2439), \\
b_{1+} &= D^1(H) + (q_{1+})[\dot{x}^1] \\
&= D^1(H) + M^2((1-x_2(t_1))Q_{22}(t_1+) - \psi_2(t_1))\psi_2(t_1) \\
&= 27.028 + 133.55 = 165.58 > 0.
\end{aligned}$$

Then the jump condition in (5.112),

$$[Q]^1 = (b_{1+})^{-1}(q_{1+})^*(q_{1+}) = \begin{pmatrix} 0 & 0 \\ 0 & [Q_{22}]^1 \end{pmatrix},$$

reduces to a jump condition for Q_{22} at t_1. However, we do not need to evaluate this jump condition explicitly because the linear equation (5.129) has a solution regardless of the value $Q_{22}(t_1-)$. Hence, we conclude from Theorem 5.27 that the numerical solution characterized by (5.124) and displayed in Figure 5.1 provides a strict bounded strong minimum.

We may hope to improve on the benefit by choosing a larger time horizon. For the final time $t_f = 1.1$, we get a bang-singular-bang maintenance control $m(t)$ as shown in Figure 5.2.

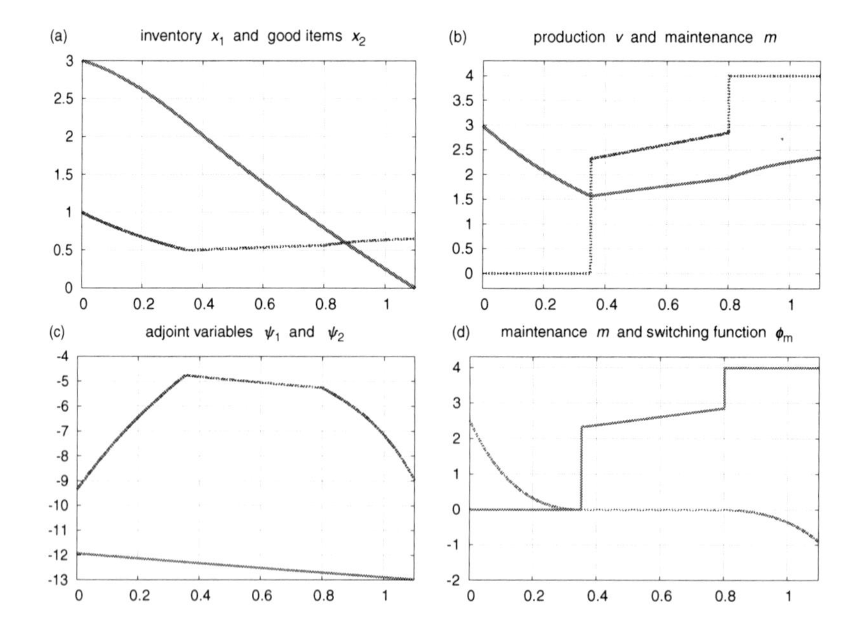

Figure 5.2. *Optimal production and maintenance, final time* $t_f = 1.1$. (a) *State variables* x_1, x_2. (b) *Regular production control* v *and bang-singular-bang maintenance control* m. (c) *Adjoint variables* ψ_1, ψ_2. (d) *Maintenance control* m *with switching function* ϕ_m.

The solution characteristics for this solution are given by

$$\begin{array}{rclrcl} J & = & 23.3567, & & & \\ x_1(t_f) & = & 0, & x_2(t_f) & = & 0.64926, \\ \psi_1(0) & = & -11.93, & \psi_2(0) & = & -9.332, \\ \psi_1(t_f) & = & -12.97, & \psi_2(t_f) & = & -8.956. \end{array} \tag{5.131}$$

Apparently, the larger time horizon $t_f = 1.1$ results in a smaller gain $J = 23.357$ compared to $J = 26.705$ for the final time $t_f = 0.9$. We are not aware of any type of sufficient optimality conditions that would apply to the extremal for $t_f = 1.1$, where one control component has a bang-singular-bang structure. Thus one is lead to ask: What is the optimal lifetime of the machine to give maximal gain? This amounts to solving the control problem (5.113)–(5.116) with *free final time* t_f. The solution is very similar to that shown in Figure 5.1. The maintenance control $m(t)$ is bang-bang with one switching time $t_1 = 0.6523$. The optimal final time $t_f = 0.8633$ gives the gain $J = 26.833$ which slightly improves on $J = 26.705$ for the final time $t_f = 0.9$.

Chapter 6

Second-Order Optimality Conditions for Bang-Bang Control

In this chapter, we investigate the pure bang-bang case, where the second-order necessary or sufficient optimality conditions amount to testing the positive (semi)definiteness of a quadratic form on a *finite-dimensional* critical cone. In Section 6.2, we deduce these conditions from the results obtained in the previous chapter. Although the quadratic conditions turned out to be finite-dimensional, the direct numerical test works only in some special cases. Therefore, in Section 6.3, we study various transformations of the quadratic form and the critical cone which are tailored to different types of control problems in practice. In particular, by a solution to a *linear matrix differential equation*, the quadratic form can be converted to perfect squares. In Section 6.5, we study second-order optimality conditions for *time-optimal control problems with control appearing linearly*. In Section 6.6, we show that an approach similar to the above mentioned Riccati equation approach is applicable for such problems. Again, the test requires us to find a solution of a linear matrix differential equation which satisfies certain jump conditions at the switching points. In Section 6.7, we discuss two numerical examples that illustrate the numerical procedure of verifying positive definiteness of the corresponding quadratic forms. In Section 6.8, following [79], we study second-order optimality conditions in a simple, but important class of time-optimal control problems for linear autonomous systems.

6.1 Bang-Bang Control Problems on Nonfixed Time Intervals

6.1.1 Optimal Control Problems with Control Appearing Linearly

We consider optimal control problems with control appearing linearly. Let $x(t) \in \mathbb{R}^{d(x)}$ denote the state variable and $u(t) \in \mathbb{R}^{d(u)}$ the control variable in the time interval $t \in \Delta = [t_0, t_f]$ with a nonfixed initial time t_0 and final time t_f. We shall refer to the following control problem (6.1)–(6.3) as the *basic bang-bang control problem*, or briefly, the *basic problem*:

$$\text{Minimize} \qquad \mathcal{J} := J(t_0, x(t_0), t_f, x(t_f)) \tag{6.1}$$

subject to the constraints

$$\dot{x}(t) = f(t,x(t),u(t)), \quad u(t) \in U, \quad (t,x(t)) \in \mathcal{Q}, \quad t_0 \le t \le t_f, \tag{6.2}$$

$$\begin{aligned} &F(t_0,x(t_0),t_f,x(t_f)) \le 0, \quad K(t_0,x(t_0),t_f,x(t_f)) = 0, \\ &(t_0,x(t_0),t_f,x(t_f)) \in \mathcal{P}, \end{aligned} \tag{6.3}$$

where the control variable appears linearly in the system dynamics,

$$f(t,x,u) = a(t,x) + B(t,x)u. \tag{6.4}$$

Here, F, K, and a are vector functions, B is a $d(x) \times d(u)$ matrix function, $\mathcal{P} \subset \mathbb{R}^{2+2d(x)}$, $\mathcal{Q} \subset \mathbb{R}^{1+d(x)}$ are open sets, and $U \subset \mathbb{R}^{d(u)}$ is a convex polyhedron. The functions J, F, and K are assumed to be twice continuously differentiable on $\mathcal{P}$, and the functions a, B are twice continuously differentiable on $\mathcal{Q}$. The dimensions of F and K are denoted by $d(F)$ and $d(K)$. We shall use the abbreviations $x_0 = x(t_0)$, $x_f = x(t_f)$, $p = (t_0, x_0, t_f, x_f)$. A trajectory $\mathcal{T} = (x(t), u(t) \mid t \in [t_0, t_f])$ is said to be *admissible* if $x(\cdot)$ is absolutely continuous, $u(\cdot)$ is measurable bounded, and the pair of functions $(x(t), u(t))$ on the interval $\Delta = [t_0, t_f]$ with the endpoints $p = (t_0, x(t_0), t_f, x(t_f))$ satisfies the constraints (6.2) and (6.3). We set $\mathcal{J}(\mathcal{T}) := J(t_0, x(t_0), t_f, x(t_f))$.

Obviously, the basic problem (6.1)–(6.3) is a special case of the general problem (5.69)–(5.72) studied in the previous chapter. This special case corresponds to the assumption that the function f does not depend on the variable v, i.e., $f = f(t,x,u)$. Under this assumption it turned out to be possible to obtain certain deeper results than in the general problem. More precisely, we formulate the necessary quadratic Condition $\mathfrak{A}$ in the problem (6.1)–(6.3), which is a simple consequence of the Condition $\mathfrak{A}$ in the general problem, whereas the correspondent sufficient quadratic Condition $\mathfrak{B}$ will be slightly simplified.

Let us give the definition of Pontryagin minimum for the basic problem. The trajectory $\mathcal{T}$ affords a *Pontryagin minimum* if there is no sequence of admissible trajectories $\mathcal{T}^n = (x^n(t), u^n(t) \mid t \in [t_0^n, t_f^n])$, $n = 1, 2, \ldots$, such that the following properties hold with $\Delta^n = [t_0^n, t_f^n]$:

(a) $\mathcal{J}(\mathcal{T}^n) < \mathcal{J}(\mathcal{T})$ for all n and $t_0^n \to t_0$, $t_f^n \to t_f$ for $n \to \infty$;

(b) $\max_{\Delta^n \cap \Delta} |x^n(t) - x(t)| \to 0$ for $n \to \infty$;

(c) $\int_{\Delta^n \cap \Delta} |u^n(t) - u(t)|\, dt \to 0$ for $n \to \infty$.

Note that for a fixed time interval Δ, a Pontryagin minimum corresponds to an L^1-*local minimum* with respect to the control variable.

6.1.2 First-Order Necessary Optimality Conditions

Let $\mathcal{T} = (x(t), u(t) \mid t \in [t_0, t_f])$ be a fixed admissible trajectory such that the control $u(\cdot)$ is a piecewise constant function on the interval $\Delta = [t_0, t_f]$. Denote by $\Theta = \{t_1, \ldots, t_s\}$, $t_0 < t_1 < \cdots < t_s < t_f$ the finite set of all discontinuity points (jump points) of the control $u(t)$. Then $\dot{x}(t)$ is a piecewise continuous function whose discontinuity points belong to Θ, and hence $x(t)$ is a piecewise smooth function on Δ. We continue to use the notation $[u]^k = u^{k+} - u^{k-}$ for the jump of function $u(t)$ at the point $t_k \in \Theta$, where $u^{k-} = u(t_k-)$, $u^{k+} = u(t_k+)$ are the left- and right-hand values of the control $u(t)$ at t_k, respectively.

Let us formulate a first-order necessary condition for optimality of the trajectory $\mathcal{T}$, that is, the Pontryagin minimum principle. The Pontryagin function has the form

$$H(t,x,u,\psi) = \psi f(t,x,u) = \psi a(t,x) + \psi B(t,x)u, \tag{6.5}$$

where ψ is a row vector of dimension $d(\psi) = d(x)$, while x, u, f, F, and K are column vectors. The factor of the control u in the Pontryagin function is the *switching function*

$$\phi(t,x,\psi) = H_u(t,x,u,\psi) = \psi B(t,x) \tag{6.6}$$

which is a row vector of dimension $d(u)$. The endpoint Lagrange function is

$$l(\alpha_0,\alpha,\beta,p) = \alpha_0 J(p) + \alpha F(p) + \beta K(p),$$

where α and β are row vectors with $d(\alpha) = d(F)$ and $d(\beta) = d(K)$, and α_0 is a number. By $\lambda = (\alpha_0,\alpha,\beta,\psi(\cdot),\psi_0(\cdot))$ we denote a tuple of Lagrange multipliers such that $\psi(\cdot) : \Delta \to (\mathbb{R}^{d(x)})^*$, $\psi_0(\cdot) : \Delta \to \mathbb{R}^1$ are continuous on Δ and continuously differentiable on each interval of the set $\Delta \setminus \Theta$.

Let M_0 be the set of the normed collections λ satisfying the minimum principle conditions for the trajectory $\mathcal{T}$:

$$\alpha_0 \geq 0, \quad \alpha \geq 0, \quad \alpha F(p) = 0, \quad \alpha_0 + \sum_{i=1}^{d(F)} \alpha_i + \sum_{j=1}^{d(K)} |\beta_j| = 1, \tag{6.7}$$

$$\dot{\psi} = -H_x, \quad \dot{\psi}_0 = -H_t \quad \forall\, t \in \Delta \setminus \Theta, \tag{6.8}$$

$$\psi(t_0) = -l_{x_0}, \quad \psi(t_f) = l_{x_f}, \quad \psi_0(t_0) = -l_{t_0}, \quad \psi_0(t_f) = l_{t_f}, \tag{6.9}$$

$$\min_{u \in U} H(t,x(t),u,\psi(t)) = H(t,x(t),u(t),\psi(t)) \quad \forall\, t \in \Delta \setminus \Theta, \tag{6.10}$$

$$H(t,x(t),u(t),\psi(t)) + \psi_0(t) = 0 \quad \forall\, t \in \Delta \setminus \Theta. \tag{6.11}$$

The derivatives l_{x_0} and l_{x_f} are taken at the point $(\alpha_0,\alpha,\beta,p)$, where $p = (t_0,x(t_0),t_f,x(t_f))$, and the derivatives H_x, H_t are evaluated at the point $(t,x(t),u(t),\psi(t))$. Again, we use the simple abbreviation (t) for indicating all arguments $(t,x(t),u(t),\psi(t))$, $t \in \Delta \setminus \Theta$.

Theorem 6.1. *If the trajectory $\mathcal{T}$ affords a Pontryagin minimum, then the set M_0 is nonempty. The set M_0 is a finite-dimensional compact set, and the projector $\lambda \mapsto (\alpha_0,\alpha,\beta)$ is injective on M_0.*

For each $\lambda \in M_0$ and $t_k \in \Theta$, let us define the quantity $D^k(H)$. Set

$$(\Delta_k H)(t) = H(t,x(t),u^{k+},\psi(t)) - H(t,x(t),u^{k-},\psi(t)) = \phi(t)[u]^k, \tag{6.12}$$

where $\phi(t) = \phi(t,x(t),\psi(t))$. For each $\lambda \in M_0$ the following equalities hold:

$$\frac{d}{dt}(\Delta_k H)\big|_{t=t_k-} = \frac{d}{dt}(\Delta_k H)\big|_{t=t_k+}, \quad k = 1,\dots,s.$$

Consequently, for each $\lambda \in M_0$ the function $(\Delta_k H)(t)$ has a derivative at the point $t_k \in \Theta$. Set

$$D^k(H) = -\frac{d}{dt}(\Delta_k H)(t_k).$$

Then, for each $\lambda \in M_0$, the minimum condition (6.10) implies the inequality

$$D^k(H) \geq 0, \quad k = 1, \ldots, s. \tag{6.13}$$

As we know, the value $D^k(H)$ can be written in the form

$$D^k(H) = -H_x^{k+} H_\psi^{k-} + H_x^{k-} H_\psi^{k+} - [H_t]^k = \dot{\psi}^{k+} \dot{x}^{k-} - \dot{\psi}^{k-} \dot{x}^{k+} + [\psi_0]^k,$$

where H_x^{k-} and H_x^{k+} are the left- and right-hand values of the function

$$H_x(t, x(t), u(t), \psi(t))$$

at t_k, respectively, $[H_t]^k$ is a jump of the function $H_t(t)$ at t_k, etc. It also follows from the above representation that we have

$$D^k(H) = -\dot{\phi}(t_k \pm)[u]^k, \tag{6.14}$$

where the values on the right-hand side agree for the derivative $\dot{\phi}(t_k+)$ from the right and the derivative $\dot{\phi}(t_k-)$ from the left. In the case of a *scalar* control u, the total derivative $\phi_t + \phi_x \dot{x} + \phi_\psi \dot{\psi}$ does not contain the control variable explicitly and hence the derivative $\dot{\phi}(t)$ is *continuous* at t_k.

Proposition 6.2. *For any $\lambda \in M_0$, we have*

$$l_{x_0} \dot{x}(t_0) + l_{t_0} = 0, \quad l_{x_f} \dot{x}(t_f) + l_{t_f} = 0. \tag{6.15}$$

Proof. The equalities (6.15) follow from the equality $\psi(t)\dot{x}(t) + \psi_0(t) = 0$ evaluated for $t = t_0$ and $t = t_f$ together with the transversality conditions

$$\psi(t_0) = -l_{x_0}, \quad \psi_0(t_0) = -l_{t_0}, \quad \psi(t_f) = l_{x_f}, \quad \psi_0(t_f) = l_{t_f}. \quad \blacksquare$$

6.1.3 Strong Minimum

As already mentioned in Section 2.7, any control problem with a cost functional in integral form $\mathcal{J} = \int_{t_0}^{t_f} f_0(t, x(t), u(t))\,dt$ can be brought to the canonical form (6.1) by introducing a new state variable y defined by the state equation $\dot{y} = f_0(t, x, u)$, $y(t_0) = 0$. This yields the cost function $\mathcal{J} = y(t_f)$. The control variable is assumed to appear linearly in the function f_0,

$$f_0(t, x, u) = a_0(t, x) + B_0(t, x)u. \tag{6.16}$$

It follows that the adjoint variable ψ^y associated with the new state variable y is given by $\psi^y(t) \equiv \alpha_0$ which yields the Pontryagin function (6.5) in the form

$$\begin{aligned} H(t, x, \psi, u) &= \alpha_0 f_0(t, x, u) + \psi f(t, x, u) \\ &= \alpha_0 a_0(t, x) + \psi a(t, x) + (\alpha_0 B_0(t, x) + \psi B(t, x))u. \end{aligned} \tag{6.17}$$

Hence, the *switching function* is given by

$$\phi(t, x, \psi) = \alpha_0 B_0(t, x) + \psi B(t, x), \quad \phi(t) = \phi(t, x(t), \psi(t)). \tag{6.18}$$

The component y was called an *unessential* component in the augmented problem. The general definition was given in Section 5.2.3: the state variable x_i is called *unessential* if the function f does not depend on x_i and if the functions F, J, and K are affine in $x_{i0} = x_i(t_0)$ and $x_{i1} = x_i(t_f)$. Let $\underline{x}$ denote the vector of all essential components of state vector x. Now we can define a strong minimum in the basic problem.

We say that the trajectory $\mathcal{T}$ affords a *strong minimum* if there is no sequence of admissible trajectories $\mathcal{T}^n = (x^n(t), u^n(t) \mid t \in [t_0^n, t_f^n])$, $n = 1, 2, \ldots$, such that

(a) $\mathcal{J}(\mathcal{T}^n) < \mathcal{J}(\mathcal{T})$;

(b) $t_0^n \to t_0$, $t_f^n \to t_f$, $x^n(t_0) \to x(t_0)$ $(n \to \infty)$;

(c) $\max_{\Delta^n \cap \Delta} |\underline{x}^n(t) - \underline{x}(t)| \to 0$ $(n \to \infty)$, where $\Delta^n = [t_0^n, t_f^n]$.

The *strict* strong minimum is defined in a similar way, with the strict inequality (a) replaced by the nonstrict one and the trajectory $\mathcal{T}^n$ required to be different from $\mathcal{T}$ for each n.

6.1.4 Bang-Bang Control

For a given extremal trajectory $\mathcal{T} = \{(x(t), u(t)) \mid t \in \Delta\}$ with a piecewise constant control $u(t)$ we say that $u(t)$ is a *strict bang-bang control* if there exists $\lambda = (\alpha_0, \alpha, \beta, \psi, \psi_0) \in M_0$ such that

$$\operatorname*{Arg\,min}_{u' \in U} \phi(t)u' = [u(t-), u(t+)], \quad t \in [t_0, t_f], \tag{6.19}$$

where $[u(t-), u(t+)]$ denotes the line segment spanned by the vectors $u(t-)$ and $u(t+)$ in $\mathbb{R}^{d(u)}$ and $\phi(t) := \phi(t, x(t), \psi(t)) = \psi(t)B(t, x(t))$. Note that $[u(t-), u(t+)]$ is a singleton $\{u(t)\}$ at each continuity point of the control $u(t)$ with $u(t)$ being a vertex of the polyhedron U. Only at the points $t_k \in \Theta$ does the line segment $[u^{k-}, u^{k+}]$ coincide with an edge of the polyhedron.

As it was already mentioned in Section 5.1.7, if the control is *scalar*, $d(u) = 1$, and $U = [u_{\min}, u_{\max}]$, then the strict bang-bang property is equivalent to $\phi(t) \neq 0$ for all $t \in \Delta \setminus \Theta$ which yields the control law

$$u(t) = \left\{ \begin{array}{ll} u_{\min} & \text{if } \phi(t) > 0 \\ u_{\max} & \text{if } \phi(t) < 0 \end{array} \right\} \quad \forall\, t \in \Delta \setminus \Theta. \tag{6.20}$$

For vector-valued control inputs, condition (6.19) imposes further restrictions. For example, if U is the unit cube in $\mathbb{R}^{d(u)}$, condition (6.19) precludes simultaneous switching of the control components; the case of simultaneous switching was studied in Felgenhauer [31]. This property holds in many examples. The condition (6.19) is indispensable in the sensitivity analysis of optimal bang-bang controls.

6.2 Quadratic Necessary and Sufficient Optimality Conditions

In this section, we shall formulate a quadratic necessary optimality condition of a Pontryagin minimum for given bang-bang control. A strengthening of this quadratic condition yields a quadratic sufficient condition for a strong minimum. These quadratic conditions are based on the properties of a quadratic form on the critical cone.

6.2.1 Critical Cone

For a given trajectory $\mathcal{T}$, we introduce the space $\mathcal{Z}(\Theta)$ and the *critical cone* $\mathcal{K} \subset \mathcal{Z}(\Theta)$. Denote by $P_\Theta C^1(\Delta,\mathbb{R}^{d(x)})$ the space of piecewise continuous functions $\bar{x}(\cdot):\Delta \to \mathbb{R}^{d(x)}$, continuously differentiable on each interval of the set $\Delta \setminus \Theta$. For each $\bar{x} \in P_\Theta C^1(\Delta,\mathbb{R}^{d(x)})$ and for $t_k \in \Theta$, we set $\bar{x}^{k-} = \bar{x}(t_k-)$, $\bar{x}^{k+} = \bar{x}(t_k+)$, $[\bar{x}]^k = \bar{x}^{k+} - \bar{x}^{k-}$. Set $\bar{z} = (\bar{t}_0,\bar{t}_f,\bar{\xi},\bar{x})$, where $\bar{t}_0,\bar{t}_f \in \mathbb{R}^1$, $\bar{\xi} \in \mathbb{R}^s$, $\bar{x} \in P_\Theta C^1(\Delta,\mathbb{R}^{d(x)})$. Thus,

$$\bar{z} \in \mathcal{Z}(\Theta) := \mathbb{R}^2 \times \mathbb{R}^s \times P_\Theta C^1(\Delta,\mathbb{R}^{d(x)}).$$

For each $\bar{z}$, we set

$$\bar{\bar{x}}_0 = \bar{x}(t_0) + \bar{t}_0\dot{x}(t_0), \quad \bar{\bar{x}}_f = \bar{x}(t_f) + \bar{t}_f\dot{x}(t_f), \quad \bar{\bar{p}} = \left(\bar{t}_0,\bar{\bar{x}}_0,\bar{t}_f,\bar{\bar{x}}_f\right). \tag{6.21}$$

The vector $\bar{\bar{p}}$ is considered as a column vector. Note that $\bar{t}_0 = 0$, respectively, $\bar{t}_f = 0$, for *fixed* initial time t_0, respectively, final time t_f. Let $I_F(p) = \{i \in \{1,\ldots,d(F)\} \mid F_i(p) = 0\}$ be the set of indices of all active endpoint inequalities $F_i(p) \le 0$ at the point $p = (t_0,x(t_0),t_f,x(t_f))$. Denote by $\mathcal{K}$ the set of all $\bar{z} \in \mathcal{Z}(\Theta)$ satisfying the following conditions:

$$J'(p)\bar{\bar{p}} \le 0, \quad F_i'(p)\bar{\bar{p}} \le 0 \;\forall\, i \in I_F(p), \quad K'(p)\bar{\bar{p}} = 0, \tag{6.22}$$

$$\dot{\bar{x}}(t) = f_x(t,x(t),u(t))\bar{x}(t), \quad [\bar{x}]^k = [\dot{x}]^k\bar{\xi}_k, \quad k = 1,\ldots,s, \tag{6.23}$$

where $p = (t_0,x(t_0),t_f,x(t_f))$. It is obvious that $\mathcal{K}$ is a convex finite-dimensional and finite-faced cone in the space $\mathcal{Z}(\Theta)$. We call it *the critical cone*. Each element $\bar{z} \in \mathcal{K}$ is uniquely defined by numbers $\bar{t}_0$, $\bar{t}_f$, a vector $\bar{\xi}$, and the initial value $\bar{x}(t_0)$ of the function $\bar{x}(t)$.

Proposition 6.3. *For any $\lambda \in M_0$ and $\bar{z} \in \mathcal{K}$, we have*

$$l_{x_0}\bar{x}(t_0) + l_{x_f}\bar{x}(t_f) = 0. \tag{6.24}$$

Proof. Integrating the equality $\psi(\dot{\bar{x}} - f_x\bar{x}) = 0$ on $[t_0,t_f]$ and using the adjoint equation $\dot{\psi} = -\psi f_x$ we obtain $\int_{t_0}^{t_f}\frac{d}{dt}(\psi\bar{x})\,dt = 0$, whence $(\psi\bar{x})|_{t_0}^{t_f} - \sum_{k=1}^{s}[\psi\bar{x}]^k = 0$. From the jump conditions $[\bar{x}]^k = [\dot{x}]^k\bar{\xi}_k$ and the equality $\psi(t)\dot{x}(t) + \psi_0(t) = 0$ it follows that $[\psi\bar{x}]^k = \psi(t_k)[\dot{x}]^k\bar{\xi}_k = [\psi\dot{x}]^k\bar{\xi}_k = -[\psi_0]^k\bar{\xi}_k = 0$ for all k. Then the equation $(\psi\bar{x})|_{t_0}^{t_f} = 0$, together with the transversality conditions $\psi(t_0) = -l_{x_0}$ and $\psi(t_f) = l_{x_f}$, implies (6.24). ∎

Proposition 6.4. *For any $\lambda \in M_0$ and $\bar{z} \in \mathcal{K}$ we have*

$$\alpha_0 J'(p)\bar{\bar{p}} + \sum_{i=1}^{s}\alpha_i F_i'(p)\bar{\bar{p}} + \beta K'(p)\bar{\bar{p}} = 0. \tag{6.25}$$

Proof. For $\lambda \in M_0$ and $\bar{z} \in \mathcal{K}$, we have, by Propositions 6.2 and 6.3,

$$\bar{t}_0(l_{x_0}\dot{x}(t_0) + l_{t_0}) + \bar{t}_f(l_{x_f}\dot{x}(t_f) + l_{t_f}) + l_{x_0}\bar{x}(t_0) + l_{x_f}\bar{x}(t_f) = 0.$$

Now using the equalities $\bar{\bar{x}}_0 = \bar{x}(t_0) + \bar{t}_0\dot{x}(t_0)$, $\bar{\bar{x}}_f = \bar{x}(t_f) + \bar{t}_f\dot{x}(t_f)$, and $\bar{\bar{p}} = \left(\bar{t}_0,\bar{\bar{x}}_0,\bar{t}_f,\bar{\bar{x}}_f\right)$, we get $l_p\bar{\bar{p}} = 0$ which is equivalent to condition (6.25). ∎

Two important properties of the critical cone follow from Proposition 6.4.

Proposition 6.5. *For any $\lambda \in M_0$ and $\bar{z} \in \mathcal{K}$, we have $\alpha_0 J'(p)\bar{\bar{p}} = 0$ and $\alpha_i F_i'(p)\bar{\bar{p}} = 0$ for all $i \in I_F(p)$.*

Proposition 6.6. *Suppose that there exist $\lambda \in M_0$ with $\alpha_0 > 0$. Then adding the equalities $\alpha_i F_i'(p)\bar{\bar{p}} = 0$ for all $i \in I_F(p)$ to the system (6.22), (6.23) defining $\mathcal{K}$, one can omit the inequality $J'(p)\bar{\bar{p}} \le 0$ in that system without affecting $\mathcal{K}$.*

Thus, $\mathcal{K}$ is defined by condition (6.23) and by the condition $\bar{\bar{p}} \in \mathcal{K}_0$, where $\mathcal{K}_0$ is the cone in $\mathbb{R}^{2d(x)+2}$ given by (6.22). But if there exists $\lambda \in M_0$ with $\alpha_0 > 0$, then we can put

$$\mathcal{K}_0 = \{\bar{\bar{p}} \in \mathbb{R}^{d(x)+2} \mid F_i'(p)\bar{\bar{p}} \le 0,\ \alpha_i F_i'(p)\bar{\bar{p}} = 0\ \forall\, i \in I_F(p),\ K'(p)\bar{\bar{p}} = 0\}. \tag{6.26}$$

If, in addition, $\alpha_i > 0$ holds for all $i \in I_F(p)$, then $\mathcal{K}_0$ is a subspace in $\mathbb{R}^{d(x)+2}$.

An explicit representation of the variations $\bar{x}(t)$ in (6.23) is obtained as follows. For each $k = 1, \dots, s$, define the vector functions $y^k(t)$ as the solutions to the system

$$\dot{y} = f_x(t)y, \quad y(t_k) = [\dot{x}]^k, \quad t \in [t_k, t_f].$$

For $t < t_k$ we put $y^k(t) = 0$ which yields the jump $[y^k]^k = [\dot{x}]^k$. Moreover, define $y^0(t)$ as the solution to the system

$$\dot{y} = f_x(t)y, \quad y(t_0) = \bar{x}(t_0) =: \bar{x}_0.$$

By the superposition principle for linear ODEs it is obvious that we have

$$\bar{x}(t) = \sum_{k=1}^{s} y^k(t)\bar{\xi}_k + y^0(t)$$

from which we obtain the representation

$$\bar{\bar{x}}_f = \sum_{k=1}^{s} y^k(t_f)\bar{\xi}_k + y^0(t_f) + \dot{x}(t_f)\bar{t}_f. \tag{6.27}$$

Furthermore, denote by $x(t; t_1, \dots, t_s)$ the solution of the state equation (6.2) using the values of the optimal bang-bang control with switching points $t_1, \dots, t_s$. It easily follows from elementary properties of ODEs that the partial derivatives of state trajectories with respect to the switching points are given by

$$\frac{\partial x}{\partial t_k}(t; t_1, \dots, t_s) = -y^k(t) \quad \text{for } t \ge t_k,\ k = 1, \dots, s. \tag{6.28}$$

This gives the following expression for $\bar{x}(t)$:

$$\bar{x}(t) = -\sum_{k=1}^{s} \frac{\partial x}{\partial t_k}(t)\bar{\xi}_k + y^0(t). \tag{6.29}$$

In a special case that frequently arises in practice, we can use these formulas to show that $\mathcal{K} = \{0\}$. This property then yields a first-order sufficient condition in view of Theorem 6.10.

Namely, consider the following problem with an integral cost functional, where the initial time $t_0 = \hat{t}_0$ is fixed, while the final time t_f is free and where the initial and final values of the state variables are given: Minimize

$$\mathcal{J} = \int_{t_0}^{t_f} f_0(t,x,u)dt \tag{6.30}$$

subject to

$$\dot{x} = f(t,x,u), \quad x(t_0) = \hat{x}_0,\ x(t_f) = \hat{x}_f, \quad u(t) \in U, \tag{6.31}$$

where f is defined by (6.4), and f_0 is defined by (6.16). In the definition of $\mathcal{K}$ we then have $\bar{t}_0 = 0$, $\bar{x}(t_0) = 0$, $\bar{\bar{x}}_f = 0$. The condition $\bar{x}(t_0) = 0$ implies that $y^0(t) \equiv 0$, whereas the condition $\bar{\bar{x}}_f = 0$ yields in view of the representation (6.27),

$$\sum_{k=1}^{s} y^k(t_f)\bar{\xi}_k + \dot{x}(t_f)\bar{t}_f = 0.$$

This equation leads to the following statement.

Proposition 6.7. *In problem* (6.30), (6.31), *assume that the* $s+1$ *vectors*

$$y^k(t_f) = -\frac{\partial x}{\partial t_k}(t_f)\ (k = 1,\dots,s),\ \dot{x}(t_f)$$

are linearly independent. Then the critical cone is $\mathcal{K} = \{0\}$.

We conclude this subsection with a special property of the critical cone for time-optimal control problems with fixed initial time and state,

$$t_f \to \min,\ \dot{x} = f(t,x,u),\ u \in U,\ t_0 = \hat{t}_0,\ x(t_0) = \hat{x}_0,\ K(x(t_f)) = 0, \tag{6.32}$$

where f is defined by (6.4). The following result will be used in Section 6.2.3; cf. Proposition 6.11.

Proposition 6.8. *Suppose that there exists* $(\psi_0, \psi) \in M_0$ *such that* $\alpha_0 > 0$. *Then* $\bar{t}_f = 0$ *holds for each* $\bar{z} = (\bar{t}_f, \bar{\xi}, \bar{x}) \in \mathcal{K}$.

Proof. For arbitrary $\lambda \in M_0$ and $\bar{z} = (\bar{t}_f, \bar{\xi}, \bar{x}) \in \mathcal{K}$ we infer from the proof of Proposition 6.3 that $\psi(t)\bar{x}(t)$ is a constant function on $[t_0, t_f]$. In view of the relations

$$\psi(t_f) = \beta K_{x_f}(x(t_f)), \qquad K_{x_f}(x(t_f))\bar{\bar{x}}_f = 0, \quad \bar{\bar{x}}_f = \bar{x}(t_f) + \dot{x}(t_f)\bar{t}_f,$$

we get

$$0 = (\psi\bar{x})(t_0) = (\psi\bar{x})(t_f) = \psi(t_f)(\bar{\bar{x}}_f - \dot{x}(t_f)\bar{t}_f) = -\psi(t_f)\dot{x}(t_f)\bar{t}_f = \psi_0(t_f)\bar{t}_f.$$

Since $\psi_0(t_f) = \alpha_0 > 0$, this relation yields $\bar{t}_f = 0$. ∎

In the case $\alpha_0 > 0$, we note as a consequence that the critical cone is a subspace defined by the conditions

$$\begin{aligned} &\dot{\bar{x}} = f_x(t)\bar{x}, \quad [\bar{x}]^k = [\dot{x}]^k\bar{\xi}_k \quad (k = 1,\dots,s), \\ &\bar{t}_0 = \bar{t}_f = 0, \quad \bar{x}(t_0) = 0, \quad K_{x_f}(x(t_f))\bar{x}(t_f) = 0. \end{aligned} \tag{6.33}$$

6.2.2 Quadratic Necessary Optimality Conditions

Let us introduce a quadratic form on the critical cone $\mathcal{K}$ defined by the conditions (6.22), (6.23). For each $\lambda \in M_0$ and $\bar{z} \in \mathcal{K}$, we set

$$\Omega(\lambda,\bar{z}) = \langle A\bar{\bar{p}},\bar{\bar{p}}\rangle + \sum_{k=1}^{s}\left(D^k(H)\bar{\xi}_k^2 + 2[H_x]^k\bar{x}_{\mathrm{av}}^k\bar{\xi}_k\right) + \int_{t_0}^{t_f}\langle H_{xx}\bar{x}(t),\bar{x}(t)\rangle\,dt, \tag{6.34}$$

where

$$\begin{aligned}\langle A\bar{\bar{p}},\bar{\bar{p}}\rangle &= \langle l_{pp}\bar{\bar{p}},\bar{\bar{p}}\rangle + 2\dot{\psi}(t_0)\bar{\bar{x}}_0\bar{t}_0 + (\dot{\psi}_0(t_0) - \dot{\psi}(t_0)\dot{x}(t_0))\bar{t}_0^2 \\ &\quad - 2\dot{\psi}(t_f)\bar{\bar{x}}_f\bar{t}_f - (\dot{\psi}_0(t_f) - \dot{\psi}(t_f)\dot{x}(t_f))\bar{t}_f^2,\end{aligned} \tag{6.35}$$

$$l_{pp} = l_{pp}(\alpha_0,\alpha,\beta,p),\ \ p = (t_0,x(t_0),t_f,x(t_f)),\ \ H_{xx} = H_{xx}(t,x(t),u(t),\psi(t)),$$

$$\bar{x}_{\mathrm{av}}^k = \frac{1}{2}(\bar{x}^{k-} + \bar{x}^{k+}).$$

Note that for a problem on a fixed time interval $[t_0,t_f]$ we have $\bar{t}_0 = \bar{t}_f = 0$ and, hence, the quadratic form (6.35) reduces to $\langle A\bar{\bar{p}},\bar{\bar{p}}\rangle = \langle l_{pp}\bar{p},\bar{p}\rangle$. The following theorem gives the main second-order necessary condition of optimality.

Theorem 6.9. *If the trajectory $\mathcal{T}$ affords a Pontryagin minimum, then the following Condition $\mathfrak{A}$ holds: The set M_0 is nonempty and* $\max_{\lambda\in M_0}\Omega(\lambda,\bar{z}) \geq 0$ *for all $\bar{z} \in \mathcal{K}$.*

6.2.3 Quadratic Sufficient Optimality Conditions

A natural strengthening of the necessary Condition $\mathfrak{A}$ turns out to be a sufficient optimality condition not only for a Pontryagin minimum, but also for a strong minimum.

Theorem 6.10. *Let the following Condition $\mathfrak{B}$ be fulfilled for the trajectory $\mathcal{T}$:*

(a) *there exists $\lambda \in M_0$ such that $D^k(H) > 0$, $k = 1,\dots,s$, and condition* (6.19) *holds (i.e., $u(t)$ is a strict bang-bang control),*

(b) $\max_{\lambda\in M_0}\Omega(\lambda,\bar{z}) > 0$ *for all $\bar{z} \in \mathcal{K}\setminus\{0\}$.*

Then $\mathcal{T}$ is a strict strong minimum.

Note that the condition (b) is automatically fulfilled, if $\mathcal{K} = \{0\}$, which gives a *first-order sufficient condition* for a strong minimum in the problem. A specific situation where $\mathcal{K} = \{0\}$ holds was described in Proposition 6.7. Also note that the condition (b) is automatically fulfilled if there exists $\lambda \in M_0$ such that

$$\Omega(\lambda,\bar{z}) > 0 \qquad \forall\,\bar{z} \in \mathcal{K}\setminus\{0\}. \tag{6.36}$$

Example: Resource allocation problem. Let $x(t)$ be the stock of a resource and let the control $u(t)$ be the investment rate. The control problem is to *maximize* the overall consumption

$$\int_0^{t_f} x(t)(1-u(t))\,dt$$

on a fixed time interval $[0, t_f]$ subject to

$$\dot{x}(t) = x(t)u(t), \quad x(0) = x_0 > 0, \quad 0 \le u(t) \le 1.$$

The Pontryagin function (6.5) for the equivalent minimization problem and the switching function are given by

$$H = \alpha_0 x(u-1) + \psi x u = -\alpha_0 x + \phi u, \quad \phi(x,\psi) = x(\alpha_0 + \psi).$$

We can put $\alpha_0 = 1$, since the terminal state $x(t_f)$ is free. A straightforward discussion of the minimum principle then shows that the optimal control has exactly one switching point at $t_1 = t_f - 1$ for $t_f > 1$,

$$u(t) = \left\{ \begin{array}{ll} 1, & 0 \le t < t_1 \\ 0, & t_1 \le t \le t_f \end{array} \right\},$$

$$(x(t), \psi(t)) = \left\{ \begin{array}{ll} (x_0 e^t, -e^{-(t-t_1)}), & 0 \le t \le t_1 \\ (x_0 e^{t_1}, t - t_f), & t_1 \le t \le t_f \end{array} \right\}.$$

The switching function is $\phi(t) = x(t)(1+\psi(t))$ which yields $\dot{\phi}(t_1) = x_0 e^{t_1} \neq 0$. Here we have $k = 1$, $[u]^1 = -1$, and thus obtain $D^1(H) = -\dot{\phi}(t_1)[u]^1 = \dot{\phi}(t_1) > 0$ in view of (6.12) and (6.14). Hence, condition (a) of Theorem 6.10 holds. Checking condition (b) is rather simple, since the quadratic form (6.34) reduces here to $\Omega(\lambda, \bar{z}) = D^1(H)\bar{\xi}_1^2$. This relation follows from $H_{xx} \equiv 0$ and $[H_x]^1 = (1+\psi(t_1))[u]^1 = 0$ and the fact that the quadratic form (6.35) vanishes. Note that the above control problem cannot be handled in the class of *convex* optimization problems. This means that the necessary conditions do not automatically imply optimality of the computed solution. ∎

We conclude this subsection with the case of a *time-optimal* control problem (6.32) with a *single switching point*, i.e., $s = 1$. Assume that $\alpha_0 > 0$ for a given $\lambda \in M_0$. Then by Proposition 6.8, we have $\bar{t}_f = 0$, and thus the critical cone is the subspace defined by (6.33). In this case, the quadratic form Ω can be computed explicitly as follows. Denote by $y(t)$, $t \in [t_1, t_f]$, the solution to the Cauchy problem

$$\dot{y} = f_x y, \quad y(t_1) = [\dot{x}]^1.$$

The following assertion is obvious: If $(\bar{\xi}, \bar{x}) \in \mathcal{K}$, then $\bar{x}(t) = 0$ for $t \in [t_0, t_1)$ and $\bar{x}(t) = y(t)\bar{\xi}$ for $t \in (t_1, t_f]$. Therefore, the inequality $K_{x_f}(x(t_f))y(t_f) \neq 0$ would imply $\mathcal{K} = \{0\}$. Consider now the case $K_{x_f}(x(t_f))y(t_f) = 0$. This condition always holds for time-optimal problems with a *scalar* function K and $\alpha_0 > 0$. Indeed, the condition $\frac{d}{dt}(\psi y) = 0$ implies $(\psi y)(t) = \text{const}$ in $[t_1, t_f]$, whence

$$(\psi y)(t_f) = (\psi y)(t_1) = \psi(t_1)[\dot{x}]^1 = \phi(t_1)[u]^1 = 0.$$

Using the transversality condition $\psi(t_f) = \beta K_{x_f}(x(t_f))$ and the inequality $\beta \neq 0$ (if $\beta = 0$, then $\psi(t_f) = 0$, and hence $\psi(t) = 0$ and $\psi_0(t) = 0$ in $[t_0, t_f]$), we see that the equality $(\psi y)(t_f) = 0$ implies the equality $K_{x_f}(x(t_f))y(t_f) = 0$.

Observe now that the cone $\mathcal{K}$ is a one-dimensional subspace, on which the quadratic form has the representation $\Omega = \rho\bar{\xi}^2$, where

$$\rho := D^1(H) - [\dot{\psi}]^1[\dot{x}]^1 + \int_{t_1}^{t_f} (y(t))^* H_{xx}(t) y(t)\, dt + (y(t_f))^* (\beta K)_{x_f x_f} y(t_f). \tag{6.37}$$

This gives the following result.

Proposition 6.11. *Suppose that we have found an extremal for the time-optimal control problem* (6.32) *that has one switching point and satisfies* $\alpha_0 > 0$ *and* $K_{x_f}(x(t_f))y(t_f) = 0$. *Then the inequality* $\rho > 0$ *with* ρ *defined in* (6.37) *is equivalent to the positive definiteness of* Ω *on* $\mathcal{K}$.

6.2.4 Proofs of Quadratic Conditions

It was already mentioned that problem (6.1)–(6.3) is a special case of the general problem (5.69)–(5.72). It is easy to check that in problem (6.1)–(6.3) we obtain the set M_0 and the critical cone $\mathcal{K}$ as the special cases of these sets in the general problem.

Let us compare the quadratic forms. It suffices to show that the endpoint quadratic form $\langle A\bar{\bar{p}}, \bar{\bar{p}}\rangle$ (see (6.35)) can be transformed into the endpoint quadratic form $\omega_e(\lambda, \bar{z})$ in (5.86) if relations (6.21) hold. Indeed, we have

$$\begin{aligned} \langle A\bar{\bar{p}}, \bar{\bar{p}}\rangle \; &:= \; \langle l_{pp}\bar{\bar{p}}, \bar{\bar{p}}\rangle + 2\dot{\psi}(t_0)\bar{\bar{x}}_0\bar{t}_0 + (\dot{\psi}_0(t_0) - \dot{\psi}(t_0)\dot{x}(t_0))\bar{t}_0^2 \\ &\quad - 2\dot{\psi}(t_f)\bar{\bar{x}}_f\bar{t}_f - (\dot{\psi}_0(t_f) - \dot{\psi}(t_f)\dot{x}(t_f))\bar{t}_f^2 \\ &= \; \langle l_{pp}\bar{\bar{p}}, \bar{\bar{p}}\rangle + 2\dot{\psi}(t_0)(\bar{x}_0 + \bar{t}_0\dot{x}(t_0))\bar{t}_0 + (\dot{\psi}_0(t_0) - \dot{\psi}(t_0)\dot{x}(t_0))\bar{t}_0^2 \\ &\quad - 2\dot{\psi}(t_f)(\bar{x}_1 + \bar{t}_f\dot{x}(t_f))\bar{t}_f - (\dot{\psi}_0(t_f) - \dot{\psi}(t_f)\dot{x}(t_f))\bar{t}_f^2 \\ &= \; \langle l_{pp}\bar{\bar{p}}, \bar{\bar{p}}\rangle - 2\dot{\psi}(t_f)\bar{x}_f\bar{t}_f - \Big(\dot{\psi}(t_f)\dot{x}(t_f) + \dot{\psi}_0(t_f)\Big)\bar{t}_f^2 \\ &\quad + 2\dot{\psi}(t_0)\bar{x}_0\bar{t}_0 + \Big(\dot{\psi}(t_0)\dot{x}(t_0) + \dot{\psi}_0(t_0)\Big)\bar{t}_0^2 =: \omega_e(\lambda, \bar{z}). \end{aligned}$$

Thus, Theorem 6.9, which gives necessary quadratic conditions in the problem (6.1)–(6.3), is a consequence of Theorem 5.22.

Now let us proceed to sufficient quadratic conditions in the same problem. Here the set M_0^+ consists of all those elements $\lambda \in M_0$ for which condition (6.19) is fulfilled, and the set $\mathrm{Leg}_+(M_0^+)$ consists of all those elements $\lambda \in M_0^+$ for which

$$D^k(H) > 0, \quad k = 1, \ldots, s. \tag{6.38}$$

Denote for brevity $\mathrm{Leg}_+(M_0^+) = \mathcal{M}$. Thus the set $\mathcal{M}$ consists of all those elements $\lambda \in M_0$ for which (6.19) and (6.38) are fulfilled. Let us also note that the strict bounded strong minimum in the problem (6.1)–(6.3) is equivalent to the strict strong minimum, since U is a compact set. Thus Theorem 5.24 implies the following result.

Theorem 6.12. *For a trajectory* $\mathcal{T}$ *in the problem* (6.1)–(6.3) *let the following Condition* $\mathfrak{B}$ *be fulfilled: The set* $\mathcal{M}$ *is nonempty and there exist a nonempty compact* $M \subset \mathcal{M}$ *and* $\varepsilon > 0$ *such that*

$$\max_{\lambda \in M} \Omega(\lambda, \bar{z}) \geq \varepsilon \bar{\gamma}(\bar{z}) \quad \forall\, \bar{z} \in \mathcal{K}, \tag{6.39}$$

where $\bar{\gamma}(\bar{z}) = \bar{t}_0^2 + \bar{t}_f^2 + \langle \bar{\xi}, \bar{\xi}\rangle + \langle \bar{x}(t_0), \bar{x}(t_0)\rangle$. *Then the trajectory* $\mathcal{T}$ *affords a strict strong minimum in this problem.*

Remarkably, the fact that the critical cone $\mathcal{K}$ in the problem (6.1)–(6.3) is finite-dimensional (since each element $\bar{z} = (\bar{t}_0, \bar{t}_f, \bar{\xi}, \bar{x})$ is uniquely defined by the parameters

$\bar{t}_0, \bar{t}_f, \bar{\xi}, \bar{x}(t_0))$ implies that condition $\mathfrak{B}$ in Theorem 6.12 is equivalent to the following, generally weaker condition.

Condition $\mathfrak{B}_0$. The set $\mathcal{M}$ is nonempty and

$$\max_{\lambda \in M_0} \Omega(\lambda, \bar{z}) > 0 \quad \forall\, \bar{z} \in \mathcal{K} \setminus \{0\}. \tag{6.40}$$

Lemma 6.13. *For a trajectory $\mathcal{T}$ in the problem* (6.1)–(6.3), *Condition $\mathfrak{B}$ is equivalent to the Condition $\mathfrak{B}_0$.*

Proof. As we pointed out, Condition $\mathfrak{B}$ always implies Condition $\mathfrak{B}_0$. We will show that, in our case, the inverse assertion also holds. Let Condition $\mathfrak{B}_0$ be fulfilled. Let $S_1(\mathcal{K})$ be the set of elements $\bar{z} \in \mathcal{K}$ satisfying the Condition $\bar{\gamma}(\bar{z}) = 1$. Then

$$\max_{\lambda \in M_0} \Omega(\lambda, \bar{z}) > 0 \quad \forall\, \bar{z} \in S_1(\mathcal{K}). \tag{6.41}$$

Recall that M_0 is a finite-dimensional compact set. It is readily verified that the relative interior of the cone $\operatorname{con} M_0$ generated by M_0 is contained in the cone $\operatorname{con} \mathcal{M}$ generated by $\mathcal{M}$, i.e., $\operatorname{reint}(\operatorname{con} M_0) \subset \operatorname{con} \mathcal{M}$. Combined with (6.41) this implies that for any element $\bar{z} \in S_1(\mathcal{K})$ there exist $\lambda \in \mathcal{M}$ and a neighborhood $\mathcal{O}_{\bar{z}} \subset S_1(\mathcal{K})$ of element $\bar{z}$ such that $\Omega(\lambda, \cdot) > 0$ on $\mathcal{O}_{\bar{z}}$. The family of neighborhoods $\{\mathcal{O}_{\bar{z}}\}$ forms an open covering of the compact set $S_1(\mathcal{K})$. Select a finite subcovering. To this subcovering there corresponds a finite compact set $M = \{\lambda_1, \dots, \lambda_r\}$ such that

$$\max_{\lambda \in M} \Omega(\lambda, \bar{z}) > 0 \quad \forall\, \bar{z} \in S_1(\mathcal{K}),$$

and consequently, due to compactness of the cross-section $S_1(\mathcal{K})$,

$$\max_{\lambda \in M} \Omega(\lambda, \bar{z}) > \varepsilon \quad \forall\, \bar{z} \in S_1(\mathcal{K}),$$

for some $\varepsilon > 0$. Hence Condition $\mathfrak{B}$ follows. ∎

Theorem 6.12 and Lemma 6.13 imply Theorem 6.10, where $\mathfrak{B} = \mathfrak{B}_0$. In what follows, for the problems of the type of basic problem (6.1)–(6.3), by Condition $\mathfrak{B}$ we will mean Condition $\mathfrak{B}_0$.

6.3 Sufficient Conditions for Positive Definiteness of the Quadratic Form Ω on the Critical Cone $\mathcal{K}$

Assume that the condition (a) of Theorem 6.10 is fulfilled for the trajectory $\mathcal{T}$. Let $\lambda \in M_0$ be a fixed element (possibly, different from that in condition (a)) and let $\Omega = \Omega(\lambda, \cdot)$ be the quadratic form (6.34) for this element. According to Theorem 6.10, the *positive definiteness of Ω on the critical cone $\mathcal{K}$ is a sufficient condition for a strict strong minimum.* Recall that $\mathcal{K}$ is defined by (6.23) and the condition $\bar{\bar{p}} \in \mathcal{K}_0$, where $\bar{\bar{p}} = (\bar{t}_0, \bar{\bar{x}}_0, \bar{t}_f, \bar{\bar{x}}_f)$, $\bar{\bar{x}}_0 = \bar{x}(t_0) + \bar{t}_0 \dot{x}(t_0)$, $\bar{\bar{x}}_f = \bar{x}(t_f) + \bar{t}_f \dot{x}(t_f)$. The cone $\mathcal{K}_0$ is defined by (6.26) in the case $\alpha_0 > 0$ and by (6.22) in the general case.

Now our aim is to find sufficient conditions for the positive definiteness of the quadratic form Ω on the cone $\mathcal{K}$. In what follows, we shall use the ideas and results presented in Chapter 4 (see also [69]).

6.3.1 Q-Transformation of Ω on $\mathcal{K}$

Let $Q(t)$ be a symmetric matrix on $[t_0, t_f]$ with piecewise continuous entries which are absolutely continuous on each interval of the set $[t_0, t_f] \setminus \Theta$. Therefore, Q may have a jump at each point $t_k \in \Theta$. For $\bar{z} \in \mathcal{K}$, formula (5.100) holds:

$$\int_{t_0}^{t_f} \frac{d}{dt}\langle Q\bar{x}, \bar{x}\rangle\, dt = \langle Q\bar{x}, \bar{x}\rangle \Big|_{t_0}^{t_f} - \sum_{k=1}^{s} [\langle Q\bar{x}, \bar{x}\rangle]^k,$$

where $[\langle Q\bar{x}, \bar{x}\rangle]^k$ is the jump of the function $\langle Q\bar{x}, \bar{x}\rangle$ at the point $t_k \in \Theta$. Using the equation $\dot{\bar{x}} = f_x \bar{x}$ with $f_x = f_x(t, x(t), u(t))$, we obtain

$$\sum_{k=1}^{s} [\langle Q\bar{x}, \bar{x}\rangle]^k + \int_{t_0}^{t_f} \langle (\dot{Q} + f_x^* Q + Q f_x)\bar{x}, \bar{x}\rangle\, dt - \langle Q\bar{x}, \bar{x}\rangle(t_f) + \langle Q\bar{x}, \bar{x}\rangle(t_0) = 0,$$

where the asterisk denotes transposition. Adding this zero form to Ω and using the equality $[H_x]^k = -[\dot{\psi}]^k$, we get

$$\begin{aligned}\Omega = {} & \langle A\bar{\bar{p}}, \bar{\bar{p}}\rangle - \langle Q\bar{x}, \bar{x}\rangle(t_f) + \langle Q\bar{x}, \bar{x}\rangle(t_0) \\ & + \sum_{k=1}^{s} \left(D^k(H)\bar{\xi}_k^2 - 2[\dot{\psi}]^k \bar{x}_{\mathrm{av}}^k \bar{\xi}_k + [\langle Q\bar{x}, \bar{x}\rangle]^k \right) \\ & + \int_{t_0}^{t_f} \langle (H_{xx} + \dot{Q} + f_x^* Q + Q f_x)\bar{x}, \bar{x}\rangle\, dt.\end{aligned} \tag{6.42}$$

We shall call this formula the *Q-transformation of Ω on $\mathcal{K}$*.

In order to eliminate the integral term in Ω, we assume that $Q(t)$ satisfies the following linear matrix differential equation:

$$\dot{Q} + f_x^* Q + Q f_x + H_{xx} = 0 \quad \text{on } [t_0, t_f] \setminus \Theta. \tag{6.43}$$

It is interesting to note that the same equation is obtained from the modified Riccati equation in Maurer and Pickenhain [73, Equation (47)], when all control variables are on the boundary of the control constraints. Using (6.43) the quadratic form (6.42) reduces to

$$\Omega = \omega_0 + \sum_{k=1}^{s} \omega_k, \tag{6.44}$$

$$\omega_k := D^k(H)\bar{\xi}_k^2 - 2[\dot{\psi}]^k \bar{x}_{\mathrm{av}}^k \bar{\xi}_k + [\langle Q\bar{x}, \bar{x}\rangle]^k, \quad k = 1, \dots, s, \tag{6.45}$$

$$\omega_0 := \langle A\bar{\bar{p}}, \bar{\bar{p}}\rangle - \langle Q\bar{x}, \bar{x}\rangle(t_f) + \langle Q\bar{x}, \bar{x}\rangle(t_0). \tag{6.46}$$

Thus, we have proved the following statement.

Proposition 6.14. *Let $Q(t)$ satisfy the linear differential equation* (6.43) *on* $[t_0,t_f]\setminus\Theta$. *Then for each $\bar{z}\in\mathcal{K}$ the representation* (6.44) *holds.*

Now our goal is to derive conditions such that $\omega_k>0$, $k=0,\dots,s$, holds on $\mathcal{K}\setminus\{0\}$. To this end, as in Section 4.1.6, we shall express ω_k via the vector $(\bar{\xi}_k,\bar{x}^{k-})$. We use the formula

$$\bar{x}^{k+}=\bar{x}^{k-}+[\dot{x}]^k\bar{\xi}_k, \tag{6.47}$$

which implies

$$\langle Q^{k+}\bar{x}^{k+},\bar{x}^{k+}\rangle=\langle Q^{k+}\bar{x}^{k-},\bar{x}^{k-}\rangle+2\langle Q^{k+}[\dot{x}]^k,\bar{x}^{k-}\rangle\bar{\xi}_k+\langle Q^{k+}[\dot{x}]^k,[\dot{x}]^k\rangle\bar{\xi}_k^2.$$

Consequently,

$$[\langle Q\bar{x},\bar{x}\rangle]^k=\langle[Q]^k\bar{x}^{k-},\bar{x}^{k-}\rangle+2\langle Q^{k+}[\dot{x}]^k,\bar{x}^{k-}\rangle\bar{\xi}_k+\langle Q^{k+}[\dot{x}]^k,[\dot{x}]^k\rangle\bar{\xi}_k^2.$$

Using this relation together with $\bar{x}^k_{\text{av}}=\bar{x}^{k-}+\frac{1}{2}[\dot{x}]^k\bar{\xi}_k$ in the definition (6.45) of ω_k, we obtain

$$\begin{aligned}\omega_k &= \left\{D^k(H)+\left(([\dot{x}]^k)^*Q^{k+}-[\dot{\psi}]^k\right)[\dot{x}]^k\right\}\bar{\xi}_k^2\\ &\quad+2\left(([\dot{x}]^k)^*Q^{k+}-[\dot{\psi}]^k\right)\bar{x}^{k-}\bar{\xi}_k+(\bar{x}^{k-})^*[Q]^k\bar{x}^{k-}.\end{aligned} \tag{6.48}$$

Here $[\dot{x}]^k$ and $\bar{x}^{k-}$ are column vectors while $([\dot{x}]^k)^*$, $(\bar{x}^{k-})^*$, and $[\dot{\psi}]^k$ are row vectors. By putting

$$q_{k+}=([\dot{x}]^k)^*Q^{k+}-[\dot{\psi}]^k,\quad b_{k+}=D^k(H)+(q_{k+})[\dot{x}]^k, \tag{6.49}$$

we get

$$\omega_k=(b_{k+})\bar{\xi}_k^2+2(q_{k+})\bar{x}^{k-}\bar{\xi}_k+(\bar{x}^{k-})^*[Q]^k\bar{x}^{k-}. \tag{6.50}$$

Note that ω_k is a quadratic form in the variables $(\bar{\xi}_k,\bar{x}^{k-})$ with the matrix

$$M_{k+}=\begin{pmatrix} b_{k+} & q_{k+}\\ (q_{k+})^* & [Q]^k\end{pmatrix}, \tag{6.51}$$

where q_{k+} is a row vector and $(q_{k+})^*$ is a column vector. Similarly, using the relation $\bar{x}^{k-}=\bar{x}^{k+}-[\dot{x}]^k\bar{\xi}_k$, we obtain

$$[\langle Q\bar{x},\bar{x}\rangle]^k=\langle[Q]^k\bar{x}^{k+},\bar{x}^{k+}\rangle+2\langle Q^{k-}[\dot{x}]^k,\bar{x}^{k+}\rangle\bar{\xi}^k-\langle Q^{k-}[\dot{x}]^k,[\dot{x}]^k\rangle\bar{\xi}_k^2.$$

This formula together with the relation $\bar{x}^k_{\text{av}}=\bar{x}^{k+}-\frac{1}{2}[\dot{x}]^k\bar{\xi}_k$ leads to the representation (cf. formula (4.160))

$$\omega_k=(b_{k-})\bar{\xi}_k^2+2(q_{k-})\bar{x}^{k+}\bar{\xi}_k+(\bar{x}^{k+})^*[Q]^k\bar{x}^{k+}, \tag{6.52}$$

where

$$q_{k-}=([\dot{x}]^k)^*Q^{k-}-[\dot{\psi}]^k,\quad b_{k-}=D^k(H)-(q_{k-})[\dot{x}]^k. \tag{6.53}$$

We consider (6.52) as a quadratic form in the variables $(\bar{\xi}_k, \bar{x}^{k+})$ with the matrix

$$M_{k-} = \begin{pmatrix} b_{k-} & q_{k-} \\ (q_{k-})^* & [Q]^k \end{pmatrix}. \tag{6.54}$$

Since the right-hand sides of equalities (6.50) and (6.52) are connected by relation (6.47), the following statement obviously holds.

Proposition 6.15. *For each $k = 1, \ldots, s$, the positive (semi)definiteness of the matrix M_{k-} is equivalent to the positive (semi)definiteness of the matrix M_{k+}.*

Now we can prove two theorems.

Theorem 6.16. *Assume that $s = 1$. Let $Q(t)$ be a solution to the linear differential equation* (6.43) *on $[t_0, t_f] \setminus \Theta$ which satisfies two conditions:*

(i) *The matrix M_{1+} is positive semidefinite;*
(ii) *the quadratic form ω_0 is positive on the cone $\mathcal{K}_0 \setminus \{0\}$.*

Then Ω is positive on $\mathcal{K} \setminus \{0\}$.

Proof. Take an arbitrary element $\bar{z} \in \mathcal{K}$. Conditions (i) and (ii) imply that $\omega_k \geq 0$ for $k = 0, 1$, and hence $\Omega = \omega_0 + \omega_1 \geq 0$ for this element. Assume now that $\Omega = 0$. Then $\omega_k = 0$ for $k = 0, 1$. By virtue of (ii) the equality $\omega_0 = 0$ implies that $\bar{t}_0 = \bar{t}_f = 0$ and $\bar{x}(t_0) = \bar{x}(t_f) = 0$. The last two equalities and the equation $\dot{\bar{x}} = f_x \bar{x}$ show that $\bar{x}(t) = 0$ in $[t_0, t_1) \cup (t_1, t_f]$. Now using formula (6.45) for $\omega_1 = 0$, as well as the conditions $D^1(H) > 0$ and $\bar{x}^{1-} = \bar{x}^{1+} = 0$, we obtain that $\bar{\xi}_1 = 0$. Consequently, we have $\bar{z} = 0$ which means that Ω is positive on $\mathcal{K} \setminus \{0\}$. ∎

Theorem 6.17. *Assume that $s \geq 2$. Let $Q(t)$ be a solution to the linear differential equation* (6.43) *on $[t_0, t_f] \setminus \Theta$ which satisfies the following conditions:*

(a) *The matrix M_{k+} is positive semidefinite for each $k = 1, \ldots, s$;*
(b) *$b_{k+} := D^k(H) + (q_{k+})[\dot{x}]^k > 0$ for each $k = 1, \ldots, s-1$;*
(c) *the quadratic form ω_0 is positive on the cone $\mathcal{K}_0 \setminus \{0\}$.*

Then Ω is positive on $\mathcal{K} \setminus \{0\}$.

Proof. Take an arbitrary element $\bar{z} \in \mathcal{K}$. Conditions (a) and (c) imply that $\omega_k \geq 0$ for $k = 0, 1, \ldots, s$, and then $\Omega \geq 0$ for this element. Assume that $\Omega = 0$. Then $\omega_k = 0$ for $k = 0, 1, \ldots, s$. By virtue of (c) the equality $\omega_0 = 0$ implies that $\bar{t}_0 = \bar{t}_f = 0$ and $\bar{x}(t_0) = \bar{x}(t_f) = 0$. The last two equalities and the equation $\dot{\bar{x}} = f_x \bar{x}$ show that $\bar{x}(t) = 0$ in $[t_0, t_1) \cup (t_s, t_f]$ and hence $\bar{x}^{1-} = \bar{x}^{s+} = 0$. The conditions $\omega_1 = 0$, $\bar{x}^{1-} = 0$, and $b_{1+} > 0$ by formula (6.50) (with $k = 1$) yield $\bar{\xi}_1 = 0$. Then $[\bar{x}]^1 = 0$ and hence $\bar{x}^{1+} = 0$. The last equality and the equation $\dot{\bar{x}} = f_x \bar{x}$ show that $\bar{x}(t) = 0$ in (t_1, t_2) and hence $\bar{x}^{2-} = 0$. Similarly, the conditions $\omega_2 = 0$, $\bar{x}^{2-} = 0$ and $b_{2+} > 0$ by formula (6.50) (with $k = 2$) imply that $\bar{\xi}_2 = 0$ and $\bar{x}(t) = 0$ in (t_2, t_3). Therefore, $\bar{x}^{3-} = 0$, etc. Continuing this process we get that $\bar{x} \equiv 0$ and $\bar{\xi}_k = 0$ for $k = 1, \ldots, s-1$. Now using formula (6.45) for $\omega_s = 0$, as well as the conditions $D^s(H) > 0$ and $\bar{x} \equiv 0$, we obtain that $\bar{\xi}_s = 0$. Consequently, $\bar{z} = 0$, and hence Ω is positive on $\mathcal{K} \setminus \{0\}$. ∎

Similarly, using representation (6.52) for ω_k we can prove the following statement.

Theorem 6.18. *Let $Q(t)$ be a solution to the linear differential equation* (6.43) *on* $[t_0, t_f] \setminus \Theta$ *which satisfies the following conditions:*

(a′) *The matrix M_{k-} is positive semidefinite for each $k = 1, \dots, s$;*

(b′) $b_{k-} := D^k(H) - (q_{k-})[\dot{x}]^k > 0$ *for each* $k = 2, \dots, s$
(if $s = 1$, then this condition is not required);

(c) *the quadratic form ω_0 is positive on the cone $\mathcal{K}_0 \setminus \{0\}$.*

Then Ω is positive on $\mathcal{K} \setminus \{0\}$.

Remark. Noble and Schättler [80] and Ledzewicz and Schättler [53] use also the linear ODE (6.43) for deriving sufficient conditions. It would be of interest to relate their approach to the results in Theorem 6.18.

6.3.2 Case of Fixed Initial Values t_0 and $x(t_0)$

Consider the problem (6.1)–(6.3) with additional constraints $t_0 = \hat{t}_0$ and $x(t_0) = \hat{x}_0$. In this case we have additional equalities in the definition of the critical cone $\mathcal{K}$: $\bar{t}_0 = 0$ and $\bar{\bar{x}}_0 := \bar{x}(t_0) + \bar{t}_0 \dot{x}(t_0) = 0$, whence $\bar{x}(t_0) = 0$. The last equality and the equation $\dot{\bar{x}} = f_x \bar{x}$ show that $\bar{x}(t) = 0$ in $[t_0, t_1)$, whence $\bar{x}^{1-} = 0$. From definitions (6.46) and (6.35) of ω_0 and $\langle A\bar{\bar{p}}, \bar{\bar{p}} \rangle$, respectively, it follows that for each $\bar{z} \in \mathcal{K}$ we have

$$\omega_0 = \langle A_1 \bar{\bar{p}}, \bar{\bar{p}} \rangle - \langle Q(t_f)(\bar{\bar{x}}_f - \bar{t}_f \dot{x}(t_f)), (\bar{\bar{x}}_f - \bar{t}_f \dot{x}(t_f)) \rangle, \tag{6.55}$$

where

$$\begin{aligned} \langle A_1 \bar{\bar{p}}, \bar{\bar{p}} \rangle = {} & l_{t_f t_f} \bar{t}_f^2 + 2 l_{t_f x_f} \bar{\bar{x}}_f \bar{t}_f + \langle l_{x_f x_f} \bar{\bar{x}}_f, \bar{\bar{x}}_f \rangle \\ & - 2 \dot{\psi}(t_f) \bar{\bar{x}}_f \bar{t}_f - (\dot{\psi}_0(t_f) - \dot{\psi}(t_f) \dot{x}(t_f)) \bar{t}_f^2. \end{aligned} \tag{6.56}$$

The equalities $\bar{t}_0 = 0$ and $\bar{\bar{x}}_0 = 0$ hold also for each element $\bar{\bar{p}}$ of the finite-dimensional and finite-faced cone $\mathcal{K}_0$, given by (6.26) for $\alpha_0 > 0$ and by (6.22) in the general case. Rewriting the terms ω_0, we get the quadratic form in the variables $(\bar{t}_f, \bar{\bar{x}}_f)$ generated by the matrix

$$B := \begin{pmatrix} B_{11} & B_{12} \\ B_{12}^* & B_{22} \end{pmatrix},$$

where

$$\begin{aligned} B_{11} &= l_{t_f t_f} + \dot{\psi}(t_f) \dot{x}(t_f) - \dot{\psi}_0(t_f) - \dot{x}(t_f)^* Q(t_f) \dot{x}(t_f), \\ B_{12} &= l_{t_f x_f} - \dot{\psi}(t_f) + \dot{x}(t_f)^* Q(t_f), \\ B_{22} &= l_{x_f x_f} - Q(t_f). \end{aligned} \tag{6.57}$$

The property $\bar{x}(t) = 0$ in $[t_0, t_1)$ for $\bar{z} \in \mathcal{K}$ allows us to refine Theorems 6.16 and 6.17.

Theorem 6.19. *Assume that the initial values $t_0 = \hat{t}_0$ and $x(t_0) = \hat{x}_0$ are fixed in the problem* (6.1)–(6.3), *and let $s = 1$. Let $Q(t)$ be a continuous solution of the linear differential equation* (6.43) *on* $[t_1, t_f]$ *which satisfies two conditions:*

(i) $b_1 := D^1(H) + \big(([\dot{x}]^1)^* Q(t_1) - [\dot{\psi}]^1\big)[\dot{x}]^1 \geq 0;$

(ii) *the quadratic form ω_0 is positive on the cone $\mathcal{K}_0 \setminus \{0\}$.*

Then Ω is positive on $\mathcal{K} \setminus \{0\}$.

Proof. Continue $Q(t)$ arbitrarily as a solution of differential equation (6.43) to the whole interval $[t_0, t_f]$ with possible jump at the point t_1. Note that the value b_1 in condition (i) is the same as the value b_{1+} for the continued solution, and hence $b_{1+} \geq 0$. Let $\bar{z} \in \mathcal{K}$, and hence $\bar{x}^{1-} = 0$. Then by (6.50) with $k = 1$ the condition $b_{1+} \geq 0$ implies the inequality $\omega_1 \geq 0$. Condition (ii) implies the inequality $\omega_0 \geq 0$. Consequently $\Omega = \omega_0 + \omega_1 \geq 0$. Further arguments are the same as in the proof of Theorem 6.16. ∎

Theorem 6.20. *Assume that the initial values $t_0 = \hat{t}_0$ and $x(t_0) = \hat{x}_0$ are fixed in the problem* (6.1)–(6.3) *and $s \geq 2$. Let $Q(t)$ be a solution of the linear differential equation* (6.43) *on $(t_1, t_f] \setminus \Theta$ which satisfies the following conditions:*

(a) *The matrix M_{k+} is positive semidefinite for each $k = 2, \ldots, s$;*

(b) *$b_{k+} := D^k(H) + (q_{k+})[\dot{x}]^k > 0$ for each $k = 1, \ldots, s-1$;*

(c) *the quadratic form ω_0 is positive on the cone $\mathcal{K}_0 \setminus \{0\}$.*

Then Ω is positive on $\mathcal{K} \setminus \{0\}$.

Proof. Again, without loss of generality, we can consider $Q(t)$ as a discontinuous solution of equation (6.43) on the whole interval $[t_0, t_f]$. Let $\bar{z} \in \mathcal{K}$. Then by (6.50) with $k = 1$ the conditions $b_{1+} > 0$ and $\bar{x}^{1-} = 0$ imply the inequality $\omega_1 \geq 0$. Further arguments are the same as in the proof of Theorem 6.17. ∎

6.3.3 Q-Transformation of Ω to Perfect Squares

As in Section 5.3.3, we shall formulate special jump conditions for the matrix Q at each point $t_k \in \Theta$. This will make it possible to transform Ω into perfect squares and thus to prove its positive definiteness on $\mathcal{K}$.

Proposition 6.21. *Suppose that*

$$b_{k+} := D^k(H) + (q_{k+})[\dot{x}]^k > 0 \tag{6.58}$$

and that Q satisfies the jump condition at t_k,

$$b_{k+}[Q]^k = (q_{k+})^*(q_{k+}), \tag{6.59}$$

where $(q_{k+})^*$ is a column vector while q_{k+} is a row vector (defined as in (6.49)*). Then ω_k can be written as the perfect square*

$$\begin{aligned} \omega_k &= (b_{k+})^{-1}\left((b_{k+})\bar{\xi}_k + (q_{k+})(\bar{x}^{k-})\right)^2 \\ &= (b_{k+})^{-1}\left(D^k(H)\bar{\xi}_k + (q_{k+})(\bar{x}^{k+})\right)^2. \end{aligned} \tag{6.60}$$

Proof. These formulas were proved in Section 4.1.6. ∎

Theorem 6.22. *Let $Q(t)$ satisfy the linear differential equation* (6.43) *on $[t_0, t_f] \setminus \Theta$, and let conditions* (6.58) *and* (6.59) *hold for each $k = 1, \ldots, s$. Let ω_0 be positive on $\mathcal{K}_0 \setminus \{0\}$. Then Ω is positive on $\mathcal{K} \setminus \{0\}$.*

Proof. By Proposition 6.21 and formulae (6.50), (6.51) the matrix M_{k+} is positive semi-definite for each $k = 1,\dots,n$. Now using Theorem 6.16 for $s = 1$ and Theorem 6.17 for $s \geq 2$, we obtain that Ω is positive on $\mathcal{K} \setminus \{0\}$. ∎

Similar assertions hold for the jump conditions that use left-hand values of Q at each point $t_k \in \Theta$. Suppose that

$$b_{k-} := D^k(H) - (q_{k-})[\dot{x}]^k > 0 \tag{6.61}$$

and that Q satisfies the jump condition at t_k

$$b_{k-}[Q]^k = (q_{k-})^*(q_{k-}). \tag{6.62}$$

Then, according to Proposition 4.30, we have

$$\begin{aligned} \omega_k &= (b_{k-})^{-1}\left((b_{k-})\bar{\xi}_k + (q_{k-})(\bar{x}^{k+})\right)^2 \\ &= (b_{k-})^{-1}\left(D^k(H)\bar{\xi}_k + (q_{k-})(\bar{x}^{k-})\right)^2. \end{aligned} \tag{6.63}$$

Using these formulas we deduce the following theorem.

Theorem 6.23. *Let $Q(t)$ satisfy the linear differential equation (6.43) on $[t_0,t_f] \setminus \Theta$, and let conditions (6.61) and (6.62) hold for each $k = 1,\dots,s$. Let ω_0 be positive on $\mathcal{K}_0 \setminus \{0\}$. Then Ω is positive on $\mathcal{K} \setminus \{0\}$.*

6.4 Example: Minimal Fuel Consumption of a Car

The following optimal control problem has been treated by Oberle and Pesch [83] as an exercise of applying the minimum principle. Consider a car whose dynamics (position x_1 and velocity x_2) are subject to friction and gravitational forces. The acceleration $u(t)$ is proportional to the fuel consumption. Thus the control problem is to minimize the total fuel consumption

$$\mathcal{J} = \int_0^{t_f} u(t)\,dt \tag{6.64}$$

in a time interval $[0,t_f]$ subject to the dynamic constraints, boundary conditions, and the control constraints

$$\dot{x}_1 = x_2, \quad \dot{x}_2 = \frac{u}{mx_2} - \alpha g - \frac{c}{m}x_2^2, \tag{6.65}$$

$$x_1(0) = 0,\ x_2(0) = 1, \quad x_1(t_f) = 10,\ x_2(t_f) = 3, \tag{6.66}$$

$$u_{\min} \leq u(t) \leq u_{\max}, \quad 0 \leq t \leq t_f. \tag{6.67}$$

The final time t_f is unspecified. The following values of the constants will be used in computations:

$$m = 4, \quad \alpha = 1, \quad g = 10, \quad c = 0.4, \quad u_{\min} = 100, \quad u_{\max} = 140.$$

In view of the integral cost criterion (6.64), we consider the Pontryagin function (Hamiltonian) (cf. (6.17)) in normalized form taking $\alpha_0 = 1$,

$$H(x_1, x_2, \psi_1, \psi_2, u) = u + \psi_1 x_2 + \psi_2 \left(\frac{u}{m x_2} - \alpha g - \frac{c}{m} x_2^2 \right). \tag{6.68}$$

The adjoint equations $\dot{\psi} = -H_x$ are

$$\dot{\psi}_1 = 0, \quad \dot{\psi}_2 = -\psi_1 + \psi_2 \left(\frac{u}{m x_2^2} + \frac{2c}{m} x_2 \right). \tag{6.69}$$

The transversality condition (6.11) evaluated at the free final time t_f yields the additional boundary condition

$$u(t_f) + 3\psi_1(t_f) + \psi_2(t_f) \left(\frac{u(t_f)}{3m} - \alpha g - \frac{9c}{m} \right) = 0. \tag{6.70}$$

The switching function

$$\phi(x, \psi) = D_u H = 1 + \frac{\psi_2}{m x_2}, \quad \phi(t) = \phi(x(t), \psi(t)),$$

determines the control law

$$u(t) = \left\{ \begin{array}{lll} u_{\min} & \text{if} & \phi(t) > 0 \\ u_{\max} & \text{if} & \phi(t) < 0 \end{array} \right\}.$$

Computations give evidence to the fact that the optimal control is bang-bang with one switching time t_1,

$$u(t) = \left\{ \begin{array}{ll} u_{\min}, & 0 \le t < t_1 \\ u_{\max}, & t_1 \le t \le t_f \end{array} \right\}.$$

We compute an extremal using the code BNDSCO of Oberle and Grimm [82] or the code NUDOCCCS of Büskens [13]. The solution is displayed in Figure 6.1. Results for the switching time t_1, final time t_f, and adjoint variables $\psi(t)$ are

$$\begin{array}{rclrcl} t_1 & = & 3.924284, & t_f & = & 4.254074, \\ \psi_1(0) & = & -42.24170, & \psi_2(0) & = & -3.876396, \\ x_1(t_1) & = & 9.086464, & x_2(t_1) & = & 2.367329, \\ \psi_1(t_f) & = & -42.24170, & \psi_2(t_f) & = & -17.31509. \end{array}$$

We will show that this trajectory satisfies the assumptions of Proposition 6.7. The critical cone is $\mathcal{K} = \{0\}$, since the computed vectors

$$\frac{\partial x}{\partial t_1}(t_f) = (-0.6326710, -0.7666666)^*, \quad \dot{x}(t_f) = (3.0, 0.7666666)^*$$

are linearly independent. Moreover, we find in view of (6.14) that

$$D^1(H) = -\dot{\phi}(t_1)[u]^1 = 0.472397 \cdot 40 > 0.$$

Theorem 6.10 then asserts that the computed bang-bang control provides a strict strong minimum.

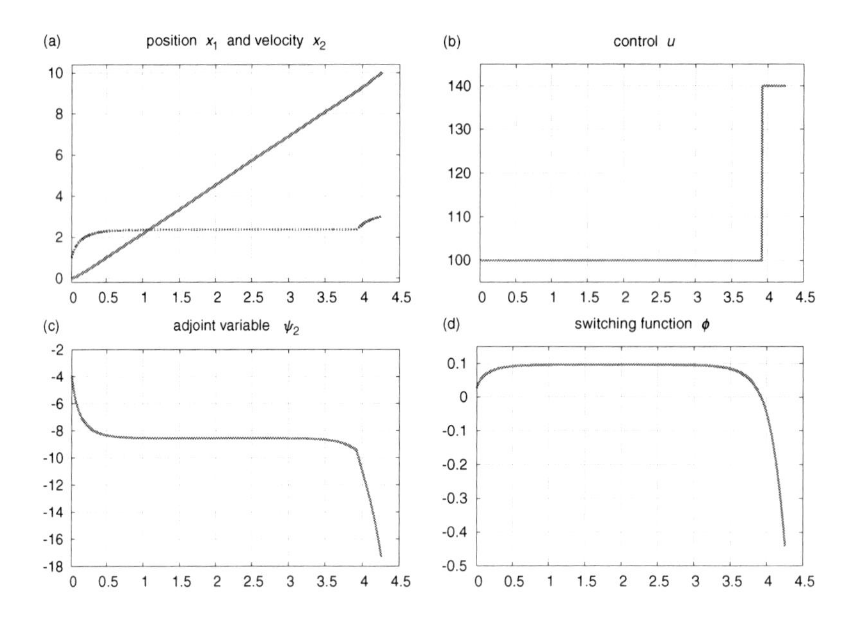

Figure 6.1. *Minimal fuel consumption of a car.* (a) *State variables* x_1, x_2. (b) *Bang-bang control* u. (c) *Adjoint variable* ψ_2. (d) *Switching function* ϕ.

6.5 Quadratic Optimality Conditions in Time-Optimal Bang-Bang Control Problems

In this section, we present the results of our paper [70].

6.5.1 Statement of the Problem; Pontryagin and Strong Minimum

We consider time-optimal control problems with control appearing linearly. Let $x(t) \in \mathbb{R}^{d(x)}$ denote the state variable and $u(t) \in \mathbb{R}^{d(u)}$ the control variable in the time interval $t \in \Delta = [0, t_f]$ with a nonfixed final time $t_f > 0$. For simplicity, the initial and terminal states are fixed in the following control problem:

$$\text{Minimize the final time} \quad t_f \tag{6.71}$$

subject to the constraints on the interval $\Delta = [0, t_f]$,

$$dx/dt = \dot{x} = f(t, x, u) = a(t, x) + B(t, x)u, \tag{6.72}$$

$$x(0) = a_0, \quad x(t_f) = a_1, \tag{6.73}$$

$$u(t) \in U, \quad (t, x(t)) \in \mathcal{Q}. \tag{6.74}$$

Here, a_0 and a_1 are given points in $\mathbb{R}^{d(x)}$, $\mathcal{Q} \subset \mathbb{R}^{1+d(x)}$ is an open set, and $U \subset \mathbb{R}^{d(u)}$ is a convex polyhedron. The functions a, B are twice continuously differentiable on $\mathcal{Q}$, with B being a $d(x) \times d(u)$ matrix function. A trajectory or control process $\mathcal{T} = \{(x(t), u(t)) \mid t \in [0, t_f]\}$ is said to be *admissible* if $x(\cdot)$ is absolutely continuous, $u(\cdot)$ is measurable and essentially bounded, and the pair of functions $(x(t), u(t))$ satisfies the constraints (6.72)–(6.74) on the interval $\Delta = [0, t_f]$. Let us define the Pontryagin and the strong minimum in the problem.

An admissible trajectory $\mathcal{T}$ is said to be a *Pontryagin minimum* if there is no sequence of admissible trajectories $\mathcal{T}^n = \{(x^n(t), u^n(t)) \mid t \in [0, t_f^n]\}$, $n = 1, 2, \ldots$, with

(a) $t_f^n < t_f$ for $n = 1, 2, \ldots$;
(b) $t_f^n \to t_f$ for $n \to \infty$;
(c) $\max_{\Delta^n} |x^n(t) - x(t)| \to 0$ for $n \to \infty$, where $\Delta^n = [0, t_f^n]$;
(d) $\int_{\Delta^n} |u^n(t) - u(t)|\, dt \to 0$ for $n \to \infty$.

An admissible trajectory $\mathcal{T}$ is said to be *a strong minimum* (respectively, *a strict strong minimum*) if there is no sequence of admissible trajectories $\mathcal{T}^n$, $n = 1, 2, \ldots$ such that

(a) $t_f^n < t_f$ ($t_f^n \le t_f$, $\mathcal{T}^n \ne \mathcal{T}$) for $n = 1, 2, \ldots$;
(b) $t_f^n \to t_f$ for $n \to \infty$;
(c) $\max_{\Delta^n} |x^n(t) - x(t)| \to 0$ for $n \to \infty$, where $\Delta^n = [0, t_f^n]$.

6.5.2 Minimum Principle

Let $\mathcal{T} = (x(t), u(t) \mid t \in [0, t_f])$ be a fixed admissible trajectory such that the control $u(\cdot)$ is a piecewise constant function on the interval $\Delta = [0, t_f]$ with finitely many points of discontinuity. Denote by $\Theta = \{t_1, \ldots, t_s\}$, $0 < t_1 < \cdots < t_s < t_f$, the finite set of all discontinuity points (jump points) of the control $u(t)$. Then $\dot{x}(t)$ is a piecewise continuous function whose discontinuity points belong to the set Θ and, thus, $x(t)$ is a piecewise smooth function on Δ. We use the notation $[u]^k = u^{k+} - u^{k-}$ for the jump of the function $u(t)$ at the point $t_k \in \Theta$, where $u^{k-} = u(t_k-)$, $u^{k+} = u(t_k+)$ are, respectively, the left- and right-hand values of the control $u(t)$ at t_k. Similarly, we denote by $[\dot{x}]^k$ the jump of the function $\dot{x}(t)$ at the same point.

Let us formulate the first-order necessary conditions of optimality for the trajectory $\mathcal{T}$, the *Pontryagin minimum principle*. To this end we introduce the *Pontryagin function* or *Hamiltonian function*

$$H(t, x, u, \psi) = \psi f(t, x, u) = \psi a(t, x) + \psi B(t, x) u, \tag{6.75}$$

where ψ is a row vector of dimension $d(x)$, while x, u, and f are column vectors. The factor of the control u in the Pontryagin function is the switching function $\phi(t, x, \psi) = \psi B(t, x)$. Consider the pair of functions $\psi_0(\cdot) : \Delta \to \mathbb{R}^1$, $\psi(\cdot) : \Delta \to \mathbb{R}^{d(x)}$, which are continuous on Δ and continuously differentiable on each interval of the set $\Delta \setminus \Theta$. Denote by M_0 the set of normed pairs of functions $(\psi_0(\cdot), \psi(\cdot))$ satisfying the conditions

$$\psi_0(t_f) \ge 0, \quad |\psi(0)| = 1, \tag{6.76}$$

$$\dot{\psi}(t) = -H_x(t, x(t), u(t), \psi(t)) \quad \forall\, t \in \Delta \setminus \Theta, \tag{6.77}$$

$$\dot{\psi}_0(t) = -H_t(t, x(t), u(t), \psi(t)) \quad \forall\, t \in \Delta \setminus \Theta, \tag{6.78}$$

$$\min_{u \in U} H(t, x(t), u, \psi(t)) = H(t, x(t), u(t), \psi(t)) \quad \forall\, t \in \Delta \setminus \Theta, \tag{6.79}$$

$$H(t, x(t), u(t), \psi(t)) + \psi_0(t) = 0 \quad \forall\, t \in \Delta \setminus \Theta. \tag{6.80}$$

Then the condition $M_0 \ne \emptyset$ is equivalent to the Pontryagin minimum principle. This is the first-order necessary condition for a Pontryagin minimum. We assume that this condition is satisfied for the trajectory $\mathcal{T}$. We say in this case that $\mathcal{T}$ is an *extremal trajectory* for the problem. The set M_0 is a finite-dimensional compact set, since in (6.76) the initial values $\psi(0)$ are assumed to belong to the unit ball of $\mathbb{R}^{d(x)}$. The case that there exists a multiplier $(\psi_0,\psi) \in M_0$ with $\psi_0(t_f) > 0$ will be called the *nondegenerate* or *normal* case. Again we use the simple abbreviation (t) for all arguments $(t,x(t),u(t),\psi(t))$, e.g., $\phi(t) = \phi(t,x(t),\psi(t))$.

Let us introduce the quantity $D^k(H)$. For $(\psi_0,\psi) \in M_0$ and $t_k \in \Theta$ consider the function

$$(\Delta_k H)(t) = H(t,x(t),u^{k+},\psi(t)) - H(t,x(t),u^{k-},\psi(t)) = \phi(t,x(t),\psi(t))[u]^k.$$

This function has a derivative at the point $t_k \in \Theta$. We set

$$D^k(H) = -\frac{d}{dt}(\Delta_k H)(t_k) = -\dot{\phi}(t_k\pm)[u]^k.$$

We know that for each $(\psi_0,\psi) \in M_0$

$$D^k(H) \geq 0 \quad \text{for } k = 1,\ldots,s. \tag{6.81}$$

We need the definition of a strict bang-bang control (see Section 6.1.4) to obtain the sufficient conditions in Theorem 6.27. For a given extremal trajectory $\mathcal{T} = \{(x(t),u(t)) \mid t \in \Delta\}$ with piecewise constant control $u(t)$ we say that $u(t)$ is a *strict bang-bang control* if there exists $(\psi_0,\psi) \in M_0$ such that

$$\operatorname*{Arg\,min}_{u' \in U} \phi(t,x(t),\psi(t))u' = [u(t-),u(t+)] \quad \forall\, t \in [t_0,t_f], \tag{6.82}$$

where

$$[u(t-),u(t+)] = \{\alpha u(t-) + (1-\alpha)u(t+) \mid 0 \leq \alpha \leq 1\}$$

denotes the line segment in $\mathbb{R}^{d(u)}$. As it was mentioned already in Section 6.1.4, if U is the unit cube in $\mathbb{R}^{d(u)}$, condition (6.82) precludes *simultaneous switching of the control components*. However, this property holds for all numerical examples in Chapter 8.

In order to formulate quadratic optimality condition for a given extremal $\mathcal{T}$ with bang-bang control $u(\cdot)$ we shall introduce the space $\mathcal{Z}(\Theta)$, the critical subspace $\mathcal{K} \subset \mathcal{Z}(\Theta)$, and the quadratic form Ω defined in $\mathcal{Z}(\Theta)$.

6.5.3 Critical Subspace

As in Section 6.2.1, we denote by $P_\Theta C^1(\Delta,\mathbb{R}^n)$ the space of piecewise continuous functions $\bar{x}(\cdot) : \Delta \to \mathbb{R}^n$, that are continuously differentiable on each interval of the set $\Delta \setminus \Theta$. For each $\bar{x} \in P_\Theta C^1(\Delta,\mathbb{R}^n)$ and for $t_k \in \Theta$, we use the abbreviation $[\bar{x}]^k = \bar{x}^{k+} - \bar{x}^{k-}$, where $\bar{x}^{k-} = \bar{x}(t_k-)$, $\bar{x}^{k+} = \bar{x}(t_k+)$. Putting

$$\bar{z} = (\bar{t}_f,\bar{\xi},\bar{x}) \quad \text{with} \quad \bar{t}_f \in \mathbb{R}^1, \quad \bar{\xi} \in \mathbb{R}^s, \quad \bar{x} \in P_\Theta C^1(\Delta,\mathbb{R}^n),$$

we have $\bar{z} \in \mathcal{Z}(\Theta) := \mathbb{R}^1 \times \mathbb{R}^s \times P_\Theta C^1(\Delta,\mathbb{R}^n)$. Denote by $\mathcal{K}$ the set of all $\bar{z} \in \mathcal{Z}(\Theta)$ satisfying the following conditions:

$$\dot{\bar{x}}(t) = f_x(t,x(t),u(t))\bar{x}(t), \quad [\bar{x}]^k = [\dot{x}]^k \bar{\xi}_k, \quad k = 1,\ldots,s, \tag{6.83}$$

$$\bar{x}(0) = 0, \quad \bar{x}(t_f) + \dot{x}(t_f)\bar{t}_f = 0. \tag{6.84}$$

Then $\mathcal{K}$ is a subspace of the space $Z(\Theta)$ which we call the *critical subspace*. Each element $\bar{z} \in \mathcal{K}$ is uniquely defined by the number $\bar{t}_f$ and the vector $\bar{\xi}$. Consequently, the subspace $\mathcal{K}$ is *finite-dimensional*.

An explicit representation of the variations $\bar{x}(t)$ in (6.83) is obtained as in Section 6.2.1. For each $k = 1,\dots,s$, define the vector functions $y^k(t)$ as the solutions to the system

$$\dot{y} = f_x(t)y, \quad y(t_k) = [\dot{x}]^k, \quad t \in [t_k, t_f]. \tag{6.85}$$

For $t < t_k$ we put $y^k(t) = 0$ which yields the jump $[y^k]^k = [\dot{x}]^k$. Then

$$\bar{x}(t) = \sum_{k=1}^{s} y^k(t)\bar{\xi}_k \tag{6.86}$$

from which we obtain the representation

$$\bar{x}(t_f) + \dot{x}(t_f)\bar{t}_f = \sum_{k=1}^{s} y^k(t_f)\bar{\xi}_k + \dot{x}(t_f)\bar{t}_f. \tag{6.87}$$

Furthermore, denote by $x(t;t_1,\dots,t_s)$ the solution of the state equation (6.72) using the optimal bang-bang control with switching points $t_1,\dots,t_s$. Then the partial derivatives of state trajectories with respect to the switching points are given by

$$\frac{\partial x}{\partial t_k}(t;t_1,\dots,t_s) = -y^k(t) \quad \text{for} \quad t \geq t_k, \quad k = 1,\dots,s. \tag{6.88}$$

This relation holds for all $t \in [0,t_f] \setminus \{t_k\}$, because for $t < t_k$ we have $\frac{\partial x}{\partial t_k}(t) = 0$ and $y^k(t) = 0$. Hence, (6.86) yields

$$\bar{x}(t) = -\sum_{k=1}^{s} \frac{\partial x}{\partial t_k}(t)\bar{\xi}_k. \tag{6.89}$$

In the nondegenerate case $\psi_0(t_f) > 0$, the critical subspace is simplified as follows.

Proposition 6.24. *If there exists* $(\psi_0, \psi) \in M_0$ *such that* $\psi_0(t_f) > 0$, *then* $\bar{t}_f = 0$ *holds for each* $\bar{z} = (\bar{t}_f, \bar{\xi}, \bar{x}) \in \mathcal{K}$.

This proposition is a straightforward consequence from Proposition 6.8. In Section 6.5.4, we shall conclude from Theorem 6.27 that the property $\mathcal{K} = \{0\}$ essentially represents a *first-order sufficient condition*. Since $\bar{x}(t_f) + \dot{x}(t_f)\bar{t}_f = 0$ by (6.84), the representations (6.86) and (6.87) and Proposition 6.24 induce the following conditions for $\mathcal{K} = \{0\}$.

Proposition 6.25. *Assume that one of the following conditions is satisfied:*

(a) *The s+1 vectors* $y^k(t_f) = \frac{\partial x}{\partial t_k}(t_f)$, $k = 1,\dots,s$, $\dot{x}(t_f)$ *are linearly independent;*

(b) *there exists* $(\psi_0, \psi) \in M_0$ *with* $\psi_0(t_f) > 0$ *and the s vectors* $y^k(t_f) = \frac{\partial x}{\partial t_k}(t_f)$, $k = 1,\dots,s$, *are linearly independent;*

(c) *there exists* $(\psi_0, \psi) \in M_0$ *with* $\psi_0(t_f) > 0$, *and the bang-bang control has exactly one switching point, i.e.,* $s = 1$.

Then the critical subspace is $\mathcal{K} = \{0\}$.

Now we discuss the case of two switching points, i.e., $s = 2$, to prepare the numerical example in Section 6.5.4. Let us assume that $\psi_0(t_f) > 0$ (for some $(\psi_0, \psi) \in M_0$) and $[\dot{x}]^1 \neq 0$, $[\dot{x}]^2 \neq 0$. By virtue of Proposition 6.24, we have $\bar{t}_f = 0$ and hence $\bar{x}(t_f) = 0$ for each element $\bar{z} \in \mathcal{K}$. Then the relations (6.84) and (6.86) yield

$$0 = \bar{x}(t_f) = y^1(t_f)\bar{\xi}_1 + y^2(t_f)\bar{\xi}_2. \tag{6.90}$$

The conditions $[\dot{x}]^1 \neq 0$ and $[\dot{x}]^2 \neq 0$ imply that $y^1(t_f) \neq 0$ and $y^2(t_f) \neq 0$, respectively. Furthermore, assume that $\mathcal{K} \neq \{0\}$. Then (6.90) shows that the nonzero vectors $y^1(t_f)$ and $y^2(t_f)$ are collinear, i.e.,

$$y^2(t_f) = \alpha y^1(t_f) \tag{6.91}$$

with some factor $\alpha \neq 0$. As a consequence, the relation $y^2(t) = \alpha y^1(t)$ is valid for all $t \in (t_2, t_f]$. In particular, we have $y^2(t_2+) = \alpha y^1(t_2)$ and thus

$$[\dot{x}]^2 = \alpha y^1(t_2) \tag{6.92}$$

which is equivalent to (6.91). In addition, the equalities (6.90) and (6.91) imply that

$$\xi_2 = -\frac{1}{\alpha}\xi_1. \tag{6.93}$$

We shall use this formula in the next section.

6.5.4 Quadratic Form

For $(\psi_0, \psi) \in M_0$ and $\bar{z} \in \mathcal{K}$, we define the functional (see formulas (6.34) and (6.35))

$$\begin{aligned}\Omega(\psi_0, \psi, \bar{z}) = & \sum_{k=1}^{s}(D^k(H)\bar{\xi}_k^2 - 2[\dot{\psi}]^k \bar{x}_{\mathrm{av}}^k \bar{\xi}_k) + \int_0^{t_f} \langle H_{xx}(t)\bar{x}(t), \bar{x}(t)\rangle\, dt \\ & -(\dot{\psi}_0(t_f) - \dot{\psi}(t_f)\dot{x}(t_f))\bar{t}_f^2.\end{aligned} \tag{6.94}$$

where $\bar{x}_{\mathrm{av}}^k := \frac{1}{2}(\bar{x}^{k-} + \bar{x}^{k+})$. Now we introduce second-order optimality conditions for bang-bang control in the problem (6.71)–(6.74). From Theorem 6.9 we easily deduce the following result.

Theorem 6.26. *Let a trajectory $\mathcal{T}$ affords a Pontryagin minimum. Then the following Condition $\mathfrak{A}$ holds for the trajectory $\mathcal{T}$: The set M_0 is nonempty and*

$$\max_{(\psi_0,\psi)\in M_0} \Omega(\psi_0, \psi, \bar{z}) \geq 0 \quad \forall\, \bar{z} \in \mathcal{K} \setminus \{0\}.$$

Similarly, from Theorem 6.10 we obtain the following theorem.

Theorem 6.27. *Let the following Condition $\mathfrak{B}$ be fulfilled for the trajectory $\mathcal{T}$:*

(a) *there exists $\lambda \in M_0$ such that $D^k(H) > 0$, $k = 1, \dots, s$, and condition* (6.82) *holds (i.e., $u(t)$ is a strict bang-bang control),*

(b) $\max_{(\psi_0,\psi)\in M_0} \Omega(\psi_0, \psi, \bar{z}) > 0$ *for all $\bar{z} \in \mathcal{K} \setminus \{0\}$.*

Then $\mathcal{T}$ is a strict strong minimum.

Remarks.

1. The sufficient Condition $\mathfrak{B}$ is a natural strengthening of the necessary Condition $\mathfrak{A}$.
2. Condition (b) is automatically fulfilled if $\mathcal{K} = \{0\}$ holds (cf. Proposition 6.25) which gives a first-order sufficient condition for a strong minimum.
3. If there exists $(\psi_0, \psi) \in M_0$ such that $\Omega(\psi_0, \psi, \bar{z}) > 0$ for all $\bar{z} \in \mathcal{K} \setminus \{0\}$, then condition (b) is obviously fulfilled.
4. For boxes $U = \{u = (u_1, \ldots, u_{d(u)}) \in \mathbb{R}^{d(u)} : u_i^{\min} \le u_i \le u_i^{\max},\ i = 1, \ldots, d(u)\}$, the condition $D^k(H) > 0$, $k = 1, \ldots, s$, is equivalent to the property $\dot{\phi}_i(t_k) \neq 0$ if t_k is a switching point of the ith control component $u_i(t)$. Note again that condition (6.82) precludes the simultaneous switching of two or more control components.
5. A further remark concerns the case that the set M_0 of Pontryagin multipliers is not a singleton. This case was illustrated in [89] by the following time-optimal control problem for a linear system:

$$\dot{x}_1 = x_2,\ \dot{x}_2 = x_3,\ \dot{x}_3 = x_4,\ \dot{x}_4 = u, \quad |u| \le 1, \quad x(0) = a,\ x(t_f) = b,$$

where $x = (x_1, x_2, x_3, x_4)$. It was shown in [89] that for some a and b there exists an extremal in this problem with two switching points of the control such that, under appropriate normalization, the set M_0 is a segment. For this extremal, the maximum of the quadratic forms Ω is positive on each nonzero element of the critical subspace, and hence the sufficient conditions of Theorem 6.27 are satisfied. But this is not true for any single quadratic form Ω (corresponding to an element of the set M_0).

6.5.5 Nondegenerate Case

Let us assume the *nondegenerate* or *normal* case that there exists $(\psi_0, \psi) \in M_0$ such that the cost function multiplier $\psi_0(t_f)$ is positive. By virtue of Proposition 6.24 we have in this case that $\bar{t}_f = 0$ for all $\bar{z} \in \mathcal{K}$. Thus the critical subspace $\mathcal{K}$ is defined by the conditions

$$\dot{\bar{x}}(t) = f_x(t)\bar{x}(t), \quad [\bar{x}]^k = [\dot{x}]^k \bar{\xi}_k \quad (k = 1, \ldots, s), \quad \bar{x}(0) = 0, \quad \bar{x}(t_f) = 0. \tag{6.95}$$

In particular, these conditions imply $\bar{x}(t) \equiv 0$ on $[0, t_1)$ and $(t_s, t_f]$. Hence we have $\bar{x}^{1-} = \bar{x}^{s+} = 0$ for all $\bar{z} \in \mathcal{K}$. Then the quadratic form (6.94) is equal to

$$\Omega(\psi, \bar{z}) = \sum_{k=1}^{s} \left(D^k(H)\bar{\xi}_k^2 + 2[H_x]^k \bar{x}_{\text{av}}^k \bar{\xi}_k \right) + \int_0^{t_f} \langle H_{xx}(t)\bar{x}(t), \bar{x}(t) \rangle \, dt. \tag{6.96}$$

This case of a time-optimal (autonomous) control problem was studied by Sarychev [104]. He used a special transformation of the problem and obtained sufficient optimality condition for the transformed problem. It is not easy, but it is possible, to reformulate his results in terms of the original problem. The comparison of both types of conditions reveals that Sarychev used the same critical subspace, but his quadratic form is a lower bound for Ω. Namely, in his quadratic form the positive term $D^k(H)\bar{\xi}_k^2$ has the factor $\frac{1}{4}$ instead of the factor 1 for the same term in Ω. Therefore, the sufficient Condition $\mathfrak{B}$ is always fulfilled whenever Sarychev's condition is fulfilled. However, there is an example of a control problem where the optimal solution satisfies Condition $\mathfrak{B}$ but does not satisfy Sarychev's condition. Finally, Sarychev proved that his condition is sufficient for an L^1-minimum with respect to the control (which is a Pontryagin minimum in this problem). In fact, it could be proved that his condition is sufficient for a strong minimum.

6.5.6 Cases of One or Two Switching Times of the Control

From Theorem 6.27 and Proposition 6.25(c), we immediately deduce sufficient conditions for a bang-bang control with one switching point. The result is used for the example in Section 6.7.1 and is also applicable to the time-optimal control of an image converter discussed by Kim et al. [47].

Theorem 6.28. *Let the following conditions be fulfilled for the trajectory* $\mathcal{T}$*:*

(a) $u(t)$ *is a bang-bang control with one switching point, i.e.,* $s = 1$*,*

(b) *there exists* $(\psi_0, \psi) \in M_0$ *such that* $D^1(H) > 0$ *and condition* (6.82) *holds (i.e.,* $u(t)$ *is a strict bang-bang control),*

(c) *there exists* $(\psi_0, \psi) \in M_0$ *with* $\psi_0(t_f) > 0$*.*

Then $\mathcal{T}$ *is a strict strong minimum.*

Now we turn our attention to the case of two switching points where $s = 2$. Assume the nondegenerate case $\psi_0(t_f) > 0$ and suppose that $[\dot{x}]^1 \neq 0$, $[\dot{x}]^1 \neq 0$ and $y^2(t_f) = \alpha y^1(t_f)$ as in (6.91). Otherwise, $\mathcal{K} = \{0\}$ holds and, hence, the first-order sufficient condition for a strong minimum is satisfied. For any element $\bar{z} \in \mathcal{K}$, we have $\bar{t}_f = 0$, $\bar{x}^{1-} = 0$, $\bar{x}^{2+} = 0$. Consequently,

$$\bar{x}^1_{\text{av}} = \frac{1}{2}[\bar{x}]^1 = \frac{1}{2}[\dot{x}]^1\bar{\xi}_1, \quad \bar{x}^2_{\text{av}} = \frac{1}{2}\bar{x}^{2-} = \frac{1}{2}y^1(t_2)\bar{\xi}_1 = \frac{1}{2\alpha}[\dot{x}]^2\bar{\xi}_1$$

in view of $\bar{x}(t) = y^1(t)\bar{\xi}_1 + y^2(t)\bar{\xi}_2$, $y^2(t_2-) = 0$ and (6.92). Using these relations in the quadratic form (6.96) together with (6.93) and the conditions $y^2(t) = 0$ for all $t < t_2$, $[H_x]^k = -[\dot{\psi}]^k$, $k = 1, 2$, we compute the quadratic form for the element of the critical subspace as

$$\begin{aligned}\Omega &= D^1(H)\bar{\xi}_1^2 + D^2(H)\bar{\xi}_2^2 - 2[\dot{\psi}]^1\bar{x}^1_{\text{av}}\bar{\xi}_1 - 2[\dot{\psi}]^2\bar{x}^2_{\text{av}}\bar{\xi}_2 + \int_{t_1}^{t_2}\langle H_{xx}\bar{x}, \bar{x}\rangle\, dt \\ &= D^1(H)\bar{\xi}_1^2 + \frac{1}{\alpha^2}D^2(H)\bar{\xi}_1^2 - [\dot{\psi}]^1[\dot{x}]^1\bar{\xi}_1^2 + \frac{1}{\alpha^2}[\dot{\psi}]^2[\dot{x}]^2\bar{\xi}_1^2 + \left(\int_{t_1}^{t_2}\langle H_{xx}y^1, y^1\rangle\, dt\right)\bar{\xi}_1^2 \\ &= \rho\bar{\xi}_1^2,\end{aligned}$$

where

$$\rho = \left(D^1(H) - [\dot{\psi}]^1[\dot{x}]^1\right) + \frac{1}{\alpha^2}\left(D^2(H) + [\dot{\psi}]^2[\dot{x}]^2\right) + \int_{t_1}^{t_2}\langle H_{xx}y^1, y^1\rangle\, dt. \tag{6.97}$$

Thus, we obtain the following proposition.

Proposition 6.29. *Assume that* $\psi_0(t_f) > 0$, $s = 2$, $[\dot{x}]^1 \neq 0$, $[\dot{x}]^2 \neq 0$, *and* $y^2(t_f) = \alpha y^1(t_f)$ *(which is equivalent to* (6.91)*) with some factor* α*. Then the condition of the positive definiteness of* Ω *on* $\mathcal{K}$ *is equivalent to the inequality* $\rho > 0$*, where* ρ *is defined as in* (6.97)*.*

6.6 Sufficient Conditions for Positive Definiteness of the Quadratic Form Ω on the Critical Subspace $\mathcal{K}$ for Time-Optimal Control Problems

In this section, we consider the nondegenerate case as in Section 6.5.5 and assume

(i) $u(t)$ is a bang-bang control with $s > 1$ switching points,

(ii) there exists $(\psi_0, \psi) \in M_0$ such that $\psi_0(t_f) > 0$ and $D^k(H) > 0, k = 1,\dots,s$.

Under these assumptions the critical subspace $\mathcal{K}$ is defined as in (6.95). Let $(\psi_0, \psi) \in M_0$ be a fixed element (possibly different from that in assumption (ii)), and denote by $\Omega = \Omega(\psi_0, \psi, \cdot)$ the quadratic form for this element. Recall that Ω is given by (6.96). According to Theorem 6.27 the positive definiteness of the quadratic form (6.96) on the subspace $\mathcal{K}$ in (6.95) is a sufficient condition for a strict strong minimum of the trajectory. Now our aim is to find conditions that guarantee the positive definiteness of Ω on $\mathcal{K}$.

6.6.1 Q-Transformation of Ω on $\mathcal{K}$

Here we shall use the same arguments as in Sections 5.3.2 and 6.3.1. Let $Q(t)$ be a symmetric matrix on $[t_1, t_s]$ with piecewise continuous entries which are absolutely continuous on each interval of the set $[t_1, t_s] \setminus \Theta$. Therefore, Q may have a jump at each point $t_k \in \Theta$ including t_1, t_s, and thus the symmetric matrices Q^{1-} and Q^{s+} are also defined. For $\bar{z} \in \mathcal{K}$, we obviously have

$$\int_{t_1}^{t_s} \frac{d}{dt} \langle Q\bar{x}, \bar{x}\rangle \, dt = \langle Q\bar{x}, \bar{x}\rangle \Big|_{t_1-}^{t_s+} - \sum_{k=1}^{s} [\langle Q\bar{x}, \bar{x}\rangle]^k,$$

where $[\langle Q\bar{x}, \bar{x}\rangle]^k$ is the jump of the function $\langle Q\bar{x}, \bar{x}\rangle$ at the point $t_k \in \Theta$. Using the conditions $\dot{\bar{x}} = f_x(t)\bar{x}$ and $\bar{x}^{1-} = \bar{x}^{s+} = 0$, we obtain

$$\sum_{k=1}^{s} [\langle Q\bar{x}, \bar{x}\rangle]^k + \int_{t_1}^{t_s} \langle (\dot{Q} + f_x^* Q + Q f_x)\bar{x}, \bar{x}\rangle \, dt = 0, \tag{6.98}$$

where the asterisk denotes transposition. Adding this zero form to Ω, we get

$$\begin{aligned} \Omega \;=\; & \sum_{k=1}^{s} \left(D^k(H)\bar{\xi}_k^2 - 2[\dot{\psi}]^k \bar{x}_{\mathrm{av}}^k \bar{\xi}_k + [\langle Q\bar{x}, \bar{x}\rangle]^k \right) \\ & + \int_{t_1}^{t_s} \langle (H_{xx} + \dot{Q} + f_x^* Q + Q f_x)\bar{x}, \bar{x}\rangle \, dt. \end{aligned} \tag{6.99}$$

We call this formula the *Q-transformation of Ω on $\mathcal{K}$*.

To eliminate the integral term in Ω, we assume that $Q(t)$ satisfies the following linear matrix differential equation:

$$\dot{Q} + f_x^* Q + Q f_x + H_{xx} = 0 \quad \text{on } [t_1, t_s] \setminus \Theta. \tag{6.100}$$

Using (6.100), the quadratic form (6.99) reduces to

$$\Omega = \sum_{k=1}^{s} \omega_k, \quad \omega_k := D^k(H)\xi_k^2 - 2[\dot{\psi}]^k \bar{x}_{\mathrm{av}}^k \xi_k + [\langle Q\bar{x}, \bar{x}\rangle]^k. \tag{6.101}$$

Thus, we have proved the following statement.

Proposition 6.30. *Let $Q(t)$ satisfy the linear differential equation* (6.100) *on* $[t_1,t_s]\setminus\Theta$. *Then for each $\bar{z}\in\mathcal{K}$ the representation* (6.101) *holds.*

Now our goal is to derive conditions such that $\omega_k>0$ holds on $\mathcal{K}\setminus\{0\}$ for $k=1,\ldots,s$. We shall use the representations of ω_k given in Section 6.3.1. According to (6.50),

$$\omega_k=\left(D^k(H)+(q_{k+})[\dot{x}]^k\right)\bar{\xi}_k^2+2(q_{k+})\bar{x}^{k-}\bar{\xi}_k+(\bar{x}^{k-})^*[Q]^k\bar{x}^{k-}, \tag{6.102}$$

where $q_{k+}=([\dot{x}]^k)^*Q^{k+}-[\dot{\psi}]^k$. We immediately see from this representation that one way to enforce $\omega_k>0$ is to impose the following conditions:

$$D^k(H)>0,\quad q_{k+}=([\dot{x}]^k)^*Q^{k+}-[\dot{\psi}]^k=0,\quad [Q]^k\geq 0. \tag{6.103}$$

In practice, however, it might be difficult to check this condition since it is necessary to satisfy the $d(x)$ equality constraints $q_{k+}=([\dot{x}]^k)^*Q^{k+}-[\dot{\psi}]^k=0$ and the inequality constraints $[Q]^k\geq 0$. It is more convenient to express ω_k as a quadratic form in the variables $(\bar{\xi}_k,\bar{x}^{k-})$ with the matrix

$$M_{k+}=\begin{pmatrix} D^k(H)+(q_{k+})[\dot{x}]^k & q_{k+} \\ (q_{k+})^* & [Q]^k \end{pmatrix}, \tag{6.104}$$

where q_{k+} is a row vector and $(q_{k+})^*$ is a column vector.

Similarly, according to (6.52), the following representation holds:

$$\omega_k=\left(D^k(H)-(q_{k-})[\dot{x}]^k\right)\bar{\xi}_k^2+2(q_{k-})\bar{x}^{k+}\bar{\xi}_k+(\bar{x}^{k+})^*[Q]^k\bar{x}^{k+}, \tag{6.105}$$

where $q_{k-}=([\dot{x}]^k)^*Q^{k-}-[\dot{\psi}]^k$. Again, we see that $\omega_k>0$ holds if we require the conditions

$$D^k(H)>0,\quad q_{k-}=([\dot{x}]^k)^*Q^{k-}-[\dot{\psi}]^k=0,\quad [Q]^k\geq 0. \tag{6.106}$$

To find a more general condition for $\omega_k>0$, we consider (6.105) as a quadratic form in the variables $(\bar{\xi}_k,\bar{x}^{k+})$ with the matrix

$$M_{k-}=\begin{pmatrix} D^k(H)-(q_{k-})[\dot{x}]^k & q_{k-} \\ (q_{k-})^* & [Q]^k \end{pmatrix}. \tag{6.107}$$

Since the right-hand sides of equalities (6.102) and (6.105) are connected by the relation $\bar{x}^{k+}=\bar{x}^{k-}+[\dot{x}]^k\bar{\xi}_k$, the following statement obviously holds.

Proposition 6.31. *For each $k=1,\ldots,s$, the positive (semi)definiteness of the matrix M_{k-} is equivalent to the positive (semi)definiteness of the matrix M_{k+}.*

Now we can prove the following theorem.

Theorem 6.32. *Let $Q(t)$ be a solution of the linear differential equation* (6.100) *on* $[t_1,t_s]\setminus\Theta$ *which satisfies the following conditions:*

(a) *the matrix* M_{k+} *is positive semidefinite for each* $k = 2,\dots,s$;
(b) $b_{k+} := D^k(H) + (q_{k+})[\dot{x}]^k > 0$ *for each* $k = 1,\dots,s-1$.

Then Ω *is positive on* $\mathcal{K} \setminus \{0\}$.

Proof. Take an arbitrary element $\bar{z} = (\bar{\xi},\bar{x}) \in \mathcal{K}$. Let us show that $\Omega \geq 0$ for this element. Condition (a) implies that $\omega_k \geq 0$ for $k = 2,\dots,s$. Condition (b) for $k = 1$ together with condition $\bar{x}^{1-} = 0$ implies that $\omega_1 \geq 0$. Consequently, $\Omega \geq 0$. Assume that $\Omega = 0$. Then $\omega_k = 0$, $k = 1,\dots,s$. The conditions $\omega_1 = 0$, $\bar{x}^{1-} = 0$, and $b_{1+} > 0$ by formula (6.102) (with $k = 1$) yield $\bar{\xi}_1 = 0$. Then $[\bar{x}]^1 = 0$ and hence $\bar{x}^{1+} = 0$. The last equality and the equation $\dot{\bar{x}} = f_x(t)\bar{x}$ show that $\bar{x}(t) = 0$ in (t_1,t_2) and hence $\bar{x}^{2-} = 0$. Similarly, the conditions $\omega_2 = 0$, $\bar{x}^{2-} = 0$, and $b_{2+} > 0$ by formula (6.102) (with $k = 2$) imply that $\bar{\xi}_2 = 0$ and $\bar{x}(t) = 0$ in (t_2,t_3). Therefore, $\bar{x}^{3-} = 0$ etc. Continuing this process we get $\bar{x} \equiv 0$ and $\bar{\xi}_k = 0$ for $k = 1,\dots,s-1$. Now using formula (6.101) for $\omega_s = 0$, as well as the conditions $D^s(H) > 0$ and $\bar{x} \equiv 0$ we obtain $\bar{\xi}_s = 0$. Consequently, we have $\bar{z} = 0$ which means that Ω is positive on $\mathcal{K} \setminus \{0\}$. ∎

Similarly, using representation (6.105) for ω_k we can prove the following statement.

Theorem 6.33. *Let* $Q(t)$ *be a solution of the linear differential equation* (6.100) *on* $[t_1,t_s] \setminus \Theta$ *which satisfies the following conditions:*

(a) *The matrix* M_{k-} *is positive semidefinite for each* $k = 1,\dots,s-1$;
(b) $b_{k-} := D^k(H) - (q_{k-})[\dot{x}]^k > 0$ *for each* $k = 2,\dots,s$.

Then Ω *is positive on* $\mathcal{K} \setminus \{0\}$.

6.6.2 Q-Transformation of Ω to Perfect Squares

Here, as in Section 6.3.3, we formulate special jump conditions for the matrix Q at each point $t_k \in \Theta$, which will make it possible to transform Ω into perfect squares and thus to prove its positive definiteness on $\mathcal{K}$. Suppose that

$$b_{k-} := D^k(H) - (q_{k-})[\dot{x}]^k > 0 \tag{6.108}$$

and that Q satisfies the jump condition at t_k,

$$b_{k-}[Q]^k = (q_{k-})^*(q_{k-}), \tag{6.109}$$

where $(q_{k-})^*$ is a column vector while q_{k-} is a row vector. Then according to (6.63),

$$\omega_k = (b_{k-})^{-1}\left((b_{k-})\bar{\xi}_k + (q_{k-})(\bar{x}^{k+})\right)^2 = (b_{k-})^{-1}\left(D^k(H)\bar{\xi}_k + (q_{k-})(\bar{x}^{k-})\right)^2. \tag{6.110}$$

Theorem 6.34. *Let* $Q(t)$ *satisfy the linear differential equation* (6.100) *on* $[t_1,t_s] \setminus \Theta$. *Let condition* (6.108) *hold for each* $k = 1,\dots,s$ *and condition* (6.109) *hold for each* $k = 1,\dots,s-1$. *Then* Ω *is positive on* $\mathcal{K} \setminus \{0\}$.

Proof. According to (6.110), the matrix M_{k-} is positive semidefinite for each $k = 1,\dots,s-1$ (cf. (6.105) and (6.107)), and hence both conditions (a) and (b) of Theorem 6.33 are fulfilled. Then by Theorem 6.33, Ω is positive on $\mathcal{K} \setminus \{0\}$. ∎

Similar assertions hold for the jump conditions that use right-hand values of Q at each point $t_k \in \Theta$. Suppose that

$$b_{k+} := D^k(H) + (q_{k+})[\dot{x}]^k > 0 \tag{6.111}$$

and that Q satisfies the jump condition at point t_k

$$b_{k+}[Q]^k = (q_{k+})^*(q_{k+}). \tag{6.112}$$

Then

$$\omega_k = (b_{k+})^{-1}\left((b_{k+})\bar{\xi}_k + (q_{k+})(\bar{x}^{k-})\right)^2 = (b_{k+})^{-1}\left(D^k(H)\bar{\xi}_k + (q_{k+})(\bar{x}^{k+})\right)^2. \tag{6.113}$$

Theorem 6.35. *Let $Q(t)$ satisfy the linear differential equation* (6.100) *on* $[t_1, t_s] \setminus \Theta$. *Let condition* (6.111) *hold for each $k = 1, \dots, s$ and condition* (6.112) *hold for each $k = 2, \dots, s$. Then Ω is positive on $\mathcal{K} \setminus \{0\}$.*

6.6.3 Case of Two Switching Times of the Control

Let $s = 2$, i.e., $\Theta = \{t_1, t_2\}$, and let $Q(t)$ be a symmetric matrix with absolutely continuous entries on $[t_1, t_2]$. Put

$$Q^k = Q(t_k), \quad q_k = ([\dot{x}]^k)^* Q^k - [\dot{\psi}]^k, \quad k = 1, 2.$$

Theorem 6.36. *Let $Q(t)$ satisfy the linear differential equation* (6.100) *on* (t_1, t_2) *and the following inequalities hold at t_1, t_2:*

$$D^1(H) + q_1[\dot{x}]^1 > 0, \quad D^2(H) - q_2[\dot{x}]^2 > 0. \tag{6.114}$$

Then Ω is positive on $\mathcal{K} \setminus \{0\}$.

Proof. In the case considered we have

$$Q^{1+} = Q^1, \quad q_{1+} = q_1, \quad Q^{2-} = Q^2, \quad q_{2-} = q_2,$$

and

$$b_{1+} := D^1(H) + q_1[\dot{x}]^1 > 0, \quad b_{2-} := D^2(H) - q_2[\dot{x}]^2 > 0. \tag{6.115}$$

Define the jumps $[Q]^1$ and $[Q]^2$ by the conditions

$$b_{1+}[Q]^1 = (q_{1+})^*(q_{1+}), \quad b_{2-}[Q]^2 = (q_{2-})^*(q_{2-}). \tag{6.116}$$

Then $[Q]^1$ and $[Q]^2$ are symmetric matrices. Put

$$Q^{1-} = Q^{1+} - [Q]^1, \quad Q^{2+} = Q^{2-} + [Q]^2.$$

Then Q^{1-} and Q^{2+} are also symmetric matrices. Thus, we obtain a symmetric matrix $Q(t)$ satisfying (6.100) on (t_1, t_2), the inequalities (6.115), and the jump conditions (6.116). By formulas (6.110) and (6.113) the terms ω_1 and ω_2 are nonnegative. In view of (6.101), we see that $\Omega = \omega_1 + \omega_2$ is nonnegative on $\mathcal{K}$. Suppose that $\Omega = 0$ for some $\bar{z} = (\xi, \bar{x}) \in \mathcal{K}$. Then $\omega_k = 0$ for $k = 1, 2$, and thus formulas (6.110) and (6.113) give

$$b_{1+}\xi_1 + (q_{1+})\bar{x}^{1-} = 0, \quad b_{2-}\xi_2 + (q_{2-})\bar{x}^{2+} = 0.$$

But $\bar{x}^{1-} = 0$ and $\bar{x}^{2+} = 0$. Consequently, $\bar{\xi}_1 = \bar{\xi}_2 = 0$, and then conditions $\bar{x}^{1-} = 0$ and $[\bar{x}]^1 = 0$ imply that $\bar{x}^{1+} = 0$. The last equality and the equation $\dot{\bar{x}} = f_x(t)\bar{x}$ imply that $\bar{x}(t) = 0$ on (t_1, t_2). Thus $\bar{x} \equiv 0$ and then $\bar{z} = 0$. We have proved that Ω is positive on $\mathcal{K} \setminus \{0\}$. ∎

6.6.4 Control System with a Constant Matrix B

In the case that $B(t,x) = B$ is a constant matrix, the adjoint equation has the form

$$\dot{\psi} = -\psi a_x,$$

which implies that

$$[\dot{\psi}]^k = 0, \quad k = 1, \ldots, s.$$

Therefore,

$$\begin{aligned} q_{k-} &= ([\dot{x}]^k)^* Q^{k-}, & q_{k+} &= ([\dot{x}]^k)^* Q^{k+}, \\ (q_{k-})^* q_{k-} &= Q^{k-}[\dot{x}]^k([\dot{x}]^k)^* Q^{k-}, & (q_{k+})^* q_{k+} &= Q^{k+}[\dot{x}]^k([\dot{x}]^k)^* Q^{k+}, \\ b_{k-} &= D^k(H) - ([\dot{x}]^k)^* Q^{k-}[\dot{x}]^k, & b_{k+} &= D^k(H) + ([\dot{x}]^k)^* Q^{k+}[\dot{x}]^k, \end{aligned}$$

where

$$D^k(H) = \dot{\psi}(t_k) B [u]^k, \quad k = 1, \ldots, s.$$

In the case of two switching points with $s = 2$, the conditions (6.114) take the form

$$D^1(H) + \langle Q^1[\dot{x}]^1, [\dot{x}]^1 \rangle\rangle > 0, \quad D^2(H) - \langle Q^2[\dot{x}]^2, [\dot{x}]^2 \rangle\rangle > 0. \tag{6.117}$$

Now assume, in addition, that u is one-dimensional and that

$$B = \beta e_n := \begin{pmatrix} 0 \\ \vdots \\ 0 \\ \beta \end{pmatrix}, \quad \beta > 0, \qquad U = [-c, c], \ c > 0.$$

In this case we get

$$[\dot{x}]^k = B[u]^k = \beta e_n [u]^k, \quad k = 1, \ldots, s,$$

and thus

$$\langle Q^k[\dot{x}]^k, [\dot{x}]^k \rangle\rangle = \beta^2 \langle Q^k e_n, e_n \rangle |[u]^k|^2 = 4\beta^2 c^2 Q_{nn}(t_k),$$

where Q_{nn} is the element of matrix

$$Q = \begin{pmatrix} Q_{11} & \ldots & Q_{1n} \\ \vdots & \vdots & \vdots \\ Q_{n1} & \ldots & Q_{nn} \end{pmatrix}.$$

Moreover, in the last case we obviously have

$$D^k(H) = 2\beta c\, |\dot{\psi}_n(t_k)|, \quad k = 1,\dots,s.$$

For $s = 2$, conditions (6.114) are thus equivalent to the estimates

$$Q_{nn}(t_1) > -\frac{|\dot{\psi}_n(t_1)|}{2\beta c}, \qquad Q_{nn}(t_2) < \frac{|\dot{\psi}_n(t_2)|}{2\beta c}. \tag{6.118}$$

6.7 Numerical Examples of Time-Optimal Control Problems

6.7.1 Time-Optimal Control of a Van der Pol Oscillator

Consider again the tunnel-diode oscillator displayed in Figure 4.1. In the control problem of the van der Pol oscillator, the state variable x_1 represents the voltage, whereas the control u is the voltage at the generator. Time-optimal solutions will be computed in two cases. First, we consider a fixed terminal state $x(t_f) = x_f$. The second case treats the nonlinear terminal constraint $x_1(t_f)^2 + x_2(t_f)^2 = r^2$ for a small $r > 0$, by which the oscillator is steered only to a neighborhood of the origin.

In the first case we consider the control problem of minimizing the final time t_f subject to the constraints

$$\dot{x}_1(t) = x_2(t),\ \dot{x}_2(t) = -x_1(t) + x_2(t)(1 - x_1^2(t)) + u(t), \tag{6.119}$$

$$x_1(0) = -0.4,\ x_2(0) = 0.6, \quad x_1(t_f) = 0.6,\ x_2(t_f) = 0.4, \tag{6.120}$$

$$|u(t)| \le 1 \quad \text{for } t \in [0, t_f]. \tag{6.121}$$

The Pontryagin function (Hamiltonian) is given by

$$H(x,u,\psi) = \psi_1 x_2 + \psi_2(-x_1 + x_2(1 - x_1^2) + u). \tag{6.122}$$

The adjoint equations $\dot{\psi} = -H_x$ are

$$\dot{\psi}_1 = \psi_2(1 + 2x_1 x_2), \quad \dot{\psi}_2 = -\psi_1 + \psi_2(x_1^2 - 1). \tag{6.123}$$

In view of the free final time we get the additional boundary condition

$$H(t_f) + \psi_0(t_f) = 0.4\psi_1(t_f) + \psi_2(t_f)(-0.344 + u(t_f)) + 1 = 0. \tag{6.124}$$

The sign of switching function $\phi(t) = \psi_2(t)$ determines the optimal control according to

$$u(t) = \left\{ \begin{array}{rl} 1 & \text{if} \quad \psi_2(t) < 0 \\ -1 & \text{if} \quad \psi_2(t) > 0 \end{array} \right\}. \tag{6.125}$$

Evaluating the derivatives of the switching function, it can easily be seen that there are no *singular arcs* with $\psi_2(\tau) \equiv 0$ holding on a time interval $[t_1, t_2]$. Nonlinear programming methods applied to the discretized control problem show that the optimal bang-bang control has two bang-bang arcs,

$$u(t) = \left\{ \begin{array}{rl} 1 & \text{for} \quad 0 \le t < t_1 \\ -1 & \text{for} \quad t_1 \le t \le t_f \end{array} \right\}, \tag{6.126}$$

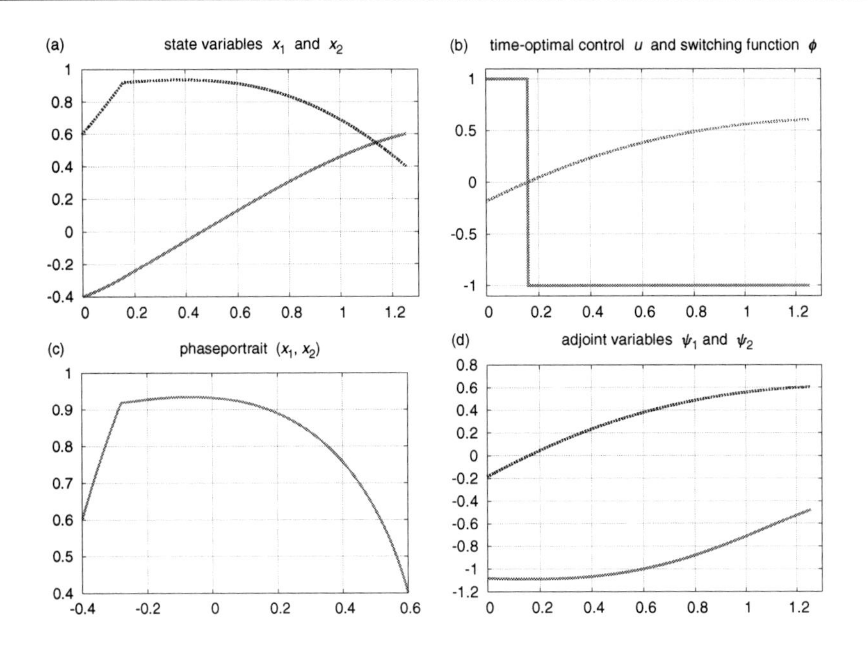

Figure 6.2. *Time-optimal solution of the van der Pol oscillator, fixed terminal state* (6.120). (a) *State variables x_1 and x_2 (dashed line).* (b) *Control u and switching function ψ_2 (dashed line).* (c) *Phase portrait (x_1, x_2).* (d) *Adjoint variables ψ_1 and ψ_2 (dashed line).*

with switching time t_1. This implies the switching condition

$$\phi(t_1) = \psi_2(t_1) = 0. \tag{6.127}$$

Hence, we must solve the boundary value problem (6.120)–(6.127). Using the code BNDSCO [82] or NUDOCCCS [13, 14], we obtain the extremal solution displayed in Figure 6.2. The optimal final time, the switching point, and some values for the adjoint variables are

$$\begin{array}{rclrcl} t_f &=& 1.2540747, & t_1 &=& 0.158320138, \\ \psi_1(0) &=& -1.0816056, & \psi_2(0) &=& -0.18436798, \\ \psi_1(t_1) &=& -1.0886321, & \psi_2(t_1) &=& 0.0, \\ \psi_1(t_f) &=& -0.47781383, & \psi_2(t_f) &=& 0.60184112. \end{array} \tag{6.128}$$

Since the bang-bang control has only one switching time, we are in the position to apply Theorem 6.27. For checking the assumptions of this theorem it remains to verify the condition $D^1(H) = |\dot{\phi}(t_1)[u]^1| > 0$. Indeed, in view of the adjoint equation (6.123) and the switching condition $\psi_2(t_1) = 0$ we get

$$D^1(H) = |\dot{\phi}(t_1)[u]^1| = 2|\psi_1(t_1)| = 2 \cdot 1.08863205 \neq 0.$$

Then Theorem 6.27 ensures that the computed solution is a strict strong minimum.

Now we treat the second case, where the two boundary conditions (6.120) are replaced by the single nonlinear condition

$$x(t_f)^2 + x_2(t_f)^2 = r^2, \quad r = 0.2. \tag{6.129}$$

Imposing this boundary condition, we aim at steering the van der Pol oscillator to a small neighborhood of the origin. The adjoint equation (6.124) remains valid. The transversality condition for the adjoint variable gives

$$\psi_1(t_f) = 2\beta x_1(t_f), \quad \psi_2(t_f) = 2\beta x_2(t_f), \quad \beta \in \mathbb{R}. \tag{6.130}$$

The boundary condition (6.11) associated with the free final time t_f yields

$$\psi_1(t_f)x_2(t_f) + \psi_2(t_f)(-x_1(t_f) + x_2(t_f)(1 - x_1(t_f)^2) + u(t_f)) + 1 = 0. \tag{6.131}$$

Again, the switching function (6.18) is given by $\phi(t) = H_u(t) = \psi_2(t)$. The structure of the bang-bang control agrees with that in (6.132),

$$u(t) = \left\{ \begin{array}{rll} 1 & \text{for} & 0 \le t < t_1 \\ -1 & \text{for} & t_1 \le t \le t_f \end{array} \right\}, \tag{6.132}$$

which yields the switching condition

$$\phi(t_1) = \psi_2(t_1) = 0. \tag{6.133}$$

Using either the boundary value solver BNDSCO of Oberle and Grimm [82] or the direct optimization routine NUDOCCCS of Büskens [13, 14], we obtain the extremal solution depicted in Figure 6.3 and the following values for the switching, final time, state, and adjoint variables:

$$\begin{array}{rcl rcl} t_1 &=& 0.7139356, & t_f &=& 2.864192, \\ \psi_1(0) &=& 0.9890682, & \psi_2(0) &=& 0.9945782, \\ x_1(t_1) &=& 1.143759, & x_2(t_1) &=& -0.5687884, \\ \psi_1(t_1) &=& 1.758128, & \psi_2(t_1) &=& 0.0, \\ x_1(t_f) &=& 0.06985245, & x_2(t_f) &=& -0.1874050, \\ \psi_1(t_f) &=& 0.4581826, & \psi_2(t_f) &=& -1.229244, \\ \beta &=& 3.279646. & & & \end{array} \tag{6.134}$$

There are two alternative ways to check sufficient conditions. We may either use Theorem 6.19 and solve the linear equation (6.43) or evaluate directly the quadratic form in Proposition 6.11. We begin by testing the assumptions of Theorem 6.19 and consider the symmetric 2×2 matrix

$$Q(t) = \begin{pmatrix} Q_{11}(t) & Q_{12}(t) \\ Q_{12}(t) & Q_{22}(t) \end{pmatrix}.$$

The linear equations $\dot{Q} = -Qf_x - f_x^* Q - H_{xx}$ in (6.100) yield the following ODEs:

$$\begin{aligned} \dot{Q}_{11} &= 2\,Q_{12}(1 + 2x_1x_2) + 2\psi_2 x_2, \\ \dot{Q}_{12} &= -Q_{11} - Q_{12}(1 - x_1^2) + Q_{22}(1 + 2x_1x_2) + 2\psi_2 x_1, \\ \dot{Q}_{22} &= -2(Q_{12} + Q_{22}(1 - x_1^2)). \end{aligned} \tag{6.135}$$

In view of Theorem 6.19 we must find a solution $Q(t)$ only in the interval $[t_1, t_f]$ such that

$$D^1(H) + (q_{k+})[\dot{x}]^1 > 0, \quad q_{k+} = ([\dot{x}]^1)^* Q(t_1) - [\dot{\psi}]^1$$

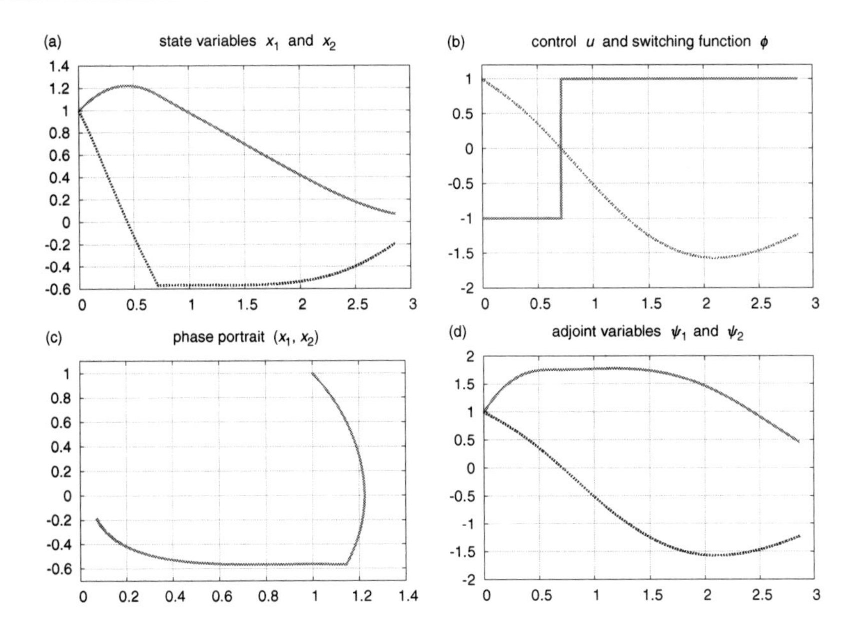

Figure 6.3. *Time-optimal solution of the van der Pol oscillator, nonlinear boundary condition* (6.129). (a) *State variables* x_1 *and* x_2 *(dashed line).* (b) *Control u and switching function* ψ_2 *(dashed line).* (c) *Phase portrait* (x_1,x_2). (d) *Adjoint variables* ψ_1 *and* ψ_2 *(dashed line).*

holds and the quadratic form ω_0 in (6.55)–(6.57) is positive definite on the cone $\mathcal{K}_0$ defined in (6.26). Since $\psi_2(t_1) = 0$ we get from (6.130),

$$D^1(H) = -\dot{\phi}(t_1)[u]^1 = 2 \cdot \psi_1(t_1) = 2 \cdot 1.758128 > 0.$$

Furthermore, from $[\dot{\psi}]^1 = 0$ we obtain the condition

$$D^1(H) + ([\dot{x}]^1)^* Q(t_1)[\dot{x}]^1 = 2 \cdot 1.758128 + 4Q_{22}(t_1) > 0,$$

which is satisfied by any initial value $Q_{22}(t_1) > -0.879064$. By Proposition 6.8, we have $\bar{t}_f = 0$ for every element $\bar{z} = (\bar{t}_f, \xi, \bar{x}) \in \mathcal{K}$. Therefore, by (6.57) we must check that the matrix $B_{22} = 2\beta I_2 - Q(t_f)$ is positive definite on the critical cone $\mathcal{K}_0$ defined in (6.26), i.e., on the cone

$$\mathcal{K}_0 = \{(\bar{t}_f, v_1, v_2) \mid \bar{t}_f = 0,\ x_1(t_f)v_1 + x_2(t_f)v_2 = 0\}.$$

Thus the variations (v_1, v_2) are related by $v_2 = -v_1 x_1(t_f)/x_2(t_f)$. Evaluating the quadratic form $\langle(2\beta I_2 - Q(t_f))(v_1, v_2), (v_1, v_2)\rangle$ with $v_2 = -v_1 x_1(t_f)/x_2(t_f)$, we arrive at the test

$$c = \left(2\beta\left(1 + \left(\frac{x_1}{x_2}\right)^2\right) - \left(Q_{11} - 2\frac{x_1}{x_2}Q_{12} + \left(\frac{x_1}{x_2}\right)^2 Q_{22}\right)\right)(t_f) > 0.$$

A straightforward integration of the ODEs (6.135) using the solution data (6.134) and the initial values $Q_{11}(t_1) = Q_{12}(t_1) = Q_{22}(t_1) = 0$ gives the numerical results

$$Q_{11}(t_f) = 0.241897, \quad Q_{12}(t_f) = -0.706142, \quad Q_{22}(t_f) = 1.163448,$$

which yield the positive value $c = 7.593456 > 0$. Thus Theorem 6.19 asserts that the bang-bang control characterized by (6.134) provides a strict strong minimum.

The alternative test for second-order sufficient conditions (SSC) is based on Proposition 6.11. The variational system $\dot{y}(t) = f_x(t)y(t)$, $y(t_1) = [\dot{x}]^1$, for the variation $y = (y_1, y_2)$ leads to the variational system

$$\begin{aligned} \dot{y}_1 &= y_2, & y_1(t_1) &= 0, \\ \dot{y}_2 &= -(1 + 2x_1 2x_2)y_1 + (1 - x_1^2)y_2, & y_2(t_1) &= 2, \end{aligned}$$

for which we compute

$$y_1(t_f) = 4.929925, \quad y_2(t_f) = 1.837486.$$

Note that the relation $K_{x_f}(x(t_f))y(t_f) = 2(x_1(t_f)y_1(t_f) + x_2(t_f)y_2(t_f)) = 0$ holds. By Proposition 6.11 we have to show that the quantity ρ in (6.37) is positive,

$$\rho = D^1(H) - [\dot{\psi}]^1[\dot{x}]^1 + \int_{t_1}^{t_f} (y(t))^* H_{xx}(t)y(t)\,dt + (y(t_f))^*(\beta K)_{x_f x_f} y(t_f) > 0.$$

Using $[\dot{\psi}]^1 = 0$ and $(y(t_f))^*(\beta K)_{x_f x_f} y(t_f) = 2\beta(y_1(t_f)^2 + y_2(t_f)^2)$, we finally obtain

$$\rho = D^1(H) + 184.550 > 0.$$

6.7.2 Time-Optimal Control of the Rayleigh Equation

In Section 4.1, the Rayleigh problem of controlling oscillations in a tunnel-diode oscillator (Figure 4.1) was considered with a regulator functional. In this section, we treat the time-optimal case of steering a given initial state to the origin in minimal time. Recall that the state variable $x_1(t) = I(t)$ denotes the electric current. The optimal control problem is to minimize the final time t_f subject to the dynamics and control constraints

$$\dot{x}_1(t) = x_2(t), \; \dot{x}_2(t) = -x_1(t) + x_2(t)(1.4 - 0.14x_2(t)^2) + u(t), \tag{6.136}$$

$$x_1(0) = x_2(0) = -5, \quad x_1(t_f) = x_2(t_f) = 0, \tag{6.137}$$

$$|u(t)| \le 4 \quad \text{for } t \in [0, t_f]. \tag{6.138}$$

Note that we have shifted the factor 4 to the control variable in the dynamics (4.134) to the control constraint (6.138). The Pontryagin function (Hamiltonian) (see (6.75)) for this problem is

$$H(x, u, \psi) = \psi_1 x_2 + \psi_2(-x_1 + x_2(1.4 - 0.14x_2^2) + u). \tag{6.139}$$

The transversality condition (6.11) yields, in view of (6.139),

$$H(t_f)+1 = \psi_2(t_f)u(t_f)+1 = 0. \tag{6.140}$$

The switching function $\phi(x,\psi) = \psi_2$ determines the optimal control

$$u(t) = \left\{ \begin{array}{rll} 4 & \text{if} & \psi_2(t) < 0 \\ -4 & \text{if} & \psi_2(t) > 0 \end{array} \right\}. \tag{6.141}$$

As for the van der Pol oscillator, it is easy to show that there are no *singular arcs* with $\psi_2(t) \equiv 0$ holding on a time interval $I_s \subset [0,t_f]$. Hence, the optimal control is bang-bang. Applying nonlinear programming methods to the suitably discretized Rayleigh problem, one realizes that the optimal control comprises the following three bang-bang arcs with two switching times t_1, t_2:

$$u(t) = \left\{ \begin{array}{rll} 4 & \text{for} & 0 \le t < t_1 \\ -4 & \text{for} & t_1 \le t < t_2 \\ 4 & \text{for} & t_2 \le t \le t_f \end{array} \right\}. \tag{6.142}$$

This control structure yields two switching conditions

$$\psi_2(t_1) = 0, \quad \psi_2(t_2) = 0. \tag{6.143}$$

Thus, we have to solve the multipoint boundary value problem comprising equations (6.136)–(6.143). The codes BNDSCO [82] and NUDOCCS [13, 14] yield the extremal depicted in Figure 6.4. The final time, the switching points, and some values for the adjoint variables are computed as

$$\begin{array}{rclrcl} t_f & = & 3.66817339, & & & \\ t_1 & = & 1.12050658, & t_2 & = & 3.31004698, \\ \psi_1(0) & = & -0.12234128, & \psi_2(0) & = & -0.08265161, \\ \psi_1(t_1) & = & -0.21521225, & \psi_1(t_2) & = & 0.89199176, \\ \psi_1(t_f) & = & 0.84276186, & \psi_2(t_f) & = & -0.25. \end{array} \tag{6.144}$$

Now we are going to show that the computed control satisfies the assumptions of Theorem 6.19 and thus provides a strict strong minimum. Consider the symmetric 2×2 matrix

$$Q(t) = \begin{pmatrix} Q_{11}(t) & Q_{12}(t) \\ Q_{12}(t) & Q_{22}(t) \end{pmatrix},$$

The linear equation (6.100), $dQ/dt = -Q\,f_x - f_x^*\,Q - H_{xx}$, leads to the following three equations:

$$\begin{aligned} \dot{Q}_{11} &= 2Q_{12}, \\ \dot{Q}_{12} &= -Q_{11} - Q_{12}(1.4 - 0.42x_2^2) + Q_{22}, \\ \dot{Q}_{22} &= -Q_{12} - Q_{22}(1.4 - 0.42x_2^2) + 0.84\psi_2 x_2. \end{aligned} \tag{6.145}$$

We must find a solution of these equations satisfying the estimates (6.114) at the switching times t_1 and t_2. From

$$D^k(H) = |\dot{\phi}(t_k)[u]^k| = 8\,|\psi_1(t_k)|, \quad k = 1,2,$$

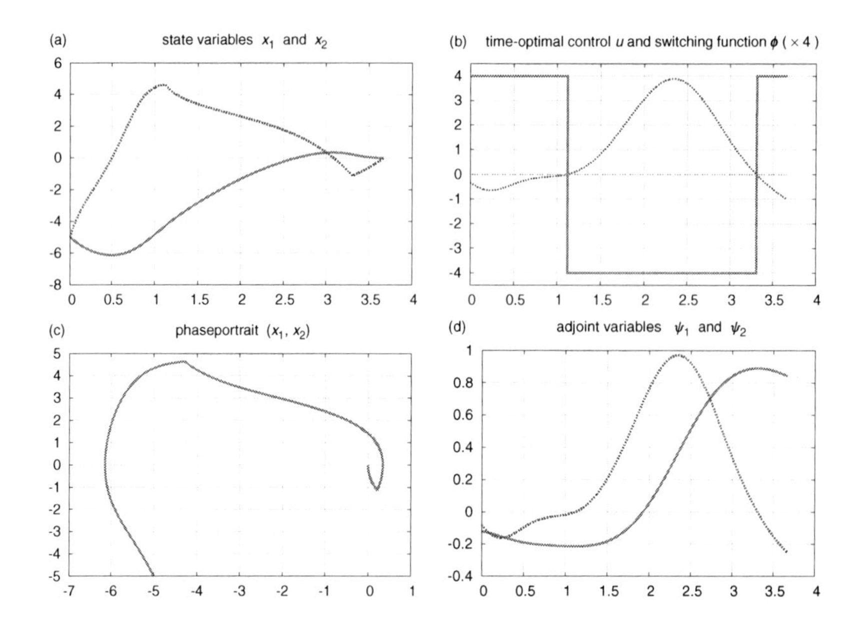

Figure 6.4. *Time-optimal control of the Rayleigh equation.* (a) *State variables* x_1 *and* x_2 *(dashed line).* (b) *Control* u *and switching function* ϕ *(dashed line).* (c) *Phase portrait* (x_1, x_2). (d) *Adjoint variables* ψ_1 *and* ψ_2 *(dashed line).*

we get

$$D^1(H) = 8 \cdot 0.21521225 = 1.7269800 > 0, \quad D^2(H) = 8 \cdot 0.89199176 = 7.1359341 > 0.$$

Furthermore, we have $[\dot{x}]^k = [u]^k(0,1)^* = (0,8)^*$ and thus obtain $\langle Q^k[\dot{x}]^k, [\dot{x}]^k \rangle = 64\, Q_{22}(t_k)$, for $k = 1,2$. The next step is to find a solution for the equations (6.145) in the interval $[t_1, t_2]$ that satisfies the inequalities

$$\begin{aligned} D^1(H) + \langle Q^1[\dot{x}]^1, [\dot{x}]^1 \rangle &= 1.7269800 + 64\, Q_{22}(t_1) > 0, \\ D^2(H) - \langle Q^2[\dot{x}]^2, [\dot{x}]^2 \rangle &= 7.1359341 - 64\, Q_{22}(t_2) > 0 \end{aligned}$$

This requires the estimates

$$Q_{22}(t_1) > -0.0269841, \quad Q_{22}(t_2) < 0.11149897. \tag{6.146}$$

These conditions can be satisfied by choosing, e.g., the following initial values at the switching time t_1:

$$Q_{11}(t_1) = 0, \quad Q_{12}(t_1) = 0.25, \quad Q_{22}(t_1) = 0.$$

Integration yields the value $Q_{22}(t_2) = -0.1677185$ which shows that the estimates (6.146) hold. Note that these estimates do not hold for the choice $Q(t_1) = 0$, since this initial value would give $Q_{22}(t_2) = 0.70592$. In summary, Theorem 6.36 asserts that the computed solution provides a strict strong minimum.

6.8 Time-Optimal Control Problems for Linear Systems with Constant Entries

In this section, we shall observe some results which were obtained in [77], [79, Part 2, Section 13], and [87].

6.8.1 Statement of the Problem, Minimum Principle, and Simple Sufficient Optimality Condition

Consider the following problem:

$$\begin{aligned} &t_f \to \min, \\ &\dot{x} = Ax + Bu, \quad x(0) = a, \; x(t_f) = b, \quad u \in U, \end{aligned} \tag{6.147}$$

where A and B are constant matrices, a and b are fixed vectors in $\mathbb{R}^{d(x)}$, and U is a convex polyhedron in $\mathbb{R}^{d(u)}$. A triple (t_f, x, u) is said to be *admissible* if $x(t)$ is a Lipschitz continuous and $u(t)$ is measurable bounded function on the interval $\Delta = [0, t_f]$ and the pair (x, u) satisfies on Δ the constraints of the problem (6.147).

Definition 6.37. We say that the admissible triple (t_f, x, u) affords *an almost global minimum* if there is no sequence of admissible triples $({t_f}^n, x^n, u^n)$, $n = 1, 2, \ldots$, such that $t_f^n < t_f$ for $n = 1, 2, \ldots$, and $t_f^n \to t_f$.

Proposition 6.38. *Suppose that, in the problem* (6.147), *there exists an* $u^* \in U$ *such that*

$$Aa + Bu^* = 0 \quad or \quad Ab + Bu^* = 0$$

(for example, let $b = 0$, $0 \in U$*). Then the almost global minimum is equivalent to the global one.*

Proof. If, for example, $Ab + Bu^* = 0$, $u^* \in U$, then any pair (x, u) admissible on $[0, t_f]$ can be extended to the right of t_f by putting $x(t) = b$, $u(t) = u^*$. ∎

Let (t_f, x, u) be an admissible triple for which the conditions of the minimum principle are fulfilled: There exists a smooth function $\psi : [0, t_f] \to \mathbb{R}^{d(x)}$ such that

$$\begin{aligned} &-\dot{\psi} = \psi A, \quad u(t) \in \operatorname{Arg\,min}_{u' \in U} (\psi(t) B u'), \\ &\psi \dot{x} = \text{const} \le 0, \quad |\psi(0)| = 1. \end{aligned} \tag{6.148}$$

These conditions follow from (6.76)–(6.80), because here $H = \psi(Ax + Bu)$, $H_t = 0$, hence $-\psi\dot{x} = \psi_0 = \text{const} =: \alpha_0 \ge 0$. Thus M_0 can be identified with the set of infinitely differentiable functions $\psi(t)$ on $[0, t_f]$ satisfying conditions (6.148).

The condition that M_0 is nonempty is necessary for a Pontryagin minimum to hold for the triple (t_f, x, u). We will refer to a triple (t_f, x, u) with nonempty M_0 as an *extremal triple*. Recall a simple *sufficient first-order condition* for an almost global minimum obtained in [79, Part 1].

Theorem 6.39 (Milyutin). *Suppose there exists* $\psi \in M_0$ *such that* $\alpha_0 := -\psi\dot{x} > 0$. *Then* (t_f, x, u) *affords an almost global minimum.*

In what follows, we assume that (t_f, x, u) is an extremal triple such that $u(t)$ is a piecewise constant function taking values in the vertices of the polyhedron U. We denote by $\Theta = \{t_1, \dots, t_s\}$ the set of discontinuity points of the control $u(t)$. Let $\psi \in M_0$. An important role in formulations of optimality conditions in problem (6.147) is played by the product $(\dot{\psi}\dot{x})(t)$.

Proposition 6.40. *The product $(\dot{\psi}\dot{x})(t)$ is a monotone nonincreasing step function with discontinuities only at the discontinuity points of the control $u(t)$.*

We now formulate yet another simple sufficient condition for the almost global minimum, obtained in [79, Part 2].

Theorem 6.41. *Suppose that there exists $\psi \in M_0$ such that the product $(\dot{\psi}\dot{x})$ fulfills one of the following two conditions: $(\dot{\psi}\dot{x})(0) < 0$, $(\dot{\psi}\dot{x})(t_f) > 0$; i.e., $(\dot{\psi}\dot{x})$ strictly retains its sign on $[0, t_f]$. Then (t_f, x, u) affords an almost global minimum.*

Theorem 6.41 implies a sufficient condition of a geometric nature.

Corollary 6.42. *Let (t_f, x, u) be an extremal triple such that for any point $t_k \in \Theta$ the vectors*

$$\dot{x}^{k-} = \dot{x}(t_k-), \quad \dot{x}^{k+} = \dot{x}(t_k+)$$

are different from zero and equally directed (so that $\dot{x}^{k-} = c_k \dot{x}^{k+}$ for some $c_k > 0$). Suppose that there exists $\psi \in M_0$ such that $(\dot{\psi}\dot{x})$ is not identically zero. Then (t_f, x, u) affords an almost global minimum.

6.8.2 Quadratic Optimality Condition

For an extremal triple (t_f, x, u) in the problem (6.147) satisfying the assumptions of Section 6.8.1 we will write the quadratic necessary Condition $\mathfrak{A}$ and the quadratic sufficient Condition $\mathfrak{B}$ using the results in Section 6.5.4.

Necessary Condition $\mathfrak{A}$. This time we begin by writing the quadratic form Ω as in (6.94). Let us show that it is completely determined by the left and right limits of the step function $\dot{\psi}\dot{x}$ at points $t_k \in \Theta$. Since $H = \psi(Ax + Bu)$ and $\dot{\psi} = -\psi A$, we have $[\dot{\psi}]^k = 0$, $k = 1, \dots, s$ and $H_{xx} = 0$. Moreover,

$$D^k(H) = -\dot{\psi}(t_k) B [u]^k = -\dot{\psi}(t_k)[\dot{x}]^k = -[\dot{\psi}\dot{x}]^k, \quad k = 1, \dots, s.$$

It follows from the condition $H_t = 0$ that $\dot{\psi}_0 = 0$. Further, the condition that $(\dot{\psi}\dot{x})$ is constant on (t_s, t_f) implies $(\dot{\psi}\dot{x})(t_f) = (\dot{\psi}\dot{x})(t_s+)$. Thus according to (6.94),

$$\Omega(\psi, \bar{z}) = -\sum_{k=1}^{s} [(\dot{\psi}\dot{x})]^k \bar{\xi}_k^2 + \bar{t}_f^2 (\dot{\psi}\dot{x})^{s+}. \tag{6.149}$$

This formula holds for any $\psi \in M_0$ and any $\bar{z} = (\bar{t}_f, \bar{\xi}, \bar{x})$ which belongs to the subspace $\mathcal{K}$ as in (6.83) and (6.84).

Let us see what is the form of $\mathcal{K}$ in the present case. Since $\bar{x}(t)$ satisfies the linear system $\dot{\bar{x}} = A\bar{x}$ on each interval $\Delta \setminus \Theta$, condition $\bar{x}(0) = 0$ in (6.84) can be replaced by $\bar{x}(t_1-) = 0$. Since $\dot{x}$ satisfies the same system, $\ddot{x} = A\dot{x}$, condition $\bar{x}(t_f) + \dot{x}(t_f)\bar{t}_f = 0$ in (6.84) can be replaced by $\bar{x}(t_s+) + \dot{x}(t_s+)\bar{t}_f = 0$. For brevity, put

$$\bar{x}(t_1-) = \bar{x}^{1-}, \quad \bar{x}(t_s+) = \bar{x}^{s+}, \quad \dot{x}(t_s+) = \dot{x}^{s+}.$$

Then by (6.83) and (6.84) the subspace $\mathcal{K}$ consists of triples $\bar{z} = (\bar{t}_f, \bar{\xi}, \bar{x})$ satisfying the conditions

$$\begin{aligned}
&\bar{t}_f \in \mathbb{R}^1, \quad \bar{\xi} \in \mathbb{R}^s, \quad \bar{x}(\cdot) \in P_\Theta C^\infty(\Delta, \mathbb{R}^n),\\
&\dot{\bar{x}} = A\bar{x}, \quad [\bar{x}]^k = [\dot{x}]^k \bar{\xi}_k, \quad k = 1, \ldots, s,\\
&\bar{x}^{1-} = 0, \quad \bar{x}^{s+} + \bar{t}_f \dot{x}^{s+} = 0,
\end{aligned}$$

where $P_\Theta C^\infty(\Delta, \mathbb{R}^n)$ is the space of piecewise continuous functions $\bar{x}(t) : \Delta \to \mathbb{R}^{d(x)}$ infinitely differentiable on each interval of the set $\Delta \setminus \Theta$. This property of $\bar{x}$ follows from the fact that on each interval of the set $\Delta \setminus \Theta$ the function $\bar{x}$ satisfies the linear system $\dot{\bar{x}} = A\bar{x}$ with constant entries.

Consider the cross-section of $\mathcal{K}$ specified by condition $\bar{t}_f = -1$. The passage to the cross-section does not weaken the quadratic necessary Condition $\mathfrak{A}$ because the functional $\max_{\psi \in M_0} \Omega(\psi, \bar{z})$ involved in it is homogeneous of degree 2 and nonnegative on any element $\bar{z} \in \mathcal{K}$ with $\bar{t}_f = 0$ (since for any $\psi \in M_0$ the inequalities $D^k(H) = -[\dot{\psi}\dot{x}]^k \geq 0, k = 1, \ldots, s$, hold).

Denote by $\mathcal{R}$ the cross-section of the subspace $\mathcal{K}$, specified by condition $\bar{t}_f = -1$. We omit the coordinate $\bar{t}_f$ in the definition of $\mathcal{R}$. Thus $\mathcal{R}$ is a set of pairs $(\bar{\xi}, \bar{x})$ such that the following conditions are fulfilled:

$$\begin{aligned}
&\bar{\xi} \in \mathbb{R}^s, \quad \bar{x}(\cdot) \in P_\Theta C^\infty(\Delta, \mathbb{R}^n),\\
&\dot{\bar{x}} = A\bar{x}, \quad [\bar{x}]^k = [\dot{x}]^k \bar{\xi}_k, \quad k = 1, \ldots, s,\\
&\bar{x}^{1-} = 0, \quad \bar{x}^{s+} = \dot{x}^{s+}.
\end{aligned}$$

For $\psi \in M_0$, $\bar{\xi} \in \mathbb{R}^s$, let

$$Q(\psi, \bar{\xi}) = -\sum_{k=1}^{s} [\dot{\psi}\dot{x}]^k \bar{\xi}_k^2 + (\dot{\psi}\dot{x})^{s+},$$

and set

$$Q_0(\bar{\xi}) = \max_{\psi \in M_0} Q(\psi, \bar{\xi}). \tag{6.150}$$

Then Theorem 6.26 implies the following theorem.

Theorem 6.43. *Let a triple (t_f, x, u) afford a Pontryagin minimum in the problem* (6.147). *Then the set M_0 as in* (6.148) *is nonempty and*

$$Q_0(\bar{\xi}) \geq 0 \quad \forall\, (\bar{\xi}, \bar{x}) \in \mathcal{R}. \tag{6.151}$$

It is clear that the set $\mathcal{R}$, in this necessary condition, can be replaced by its projection under the mapping $(\bar{\xi}, \bar{x}) \to \bar{\xi}$. Denote this projection by Ξ and find out what conditions specify it. Conditions $\dot{\bar{x}} = A\bar{x}$ and $\ddot{x} = A\dot{x}$ imply

$$\bar{x}(t) = e^{At}\bar{c}(t), \quad \dot{x}(t) = e^{At}c(t),$$

where $\bar{c}(t)$ and $c(t)$ are step functions whose discontinuity points are contained in Θ. Conditions

$$\bar{x}^{1-} = 0, \quad \bar{x}^{s+} = \dot{x}^{s+}, \quad [\bar{x}]^k = [\dot{x}]^k \bar{\xi}_k, \quad k = 1,\dots,s,$$

imply

$$\bar{c}^{1-} = 0, \quad \bar{c}^{s+} = c^{s+}, \quad [\bar{c}]^k = [c]^k \bar{\xi}_k, \quad k = 1,\dots,s.$$

Therefore

$$c^{s+} = \bar{c}^{s+} = \sum_{k=1}^{s} [\bar{c}]^k = \sum_{k=1}^{s} [c]^k \bar{\xi}_k.$$

It is easily seen that the conditions

$$\sum_{k=1}^{s} [c]^k \bar{\xi}_k = c^{s+}$$

determine the projection of $\mathcal{R}$ under the mapping $(\bar{\xi}, \bar{x}) \to \bar{\xi}$, which we denote by Ξ. Since

$$[c]^k = e^{-At_k}[\dot{x}]^k \;\forall\, k, \qquad c^{s+} = e^{-At_s}\dot{x}^{s+},$$

the set Ξ is determined by the condition

$$\sum_{k=1}^{s} e^{-At_k}[\dot{x}]^k \bar{\xi}_k = e^{-At_s}\dot{x}^{s+},$$

which after multiplication by e^{At_s} from the left takes the final form

$$\sum_{k=1}^{s} e^{A(t_s - t_k)}[\dot{x}]^k \bar{\xi}_k = \dot{x}^{s+}. \tag{6.152}$$

Hence Ξ is the set of vectors $\bar{\xi} \in \mathbb{R}^s$ satisfying the system of algebraic equations (6.152). Thus Theorem 6.43 implies the following theorem.

Theorem 6.44. *Let a triple* (t_f, x, u) *afford a Pontryagin minimum in the problem* (6.147). *Then the set* M_0 *as in* (6.148) *is nonempty and*

$$Q_0(\bar{\xi}) \geq 0 \quad \forall\, \bar{\xi} \in \Xi. \tag{6.153}$$

It makes sense to use necessary condition (6.153) for investigation of only those extremals (t_f, x, u) for which

$$\alpha_0 := -\psi\dot{x} = 0 \quad \forall\, \psi \in M_0, \tag{6.154}$$

because otherwise, by Theorem 6.39, (t_f, x, u) affords an almost global minimum in the problem. Condition (6.154) guarantees that the set Ξ is nonempty (see Theorem 13.7 in [79, p. 322]). Theorem 6.44 implies a simple consequence of a geometric nature.

Corollary 6.45. *Suppose that (t_f, x, u) affords a Pontryagin minimum in the problem* (6.147). *Let the vectors $\dot{x}^{k-}$ and $\dot{x}^{k+}$ be different from zero and collinear for some $t_k \in \Theta$, and let the jump $[\dot{\psi}\dot{x}]^k$ of the product $\dot{\psi}\dot{x}$ at the point t_k be different from zero for any $\psi \in M_0$. Then the vectors $\dot{x}^{k-}$ and $\dot{x}^{k+}$ are equally directed.*

The proof is given in [79].

Sufficient Condition $\mathfrak{B}$. Let (t_f, x, u) be an admissible triple in the problem (6.147) satisfying the assumptions of Section 6.8.1. As in Section 6.2.4, denote by $\mathcal{M}$ the set of functions $\psi \in M_0$ satisfying the two conditions

$$-[\dot{\psi}\dot{x}]^k > 0, \quad k = 1, \ldots, s, \tag{6.155}$$

$$\operatorname*{Arg\,min}_{u' \in U} \psi(t)Bu' = [u(t-), u(t+)] \quad \forall\, t \in [0, t_f]. \tag{6.156}$$

Obviously, the interval $[u(t-), u(t+)]$ is a singleton for all $t \in [0, t_f] \setminus \Theta$.

Theorem 6.46. *For an admissible triple (t_f, x, u) satisfying the assumptions of Section* 6.8.1, *let the set $\mathcal{M}$ be nonempty and*

$$Q_0(\bar{\xi}) > 0 \quad \forall\, \bar{\xi} \in \Xi. \tag{6.157}$$

Then (t_f, x, u) affords a strict almost global minimum in the problem (6.147).

Proof. Condition (6.157) implies that

$$\max_{\psi \in M_0} \Omega(\lambda, \bar{z}) > 0 \tag{6.158}$$

for all $\bar{z} \in \mathcal{K}$ such that $\bar{t}_f \neq 0$. Consider an element $\bar{z} \in \mathcal{K} \setminus \{0\}$ such that $\bar{t}_f = 0$. For this element, $\bar{\xi} \neq 0$, since otherwise $\bar{z} = 0$. Take an arbitrary element $\psi \in \mathcal{M}$ and put $q = -\dot{\psi}\dot{x}$. Then $[q]^k > 0$, $k = 1, \ldots, s$. Hence

$$\Omega(\lambda, \bar{z}) = \sum_{k=1}^{s} [q]^k \bar{\xi}_k^2 > 0.$$

Thus the inequality (6.158) holds for all $\bar{z} \in \mathcal{K} \setminus \{0\}$. Therefore by Theorem 6.27, (t_f, x, u) affords a strict strong minimum in the problem (6.147). Moreover, condition $\mathcal{M} \neq \emptyset$ in the problem (6.147) implies that the strict strong minimum is equivalent to the strict almost global minimum. The last assertion is a consequence of Proposition 13.4 and Lemma 13.1 in [79]. ∎

6.8.3 Example

Consider the problem

$$t_f \to \min, \quad \dot{x}_1 = x_2, \quad \dot{x}_2 = u, \quad |u| \leq 1, \quad x(0) = a, \quad x(t_f) = b, \tag{6.159}$$

where $x = (x_1, x_2) \in \mathbb{R}^2$, $u \in \mathbb{R}^1$. The minimum principle conditions for this problem are as follows:

$$\begin{aligned} &\dot{\psi}_1 = 0, \quad \dot{\psi}_2 = -\psi_1, \quad u = -\operatorname{sgn}\psi_2, \\ &\psi_1 x_2 + \psi_2 u = -\alpha_0 \le 0, \quad |\psi(0)| = 1, \end{aligned} \tag{6.160}$$

where $\psi = (\psi_1, \psi_2)$. This implies that $\psi_2(t)$ is a linear function, and $u(t)$ is a step function having at most one switching and taking values ± 1.

The system corresponding to $u = 1$ is $\dot{x}_1 = x_2$, $\dot{x}_2 = 1$, and omitting t we have

$$x_1 = \frac{1}{2}x_2^2 + C. \tag{6.161}$$

The system corresponding to $u = -1$ is $\dot{x}_1 = x_2$, $\dot{x}_2 = -1$, whence

$$x_1 = -\frac{1}{2}x_2^2 + C. \tag{6.162}$$

Through each point $x = (x_1, x_2)$ of the state plane there pass one curve of the family (6.161) and one curve of the family (6.162). The condition $\psi_1 x_2 + \psi_2 u = -\alpha_0 \le 0$ implies that the passage from a curve of the family (6.161) to a curve of the family (6.162) is only possible in the upper half-plane $x_2 \ge 0$, and the passage from a curve of the family (6.162) to a curve of the family (6.161) in the lower half-plane $x_2 \le 0$. The direction of movement is such that along the curves of the family (6.161) the point can go to $(+\infty, +\infty)$, and along the curves of the family (6.162) to $(-\infty, -\infty)$. This means that any two points can be joined by the extremals in no more than two ways. By Theorem 6.39 each extremal with

$$\alpha_0 := -\dot{\psi}\dot{x} = -\psi_1 x_2 - \psi_2 u > 0$$

yields an almost global minimum.

Consider extremals with switching and $\alpha_0 = 0$. These are extremals with $x_2 = 0$ at the switching time and $\dot{x}^{1-} = (0, u^{1-})$, $\dot{x}^{1+} = (0, u^{1+})$. Since $u^{1-} = -u^{1+} \ne 0$, the vectors $\dot{x}^{1-}$ and $\dot{x}^{1+}$ are different from zero and directed in an opposite way. The set M_0 consists of a single pair $\psi = (\psi_1, \psi_2)$, and

$$\dot{\psi}\dot{x} = -\psi_1 u, \quad [\dot{\psi}\dot{x}]^1 = -\psi_1(t_1)[u]^1 \ne 0. \tag{6.163}$$

According to Corollary 6.45 this extremal does not yield a Pontryagin minimum because it is necessary for Pontryagin minimum that the collinear vectors $\dot{x}^{1-}$ and $\dot{x}^{1+}$ are equally directed. In this case the necessary second-order condition fails.

Thus in this problem an extremal with a switching affords an almost global minimum iff $\alpha_0 > 0$ for this extremal; i.e., the point $x = (x_1, x_2)$ at the switching time does not lie on the x_1 axis in the state plane.

For an extremal without switchings there always exists $\psi \in M_0$ such that $\alpha_0 > 0$. Indeed, for this extremal one can set $\psi_1 = 0, \psi_2 = -u$. Then $\alpha_0 = -\psi_2 u = 1$, and all conditions of the minimum principle are fulfilled.

More interesting examples of investigation of extremals in time-optimal control problems for linear systems with constant entries are given in [79, Part 2, Section 14]. The time-optimal control in a simplified model of a container crane (ore unloader) was discussed in [63, Section 5]. The optimality of the time-optimal control with three switching times follows from Theorem 6.39.

Chapter 7

Bang-Bang Control Problem and Its Induced Optimization Problem

We continue our investigation of the pure bang-bang case. As it was mentioned in introduction, second-order sufficient optimality conditions for bang-bang controls had been derived in the literature in two different forms. The first form was discussed in the last chapter. The second one is due to Agrachev, Stefani, and Zezza [1], who first reduced the bang-bang control problem to a finite-dimensional optimization problem and then showed that well-known sufficient optimality conditions for this optimization problem supplemented by the strict bang-bang property furnish sufficient conditions for the bang-bang control problem. The bang-bang control problem, considered in this chapter, is more general than that in [1]. Following [99, 100], we establish the equivalence of both forms of second-order conditions for this problem.

7.1 Main Results

7.1.1 Induced Optimization Problem

Again, let $\hat{\mathcal{T}} = (\hat{x}(t), \hat{u}(t) \mid t \in [\hat{t}_0, \hat{t}_f])$ be an admissible trajectory for the basic problem (6.1)–(6.3). We denote by $V = \operatorname{ex} U$ the set of vertices of the polyhedron U. Assume that $\hat{u}(t)$ is a bang-bang control in $\hat{\Delta} = [\hat{t}_0, \hat{t}_f]$ taking values in the set V,

$$\hat{u}(t) = u^k \in V \text{ for } t \in (\hat{t}_{k-1}, \hat{t}_k), \quad k = 1, \dots, s+1,$$

where $\hat{t}_{s+1} = \hat{t}_f$. Thus, $\hat{\Theta} = \{\hat{t}_1, \dots, \hat{t}_s\}$ is the set of switching points of the control $\hat{u}(\cdot)$ with $\hat{t}_k < \hat{t}_{k+1}$ for $k = 0, 1, \dots, s$. Assume now that the set M_0 of multipliers is nonempty for the trajectory $\hat{\mathcal{T}}$. Put

$$\hat{x}(\hat{t}_0) = \hat{x}_0, \qquad \hat{\theta} = (\hat{t}_1, \dots, \hat{t}_s), \qquad \hat{\zeta} = (\hat{t}_0, \hat{t}_f, \hat{x}_0, \hat{\theta}). \tag{7.1}$$

Then $\hat{\theta} \in \mathbb{R}^s$, $\hat{\zeta} \in \mathbb{R}^2 \times \mathbb{R}^n \times \mathbb{R}^s$, where $n = d(x)$.

Take a small neighborhood $\mathcal{V}$ of the point $\hat{\zeta}$ in $\mathbb{R}^2 \times \mathbb{R}^n \times \mathbb{R}^s$, and let

$$\zeta = (t_0, t_f, x_0, \theta) \in \mathcal{V},$$

where $\theta = (t_1,\dots,t_s)$ satisfies $t_0 < t_1 < t_2 < \cdots < t_s < t_f$. Define the function $u(t;\theta)$ by the condition

$$u(t;\theta) = u^k \text{ for } t \in (t_{k-1},t_k), \quad k = 1,\dots,s+1, \tag{7.2}$$

where $t_{s+1} = t_f$. The values $u(t_k;\theta)$, $k = 1,\dots,s$, may be chosen in U arbitrarily. For definiteness, define them by the condition of continuity of the control from the left: $u(t_k;\theta) = u(t_k-;\theta)$, $k = 1,\dots,s$.

Let $x(t;t_0,x_0,\theta)$ be the solution of the initial value problem (IVP),

$$\dot{x} = f(t,x,u(t;\theta)), \quad t \in [t_0,t_f], \quad x(t_0) = x_0. \tag{7.3}$$

For each $\zeta \in \mathcal{V}$ this solution exists if the neighborhood $\mathcal{V}$ of the point $\hat{\zeta}$ is sufficiently small. We obviously have

$$x(t;\hat{t}_0,\hat{x}_0,\hat{\theta}) = \hat{x}(t), \quad t \in \hat{\Delta}, \quad u(t;\hat{\theta}) = \hat{u}(t), \quad t \in \hat{\Delta} \setminus \hat{\Theta}.$$

Consider now the following finite-dimensional optimization problem in the space $\mathbb{R}^2 \times \mathbb{R}^n \times \mathbb{R}^s$ of the variables $\zeta = (t_0,t_f,x_0,\theta)$:

$$\begin{aligned} \mathcal{F}_0(\zeta) &:= J(t_0,x_0,t_f,x(t_f;t_0,x_0,\theta)) \to \min, \\ \mathcal{F}(\zeta) &:= F(t_0,x_0,t_f,x(t_f;t_0,x_0,\theta)) \le 0, \\ \mathcal{G}(\zeta) &:= K(t_0,x_0,t_f,x(t_f;t_0,x_0,\theta)) = 0. \end{aligned} \tag{7.4}$$

We call (7.4) the *Induced Optimization Problem* (IOP) or simply *Induced Problem*, which represents an extension of the IOP introduced in Agrachev, Stefani, and Zezza [1]. The following assertion is almost obvious.

Theorem 7.1. *Let the trajectory $\hat{\mathcal{T}}$ be a Pontryagin local minimum for the basic control problem* (6.1)–(6.3). *Then the point $\hat{\zeta}$ is a local minimum for the IOP* (7.4), *and hence it satisfies first- and second-order necessary conditions for this problem.*

Proof. Assume that $\hat{\zeta}$ is not a local minimum in problem (7.4). Then there exists a sequence of admissible points $\zeta^\nu = (t_0^\nu,t_1^\nu,x_0^\nu,\theta^\nu)$ in problem (7.4) such that $\zeta^\nu \to \hat{\zeta}$ for $\nu \to \infty$ and $\mathcal{F}_0(\zeta^\nu) < \mathcal{F}_0(\hat{\zeta})$. Take the corresponding sequence of admissible trajectories

$$\mathcal{T}^\nu = \{x(t;t_0^\nu,x_0^\nu,\theta^\nu),u(t;\theta^\nu) \mid t \in [t_0^\nu,t_f^\nu]\}$$

in problem (6.1)–(6.3). Then the conditions $t_0^\nu \to \hat{t}_0$, $t_f^\nu \to \hat{t}_f$, $x_0^\nu \to \hat{x}_0$, $\theta^\nu \to \hat{\theta}$ imply that

$$\int_{\Delta^\nu \cap \hat{\Delta}} |u(t;\theta^\nu) - \hat{u}(t)|\,dt \to 0, \quad \max_{\Delta^\nu \cap \hat{\Delta}} |x(t;t_0^\nu,x_0^\nu,\theta^\nu) - \hat{x}(t)| \to 0,$$

where $\Delta^\nu = [t_0^\nu,t_f^\nu]$. Moreover, $\mathcal{J}(\mathcal{T}^\nu) = \mathcal{F}_0(\zeta^\nu) < \mathcal{F}_0(\hat{\zeta}) = \mathcal{J}(\hat{\mathcal{T}})$. It means that the trajectory $\hat{\mathcal{T}}$ is not a Pontryagin local minimum for the basic problem (6.1)–(6.3). ∎

We shall clarify a relationship between the second-order conditions for the IOP (7.4) at the point $\hat{\zeta}$ and those in the basic bang-bang control problem (6.1)–(6.3) for the trajectory $\hat{\mathcal{T}}$.

We shall state that there is a one-to-one correspondence between Lagrange multipliers in these problems and a one-to-one correspondence between elements of the critical cones. Moreover, for corresponding Lagrange multipliers, the quadratic forms in these problems take equal values on the corresponding elements of the critical cones. This will allow us to express the necessary and sufficient quadratic optimality conditions for bang-bang control, formulated in Theorems 6.9 and 6.10, in terms of the IOP (7.4). In particular, we thus establish the equivalence between our quadratic sufficient conditions and those due to Agrachev, Stefani, and Zezza [1].

First, for convenience we recall second-order necessary and sufficient conditions for a smooth finite-dimensional optimization problem with inequality- and equality-type constraints (see Section 1.3.1). Consider a problem in $\mathbb{R}^n$,

$$f_0(x) \to \min, \qquad f_i(x) \le 0 \ (i = 1,\dots,k), \quad g_j(x) = 0 \ (j = 1,\dots,m), \tag{7.5}$$

where $f_0,\dots,f_k,g_1,\dots,g_m$ are C^2-functions in $\mathbb{R}^n$. Let $\hat{x}$ be an admissible point in this problem. Define, at this point, the set of normalized vectors $\mu = (\alpha_0,\dots,\alpha_k,\beta_1,\dots,\beta_m)$ of Lagrange multipliers

$$\begin{aligned} \Lambda_0 \ = \ \Big\{ &\mu \in \mathbb{R}^{k+1+m} \mid \alpha_i \ge 0 \ (i = 0,\dots,k), \quad \alpha_i f_i(\hat{x}) = 0 \ (i = 1,\dots,k), \\ &\sum_{i=0}^{k} \alpha_i + \sum_{j=1}^{m} |\beta_j| = 1; \quad L_x(\mu,\hat{x}) = 0 \Big\}, \end{aligned}$$

where $L(\mu,x) = \sum_{i=0}^{k} \alpha_i f_i(x) + \sum_{j=1}^{m} \beta_j g_j(x)$ is the Lagrange function. Define the set of active indices $I = \{i \in \{1,\dots,k\} \mid f_i(\hat{x}) = 0\}$ and the *critical cone*

$$\mathcal{K}_0 = \{\bar{x} \mid f_0'(\hat{x})\bar{x} \le 0, \quad f_i'(\hat{x})\bar{x} \le 0, \ i \in I, \quad g_j'(\hat{x})\bar{x} = 0, \ j = 1,\dots,m\}.$$

Theorem 7.2. *Let $\hat{x}$ be a local minimum in problem* (7.5). *Then, at this point, the set Λ_0 is nonempty and the following inequality holds:*

$$\max_{\mu \in \Lambda_0} \langle L_{xx}(\mu,\hat{x})\bar{x},\bar{x}\rangle \ge 0 \quad \forall\, \bar{x} \in \mathcal{K}_0.$$

Theorem 7.3. *Let the set Λ_0 be nonempty at the point $\hat{x}$ and let*

$$\max_{\mu \in \Lambda_0} \langle L_{xx}(\mu,\hat{x})\bar{x},\bar{x}\rangle > 0 \quad \forall\, \bar{x} \in \mathcal{K}_0 \setminus \{0\}.$$

Then $\hat{x}$ is a local minimum in problem (7.5).

These conditions were obtained by Levitin, Milyutin, and Osmolovskii [54, 55]; cf. also Ben-Tal and Zowe [4].

7.1.2 Relationship between Second-Order Conditions for the Basic and Induced Optimization Problem

Let $\hat{\mathcal{T}} = (\hat{x}(t),\hat{u}(t) \mid t \in [\hat{t}_0,\hat{t}_f])$ be an admissible trajectory in the basic problem with the properties assumed in Section 7.1.1, and let $\hat{\zeta} = (\hat{t}_0,\hat{t}_f,\hat{x}_0,\hat{\theta})$ be the corresponding admissible point in the Induced Problem.

Lagrange multipliers. Let us define the set $\Lambda_0 \subset \mathbb{R}^{1+d(F)+d(K)}$ of the triples $\mu = (\alpha_0, \alpha, \beta)$ of normalized Lagrange multipliers at the point $\hat{\zeta}$ for the Induced Problem. The Lagrange function for the Induced Problem is

$$\begin{aligned} L(\mu,\zeta) &= L(\mu,t_0,t_f,x_0,\theta) = \alpha_0 J(t_0,x_0,t_f,x(t_f;t_0,x_0,\theta)) \\ &\quad + \alpha F(t_0,x_0,t_f,x(t_f;t_0,x_0,\theta)) + \beta K(t_0,x_0,t_f,x(t_f;t_0,x_0,\theta)) \\ &= l(\mu,t_0,x_0,t_f,x(t_f;t_0,x_0,\theta)), \end{aligned} \tag{7.6}$$

where $l = \alpha_0 J + \alpha F + \beta K$. By definition, Λ_0 is the set of multipliers $\mu = (\alpha_0, \alpha, \beta)$ such that

$$\alpha_0 \geq 0, \quad \alpha \geq 0, \quad \alpha_0 + |\alpha| + |\beta| = 1, \quad \alpha F(\hat{p}) = 0, \quad L_\zeta(\mu, \hat{\zeta}) = 0, \tag{7.7}$$

where $\hat{p} = (\hat{t}_0, \hat{x}_0, \hat{t}_f, \hat{x}_f)$, $\hat{x}_0 = \hat{x}(\hat{t}_0)$, $\hat{x}_f = \hat{x}(\hat{t}_f) = x(\hat{t}_f; \hat{t}_0, \hat{x}_0, \hat{\theta})$. Now, let us define the corresponding set of normalized Lagrange multipliers for the trajectory $\hat{\mathcal{T}}$ in the basic problem. Denote by Λ the set of multipliers $\lambda = (\alpha_0, \alpha, \beta, \psi, \psi_0)$ such that

$$\begin{aligned} &\alpha_0 \geq 0, \quad \alpha \geq 0, \quad \alpha_0 + |\alpha| + |\beta| = 1, \quad \alpha F(\hat{p}) = 0, \\ &-\dot{\psi}(t) = \psi(t) f_x(t, \hat{x}(t), \hat{u}(t)), \quad -\dot{\psi}_0(t) = \psi(t) f_t(t, \hat{x}(t), \hat{u}(t)), \\ &\psi(\hat{t}_0) = -l_{x_0}(\mu, \hat{p}), \quad \psi(\hat{t}_f) = l_{x_f}(\mu, \hat{p}), \\ &\psi_0(\hat{t}_0) = -l_{t_0}(\mu, \hat{p}), \quad \psi_0(\hat{t}_f) = l_{t_f}(\mu, \hat{p}), \\ &\psi(t) f(t, \hat{x}(t), \hat{u}(t)) + \psi_0(t) = 0 \quad \forall\, t \in \hat{\Delta} \setminus \hat{\Theta}, \end{aligned} \tag{7.8}$$

where $\hat{\Delta} = [\hat{t}_0, \hat{t}_f]$ and $\hat{\Theta} = \{\hat{t}_1, \dots, \hat{t}_s\}$.

Proposition 7.4. *The projector*

$$\pi_0 : (\alpha_0, \alpha, \beta, \psi, \psi_0) \to (\alpha_0, \alpha, \beta) \tag{7.9}$$

maps one-to-one the set Λ onto the set Λ_0.

Let us define the inverse mapping. Take an arbitrary multiplier $\mu = (\alpha_0, \alpha, \beta) \in \Lambda_0$. This tuple defines the gradient $l_{x_f}(\mu, \hat{p})$, and hence the system

$$-\dot{\psi} = \psi f_x(t, \hat{x}(t), \hat{u}(t)), \quad \psi(\hat{t}_f) = l_{x_f}(\mu, \hat{p}) \tag{7.10}$$

defines $\psi(t)$. Define $\psi_0(t)$ by the equality

$$\psi(t) f(t, \hat{x}(t), \hat{u}(t)) + \psi_0(t) = 0. \tag{7.11}$$

Proposition 7.5. *The inverse mapping*

$$\pi_0^{-1} : (\alpha_0, \alpha, \beta) \in \Lambda_0 \to (\alpha_0, \alpha, \beta, \psi, \psi_0) \in \Lambda \tag{7.12}$$

is defined by formulas (7.10) *and* (7.11).

We note that $M_0 \subset \Lambda$ holds, because the system of conditions (6.7)–(6.9) and (6.11) is equivalent to system (7.8). But it may happen that $M_0 \neq \Lambda$, since in the definition of Λ there is no requirement that its elements satisfy minimum condition (6.10). Let us denote $\Lambda_0^{MP} := \pi_0(M_0)$, where *MP* stands for Minimum Principle.

We say that multipliers $\mu = (\alpha_0, \alpha, \beta)$ and $\lambda = (\alpha_0, \alpha, \beta, \psi, \psi_0)$ *correspond to each other* if they have the same components α_0, α and β, i.e.,

$$\pi_0(\alpha_0, \alpha, \beta, \psi, \psi_0) = (\alpha_0, \alpha, \beta).$$

Critical cones. We denote by $\mathcal{K}_0$ the critical cone at the point $\hat{\zeta}$ in the Induced Problem. Thus, $\mathcal{K}_0$ is the set of collections $\bar{\zeta} = (\bar{t}_0, \bar{t}_f, \bar{x}_0, \bar{\theta})$ such that

$$\mathcal{F}_0'(\hat{\zeta})\bar{\zeta} \leq 0, \quad \mathcal{F}_i'(\hat{\zeta})\bar{\zeta} \leq 0,\ i \in I, \quad \mathcal{G}'(\hat{\zeta})\bar{\zeta} = 0, \tag{7.13}$$

where I is the set of indices of the inequality constraints active at the point $\hat{\zeta}$. Let $\mathcal{K}$ be the critical cone for the trajectory $\hat{\mathcal{T}}$ in the basic problem, i.e., the set of all tuples $\bar{z} = (\bar{t}_0, \bar{t}_f, \bar{\xi}, \bar{x}) \in Z(\hat{\Theta})$, satisfying conditions (6.21)–(6.23).

Proposition 7.6. *The operator* $\pi_1 : (\bar{t}_0, \bar{t}_f, \bar{\xi}, \bar{x}) \to (\bar{t}_0, \bar{t}_f, \bar{x}_0, \bar{\theta})$ *defined by*

$$\bar{\theta} = -\bar{\xi}, \quad \bar{x}_0 = \bar{x}(\hat{t}_0) \tag{7.14}$$

is a one-to-one mapping of the critical cone $\mathcal{K}$ *(for the trajectory* $\hat{\mathcal{T}}$ *in the basic problem) onto the critical cone* $\mathcal{K}_0$ *(at the point* $\hat{\zeta}$ *in the Induced Problem).*

We say that the elements $\bar{\zeta} = (\bar{t}_0, \bar{t}_f, \bar{x}_0, \bar{\theta}) \in \mathcal{K}_0$ and $\bar{z} = (\bar{t}_0, \bar{t}_f, \bar{\xi}, \bar{x}) \in \mathcal{K}$ *correspond to each other* if $\bar{\theta} = -\bar{\xi}$ and $\bar{x}_0 = \bar{x}(\hat{t}_0)$, i.e., $\pi_1(\bar{t}_0, \bar{t}_f, \bar{\xi}, \bar{x}) = (\bar{t}_0, \bar{t}_f, \bar{x}_0, \bar{\theta})$.

Now we give explicit formulas for the inverse mapping for π_1. Let $V(t)$ be an $n \times n$ matrix-valued function ($n = d(x)$) which is absolutely continuous in $\hat{\Delta} = [\hat{t}_0, \hat{t}_f]$ and satisfies the system

$$\dot{V}(t) = f_x(t, \hat{x}(t), \hat{u}(t))V(t), \quad V(\hat{t}_0) = E, \tag{7.15}$$

where E is the identity matrix. For each $k = 1, \ldots, s$ denote by $y^k(t)$ the n-dimensional vector function which is equal to zero in $[\hat{t}_0, \hat{t}_k)$, and in $[\hat{t}_k, \hat{t}_f]$ it is the solution to the IVP

$$\dot{y}^k = f_x(t, \hat{x}(t), \hat{u}(t))y^k, \quad y^k(\hat{t}_k) = -[\dot{\hat{x}}]^k. \tag{7.16}$$

Hence y^k is a piecewise continuous function with one jump $[y^k]^k = -[\dot{\hat{x}}]^k$ at $\hat{t}_k$.

Proposition 7.7. *The inverse mapping* $\pi_1^{-1} : (\bar{t}_0, \bar{t}_f, \bar{x}_0, \bar{\theta}) \in \mathcal{K}_0 \to (\bar{t}_0, \bar{t}_f, \bar{\xi}, \bar{x}) \in \mathcal{K}$ *is given by the formulas*

$$\bar{\xi} = -\bar{\theta}, \qquad \bar{x}(t) = V(t)\left(\bar{x}_0 - \dot{\hat{x}}(\hat{t}_0)\bar{t}_0\right) + \sum_{k=1}^{s} y^k(t)\bar{t}_k, \tag{7.17}$$

where $\bar{t}_k$ *is the kth component of the vector* $\bar{\theta}$.

Quadratic forms. For $\mu \in \Lambda_0$ the quadratic form of the IOP is equal to

$$\langle L_{\zeta\zeta}(\mu, \hat{\zeta})\bar{\zeta}, \bar{\zeta}\rangle.$$

The main result of this section is the following.

Theorem 7.8. *Let the Lagrange multipliers*

$$\mu = (\alpha_0, \alpha, \beta) \in \Lambda_0^{MP} \quad \text{and} \quad \lambda = (\alpha_0, \alpha, \beta, \psi, \psi_0) \in M_0$$

correspond to each other, i.e., $\pi_0 \lambda = \mu$, and let the elements of the critical cones $\bar{\zeta} = (\bar{t}_0, \bar{t}_f, \bar{x}_0, \bar{\theta}) \in \mathcal{K}_0$ and $\bar{z} = (\bar{t}_0, \bar{t}_f, \bar{\xi}, \bar{x}) \in \mathcal{K}$ correspond to each other, i.e., $\pi_1 \bar{z} = \bar{\zeta}$. Then the quadratic forms in the basic and induced problems take equal values: $\langle L_{\zeta\zeta}(\mu, \hat{\zeta})\bar{\zeta}, \bar{\zeta} \rangle = \Omega(\lambda, \bar{z})$. Consequently,

$$\max_{\mu \in \Lambda_0^{MP}} \langle L_{\zeta\zeta}(\mu, \hat{\zeta})\bar{\zeta}, \bar{\zeta} \rangle = \max_{\lambda \in M_0} \Omega(\lambda, \bar{z})$$

for each pair of elements of the critical cones $\bar{\zeta} \in \mathcal{K}_0$ and $\bar{z} \in \mathcal{K}$ such that $\pi_1 \bar{z} = \bar{\zeta}$.

Theorems 6.9 and 7.8 and Proposition 7.6 imply the following second-order *necessary optimality condition* for the basic problem.

Theorem 7.9. *If the trajectory $\hat{\mathcal{T}}$ affords a Pontryagin minimum in the basic problem, then the following Condition $\mathfrak{A}_0$ holds: The set M_0 is nonempty and*

$$\max_{\mu \in \Lambda_0^{MP}} \langle L_{\zeta\zeta}(\mu, \hat{\zeta})\bar{\zeta}, \bar{\zeta} \rangle \geq 0 \qquad \forall\, \bar{\zeta} \in \mathcal{K}_0.$$

Theorems 6.10 and 7.8 and Proposition 7.6 imply the following second-order *sufficient optimality condition* for the basic control problem.

Theorem 7.10. *Let the following Condition $\mathfrak{B}_0$ be fulfilled for an admissible trajectory $\hat{\mathcal{T}}$ in the basic problem:*

(a) *$\hat{u}(t)$ is a bang-bang control taking values in the set $V = \operatorname{ex} U$,*

(b) *the set M_0 is nonempty, and there exists $\lambda \in M_0$ such that $D^k(H) > 0$, $k = 1, \ldots, s$, and condition (6.19) holds (hence, $u(t)$ is a strict bang-bang control),*

(c) *$\max_{\mu \in \Lambda_0^{MP}} \langle L_{\zeta\zeta}(\mu, \hat{\zeta})\bar{\zeta}, \bar{\zeta} \rangle > 0$ for all $\bar{\zeta} \in \mathcal{K}_0 \setminus \{0\}$.*

Then $\hat{\mathcal{T}}$ is a strict strong minimum.

Theorem 7.10 is a generalization of sufficient optimality conditions for bang-bang controls obtained in Agrachev, Stefani, and Zezza [1]. The detailed proofs of the preceding theorems will be given in the following sections. Let us point out that the proofs reveals the useful fact that all elements of the Hessian $L_{\zeta\zeta}(\mu, \hat{\zeta})$ can be computed explicitly on the basis of the transition matrix $V(\hat{t}_f)$ in (7.15) and of the *first-order* variations y^k defined by (7.16). We shall need formulas for all first-order partial derivatives of the function $x(t_f; t_0, x_0, \theta)$. We shall make extensive use of the variational system

$$\dot{V} = f_x(t, x(t; t_0, x_0, \theta), u(t; \theta))\, V, \quad V(t_0) = E, \tag{7.18}$$

where E is the identity matrix. The solution $V(t)$ is an $n \times n$ matrix-valued function ($n = d(x)$) which is absolutely continuous in $\Delta = [t_0, t_f]$. The solution of (7.18) is denoted by $V(t; t_0, x_0, \theta)$. Along the reference trajectory $\hat{x}(t), \hat{u}(t)$, i.e., for $\zeta = \hat{\zeta}$, we shall use the notation $V(t)$ for simplicity.

7.2 First-Order Derivatives of $x(t_f;t_0,x_0,\theta)$ with Respect to t_0, t_f, x_0, and θ. Lagrange Multipliers and Critical Cones

Let $x(t;t_0,x_0,\theta)$ be the solution of the IVP (7.3) and put

$$g(\zeta) = g(t_0,t_f,x_0,\theta) := x(t_f;t_0,x_0,\theta). \tag{7.19}$$

Under our assumptions, the operator $g : \mathcal{V} \to \mathbb{R}^n$ is well defined and C^2-smooth if the neighborhood $\mathcal{V}$ of the point $\hat{\zeta}$ is sufficiently small. In this section, we shall derive the first-order partial derivatives of $g(t_0,t_f,x_0,\theta)$ with respect to t_0, t_f, x_0, and θ at the point $\hat{\zeta}$. We shall use well-known results from the theory of ODEs about differentiation of solutions to ODEs with respect to parameters and initial values. In what follows, it will be convenient to drop those arguments in $x(t;t_0,x_0,\theta)$, $u(t,\theta)$, $V(t;t_0,x_0,\theta)$, etc., that are kept fixed.

7.2.1 Derivative $\partial x/\partial x_0$

Let us fix θ and t_0. The following result is well known in the theory of ODEs.

Proposition 7.11. *We have*

$$\frac{\partial x(t;x_0)}{\partial x_0} = V(t;x_0), \tag{7.20}$$

where the matrix-valued function $V(t;x_0)$ is the solution to the IVP (7.18), *i.e.,*

$$\dot{V} = f_x(t,x(t),u(t))V, \qquad V|_{t=t_0} = E, \tag{7.21}$$

where $x(t) = x(t;x_0)$, $\dot{V} = \frac{\partial V}{\partial t}$.

Consequently, we have

$$g_{x_0}(\hat{\zeta}) := \frac{\partial x(\hat{t}_f;\hat{t}_0,\hat{x}_0,\hat{\theta})}{\partial x_0} = V(\hat{t}_f), \tag{7.22}$$

where $V(t)$ satisfies IVP (7.21) along the trajectory $(\hat{x}(t),\hat{u}(t))$, $t \in [\hat{t}_0,\hat{t}_f]$.

7.2.2 Derivatives $\partial x/\partial t_0$ and $\partial x/\partial t_f$

Fix x_0 and θ and put

$$w(t;t_0) = \frac{\partial x(t;t_0)}{\partial t_0}.$$

Proposition 7.12. *The vector function $w(t;t_0)$ is the solution to the IVP*

$$\dot{w} = f_x(t,x(t),u(t))w, \qquad w|_{t=t_0} = -\dot{x}(t_0), \tag{7.23}$$

where $x(t) = x(t;t_0)$, $\dot{w} = \frac{\partial w}{\partial t}$. Therefore, we have $w(t;t_0) = -V(t;t_0)\dot{x}(t_0)$, where the matrix-valued function $V(t;t_0)$ is the solution to the IVP (7.18).

Hence, we obtain

$$g_{t_0}(\hat{\zeta}) := \frac{\partial x(\hat{t}_f;\hat{t}_0,\hat{x}_0,\hat{\theta})}{\partial t_0} = -V(\hat{t}_f)\dot{\hat{x}}(\hat{t}_0). \tag{7.24}$$

Obviously, we have

$$g_{t_f}(\hat{\zeta}) := \frac{\partial x(\hat{t}_f;\hat{t}_0,\hat{x}_0,\hat{\theta})}{\partial t_f} = \dot{\hat{x}}(\hat{t}_f). \tag{7.25}$$

7.2.3 Derivatives $\partial x/\partial t_k$

Fix t_0 and x_0. Take some k and fix t_j for all $j \neq k$. Put

$$y^k(t;t_k) = \frac{\partial x(t;t_k)}{\partial t_k}$$

and denote by $\dot{y}^k$ the derivative of y^k with respect to t.

Proposition 7.13. *For $t \geq t_k$ the function $y^k(t;t_k)$ is the solution to the IVP*

$$\dot{y}^k = f_x(t,x(t;t_k),u(t;t_k))\,y^k, \quad y^k|_{t=t_k} = -[f]^k, \tag{7.26}$$

where $[f]^k = f(t_k,x(t_k;t_k),u^{k+1}) - f(t_k,x(t_k;t_k),u^k)$ is the jump of the function $f(t,x(t;t_k), u(t;t_k))$ at the point t_k. For $t < t_k$, we have $y^k(t;t_k) = 0$. Thus, $[y]^k = -[f]^k$, where $[y]^k = y(t_k+;t_k) - y(t_k-;t_k)$ is the jump of the function $y^k(t;t_k)$ at the point t_k.

Proof. Let us sketch how to obtain the representation (7.26). For $t \geq t_k$ the trajectory $x(t;t_k)$ satisfies the integral equation

$$x(t;t_k) = x(t_k-;t_k) + \int_{t_k+}^{t} f(h,x(h;t_k),u(h,t_k))\,dh.$$

By differentiating this equation with respect to t_k, we obtain

$$y^k(t;t_k) = \dot{x}(t_k-;t_k) - \dot{x}(t_k+;t_k) + \int_{t_k+}^{t} f_x(h,x(h;t_k),u(h,t_k))y^k(h;t_k)\,dh,$$

from which we get $y^k|_{t=t_k} = -[f]^k$ and the variational equation in (7.26). ∎

In particular, we obtain

$$g_{t_k}(\hat{\zeta}) := \frac{\partial x(\hat{t}_f;\hat{t}_0,\hat{x}_0,\hat{\theta})}{\partial t_k} = y^k(\hat{t}_f). \tag{7.27}$$

7.2.4 Comparison of Lagrange Multipliers

Here, we prove Propositions 7.4 and 7.5. Consider the Lagrangian (7.6) with a multiplier $\mu = (\alpha_0,\alpha,\beta) \in \Lambda_0$, where Λ_0 is the set (7.7) of normalized Lagrange multipliers at the point $\hat{\zeta}$ in the Induced Problem (7.4). Define the absolutely continuous function $\psi(t)$ and the function $\psi_0(t)$ by equation (7.10) and (7.11), respectively. We will show that the function

$\psi_0(t)$ is absolutely continuous and the collection $\lambda = (\alpha_0,\alpha,\beta,\psi,\psi_0)$ satisfies all conditions in (7.8) and hence belongs to the set Λ. The conditions

$$\alpha_0 \geq 0,\ \alpha \geq 0, \quad \alpha_0 + |\alpha| + |\beta| = 1, \quad \alpha F(\hat{p}) = 0$$

in the definitions of Λ_0 and Λ are identical. Hence, we must analyze the equation $L_\zeta(\mu,\hat{\zeta}) = 0$ in the definition of Λ_0 which is equivalent to the system

$$\begin{aligned}
L_{t_0}(\mu,\hat{\zeta}) &= l_{t_0}(\hat{p}) + l_{x_f}(\hat{p}) g_{t_0}(\hat{\zeta}) = 0,\\
L_{t_f}(\mu,\hat{\zeta}) &= l_{t_f}(\hat{p}) + l_{x_f}(\hat{p}) g_{t_f}(\hat{\zeta}) = 0,\\
L_{x_0}(\mu,\hat{\zeta}) &= l_{x_0}(\hat{p}) + l_{x_f}(\hat{p}) g_{x_0}(\hat{\zeta}) = 0,\\
L_{t_k}(\mu,\hat{\zeta}) &= l_{x_f}(\hat{p}) g_{t_k}(\hat{\zeta}) = 0, \quad k = 1,\ldots,s.
\end{aligned}$$

Using the equality $l_{x_f}(\hat{p}) = \psi(\hat{t}_f)$ and formulas (7.24), (7.25), (7.22), (7.27) for the derivatives of g with respect to t_0, t_f, x_0, t_k, at the point $\hat{\zeta}$, we get

$$L_{t_0}(\mu,\hat{\zeta}) = l_{t_0}(\hat{p}) - \psi(\hat{t}_f) V(\hat{t}_f) \dot{\hat{x}}(\hat{t}_0) = 0, \tag{7.28}$$
$$L_{t_f}(\mu,\hat{\zeta}) = l_{t_f}(\hat{p}) + \psi(\hat{t}_f) \dot{\hat{x}}(\hat{t}_f) = 0, \tag{7.29}$$
$$L_{x_0}(\mu,\hat{\zeta}) = l_{x_0}(\hat{p}) + \psi(\hat{t}_f) V(\hat{t}_f) = 0, \tag{7.30}$$
$$L_{t_k}(\mu,\hat{\zeta}) = \psi(\hat{t}_f) y^k(\hat{t}_f) = 0, \quad k = 1,\ldots,s. \tag{7.31}$$

Analysis of (7.28). The $n \times n$ matrix-value function $V(t)$ satisfies the equation

$$\dot{V} = f_x V, \qquad V(\hat{t}_0) = E$$

with $f_x = f_x(t,\hat{x}(t),\hat{u}(t))$. Then, $\Psi(t) := V^{-1}(t)$ is the solution to the adjoint equation

$$-\dot{\Psi} = \Psi f_x, \qquad \Psi(\hat{t}_0) = E.$$

Consequently, $\psi(\hat{t}_f) = \psi(\hat{t}_0)\Psi(\hat{t}_f) = \psi(\hat{t}_0)V^{-1}(\hat{t}_f)$. Using these relations in (7.28), we get

$$l_{t_0}(\hat{p}) - \psi(\hat{t}_0)\dot{\hat{x}}(\hat{t}_0) = 0.$$

By virtue of (7.11), we have $\psi(\hat{t}_0)\dot{\hat{x}}(\hat{t}_0) = -\psi_0(\hat{t}_0)$. Hence, (7.28) is equivalent to the transversality condition for ψ_0 at the point $\hat{t}_0$:

$$l_{t_0}(\hat{p}) + \psi_0(\hat{t}_0) = 0.$$

Analysis of (7.29). Since $\psi(\hat{t}_f)\dot{\hat{x}}(\hat{t}_f) = -\psi_0(\hat{t}_f)$ holds, (7.29) is equivalent to the transversality condition for ψ_0 at the point $\hat{t}_f$:

$$l_{t_f}(\hat{p}) - \psi_0(\hat{t}_f) = 0.$$

Analysis of (7.30). Since $\psi(\hat{t}_f) = \psi(\hat{t}_0)V^{-1}(\hat{t}_f)$, equality (7.30) is equivalent to the transversality condition for ψ at the point $\hat{t}_0$:

$$l_{x_0}(\hat{p}) + \psi(\hat{t}_0) = 0.$$

Analysis of (7.31). We need the following result.

Proposition 7.14. *Let the absolutely continuous function y be a solution to the system $\dot{y} = f_x y$ on an interval Δ, and let the absolutely continuous function ψ be a solution to the adjoint system $-\dot{\psi} = \psi f_x$ on the same interval, where $f_x = f_x(t,\hat{x}(t),\hat{u}(t))$. Then $\psi(t)y(t) \equiv \text{const}$ on Δ.*

Proof. We have $\frac{d}{dt}(\psi y) = \dot{\psi} y + \psi \dot{y} = -\psi f_x y + \psi f_x y = 0$. ∎

It follows from this proposition and (7.26) that for $k = 1,\ldots,s$,

$$\psi(\hat{t}_f)y^k(\hat{t}_f) = \psi(\hat{t}_k)y^k(\hat{t}_k+0) = \psi(\hat{t}_k)[y^k]^k = -\psi(\hat{t}_k)[\dot{\hat{x}}]^k = -[\psi\dot{\hat{x}}]^k = [\psi_0]^k.$$

Therefore, (7.31) is equivalent to the conditions

$$[\psi_0]^k = 0, \quad k = 1,\ldots,s,$$

which means that ψ_0 is continuous at each point $\hat{t}_k$, $k = 1,\ldots,s$, and hence absolutely continuous on $\hat{\Delta} = [\hat{t}_0,\hat{t}_f]$. Moreover, it follows from $0 = [\psi_0]^k = -\psi(\hat{t}_k)[\dot{\hat{x}}]^k$ that

$$\phi(\hat{t}_k)[\hat{u}]^k = 0, \tag{7.32}$$

where $\phi(t) = \psi(t)B(t,\hat{x}(t))$ denotes the switching function. Finally, differentiating (7.11) with respect to t, we get

$$-\psi f_x\dot{\hat{x}} + \psi f_t + \psi f_x\dot{\hat{x}} + \dot{\psi}_0 = 0, \quad \text{i.e.,} \; -\dot{\psi}_0 = \psi f_t.$$

Thus, we have proved that $\lambda = (\alpha_0,\alpha,\beta,\psi,\psi_0) \in \Lambda$. Conversely, if $(\alpha_0,\alpha,\beta,\psi) \in \Lambda$, then one can show similarly that $(\alpha_0,\alpha,\beta) \in \Lambda_0$. Moreover, it is obvious that the projector (7.9) is injective on Λ_0, because ψ and ψ_0 are defined uniquely by condition (7.10) and (7.11), respectively.

7.2.5 Comparison of the Critical Cones

Take an element $\bar{\zeta} = (\bar{t}_0,\bar{t}_f,\bar{x}_0,\bar{\theta})$ of the critical cone $\mathcal{K}_0$ (see (7.13)) at the point $\hat{\zeta}$ in the Induced Problem:

$$\mathcal{F}_0'(\hat{\zeta})\bar{\zeta} \le 0, \quad \mathcal{F}_i'(\hat{\zeta})\bar{\zeta} \le 0, \; i \in I, \quad \mathcal{G}'(\hat{\zeta})\bar{\zeta} = 0.$$

Define $\bar{\xi}$ and $\bar{x}$ by formulas (7.17),

$$\bar{\xi} = -\bar{\theta}, \quad \bar{x}(t) = V(t)\Big(\bar{x}_0 - \dot{\hat{x}}(\hat{t}_0)\bar{t}_0\Big) + \sum_{k=1}^{s} y^k(t)\bar{t}_k,$$

where $\bar{t}_k$ is the kth component of the vector $\bar{\theta}$, and put $\bar{z} = (\bar{t}_0,\bar{t}_f,\bar{\xi},\bar{x})$. We shall show that $\bar{z}$ is an element of the critical cone $\mathcal{K}$ (Equations (6.22) and (6.23)) for the trajectory

$\hat{\mathcal{T}} = \{(\hat{x}(t),\hat{u}(t)) \mid t \in [\hat{t}_0,\hat{t}_f]\}$ in the basic problem. Consider the first inequality $\mathcal{F}_0'(\hat{\zeta})\bar{\zeta} \le 0$, where $\mathcal{F}_0(\zeta) := J(t_0,x_0,t_f,x(t_f;t_0,x_0,\theta))$. We obviously have

$$\begin{aligned}\mathcal{F}_0'(\hat{\zeta})\bar{\zeta} &= (J_{t_0}(\hat{p}) + J_{x_f}(\hat{p})g_{t_0}(\hat{\zeta}))\bar{t}_0 + (J_{t_f}(\hat{p}) + J_{x_f}(\hat{p})g_{t_f}(\hat{\zeta}))\bar{t}_f \\ &\quad + (J_{x_0}(\hat{p}) + J_{x_f}(\hat{p})g_{x_0}(\hat{\zeta}))\bar{x}_0 + \sum_{k=1}^{s} J_{x_f}(\hat{p})g_{t_k}(\hat{\zeta})\bar{\theta}_k,\end{aligned}$$

where $\bar{\theta}_k = \bar{t}_k$ is the kth component of the vector $\bar{\theta}$. Using formulas (7.24), (7.25), (7.22), (7.27) for the derivatives of g with respect to t_0, t_f, x_0, t_k, at the point $\hat{\zeta}$, we get

$$\begin{aligned}\mathcal{F}_0'(\hat{\zeta})\bar{\zeta} &= (J_{t_0}(\hat{p}) - J_{x_f}(\hat{p})V(t_f)\dot{\hat{x}}(\hat{t}_0))\bar{t}_0 + (J_{t_f}(\hat{p}) + J_{x_f}(\hat{p})\dot{\hat{x}}(\hat{t}_f))\bar{t}_f \\ &\quad + (J_{x_0}(\hat{p}) + J_{x_f}(\hat{p})V(\hat{t}_f))\bar{x}_0 + \sum_{k=1}^{s} J_{x_f}(\hat{p})y^k(\hat{t}_f)\bar{\theta}_k.\end{aligned}$$

Hence, the inequality $\mathcal{F}_0'(\hat{\zeta})\bar{\zeta} \le 0$ is equivalent to the inequality

$$\begin{aligned}&J_{t_0}(\hat{p})\bar{t}_0 + J_{t_f}(\hat{p})\bar{t}_f + J_{x_0}(\hat{p})\bar{x}_0 \\ &\quad + J_{x_f}(\hat{p})\left(V(t_f)(\bar{x}_0 - \dot{\hat{x}}(\hat{t}_0)\bar{t}_0) + \sum_{k=1}^{s} y^k(\hat{t}_f)\bar{\theta}_k + \dot{\hat{x}}(\hat{t}_f)\bar{t}_f\right) \le 0.\end{aligned}$$

It follows from the definition (7.17) of $\bar{x}$ that

$$\bar{\bar{x}}_0 := \bar{x}(\hat{t}_0) + \dot{\hat{x}}(\hat{t}_0)\bar{t}_0 = \bar{x}_0, \tag{7.33}$$

since $V(\hat{t}_0) = E$, and $y^k(\hat{t}_0) = 0$, $k = 1,\dots,s$. Moreover, using the same definition, we get

$$\bar{\bar{x}}_f := \bar{x}(\hat{t}_f) + \dot{\hat{x}}(\hat{t}_f)\bar{t}_f = V(\hat{t}_f)(\bar{x}_0 - \dot{\hat{x}}(\hat{t}_0)\bar{t}_0) + \sum_{k=1}^{s} y^k(\hat{t}_f)\bar{t}_k + \dot{\hat{x}}(\hat{t}_f)\bar{t}_f. \tag{7.34}$$

Thus, the inequality $\mathcal{F}_0'(\hat{\zeta})\bar{\zeta} \le 0$ is equivalent to the inequality

$$J_{t_0}(\hat{p})\bar{t}_0 + J_{t_f}(\hat{p})\bar{t}_f + J_{x_0}(\hat{p})\bar{\bar{x}}_0 + J_{x_f}(\hat{p})\bar{\bar{x}}_f \le 0,$$

or briefly,

$$J'(\hat{p})\bar{\bar{p}} \le 0,$$

where $\bar{\bar{p}} = (\bar{t}_0, \bar{\bar{x}}_0, \bar{t}_f, \bar{\bar{x}}_f)$; see equation (6.21).

Similarly, the inequalities $\mathcal{F}_i'(\hat{\zeta})\bar{\zeta} \le 0$ for all $i \in I$ and the equality $\mathcal{G}'(\hat{\zeta})\bar{\zeta} = 0$ in the definition of $\mathcal{K}_0$ are equivalent to the inequalities (respectively, equalities)

$$F_i'(\hat{p})\bar{\bar{p}} \le 0, \quad i \in I, \quad K'(\hat{p})\bar{\bar{p}} = 0,$$

in the definition of $\mathcal{K}$; cf. (6.22).

Since $\dot{V} = f_x(t,\hat{x}(t),\hat{u}(t))V$ and $\dot{y}^k = f_x(t,\hat{x}(t),\hat{u}(t))y^k$, $k = 1,\dots,s$, it follows from definition (7.17) that $\bar{x}$ is a solution to the same linear system

$$\dot{\bar{x}} = f_x(t,\hat{x}(t),\hat{u}(t))\bar{x}.$$

Finally, recall from (7.26) that for each $k = 1,\dots,s$ the function $y^k(t)$ is piecewise continuous with only one jump $[y^k]^k = -[\dot{\hat{x}}]^k$ at the point $\hat{t}_k$ and is absolutely continuous on each of the half-open intervals $[\hat{t}_0,\hat{t}_k)$ and $(\hat{t}_k,\hat{t}_f]$. Moreover, the function $V(t)$ is absolutely continuous in $[\hat{t}_0,\hat{t}_f]$. Hence, $\bar{x}(t)$ is a piecewise continuous function which is absolutely continuous on each interval of the set $[\hat{t}_0,\hat{t}_f]\setminus\hat{\Theta}$ and satisfies the jump conditions

$$[\bar{x}]^k = [\dot{\hat{x}}]^k\bar{\xi}_k, \quad \bar{\xi}_k = -\bar{t}_k, \quad k = 1,\dots,s.$$

Thus, we have proved that $\bar{z} = (\bar{t}_0,\bar{t}_f,\bar{\xi},\bar{x})$ is an element of the critical cone $\mathcal{K}$. Similarly, one can show that if $\bar{z} = (\bar{t}_0,\bar{t}_f,\bar{\xi},\bar{x}) \in \mathcal{K}$, then putting $\bar{x}_0 = \bar{x}(\hat{t}_0)$ and $\bar{\theta} = -\bar{\xi}$, we obtain the element $\bar{\zeta} = (\bar{t}_0,\bar{t}_f,\bar{x}_0,\bar{\theta})$ of the critical cone $\mathcal{K}_0$.

7.3 Second-Order Derivatives of $x(t_f;t_0,x_0,\theta)$ with Respect to t_0, t_f, x_0, and θ

In this section we shall give formulas for all second-order partial derivatives of the functions

$$x(t;t_0,x_0,\theta) \quad \text{and} \quad g(\zeta) = g(t_0,t_f,x_0,\theta) := x(t_f;t_0,x_0,\theta)$$

at the point $\hat{\zeta}$. We are not sure whether all of them are known; therefore we shall also sketch the proofs. Here $x(t;t_0,x_0,\theta)$ is the solution to IVP (7.3). Denote by $g_k(\zeta) := x_k(t_f;t_0,x_0,\theta)$ the kth component of the function g.

7.3.1 Derivatives $(g_k)_{x_0x_0}$

Let $x(t;x_0)$ be the solution to the IVP (7.3) with fixed t_0 and θ, and let $x_k(t;x_0)$ be its kth component. For $k = 1,\dots,n$, we define the $n\times n$ matrix

$$W^k(t;x_0) := \frac{\partial^2 x_k(t;x_0)}{\partial x_0 \partial x_0} \quad \text{with entries} \quad w_{ij}^k(t;x_0) = \frac{\partial^2 x_k(t;x_0)}{\partial x_{0i}\partial x_{0j}},$$

where x_{0i} is the ith component of the column vector $x_0 \in \mathbb{R}^n$.

Proposition 7.15. *The matrix-valued functions $W^k(t;x_0)$, $k = 1,\dots,n$, satisfy the IVPs*

$$\dot{W}^k = V^* f_{kxx} V + \sum_{r=1}^{n} f_{kx_r} W^r, \qquad W^k|_{t=t_0} = O, \quad k = 1,\dots,n, \tag{7.35}$$

where $\dot{W}^k = \frac{\partial W^k}{\partial t}$, O is the zero matrix, f_k is the kth component of the vector function f, and

$$f_{kx_r} = \frac{\partial f_k(t,x(t;x_0),u(t))}{\partial x_r}, \qquad f_{kxx} = \frac{\partial^2 f_k(t,x(t;x_0),u(t))}{\partial x \partial x}$$

are its partial derivatives at the point $(t,x(t;x_0),u(t))$ for $t \in [t_0,t_f]$.

Proof. For notational convenience, we use the function $\varphi(t,x) := f(t,x,u(t))$. By Proposition 7.11, the matrix-valued function $V(t;x_0) = \frac{\partial x(t;x_0)}{\partial x_0}$ with entries $v_{ij}(t;x_0) = \frac{\partial x_i(t;x_0)}{\partial x_{0j}}$ is the solution to the IVP (7.18). Consequently, its entries satisfy the equations

$$\frac{\partial \dot{x}_k(t;x_0)}{\partial x_{0i}} = \sum_r \varphi_{kx_r}(t,x(t;x_0))\frac{\partial x_r(t;x_0)}{\partial x_{0i}}$$

$$\frac{\partial x_k(t_0;x_0)}{\partial x_{0i}} = e_{ki}, \quad k,i = 1,\dots,n,$$

where e_{ki} are the elements of the identity matrix E. By differentiating these equations with respect to x_{0j}, we get

$$\frac{\partial^2 \dot{x}_k(t;x_0)}{\partial x_{0i}\partial x_{0j}} = \sum_r \left(\varphi_{kx_r}(t,x(t;x_0))\right)_{x_{0j}} \frac{\partial x_r(t;x_0)}{\partial x_{0i}} + \sum_r \varphi_{kx_r}(t,x(t;x_0))\frac{\partial^2 x_r(t;x_0)}{\partial x_{0i}\partial x_{0j}}, \tag{7.36}$$

$$\frac{\partial^2 x_k(t_0;x_0)}{\partial x_{0i}\partial x_{0j}} = 0, \quad k,i,j = 1,\dots,n. \tag{7.37}$$

Transforming the first sum in the right-hand side of (7.36), we get

$$\begin{aligned}
&\sum_r \left(\varphi_{kx_r}(t,x(t;x_0))\right)_{x_{0j}} \frac{\partial x_r(t;x_0)}{\partial x_{0i}} \\
&\quad = \sum_r\sum_s \varphi_{kx_rx_s}(t,x(t;x_0))\frac{\partial x_s(t;x_0)}{\partial x_{0j}} \cdot \frac{\partial x_r(t;x_0)}{\partial x_{0i}} \\
&\quad = \left(V^*\varphi_{kxx}(t,x(t;x_0))V\right)_{ij}, \quad k,i,j = 1,\dots,n,
\end{aligned}$$

where $(A)_{ij}$ denotes the element a_{ij} of a matrix A, and A^* denotes the transposed matrix. Thus, (7.36) and (7.37) imply (7.35). ■

It follows from Proposition 7.15 that

$$(g_k)_{x_0x_0}(\hat{\zeta}) := \frac{\partial^2 x_k(\hat{t}_f;\hat{t}_0,\hat{x}_0,\hat{\theta})}{\partial x_0\partial x_0} = W^k(\hat{t}_f), \quad k = 1,\dots,n, \tag{7.38}$$

where the matrix-valued functions $W^k(t)$, $k = 1,\dots,n$, satisfy the IVPs (7.35) along the reference trajectory $(\hat{x}(t),\hat{u}(t))$.

7.3.2 Mixed Derivatives $g_{x_0t_k}$

Let $s = 1$ for notational convenience, and thus $\theta = t_1$. Fix t_0 and consider the functions

$$V(t;x_0,\theta) = \frac{\partial x(t;x_0,\theta)}{\partial x_0}, \quad y(t;x_0,\theta) = \frac{\partial x(t;x_0,\theta)}{\partial \theta},$$

$$R(t;x_0,\theta) = \frac{\partial V(t;x_0,\theta)}{\partial \theta} = \frac{\partial^2 x(t;x_0,\theta)}{\partial x_0\partial \theta},$$

$$\dot{V}(t;x_0,\theta) = \frac{\partial V(t;x_0,\theta)}{\partial t}, \quad \dot{R}(t;x_0,\theta) = \frac{\partial R(t;x_0,\theta)}{\partial t}.$$

Then V, $\dot{V}$ and R, $\dot{R}$ are $n \times n$ matrix-valued functions, and y is a vector function of dimension n.

Proposition 7.16. *For $t \geq \theta$, the function $R(t;x_0,\theta)$ is the solution to the IVP*

$$\dot{R} = (y^* f_{xx})V + f_x R, \quad R(\theta;x_0,\theta) = -[f_x]V(\theta;x_0,\theta), \tag{7.39}$$

where f_x and f_{xx} are taken along the trajectory $(t, x(t;x_0,\theta), u(t,\theta))$, $t \in [t_0,t_f]$. Here, by definition, $(y^* f_{xx})$ is an $n \times n$ matrix with entries

$$(y^* f_{xx})_{k,j} = \sum_{i=1}^{n} \frac{\partial^2 f_k}{\partial x_i \partial x_j} y_i \tag{7.40}$$

in the kth row and jth column, and

$$[f_x] = f_x(\theta, x(\theta;x_0,\theta), u^2) - f_x(\theta, x(\theta;x_0,\theta), u^1)$$

is the jump of the function $f_x(\cdot, x(\cdot;x_0,\theta), u(\cdot,\theta))$ at the point θ. For $t < \theta$, we have $R(t;x_0,\theta) = 0$.

Proof. According to Proposition 7.11 the matrix-valued function V is the solution to the system

$$\dot{V}(t;x_0,\theta) = f_x(t, x(t;x_0,\theta), u(t;\theta))V(t;x_0,\theta). \tag{7.41}$$

By differentiating this equality with respect to θ, we get the equation

$$\frac{\partial \dot{V}}{\partial \theta} = \sum_i (f_x V)'_{x_i} \frac{\partial x_i}{\partial \theta} + f_x \frac{\partial V}{\partial \theta},$$

which is equivalent to

$$\dot{R} = \sum_i (f_x V)_{x_i} y_i + f_x R. \tag{7.42}$$

Upon defining

$$A = \sum_i (f_x V)_{x_i} y_i,$$

the element in the rth row and sth column of the matrix A is equal to

$$\begin{aligned} a_{rs} &= \sum_i ((f_x V)_{rs})_{x_i} y_i = \sum_i \left(\sum_j f_{rx_j} v_{js} \right)_{x_i} y_i \\ &= \sum_i \sum_j y_i f_{rx_i x_j} v_{js} = \sum_j \left(\sum_i y_i f_{rx_i x_j} \right) v_{js} \\ &= \sum_j \left(y^* f_{xx}\right)_{rj} v_{js} = \left((y^* f_{xx})V\right)_{rs}, \end{aligned}$$

where v_{js} is the element in the jth row and sth column of the matrix V. Hence we have $A = (y^* f_{xx})V$ and see that (7.42) is equivalent to (7.39). The initial condition in (7.39), which is similar to the initial condition (7.26) in Proposition 7.13, follows from (7.41) (see the proof of Proposition 7.13). The condition $R(t;x_0,\theta) = 0$ for $t < \theta$ is obvious. ∎

Proposition 7.16 yields

$$g_{x_0 t_k}(\hat{\zeta}) := \frac{\partial^2 x(\hat{t}_f;\hat{t}_0,\hat{x}_0,\hat{\theta})}{\partial x_0 \partial t_k} = R^k(\hat{t}_f), \tag{7.43}$$

where the matrix-valued function $R^k(t)$ satisfies the IVP

$$\begin{aligned} \dot{R}^k(t) &= \left(y^k(t)^* f_{xx}(t,\hat{x}(t),\hat{u}(t))\right) V(t) + f_x(t,\hat{x}(t),\hat{u}(t))R^k(t), \quad t \in [\hat{t}_k,\hat{t}_f], \\ R^k(\hat{t}_k) &= -[f_x]^k V(\hat{t}_k). \end{aligned} \tag{7.44}$$

Here, $V(t)$ is the solution to the IVP (7.18), $y^k(t)$ is the solution to the IVP (7.26) (for $t_0 = \hat{t}_0$, $x_0 = \hat{x}_0$, $\theta = \hat{\theta}$), and $[f_x]^k = f(\hat{t}_k,\hat{x}(\hat{t}_k),\hat{u}(\hat{t}_k+)) - f(\hat{t}_k,\hat{x}(\hat{t}_k),\hat{u}(\hat{t}_k-))$, $k = 1,\dots,s$.

7.3.3 Derivatives $g_{t_k t_k}$

Again, for simplicity let $s = 1$. Fix t_0 and x_0 and put

$$\begin{aligned} y(t;\theta) &= \frac{\partial x(t;\theta)}{\partial \theta}, \quad z(t;\theta) = \frac{\partial y(t;\theta)}{\partial \theta} = \frac{\partial^2 x(t;\theta)}{\partial \theta^2}, \\ \dot{y}(t;\theta) &= \frac{\partial y(t;\theta)}{\partial t}, \quad \dot{z}(t;\theta) = \frac{\partial z(t;\theta)}{\partial t}. \end{aligned}$$

Then y, $\dot{y}$ and z, $\dot{z}$ are vector functions of dimension n.

Proposition 7.17. *For $t \ge \theta$ the function $z(t;\theta)$ is the solution to the system*

$$\dot{z} = f_x z + y^* f_{xx} y \tag{7.45}$$

with the initial condition at the point $t = \theta$,

$$z(\theta;\theta) + \dot{y}(\theta+;\theta) = -[f_t] - [f_x](\dot{x}(\theta+;\theta) + y(\theta;\theta)). \tag{7.46}$$

In (7.45), f_x and f_{xx} are taken along the trajectory $(t,x(t;\theta),u(t;\theta))$, $t \in [t_0,t_f]$, and $y^* f_{xx} y$ is a vector with elements

$$(y^* f_{xx} y)_k = y^* f_{kxx} y = \sum_{i,j=1}^{n} \frac{\partial^2 f_k}{\partial x_i \partial x_j} y_i y_j, \quad k = 1,\dots,n.$$

In (7.46), the expressions

$$\begin{aligned} [f_t] &= f_t(\theta,x(\theta;\theta),u^2) - f_t(\theta,x(\theta;\theta),u^1), \\ [f_x] &= f_x(\theta,x(\theta;\theta),u^2) - f_x(\theta,x(\theta;\theta),u^1) \end{aligned}$$

are the jumps of the derivatives $f_t(t,x(t;\theta),u(t;\theta))$ and $f_x(t,x(t;\theta),u(t;\theta))$ at the point θ ($u^2 = u(\theta+;\theta)$, $u^1 = u(\theta-;\theta)$). For $t < \theta$, we have $z(t;\theta) = 0$.

Proof. By Proposition 7.13, for $t \geq \theta$ the function $y(t;\theta)$ is the solution to the IVP

$$\dot{y}(t;\theta) = f_x(t,x(t;\theta),u(t;\theta))y(t;\theta),$$
$$y(\theta;\theta) = -(f(\theta,x(\theta;\theta),u^2) - f(\theta,x(\theta;\theta),u^1)).$$

By differentiating these equalities with respect to θ at the points θ and $\theta+$, we obtain (7.45) and (7.46). For $t < \theta$ we have $y = 0$ and hence $z = 0$. ∎

For the solution $x(t;t_0,x_0,\theta)$ to the IVP (7.3) with an arbitrary s, it follows from Proposition 7.17 that

$$g_{t_k t_k}(\hat{\zeta}) := \frac{\partial^2 x(\hat{t}_f;\hat{t}_0,\hat{x}_0,\hat{\theta})}{\partial t_k \partial t_k} = z^{kk}(\hat{t}_f), \quad k = 1,\ldots,s, \tag{7.47}$$

where for $t \geq \hat{t}_k$ the vector function $z^{kk}(t)$ satisfies the equation

$$\dot{z}^{kk}(t) = f_x(t,\hat{x}(t),\hat{u}(t))z^{kk}(t) + y^k(t)^* f_{xx}(t,\hat{x}(t),\hat{u}(t))y^k(t) \tag{7.48}$$

with the initial condition at the point $t = \hat{t}_k$:

$$z^{kk}(\hat{t}_k) + \dot{y}^k(\hat{t}_k+) = -[f_t]^k - [f_x]^k(\dot{\hat{x}}(\hat{t}_k+) + y^k(\hat{t}_k)). \tag{7.49}$$

Here, for $t \geq \hat{t}_k$, the function $y^k(t)$ is the solution to the IVP (7.26), and $y^k(t) = 0$ for $t < \hat{t}_k$, $k = 1,\ldots,s$. Furthermore, by definition, $[f_t]^k = f_t(\hat{t}_k,\hat{x}(\hat{t}_k),\hat{u}(\hat{t}_k+)) - f_t(\hat{t}_k,\hat{x}(\hat{t}_k),\hat{u}(\hat{t}_k-))$ and $[f_x]^k = f_x(\hat{t}_k,\hat{x}(\hat{t}_k),\hat{u}(\hat{t}_k+)) - f_x(\hat{t}_k,\hat{x}(\hat{t}_k),\hat{u}(\hat{t}_k-))$ are the jumps of the derivatives $f_t(t,\hat{x}(t),\hat{u}(t))$ and $f_x(t,\hat{x}(t),\hat{u}(t))$ at the point $\hat{t}_k$. For $t < \hat{t}_k$ we put $z^{kk}(t) = 0$, $k = 1,\ldots,s$.

7.3.4 Mixed Derivatives $g_{t_k t_j}$

For simplicity, let $s = 2$, $\theta = (t_1,t_2)$, and $t_0 < t_1 < t_2 < t_f$. Fix x_0 and t_0 and put

$$y^k(t;\theta) = \frac{\partial x(t;\theta)}{\partial t_k}, \quad k = 1,2, \quad z^{12}(t;\theta) = \frac{\partial y^1(t;\theta)}{\partial t_2} = \frac{\partial^2 x(t;\theta)}{\partial t_1 \partial t_2},$$
$$\dot{y}^k(t;\theta) = \frac{\partial y^k(t;\theta)}{\partial t}, \quad k = 1,2, \quad \dot{z}^{12}(t;\theta) = \frac{\partial z^{12}(t;\theta)}{\partial t}.$$

Then y^k, $\dot{y}^k$, $k = 1,2$, and z^{12}, $\dot{z}^{12}$ are vector functions of dimension n.

Proposition 7.18. *For $t \geq t_2$ the function $z^{12}(t;\theta)$ is the solution to the system*

$$\dot{z}^{12} = f_x z^{12} + (y^1)^* f_{xx} y^2 \tag{7.50}$$

with the initial condition at the point $t = t_2$,

$$z^{12}(t_2;\theta) = -[\dot{y}^1]^2. \tag{7.51}$$

In (7.50), f_x and f_{xx} are taken along the trajectory $(t,x(t;\theta),u(t;\theta))$, $t\in[t_0,t_f]$, and $(y^1)^* f_{xx}y^2$ is a vector with elements

$$((y^1)^* f_{xx}y^2)_k = (y^1)^* f_{kxx}y^2 = \sum_{i,j=1}^{n} \frac{\partial^2 f_k}{\partial x_i \partial x_j} y_i^1 y_j^2, \quad k=1,\dots,n.$$

In (7.51) we have $[\dot y^1]^2 = [f_x]^2 y^1(t_2;\theta)$, where

$$[f_x]^2 = f_x(t_2,x(t_2;\theta),u^3) - f_x(t_2,x(t_2;\theta),u^2).$$

For $t<t_2$ we have $z^{12}(t;\theta)=0$.

Proof. By Proposition 7.13, for $t\ge t_1$ the function $y^1(t;\theta)$ is a solution to the equation

$$\dot y^1(t;\theta) = f_x(t,x(t;\theta),u(t;\theta))y^1(t;\theta),$$

where $y^1(t;\theta)=0$ for $t<t_1$. Differentiating this equation with respect to t_2, we see that for $t\ge t_2$, the function $z^{12}(t;\theta) = \frac{\partial y^1(t;\theta)}{\partial t_2}$ is a solution to system (7.50). The initial condition (7.51) is similar to the initial condition (7.26) in Proposition 7.13. For $t<t_2$, we obviously have $z^{12}(t;\theta)=0$. ∎

For the solution $x(t;t_0,x_0,\theta)$ of IVP (7.3) and for $t_k<t_j$ $(k,j=1,\dots,s)$, it follows from Proposition 7.18 that

$$g_{t_k t_j}(\hat\zeta) := \frac{\partial^2 x(\hat t_f;\hat t_0,\hat x_0,\hat\theta)}{\partial t_k \partial t_j} = z^{kj}(\hat t_f), \tag{7.52}$$

where for $t\ge \hat t_j$ the vector function $z^{kj}(t)$ is the solution to the equation

$$\dot z^{kj}(t) = f_x(t,\hat x(t),\hat u(t))z^{kj}(t) + y^k(t)^* f_{xx}(t,\hat x(t),\hat u(t))y^j(t) \tag{7.53}$$

satisfying the initial condition

$$z^{kj}(\hat t_j) = -[\dot y^k]^j = -[f_x]^j y^k(\hat t_j). \tag{7.54}$$

Here, for $t\ge \hat t_k$, the function $y^k(t)$ is the solution to the IVP (7.26), while $y^k(t)=0$ holds for $t<\hat t_k$, $k=1,\dots,s$. By definition, $[\dot y^k]^j = \dot y^k(\hat t_j+) - \dot y^k(\hat t_j-)$ and $[f_x]^j = f_x(\hat t_j,\hat x(\hat t_j),\hat u(\hat t_j+)) - f_x(\hat t_j,\hat x(\hat t_j),\hat u(\hat t_j-))$ are the jumps of the derivatives $\dot y^k(t)$ and $f_x(t,\hat x(t),\hat u(t))$, respectively, at the point $\hat t_j$. For $t<\hat t_j$ we put $z^{kj}(t)=0$.

7.3.5 Derivatives $g_{t_0t_0}$, $g_{t_0t_f}$, and $g_{t_ft_f}$

Here, we fix x_0 and θ and study the functions

$$w(t;t_0) = \frac{\partial x(t;t_0)}{\partial t_0}, \quad q(t;t_0) = \frac{\partial w(t;t_0)}{\partial t_0} = \frac{\partial^2 x(t;t_0)}{\partial t_0^2},$$

$$\dot w(t;t_0) = \frac{\partial w(t;t_0)}{\partial t}, \quad \dot q(t;t_0) = \frac{\partial q(t;t_0)}{\partial t}, \quad \ddot x(t;t_0) = \frac{\partial^2 x(t;t_0,)}{\partial t^2}.$$

Proposition 7.19. *The function $q(t;t_0)$ is the solution to the system*

$$\dot{q} = f_x q + w^* f_{xx} w, \quad t \in [t_0, t_f] \tag{7.55}$$

satisfying the initial condition at the point $t = t_0$,

$$\ddot{x}(t_0;t_0) + 2\dot{w}(t_0;t_0) + q(t_0;t_0) = 0. \tag{7.56}$$

In (7.55), *f_x and f_{xx} are taken along the trajectory $(t, x(t;t_0), u(t))$, $t \in [t_0, t_f]$, and $w^* f_{xx} w$ is a vector with elements*

$$(w^* f_{xx} w)_k = w^* f_{kxx} w = \sum_{i,j=1}^{n} \frac{\partial^2 f_k}{\partial x_i \partial x_j} w_i w_j, \quad k = 1, \ldots, n.$$

Proof. By Proposition 7.12 we have

$$\dot{w}(t;t_0) = f_x(t, x(t;t_0)) w(t;t_0), \quad \dot{x}(t_0;t_0) + w(t_0;t_0) = 0.$$

Differentiating these equalities with respect to t_0, we obtain (7.55) and (7.56). ∎

From Proposition 7.19 it follows that

$$g_{t_0 t_0}(\hat{\zeta}) := \frac{\partial^2 x(\hat{t}_f; \hat{t}_0, \hat{x}_0, \hat{\theta})}{\partial t_0^2} = q(\hat{t}_f), \tag{7.57}$$

where the vector function $q(t)$ is the solution to the equation

$$\dot{q}(t) = f_x(t, \hat{x}(t), \hat{u}(t)) q(t) + w^*(t) f_{xx}(t, \hat{x}(t), \hat{u}(t)) w(t) \tag{7.58}$$

satisfying the initial condition

$$\ddot{\hat{x}}(\hat{t}_0) + 2\dot{w}(\hat{t}_0) + q(\hat{t}_0) = 0. \tag{7.59}$$

Since $w(t) = -V(t)\dot{\hat{x}}(\hat{t}_0)$ in view of Proposition 7.12, $\dot{V} = f_x V$, and $V(\hat{t}_0) = E$, we obtain

$$\dot{w}(\hat{t}_0) = -\dot{V}(\hat{t}_0)\dot{\hat{x}}(\hat{t}_0) = -f_x(\hat{t}_0, \hat{x}(\hat{t}_0), \hat{u}(\hat{t}_0))\dot{\hat{x}}(\hat{t}_0).$$

Thus, the initial condition (7.59) is equivalent to

$$\ddot{\hat{x}}(\hat{t}_0) - 2 f_x(\hat{t}_0, \hat{x}(\hat{t}_0), \hat{u}(\hat{t}_0))\dot{\hat{x}}(\hat{t}_0) + q(\hat{t}_0) = 0. \tag{7.60}$$

From (7.24) it follows that

$$\begin{aligned} g_{t_0 t_f}(\hat{\zeta}) &:= \frac{\partial^2 x(\hat{t}_f; \hat{t}_0, \hat{x}_0, \hat{\theta})}{\partial t_0 \partial t_f} \\ &= -\dot{V}(\hat{t}_f)\dot{\hat{x}}(\hat{t}_0) = -f_x(\hat{t}_f, \hat{x}(\hat{t}_f), \hat{u}(\hat{t}_f)) V(\hat{t}_f)\dot{\hat{x}}(\hat{t}_0). \end{aligned} \tag{7.61}$$

Formula (7.25) implies that

$$g_{t_f t_f}(\hat{\zeta}) := \frac{\partial^2 x(\hat{t}_f; \hat{t}_0, \hat{x}_0, \hat{\theta})}{\partial t_f^2} = \ddot{\hat{x}}(\hat{t}_f). \tag{7.62}$$

7.3.6 Derivatives $g_{x_0t_f}$ and $g_{t_kt_f}$

Formula (7.22) implies that

$$g_{x_0t_f}(\hat{\zeta}) := \frac{\partial^2 x(\hat{t}_f;\hat{t}_0,\hat{x}_0,\hat{\theta})}{\partial x_0 \partial t_f} = \dot{V}(\hat{t}_f), \tag{7.63}$$

where $V(t)$ is the solution to the IVP (7.18). From (7.27) it follows that

$$g_{t_kt_f}(\hat{\zeta}) := \frac{\partial^2 x(\hat{t}_f;\hat{t}_0,\hat{x}_0,\hat{\theta})}{\partial t_k \partial t_f} = \dot{y}^k(\hat{t}_f), \quad k = 1,\dots,s, \tag{7.64}$$

where $y^k(t)$ is the solution to the IVP (7.26).

7.3.7 Derivative $g_{x_0t_0}$

Let us fix θ and consider

$$V(t;t_0,x_0) = \frac{\partial x(t;t_0,x_0)}{\partial x_0}, \quad S(t;t_0,x_0) = \frac{\partial V(t;t_0,x_0)}{\partial t_0} = \frac{\partial^2 x(t;t_0,x_0)}{\partial x_0 \partial t_0},$$
$$\dot{V}(t;t_0,x_0) = \frac{\partial V(t;t_0,x_0)}{\partial t}, \quad \dot{S}(t;t_0,x_0) = \frac{\partial S(t;t_0,x_0)}{\partial t}.$$

Proposition 7.20. *The elements $s_{ij}(t;t_0,x_0)$ of the matrix $S(t;t_0,x_0)$ satisfy the system*

$$\dot{s}_{ij} = -e_j^* V^*(f_i)_{xx} V \dot{x}(t_0) + f_{ix} S e_j, \quad i,j = 1,\dots,n, \tag{7.65}$$

and the matrix S itself satisfies the initial condition at the point $t = t_0$,

$$S(t_0;t_0,x_0) + \dot{V}(t_0;t_0,x_0) = 0. \tag{7.66}$$

In (7.65), the derivatives f_x and f_{xx} are taken along the trajectory $(t,x(t;t_0,x_0),u(t))$, $t \in [t_0,t_f]$, e_j is the jth column of the identity matrix E, and, by definition, $\dot{x}(t_0) = \dot{x}(t_0;t_0,x_0)$.

Proof. By Proposition 7.11,

$$\dot{V}(t;t_0,x_0) = f_x(t,x(t;t_0,x_0),u(t))V(t;t_0,x_0), \quad V(t_0;t_0,x_0) = E. \tag{7.67}$$

The first equality in (7.67) is equivalent to

$$\dot{v}_{ij}(t;t_0,x_0) = f_{ix}(t,x(t;t_0,x_0),u(t))V(t,t_0)e_j, \quad i,j = 1,\dots,n.$$

By differentiating these equalities with respect to t_0 and using Proposition 7.12, we obtain (7.65). Differentiating the second equality in (7.67) with respect to t_0 yields (7.66). ∎

Proposition 7.20 implies that

$$g_{x_0t_0}(\hat{\zeta}) := \frac{\partial^2 x(\hat{t}_f;\hat{t}_0,\hat{x}_0,\hat{\theta})}{\partial x_0 \partial t_0} = S(\hat{t}_f), \tag{7.68}$$

where the elements $s_{ij}(t)$ of the matrix $S(t)$ satisfy the system

$$\dot{s}_{ij}(t) = e_j^* V^*(t)(f_i)_{xx}(t,\hat{x}(t),\hat{u}(t))V(t)\dot{\hat{x}}(\hat{t}_0) + f_{ix}(t,\hat{x}(t),\hat{u}(t))S(t)e_j,$$
$$i,j = 1,\dots,n. \tag{7.69}$$

Here, $V(t)$ is the solution to the IVP (7.18), and the matrix $S(t)$ itself satisfies the initial condition at the point $t = \hat{t}_0$,

$$S(\hat{t}_0) + \dot{V}(\hat{t}_0) = 0. \tag{7.70}$$

7.3.8 Derivative $g_{t_k t_0}$

Consider again the case $s = 1$ with $\theta = t_1$ and define

$$y(t;t_0,\theta) = \frac{\partial x(t;t_0,\theta)}{\partial \theta}, \quad r(t;t_0,\theta) = \frac{\partial y(t;t_0,\theta)}{\partial t_0} = \frac{\partial^2 x(t;t_0,\theta)}{\partial t_0 \partial \theta},$$
$$\dot{y}(t;t_0,\theta) = \frac{\partial y(t;t_0,\theta)}{\partial t}, \quad \dot{r}(t;t_0,\theta) = \frac{\partial r(t,t_0,\theta)}{\partial t},$$
$$\dot{x}(t;t_0,\theta) = \frac{\partial x(t;t_0,\theta)}{\partial t}, \quad V(t;t_0,\theta) = \frac{\partial x(t;t_0,\theta)}{\partial x_0}.$$

Proposition 7.21. *For $t \geq \theta$, the function $r(t;t_0,\theta)$ is the solution to the IVP*

$$\dot{r} = f_x r - y^* f_{xx} V \dot{x}(t_0), \quad r|_{t=\theta} = [f_x]V(\theta)\dot{x}(t_0), \tag{7.71}$$

where $y^* f_{xx} V \dot{x}(t_0)$ is the vector with elements $(y^* f_{xx} V \dot{x}(t_0))_i = y^* f_{ixx} V \dot{x}(t_0)$, $i = 1,\dots,n$, $V(\theta) = V(\theta;t_0,\theta)$, and

$$[f_x] = f_x(\theta, x(\theta;t_0,\theta), u^2) - f_x(\theta, x(\theta;t_0,\theta), u^1)$$

is the jump of the derivative $f_x(t,x(t;t_0,\theta),u(t;\theta))$ at the point θ. The derivatives f_x and f_{xx} are taken along the trajectory $(t,x(t;t_0,\theta),u(t;\theta))$, $t \in [\theta,t_f]$. For $t < \theta$ we have $r(t;t_0,\theta) = 0$. Then the jump of the function $r(t;t_0,\theta)$ at the point $t = \theta$ is given by $[r] = [f_x]V(\theta)\dot{x}(t_0)$.

Proof. By Proposition 7.13 we have $y(t;t_0,\theta) = 0$ for $t < \theta$ and hence $r(t;t_0,\theta) = 0$ for $t < \theta$. According to the same proposition, for $t \geq \theta$ the function $y(t;t_0,\theta)$ satisfies the equation

$$\dot{y}(t;t_0,\theta) = f_x(t,x(t;t_0,\theta),u(t;\theta))y(t;t_0,\theta).$$

Differentiating this equation with respect to t_0, we get

$$\dot{r} = f_x r + y^* f_{xx} \frac{\partial x}{\partial t_0}.$$

According to Proposition 7.12,

$$\frac{\partial x(t;t_0,\theta)}{\partial t_0} = -V(t;t_0,\theta)\dot{x}(t_0),$$

where $\dot{x}(t_0) = \dot{x}(t_0; t_0, \theta)$. This yields

$$\dot{r} = f_x r - y^* f_{xx} V \dot{x}(t_0).$$

By Proposition 7.13, the following initial condition holds at the point $t = \theta$:

$$y(\theta; t_0, \theta) = -(f(\theta, x(\theta; t_0, \theta), u^2) - f(\theta, x(\theta; t_0, \theta), u^1)).$$

Differentiating this condition with respect to t_0, we get

$$r|_{t=\theta} = -[f_x] \frac{\partial x}{\partial t_0}|_{t=\theta} = [f_x] V(\theta) \dot{x}(t_0),$$

where $V(\theta) = V(\theta; t_0, \theta)$. ∎

It follows from Proposition 7.21 that for each $k = 1, \ldots, s$,

$$g_{t_k t_0}(\hat{\zeta}) := \frac{\partial^2 x(\hat{t}_f; \hat{t}_0, \hat{x}_0, \hat{\theta})}{\partial t_k \partial t_0} = r^k(\hat{t}_f), \tag{7.72}$$

where the function $r^k(t)$ is the solution to the system

$$\dot{r}^k(t) = f_x(t, \hat{x}(t), \hat{u}(t)) r^k(t) - (y^k(t))^* f_{xx}(t, \hat{x}(t), \hat{u}(t)) V(t) \dot{\hat{x}}(\hat{t}_0) \tag{7.73}$$

and satisfies the initial condition at the point $t = \hat{t}_k$,

$$r^k(\hat{t}_k) = [f_x]^k V(\hat{t}_k) \dot{\hat{x}}(\hat{t}_0). \tag{7.74}$$

Here $V(t)$ is the solution to the IVP (7.18) and $y^k(t)$ is the solution to the IVP (7.26). The vector $(y^k)^* f_{xx} V \dot{\hat{x}}(\hat{t}_0)$ has components

$$((y^k)^* f_{xx} V \dot{\hat{x}}(\hat{t}_0))_j = (y^k)^* f_{jxx} V \dot{\hat{x}}(\hat{t}_0), \quad j = 1, \ldots, n.$$

7.4 Explicit Representation of the Quadratic Form for the Induced Optimization Problem

Let the Lagrange multipliers $\mu = (\alpha_0, \alpha, \beta) \in \Lambda_0$ and $\lambda = (\alpha_0, \alpha, \beta, \psi, \psi_0) \in \Lambda$ correspond to each other, i.e., let $\pi_0 \lambda = \mu$ hold; see Proposition 7.4. For any $\bar{\zeta} = (\bar{t}_0, \bar{t}_f, \bar{x}_0, \bar{\theta}) \in \mathbb{R}^{2+n+s}$, let us find an explicit representation for the quadratic form $\langle L_{\zeta\zeta}(\mu, \hat{\zeta})\bar{\zeta}, \bar{\zeta} \rangle$. By definition,

$$\begin{aligned}
\langle L_{\zeta\zeta}(\mu, \hat{\zeta})\bar{\zeta}, \bar{\zeta} \rangle = {} & \langle L_{x_0 x_0} \bar{x}_0, \bar{x}_0 \rangle + 2 \sum_{k=1}^{s} L_{x_0 t_k} \bar{x}_0 \bar{t}_k + \sum_{k,j=1}^{s} L_{t_k t_j} \bar{t}_k \bar{t}_j \\
& + 2 L_{x_0 t_f} \bar{x}_0 \bar{t}_f + 2 \sum_{k=1}^{s} L_{t_k t_f} \bar{t}_k \bar{t}_f + L_{t_f t_f} \bar{t}_f^2 \\
& + 2 L_{x_0 t_0} \bar{x}_0 \bar{t}_0 + 2 \sum_{k=1}^{s} L_{t_0 t_k} \bar{t}_0 \bar{t}_k + 2 L_{t_0 t_f} \bar{t}_0 \bar{t}_f + L_{t_0 t_0} \bar{t}_0^2.
\end{aligned} \tag{7.75}$$

All derivatives in formula (7.75) are taken at the point $(\mu, \hat{\zeta})$. Now we shall calculate these derivatives. Recall the definition (7.6) of the Lagrangian,

$$L(\mu,\zeta) = L(\mu,t_0,t_f,x_0,\theta) = l(\mu,t_0,x_0,t_f,x(t_f;t_0,x_0,\theta)). \tag{7.76}$$

Note that all functions V, W^k, y^k, z^{kj}, S, R^k, q, w, r^k, introduced in Sections 7.1.2 and 7.3, depend now on t, t_0, x_0, and θ. For simplicity, we put $V(t) = V(t;\hat{t}_0,\hat{x}_0,\hat{\theta})$, etc.

7.4.1 Derivative $L_{x_0x_0}$

Using Proposition 7.11, we get

$$\begin{aligned}\left(\frac{\partial}{\partial x_0} l(t_0,x_0,t_f,x(t_f;t_0,x_0,\theta))\right)\bar{x}_0 &= l_{x_0}(t_0,x_0,t_f,x(t_f;t_0,x_0,\theta))\bar{x}_0 \\ &\quad + l_{x_f}(t_0,x_0,t_f,x(t_f;t_0,x_0,\theta))V(t_f;t_0,x_0,\theta)\bar{x}_0. \end{aligned}\tag{7.77}$$

Let us find the derivative of this function with respect to x_0. We have

$$\begin{aligned}\frac{\partial}{\partial x_0}\Big(l_{x_0}(t_0,x_0,t_f,x(t_f;t_0,x_0,\theta))\bar{x}_0\Big) &= \bar{x}_0^* l_{x_0x_0}(t_0,x_0,t_f,x(t_f;t_0,x_0,\theta)) \\ &\quad + \bar{x}_0^* l_{x_0x_f}(t_0,x_0,t_f,x(t_f;t_0,x_0,\theta))V(t_f;t_0,x_0,\theta), \end{aligned}\tag{7.78}$$

and

$$\begin{aligned}&\frac{\partial}{\partial x_0}\Big(l_{x_f}(t_0,x_0,t_f,x(t_f;t_0,x_0,\theta))V(t_f;t_0,x_0,\theta)\bar{x}_0\Big) \\ &\quad = \bar{x}_0^* V^*(t_f;t_0,x_0,\theta)\Big(l_{x_fx_0}(t_0,x_0,t_f,x(t_f;t_0,x_0,\theta)) \\ &\qquad\qquad + l_{x_fx_f}(t_0,x_0,t_f,x(t_f;t_0,x_0,\theta))V(t_f;t_0,x_0,\theta)\Big) \\ &\quad + l_{x_f}(t_0,x_0,t_f,x(t_f;t_0,x_0,\theta))\frac{\partial}{\partial x_0}V(t_f;t_0,x_0,\theta)\bar{x}_0. \end{aligned}\tag{7.79}$$

From (7.77)–(7.79) and the transversality condition $l_{x_f}(\hat{p}) = \psi(\hat{t}_f)$ it follows that at the point $\hat{\zeta}$, we have

$$\begin{aligned}\langle L_{x_0x_0}\bar{x}_0,\bar{x}_0\rangle &= \bar{x}_0^* l_{x_0x_0}(\hat{p})\bar{x}_0 + 2\bar{x}_0^* l_{x_0x_f}(\hat{p})V(\hat{t}_f)\bar{x}_0 + \bar{x}_0^* V^*(\hat{t}_f) l_{x_fx_f}(\hat{p})V(\hat{t}_f)\bar{x}_0 \\ &\quad + \left\{\psi(t_f)\frac{\partial}{\partial x_0}(V(t_f;t_0,x_0,\theta)\bar{x}_0)\bar{x}_0\right\}\Bigg|_{\zeta=\hat{\zeta}}. \end{aligned}\tag{7.80}$$

Let us calculate the last term in this formula.

Proposition 7.22. *The following equality holds:*

$$\psi(t_f)\frac{\partial}{\partial x_0}\big(V(t_f;t_0,x_0,\theta)\bar{x}_0\big)\bar{x}_0 = \bar{x}_0^*\left(\sum_k \psi_k(t_f)W^k(t_f;t_0,x_0,\theta)\right)\bar{x}_0. \tag{7.81}$$

Proof. For brevity, put $\psi(t_f) = \psi$, $V(t_f;t_0,x_0,\theta) = V$, $W(t_f;t_0,x_0,\theta) = W$. Then we have

$$\begin{aligned}
\psi\frac{\partial}{\partial x_0}(V\bar{x}_0)\bar{x}_0 &= \psi\frac{\partial}{\partial x_0}\left(\frac{\partial x}{\partial x_0}\bar{x}_0\right)\bar{x}_0 = \psi\frac{\partial}{\partial x_0}\left(\sum_i\frac{\partial x}{\partial x_{0i}}\bar{x}_{0i}\right)\bar{x}_0 \\
&= \psi\sum_j\sum_i\frac{\partial^2 x}{\partial x_{0i}\partial x_{0j}}\bar{x}_{0i}\bar{x}_{0j} = \sum_k\sum_j\sum_i\psi_k\frac{\partial^2 x_k}{\partial x_{0i}\partial x_{0j}}\bar{x}_{0i}\bar{x}_{0j} \\
&= \sum_i\sum_j\left(\sum_k\psi_k\frac{\partial^2 x_k}{\partial x_{0i}\partial x_{0j}}\right)\bar{x}_{0i}\bar{x}_{0j} = \bar{x}_0^*\left(\sum_k\psi_k(t_f)W^k\right)\bar{x}_0. \qquad\blacksquare
\end{aligned}$$

Proposition 7.23. *For $\zeta = \hat{\zeta}$, the following equality holds:*

$$\frac{d}{dt}\left(\sum_k\psi_k W^k\right) = V^* H_{xx} V, \tag{7.82}$$

where $H = \psi f(t,x,u)$, $H_{xx} = H_{xx}(t,\hat{x}(t),\psi(t),\hat{u}(t))$.

Proof. According to Proposition 7.15, we have

$$\dot{W}^k = V^* f_{kxx} V + \sum_r f_{kx_r} W^r, \quad k = 1,\dots,n. \tag{7.83}$$

Using these equations together with the adjoint equation $-\dot{\psi} = \psi f_x$, we obtain

$$\begin{aligned}
\frac{d}{dt}\left(\sum_k\psi_k W^k\right) &= \sum_k\dot{\psi}_k W^k + \sum_k\psi_k\dot{W}^k \\
&= -\sum_k\psi f_{x_k}W^k + \sum_k\psi_k\left(V^* f_{kxx}V + \sum_r f_{kx_r}W^r\right) \\
&= -\sum_k\psi f_{x_k}W^k + \sum_k V^*(\psi_k f_{kxx})V + \sum_k\psi_k\sum_r f_{kx_r}W^r \\
&= -\sum_r\psi f_{x_r}W^r + V^*\left(\sum_k\psi_k f_{kxx}\right)V + \sum_r\left(\sum_k\psi_k f_{kx_r}\right)W^r \\
&= -\sum_r\psi f_{x_r}W^r + V^*(\psi f_{xx})V + \sum_r\psi f_{x_r}W^r = V^* H_{xx} V. \qquad\blacksquare
\end{aligned}$$

Now we can prove the following assertion.

Proposition 7.24. *The following formula holds:*

$$\left.\left\{\psi(t_f)\frac{\partial}{\partial x_0}\left(V(t_f;t_0,x_0,\theta)\bar{x}_0\right)\bar{x}_0\right\}\right|_{\hat{\zeta}} = \int_{\hat{t}_0}^{\hat{t}_f}(V(t)\bar{x}_0)^* H_{xx}(t,\hat{x}(t),\hat{u}(t),\psi(t))V(t)\bar{x}_0\,dt. \tag{7.84}$$

Proof. Using Propositions 7.22 and 7.23 and the initial conditions $W^k(\hat{t}_0) = 0$ for $k = 1,\dots,n$, we get

$$\begin{aligned}
&\left\{\psi(t_f)\frac{\partial}{\partial x_0}\left(V(t_f;t_0,x_0,\theta)\bar{x}_0\right)\bar{x}_0\right\}\Bigg|_{\hat{\zeta}} \\
&\quad = \bar{x}_0^*\left(\sum_k \psi_k(\hat{t}_f)W^k(\hat{t}_f)\right)\bar{x}_0 = \bar{x}_0^*\left(\sum_k \psi_k(t)W^k(t)\right)\bar{x}_0\Bigg|_{\hat{t}_0}^{\hat{t}_f} \\
&\quad = \int_{\hat{t}_0}^{\hat{t}_f} \bar{x}_0^*\frac{d}{dt}\left(\sum_k \psi_k W^k\right)\bar{x}_0\,dt = \int_{\hat{t}_0}^{\hat{t}_f} \bar{x}_0^* V^* H_{xx} V\bar{x}_0\,dt \\
&\quad = \int_{\hat{t}_0}^{\hat{t}_f} (V\bar{x}_0)^* H_{xx}(V\bar{x}_0)\,dt. \qquad \blacksquare
\end{aligned}$$

In view of formulas (7.80) and (7.84), we obtain

$$\begin{aligned}
\langle L_{x_0x_0}\bar{x}_0,\bar{x}_0\rangle &= \bar{x}_0^* l_{x_0x_0}(\hat{p})\bar{x}_0 \\
&\quad +2\bar{x}_0^* l_{x_0x_f}(\hat{p})V(\hat{t}_f)\bar{x}_0 + (V(\hat{t}_f)\bar{x}_0)^* l_{x_fx_f}(\hat{p})V(\hat{t}_f)\bar{x}_0 \\
&\quad + \int_{\hat{t}_0}^{\hat{t}_f} (V(t)\bar{x}_0)^* H_{xx}(t,\hat{x}(t),\psi(t),\hat{u}(t))V(t)\bar{x}_0\,dt.
\end{aligned} \tag{7.85}$$

7.4.2 Derivative $L_{x_0t_k}$

Differentiating (7.77) with respect to t_k and using Propositions 7.13 and 7.16, we get

$$\begin{aligned}
&\frac{\partial^2}{\partial x_0\partial t_k} l(t_0,x_0,t_f,x(t_f;t_0,x_0,\theta))\bar{x}_0 \\
&\quad = \frac{\partial}{\partial t_k} l_{x_0}(t_0,x_0,t_f,x(t_f;t_0,x_0,\theta))\bar{x}_0 \\
&\qquad + \frac{\partial}{\partial t_k}\left(l_{x_f}(t_0,x_0,t_f,x(t_f;t_0,x_0,\theta))V(t_f;t_0,x_0,\theta)\bar{x}_0\right) \\
&\quad = \bar{x}_0^* l_{x_0x_f}(t_0,x_0,t_f,x(t_f;t_0,x_0,\theta))\frac{\partial x(t_f;t_0,x_0,\theta)}{\partial t_k} \\
&\qquad + \left(\frac{\partial}{\partial t_k} l_{x_f}(t_0,x_0,t_f,x(t_f;t_0,x_0,\theta))\right)V(t_f;t_0,x_0,\theta)\bar{x}_0 \\
&\qquad + l_{x_f}(t_0,x_0,t_f,x(t_f;t_0,x_0,\theta))\frac{\partial V(t_f;t_0,x_0,\theta)}{\partial t_k}\bar{x}_0 \\
&\quad = \bar{x}_0^* l_{x_0x_f}(t_0,x_0,t_f,x(t_f;t_0,x_0,\theta))y^k(t_f;t_0,x_0,\theta) \\
&\qquad + (V(t_f;t_0,x_0,\theta)\bar{x}_0)^* l_{x_fx_f}(t_0,x_0,t_f,x(t_f;t_0,x_0,\theta))y^k(t_f;t_0,x_0,\theta) \\
&\qquad + l_{x_f}(t_0,x_0,t_f,x(t_f;t_0,x_0,\theta))R^k(t_f;t_0,x_0,\theta))\bar{x}_0.
\end{aligned} \tag{7.86}$$

Hence at the point $\zeta = \hat{\zeta}$, we have

$$L_{x_0t_k}\bar{x}_0\bar{t}_k = \bar{x}_0^* l_{x_0x_f}(\hat{p})y^k(\hat{t}_f)\bar{t}_k + \left(V(\hat{t}_f)\bar{x}_0\right)^* l_{x_fx_f}(\hat{p})y^k(\hat{t}_f)\bar{t}_k + \psi(\hat{t}_f)R^k(\hat{t}_f)\bar{x}_0\bar{t}_k. \tag{7.87}$$

Let us transform the last term.

Proposition 7.25. *The following formula holds:*

$$\psi(\hat{t}_f)R^k(\hat{t}_f)\bar{x}_0\bar{t}_k = -[H_x]^k V(\hat{t}_k)\bar{x}_0\bar{t}_k + \int_{\hat{t}_k}^{\hat{t}_f} \langle H_{xx} y^k \bar{t}_k, V\bar{x}_0\rangle \, dt. \tag{7.88}$$

Proof. Using equation (7.44) and the adjoint equation $-\dot{\psi} = \psi f_x$, we get for $t \in [\hat{t}_k, \hat{t}_f]$,

$$\begin{aligned} \frac{d}{dt}(\psi R^k) &= \dot{\psi} R^k + \psi \dot{R}^k = -\psi f_x R^k + \psi\left(((y^k)^* f_{xx})V + f_x R^k\right) \\ &= \sum_j \psi_j (y^k)^* f_{jxx} V = (y^k)^* \sum_j \psi_j f_{jxx} V = (y^k)^* H_{xx} V, \end{aligned}$$

where H_{xx} is taken along the trajectory $(t, \hat{x}(t), \psi(t), \hat{u}(t))$. Consequently,

$$\psi(\hat{t}_f)R^k(\hat{t}_f) = \psi(\hat{t}_k)R^k(\hat{t}_k) + \int_{\hat{t}_k}^{\hat{t}_f} (y^k)^* H_{xx} V \, dt.$$

Using the initial condition (7.44) for R^k at $\hat{t}_k$, we get

$$\psi(\hat{t}_f)R^k(\hat{t}_f) = -\psi(\hat{t}_k)[f_x]^k V(\hat{t}_k) + \int_{\hat{t}_k}^{\hat{t}_f} (y^k)^* H_{xx} V \, dt.$$

Hence,

$$\psi(\hat{t}_f)R^k(\hat{t}_f)\bar{x}_0\bar{t}_k = -[H_x]^k V(\hat{t}_k)\bar{x}_0\bar{t}_k + \int_{\hat{t}_k}^{\hat{t}_f} \langle H_{xx} y^k \bar{t}_k, V\bar{x}_0\rangle \, dt. \qquad \blacksquare$$

Formulas (7.87) and (7.88) and the condition $y^k(t) = 0$ for $t < \hat{t}_k$ imply the equality

$$\begin{aligned} L_{x_0 t_k}\bar{x}_0\bar{t}_k &= \bar{x}_0^* l_{x_0 x_f}(\hat{p}) y^k(\hat{t}_f)\bar{t}_k + (V(\hat{t}_f)\bar{x}_0)^* l_{x_f x_f}(\hat{p}) y^k(\hat{t}_f)\bar{t}_k \\ &\quad - [H_x]^k V(\hat{t}_k)\bar{x}_0\bar{t}_k + \textstyle\int_{\hat{t}_0}^{\hat{t}_f} \langle H_{xx} y^k \bar{t}_k, V\bar{x}_0\rangle \, dt. \end{aligned}$$

7.4.3 Derivative $L_{t_k t_k}$

Using the notation $\frac{\partial x}{\partial t_k} = y^k$ from Proposition 7.13, we get

$$\frac{\partial}{\partial t_k} l(t_0, x_0, t_f, x(t_f; t_0, x_0, \theta)) = l_{x_f}(t_0, x_0, t_f, x(t_f; t_0, x_0, \theta)) y^k(t_f; t_0, x_0, \theta)). \tag{7.89}$$

Now, using the notation $\frac{\partial y^k}{\partial t_k} = z^{kk}$ as in Proposition 7.17, we obtain

$$\begin{aligned} &\frac{\partial^2}{\partial t_k^2} l(t_0, x_0, t_f, x(t_f; t_0, x_0, \theta)) \\ &\quad = \left(\frac{\partial}{\partial t_k} l_{x_f}(t_0, x_0, t_f, x(t_f; t_0, x_0, \theta))\right) y^k(t_f; t_0, x_0, \theta) \\ &\qquad + l_{x_f}(t_0, x_0, t_f, x(t_f; t_0, x_0, \theta)) \frac{\partial}{\partial t_k} y^k(t_f; t_0, x_0, \theta) \\ &\quad = \langle l_{x_f x_f}(t_0, x_0, t_f, x(t_f; t_0, x_0, \theta)) y^k(t_f; t_0, x_0, \theta), y^k(t_f; t_0, x_0, \theta)\rangle \\ &\qquad + l_{x_f}(t_0, x_0, t_f, x(t_f; t_0, x_0, \theta)) z^{kk}(t_f; t_0, x_0, \theta), \end{aligned} \tag{7.90}$$

and thus,

$$\begin{aligned} L_{t_k t_k} &= \frac{\partial^2}{\partial t_k^2} l(t_0, x_0, t_f, x(t_f; t_0, x_0, \theta)) \bigg|_{\zeta=\hat{\zeta}} \\ &= \langle l_{x_f x_f}(\hat{p}) y^k(\hat{t}_f), y^k(\hat{t}_f) \rangle + l_{x_f}(\hat{p}) z^{kk}(\hat{t}_f). \end{aligned}$$

Let us rewrite the last term in this formula. The transversality condition $l_{x_f} = \psi(\hat{t}_f)$ implies

$$l_{x_f}(\hat{p}) z^{kk}(\hat{t}_f) = \psi(\hat{t}_f) z^{kk}(\hat{t}_f) = \int_{\hat{t}_k}^{\hat{t}_f} \frac{d}{dt}(\psi z^{kk})\, dt + \psi(\hat{t}_k) z^{kk}(\hat{t}_k). \tag{7.91}$$

By formula (7.48), we have

$$\dot{z}^{kk} = f_x z^{kk} + (y^k)^* f_{xx} y^k, \qquad t \geq \hat{t}_k.$$

Using this equation together with the adjoint equation $-\dot{\psi} = \psi f_x$, we get

$$\frac{d}{dt}(\psi z^{kk}) = \dot{\psi} z^{kk} + \psi \dot{z}^{kk} = -\psi f_x z^{kk} + \psi f_x z^{kk} + \sum_j \psi_j ((y^k)^* f_{jxx} y^k) = (y^k)^* H_{xx} y^k, \tag{7.92}$$

and thus

$$l_{x_f}(\hat{p}) z^{kk}(\hat{t}_f) = \int_{\hat{t}_k}^{\hat{t}_f} (y^k)^* H_{xx} y^k\, dt + \psi(\hat{t}_k) z^{kk}(\hat{t}_k). \tag{7.93}$$

We shall transform the last term in (7.93) using the relations

$$\begin{aligned} &(\Delta_k H)(t) = H(t, \hat{x}(t), \hat{u}^{k+}, \psi(t)) - H(t, \hat{x}(t), \hat{u}^{k-}, \psi(t)), \\ &D^k(H) = -\frac{d}{dt}(\Delta_k H)\Big|_{t=t_k+} = -[H_t]^k - [H_x]^k \dot{\hat{x}}(\hat{t}_k+) - \dot{\psi}(\hat{t}_k+)[H_\psi]^k \end{aligned} \tag{7.94}$$

(see Section 5.2.2).

Proposition 7.26. *The following equality holds:*

$$\psi(\hat{t}_k) z^{kk}(\hat{t}_k) = D^k(H) - [H_x]^k [y^k]^k. \tag{7.95}$$

Proof. Multiplying the initial condition (7.49) for z^{kk} at the point $t = \hat{t}_k$ by $\psi(\hat{t}_k)$, we get

$$\psi(\hat{t}_k) z^{kk}(\hat{t}_k) + \psi(\hat{t}_k) \dot{y}^k(\hat{t}_k+) = -\psi(\hat{t}_k)[f_t]^k - \psi(\hat{t}_k)[f_x]^k \left(\dot{\hat{x}}(\hat{t}_k+) + y^k(\hat{t}_k) \right). \tag{7.96}$$

Here, we obviously have the relations $\psi(\hat{t}_k)[f_t]^k = [H_t]^k$, $\psi(\hat{t}_k)[f_x]^k = [H_x]^k$, and $y^k(\hat{t}_k) = [y^k]^k$. Moreover, equation (7.26) for y^k together with the adjoint equation $-\dot{\psi} = \psi f_x$ implies that $\psi \dot{y}^k = \psi f_x y^k = -\dot{\psi} y^k$. Hence, in view of the initial condition (7.26) for y^k, we find

$$\psi(\hat{t}_k) \dot{y}^k(\hat{t}_k+) = -\dot{\psi}(\hat{t}_k+) y^k(\hat{t}_k) = \dot{\psi}(\hat{t}_k+)[f]^k = \dot{\psi}(\hat{t}_k+)[H_\psi]^k.$$

Thus, (7.96) and (7.94) imply (7.95). ∎

From the relations (7.91), (7.93), and (7.95) and the equality $y^k(t) = 0$ for $t < \hat{t}_k$, it follows that

$$L_{t_k t_k} \bar{t}_k^2 = \langle l_{x_f x_f}(\hat{p}) y^k(\hat{t}_f)\bar{t}_k, y^k(\hat{t}_f)\bar{t}_k\rangle + \int_{\hat{t}_0}^{\hat{t}_f} (y^k \bar{t}_k)^* H_{xx} y^k \bar{t}_k \, dt$$
$$+ D^k(H)\bar{t}_k^2 - [H_x]^k [y^k]^k \bar{t}_k^2, \quad k = 1, \ldots, s. \tag{7.97}$$

7.4.4 Derivative $L_{t_k t_j}$

Note that $L_{t_k t_j} = L_{t_j t_k}$ for all k, j. Therefore,

$$\sum_{k,j=1}^{s} L_{t_k t_j} \bar{t}_k \bar{t}_j = \sum_{k=1}^{s} L_{t_k t_k} \bar{t}_k^2 + 2 \sum_{k<j} L_{t_k t_j} \bar{t}_k \bar{t}_j. \tag{7.98}$$

Let us calculate $L_{t_k t_j}$ for $k < j$. Differentiating (7.89) with respect to t_j, we get

$$\frac{\partial^2}{\partial t_k \partial t_j} l(t_0, x_0, t_f, x(t_f; t_0, x_0, \theta))$$
$$= \left(\frac{\partial}{\partial t_j} l_{x_f}(t_0, x_0, t_f, x(t_f; t_0, x_0, \theta)) \right) y^k(t_f; t_0, x_0, \theta)$$
$$+ l_{x_f}(t_0, x_0, t_f, x(t_f; t_0, x_0, \theta)) \frac{\partial}{\partial t_j} y^k(t_f; t_0, x_0, \theta)$$
$$= \langle l_{x_f x_f}(t_0, x_0, t_f, x(t_f; t_0, x_0, \theta)) y^k(t_f; t_0, x_0, \theta), y^j(t_f; t_0, x_0, \theta)\rangle$$
$$+ l_{x_f}(t_0, x_0, t_f, x(t_f; t_0, x_0, \theta)) z^{kj}(t_f; t_0, x_0, \theta). \tag{7.99}$$

Thus,

$$L_{t_k t_j} = \frac{\partial^2}{\partial t_k \partial t_j} l(t_0, x_0, t_f, x(t_f; t_0, x_0, \theta))|_{\zeta = \hat{\zeta}}$$
$$= \langle l_{x_f x_f}(\hat{p}) y^k(\hat{t}_f), y^j(\hat{t}_f)\rangle + l_{x_f}(\hat{p}) z^{kj}(\hat{t}_f). \tag{7.100}$$

We can rewrite the last term in this formula as

$$l_{x_f}(\hat{p}) z^{kj}(\hat{t}_f) = \psi(\hat{t}_f) z^{kj}(\hat{t}_f) = \int_{\hat{t}_j}^{\hat{t}_f} \frac{d}{dt}(\psi z^{kj}) \, dt + \psi(\hat{t}_j) z^{kj}(\hat{t}_j).$$

By formula (7.53), $\dot{z}^{kk} = f_x z^{kk} + (y^k)^* f_{xx} y^j$ for $t \geq \hat{t}_j$. Similarly to (7.92), we get $\frac{d}{dt}(\psi z^{kj}) = (y^k)^* H_{xx} y^j$ and thus obtain

$$l_{x_f}(\hat{p}) z^{kj}(\hat{t}_f) = \int_{\hat{t}_j}^{\hat{t}_f} (y^k)^* H_{xx} y^j \, dt + \psi(\hat{t}_j) z^{kj}(\hat{t}_j). \tag{7.101}$$

Since $y^j(t) = 0$ for $t < \hat{t}_j$, we have

$$\int_{\hat{t}_j}^{\hat{t}_f} (y^k)^* H_{xx} y^j \, dt = \int_{\hat{t}_0}^{\hat{t}_f} (y^k)^* H_{xx} y^j \, dt. \tag{7.102}$$

Using the initial condition (7.54) for z^{kj} at the point $\hat{\theta}^j$, we get

$$\psi(\hat{t}_j)z^{kj}(\hat{t}_j) = -\psi(\hat{t}_j)[f_x]^j y^k(\hat{t}_j) = -[H_x]^j y^k(\hat{t}_j). \tag{7.103}$$

Formulas (7.100)–(7.103) imply the following representation for all $k < j$:

$$L_{t_k t_j}\bar{t}_k\bar{t}_j = \langle l_{x_f x_f}(\hat{p})y^k(\hat{t}_f)\bar{t}_k, y^j(\hat{t}_f)\bar{t}_j\rangle + \int_{\hat{t}_0}^{\hat{t}_f} (y^k\bar{t}_k)^* H_{xx} y^j \bar{t}_j \, dt - [H_x]^j y^k(\hat{t}_j)\bar{t}_k\bar{t}_j. \tag{7.104}$$

7.4.5 Derivative $L_{x_0 t_f}$

Using Proposition 7.11, we get

$$\frac{\partial^2}{\partial x_0 \partial t_f} l(t_0, x_0, t_f, x(t_f; t_0, x_0, \theta)) = \frac{\partial}{\partial t_f}\{l_{x_0} + l_{x_f} V\}|_{t=t_f}$$
$$= (l_{x_0 t_f} + l_{x_0 x_f}\dot{x} + \frac{\partial l_{x_f}}{\partial t_f} V + l_{x_f}\dot{V})|_{t=t_f}$$
$$= (l_{x_0 t_f} + l_{x_0 x_f}\dot{x} + (l_{x_f t_f} + l_{x_f x_f}\dot{x})V + l_{x_f} f_x V)|_{t=t_f}.$$

Again, we transform the last term in this formula at the point $\zeta = \hat{\zeta}$. Using the adjoint equation $-\dot{\psi} = \psi f_x$ and the transversality condition $\psi(\hat{t}_f) = l_{x_f}$, we get

$$l_{x_f} f_x V|_{t=\hat{t}_f} = \psi f_x V|_{t=\hat{t}_f} = -\dot{\psi}(\hat{t}_f)V(\hat{t}_f).$$

Consequently,

$$\begin{aligned} L_{x_0 t_f}\bar{x}_0\bar{t}_f = {}& l_{x_0 t_f}\bar{x}_0\bar{t}_f + \langle l_{x_0 x_f}\dot{\hat{x}}(\hat{t}_f)\bar{t}_f, \bar{x}_0\rangle + l_{x_f t_f} V(\hat{t}_f)\bar{x}_0\bar{t}_f \\ & + \langle l_{x_f x_f}\dot{\hat{x}}(\hat{t}_f)\bar{t}_f, V(\hat{t}_f)\bar{x}_0\rangle - \dot{\psi}(\hat{t}_f)V(\hat{t}_f)\bar{x}_0\bar{t}_f. \end{aligned} \tag{7.105}$$

7.4.6 Derivative $L_{t_k t_f}$

Using the notation $\frac{\partial x}{\partial t_k} = y^k$ and Proposition 7.13, we get

$$\frac{\partial^2}{\partial t_k \partial t_f} l(t_0, x_0, t_f, x(t_f; t_0, x_0, \theta)) = \frac{\partial}{\partial t_f}\{l_{x_f} y^k\}|_{t=t_f}$$
$$= \{(l_{x_f x_f}\dot{x} + l_{x_f t_f})y^k + l_{x_f}\dot{y}^k\}|_{t=t_f}$$
$$= \{l_{x_f x_f}\dot{x} y^k + l_{x_f t_f} y^k + l_{x_f} f_x y^k\}|_{t=t_f}.$$

We evaluate the last term in this formula at the point $\zeta = \hat{\zeta}$ using the adjoint equation $-\dot{\psi} = \psi f_x$ and the transversality condition $\psi(\hat{t}_f) = l_{x_f}$:

$$l_{x_f} f_x y^k|_{t=\hat{t}_f} = \psi f_x y^k|_{t=\hat{t}_f} = -\dot{\psi}(\hat{t}_f)y^k(\hat{t}_f).$$

Therefore,

$$L_{t_k t_f}\bar{t}_k\bar{t}_f = \langle l_{x_f x_f}\dot{\hat{x}}(\hat{t}_f)\bar{t}_f, y^k(\hat{t}_f)\bar{t}_k\rangle + l_{x_f t_f} y^k(\hat{t}_f)\bar{t}_k\bar{t}_f - \dot{\psi}(\hat{t}_f)y^k(\hat{t}_f)\bar{t}_k\bar{t}_f. \tag{7.106}$$

7.4.7 Derivative $L_{t_f t_f}$

We have

$$\frac{\partial^2}{\partial t_f^2} l(t_0, x_0, t_f, x(t_f; t_0, x_0, \theta)) = \frac{\partial}{\partial t_f}\{l_{t_f} + l_{x_f}\dot{x}\}|_{t=t_f}$$
$$= \{(l_{t_f t_f} + l_{t_f x_f}\dot{x}) + (l_{x_f t_f} + l_{x_f x_f}\dot{x})\dot{x} + l_{x_f}\ddot{x}\}\Big|_{t=t_f},$$

which gives

$$L_{t_f t_f} = l_{t_f t_f} + 2 l_{t_f x_f}\dot{\hat{x}}(\hat{t}_f) + \langle l_{x_f x_f}\dot{\hat{x}}(\hat{t}_f), \dot{\hat{x}}(\hat{t}_f)\rangle + \psi(\hat{t}_f)\ddot{\hat{x}}(\hat{t}_f). \tag{7.107}$$

Let us transform the last term. Equation (6.11) in the definition of M_0 is equivalent to the relation $\psi\dot{\hat{x}} + \psi_0 = 0$. Differentiating this equation with respect to t, we get

$$\dot{\psi}\dot{\hat{x}} + \psi\ddot{\hat{x}} + \dot{\psi}_0 = 0. \tag{7.108}$$

Hence, formula (7.107) implies the equality

$$L_{t_f t_f}\bar{t}_f^2 = l_{t_f t_f}\bar{t}_f^2 + 2 l_{t_f x_f}\dot{\hat{x}}(\hat{t}_f)\bar{t}_f^2 + \langle l_{x_f x_f}\dot{\hat{x}}(\hat{t}_f)\bar{t}_f, \dot{\hat{x}}(\hat{t}_f)\bar{t}_f\rangle - (\dot{\psi}(\hat{t}_f)\dot{\hat{x}}(\hat{t}_f) + \dot{\psi}_0(\hat{t}_f))\bar{t}_f^2.$$

7.4.8 Derivative $L_{x_0 t_0}$

In view of the relation $\frac{\partial x}{\partial x_0} = V$, we obtain

$$\frac{\partial^2}{\partial x_0 \partial t_0} l(t_0, x_0, t_f, x(t_f; t_0, x_0, \theta)) = \frac{\partial}{\partial t_0}\{l_{x_0} + l_{x_f} V\}|_{t=t_f}$$
$$= \left\{ l_{x_0 t_0} + l_{x_0 x_f}\frac{\partial x}{\partial t_0} + \left(l_{x_f t_0} + l_{x_f x_f}\frac{\partial x}{\partial t_0}\right)V + l_{x_f}\frac{\partial V}{\partial t_0}\right\}\Big|_{t=t_f}.$$

Now, using the transversality condition $l_{x_f} = \psi(\hat{t}_f)$, formula (7.24), and the notation $\frac{\partial V}{\partial t_0} = S$, we get

$$L_{x_0 t_0} = l_{x_0 t_0} - l_{x_0 x_f} V(\hat{t}_f)\dot{\hat{x}}(\hat{t}_0) + l_{x_f t_0} V(\hat{t}_f) - \dot{\hat{x}}(\hat{t}_0)^* V(\hat{t}_f)^* l_{x_f x_f} V(\hat{t}_f) + \psi(\hat{t}_f) S(\hat{t}_f).$$

The transformation of the last term in this formula proceeds as follows. Using the adjoint equation for ψ and the system (7.69) for S, we obtain the equation

$$\frac{d}{dt}(\psi S) = \dot{\psi} S + \psi \dot{S} = -\psi f_x S + \psi f_x S - \dot{\hat{x}}(\hat{t}_0)^* V^* \sum_i \psi_i f_{ixx} V$$
$$= -\dot{\hat{x}}(\hat{t}_0)^* V^* H_{xx} V,$$

which yields

$$\psi(\hat{t}_f) S(\hat{t}_f) = -\int_{\hat{t}_0}^{\hat{t}_f} \dot{\hat{x}}(\hat{t}_0)^* V^* H_{xx} V\, dt + \psi(\hat{t}_0) S(\hat{t}_0). \tag{7.109}$$

Using now the initial condition (7.70) for S at the point $t = \hat{t}_0$ and the equation $\dot{V} = f_x V$, we get

$$(\psi S)|_{\hat{t}_0} = -(\psi \dot{V})|_{\hat{t}_0} = -(\psi f_x V)|_{\hat{t}_0} = (\dot{\psi} V)|_{\hat{t}_0} = \dot{\psi}(\hat{t}_0), \tag{7.110}$$

since $V(\hat{t}_0) = E$. Formulas (7.109) and (7.110) then imply the equality

$$\begin{aligned} L_{x_0 t_0} &= l_{x_0 t_0} - l_{x_0 x_f} V(\hat{t}_f)\dot{\hat{x}}(\hat{t}_0) + l_{x_f t_0} V(\hat{t}_f) - \dot{\hat{x}}(\hat{t}_0)^* V(\hat{t}_f)^* l_{x_f x_f} V(\hat{t}_f) \\ &\quad - \int_{\hat{t}_0}^{\hat{t}_f} \dot{\hat{x}}(\hat{t}_0)^* V^* H_{xx} V\, dt + \dot{\psi}(\hat{t}_0). \end{aligned} \tag{7.111}$$

Therefore,

$$\begin{aligned} L_{x_0 t_0} \bar{x}_0 \bar{t}_0 &= l_{x_0 t_0} \bar{x}_0 \bar{t}_0 - \langle l_{x_0 x_f} V(\hat{t}_f)\dot{\hat{x}}(\hat{t}_0)\bar{t}_0, \bar{x}_0 \rangle + l_{x_f t_0} V(\hat{t}_f)\bar{x}_0 \bar{t}_0 \\ &\quad - \langle l_{x_f x_f} V(\hat{t}_f)\bar{x}_0, V(\hat{t}_f)\dot{\hat{x}}(\hat{t}_0)\bar{t}_0 \rangle + \dot{\psi}(\hat{t}_0)\bar{x}_0 \bar{t}_0 \\ &\quad - \int_{\hat{t}_0}^{\hat{t}_f} \langle H_{xx} V \bar{x}_0, V \dot{\hat{x}}(\hat{t}_0)\bar{t}_0 \rangle\, dt. \end{aligned} \tag{7.112}$$

7.4.9 Derivative $L_{t_k t_0}$

Using the notation $\frac{\partial x}{\partial t_0} = w$, $\frac{\partial x}{\partial t_k} = y^k$, and $\frac{\partial y_k}{\partial t_0} = r^k$, we obtain

$$\begin{aligned} \frac{\partial^2}{\partial t_k \partial t_0} l(t_0, x_0, t_f, x(t_f; t_0, x_0, \theta)) &= \frac{\partial}{\partial t_0} \{ l_{x_f} y^k \}|_{t=t_f} \\ &= \left\{ l_{x_f t_0} y^k + (y^k)^* l_{x_f x_f} \frac{\partial x}{\partial t_0} + l_{x_f} \frac{\partial y^k}{\partial t_0} \right\} \bigg|_{t=t_f} \\ &= \{ l_{x_f t_0} y^k + (y^k)^* l_{x_f x_f} w + l_{x_f} r^k \}|_{t=t_f}. \end{aligned}$$

According to condition (7.24) we have $w|_{t=\hat{t}_f} = -V(\hat{t}_f)\dot{\hat{x}}(\hat{t}_0)$. Using this condition together with the transversality condition $l_{x_f} = \psi(\hat{t}_f)$, we find

$$L_{t_k t_0} = l_{x_f t_0} y^k(\hat{t}_f) - (y^k(\hat{t}_f))^* l_{x_f x_f} V(\hat{t}_f)\dot{\hat{x}}(\hat{t}_0) + \psi(\hat{t}_f) r^k(\hat{t}_f). \tag{7.113}$$

Let us transform the last term in this formula. Using the adjoint equation for ψ and the system (7.73) for r^k, we get for $t \geq \hat{t}_k$:

$$\begin{aligned} \frac{d}{dt}(\psi r^k) &= \dot{\psi} r^k + \psi \dot{r}^k = -\psi f_x r^k + \psi f_x r^k - (y_i)^* \sum_j \psi_j f_{jxx} V \dot{\hat{x}}(\hat{t}_0) \\ &= -(y_k)^* H_{xx} V \dot{\hat{x}}(\hat{t}_0). \end{aligned}$$

It follows that

$$\psi(\hat{t}_f) r^k(\hat{t}_f) = -\int_{\hat{t}_k}^{\hat{t}_f} (y_k)^* H_{xx} V \dot{\hat{x}}(\hat{t}_0)\, dt + \psi(\hat{t}_k) r^k(\hat{t}_k). \tag{7.114}$$

The initial condition (7.74) for r^k at the point $\hat{t}_k$ then yields

$$\psi(\hat{t}_k)r^k(\hat{t}_k) = \psi(\hat{t}_k)[f_x]^k V(\hat{t}_k)\dot{\hat{x}}(\hat{t}_0) = [H_x]^k V(\hat{t}_k)\dot{\hat{x}}(\hat{t}_0). \tag{7.115}$$

Formulas (7.113)–(7.115) and the condition $y^k(t) = 0$ for $t < \hat{t}_k$ then imply the equality

$$\begin{aligned} L_{t_k t_0} &= l_{x_f t_0} y^k(\hat{t}_f) - (y^k(\hat{t}_f))^* l_{x_f x_f} V(\hat{t}_f)\dot{\hat{x}}(\hat{t}_0) \\ &\quad + [H_x]^k V(\hat{t}_k)\dot{\hat{x}}(\hat{t}_0) - \int_{\hat{t}_0}^{\hat{t}_f} (y_k)^* H_{xx} V \dot{\hat{x}}(\hat{t}_0)\,dt. \end{aligned} \tag{7.116}$$

Hence,

$$\begin{aligned} L_{t_k t_0}\bar{t}_k\bar{t}_0 &= l_{x_f t_0} y^k(\hat{t}_f)\bar{t}_k\bar{t}_0 - (y^k(\hat{t}_f)\bar{t}_k)^* l_{x_f x_f} V(\hat{t}_f)\dot{\hat{x}}(\hat{t}_0)\bar{t}_0 \\ &\quad + [H_x]^k V(\hat{t}_k)\dot{\hat{x}}(\hat{t}_0)\bar{t}_0\bar{t}_k - \int_{\hat{t}_0}^{\hat{t}_f} (y^k\bar{t}_k)^* H_{xx} V \dot{\hat{x}}(\hat{t}_0)\bar{t}_0\,dt. \end{aligned} \tag{7.117}$$

7.4.10 Derivative $L_{t_f t_0}$

We have

$$\begin{aligned} \frac{\partial^2}{\partial t_f \partial t_0} l(t_0, x_0, t_f, x(t_f; t_0, x_0, \theta)) &= \frac{\partial}{\partial t_0}\{l_{t_f} + l_{x_f}\dot{x}\}|_{t=t_f} \\ &= \left\{ l_{t_f t_0} + l_{t_f x_f}\frac{\partial x}{\partial t_0} + \left(\frac{\partial}{\partial t_0} l_{x_f}\right)\dot{x} + l_{x_f}\frac{\partial}{\partial t_0}\dot{x} \right\}\bigg|_{t=t_f}. \end{aligned}$$

Using the equalities

$$\frac{\partial}{\partial t_0} l_{x_f} = l_{x_f t_0} + l_{x_f x_f}\frac{\partial x}{\partial t_0}, \quad \frac{\partial x}{\partial t_0} = -V\dot{x}(t_0),$$

we get

$$\begin{aligned} \frac{\partial^2}{\partial t_f \partial t_0} l(t_0, x_0, t_f, x(t_f; t_0, x_0, \theta)) &= l_{t_f t_0} - l_{t_f x_f} V\dot{x}(t_0) + l_{x_f t_0}\dot{x}(t_f) \\ &\quad - (V(t_f)\dot{x}(t_0))^* l_{x_f x_f}\dot{x}(t_f) + l_{x_f}\frac{\partial}{\partial t_0}\dot{x}|_{t=t_f}. \end{aligned} \tag{7.118}$$

Let us calculate the last term. Differentiating the equation

$$\dot{x}(t; t_0, x_0, \theta) = f(t, x(t; t_0, x_0, \theta), u(t; \theta))$$

with respect to t_0, we get

$$\frac{\partial}{\partial t_0}\dot{x} = f_x \frac{\partial}{\partial t_0} x = -f_x V\dot{x}(t_0).$$

Consequently, at the point $\zeta = \hat{\zeta}$, we obtain

$$l_{x_f}\frac{\partial}{\partial t_0}\dot{x}\bigg|_{t=\hat{t}_f} = \left\{\psi\frac{\partial}{\partial t_0}\dot{x}\right\}\bigg|_{t=\hat{t}_f} = \{-\psi f_x V\dot{\hat{x}}(\hat{t}_0)\}|_{t=\hat{t}_f} = \dot{\psi}(\hat{t}_f)V(\hat{t}_f)\dot{\hat{x}}(\hat{t}_0).$$

Using this equality in (7.118), we get at the point $\zeta = \hat{\zeta}$,

$$L_{t_f t_0} = l_{t_f t_0} - l_{t_f x_f} V(\hat{t}_f)\dot{\hat{x}}(\hat{t}_0) + l_{x_f t_0}\dot{\hat{x}}(\hat{t}_f) - \langle l_{x_f x_f} V(\hat{t}_f)\dot{\hat{x}}(\hat{t}_0), \dot{\hat{x}}(\hat{t}_f)\rangle + \dot{\psi}(\hat{t}_f)V(\hat{t}_f)\dot{\hat{x}}(\hat{t}_0), \tag{7.119}$$

which yields

$$L_{t_f t_0}\bar{t}_f\bar{t}_0 = l_{t_f t_0}\bar{t}_f\bar{t}_0 - l_{t_f x_f}(V(\hat{t}_f)\dot{\hat{x}}(\hat{t}_0)\bar{t}_0)\bar{t}_f + l_{x_f t_0}(\dot{\hat{x}}(\hat{t}_f)\bar{t}_f)\bar{t}_0 - \langle l_{x_f x_f} V(\hat{t}_f)\dot{\hat{x}}(\hat{t}_0)\bar{t}_0, \dot{\hat{x}}(\hat{t}_f)\bar{t}_f\rangle + \dot{\psi}(\hat{t}_f)(V(\hat{t}_f)\dot{\hat{x}}(\hat{t}_0)\bar{t}_0)\bar{t}_f. \tag{7.120}$$

7.4.11 Derivative $L_{t_0 t_0}$

We have

$$\begin{aligned}
\frac{\partial^2}{\partial t_0^2} l(t_0, x_0, t_f, x(t_f; t_0, x_0, \theta)) &= \frac{\partial}{\partial t_0}\left\{ l_{t_0} + l_{x_f}\frac{\partial x}{\partial t_0}\right\}\Big|_{t=t_f} \\
&= \left\{ l_{t_0 t_0} + l_{t_0 x_f}\frac{\partial x}{\partial t_0} + \left(l_{x_f t_0} + l_{x_f x_f}\frac{\partial x}{\partial t_0}\right)\frac{\partial x}{\partial t_0} + l_{x_f}\frac{\partial^2 x}{\partial t_0^2}\right\}\Big|_{t=t_f} \\
&= \left\{ l_{t_0 t_0} + 2 l_{t_0 x_f}\frac{\partial x}{\partial t_0} + \left\langle l_{x_f x_f}\frac{\partial x}{\partial t_0}, \frac{\partial x}{\partial t_0}\right\rangle + l_{x_f}\frac{\partial^2 x}{\partial t_0^2}\right\}\Big|_{t=t_f} \\
&= \left\{ l_{t_0 t_0} + 2 l_{t_0 x_f} w + \langle l_{x_f x_f} w, w\rangle + l_{x_f} q\right\}\big|_{t=t_f},
\end{aligned} \tag{7.121}$$

where

$$w = \frac{\partial x}{\partial t_0}, \quad q = \frac{\partial w}{\partial t_0} = \frac{\partial^2 x}{\partial t_0^2}.$$

The transversality condition $l_{x_f} = \psi(\hat{t}_f)$ yields

$$L_{t_0 t_0} = l_{t_0 t_0} + 2 l_{t_0 x_f} w(\hat{t}_f) + \langle l_{x_f x_f} w(\hat{t}_f), w(\hat{t}_f)\rangle + \psi(\hat{t}_f) q(\hat{t}_f). \tag{7.122}$$

Let us transform the last term using the adjoint equation for ψ and the system (7.58) for q:

$$\frac{d}{dt}(\psi q) = \dot{\psi} q + \psi \dot{q} = -\psi f_x q + \psi f_x q + \sum_j \psi_j (w^* f_{jxx} w) = w^* H_{xx} w.$$

Also, using the equality $w = -V\dot{\hat{x}}(\hat{t}_0)$, we obtain

$$\begin{aligned}
\psi(\hat{t}_f) q(\hat{t}_f) &= \psi(\hat{t}_0) q(\hat{t}_0) + \int_{\hat{t}_0}^{\hat{t}_f} w^* H_{xx} w \, dt \\
&= \psi(\hat{t}_0) q(\hat{t}_0) + \int_{\hat{t}_0}^{\hat{t}_f} \langle H_{xx} V\dot{\hat{x}}(\hat{t}_0), V\dot{\hat{x}}(\hat{t}_0)\rangle \, dt.
\end{aligned} \tag{7.123}$$

The initial condition (7.59) for q then implies

$$\psi(\hat{t}_0)q(\hat{t}_0) = -\psi(\hat{t}_0)\ddot{\hat{x}}(\hat{t}_0) - 2\psi(\hat{t}_0)\dot{w}(\hat{t}_0). \tag{7.124}$$

From the equation $\dot{w} = f_x w$ (see Proposition 7.12), the adjoint equation $-\dot{\psi} = \psi f_x$, and the formula $w = -V\dot{\hat{x}}(\hat{t}_0)$, it follows that

$$-\psi\dot{w} = -\psi f_x w = \dot{\psi} w = -\dot{\psi} V\dot{\hat{x}}(\hat{t}_0).$$

Since $V(\hat{t}_0) = E$, we obtain

$$\psi(\hat{t}_0)\dot{w}(\hat{t}_0) = \dot{\psi}(\hat{t}_0)\dot{\hat{x}}(\hat{t}_0). \tag{7.125}$$

Moreover, by formula (7.108) we have

$$-\psi\ddot{\hat{x}} = \dot{\psi}\dot{\hat{x}} + \dot{\psi}_0. \tag{7.126}$$

Formulas (7.124)–(7.126) imply

$$\psi(\hat{t}_0)q(\hat{t}_0) = \dot{\psi}_0(\hat{t}_0) - \dot{\psi}(\hat{t}_0)\dot{\hat{x}}(\hat{t}_0). \tag{7.127}$$

Combining formulas (7.122), (7.123), and (7.127), we obtain

$$\begin{aligned} L_{t_0 t_0} &= l_{t_0 t_0} - 2 l_{t_0 x_f} V(\hat{t}_f)\dot{\hat{x}}(\hat{t}_0) + \langle l_{x_f x_f} V(\hat{t}_f)\dot{\hat{x}}(\hat{t}_0), V(\hat{t}_f)\dot{\hat{x}}(\hat{t}_0)\rangle \\ &\quad + \dot{\psi}_0(\hat{t}_0) - \dot{\psi}(\hat{t}_0)\dot{\hat{x}}(\hat{t}_0) + \int_{\hat{t}_0}^{\hat{t}_f} \langle H_{xx} V(\hat{t}_f)\dot{\hat{x}}(\hat{t}_0), V(\hat{t}_f)\dot{\hat{x}}(\hat{t}_0)\rangle\, dt. \end{aligned} \tag{7.128}$$

Thus we have found the representation

$$\begin{aligned} L_{t_0 t_0}\bar{t}_0^2 &= l_{t_0 t_0}\bar{t}_0^2 - 2 l_{t_0 x_f} V(\hat{t}_f)\dot{\hat{x}}(\hat{t}_0)\bar{t}_0^2 + \langle l_{x_f x_f} V(\hat{t}_f)\dot{\hat{x}}(\hat{t}_0)\bar{t}_0, V(\hat{t}_f)\dot{\hat{x}}(\hat{t}_0)\bar{t}_0\rangle \\ &\quad + \dot{\psi}_0(\hat{t}_0)\bar{t}_0^2 - \dot{\psi}(\hat{t}_0)\dot{\hat{x}}(\hat{t}_0)\bar{t}_0^2 + \int_{\hat{t}_0}^{\hat{t}_f} \langle H_{xx} V(\hat{t}_f)\dot{\hat{x}}(\hat{t}_0)\bar{t}_0, V(\hat{t}_f)\dot{\hat{x}}(\hat{t}_0)\bar{t}_0\rangle\, dt. \end{aligned} \tag{7.129}$$

7.4.12 Representation of the Quadratic Form $\langle L_{\zeta\zeta}\bar{\zeta},\bar{\zeta}\rangle$

Combining all results and formulas in the preceding sections, we have proved the following theorem.

Theorem 7.27. *Let the Lagrange multipliers $\mu = (\alpha_0,\alpha,\beta) \in \Lambda_0$ and $\lambda = (\alpha_0,\alpha,\beta,\psi,\psi_0) \in \Lambda$ correspond to each other, i.e., let $\pi_0\lambda = \mu$ hold; see Proposition* 7.4. *Then, for any $\bar{\zeta} = (\bar{t}_0,\bar{t}_f,\bar{x}_0,\bar{\theta}) \in \mathbb{R}^{2+n+s}$, formulas* (7.75), (7.85), (7.89), (7.97), (7.98), (7.104)–(7.106), (7.109), (7.112), (7.117), (7.120), *and* (7.129) *hold, where the matrix $V(t)$ is the solution to the IVP* (7.18) *and the function y^k is the solution to the IVP* (7.26) *for each $k = 1,\ldots,s$.*

Thus we have obtained the following explicit and massive representation of the quadratic form in the IOP:

$$
\begin{aligned}
\langle L_{\zeta\zeta}\bar{\zeta},\bar{\zeta}\rangle &= \langle L_{\zeta\zeta}(\mu,\hat{\zeta})\bar{\zeta},\bar{\zeta}\rangle \qquad (7.130)\\
&= \langle L_{x_0x_0}\bar{x}_0,\bar{x}_0\rangle + 2\sum_{k=1}^{s} L_{x_0t_k}\bar{x}_0\bar{t}_k + \sum_{k=1}^{s} L_{t_kt_k}\bar{t}_k^2 + 2\sum_{k<j}^{s} L_{t_kt_j}\bar{t}_k\bar{t}_j\\
&\quad + 2L_{x_0t_f}\bar{x}_0\bar{t}_f + 2\sum_{k=1}^{s} L_{t_kt_f}\bar{t}_k\bar{t}_f + L_{t_ft_f}\bar{t}_f^2\\
&\quad + 2L_{x_0t_0}\bar{x}_0\bar{t}_0 + 2\sum_{k=1}^{s} L_{t_0t_k}\bar{t}_0\bar{t}_k + 2L_{t_0t_f}\bar{t}_0\bar{t}_f + L_{t_0t_0}\bar{t}_0^2\\
&= \bar{x}_0^* l_{x_0x_0}\bar{x}_0 + 2\bar{x}_0^* l_{x_0x_f} V(\hat{t}_f)\bar{x}_0 + (V(\hat{t}_f)\bar{x}_0)^* l_{x_fx_f} V(\hat{t}_f)\bar{x}_0\\
&\quad + \int_{\hat{t}_0}^{\hat{t}_f} (V\bar{x}_0)^* H_{xx} V\bar{x}_0\, dt\\
&\quad + \sum_{k=1}^{s} 2\bar{x}_0^* l_{x_0x_f} y^k(\hat{t}_f)\bar{t}_k + \sum_{k=1}^{s} 2(V(\hat{t}_f)\bar{x}_0)^* l_{x_fx_f} y^k(\hat{t}_f)\bar{t}_k\\
&\quad - \sum_{k=1}^{s} 2[H_x]^k V(\hat{t}_k)\bar{x}_0\bar{t}_k + \sum_{k=1}^{s}\int_{\hat{t}_0}^{\hat{t}_f} 2\langle H_{xx} y^k\bar{t}_k, V\bar{x}_0\rangle\, dt\\
&\quad + \sum_{k=1}^{s}\langle l_{x_fx_f} y^k(\hat{t}_f)\bar{t}_k, y^k(\hat{t}_f)\bar{t}_k\rangle + \sum_{k=1}^{s}\int_{\hat{t}_0}^{\hat{t}_f} (y^k\bar{t}_k)^* H_{xx} y^k\bar{t}_k\, dt\\
&\quad + \sum_{k=1}^{s} D^k(H)\bar{t}_k^2 - \sum_{k=1}^{s}[H_x]^k[y^k]^k\bar{t}_k^2 + \sum_{k<j} 2\langle l_{x_fx_f} y^k(\hat{t}_f)\bar{t}_k, y^j(\hat{t}_f)\bar{t}_j\rangle\\
&\quad + \sum_{k<j}\int_{\hat{t}_0}^{\hat{t}_f} 2(y^k\bar{t}_k)^* H_{xx} y^j\bar{t}_j\, dt - \sum_{k<j} 2[H_x]^j y^k(\hat{t}_j)\bar{t}_k\bar{t}_j\\
&\quad + 2l_{x_0t_f}\bar{x}_0\bar{t}_f + 2\langle l_{x_0x_f}\dot{\hat{x}}(\hat{t}_f)\bar{t}_f, \bar{x}_0\rangle + 2l_{x_ft_f} V(\hat{t}_f)\bar{x}_0\bar{t}_f\\
&\quad + 2\langle l_{x_fx_f}\dot{\hat{x}}(\hat{t}_f)\bar{t}_f, V(\hat{t}_f)\bar{x}_0\rangle - 2\dot{\psi}(\hat{t}_f)V(\hat{t}_f)\bar{x}_0\bar{t}_f\\
&\quad + \sum_{k=1}^{s} 2\langle l_{x_fx_f}\dot{\hat{x}}(\hat{t}_f)\bar{t}_f, y^k(\hat{t}_f)\bar{t}_k\rangle + \sum_{k=1}^{s} 2l_{x_ft_f} y^k(\hat{t}_f)\bar{t}_k\bar{t}_f\\
&\quad - \sum_{k=1}^{s} 2\dot{\psi}(\hat{t}_f) y^k(\hat{t}_f)\bar{t}_k\bar{t}_f\\
&\quad + l_{t_ft_f}\bar{t}_f^2 + 2l_{t_fx_f}\dot{\hat{x}}(\hat{t}_f)\bar{t}_f^2 + \langle l_{x_fx_f}\dot{\hat{x}}(\hat{t}_f)\bar{t}_f, \dot{\hat{x}}(\hat{t}_f)\bar{t}_f\rangle\\
&\quad - (\dot{\psi}(\hat{t}_f)\dot{\hat{x}}(\hat{t}_f) + \dot{\psi}_0(\hat{t}_f))\bar{t}_f^2\\
&\quad + 2l_{x_0t_0}\bar{x}_0\bar{t}_0 - 2\langle l_{x_0x_f} V(\hat{t}_f)\dot{\hat{x}}(\hat{t}_0)\bar{t}_0, \bar{x}_0\rangle + 2l_{x_ft_0} V(\hat{t}_f)\bar{x}_0\bar{t}_0
\end{aligned}
$$

$$
\begin{aligned}
&-2\langle l_{x_f x_f} V(\hat{t}_f)\bar{x}_0, V(\hat{t}_f)\dot{\hat{x}}(\hat{t}_0)\bar{t}_0\rangle + 2\dot{\psi}(\hat{t}_0)\bar{x}_0\bar{t}_0 \\
&-\int_{\hat{t}_0}^{\hat{t}_f} 2\langle H_{xx} V\bar{x}_0, V\dot{\hat{x}}(\hat{t}_0)\bar{t}_0\rangle\, dt \\
&+\sum_{k=1}^{s} 2l_{x_f t_0} y^k(\hat{t}_f)\bar{t}_k\bar{t}_0 - \sum_{k=1}^{s} 2(y^k(\hat{t}_f)\bar{t}_k)^* l_{x_f x_f} V(\hat{t}_f)\dot{\hat{x}}(\hat{t}_0)\bar{t}_0 \\
&+\sum_{k=1}^{s} 2[H_x]^k V(\hat{t}_k)\dot{\hat{x}}(\hat{t}_0)\bar{t}_0\bar{t}_k - \sum_{k=1}^{s}\int_{\hat{t}_0}^{\hat{t}_f} 2(y^k\bar{t}_k)^* H_{xx} V\dot{\hat{x}}(\hat{t}_0)\bar{t}_0\, dt \\
&+2l_{t_f t_0}\bar{t}_f\bar{t}_0 - 2l_{t_f x_f}(V(\hat{t}_f)\dot{\hat{x}}(\hat{t}_0)\bar{t}_0)\bar{t}_f + 2l_{x_f t_0}(\dot{\hat{x}}(\hat{t}_f)\bar{t}_f)\bar{t}_0 \\
&-2\langle l_{x_f x_f} V(\hat{t}_f)\dot{\hat{x}}(\hat{t}_0)\bar{t}_0, \dot{\hat{x}}(\hat{t}_f)\bar{t}_f\rangle + 2\dot{\psi}(\hat{t}_f)(V(\hat{t}_f)\dot{\hat{x}}(\hat{t}_0)\bar{t}_0)\bar{t}_f \\
&+l_{t_0 t_0}\bar{t}_0^2 - 2l_{t_0 x_f} V(\hat{t}_f)\dot{\hat{x}}(\hat{t}_0)\bar{t}_0^2 + \langle l_{x_f x_f} V(\hat{t}_f)\dot{\hat{x}}(\hat{t}_0)\bar{t}_0, V(\hat{t}_f)\dot{\hat{x}}(\hat{t}_0)\bar{t}_0\rangle \\
&+\dot{\psi}_0(\hat{t}_0)\bar{t}_0^2 - \dot{\psi}(\hat{t}_0)\dot{\hat{x}}(\hat{t}_0)\bar{t}_0^2 + \int_{\hat{t}_0}^{\hat{t}_f}\langle H_{xx} V\dot{\hat{x}}(\hat{t}_0)\bar{t}_0, V\dot{\hat{x}}(\hat{t}_0)\bar{t}_0\rangle\, dt.
\end{aligned}
$$

Again, we wish to emphasize that this explicit representation involves only *first-order* variations y^k and V of the trajectories $x(t;t_0,x_0,\theta)$.

7.5 Equivalence of the Quadratic Forms in the Basic and Induced Optimization Problem

In this section, we shall prove Theorem 7.8 which is the main result of this chapter. Let the Lagrange multipliers $\mu = (\alpha_0,\alpha,\beta) \in \Lambda_0$ and $\lambda = (\alpha_0,\alpha,\beta,\psi,\psi_0) \in \Lambda$ correspond to each other, and take any $\bar{\zeta} = (\bar{t}_0,\bar{t}_f,\bar{x}_0,\bar{\theta}) \in \mathbb{R}^{2+n+s}$. Consider the representation (7.130) of the quadratic form $\langle L_{\zeta\zeta}\bar{\zeta},\bar{\zeta}\rangle$, which is far from revealing the equivalence of the quadratic forms for the basic control problem and the IOP. However, we show now that by a careful regrouping of the terms in (7.130) we shall arrive at the desired equivalence. The quadratic form (7.130) contains terms of the following types.

Type (a): Positive terms with coefficients $D^k(H)$ multiplied by the variation of the switching time $\bar{t}_k$ squared,

$$a := \sum_{k=1}^{s} D^k(H)\bar{t}_k^2. \tag{7.131}$$

Type (b): Mixed terms with $[H_x]^k$ connected with the variation $\bar{t}_k$,

$$
\begin{aligned}
b := -\sum_{k=1}^{s} 2[H_x]^k V(\hat{t}_k)\bar{x}_0\bar{t}_k - \sum_{k=1}^{s}[H_x]^k[y^k]^k\bar{t}_k^2 \\
-\sum_{k<j} 2[H_x]^j y^k(\hat{t}_j)\bar{t}_k\bar{t}_j + \sum_{k=1}^{s} 2[H_x]^k V(\hat{t}_k)\dot{\hat{x}}(\hat{t}_0)\bar{t}_0\bar{t}_k.
\end{aligned} \tag{7.132}
$$

Since

$$\sum_{k<j}[H_x]^j y^k(\hat{t}_j)\bar{t}_k\bar{t}_j = \sum_{j<k}[H_x]^k y^j(\hat{t}_k)\bar{t}_k\bar{t}_j = \sum_{k=1}^{s}\sum_{j=1}^{k-1}[H_x]^k y^j(\hat{t}_k)\bar{t}_k\bar{t}_j,$$

we get from (7.132),

$$b = -\sum_{k=1}^{s} 2[H_x]^k\left(V(\hat{t}_k)\bar{x}_0 + \frac{1}{2}[y^k]^k\bar{t}_k + \sum_{j=1}^{k-1} y^j(\hat{t}_k)\bar{t}_j - V(\hat{t}_k)\dot{\hat{x}}(\hat{t}_0)\bar{t}_0\right)\bar{t}_k. \tag{7.133}$$

According to (7.17) put

$$\bar{x}(t) = V(t)\bar{x}_0 + \sum_{k=1}^{s} y^k(t)\bar{t}_k - V(t)\dot{\hat{x}}(\hat{t}_0)\bar{t}_0. \tag{7.134}$$

Then we have

$$\bar{x}(\hat{t}_k-) = V(\hat{t}_k)\bar{x}_0 + \sum_{j=1}^{k-1} y^j(\hat{t}_k)\bar{t}_j - V(\hat{t}_k)\dot{\hat{x}}(\hat{t}_0)\bar{t}_0,$$

since $y^j(\hat{t}_k-) = y^j(\hat{t}_k) = 0$ for $j > k$ and $y^k(\hat{t}_k-) = 0$. Moreover, the jump of $\bar{x}(t)$ at the point $\hat{t}_k$ is equal to the jump of $y^k(t)\bar{t}_k$ at the same point, i.e., $[\bar{x}]^k = [y^k]^k\bar{t}_k$. Therefore,

$$\begin{aligned} &V(\hat{t}_k)\bar{x}_0 + \frac{1}{2}[y^k]^k\bar{t}_k + \sum_{j=1}^{k-1} y^j(\hat{t}_k)\bar{t}_j - V(\hat{t}_k)\dot{\hat{x}}(\hat{t}_0)\bar{t}_0 \\ &= \bar{x}(\hat{t}_k-) + \frac{1}{2}[\bar{x}]^k = \frac{1}{2}(\bar{x}(\hat{t}_k-) + \bar{x}(\hat{t}_k+)) = \bar{x}^k_{\mathrm{av}}. \end{aligned}$$

Thus, we get

$$b = -\sum_{k=1}^{s} 2[H_x]^k\bar{x}^k_{\mathrm{av}}\bar{t}_k. \tag{7.135}$$

Type (c): Integral terms

$$\begin{aligned} c := &\int_{\hat{t}_0}^{\hat{t}_f} (V\bar{x}_0)^* H_{xx} V\bar{x}_0\,dt + \sum_{k=1}^{s}\int_{\hat{t}_0}^{\hat{t}_f} 2\langle H_{xx}y^k\bar{t}_k, V\bar{x}_0\rangle\,dt \\ &+ \sum_{k=1}^{s}\int_{\hat{t}_0}^{\hat{t}_f} (y^k\bar{t}_k)^* H_{xx}y^k\bar{t}_k\,dt + \sum_{k<j}\int_{\hat{t}_0}^{\hat{t}_f} 2(y^k\bar{t}_k)^* H_{xx}y^j\bar{t}_j\,dt \\ &- \int_{\hat{t}_0}^{\hat{t}_f} 2\langle H_{xx}V\bar{x}_0, V\dot{\hat{x}}(\hat{t}_0)\bar{t}_0\rangle\,dt - \sum_{k=1}^{s}\int_{\hat{t}_0}^{\hat{t}_f} 2(y^k\bar{t}_k)^* H_{xx}V\dot{\hat{x}}(\hat{t}_0)\bar{t}_0\,dt \\ &+ \int_{\hat{t}_0}^{\hat{t}_f} \langle H_{xx}V\dot{\hat{x}}(\hat{t}_0)\bar{t}_0, V\dot{\hat{x}}(\hat{t}_0)\bar{t}_0\rangle\,dt. \end{aligned} \tag{7.136}$$

Obviously, this sum can be transformed into a *perfect square*:

$$c = \int_{\hat{t}_0}^{\hat{t}_f} \left\langle H_{xx}(V\bar{x}_0 + \sum_{k=1}^{s} y^k \bar{t}_k - V\dot{\hat{x}}(\hat{t}_0)\bar{t}_0), V\bar{x}_0 + \sum_{k=1}^{s} y^k \bar{t}_k - V\dot{\hat{x}}(\hat{t}_0)\bar{t}_0 \right\rangle dt$$

$$= \int_{\hat{t}_0}^{\hat{t}_f} \langle H_{xx}\bar{x}, \bar{x}\rangle \, dt. \tag{7.137}$$

Type (d): Endpoints terms. We shall divide them into several groups.

Group (d1): This group contains the terms with second-order derivatives of the endpoint Lagrangian l with respect to t_0, x_0, and t_f:

$$d_1 := \bar{x}_0^* l_{x_0x_0}\bar{x}_0 + 2l_{x_0t_f}\bar{x}_0\bar{t}_f + l_{t_ft_f}\bar{t}_f^2 + 2l_{x_0t_0}\bar{x}_0\bar{t}_0 + 2l_{t_ft_0}\bar{t}_f\bar{t}_0 + l_{t_0t_0}\bar{t}_0^2. \tag{7.138}$$

Group (d2): We collect the terms with $l_{t_0x_f}$:

$$\begin{aligned} d_2 :=\; & 2l_{x_ft_0}V(\hat{t}_f)\bar{x}_0\bar{t}_0 + \sum_{k=1}^{s} 2l_{x_ft_0}y^k(\hat{t}_f)\bar{t}_k\bar{t}_0 \\ & + 2l_{x_ft_0}\dot{\hat{x}}(\hat{t}_f)\bar{t}_f\bar{t}_0 - 2l_{t_0x_f}V(\hat{t}_f)\dot{\hat{x}}(\hat{t}_0)\bar{t}_0^2 \\ =\; & 2l_{x_ft_0}\left(V(\hat{t}_f)\bar{x}_0 + \sum_{k=1}^{s} y^k(\hat{t}_f)\bar{t}_k + \dot{\hat{x}}(\hat{t}_f)\bar{t}_f - V(\hat{t}_f)\dot{\hat{x}}(\hat{t}_0)\bar{t}_0\right)\bar{t}_0, \\ =\; & 2l_{x_ft_0}\bar{\bar{x}}_f\bar{t}_0, \end{aligned} \tag{7.139}$$

where, in view of (7.34),

$$\bar{\bar{x}}_f := V(\hat{t}_f)\bar{x}_0 + \sum_{k=1}^{s} y^k(\hat{t}_f)\bar{t}_k + \dot{\hat{x}}(\hat{t}_f)\bar{t}_f - V(\hat{t}_f)\dot{\hat{x}}(\hat{t}_0)\bar{t}_0 = \bar{x}(\hat{t}_f) + \dot{\hat{x}}(\hat{t}_f)\bar{t}_f. \tag{7.140}$$

Group (d3): Consider the terms with $l_{x_0x_f}$:

$$\begin{aligned} d_3 :=\; & 2\bar{x}_0^* l_{x_0x_f}V(\hat{t}_f)\bar{x}_0 + \sum_{k=1}^{s} 2\bar{x}_0^* l_{x_0x_f}y^k(\hat{t}_f)\bar{t}_k \\ & + 2\langle l_{x_0x_f}\dot{\hat{x}}(\hat{t}_f)\bar{t}_f, \bar{x}_0\rangle - 2\langle l_{x_0x_f}V(\hat{t}_f)\dot{\hat{x}}(\hat{t}_0)\bar{t}_0, \bar{x}_0\rangle \\ =\; & 2\left\langle l_{x_0x_f}\left(V(\hat{t}_f)\bar{x}_0 + \sum_{k=1}^{s} y^k(\hat{t}_f)\bar{t}_k + \dot{\hat{x}}(\hat{t}_f)\bar{t}_f - V(\hat{t}_f)\dot{\hat{x}}(\hat{t}_0)\bar{t}_0\right), \bar{x}_0\right\rangle \\ =\; & 2\langle l_{x_0x_f}\bar{\bar{x}}_f, \bar{x}_0\rangle. \end{aligned} \tag{7.141}$$

Group (d4): This group contains all terms with $l_{t_fx_f}$:

$$\begin{aligned} d_4 :=\; & 2l_{x_ft_f}V(\hat{t}_f)\bar{x}_0\bar{t}_f + \sum_{k=1}^{s} 2l_{x_ft_f}y^k(\hat{t}_f)\bar{t}_k\bar{t}_f \\ & + 2l_{t_fx_f}\dot{\hat{x}}(\hat{t}_f)\bar{t}_f^2 - 2l_{t_fx_f}(V(\hat{t}_f)\dot{\hat{x}}(\hat{t}_0)\bar{t}_0)\bar{t}_f \\ =\; & 2l_{x_ft_f}\left(V(\hat{t}_f)\bar{x}_0 + \sum_{k=1}^{s} y^k(\hat{t}_f)\bar{t}_k + \dot{\hat{x}}(\hat{t}_f)\bar{t}_f - V(\hat{t}_f)\dot{\hat{x}}(\hat{t}_0)\bar{t}_0\right)\bar{t}_f \\ =\; & 2l_{x_ft_f}\bar{\bar{x}}_f\bar{t}_f. \end{aligned} \tag{7.142}$$

Group (d5): We collect all terms containing $l_{x_f x_f}$:

$$
\begin{aligned}
d_5 := {} & (V(\hat{t}_f)\bar{x}_0)^* l_{x_f x_f} V(\hat{t}_f)\bar{x}_0 + \sum_{k=1}^{s} 2(V(\hat{t}_f)\bar{x}_0)^* l_{x_f x_f} y^k(\hat{t}_f)\bar{t}_k \\
& + \sum_{k=1}^{s} \langle l_{x_f x_f} y^k(\hat{t}_f)\bar{t}_k, y^k(\hat{t}_f)\bar{t}_k \rangle + \sum_{k<j} 2\langle l_{x_f x_f} y^k(\hat{t}_f)\bar{t}_k, y^j(\hat{t}_f)\bar{t}_j \rangle \\
& + 2\langle l_{x_f x_f} \dot{\hat{x}}(\hat{t}_f)\bar{t}_f, V(\hat{t}_f)\bar{x}_0 \rangle + \sum_{k=1}^{s} 2\langle l_{x_f x_f} \dot{\hat{x}}(\hat{t}_f)\bar{t}_f, y^k(\hat{t}_f)\bar{t}_k \rangle \\
& + \langle l_{x_f x_f} \dot{\hat{x}}(\hat{t}_f)\bar{t}_f, \dot{\hat{x}}(\hat{t}_f)\bar{t}_f \rangle - 2\langle l_{x_f x_f} V(\hat{t}_f)\bar{x}_0, V(\hat{t}_f)\dot{\hat{x}}(\hat{t}_0)\bar{t}_0 \rangle \\
& - \sum_{k=1}^{s} 2(y^k(\hat{t}_f)\bar{t}_k)^* l_{x_f x_f} V(\hat{t}_f)\dot{\hat{x}}(\hat{t}_0)\bar{t}_0 - 2\langle l_{x_f x_f} V(\hat{t}_f)\dot{\hat{x}}(\hat{t}_0)\bar{t}_0, \dot{\hat{x}}(\hat{t}_f)\bar{t}_f \rangle \\
& + \langle l_{x_f x_f} V(\hat{t}_f)\dot{\hat{x}}(\hat{t}_0)\bar{t}_0, V(\hat{t}_f)\dot{\hat{x}}(\hat{t}_0)\bar{t}_0 \rangle.
\end{aligned} \tag{7.143}
$$

One can easily check that this sum can be transformed into the *perfect square*

$$
\begin{aligned}
d_5 := {} & \Big\langle l_{x_f x_f} \Big(V(\hat{t}_f)\bar{x}_0 + \sum_{k=1}^{s} y^k(\hat{t}_f)\bar{t}_k + \dot{\hat{x}}(\hat{t}_f)\bar{t}_f - V(\hat{t}_f)\dot{\hat{x}}(\hat{t}_0)\bar{t}_0 \Big), \\
& V(\hat{t}_f)\bar{x}_0 + \sum_{k=1}^{s} y^k(\hat{t}_f)\bar{t}_k + \dot{\hat{x}}(\hat{t}_f)\bar{t}_f - V(\hat{t}_f)\dot{\hat{x}}(\hat{t}_0)\bar{t}_0 \Big\rangle \\
= {} & \langle l_{x_f x_f} \bar{\bar{x}}_f, \bar{\bar{x}}_f \rangle.
\end{aligned} \tag{7.144}
$$

Group (d6): Terms with $\dot{\psi}(\hat{t}_0)$ and $\dot{\psi}_0(\hat{t}_0)$:

$$
d_6 := 2\dot{\psi}(\hat{t}_0)\bar{x}_0\bar{t}_0 + (\dot{\psi}_0(\hat{t}_0) - \dot{\psi}(\hat{t}_0)\dot{\hat{x}}(\hat{t}_0))\bar{t}_0^2. \tag{7.145}
$$

Group (d7): Terms with $\dot{\psi}(\hat{t}_f)$ and $\dot{\psi}_0(\hat{t}_f)$:

$$
\begin{aligned}
d_7 := {} & -2\dot{\psi}(\hat{t}_f)V(\hat{t}_f)\bar{x}_0\bar{t}_f - \sum_{k=1}^{s} 2\dot{\psi}(\hat{t}_f)y^k(\hat{t}_f)\bar{t}_k\bar{t}_f \\
& - (\dot{\psi}(\hat{t}_f)\dot{\hat{x}}(\hat{t}_f) + \dot{\psi}_0(\hat{t}_f))\bar{t}_f^2 + 2\dot{\psi}(\hat{t}_f)(V(\hat{t}_f)\dot{\hat{x}}(\hat{t}_0)\bar{t}_0)\bar{t}_f \\
= {} & -2\dot{\psi}(\hat{t}_f)\Big(V(\hat{t}_f)\bar{x}_0 + \sum_{k=1}^{s} y^k(\hat{t}_f)\bar{t}_k - V(\hat{t}_f)\dot{\hat{x}}(\hat{t}_0)\bar{t}_0 \Big)\bar{t}_f \\
& - (\dot{\psi}(\hat{t}_f)\dot{\hat{x}}(\hat{t}_f) + \dot{\psi}_0(\hat{t}_f))\bar{t}_f^2 \\
= {} & -2\dot{\psi}(\hat{t}_f)\bar{x}(\hat{t}_f)\bar{t}_f - (\dot{\psi}(\hat{t}_f)\dot{\hat{x}}(\hat{t}_f) + \dot{\psi}_0(\hat{t}_f))\bar{t}_f^2.
\end{aligned} \tag{7.146}
$$

Using the equality $\bar{\bar{x}}_f = \bar{x}(\hat{t}_f) + \dot{\hat{x}}(\hat{t}_f)\bar{t}_f$ in (7.146), we obtain

$$
d_7 = -2\dot{\psi}(\hat{t}_f)\bar{\bar{x}}_f\bar{t}_f - (\dot{\psi}_0(\hat{t}_f) - \dot{\psi}(\hat{t}_f)\dot{\hat{x}}(\hat{t}_f))\bar{t}_f^2. \tag{7.147}
$$

This completes the whole list of all terms in the quadratic form associated with the IOP. Hence, we have

$$\langle L_{\zeta\zeta}\bar{\zeta},\bar{\zeta}\rangle = a+b+c+d, \quad d=\sum_{i=1}^{7} d_i .$$

We thus have found the following representation of this quadratic form; see formulas (7.131) for a, (7.135) for b, and (7.137) for c:

$$\langle L_{\zeta\zeta}\bar{\zeta},\bar{\zeta}\rangle = \sum_{k=1}^{s} D^k(H)\bar{t}_k^2 - \sum_{k=1}^{s} 2[H_x]^k \bar{x}_{\mathrm{av}}^k \bar{t}_k + \int_{\hat{t}_0}^{\hat{t}_f} \langle H_{xx}\bar{x},\bar{x}\rangle + d, \tag{7.148}$$

where according to formulas (7.138), (7.139), (7.141), (7.142), (7.144), (7.145), (7.147) for $d_1,\ldots,d_7$, respectively,

$$\begin{aligned} d = {} & \langle l_{x_0x_0}\bar{x}_0,\bar{x}_0\rangle + 2l_{x_0t_f}\bar{x}_0\bar{t}_f + l_{t_ft_f}\bar{t}_f^2 + 2l_{x_0t_0}\bar{x}_0\bar{t}_0 + 2l_{t_ft_0}\bar{t}_f\bar{t}_0 + l_{t_0t_0}\bar{t}_0^2 + 2l_{x_ft_0}\bar{\bar{x}}_f\bar{t}_0 \\ & + 2\langle l_{x_0x_f}\bar{\bar{x}}_f,\bar{x}_0\rangle + 2l_{x_ft_f}\bar{\bar{x}}_f\bar{t}_f + \langle l_{x_fx_f}\bar{\bar{x}}_f,\bar{\bar{x}}_f\rangle + 2\dot{\psi}(\hat{t}_0)\bar{x}_0\bar{t}_0 + (\dot{\psi}_0(\hat{t}_0) - \dot{\psi}(\hat{t}_0)\dot{\hat{x}}(\hat{t}_0))\bar{t}_0^2 \\ & - 2\dot{\psi}(\hat{t}_f)\bar{\bar{x}}_f\bar{t}_f - (\dot{\psi}_0(\hat{t}_f) - \dot{\psi}(\hat{t}_f)\dot{\hat{x}}(\hat{t}_f))\bar{t}_f^2. \end{aligned} \tag{7.149}$$

In (7.148) and (7.149) the function $\bar{x}(t)$ and the vector $\bar{\bar{x}}_f$ are defined by (7.134) and (7.140), respectively. Note that in (7.149),

$$\begin{aligned} & \langle l_{x_0x_0}\bar{x}_0,\bar{x}_0\rangle + 2l_{x_0t_f}\bar{x}_0\bar{t}_f + l_{t_ft_f}\bar{t}_f^2 + 2l_{x_0t_0}\bar{x}_0\bar{t}_0 + 2l_{t_ft_0}\bar{t}_f\bar{t}_0 + l_{t_0t_0}\bar{t}_0^2 + 2l_{x_ft_0}\bar{\bar{x}}_f\bar{t}_0 \\ & \quad + 2\langle l_{x_0x_f}\bar{\bar{x}}_f,\bar{x}_0\rangle + 2l_{x_ft_f}\bar{\bar{x}}_f\bar{t}_f + \langle l_{x_fx_f}\bar{\bar{x}}_f,\bar{\bar{x}}_f\rangle = \langle l_{pp}\bar{\bar{p}},\bar{\bar{p}}\rangle, \end{aligned} \tag{7.150}$$

where, by definition,

$$\bar{\bar{p}} = (\bar{t}_0,\bar{x}_0,\bar{t}_f,\bar{\bar{x}}_f). \tag{7.151}$$

Finally, we get

$$\begin{aligned} d = {} & \langle l_{pp}\bar{\bar{p}},\bar{\bar{p}}\rangle + 2\dot{\psi}(\hat{t}_0)\bar{x}_0\bar{t}_0 + (\dot{\psi}_0(\hat{t}_0) - \dot{\psi}(\hat{t}_0)\dot{\hat{x}}(\hat{t}_0))\bar{t}_0^2 \\ & - 2\dot{\psi}(\hat{t}_f)\bar{\bar{x}}_1\bar{t}_f - (\dot{\psi}_0(\hat{t}_f) - \dot{\psi}(\hat{t}_f)\dot{\hat{x}}(\hat{t}_f))\bar{t}_f^2. \end{aligned} \tag{7.152}$$

Thus, we have proved the following result.

Theorem 7.28. *Let the Lagrange multipliers*

$$\mu = (\alpha_0,\alpha,\beta) \in \Lambda_0 \quad \text{and} \quad \lambda = (\alpha_0,\alpha,\beta,\psi,\psi_0) \in \Lambda$$

correspond to each other, i.e., let $\pi_0\lambda = \mu$ hold. Then for any $\bar{\zeta} = (\bar{t}_0,\bar{t}_f,\bar{x}_0,\bar{\theta}) \in \mathbb{R}^{2+n+s}$ the quadratic form $\langle L_{\zeta\zeta}\bar{\zeta},\bar{\zeta}\rangle$ has the representation (7.148)–(7.152), *where the vector function $\bar{x}(t)$ and the vector $\bar{\bar{x}}_f$ are defined by* (7.134) *and* (7.140). *The matrix-valued function $V(t)$ is the solution to the IVP* (7.18), *and, for each $k = 1,\ldots,s$, the vector function y^k is the solution to the IVP* (7.26).

Finally, we have arrived at the main result of this section.

Theorem 7.29. *Let* $\lambda = (\alpha_0, \alpha, \beta, \psi, \psi_0) \in \Lambda$ *and* $\bar{\zeta} = (\bar{t}_0, \bar{t}_f, \bar{x}_0, \bar{\theta}) \in \mathbb{R}^{2+n+s}$. *Put* $\mu = (\alpha_0, \alpha, \beta)$, *i.e., let* $\pi_0 \lambda = \mu \in \Lambda_0$ *hold; see Proposition 7.4. Define the function* $\bar{x}(t)$ *by formula* (7.134). *Put* $\bar{\xi} = -\bar{\theta}$ *and* $\bar{z} = (\bar{t}_0, \bar{t}_f, \bar{\xi}, \bar{x})$, *which means* $\pi_1 \bar{z} = \bar{\zeta}$; *see Propositions 7.6 and 7.7. Then the following equality holds:*

$$\langle L_{\zeta\zeta}(\mu, \hat{\zeta})\bar{\zeta}, \bar{\zeta} \rangle = \Omega(\lambda, \bar{z}), \tag{7.153}$$

where $\Omega(\lambda, \bar{z})$ *is defined by formulas* (6.34) *and* (6.35).

Proof. By Theorem 7.28, the equalities (7.148)–(7.152) hold. In view of equation (6.21), put

$$\bar{\bar{x}}_0 = \bar{x}(\hat{t}_0) + \bar{t}_0 \dot{\hat{x}}(\hat{t}_0) = \left(V(\hat{t}_0)\bar{x}_0 + \sum_{k=1}^{s} y^k(\hat{t}_0)\bar{t}_k - V(\hat{t}_0)\dot{\hat{x}}(\hat{t}_0)\bar{t}_0 \right) + \bar{t}_0 \dot{\hat{x}}(\hat{t}_0).$$

Since $y^k(\hat{t}_0) = 0$ for $k = 1, \ldots, s$ and $V(\hat{t}_0) = E$, it follows that $\bar{\bar{x}}_0 = \bar{x}_0$. Consequently, the vector $\bar{\bar{p}}$, which was defined by (6.21) as $(\bar{t}_0, \bar{\bar{x}}_0, \bar{t}_f, \bar{\bar{x}}_f)$, coincides with the vector $\bar{\bar{p}}$, defined in this section by (7.151). Hence, the endpoint quadratic form d in (7.152) and the endpoint quadratic form $\langle A\bar{\bar{p}}, \bar{\bar{p}} \rangle$ in (6.35) take equal values, $d = \langle A\bar{\bar{p}}, \bar{\bar{p}} \rangle$. Moreover, the integral terms $\int_{\hat{t}_0}^{\hat{t}_f} \langle H_{xx}\bar{x}, \bar{x} \rangle \, dt$ in the representation (7.148) of the form $\langle L_{\zeta\zeta}\bar{\zeta}, \bar{\zeta} \rangle$ and those in the representation (6.34) of the form Ω coincide, and

$$\sum_{k=1}^{s} \left(D^k(H)\bar{\xi}_k^2 + 2[H_x]^k \bar{x}_{\mathrm{av}}^k \bar{\xi}_k \right) = \sum_{k=1}^{s} D^k(H)\bar{t}_k^2 - \sum_{k=1}^{s} 2[H_x]^k \bar{x}_{\mathrm{av}}^k \bar{t}_k,$$

because $\bar{\xi}_k = -\bar{t}_k$, $k = 1, \ldots, s$. Thus, the representation (7.148) of the form $\langle L_{\zeta\zeta}\bar{\zeta}, \bar{\zeta} \rangle$ implies the equality (7.153) of both forms. ∎

Theorem 7.8, which is the main result of this chapter, then follows from Theorem 7.29.

Remark. Theorems 7.8 and 7.29 pave the way for the sensitivity analysis of parametric bang-bang control problems. Since the IOP is finite-dimensional, the SSC imply that we may take advantage of the well-known sensitivity results by Fiacco [32] to obtain solution differentiability for the IOP. The strict bang-bang property then ensures solution differentiability for the parametric bang-bang control problems; cf. Kim and Maurer [48] and Felgenhauer [31].

Chapter 8

Numerical Methods for Solving the Induced Optimization Problem and Applications

8.1 The Arc-Parametrization Method

8.1.1 Brief Survey on Numerical Methods for Optimal Control Problems

It has become customary to divide numerical methods for solving optimal control problems into two main classes: *indirect methods* and *direct methods*. *Indirect methods* take into account the full set of necessary optimality conditions of the Minimum Principle which gives rise to a *multipoint boundary value problem* (MPBVP) for state and adjoint variables. Shooting methods provide a powerful approach to solving an MPBVP; cf. articles from the German School [18, 19, 61, 64, 65, 81, 101, 102, 106] and French School [9, 10, 60]. E.g., the code BNDSCO by Oberle and Grimm [82] represents a very efficient implementation of shooting methods. Indirect methods produce highly accurate solutions but suffer from the drawback that the control structure, i.e., the sequence and number of bang-bang and singular arcs or constrained arcs, must be known a priori. Moreover, shooting methods need rather accurate initial estimates of the adjoint variables which in general are tedious to obtain.

In *direct methods*, the optimal control problem is transcribed into a nonlinear programming problem (NLP) by suitable discretization techniques; cf. Betts [5], Bock and Plitt [7], and Bueskens [13, 14]. Direct methods dispense with adjoint variables which are computed a posteriori from Lagrange multipliers of the NLP. These methods have shown to be very robust and are mostly capable of determining the correct control structure without assuming an a priori knowledge of the structure.

In all examples and applications in this chapter, optimal bang-bang and singular controls were computed in two steps. First, direct methods are applied to determine the correct optimal structure and obtain good estimates for the switching times. In a second step, the Induced Optimization Problem (IOP) in (7.4) of Section 7.1.1 is solved by using the arc-parametrization method developed in [44, 45, 66]. To prove optimality of the computed extremal, we then verify that the second-order sufficient conditions (SSC) in Theorem 7.10 hold.

The next section presents the arc-parametrization method by which the arc-lengths of bang-bang arcs are optimized instead of the switching times. Section 8.1.3 describes an extension of the arc-parametrization method to control problems, where the control is piecewise defined by feedback functions. Numerical examples for bang-bang controls are presented in Sections 8.2–8.6, while Section 8.7 exhibits a bang-singular control for a van der Pol oscillator with a regulator functional.

8.1.2 The Arc-Parametrization Method for Solving the Induced Optimization Problem

This section is based on the article of Maurer et al. [66] but uses a slightly different terminology. We consider the optimal control problem (6.1)–(6.3). To simplify the exposition we assume that the initial time t_0 is fixed and that the mixed boundary conditions are given as *equality constraints*. Hence, we study the following optimal control problem:

$$\text{Minimize} \qquad J(x(t_0),t_f,x(t_f)) \tag{8.1}$$

subject to the constraints

$$\dot{x}(t) = f(t,x(t),u(t)), \quad u(t) \in U, \quad t_0 \le t \le t_f, \tag{8.2}$$

$$K(x(t_0),t_f,x(t_f)) = 0. \tag{8.3}$$

The control variable appears linearly in the system dynamics,

$$f(t,x,u) = a(t,x) + B(t,x)u. \tag{8.4}$$

The control set $U \subset \mathbb{R}^{d(u)}$ is a convex polyhedron with $V = \operatorname{ex} U$ denoting the set of vertices.

Let us recall the IOP in (7.4). Assume that $\hat{u}(t)$ is a bang-bang control in $\hat{\Delta} = [t_0, \hat{t}_f]$ with switching points $\hat{t}_k \in (t_0, \hat{t}_f)$ and values u^k in the set $V = \operatorname{ex} U$,

$$\hat{u}(t) = u^k \in V \text{ for } t \in (\hat{t}_{k-1}, \hat{t}_k), \quad k = 1, \dots, s+1,$$

where $\hat{t}_0 = t_0$, $\hat{t}_{s+1} = \hat{t}_f$. Thus, $\hat{\Theta} = \{\hat{t}_1, \dots, \hat{t}_s\}$ is the set of switching points of the control $\hat{u}(\cdot)$ with $\hat{t}_k < \hat{t}_{k+1}$ for $k = 0, 1, \dots, s$. Put

$$\hat{x}(t_0) = \hat{x}_0 \in \mathbb{R}^n, \quad \hat{\theta} = (\hat{t}_1, \dots, \hat{t}_s) \in \mathbb{R}^s, \quad \hat{\zeta} = (\hat{x}_0, \hat{\theta}, \hat{t}_f) \in \mathbb{R}^n \times \mathbb{R}^{s+1}. \tag{8.5}$$

For convenience, the sequence of components of the vector $\hat{\zeta}$ in definition (7.1) has been modified. Take a small neighborhood $\mathcal{V}$ of the point $\hat{\zeta}$ and let

$$\zeta = (x_0, \theta, t_{s+1}) \in \mathcal{V}, \quad \theta = (t_1, \dots, t_s), \quad t_{s+1} = t_f,$$

where the switching times satisfy $t_0 < t_1 < t_2 < \cdots < t_s < t_{s+1} = t_f$. Define the function $u(t;\theta)$ by the condition

$$u(t;\theta) = u^k \text{ for } t \in (t_{k-1}, t_k), \quad k = 1, \dots, s+1. \tag{8.6}$$

The values $u(t_k;\theta)$, $k = 1, \dots, s$, may be chosen in U arbitrarily. For definiteness, define them by the condition of continuity of the control from the left, $u(t_k;\theta) = u(t_k-;\theta)$ for $k = 1, \dots, s$, and let $u(0;\theta) = u^1$. Denote by $x(t;x_0,\theta)$ the absolutely continuous solution of the Initial Value Problem (IVP)

$$\dot{x} = f(t,x,u(t;\theta)), \quad t \in [t_0, t_f], \quad x(t_0) = x_0. \tag{8.7}$$

For each $\zeta \in \mathcal{V}$ this solution exists in a sufficiently small neighborhood $\mathcal{V}$ of the point $\hat{\zeta}$. Obviously, we have

$$x(t;\hat{x}_0,\hat{\theta}) = \hat{x}(t), \quad t \in \hat{\Delta}, \quad u(t;\hat{\theta}) = \hat{u}(t), \quad t \in \hat{\Delta} \setminus \hat{\Theta}.$$

Consider now the following Induced Optimization Problem (IOP) in the space $\mathbb{R}^n \times \mathbb{R}^{s+1}$ of variables $\zeta = (x_0, \theta, t_{s+1})$:

$$\begin{aligned} \mathcal{F}_0(\zeta) &:= J(x_0, t_{s+1}, x(t_{s+1}; x_0, \theta)) \to \min, \\ \mathcal{G}(\zeta) &:= K(x_0, t_{s+1}, x(t_{s+1}; x_0, \theta)) = 0. \end{aligned} \tag{8.8}$$

To get uniqueness of the Lagrange multiplier for the equality constraint we assume that the following *regularity (normality) condition* holds:

$$\operatorname{rank}(\mathcal{G}_\zeta(\hat{\zeta})) = d(K). \tag{8.9}$$

The Lagrange function (7.6) in normalized form is given by

$$L(\beta, \zeta) = \mathcal{F}_0(\zeta) + \beta\, \mathcal{G}(\zeta), \quad \beta \in \mathbb{R}^{d(K)}. \tag{8.10}$$

The critical cone $\mathcal{K}_0$ in (7.13) then reduces to the subspace

$$\mathcal{K}_0 = \{\bar{\zeta} \in \mathbb{R}^n \times \mathbb{R}^{s+1} \mid \mathcal{G}_\zeta(\hat{\zeta})\bar{\zeta} = 0\}. \tag{8.11}$$

Assuming regularity (8.9), the SSC for the IOP reduce to the condition that there exist $\hat{\beta} \in \mathbb{R}^{d(K)}$ with

$$L_\zeta(\hat{\beta}, \hat{\zeta}) = (\mathcal{F}_0)_\zeta(\hat{\zeta}) + \hat{\beta}\, \mathcal{G}_\zeta(\hat{\zeta}) = 0, \tag{8.12}$$

$$\langle L_{\zeta\zeta}(\hat{\beta}, \hat{\zeta})\bar{\zeta}, \bar{\zeta}\rangle > 0 \quad \forall\, \bar{\zeta} \in \mathcal{K}_0 \setminus \{0\}. \tag{8.13}$$

From a numerical point of view it is not convenient to optimize the switching times t_k ($k = 1, \dots, s$) and terminal time $t_{s+1} = t_f$ directly. Instead, as suggested in [44, 45, 66] one computes the *arc durations* or *arc lengths*

$$\tau_k := t_k - t_{k-1}, \quad k = 1, \dots, s, s+1, \tag{8.14}$$

of the bang-bang arcs. Hence, the final time t_f can be expressed by the arc lengths as

$$t_f = t_0 + \sum_{k=1}^{s+1} \tau_k. \tag{8.15}$$

Next, we replace the optimization variable $\zeta = (x_0, t_1, \dots, t_s, t_{s+1})$ by the optimization variable

$$z := (x_0, \tau_1, \dots, \tau_s, \tau_{s+1}) \in \mathbb{R}^n \times \mathbb{R}^{s+1}, \quad \tau_k := t_k - t_{k-1}. \tag{8.16}$$

The variables ζ and z are related by the following linear transformation involving the regular $(n+s+1) \times (n+s+1)$ matrix R:

$$\begin{gathered} z = R\zeta, \quad R = \begin{pmatrix} I_n & \mathbf{0} \\ \mathbf{0} & S \end{pmatrix}, \qquad \zeta = R^{-1} z, \quad R^{-1} = \begin{pmatrix} I_n & \mathbf{0} \\ \mathbf{0} & S^{-1} \end{pmatrix}, \\ S = \begin{pmatrix} 1 & 0 & \dots & 0 \\ -1 & 1 & \ddots & \vdots \\ & \ddots & \ddots & 0 \\ 0 & & -1 & 1 \end{pmatrix}, \qquad S^{-1} = \begin{pmatrix} 1 & 0 & \dots & 0 \\ 1 & 1 & \ddots & \vdots \\ \vdots & \ddots & \ddots & 0 \\ 1 & \dots & 1 & 1 \end{pmatrix}. \end{gathered} \tag{8.17}$$

Denoting the solution to the equations (8.7) by $x(t;z)$, the IOP (8.8) obviously is equivalent to the following IOP with t_f defined by (8.15):

$$\begin{aligned} \mathcal{F}_0(z) &:= J(x_0, t_f, x(t_f; x_0, \tau_1, \ldots, \tau_s)) \to \min, \\ \mathcal{G}(z) &:= K(x_0, t_f, x(t_f; x_0, \tau_1, \ldots, \tau_s)) = 0. \end{aligned} \tag{8.18}$$

This approach is called the *arc-parametrization method.* The Lagrangian for this problem is given in normal form by

$$L(\rho, z) = \mathcal{F}_0(z) + \rho\, \mathcal{G}(z). \tag{8.19}$$

It is easy to see that the Lagrange multiplier ρ agrees with the multiplier β in the Lagrangian (8.10). Furthermore, the SSC for the optimization problems (8.8), respectively, (8.18), are equivalent. This immediately follows from the fact that the Jacobian and the Hessian for both optimization problems are related through

$$\mathcal{K}_\zeta = \mathcal{K}_z R, \quad L_\zeta = L_z R, \quad L_{\zeta\zeta} = R^* L_{zz} R.$$

Thus we can express the positive definiteness condition (8.13) evaluated for the variable z as

$$\langle L_{zz}(\hat{\beta}, \hat{z})\bar{z}, \bar{z} \rangle > 0 \quad \forall\, \bar{z} \in (\mathbb{R}^n \times \mathbb{R}^{s+1}) \setminus \{0\},\ \mathcal{G}_{\hat{z}}(\hat{z})\bar{z} = 0. \tag{8.20}$$

This condition is equivalent to the property that the so-called *reduced Hessian* is positive definite. Let N be the $(n_z \times (n_z - d(K)))$ matrix, $n_z = n + s + 1$, with full column rank $n_z - d(K)$, whose columns span the kernel of $\mathcal{G}_z(\hat{z})$. Then condition (8.20) is reformulated as

$$N^* L_{zz}(\hat{\beta}, \hat{z}) N > 0 \quad \text{(positive definite)}. \tag{8.21}$$

The computational method for determining the optimal vector $\hat{z} \in \mathbb{R}^{n+s+1}$ is based on a *multiprocess approach* proposed in [44, 45, 66]. The time interval $[t_{k-1}, t_k]$ is mapped to the fixed interval $I_k = [\frac{k-1}{s+1}, \frac{k}{s+1}]$ by the linear transformation

$$t = a_k + b_k r, \quad a_k = t_k - k\tau_k, \quad b_k = (s+1)\tau_k, \quad r \in I_k = \left[\frac{k-1}{s+1}, \frac{k}{s+1}\right], \tag{8.22}$$

where r denotes the running time. Identifying $x(r) \cong x(a_k + b_k \cdot r) = x(t)$ in the relevant intervals, we obtain the transformed dynamic system

$$\frac{dx}{dr} = \frac{dx}{dt} \cdot \frac{dt}{dr} = (s+1)\tau_k\, f(a_k + b_k r, x(r), u^k) \quad \text{for} \quad r \in I_k. \tag{8.23}$$

By way of concatenation of the solutions on the intervals I_k, we obtain an absolutely continuous solution $x(r) = x(r; \tau_1, \ldots, \tau_s)$ for $r \in [0,1]$. Thus we are confronted with the task of solving the IOP

$$\begin{aligned} \mathcal{F}_0(z) &:= J(x_0, t_f, x(1; x_0, \tau_1, \ldots, \tau_s)) \to \min, \quad \left(t_f = t_0 + \sum_{k=1}^{s+1} \tau_k \right), \\ \mathcal{G}(z) &:= K(x_0, t_f, x(1; x_0, \tau_1, \ldots, \tau_s)) = 0. \end{aligned} \tag{8.24}$$

This approach can be conveniently implemented using the routine NUDOCCCS developed by Büskens [13]. In this way, we can also take advantage of the fact that NUDOCCCS provides the Jacobian of the equality constraints and the Hessian of the Lagrangian, which are needed in the check of the second-order condition (8.13), respectively, the positive definiteness of the reduced Hessian (8.21). Moreover, this code allows for the computation of *parametric sensitivity derivatives*; cf. [48, 74, 23].

8.1.3 Extension of the Arc-Parametrization Method to Piecewise Feedback Control

The purpose of this section is to extend the IOPs (8.18) and (8.24) to the situation in which the control is piecewise defined by *feedback functions* and not only by a constant vector $u^k \in V = \operatorname{ex} U$. We consider the control problem (8.1)–(8.3) with an arbitrary dynamical system

$$\dot{x}(t) = f(t, x(t), u(t)), \quad t_0 \le t \le t_f, \tag{8.25}$$

and fixed initial time t_0. Instead of considering a bang-bang control $u(t)$ with s switching times $t_0 = t_0 < t_1 < \cdots < t_s < t_{s+1} = t_f$ and constant values

$$u(t) = u^k \quad \text{for} \quad t \in (t_{k-1}, t_k),$$

we assume that there exist continuous functions $u^k : D \to \mathbb{R}^{d(u)}$, where $D \subset \mathbb{R} \times \mathbb{R}^n$ is open, with

$$u(t) = u^k(t, x(t)) \quad \text{for} \quad t \in (t_{k-1}, t_k). \tag{8.26}$$

Such functions are provided, e.g., by singular controls in feedback form or boundary controls in the presence of state constraints, or may be simply viewed as suitable *feedback approximations of an optimal control.*

The vector of switching times is denoted by $\theta = (t_1, \ldots, t_s)$. Let $x(t; x_0, \theta)$ be the absolutely continuous solution of the piecewise defined equations

$$\dot{x}(t) = f(t, x(t), u^k(t, x(t))) \quad \text{for} \quad t \in (t_{k-1}, t_k) \quad (k = 1, \ldots, s+1) \tag{8.27}$$

with given initial value $x(t_0; x_0, \theta) = x_0$. The *IOP* with the optimization variable

$$\zeta = (x_0, \theta, t_f) = (x_0, t_1, \ldots, t_s, t_{s+1}) \in \mathbb{R}^n \times \mathbb{R}^{s+1}$$

agrees with (8.8):

$$\begin{aligned} \mathcal{F}_0(\zeta) \ &:= J(x_0, t_{s+1}, x(t_{s+1}; x_0, \theta)) \to \min, \\ \mathcal{G}(\zeta) \ &:= K(x_0, t_{s+1}, x(t_{s+1}; x_0, \theta)) = 0. \end{aligned} \tag{8.28}$$

The *arc-parametrization method* consists of optimizing x_0 and the arc lengths

$$\tau_k := t_k - t_{k-1}, \quad k = 1, \ldots, s+1.$$

Invoking the linear time transformation (8.22) for mapping the time interval $[t_{k-1}, t_k]$ to the fixed interval $I_k = [\frac{k-1}{s+1}, \frac{k}{s+1}]$,

$$t = a_k + b_k r, \quad a_k = t_k - k\tau_k, \quad b_k = (s+1)\tau_k, \quad r \in I_k,$$

the dynamic system is piecewise defined by

$$dx/dr = (s+1)\tau_i \, f(a_k + b_k r, x(r), u^k(a_k + b_k r, x(r)) \quad \text{for} \quad r \in I_k. \tag{8.29}$$

Therefore, the implementation of the arc-parametrization method using the routine NUDOCCCS [13] requires only a minor modification of the dynamic system. Applications of this method to bang-singular controls may be found in [50, 51] and to state constrained problems in [74]. An example for a bang-singular control will be given in Section 8.7.

8.2 Time-Optimal Control of the Rayleigh Equation Revisited

We revisit the problem of time-optimal control of the Rayleigh equation from Section 6.4 and show that the sufficient conditions in Theorem 7.10 can be easily verified numerically on the basis of the IOP (8.8) or (8.18). The control problem is to minimize the final time t_f subject to

$$\dot{x}_1(t) = x_2(t), \quad \dot{x}_2(t) = -x_1(t) + x_2(t)(1.4 - 0.14x_2(t)^2) + u(t), \tag{8.30}$$

$$x_1(0) = x_2(0) = -5, \quad x_1(t_f) = x_2(t_f) = 0, \tag{8.31}$$

$$|u(t)| \le 4 \quad \text{for } t \in [0, t_f]. \tag{8.32}$$

The Pontryagin function (Hamiltonian)

$$H(x, \psi, u) = \psi_1 x_2 + \psi_2(-x_1 + x_2(1.4 - 0.14x_2^2) + u) \tag{8.33}$$

yields the adjoint equations

$$\dot{\psi}_1 = \psi_2, \quad \dot{\psi}_2 = \psi_1 + \psi_2(1.4 - 0.42x_2^2). \tag{8.34}$$

The transversality conditions (6.9) gives the relation

$$\psi_2(t_f)u(t_f) + 1 = 0. \tag{8.35}$$

The switching function

$$\phi(t) = H_u(t) = \psi_2(t) \tag{8.36}$$

determines the optimal control via the minimum condition as

$$u(t) = \left\{ \begin{array}{rl} 4 & \text{if} \quad \psi_2(t) < 0 \\ -4 & \text{if} \quad \psi_2(t) > 0 \end{array} \right\}. \tag{8.37}$$

In section 6.4 it was found that the optimal control is composed of three bang-bang arcs,

$$u(t) = \left\{ \begin{array}{rll} 4 & \text{for} & 0 \le t < t_1 \\ -4 & \text{for} & t_1 \le t < t_2 \\ 4 & \text{for} & t_2 \le t \le t_3 = t_f \end{array} \right\}. \tag{8.38}$$

This implies the two switching conditions $\psi_2(t_1) = 0$ and $\psi_2(t_2) = 0$. Hence, the optimization vector for the IOP (8.18) is given by

$$z = (\tau_1, \tau_2, \tau_3), \quad \tau_1 = t_1, \quad \tau_2 = t_2 - t_1, \quad \tau_3 = t_f - t_2.$$

The code NUDOCCCS gives the following numerical results for the arc lengths, switching times, and adjoint variables:

$$\begin{array}{rclrclrcl} \tau_1 = t_1 & = & 1.12051, & \tau_2 & = & 2.18954, & t_2 & = & 3.31005, \\ \tau_3 & = & 0.35813, & t_f & = & 3.66817, & & & \\ \psi_1(0) & = & -0.122341, & \psi_2(0) & = & -0.082652, & & & \\ \psi_1(t_1) & = & -0.215212, & \psi_1(t_2) & = & 0.891992, & & & \\ \psi_1(t_f) & = & 0.842762, & \psi_2(t_f) & = & -0.25, & \beta & = & \psi(t_f). \end{array} \tag{8.39}$$

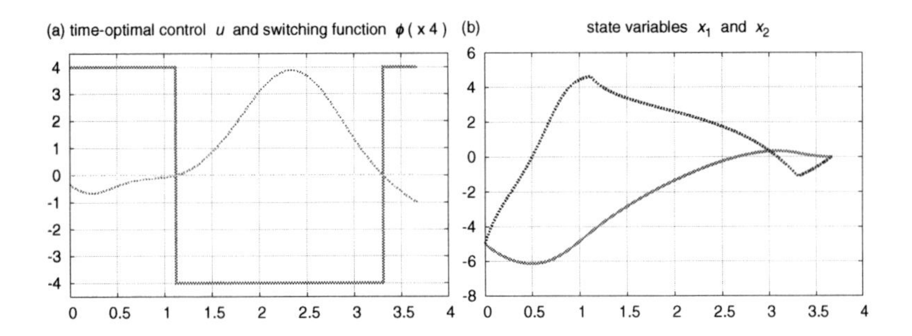

Figure 8.1. *Time-optimal control of the Rayleigh equation with boundary conditions* (8.31). (a) *Bang-bang control and scaled switching function* $\Phi(\times 4)$, (b) *State variables* x_1, *and* x_2.

The corresponding time-optimal bang-bang control with two switches and the state variables are shown in Figure 8.1.

We have already shown in Section 6.4 that the control u in (8.38) enjoys the strict bang-bang property and that the estimates $D^k(H) > 0$, $k = 1, 2$, in (6.114) are satisfied. For the terminal conditions (8.31), the Jacobian is the (2×3) matrix

$$\mathcal{G}_z(\hat{z}) = \begin{pmatrix} -4.53176 & -3.44715 & 0.0 \\ -11.2768 & -7.62049 & 4.0 \end{pmatrix}$$

which is of rank 2. The Hessian of the Lagrangian is the (3×3) matrix

$$L_{zz}(\hat{\beta}, \hat{z}) = \begin{pmatrix} -10.3713 & -8.35359 & -6.68969 \\ -8.35359 & -5.75137 & -4.61687 \\ -6.68969 & -4.61687 & 1.97104 \end{pmatrix}.$$

Note that this Hessian is *not* positive definite. However, the projected Hessian (8.21) is the positive number

$$N^* \tilde{L}_{zz}(\hat{z}, \hat{\beta}) N = 0.515518,$$

which shows that the second-order test (8.20) holds. Hence, the extremal characterized by the data (8.39) provides a strict strong minimum.

Now we consider a modified control problem, where the two terminal conditions $x_1(t_f) = x_2(t_f) = 0$ are substituted by the scalar terminal condition

$$x_1(t_f)^2 + x_2(t_f)^2 = 0.25. \tag{8.40}$$

The Hamiltonian (8.33) and the adjoint equations (8.34) remain the same. The transversality condition (6.9) yields

$$\lambda_i(t_f) = 2\beta x_i(t_f) \quad (i = 1, 2), \quad \beta \in \mathbb{R}. \tag{8.41}$$

The transversality condition for the free final time is $H(t) + 1 \equiv 0$. It turns out that the control is bang-bang with only one switching point t_1 in contrast to the control structure (8.38),

$$u(t) = \begin{Bmatrix} 4 & \text{for} & 0 \le t < t_1 \\ -4 & \text{for} & t_1 \le t \le t_f \end{Bmatrix}. \tag{8.42}$$

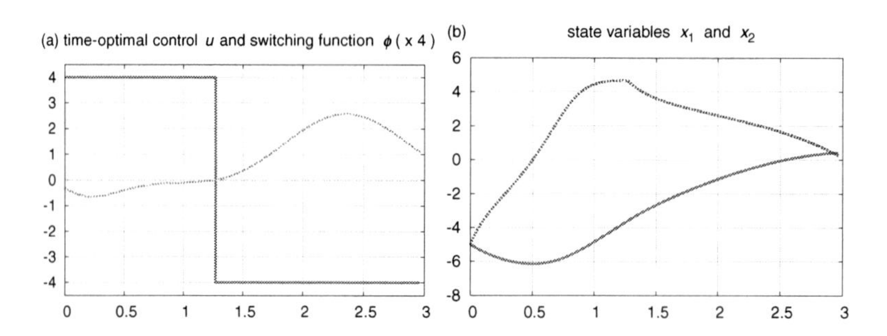

Figure 8.2. *Time-optimal control of the Rayleigh equation with boundary condition* (8.40). (a) *Bang-bang control u and scaled switching function φ (dashed line).* (b) *State variables* x_1, x_2.

Hence, the optimization vector is

$$z = (\tau_1, \tau_2), \quad \tau_1 = t_1, \quad \tau_2 = t_f - t_1.$$

Using the code NUDOCCCS, we obtain the following numerical results for the arc lengths and adjoint variables:

$$\begin{array}{rclrclrcl} t_1 & = & 1.27149, & \tau_2 & = & 1.69227, & t_f & = & 2.96377, \\ \psi_1(0) & = & -0.117316, & \psi_2(0) & = & -0.0813638, & & & \\ \psi_1(t_1) & = & -0.213831, & \psi_2(t_1) & = & 0.0, & & & \\ x_1(t_f) & = & 0.426176, & x_2(t_f) & = & 0.261484, & & & \\ \psi_1(t_f) & = & 0.448201, & \psi_2(t_f) & = & 0.274997, & \beta & = & 0.525839. \end{array} \tag{8.43}$$

Figure 8.2 displays the time-optimal bang-bang control with only one switch and the two state variables.

The relations $\dot{\phi}(t_1) = \dot{\psi}_2(t_1) = \psi_1(t_1)$ and $[u]^1 = 8$ yield

$$D^1(H) = -8 \cdot \dot{\phi}(t_1) = 8 \cdot 0.213831 > 0.$$

For the scalar terminal condition (8.40), the Jacobian is the nonzero row vector

$$\mathcal{G}_z(\hat{z}) = (-1.90175, \ -1.90173),$$

while the Hessian of the Lagrangian is the positive definite (2×2) matrix

$$L_{zz}(\hat{\beta}, \hat{z}) = \begin{pmatrix} 28.1299 & 19.0384 \\ 19.0384 & 14.0048 \end{pmatrix}.$$

Hence, the second-order conditions (8.20), respectively, the second-order conditions in Theorem 7.10, hold, which shows that the extremal characterized by the data (8.43) furnishes a strict strong minimum.

8.3 Time-Optimal Control of a Two-Link Robot

The control of two-link robots has been the subject of various articles; cf., e.g., [21, 35, 37, 81]. In these papers, optimal control policies are determined solely on the basis of first-order

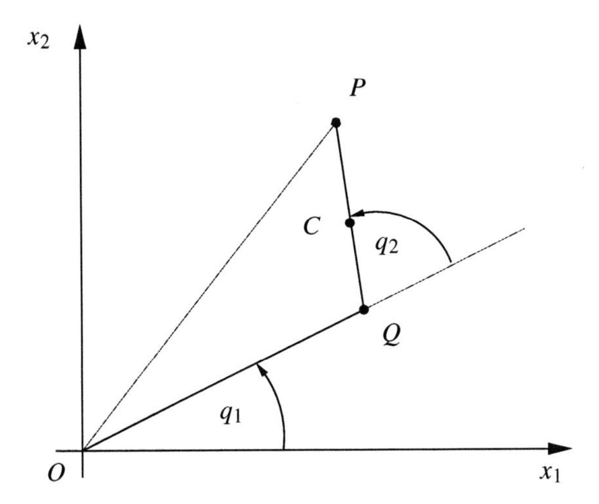

Figure 8.3. *Two-link robot* [67]*: upper arm $\overline{OQ}$, lower arm $\overline{OP}$, and angles q_1 and q_2.*

necessary conditions, since sufficient conditions were not available. In this section, we show that SSC hold for both types of robots considered in [21, 37, 81].

First, we study the robot model considered in Chernousko et al. [21]. Göllmann [37] has shown that the optimal control candidate presented in [21] is not optimal, since the sign conditions of the switching functions do not comply with the Minimum Principle. Figure 8.3 represents the two-link robot schematically. The state variables are the angles q_1 and q_2. The parameters I_1 and I_2 are the moments of inertia of the upper arm $\overline{OQ}$ and the lower arm $\overline{QP}$ with respect to the points O and Q. Further, let m_2 be the mass of the lower arm, $L_1 = |\overline{OQ}|$ the length of the upper arm, and $L_1 = |\overline{QC}|$ the distance between the second link Q and the center of gravity C of the lower arm. With the abbreviations

$$\begin{aligned} A &= I_1 + m_2 L_1^2 + I_2 + 2 m_2 L_1 L \cos q_2, & B &= I_2 + m_2 L_1 L \cos q_2, \\ R_1 &= u_1 + m_2 L_1 L (2\dot{q}_1 + \dot{q}_2)\dot{q}_2 \sin q_2, & R_2 &= u_2 - m_2 L_1 L \dot{q}_1^2 \sin q_2, \\ D &= I_2, & \Delta &= AD - B^2, \end{aligned} \tag{8.44}$$

the dynamics of the two-link robot can be described by the ODE system

$$\begin{aligned} \dot{q}_1 &= \omega_1, & \dot{\omega}_1 &= \frac{1}{\Delta}(DR_1 - BR_2), \\ \dot{q}_2 &= \omega_2, & \dot{\omega}_2 &= \frac{1}{\Delta}(AR_2 - BR_1), \end{aligned} \tag{8.45}$$

where ω_1 and ω_2 are the angular velocities. The torques u_1 and u_2 in the two links represent the two control variables. The control problem consists of steering the robot from a given initial position to a terminal position in minimal final time t_f,

$$\begin{aligned} q_1(0) &= 0, & q_2(0) &= 0, & \omega_1(0) &= 0, & \omega_2(0) &= 0, \\ q_1(t_f) &= -0.44, & q_2(t_f) &= 1.83, & \omega_1(t_f) &= 0, & \omega_2(t_f) &= 0. \end{aligned} \tag{8.46}$$

Both control components are bounded by

$$|u_1(t)| \le 2, \qquad |u_2(t)| \le 1, \qquad t \in [0, t_f]. \tag{8.47}$$

The Pontryagin function (Hamiltonian) is

$$H = \psi_1\omega_1 + \psi_2\omega_2 + \frac{\psi_3}{\Delta}(DR_1(u_1) - BR_2(u_2)) + \frac{\psi_4}{\Delta}(AR_2(u_2) - BR_1(u_1)). \quad (8.48)$$

The adjoint equations are rather complicated and are not given here explicitly. The switching functions are

$$\phi_1(t) = H_{u_1}(t) = \frac{\psi_3}{\Delta}D - \frac{\psi_4}{\Delta}B, \quad \phi_2(t) = H_{u_2}(t) = \frac{\psi_4}{\Delta}A - \frac{\psi_3}{\Delta}B. \quad (8.49)$$

For the parameter values

$$L_1 = 1, \quad L = 0.5, \quad m_2 = 10, \quad I_1 = I_2 = \frac{10}{3},$$

Göllmann [37] has found the following control structure with four bang-bang arcs:

$$u(t) = (u_1(t), u_2(t)) = \left\{ \begin{array}{cc} (-2,1), & 0 \le t < t_1 \\ (2,1), & t_1 \le t < t_2 \\ (2,-1), & t_2 \le t < t_3 \\ (-2,-1), & t_3 \le t \le t_f \end{array} \right\}, \quad 0 < t_1 < t_2 < t_3 < t_f. \quad (8.50)$$

This control structure differs substantially from that in Chernousko et al. [21] which violates the switching conditions. Obviously, the bang-bang control (8.50) satisfies the assumption that only one control component switches at a time. Since the initial point $(q_1(0), q_2(0), \omega_1(0), \omega_2(0))$ is specified, the optimization variable in the IOP (8.18) is

$$z = (\tau_1, \tau_2, \tau_3, \tau_4), \quad \tau_1 = t_1, \ \tau_2 = t_2 - t_1, \quad \tau_3 = t_3 - t_2, \quad \tau_4 = t_f - t_3.$$

Using the code NUDOCCCS, we compute the following arc durations and switching times:

$$\begin{array}{lll} t_1 = 0.7677893, & \tau_2 = 0.3358820, & t_2 = 1.1036713, \\ \tau_3 = 1.2626739, & t_3 = 2.3663452, & \tau_4 = 0.8307667, \\ t_f = 3.1971119. & & \end{array} \quad (8.51)$$

Numerical values for the adjoint functions are also provided by the code NUDOCCCS, e.g., the initial values

$$\begin{array}{lll} \psi_1(0) = -1.56972, & \psi_2(0) = -0.917955, \\ \psi_3(0) = -2.90537, & \psi_4(0) = -1.45440. \end{array} \quad (8.52)$$

Figure 8.4 shows that the switching functions ϕ_1 and ϕ_2 comply with the minimum condition and that the strict bang-bang property (6.19) and the inequalities (6.14) are satisfied:

$$\begin{array}{l} \phi_1(t) \ne 0 \quad \text{for} \quad t \ne t_1, t_3, \quad \phi_2(t) \ne 0 \quad \text{for} \quad t \ne t_2, \\ \dot{\phi}_1(t_1) < 0, \quad \dot{\phi}_1(t_3) > 0, \quad \dot{\phi}_2(t_2) > 0. \end{array}$$

For the terminal conditions (8.46) we compute the Jacobian

$$\mathcal{G}_z(\hat{z}) = \begin{pmatrix} -0.751043 & 0.0351060 & 0.258904 & 0 \\ 3.76119 & 1.84929 & -0.204170 & 0 \\ -0.326347 & 0.0770047 & 0.212722 & -0.107819 \\ 1.26849 & 0.445447 & -0.487447 & -0.233634 \end{pmatrix}.$$

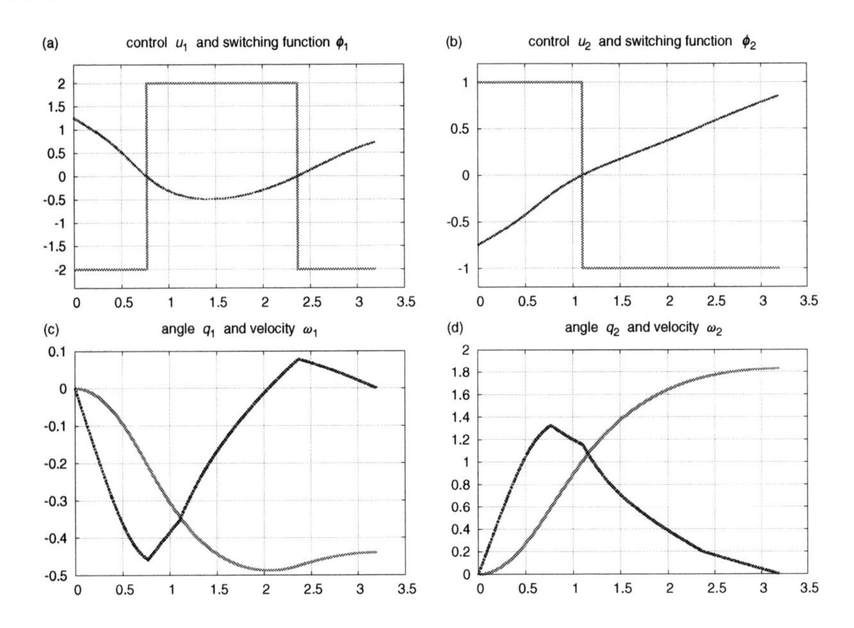

Figure 8.4. *Control of the two-link robot* (8.44)–(8.47). (a) *Control* u_1 *and scaled switching function* ϕ_1 *(dashed line).* (b) *Control* u_2 *and scaled switching function* ϕ_2 *(dashed line).* (c) *Angle* q_1 *and velocity* ω_1. (d) *Angle* q_2 *and velocity* ω_2.

This square matrix has full-rank in view of

$$\det \mathcal{G}_z(\hat{z}) = 0.0766524 \neq 0,$$

which means that the positive definiteness condition (8.13) trivially holds. We have thus verified *first-order* sufficient conditions showing that the extremal solution given by (8.50)–(8.52) provides a strict strong minimum.

In the model treated above, some parameters such as the mass of the upper arm and the mass of a load at the end of the lower arm appear implicitly in the system equations. The mass m_1 of the upper arm is included in the moment of inertia I_2, and the mass M of a load in the point P can be added to the mass m_2, where the point C and therefore the length L have to be adjusted. The length L_2 of the lower arm is incorporated in the parameter L.

The *second robot model* that we are going to discuss is taken from Geering et al. [35] and Oberle [81]. Here, every physical parameter enters the system equation explicitly. The dynamic system is as follows:

$$\begin{aligned} \dot{q}_1 &= \omega_1, & \dot{\omega}_1 &= \frac{1}{\Delta}(AI_{22} - BI_{12}\cos q_2), \\ \dot{q}_2 &= \omega_2 - \omega_1, & \dot{\omega}_2 &= \frac{1}{\Delta}(BI_{11} - AI_{12}\cos q_2), \end{aligned} \tag{8.53}$$

where we have used the abbreviations

$$\begin{aligned} A &= I_{12}\omega_2^2 \sin q_2 + u_1 - u_2, & B &= -I_{12}\omega_1^2 \sin q_2 + u_2, \\ \Delta &= I_{11}I_{22} - I_{12}^2\cos^2 q_2, & I_{11} &= I_1 + (m_2 + M)L_1^2, \\ I_{12} &= m_2 L L_1 + M L_1 L_2, & I_{22} &= I_2 + I_3 + M L_2^2. \end{aligned} \tag{8.54}$$

I_3 denotes the moment of inertia of the load with respect to the point P, and ω_2 is now the angular velocity of the angle q_1+q_2. For simplicity, we set $I_3=0$. Again, the torques u_1 and u_2 in the two links are used as control variables by which the robot is steered from a given initial position to a nonfixed end position in minimal final time t_f,

$$\begin{array}{rclrcl} q_1(0) &=& 0, & \sqrt{(x_1(t_f)-x_1(0))^2+(x_2(t_f)-x_2(0))^2} &=& r,\\ q_2(0) &=& 0, & q_2(t_f) &=& 0,\\ \omega_1(0) &=& 0, & \omega_1(t_f) &=& 0,\\ \omega_2(0) &=& 0, & \omega_2(t_f) &=& 0, \end{array} \tag{8.55}$$

where $(x_1(t),x_2(t))$ are the Cartesian coordinates of the point P,

$$\begin{array}{rcl} x_1(t) &=& L_1\cos q_1(t)+L_2\cos(q_1(t)+q_2(t)),\\ x_2(t) &=& L_1\sin q_1(t)+L_2\sin(q_1(t)+q_2(t)). \end{array} \tag{8.56}$$

The initial point $(x_1(0),x_2(0))=(2,0)$ is fixed. Both control components are bounded,

$$|u_1(t)|\le 1, \qquad |u_2(t)|\le 1, \qquad t\in[0,t_f]. \tag{8.57}$$

The Hamilton–Pontryagin function is given by

$$\begin{array}{rcl} H &=& \psi_1\omega_1+\psi_2(\omega_2-\omega_1)+\dfrac{\psi_3}{\Delta}(A(u_1,u_2)I_{22}-B(u_2)I_{12}\cos q_2)\\ && +\dfrac{\psi_4}{\Delta}(B(u_2)I_{11}-A(u_1,u_2)I_{12}\cos q_2). \end{array} \tag{8.58}$$

The switching functions are computed as

$$\begin{array}{l} \phi_1 = H_{u_1} = \dfrac{1}{\Delta}(\psi_3 I_{22}-\psi_4 I_{12}\cos q_2),\\[2ex] \phi_2 = H_{u_2} = \dfrac{1}{\Delta}(\psi_3(-I_{22}-I_{12}\cos q_2)+\psi_4(I_{11}+I_{12}\cos q_2)). \end{array} \tag{8.59}$$

For the parameter values

$$L_1=L_2=1, \quad L=0.5, \quad m_1=m_2=M=1, \quad I_1=I_2=\frac{1}{3}, \quad r=3,$$

we will show that the optimal control has the following structure with five bang-bang arcs:

$$u(t)=(u_1(t),u_2(t))=\left\{\begin{array}{rl} (-1,1) & \text{for} \quad 0\le t<t_1\\ (-1,-1) & \text{for} \quad t_1\le t<t_2\\ (1,-1) & \text{for} \quad t_2\le t<t_3\\ (1,1) & \text{for} \quad t_3\le t<t_4\\ (-1,1) & \text{for} \quad t_4\le t\le t_f \end{array}\right\}, \tag{8.60}$$

where $0=t_0<t_1<t_2<t_3<t_4<t_5=t_f$. Since the initial point $(q_1(0),q_2(0),\omega_1(0),\omega_2(0))$ is specified, the optimization variable in the optimization problem (8.8), respectively, (8.18), is

$$z=(\tau_1,\tau_2,\tau_3,\tau_4,\tau_5), \quad \tau_k=t_k-t_{k-1},\ k=1,\dots,5.$$

The code NUDOCCCS yields the arc durations and switching times

$$\begin{array}{rclrclrcl} t_1 &=& 0.546174, & \tau_2 &=& 1.21351, & t_2 &=& 1.75968,\\ \tau_3 &=& 1.03867, & t_3 &=& 2.79835, & \tau_4 &=& 0.906039,\\ t_4 &=& 3.70439, & \tau_5 &=& 0.185023, & t_f &=& 3.889409, \end{array} \tag{8.61}$$

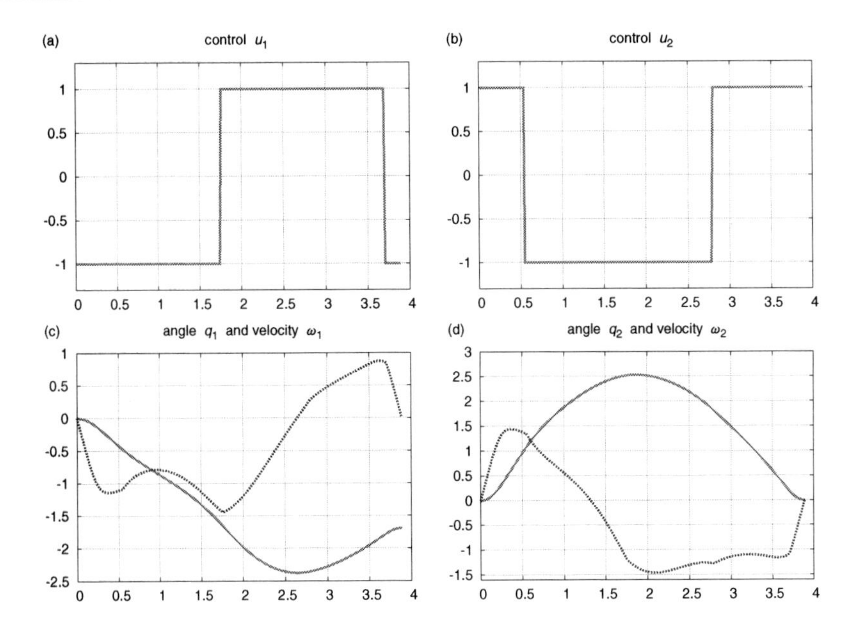

Figure 8.5. *Control of the two-link robot* (8.53)–(8.57). (a) *Control* u_1. (b) *Control* u_2. (c) *Angle* q_1 *and velocity* ω_1. (d) *Angle* q_2 *and velocity* ω_2 [17].

as well as the initial values of the adjoint variables,

$$\begin{array}{rclrcl} \psi_1(0) &=& 0.184172, & \psi_2(0) &=& -0.011125, \\ \psi_3(0) &=& 1.482636, & \psi_4(0) &=& 0.997367. \end{array} \tag{8.62}$$

The two bang-bang control components as well as the four state variables are shown in Figure 8.5. The strict bang-bang property (6.19) and the inequalities (6.14) hold in view of

$$\begin{array}{ll} \phi_1(t) \neq 0 \quad \text{for} \quad t \neq t_2, t_4, & \phi_2(t) \neq 0 \quad \text{for} \quad t \neq t_1, t_3, \\ \dot{\phi}_1(t_2) < 0, \quad \dot{\phi}_1(t_4) > 0, & \dot{\phi}_2(t_1) > 0, \quad \dot{\phi}_2(t_3) < 0. \end{array}$$

For the terminal conditions in (8.55), the Jacobian in the optimization problem is computed as the (4×5) matrix

$$\mathcal{G}_z(\hat{z}) = \begin{pmatrix} -10.8575 & -12.7462 & -5.88332 & -1.14995 & 0 \\ 0.199280 & -2.71051 & -1.45055 & -1.91476 & -4.83871 \\ -0.622556 & 3.31422 & 2.31545 & 2.94349 & 6.19355 \\ 9.36085 & 3.03934 & 0.484459 & 0.0405811 & 0 \end{pmatrix},$$

which has full rank. The Hessian of the Lagrangian is given by

$$L_{zz}(\hat{\beta}, \hat{z}) = \begin{pmatrix} 71.1424 & 90.7613 & 42.1301 & 8.49889 & -0.0518216 \\ 90.7613 & 112.544 & 51.3129 & 10.7691 & 0.149854 \\ 42.1301 & 51.3129 & 23.9633 & 5.12403 & 0.138604 \\ 8.49889 & 10.7691 & 5.12403 & 1.49988 & 0.170781 \\ -0.0518216 & 0.149854 & 0.138604 & 0.170781 & 0.297359 \end{pmatrix}.$$

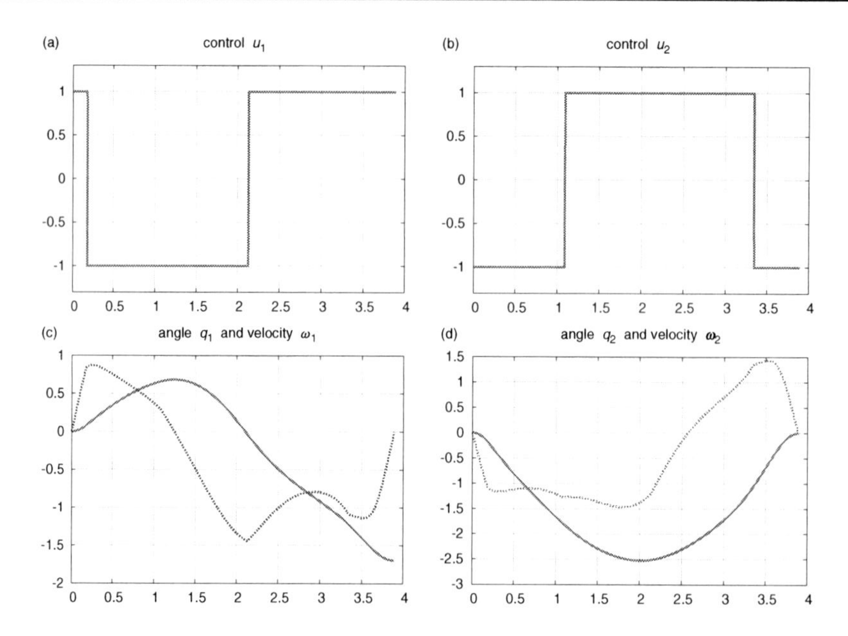

Figure 8.6. *Control of the two-link robot* (8.53)–(8.57): *Second solution.* (a) *Control u_1.* (b) *Control u_2.* (c) *Angle q_1 and velocity ω_1.* (d) *Angle q_2 and velocity ω_2.*

This yields the projected Hessian (8.21) as the positive number

$$N^* L_{zz}(\hat{\beta},\hat{z})N = 0.326929.$$

Hence, all conditions in Theorem 7.10 are satisfied, and thus the extremal (8.60)–(8.62) yields a strict strong minimum.

It is interesting to note that there exists a *second local minimum* with the same terminal time $t_f = 3.88941$. Though the control has also five bang-bang arcs, the control structure is substantially different from that in (8.60),

$$u(t) = (u_1(t), u_2(t)) = \left\{ \begin{array}{cc} (1,-1), & 0 \le t < t_1 \\ (-1,-1), & t_1 \le t < t_2 \\ (-1,1), & t_2 \le t < t_3 \\ (1,1), & t_3 \le t < t_4 \\ (1,-1), & t_4 \le t \le t_f \end{array} \right\}, \tag{8.63}$$

where $0 < t_1 < t_2 < t_3 < t_4 < t_5 = t_f$. Figure 8.6 displays the second time-optimal solution: the bang-bang controls and state variables. The code NUDOCCCS determines the switching times

$$\begin{array}{lll} t_1 = 0.1850163, & t_2 = 1.091075, & t_3 = 2.129721, \\ t_4 = 3.343237, & t_f = 3.889409. & \end{array} \tag{8.64}$$

for which the strict bang-bang property (6.19) and the inequalities (6.14) hold, i.e., $D^k(H) > 0$ for $k = 1,2,3,4$. Moreover, computations show that $\operatorname{rank}(G_z(\hat{z})) = 4$ and the projected Hessian of the Lagrangian (8.21) is the positive number

$$N^* L_{zz}(\hat{\beta},\hat{z})N = 0.326929.$$

It is remarkable that this value is identical to the value of the projected Hessian for the first local minimum. Therefore, also for the second solution we have verified that all conditions in Theorem 7.10 hold, and thus the extremal (8.63), (8.64) is a strict strong minimum. The phenomenon of multiple local solutions all with the same minimal time t_f has also been observed by Betts [5, Example 6.8 (Reorientation of a rigid body)].

8.4 Time-Optimal Control of a Single Mode Semiconductor Laser

In [46] we studied the optimal control for two classes of laser, a so-called class B laser and a semiconductor laser. In this section, we present only the semiconductor laser, whose dynamical model has been derived in [20, 29]. The primary goal is to control the transition between the initial stationary state, the switch-off state, and the terminal stationary state, the switch-on state. The control variable is the electric current injected into the laser by which the laser output power is modified.

In response to an abrupt switch from a low value of the current (initial state) to a high value (terminal state), the system responds with damped oscillations (cf. Figure 8.8) which are a great nuisance in several laser applications. We will show that by injecting an appropriate bang-bang current (control) the oscillations can be completely removed while simultaneously shortening the transition time. Semiconductor lasers are a case where the removal of the oscillations can be particularly beneficial, since in telecommunications one would like to be able to obtain the fastest possible response to the driving, with the cleanest and most direct transition, to maximize the data transmission rate and efficiency.

We consider the dynamics of a standard single-mode laser model [20], reduced to single-mode form [29]. In this model, $S(t)$ represents the normalized photon number and $N(t)$ the carrier density, $I(t)$ is the injected current (control) that is used to steer the transition between different laser power levels.

$$\begin{aligned} \dot{S} &= -\frac{S}{t_p} + \Gamma G(N,S)S + \beta BN(N+P_0), \\ \dot{N} &= \frac{I(t)}{q} - R(N) - \Gamma G(N,S)S. \end{aligned} \tag{8.65}$$

The process is considered in a time interval $t \in [0, t_f]$ with terminal time $t_f > 0$. The gain function $G(N,S)$ and recombination function $R(N)$ are given by

$$\begin{aligned} G(N,S) &= G_p(N - N_{tr})(1 - \epsilon S), \\ R(N) &= AN + BN(N+P_0) + CN(N+P_0)^2. \end{aligned} \tag{8.66}$$

The parameters have the following meaning: t_p, cavity lifetime of the photon; Γ, cavity confinement factor; β, coefficient that weights the (average) amount of spontaneous emission coupled into the lasing mode; B, incoherent band-to-band recombination coefficient; P_0, carrier number without injection; q, carrier charge; G_p, gain term; ϵ, gain compression factor; N_{tr}, number of carriers at transparency. All parameter values are given in Table 8.1. The following bounds are imposed for the injected current:

$$I_{\min} \le I(t) \le I_{\max} \qquad \forall\, t \in [0, t_f], \tag{8.67}$$

Table 8.1. *List of parameters from* [29]. *The time unit is a picosecond* [ps] = [10^{-12} s].

t_p	2.072×10^{-12} s	G_p	2.628×10^4 s^{-1}	Γ	0.3
ϵ	9.6×10^{-8}	N_{tr}	7.8×10^7	β	1.735×10^{-4}
P_0	1.5×10^7	q	1.60219×10^{-19} C	A	1×10^8 s^{-1}
B	2.788 s^{-1}	C	7.3×10^{-9} s^{-1}	I_0	20.5 mA
$I_{\min}$	2.0 mA	$I_{\max}$	67.5 mA	I_∞	42.5 mA

where $0 \le I_{\min} < I_{\max}$. To define appropriate initial and terminal values for $S(t)$ and $N(t)$, we choose two values I_0 and I_∞ with $I_{\min} < I_0 < I_\infty < I_{\max}$. Then inserting the *constant* control functions $I(t) \equiv I_0$ and $I(t) \equiv I_\infty$ into the dynamics (8.65), one can show that there exist two asymptotically stable stationary points (S_0, N_0) and (S_f, N_f) with $\dot{S} = \dot{N} = 0$ such that

$$\begin{array}{llll} (S(t),N(t)) \to (S_0,N_0) & \text{for} \quad t \to \infty & \text{and} & I(t) \equiv I_0, \\ (S(t),N(t)) \to (S_f,N_f) & \text{for} \quad t \to \infty & \text{and} & I(t) \equiv I_\infty. \end{array}$$

Hence, we shall impose the following initial and terminal conditions for the control process (8.65):

$$S(0) = S_0,\ N(0) = N_0 \quad \text{and} \quad S(t_f) = S_f,\ N(t_f) = N_f. \tag{8.68}$$

When controlling the process by the function $I(t)$, one goal is to determine a control function by which the terminal stationary point is reached in a *finite time* $t_f > 0$. But we can set a higher goal by considering the following time-optimal control problem:

$$\text{Minimize the final time } t_f \tag{8.69}$$

subject to the dynamic constraints and boundary conditions (8.65)–(8.68). For computation, we shall use the nominal parameters from [29] (see Table 8.1).

For these parameters, the stationary points, respectively, initial and terminal values, in (8.68) are computed in normalized units as

$$\begin{array}{ll} S_0 = 0.6119512914 \times 10^5, & N_0 = 1.3955581328 \times 10^8, \\ S_f = 3.4063069073 \times 10^5, & N_f = 1.4128116637 \times 10^8. \end{array} \tag{8.70}$$

The Hamilton–Pontryagin function is given by

$$\begin{aligned} H(S,N,\psi_S,\psi_N,I) &= \psi_S\left(-\frac{S}{t_p} + \Gamma G(N,S)S + \beta BN(N+P_0)\right) \\ &\quad + \psi_N\left(\frac{I}{q} - R(N) - \Gamma G(N,S)S\right), \end{aligned} \tag{8.71}$$

where the adjoint variables (ψ_S, ψ_N) satisfy the adjoint equations

$$\begin{aligned} \dot{\psi}_S = -H_S &= \psi_S\left(\frac{1}{t_p} - \Gamma G(N,S) + \Gamma G_p(N-N_{tr})\epsilon S\right) \\ &\quad + \psi_N\left(\Gamma G(N,S) - \Gamma G_p(N-N_{tr})\epsilon S\right), \\ \dot{\psi}_N = -H_N &= -\psi_S(G_p(1-\epsilon S)S + \beta B(2N+P_0)) + \psi_N(A + B(2N+P_0) \\ &\quad + C(3N^2 + 4NP_0 + P_0^2) + \Gamma G_p(1-\epsilon S)S). \end{aligned} \tag{8.72}$$

The switching function becomes

$$\phi(t) = H_I(t) = \psi_N(t)/q. \tag{8.73}$$

The minimization of the Hamiltonian (8.71) yields the following characterization of *bang-bang controls*:

$$I(t) = \begin{cases} I_{\min} & \text{if} \quad \psi_N(t) > 0, \\ I_{\max} & \text{if} \quad \psi_N(t) < 0. \end{cases} \tag{8.74}$$

In this problem, a singular control satisfying the condition $\psi_N(t) \equiv 0$ for all $t \in [t_1, t_2] \subset [0, t_f]$, $t_1 < t_2$, cannot be excluded a priori. However, the direct optimization approach yields the following bang-bang control with only one switching time t_1,

$$I(t) = \begin{cases} I_{\max} & \text{if} \quad 0 \le t < t_1, \\ I_{\min} & \text{if} \quad t_1 \le t \le t_f. \end{cases} \tag{8.75}$$

In view of the control law (8.74), we get the switching condition $\psi_N(t_1) = 0$. Moreover, since the final time t_f is free and the control problem is autonomous, we obtain the additional boundary condition for a normal trajectory,

$$H(S(t_f), N(t_f), \psi_S(t_f), \psi_N(t_f), I(t_f)) + 1 = 0. \tag{8.76}$$

The optimization variable in the IOP (8.18) is

$$z = (\tau_1, \tau_2), \quad \tau_1 = t_1, \quad \tau_2 = t_f - t_1. \tag{8.77}$$

It is noteworthy that the IOP (8.18) reduces to solving an implicitly defined nonlinear equation: determine two variables τ_1, τ_2 such that the two boundary conditions $S(t_f) = S_f$ and $N(t_f) = N_f$ in (8.70) are satisfied. Thus solving the IOP is equivalent to applying a Newton-type method to the system of equations. We obtain the following switching time, terminal time, and initial values of adjoint variables:

$$\begin{array}{rclrcl} t_1 & = & 29.52274, & t_f & = & 56.89444 \text{ ps}, \\ \psi_S(0) & = & -21.6227, & \psi_N(0) & = & -340.892, \\ \psi_S(t_f) & = & -4.6956, & \psi_N(t_f) & = & 395.60. \end{array} \tag{8.78}$$

The corresponding control and (normalized) state functions as well as adjoint variables are shown in Figure 8.7. Note that the constant control $I(t) \equiv I_\infty$ has to be applied for $t \ge t_f$ in order to fix the system at the terminal stationary point (S_f, N_f). Since the bang-bang control $I(t)$ has only one switch, Proposition 6.25 asserts that the computed extremal furnishes a strict strong minimum. The computed trajectory is normal, because the adjoint variables satisfy the necessary condition (6.7)–(6.11) with $\alpha_0 = 1$. Moreover, the graph of $\psi_N(t)$ in Figure 8.7 shows that the strict bang-bang property and $D^1(H) > 0$ in (6.14) hold in view of

$$\phi(t) < 0 \quad \forall\, 0 \le t < t_1, \quad \phi(t) > 0 \quad \forall\, t_1 < t \le t_f, \quad \dot{\phi}(t_1) > 0.$$

These conditions provide first-order sufficient conditions. Alternatively, we can use Theorem 7.10 for proving optimality. The critical cone is the zero element, since the computed 2×2 Jacobian matrix

$$g_z(\hat{z}) = \begin{pmatrix} 0.199855 & -0.000155599 \\ 0.0 & -0.00252779 \end{pmatrix}$$

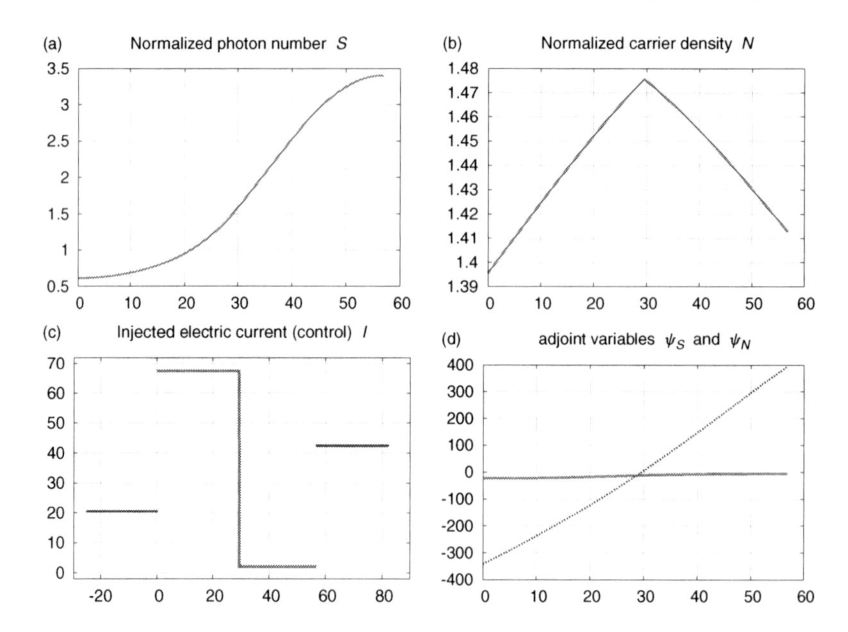

Figure 8.7. *Time-optimal control of a semiconductor laser.* (a) *Normalized photon density* $S(t) \times 10^{-5}$. (b) *Normalized photon density* $N(t) \times 10^{-8}$. (c) *Electric current (control)* $I(t)$ *with* $I(t) = I_0 = 20.5$ *for* $t < 0$ *and* $I(t) = I_\infty = 42.5$ *for* $t > t_f$. (d) *Adjoint variables* $\psi_S(t)$, $\psi_N(t)$.

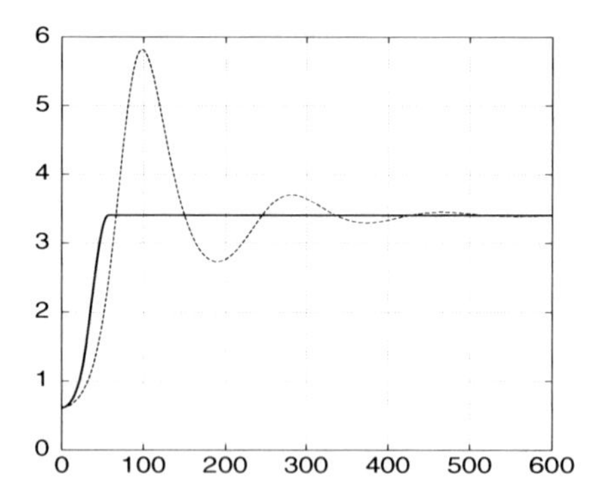

Figure 8.8. *Normalized photon number* $S(t)$ *for* $I(t) \equiv 42.5$ *mA and optimal* $I(t)$ [46].

is regular. Comparing the optimal control approach with the topological phase-space technique proposed in [29], we recognize that the control structure (8.75) constitutes a translation of the latter technique into rigorous mathematical terms.

The comparison in Figure 8.8 between the uncontrolled and optimally controlled laser shows the strength of the optimal control approach: the damped oscillations have been completely eliminated and a substantial shortening of the transient time has been achieved. Indeed, it is surprising that such a dramatic improvement is caused by the simple control strategy (8.75) adopted here.

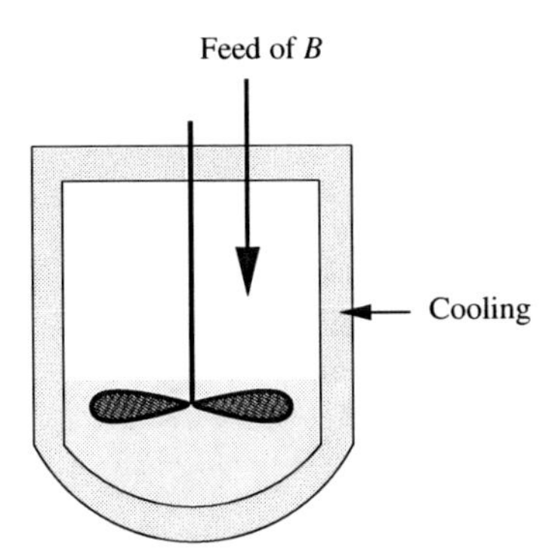

Figure 8.9. *Schematics of a batch-reaction with two control variables.*

It is worth stressing the improvement that the optimal control approach has brought to the problem. The disappearance of the damped oscillations allows the laser to attain its final state in a *finite* time rather than asymptotically. In practical terms, one can set a threshold value, δ, around the asymptotic level, S_∞, and consider the state attained once $S_\infty - \delta < S(t) < S_\infty + \delta$ holds. The parameter δ can be determined by the amount of noise present in the system, which we do not consider in our analysis. Visually, this operation corresponds to saying that the damped oscillation and asymptotic state are indistinguishable below a certain level of detail, e.g., if $t > 500$ ps as in Figure 8.8.

With this convention, one can give a quantitative estimate of the amount of improvement introduced by the control function: the asymptotic state is attained at $t \approx 50$ ps, even before the first crossing of the same level which occurs at $t \approx 70$ ps (dashed line in Figure 8.8). The improvement is of the order of a factor 10!

8.5 Optimal Control of a Batch-Reactor

The following optimal control problem for a batch-reactor is taken from [17, 110, 66]. Consider a chemical reaction

$$A + B \to C$$

and its side reaction

$$B + C \to D$$

which are assumed to be strongly exothermic. Thus, direct mixing of the entire necessary amounts of the reactants must be avoided.

The reactant A is charged in the reactor vessel, which is fitted with a cooling jacket to remove the generated heat of the reaction, while the reactant B is added. These reactions result in the product C and the undesired byproduct D.

The vector of state variables is denoted by

$$x = (M_A, M_B, M_C, M_D, H) \in \mathbb{R}^5,$$

where $M_i(t)$ [mol] and $C_i(t)$ [mol/m^3] stand for the molar holdups and the molar concentrations of the components $i = A, B, C, D$, respectively. $H(t)$ [MJ] denotes the total energy holdup, $T_R(t)$ [K] the reactor temperature and $V(t)$ [m^3] the volume of liquid in the system. The two-dimensional control vector is given by

$$u = (F_B, Q) \in \mathbb{R}^2,$$

where $F_B(t)$ [mol/s] controls the feed rate of the component B while $Q(t)$ [kW] controls the cooling load. The objective is to determine a control u that *maximizes* the molar holdup of the component C. Hence, the performance index is

$$J_1(x(t_f)) = -M_C(t_f) \tag{8.79}$$

which has to be *minimized* subject to the dynamical equations

$$\begin{array}{rclrcl} \dot{M}_A & = & -V \cdot r_1, & \dot{M}_B & = & F_B - V \cdot (r_1 + r_2), \\ \dot{M}_C & = & V \cdot (r_1 - r_2), & \dot{M}_D & = & V \cdot r_2, \\ \dot{H} & = & F_B \cdot h_f - Q - V \cdot (r_1 \cdot \Delta H_1 + r_2 \cdot \Delta H_2). & & & \end{array} \tag{8.80}$$

Here, r_j denote the reaction rates and k_j the corresponding Arrhenius rate constants of both reactions ($j = 1, 2$):

$$\begin{array}{lllllll} \mathbf{A+B} \rightarrow \mathbf{C}: & r_1 & = & k_1 \cdot C_A \cdot C_B & \text{with} & k_1 & = & A_1 \cdot e^{-E_1/T_R}, \\ \mathbf{C+B} \rightarrow \mathbf{D}: & r_2 & = & k_2 \cdot C_B \cdot C_C & \text{with} & k_2 & = & A_2 \cdot e^{-E_2/T_R}, \end{array} \tag{8.81}$$

where functions are defined by

$$\begin{array}{rcl} S & = & \sum\limits_{i=A,B,C,D} M_i \cdot \alpha_i, \qquad W = \sum\limits_{i=A,B,C,D} M_i \cdot \beta_i, \\ T_R & = & \dfrac{1}{W} \cdot \left(-S + \sqrt{(W \cdot T_{\text{ref}} + S)^2 + 2 \cdot W \cdot H} \right), \\ C_i & = & \dfrac{M_i}{V} \quad (i = A, B, C, D), \qquad V = \sum\limits_{i=A,B,C,D} \dfrac{M_i}{\rho_i}. \end{array} \tag{8.82}$$

The reference temperature for the enthalpy calculations is $T_{\text{ref}} = 298$ K and the specific molar enthalpy of the reactor feed stream is $h_f = 20$ kJ/mol. Initial values are given for all state variables,

$$M_A(0) = 9000, \qquad M_i(0) = 0 \quad (i = B, C, D), \qquad H(0) = 152509.97, \tag{8.83}$$

while there is only one terminal constraint

$$T_R(t_f) = 300 \tag{8.84}$$

with T_R defined as in (8.82). The control vector $u = (F_B, Q)$ appears linearly in the control system (8.80) and is bounded by

$$0 \leq F_B(t) \leq 10 \quad \text{and} \quad 0 \leq Q(t) \leq 1000 \quad \forall\, t \in [0, t_f]. \tag{8.85}$$

The reaction and component data appearing in (8.80)–(8.82) are given in Table 8.2.

Table 8.2. *Reaction and component data.*

Notation	Reactions		Meaning
	$j=1$	$j=2$	
$A_j\left[\frac{\text{m}^3}{\text{mol}\cdot\text{s}}\right]$	0.008	0.002	Preexponential Arrhenius constants
E_j [K]	$3000\cdot p$	2400	Activation energies
$\Delta H_j\left[\frac{\text{kJ}}{\text{mol}}\right]$	−100	−75	Enthalpies

Notation	Components				Meaning
	$i=A$	$i=B$	$i=C$	$i=D$	
$\rho_i\left[\frac{\text{mol}}{\text{m}^3}\right]$	11250	16000	10400	10000	Molar density of pure component i
$\alpha_i\left[\frac{\text{kJ}}{\text{mol}\cdot\text{K}}\right]$	0.1723	0.2	0.16	0.155	Coefficient of the linear (α_i)
$\beta_i\left[\frac{\text{kJ}}{\text{mol}\cdot\text{K}^2}\right]$	0.000474	0.0005	0.00055	0.000323	and quadratic (β_i) term in the pure component specific enthalpy expression

Calculations show that for increasing t_f the switching structure gets more and more complex. However, the total profit of $M_C(t_f)$ is nearly constant if t_f is greater than a certain value, $t_f \approx 1600$. For these values one obtains singular controls. We choose the final time $t_f = 1450$ and will show that the optimal control has the following bang-bang structure with $0 < t_1 < t_2 < t_f$:

$$u(t) = (F_B(t), Q(t)) = \left\{ \begin{array}{lll} (10,0) & \text{for} & 0 \le t < t_1 \\ (10,1000) & \text{for} & t_1 \le t < t_2 \\ (0,1000) & \text{for} & t_2 \le t \le t_f \end{array} \right\}. \tag{8.86}$$

Since the initial point $x(0)$ is specified and the final time t_f is fixed, the optimization variable in the IOP (8.18) is given by

$$z = (\tau_1, \tau_2), \quad \tau_1 = t_1, \quad \tau_2 = t_2 - t_1.$$

Then the arc-length of the terminal bang-bang arc is $\tau_3 = 1450 - \tau_1 - \tau_2$. The code NUDOCCCS yields the following arc-lengths and switching times:

$$\begin{array}{lll} J_1(x(t_f)) = 3555.292, & t_1 = 433.698, & \tau_2 = 333.575, \\ & t_2 = 767.273, & \tau_3 = 1450 - t_2 = 682.727. \end{array} \tag{8.87}$$

We note that for this control the state constraint $T_R(t) \le 520$ imposed in [17] does not become active. The adjoint equations are rather complicated and are not given here explicitly. The code NUDOCCCS also provides the adjoint functions, e.g., the initial values

$$\begin{array}{rclrcl} \psi_{M_A}(0) & = & -0.0299034, & \psi_{M_B}(0) & = & 0.0433083, \\ \psi_{M_C}(0) & = & -2.83475, & \psi_{M_D}(0) & = & -0.10494943, \\ \psi_H(0) & = & 0.00192489. & & & \end{array} \tag{8.88}$$

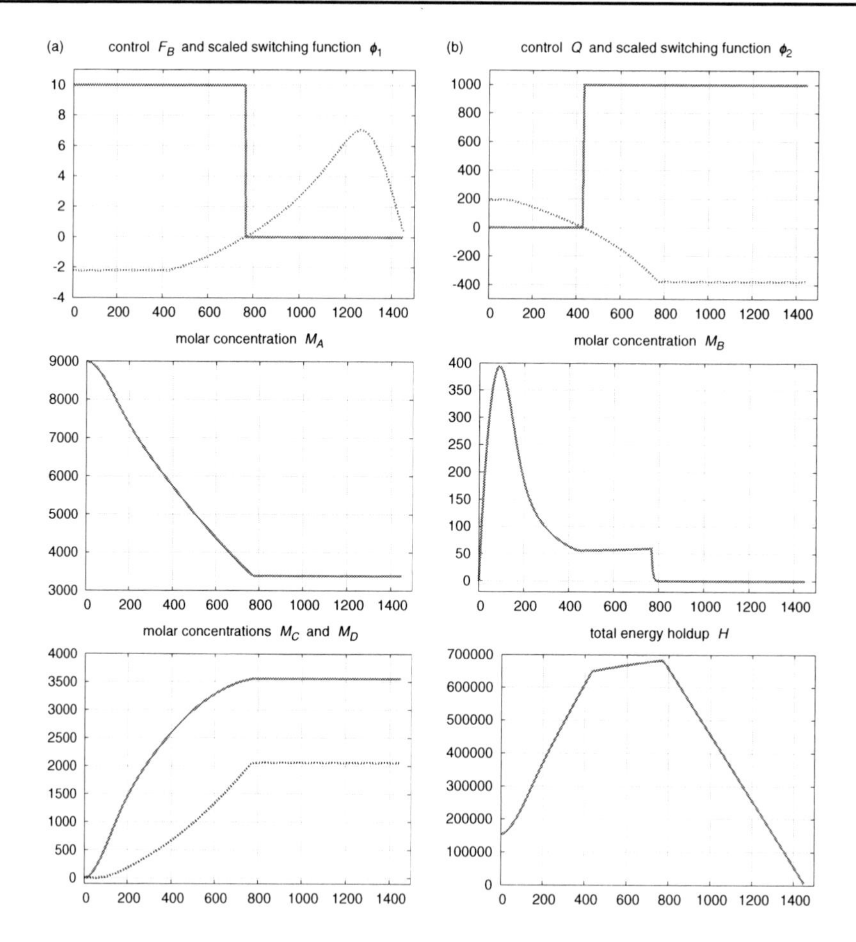

Figure 8.10. *Control of a batch reactor with functional* (8.79). *Top row: Control* $u = (F_B, Q)$ *and scaled switching functions. Middle row: Molar concentrations* M_A *and* M_B. *Bottom row: Molar concentrations* (M_C, M_D) *and energy holdup* H.

Figure 8.10 (top row) clearly shows that the strict bang-bang property holds with

$$\dot{\phi}_2(t_1) < 0, \quad \dot{\phi}_1(t_2) > 0.$$

The Jacobian of the scalar terminal condition (8.84) is computed as

$$\mathcal{G}_z(\hat{z}) = (0.764966,\ 0.396419),$$

while the Hessian of the Lagrangian is the positive matrix

$$L_{zz}(\hat{\beta}, \hat{z}) = \begin{pmatrix} 0.00704858 & 0.00555929 \\ 0.00555929 & 0.00742375 \end{pmatrix}.$$

Thus, the second-order condition (8.20) is satisfied and Theorem 7.10 tells us that the solution (8.86)–(8.88) provides a strict strong minimum.

Let us now change the cost functional (8.79) and maximize the average gain of the component C in time, i.e.,

$$\text{minimize} \quad J_2(t_f, x(t_f)) = -\frac{M_C(t_f)}{t_f}, \tag{8.89}$$

where the final time t_f is free. We will show that the bang-bang control

$$u(t) = (F_B(t), Q(t)) = \left\{ \begin{array}{lll} (10, 1000) & \text{for} & 0 \le t < t_1 \\ (0, 1000) & \text{for} & t_1 \le t \le t_f \end{array} \right\} \tag{8.90}$$

with only one switching point $0 < t_1 < t_f$ of the control $u_1(t) = F_B(t)$ is optimal. Since the initial point $x(0)$ is specified, the optimization variable in the IOP (8.18) is

$$z = (\tau_1, \tau_2), \quad \tau_1 = t_1, \quad \tau_2 = t_f - t_1.$$

Using the code NUDOCCCS, we obtain the switching and terminal times

$$\begin{array}{ll} J_2(t_f, x(t_f)) = 3.877103, & t_1 = 285.519, \\ \tau_2 = 171.399, & t_f = 456.918. \end{array} \tag{8.91}$$

and initial values of the adjoint variables

$$\begin{array}{rclrcl} \psi_{M_A}(0) & = & -0.9932 \cdot 10^{-4}, & \psi_{M_B}(0) & = & -0.61036 \cdot 10^{-3}, \\ \psi_{M_C}(0) & = & -0.108578 \cdot 10^{-2}, & \psi_{M_D}(0) & = & 0.9495 \cdot 10^{-4}, \\ \psi_H(0) & = & 0.298 \cdot 10^{-5}. & & & \end{array} \tag{8.92}$$

We may conclude from Figure 8.11 that $\dot{\phi}_1(t_1) > 0$ holds, while the switching function for the control $u_2(t) = Q(t) = 1000$ satisfies $\phi_2(t) < 0$ on $[0, t_f]$. The Jacobian for the scalar terminal condition (8.84) is

$$\mathcal{G}_z(\hat{z}) = (0.054340, \ -0.31479),$$

while the Hessian of the Lagrangian is the positive definite matrix

$$L_{zz}(\hat{\beta}, \hat{z}) = \begin{pmatrix} 0.37947 & -0.11213 \\ -0.11213 & 0.37748 \end{pmatrix} \cdot 10^{-4}.$$

Hence the SSC (8.20) hold and, again, Theorem 7.10 asserts that the solution (8.90)– (8.92) is a strict strong minimum.

8.6 Optimal Production and Maintenance with L^1-Functional

In section 5.4, we studied the optimal control model presented in Cho, Abad, and Parlar [22], where optimal production and maintenance policies are determined simultaneously. The cost functional was quadratic with respect to the production control u, which enhances a *continuous* control. In this section, we consider the case when the production control enters the cost functional linearly. In this model, $x = x_1$ denotes the inventory level, $y = x_2$

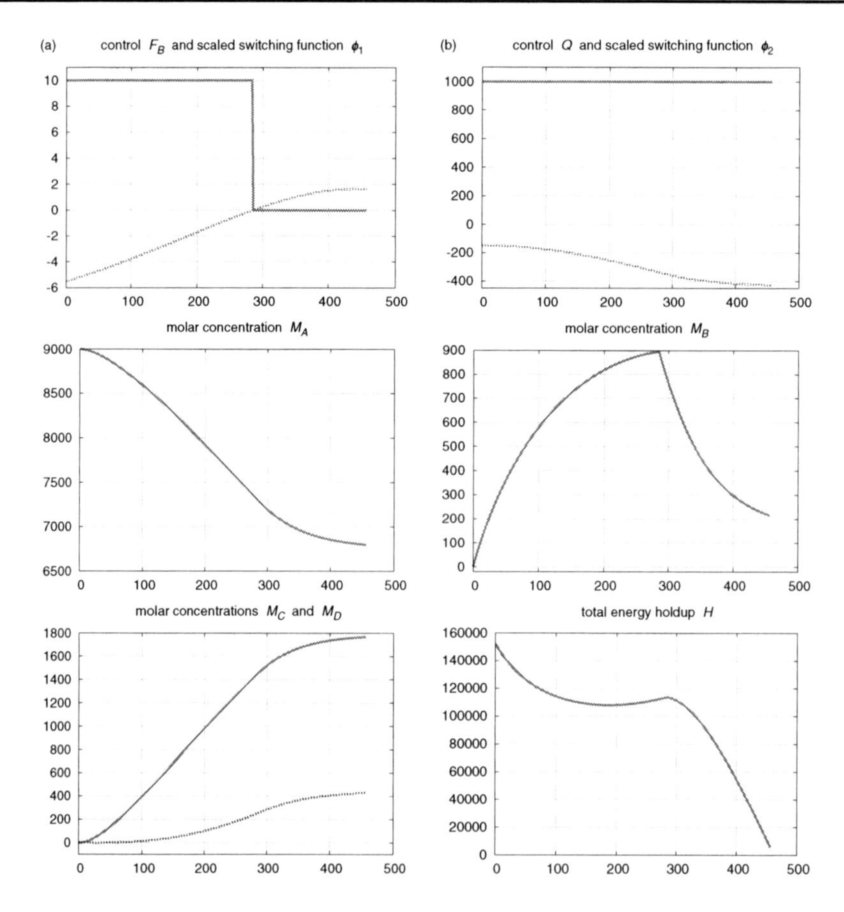

Figure 8.11. *Control of a batch reactor with functional* (8.89). *Top row: Control* $u = (F_B, Q)$ *and scaled switching functions. Middle row: Molar concentrations* M_A, M_B. *Bottom row: Molar concentrations* (M_C, M_D) *and energy holdup* H.

the proportion of good units of end items produced, v is the scheduled production rate (control), and m denotes the preventive maintenance rate (control). The parameters are $\alpha = 2$ obsolescence rate, $s = 4$ demand rate, and $\rho = 0.1$ discount rate. The control bounds and weights in the cost functional will be specified below. The dynamics of the process is given by

$$\begin{aligned} \dot{x}(t) &= y(t)v(t) - 4, && x(0) = 3, \quad x(t_f) = 0, \\ \dot{y}(t) &= -2y(t) + (1 - y(t))m(t), && y(0) = 1, \end{aligned} \tag{8.93}$$

with the following bounds on the control variables:

$$0 \le v(t) \le 3, \quad 0 \le m(t) \le 4 \quad \text{for } 0 \le t \le t_f. \tag{8.94}$$

Recall also that the terminal condition $x(t_f) = 0$ implies the nonnegativity condition $x(t) \ge 0$ for all $t \in [0, t_f]$. Now we choose the following L^1-functional: Maximize the total discounted profit and salvage value of $y(t_f)$,

$$\begin{aligned} J(x,y,m,v) = &\int_0^{t_f} [8s - x(t) - 4\,v(t) - 2.5\,m(t)]e^{-0.1\cdot t}\,dt \\ &+ 10\,y(t_f)e^{-0.1\cdot t_f} \quad (t_f = 1) \end{aligned} \tag{8.95}$$

under the constraints (8.93) and (8.94). Though the objective has to be *maximized*, we discuss the necessary conditions on the basis of the *Minimum Principle* which has been used throughout the book. Again, we use the standard Hamilton–Pontryagin function instead of the *current value* Hamiltonian:

$$\begin{aligned} H(t,x,y,\psi_1,\psi_2,m,v) = &\, e^{-0.1\cdot t}(-32 + x + 4v + 2.5\,m) \\ &+ \psi_1(yv - 4) + \psi_2(-2y + (1-y)m), \end{aligned} \tag{8.96}$$

where ψ_1, ψ_2 denote the adjoint variables. The adjoint equations and transversality condition yield, in view of the terminal constraint $x_1(t_f) = 0$ and the salvage term in the cost functional,

$$\begin{aligned} \dot\psi_1 &= -e^{-0.1}, & \psi_1(t_f) &= \nu, \\ \dot\psi_2 &= -\psi_1 v + \psi_2(2+m), & \psi_2(t_f) &= -10e^{-0.1}, \end{aligned} \tag{8.97}$$

where $\nu \in \mathbb{R}$ is an unknown multiplier. The switching functions are given by

$$\phi_m = H_m = 2.5\,e^{-0.1\cdot t} + \psi_2(1-y), \quad \phi_v = H_v = 4\,e^{-0.1\cdot t} + \psi_1 y,$$

which determines the controls according to

$$v(t) = \left\{ \begin{array}{lll} 3 & \text{if} & \phi_v(t) < 0 \\ 0 & \text{if} & \phi_v(t) > 0 \end{array} \right\}, \qquad m(t) = \left\{ \begin{array}{lll} 4 & \text{if} & \phi_m(t) < 0 \\ 0 & \text{if} & \phi_m(t) > 0 \end{array} \right\}. \tag{8.98}$$

Singular controls for which either $\phi_v(t) = 0$ or $\phi_m(t) = 0$ holds on an interval $I \subset [0, t_f]$ cannot be excluded a priori. However, for the data and the final time $t_f = 1$ chosen here, the application of direct optimization methods [5, 13, 14] reveals the following control structure with four bang-bang arcs:

$$(v(t), m(t)) = \left\{ \begin{array}{lll} (3,0) & \text{for} & 0 \le t < t_1 \\ (0,0) & \text{for} & t_1 \le t < t_2 \\ (0,4) & \text{for} & t_2 \le t < t_3 \\ (3,4) & \text{for} & t_3 \le t \le t_f = 1 \end{array} \right\}. \tag{8.99}$$

The optimization vector for the IOP (8.18) is

$$z = (\tau_1, \tau_2, \tau_3), \quad \tau_1 = t_1, \quad \tau_2 = t_2 - t_1, \quad \tau_3 = t_3 - t_2.$$

Therefore, the terminal arc-length is given by $\tau_4 = t_f - (\tau_1 + \tau_2 + \tau_3)$. The code NUDOCCCS yields the following numerical results for the arc-lengths and adjoint variables:

$$\begin{array}{rclrclrcl} t_1 &=& 0.346533, & \tau_2 &=& 0.380525, & t_2 &=& 0.727058, \\ \tau_3 &=& 0.114494, & t_3 &=& 0.841552, & t_f &=& 1.0, \\ J &=& 25.7969, & \psi(t_f) &=& -0.833792, & \psi_2(t_f) &=& -0.904837. \end{array} \tag{8.100}$$

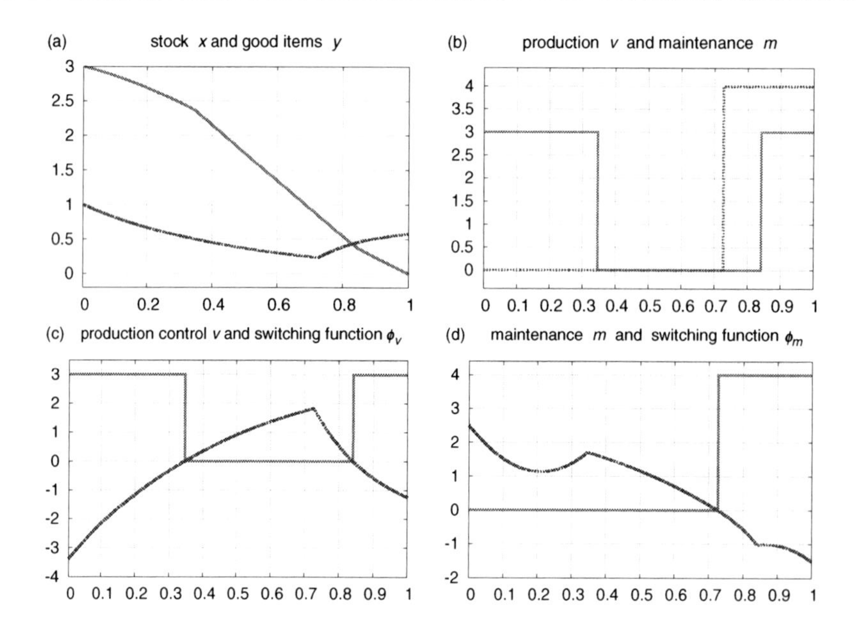

Figure 8.12. *Optimal production and maintenance with L^1-functional* (8.95). (a) *State variables x and y.* (b) *Control variables v and m.* (c), (d) *Control variables and switching functions.*

Figure 8.12 clearly indicates that the strict bang-bang property (6.19) holds, since in particular we have

$$\dot{\phi}_v(t_1) > 0, \quad \dot{\phi}_v(t_3) < 0, \quad \dot{\phi}_m(t_2) < 0.$$

For the scalar terminal condition $x(t_f) = 0$, the Jacobian is the nonzero row vector

$$\mathcal{G}_z(\hat{z}) = (-0.319377, \ -1.81950, \ -1.35638).$$

The Hessian of the Lagrangian is computed as the (3×3) matrix

$$L_{zz}(\hat{\beta}, \hat{z}) = \begin{pmatrix} 41.6187 & 21.0442 & -3.43687 \\ 21.0442 & 21.0442 & -3.43687 \\ -3.43687 & -3.43687 & 34.4731 \end{pmatrix},$$

from which the reduced Hessian (8.20) is obtained as the positive definite (2×2) matrix

$$N^* L_{zz}(\hat{\beta}, \hat{z}) N = \begin{pmatrix} 20.4789 & 6.61585 \\ 6.61585 & 49.6602 \end{pmatrix}.$$

Hence the second-order test (8.20) holds, which ensures that the control (8.99) with the data (8.100) yields a strict strong minimum.

8.7 Van der Pol Oscillator with Bang-Singular Control

The following example with a bang-singular control is taken from Vossen [111, 112]. The optimal control is a concatenation of two bang-bang arcs and one terminal singular arc. The singular control is given by a feedback expression which allows us to optimize switching times directly using the arc-parametrization method presented in Section 8.1.3; cf. also [111, 112]. We consider again the dynamic model of a Van der Pol oscillator introduced in Section 6.7.1. The aim is to minimize the regulator functional

$$J(x,u) = \frac{1}{2}\int_0^{t_f} (x_1(t)^2 + x_2(t)^2)\,dt \quad (t_f = 4) \tag{8.101}$$

subject to the dynamics, boundary conditions, and control constraints

$$\begin{aligned} &\dot{x}_1(t) = x_2(t), && x_1(0) = 0, \\ &\dot{x}_2(t) = -x_1(t) + x_2(t)(1 - x_1(t)^2) + u(t), && x_2(0) = 1, \\ &|u(t)| \le 1 \quad \text{for } t \in [0, t_f]. \end{aligned} \tag{8.102}$$

The Hamilton–Pontryagin function

$$H(x,u,\psi) = \psi_1 x_2 + \psi_2(-x_1 + x_2(1 - x_1^2) + u) \tag{8.103}$$

yields the adjoint equations and transversality conditions

$$\begin{aligned} \dot{\psi}_1 &= -x_1 + \psi_2(1 + 2x_1x_2), \quad && \psi(t_f) = 0, \\ \dot{\psi}_2 &= -x_2 - \psi - \psi_2(1 - x_1^2), && \psi(t_f) = 0. \end{aligned} \tag{8.104}$$

The switching function

$$\phi(t) = H_u(t) = \psi_2(t) \tag{8.105}$$

determines the bang-bang controls by

$$u(t) = \left\{ \begin{array}{rl} 1 & \text{if} \quad \psi_2(t) < 0 \\ -1 & \text{if} \quad \psi_2(t) > 0 \end{array} \right\}. \tag{8.106}$$

A singular control on an interval $I \subset [0, t_f]$ can be computed from the relations

$$\phi = \psi_2 = 0, \quad \dot{\phi} = \dot{\psi}_2 = -x_2 - \psi_1 = 0, \quad \ddot{\phi} = 2x_1 - x_2(1 - x_1^2) - u = 0,$$

which give the feedback expression

$$u = u_{\text{sing}}(x) = 2x_1 - x_2(1 - x_1^2). \tag{8.107}$$

It follows that the order of a singular arc is $q = 1$; cf. the definition of the order in [49]. Moreover, the strict Generalized Legendre Condition holds:

$$(-1)^q \frac{\partial}{\partial u}\left[\frac{d^2}{dt^2} H_u\right] = -\frac{\partial}{\partial u}\ddot{\phi} = 1 > 0.$$

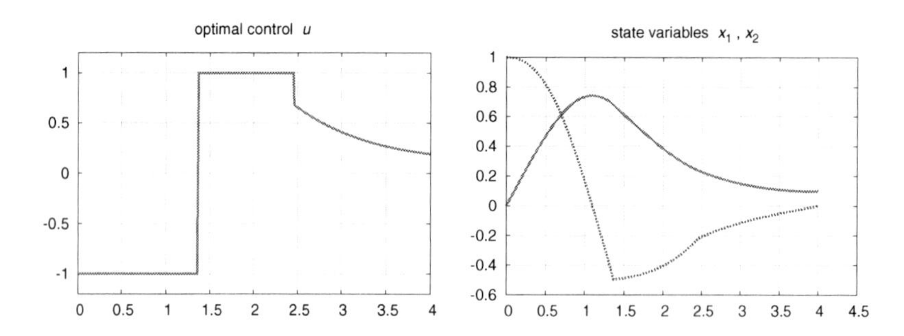

Figure 8.13. *Control of the van der Pol oscillator with regulator functional.* (a) *Bang-singular control u.* (b) *State variables x_1 and x_2.*

The application of direct optimization methods yields the following control structure with two bang-bang arcs and a terminal singular arc:

$$u(t) = \left\{ \begin{array}{lll} -1 & \text{for} & 0 \le t < t_1 \\ 1 & \text{for} & t_1 \le t < t_2 \\ 2x_1(t) - x_2(t)(1 - x_1(t)^2) & \text{for} & t_2 \le t \le t_f = 4 \end{array} \right\}. \tag{8.108}$$

Hence, the feedback functions in (8.26) are given by

$$u^1(x) = -1, \quad u^2(x) = 1, \quad u^3(x) = 2x_1 - x_2(1 - x_1^2),$$

which are inserted into the dynamic equations (8.102). Therefore, the optimization vector for the IOP (8.28) is given by

$$z = (\tau_1, \tau_2), \quad \tau_1 = t_1, \quad \tau_2 = t_2 - t_1.$$

This yields the terminal arc length $\tau_3 = 4 - \tau_1 - \tau_2$. The code NUDOCCCS provides the following numerical results for the arc-lengths, switching times, and adjoint variables:

$$\begin{array}{rclrclrcl} t_1 & = & 1.36674, & \tau_2 & = & 0.109404, & t_2 & = & 2.46078, \\ J(x,u) & = & 0.757618, & \psi_1(0) & = & -0.495815, & \psi_2(0) & = & 2.58862. \end{array} \tag{8.109}$$

The corresponding control and state variables are shown in Figure 8.13. The code NUDOCCCS furnishes the Hessian of the Lagrangian as the positive definite matrix

$$L_{zz}(\hat{z}) = \begin{pmatrix} 194.93 & -9.9707 \\ -9.9707 & 0.56653 \end{pmatrix}.$$

At this stage we have only found a strict local minimum of the IOP (8.28). Recently, SSC for a class of bang-singular controls were obtained by Aronna et al. [2]. Combining these new results with the Riccati approach in Vossen [111] for testing the positive definiteness of quadratic forms, one can now verify the SSC for the extremal solutions (8.108) and (8.109). Here, the following property of the switching function is important:

$$\begin{array}{lll} \phi(t) > 0 \quad \text{for } 0 \le t < t_1, & \dot{\phi}(t_1) < 0, & \phi(t) < 0 \text{ for } t_1 < t < t_2, \\ \phi(t_2) = \dot{\phi}(t_2) = 0, & \ddot{\phi}(t_2^-) < 0, & \phi(t) = 0 \text{ for } t_2 < t < t_f. \end{array}$$

Bibliography

[1] A. A. AGRACHEV, G. STEFANI, AND P. ZEZZA, *Strong optimality for a bang-bang trajectory*, SIAM Journal on Control and Optimization, **41**, pp. 991–1014 (2002). [5, 299, 300, 301, 304]

[2] M. S. ARONNA, J. F. BONNANS, A.V. DMITRUK, AND P. A. LOTITO, *Quadratic conditions for bang-singular extremals*, to appear in Numerical Algebra, Control and Optimization, **2**, pp. 511–546 (2012). [366]

[3] D. AUGUSTIN AND H. MAURER, *Sensitivity analysis and real-time control of a container crane under state constraints*, In: *Online Optimization of Large Scale Systems* (M. Grötschel, S. O. Krumke, J. Rambau, eds.), pp. 69–82, Springer, Berlin, 2001. [219]

[4] A. BEN-TAL AND J. ZOWE, *Second-order optimality conditions for the L_1-minimization problem*, Applied Mathematics and Optimization, **13**, pp. 45–48 (1985). [301]

[5] J. T. BETTS, *Practical Methods for Optimal Control and Estimation Using Nonlinear Programming*, 2nd ed., Advances in Design and Control **19**, SIAM, Philadelphia, 2010. [250, 339, 353, 363]

[6] G. A. BLISS, *Lectures on the Calculus of Variations*, University of Chicago Press, Chicago, 1946. [24]

[7] H. G. BOCK AND K. J. PLITT, *A multiple shooting algorithm for direct solution of optimal control problems*, In: Proceedings of the 9th IFAC World Congress, Budapest, pp. 243–247, 1984. [339]

[8] J. F. BONNANS AND N. P. OSMOLOVSKII, *Characterization of a local quadratic growth of the Hamiltonian for control constrained optimal control problems*, Dynamics of Continuous, Discrete and Impulsive Systems, Series B: Applications and Algorithms, **19**, pp. 1–16 (2012). [177]

[9] B. BONNARD, J.-B. CAILLAU, AND E. TRÉLAT, *Second order optimality conditions in the smooth case and applications in optimal control*, ESAIM Control, Optimization and Calculus of Variations, **13**, pp. 207–236 (2007). [339]

[10] B. BONNARD, J.-B. CAILLAU, AND E. TRÉLAT, *Computation of conjugate times in smooth optimal control: The COTCOT algorithm*, In: Proceedings of the 44th

IEEE Conference on Decision and Control and European Control Conference 2005, pp. 929–933, Sevilla, December 2005. [339]

[11] A. BRESSAN, *A high order test for optimality of bang-bang controls*, SIAM Journal on Control and Optimization, **23**, pp. 38–48 (1985).

[12] A. E. BRYSON AND Y. C. HO, *Applied Optimal Control*, Revised Printing, Hemisphere Publishing Corporation, New York, 1975. [219]

[13] C. BÜSKENS, *Optimierungsmethoden und Sensitivitätsanalyse für optimale Steuerprozesse mit Steuer- und Zustands-Beschränkungen*, Dissertation Institut für Numerische Mathematik, Universität Münster, Germany (1998). [6, 219, 273, 287, 288, 291, 339, 342, 343, 363]

[14] C. BÜSKENS AND H. MAURER, *SQP-methods for solving optimal control problems with control and state constraints: Adjoint variables, sensitivity analysis and real-time control*, Journal of Computational and Applied Mathematics, **120**, pp. 85–108 (2000). [6, 219, 250, 287, 288, 291, 339, 363]

[15] C. BÜSKENS AND H. MAURER, *Sensitivity analysis and real-time optimization of parametric nonlinear programming problems*, In: Online Optimization of Large Scale Systems (M. Grötschel, S. O. Krumke, J. Rambau, eds.), pp. 3–16, Springer-Verlag, Berlin, 2001. [219]

[16] C. BÜSKENS AND H. MAURER, *Sensitivity analysis and real-time control of parametric optimal control problems using nonlinear programming methods*, In: Online Optimization of Large Scale Systems (M. Grötschel, S. O. Krumke, J. Rambau, eds.), pp. 57–68, Springer-Verlag, Berlin, 2001. [219]

[17] C. BÜSKENS, H. J. PESCH, AND S. WINDERL, *Real-time solutions of bang-bang and singular optimal control problems*, In: *Online Optimization of Large Scale Systems* (M. Grötschel et al., eds.), pp. 129–142, Springer-Verlag, Berlin, 2001. [xii, 351, 357, 359]

[18] R. BULIRSCH, *Die Mehrzielmethode zur numerischen Lösung von nichtlinearen Randwertproblemen und Aufgaben der optimalen Steuerung*, Report of the Carl-Cranz-Gesellschaft, Oberpfaffenhofen, Germany, 1971. [339]

[19] R. BULIRSCH, F. MONTRONE, AND H. J. PESCH, *Abort landing in the presence of windshear as a minimax optimal control problem. Part 2: Multiple shooting and homotopy*, Journal of Optimization Theory and Applications, **70**, pp. 223–254 (1991). [339]

[20] D. M. BYRNE, *Accurate simulation of multifrequency semiconductor laser dynamics under gigabit-per-second modulation*, Journal of Lightwave Technology, **10**, pp. 1086–1096 (1992). [353]

[21] F. L. CHERNOUSKO, L. D. AKULENKO, AND N. N. BOLOTNIK, *Time-optimal control for robotic manipulators*, Optimal Control Applications & Methods, **10**, pp. 293–311 (1989). [346, 347, 348]

[22] D. I. Cho, P. L. Abad, and M. Parlar, *Optimal production and maintenance decisions when a system experiences age-dependent deterioration*, Optimal Control Applications & Methods, **14**, pp. 153–167 (1993). [4, 248, 249, 250, 361]

[23] B. Christiansen, H. Maurer, O. Zirn, *Optimal control of servo actuators with flexible load and Coulombic friction*, European Journal of Control, **17**, pp. 1–11 (2011). [342]

[24] B. Christiansen, H. Maurer, O. Zirn, *Optimal control of machine tool manipulators*, In: Recent Advances in Optimization and Its Applications in Engineering: The 14th Belgian-French-German Conference on Optimization, Leuven, September 2009 (M. Diehl, F. Glineur, E. Jarlebring, W. Michels, eds.), pp. 451–460, Springer, Berlin, 2010.

[25] A. V. Dmitruk, *Euler–Jacobi equation in the calculus of variations*, Mat. Zametki, **20**, pp. 847–858 (1976). [4, 191, 197]

[26] A. V. Dmitruk, *Jacobi-type conditions for the Bolza problem with inequalities*, Mat. Zametki, **35**, pp. 813–827 (1984). [4, 191, 193]

[27] A. V. Dmitruk, *Quadratic conditions for the Pontryagin minimum in a control-linear optimal control problem. I. Decoding theorem*, Izv. Akad. Nauk SSSR Ser. Mat., **50**, pp. 284–312 (1986). [68]

[28] A. V. Dmitruk, A. A. Milyutin, and N. P. Osmolovskii, *Lyusternik's theorem and extremum theory*, Usp. Mat. Nauk, **35**, pp. 11–46 (1980); English translation: Russian Math. Surveys, **35**, pp. 6, 11–51 (1980). [132]

[29] N. Dokhane and G. L. Lippi, *Chirp reduction in semiconductor lasers through injection current patterning*, Applied Physics Letters, **78**, pp. 3938–3940 (2001). [353, 354, 356]

[30] A. Ya. Dubovitskii and A. A. Milyutin, *Theory of the maximum principle*, In: Methods of Extremal Problem Theory in Economics [in Russian], Nauka, Moscow, pp. 6–47 (1981). [166, 167]

[31] U. Felgenhauer, *On stability of bang-bang type controls*, SIAM Journal on Control and Optimization, **41**, pp. 1843–1867 (2003). [259, 338]

[32] A.V. Fiacco, *Introduction to Sensitivity and Stability Analysis in Nonlinear Programming*, Mathematics in Science and Engineering **165**, Academic Press, New York, 1983. [338]

[33] R., Fourer, D. M. Gay, and B. W. Kernighan, *AMPL: A Modeling Language for Mathematical Programming*, Duxbury Press, Brooks-Cole Publishing Company, 1993. [250]

[34] R. Gabasov and F. Kirillova, *The Qualitative Theory of Optimal Processes*, Marcel Dekker Inc., New York, Basel, 1976.

[35] H. P. GEERING, L. GUZELLA, S. A. R. HEPNER, AND CH. ONDER, *Time-optimal motions of robots in assembly tasks*, IEEE Transactions on Automatic Control, **AC-31**, pp. 512–518 (1986). [346, 349]

[36] E. G. GILBERT AND D. S. BERNSTEIN, *Second-order necessary conditions of optimal control: Accessory-problem results without normality conditions*, Journal of Optimization Theory Applications, **41**, pp. 75–106 (1983).

[37] L. GÖLLMANN, *Numerische Berechnung zeitoptimaler Trajektorien für zweigliedrige Roboterarme*, Diploma thesis, Institut für Numerische Mathematik, Universität Münster, 1991. [346, 347, 348]

[38] B. S. GOH, *Necessary conditions for singular extremals involving multiple control variables*, SIAM Journal on Control, **4**, pp. 716–731 (1966).

[39] R. HENRION, *La Théorie de la Variation Seconde et ses Applications en Commande Optimale*, Academie Royal de Belgique, Bruxelles: Palais des Academies (1979). [1, 32]

[40] M. HESTENS, *Calculus of Variations and Optimal Control Theory*, John Wiley, New York, 1966. [225]

[41] A. J. HOFFMAN, *On approximate solutions of systems of linear inequalities*, Journal of Research of the National Bureau of Standards, **49**, pp. 263–265 (1952). [16]

[42] A. D. IOFFE AND V. M. TIKHOMIROV, *Theory of Extremal Problems* [in Russian], Nauka, Moscow (1974). Also published in German as *Theorie der Extremalaufgaben*, VEB Deutscher Verlag der Wissenschaften, Berlin, 1979. [191]

[43] D. H. JACOBSON AND D.Q. MAYNE, *Differential Dynamic Programming*, American Elsevier Publishing Company Inc., New York, 1970.

[44] Y. KAYA AND J. L. NOAKES, *Computations and time-optimal controls*, Optimal Control Applications and Methods, **17**, pp. 171–185 (1996). [6, 339, 341, 342]

[45] C. Y. KAYA AND J. L. NOAKES, *Computational method for time-optimal switching control*, Journal of Optimization Theory and Applications, **117**, pp. 69–92 (2003). [6, 339, 341, 342]

[46] J.-H. R. KIM, G. L. LIPPI, AND H. MAURER, *Minimizing the transition time in lasers by optimal control methods. Single mode semiconductor lasers with homogeneous transverse profile*, Physica D (Nonlinear Phenomena) **191**, pp. 238–260 (2004). [xii, 353, 356]

[47] J.-H. R. KIM, H. MAURER, YU. A. ASTROV, M. BODE, AND H.-G. PURWINS, *High-speed switch-on of a semiconductor gas discharge image converter using optimal control methods*, Journal of Computational Physics, **170**, pp. 395–414 (2001). [280]

[48] J.-H. R. KIM AND H. MAURER, *Sensitivity analysis of optimal control problems with bang-bang controls*, In: Proceedings of the 42nd IEEE Conference on Decision and Control, Maui, Dec. 9–12, 2003, IEEE Control Society, Washington, DC, pp. 3281–3286, 2003. [338, 342]

[49] A. J. KRENER, *The high order maximum principle, and its application to singular extremals*, SIAM Journal on Control and Optimization, **15**, pp. 256–293 (1977). [365]

[50] U. LEDZEWICZ, H. MAURER, AND H. SCHÄTTLER, *Optimal and suboptimal protocols for a mathematical model for tumor anti-angiogenesis in combination with chemotherapy*, Mathematical Biosciences and Engineering, **8**, pp. 307–328 (2011). [343]

[51] U. LEDZEWICZ, H. MAURER, AND H. SCHÄTTLER, *On optimal delivery of combination therapy for tumors*, Mathematical Biosciences, **22**, pp. 13–26 (2009). [343]

[52] U. LEDZEWICZ, J. MARRIOTT, H. MAURER, AND H. SCHÄTTLER, *The scheduling of angiogenic inhibitors minimizing tumor volume*, Journal of Medical Informatics and Technologies, **12**, pp. 23–28 (2008).

[53] U. LEDZEWICZ AND H. SCHÄTTLER, *Optimal bang-bang controls for a 2-compartment model in cancer chemotherapy*, Journal of Optimization Theory and Applications, **114**, pp. 609–637 (2002). [270]

[54] E. S. LEVITIN, A. A. MILYUTIN, AND N. P. OSMOLOVSKII, *On local minimum conditions in problems with constraints*, In: Mathematical Economics and Functional Analysis [in Russian], Nauka, Moscow, pp. 139–202 (1974). [3, 301]

[55] E. S. LEVITIN, A. A. MILYUTIN, AND N. P. OSMOLOVSKII, *Higher-order local minimum conditions in problems with constraints*, Uspehi Mat. Nauk, **33**, pp. 85–148 (1978). [3, 21, 174, 301]

[56] E. S. LEVITIN, A. A. MILYUTIN, AND N. P. OSMOLOVSKII, *Theory of higher-order conditions in smooth constrained extremal problems*, In: Theoretical and Applied Optimal Control Problems [in Russian], Nauka, Novosibirsk, pp. 4–40, 246 (1985). [3]

[57] K. MALANOWSKI, C. BÜSKENS, AND H. MAURER, *Convergence of approximations to nonlinear optimal control problems*, In: Mathematical Programming with Data Perturbations (A.V. Fiacco, ed.), Lecture Notes in Pure and Applied Mathematics, Vol. **195**, pp. 253–284, Marcel-Dekker, Inc., New York, 1997.

[58] K. MALANOWSKI AND H. MAURER, *Sensitivity analysis for parametric control problems with control-state constraints*, Computational Optimization and Applications, **5**, pp. 253–283 (1996). [219]

[59] K. MALANOWSKI AND H. MAURER, *Sensitivity analysis for state constrained optimal control problems*, Discrete and Continuous Dynamical Systems, **4**, pp. 241–272 (1998). [219]

[60] P. MARTINON AND J. GERGAUD, *Using switching detection and variational equations for the shooting method*, Optimal Control Applications and Methods, **28**, pp. 95–116 (2007). [339]

[61] H. MAURER, *Numerical solution of singular control problems using multiple shooting techniques*, Journal of Optimization Theory and Applications, **18**, pp. 235–257 (1976). [339]

[62] H. MAURER, *First and second order sufficient optimality conditions in mathematical programming and optimal control,* Mathematical Programming Study, **14**, pp. 163–177 (1981). [25]

[63] H. MAURER, *Differential stability in optimal control problems*, Applied Mathematics and Optimization, **5**, pp. 283–295 (1979). [298]

[64] H. MAURER AND D. AUGUSTIN, *Second order sufficient conditions and sensitivity analysis for the controlled Rayleigh problem*, In: Parametric Optimization and Related Topics IV (J. Guddat et al., eds.) Lang, Frankfurt am Main, pp. 245–259 (1997). [219, 339]

[65] H. MAURER AND D. AUGUSTIN, *Sensitivity analysis and real-time control of parametric optimal control problems using boundary value methods*, In: Online Optimization of Large Scale Systems (M. Grötschel et al., eds.), pp. 17–55, Springer Verlag, Berlin, 2001. [203, 219, 339]

[66] H. MAURER, C. BÜSKENS, J.-H. R. KIM, AND Y. KAYA, *Optimization methods for the verification of second-order sufficient conditions for bang-bang controls*, Optimal Control Methods and Applications, **26**, pp. 129–156 (2005). [6, 339, 340, 341, 342, 357]

[67] H. MAURER, J.-H. R. KIM, AND G. VOSSEN, *On a state-constrained control problem in optimal production and maintenance*, In: Optimal Control and Dynamic Games, Applications in Finance, Management Science and Economics (C. Deissenberg, R. F. Hartl, eds.), pp. 289–308, Springer Verlag, 2005. [xii, 5, 223, 248, 249, 250, 347]

[68] H. MAURER AND H. J. OBERLE, *Second order sufficient conditions for optimal control problems with free final time: The Riccati approach*, SIAM Journal on Control and Optimization, **41**, pp. 380–403 (2002). [219]

[69] H. MAURER AND N. P. OSMOLOVSKII, *Second order optimality conditions for bang-bang control problems*, Control and Cybernetics, **32**, pp. 555–584 (2003). [267]

[70] H. MAURER AND N. P. OSMOLOVSKII, *Second order sufficient conditions for time-optimal bang-bang control*, SIAM Journal on Control and Optimization, **42**, pp. 2239–2263 (2004). [274]

[71] H. MAURER AND H. J. PESCH, *Solution differentiability for nonlinear parametric control problems*, SIAM Journal on Control and Optimization, **32**, pp. 1542–1554 (1994). [205, 219]

[72] H. MAURER AND H. J. PESCH, *Solution differentiability for nonlinear parametric control problems with control-state constraints*, Journal of Optimization Theory and Applications, **86**, pp. 285–309 (1995). [219]

[73] H. MAURER AND S. PICKENHAIN, *Second order sufficient conditions for optimal control problems with mixed control-state constraints*, Journal of Optimization Theory and Applications, **86** (1995), pp. 649–667 (1995). [219, 267]

[74] H. MAURER AND G. VOSSEN, *Sufficient conditions and sensitivity analysis for bang-bang control problems with state constraints,* In: Proceedings of the 23rd IFIP Conference on System Modeling and Optimization, Cracow, Poland (A. Korytowski, M. Szymkat, eds.), pp. 82–99, Springer Verlag, Berlin, 2009. [342, 343]

[75] H. MAURER AND J. ZOWE, *First and second-order necessary and sufficient conditions for infinite-dimensional programming problems*, Mathematical Programming, **16**, pp. 98–110 (1979). [25]

[76] A. A. MILYUTIN, *Maximum Principle in the General Optimal Control Problem* [in Russian], Fizmatlit, Moscow (2001). [179]

[77] A. A. MILYUTIN, A. E. ILYUTOVICH, N. P. OSMOLOVSKII, AND S. V. CHUKANOV, *Optimal Control in Linear Systems* [in Russian], Nauka, Moscow (1993). [293]

[78] A. A. MILYUTIN AND N. P. OSMOLOVSKII, (1) *Higher-order minimum conditions on a set of sequences in the abstract problem with inequality-type constraints;* (2) *Higher-order minimum conditions on a set of sequences in the abstract problem with inequality- and equality-type constraints;* (3) *Higher-order minimum conditions with respect to a subsystem in the abstract minimization problems on a set of sequences*, In: Optimality of Control Dynamical Systems [in Russian], No. 14, All-Union Institute of System Studies, Moscow, pp. 68–95 (1990); English translation: Computational Mathematics and Modeling, **4**, pp. 393–400 (1993). [9, 16]

[79] A. A. MILYUTIN AND N. P. OSMOLOVSKII, *Calculus of Variations and Optimal Control*, Translations of Mathematical Monographs, Vol. **180**, American Mathematical Society, Providence, 1998. [1, 2, 3, 4, 5, 30, 32, 71, 102, 103, 104, 105, 106, 107, 108, 109, 115, 127, 128, 130, 131, 207, 225, 229, 232, 255, 293, 294, 296, 297, 298]

[80] J. NOBLE AND H. SCHÄTTLER, *Sufficient conditions for relative minima of broken extremals in optimal control theory*, Journal of Mathematical Analysis and Applications, **269**, pp. 98–128 (2002). [270]

[81] H. J. OBERLE, *Numerical computation of singular control functions for a two-link robot arm*, In: Optimal Control, Lecture Notes in Control and Information Sciences, **95**, pp. 244–253 (1987). [339, 346, 347, 349]

[82] H. J. OBERLE AND W. GRIMM, *BNDSCO—A program for the numerical solution of optimal control problems*, Institute for Flight Systems Dynamics, DLR, Oberpfaffenhofen, Germany, Internal Report No. 515–89/22, 1989. [204, 273, 287, 288, 291, 339]

[83] H. J. OBERLE AND H. J. PESCH, *Private communication*, 2000. [272]

[84] N. P. OSMOLOVSKII, *Second-order weak local minimum conditions in an optimal control problem (necessity, sufficiency)*, Dokl. Akad. Nauk SSSR, **225**, pp. 259–262 (1975); English translation: Soviet Math. Dokl., **15**, pp. 1480–1484 (1975). [25]

[85] N. P. OSMOLOVSKII, *High-order necessary and sufficient conditions for Pontryagin and bounded strong minima in the optimal control problems*, Dokl. Akad. Nauk SSSR Ser. Cybernetics and Control Theory, **303**, pp. 1052–1056 (1988); English translation: Soviet Physics Doklady, **33**, pp. 883–885 (1988). [1, 2]

[86] N. P., OSMOLOVSKII, *Theory of higher-order conditions in optimal control*, Doctoral Thesis, MISI (Moscow Civil Engineering Institute), Moscow, (1988). [1, 68, 173, 175]

[87] N. P. OSMOLOVSKII, *Second-order conditions in a time-optimal control problem for linear systems*, In: System Modelling and Optimization, Notes in Control and Information Sciences, **143**, pp. 368–376 (1989). [293]

[88] N. P. OSMOLOVSKII, *Quadratic conditions for nonsingular extremals in optimal control (Theory)*, Russian Journal of Mathematical Physics, **2**, pp. 487–516 (1995). [2]

[89] N. P. OSMOLOVSKII, *Quadratic conditions for nonsingular extremals in optimal control (examples)*, Russian Journal of Mathematical Physics, **5**, pp. 487–516 (1998). [279]

[90] N. P. OSMOLOVSKII, *Second-order conditions for broken extremals*, In: Calculus of Variations and Optimal Control, Boca Raton, FL, pp. 198–216 (2000). [2]

[91] N. P. OSMOLOVSKII, *Second-order sufficient conditions for an extremum in optimal control*, Control and Cybernetics, **31**, pp. 803–831 (2002).

[92] N. P. OSMOLOVSKII, *Quadratic optimality conditions for broken extremals in the general problem of calculus of variations*, Journal of Mathematical Sciences, **123**, pp. 3987–4122 (2004). [27, 68, 127]

[93] N. P. OSMOLOVSKII, *Second order conditions in optimal control problems with mixed equality-type constraints on a variable time interval*, Control and Cybernetics, **38**, pp. 1535–1556 (2009). [150]

[94] N. P. OSMOLOVSKII, *Sufficient quadratic conditions of extremum for discontinuous controls in optimal control problem with mixed constraints*, Journal of Mathematical Sciences, **173**, pp. 1–106 (2011). [4, 175]

[95] N. P. OSMOLOVSKII, *Necessary quadratic conditions of extremum for discontinuous controls in optimal control problem with mixed constraints*, Journal of Mathematical Sciences, **183**, pp. 435–576 (2012). [4, 68, 173, 175]

[96] N. P. OSMOLOVSKII AND F. LEMPIO, *Jacobi conditions and the Riccati equation for a broken extremal*, Journal of Mathematical Sciences, **100**, pp. 2572–2592 (2000). [2, 4, 212]

[97] N. P. OSMOLOVSKII AND F. LEMPIO, *Transformation of quadratic forms to perfect squares for broken extremals*, Journal of Set-Valued Analysis, **10**, pp. 209–223 (2002). [2, 4]

[98] N. P. OSMOLOVSKII AND H. MAURER, *Second order sufficient optimality conditions for a control problem with continuous and bang-bang control components: Riccati approach,* In: Proceedings of the 23rd IFIP Conference on System Modeling and Optimization, Cracow, Poland (A. Korytowski, M. Szymkat, eds.), pp. 411–429, Springer Verlag, Berlin, 2009. [223, 247]

[99] N. P. OSMOLOVSKII AND H. MAURER, *Equivalence of second order optimality conditions for bang-bang control problems. Part 1: Main results*, Control and Cybernetics, **34**, pp. 927–950 (2005). [299]

[100] N. P. OSMOLOVSKII AND H. MAURER, *Equivalence of second order optimality conditions for bang-bang control problems. Part 2: Proofs, variational derivatives and representations*, Control and Cybernetics, **36**, pp. 5–45 (2007). [299]

[101] H. J. PESCH, *Real-time computation of feedback controls for constrained optimal control problems, Part 1: Neighbouring extremals*, Optimal Control Applications and Methods, **10**, pp. 129–145 (1989). [219, 339]

[102] H. J. PESCH, *Real-time computation of feedback controls for constrained optimal control problems, Part 2: A correction method based on multiple shooting*, Optimal Control Applications and Methods, **10**, pp. 147–171 (1989). [339]

[103] L. S. PONTRYAGIN, V. G. BOLTYANSKI, R. V. GRAMKRELIDZE, AND E. F. MISCENKO, *The Mathematical Theory of Optimal Processes* [in Russian], Fitzmatgiz, Moscow; English translation: Pergamon Press, New York, 1964. [225]

[104] A. V. SARYCHEV, *First- and second-order sufficient optimality conditions for bang-bang controls*, SIAM Journal on Control Optimization, **35**, pp. 315–340 (1997). [279]

[105] È. È. SHNOL′, *On the degeneration in the simplest problem of the calculus of variations*, Mat. Zametki, **24** (1978).

[106] J. STOER AND R. BULIRSCH, *Introduction to Numerical Analysis*, 2nd ed., Springer-Verlag, Berlin, 1992. [206, 339]

[107] H. J. SUSSMANN, *The structure of time-optimal trajectories for single-input systems in the plane: The C^∞ nonsingular case*, SIAM Journal on Control and Optimization, **25**, pp. 433–465 (1987). [1]

[108] H. J. SUSSMANN, *The structure of time-optimal trajectories for single-input systems in the plane: The general real analytic case*, SIAM Journal on Control and Optimization, **25**, pp. 868–904 (1987).

[109] M. G. TAGIEV, *Necessary and sufficient strong extremum condition in a degenerate problem of the calculus of variations*, Uspehi. Mat. Nauk, **34**, pp. 211–212 (1979).

[110] V. S. VASSILIADIS, R. W. SARGENT, AND C. C. PANTELIDES, *Solution of a class of multistage dynamic optimization problems. 2. Problems with path constraints*, Industrial & Engineering Chemistry Research, **33**, pp. 2123–2133 (1994). [357]

[111] G. VOSSEN, *Numerische Lösungsmethoden, hinreichende Optimalitätsbedingungen und Sensitivitätsanalyse für optimale bang-bang und singuläre Steuerungen*, Dissertation, Institut für Numerische und Angewandte Mathematik, Universität Münster, Münster, Germany, 2005. [6, 365, 366]

[112] G. VOSSEN, *Switching time optimization for bang-bang and singular controls*, Journal of Optimization Theory and Applications, **144**, pp. 409–429 (2010). [6, 365]

[113] G. VOSSEN AND H. MAURER, *On L^1-minimization in optimal control and applications to robotics*, Optimal Control Applications and Methods, **27**, pp. 301–321 (2006). [6]

[114] A. WÄCHTER AND L. T. BIEGLER, *On the implementation of an interior-point filter line-search algorithm for large-scale nonlinear programming*, Mathematical Programming, **106**, pp. 25–57 (2006); cf. IPOPT home page (C. Laird and A. Wächter): https://projects.coin-or.org/Ipopt. [250]

[115] J. WARGA, *A second-order condition that strengthens Pontryagin's maximum principle*, Journal of Differential Equations, **28**, pp. 284–307 (1979).

[116] V. ZEIDAN, *Sufficiency criteria via focal points and via coupled points*, SIAM Journal of Control Optimization, **30**, pp. 82–98 (1992). [219]

[117] V. ZEIDAN, *The Riccati equation for optimal control problems with mixed state-control constraints: Necessity and sufficiency*, SIAM Journal on Control and Optimization, **32**, pp. 1297–1321 (1994). [219]

Index